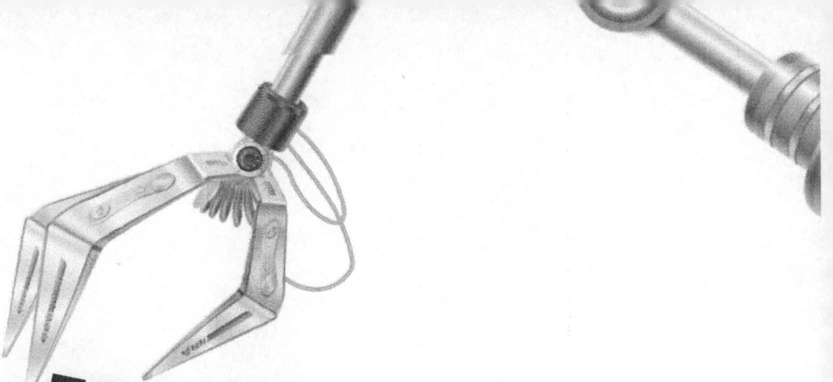

PROBLEMS IN PHYSICS

Electrostatics & Current Electricity

JEE MAIN & ADVANCED

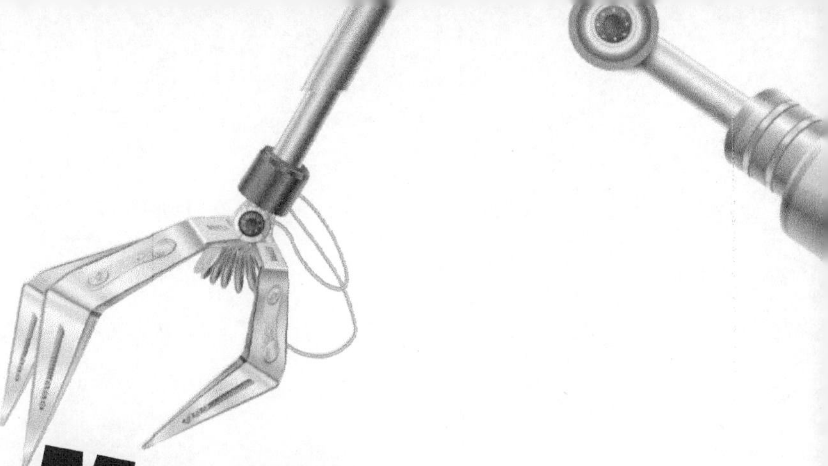

PROBLEMS IN PHYSICS

Electrostatics & Current Electricity

JEE MAIN & ADVANCED

Topically Divided 2000+ Problems
Grouped into 3 Levels

Chapters Covered

- **Electric Field & Potential**
- **Electric Dipole & Flux**
- **Current Electricity**
- **Capacitance & Dielectrics**
- ***R-C* Circuit**

PRADEEP BENIWAL
IIT Kharagpur

arihant
ARIHANT PRAKASHAN (Series), MEERUT

arihant

ARIHANT PRAKASHAN (Series), MEERUT

Administrative & Production Offices

Regd. Office
'Ramchhaya' 4577/15, Agarwal Road, Darya Ganj, New Delhi -110002
Tele: 011- 47630600, 43518550

Head Office
Kalindi, TP Nagar, Meerut (UP) - 250002
Tel: 0121-7156203, 7156204

Sales & Support Offices
Agra, Ahmedabad, Bengaluru, Bareilly, Chennai, Delhi, Guwahati, Hyderabad, Jaipur, Jhansi, Kolkata, Lucknow, Nagpur & Pune.

ISBN : 978-93-26196-13-0

PO No : TXT-XX-XXXXXXX-X-XX

Published by Arihant Publications (India) Ltd.

For further information about the books published by Arihant, log on to www.arihantbooks.com or e-mail at info@arihantbooks.com

Follow us on

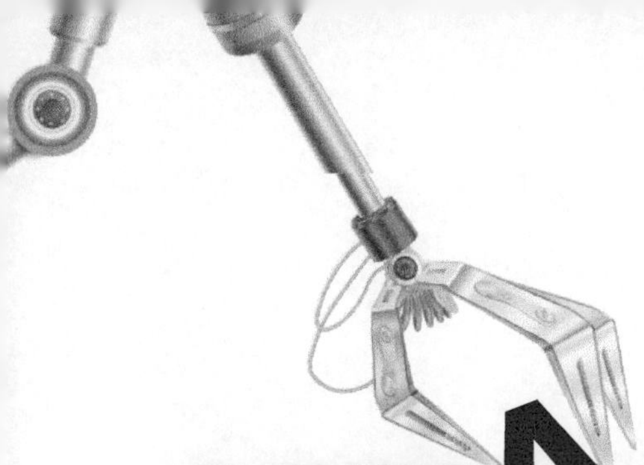

PREFACE

Physics is a subject that requires a lot of practice to master it and the best way to practice Physics is to solve more & more problems right from the beginner's level to the most complex ones. That's why the search for a good Physics Problems book has always been the top notch task of each good student.

During the last 15 Years while teaching & mentoring thousands of students preparing for JEE Main, JEE Advanced, KVPY, Olympiads etc. I always felt the need of such a book that can offer quality problems of Physics for all the topics of different chapters. There are many problems books available in the market but unfortunately none of them serve the purpose perfectly.

To fulfil the need of a good Physics Problems book, I have prepared the book 'Problems in Physics Electrostatics & Current Electricity. All the problems given in this book are arranged conceptually and systematically in order to test the conceptual understanding & analytical ability of the students.

All the problems given in each chapter of this book have been developed with the utmost care and arranged in a way to enhance the problem-solving skills of the students & to boost their confidence to tackle complex JEE Problems.

The Points given below will help you to learn the basic structure of the book;

- Each chapter is divided into three major heads; **Level 1** 'Starter Level', **Level 2** 'JEE Main Level' & **Level 3** 'JEE Advanced Level'.
- **Level 1 'Starter Level'** consists of preliminary questions that test basic concepts. After completing this level student will have a solid understanding of fundamentals of the chapter necessary to solve the problems.
- **Level 2 'JEE Main Level'** has single choice correct questions and numerical value questions aimed especially for JEE Main.
- **Level 3 'JEE Advanced Level'** consists of various types of questions; MCQ with one or more than one correct answer, Comprehension Based Questions, Match the Columns and Numerical Value Questions as per the latest JEE Advanced Pattern.
- The problems given in JEE Advanced Level are made to be conceptually challenging resulting in practical and relevant practice for JEE Advanced.
- **Past Exams' Questions** of JEE Main & JEE Advanced are also covered in each chapter.
- **Detailed solutions to all the problems have been made available through QR Code just to help the students in case they stuck somewhere while solving some questions.**

At the end, I can say confidently that if you rigorously practice the problems given in this book then surely you will get an edge in Physics and will increase your chance of getting good JEE Score.

However, I have taken the utmost care while writing this book but if any error or whatsoever has skipped my attention then please bring it to my notice so that it can be rectified. I would also like to appeal all my fellow Physics Teachers to have a look at the book and share their valuable suggestions for its further improvement.

Pradeep Beniwal

B.Tech, IIT Kharagpur

problemsinphysics@yahoo.com

CONTENTS

CHAPTER

01

ELECTRIC FIELD

LEVEL 01 \ Starter Level

1.1 Electric Charge

1 State True or False.

(i) Charge cannot exist without mass though mass can exist without charge.

(ii) Charge and mass both are conserved for an isolated system.

(iii) Electric charge of a body is always quantised.

(iv) Charge is relativistically invariant.

(v) An accelerated charged particle emits electromagnetic radiations in addition to producing electric and magnetic fields.

(vi) Relative permittivity of a medium lies between 0 and ∞.

(vii) A charged body and a neutral body cannot attract each other.

(viii) Repulsion is a sure test of electrification.

(ix) If two identical conductors having charges $+4e$ and $-2e$ are connected through a metallic wire, then charge on each conductor would be $+e$.

(x) When a positively charged rod is kept near a neutral conductor, they attract each other and when a negatively charged rod is kept near a neutral conductor, they repel each other.

2 A copper sphere contains about 2×10^{22} atoms. The charge on the nucleus of each atom is $+29e$. What fraction of the electrons must be removed from the sphere to give it a charge of $+2\mu C$?

3 A polythene piece rubbed with wool is found to have a negative charge of 3×10^{-7} C.

(i) Estimate the number of electrons transferred (from which to which?).

(ii) Is there a transfer of mass from wool to polythene?

4 Which of the following charges cannot exist?

(i) 1.6×10^{-18} C (ii) 1.6×10^{-20} C (iii) 0.001 C (iv) $+2.5\ e$

(v) $-3.2\ e$

5 Two identical conducting spheres are charged by induction and then separated by a large distance; sphere-1 has charge $+Q$ and sphere-2 has charge $-Q$. A third sphere is initially uncharged. If sphere-3 is touched to sphere-1 and separated and then touched to sphere-2 and separated, what is the final charge on each of the three spheres?

1.2 Coulomb's Law

6 Two neutral objects are rubbed against each other and the electrostatic force of attraction between them is 3.6 N, when they are separated by a distance of 10 cm. How many electrons are transferred from one object to the other during rubbing?

7 Two equally charged identical metal sphere A and B repel each other with a force 20N. Another identical uncharged sphere C is touched to B and then placed at the mid-point between A and B. What is the net electric force on C?

8 Two point charges q_1 and q_2 repel each other with force F, when kept certain distance apart in air. If now both the charges are kept in water at the same separation,

(i) how much force q_1 and q_2 will exert on each other?

(ii) how much force each of them will experience?

(iii) how much force water will exert on each charge? (Relative permittivity of water = 80)

9 Two point charges $q_1 = 20\mu C$ and $q_2 = 25\mu C$ are placed at $(-1, 1, 1)$ m and $(3, 1, -2)$ m, with respect to a coordinate system. Find the electrostatic force on q_2.

10 A particle of mass m carrying charge q_1 is revolving around a fixed charge $-q_2$ in a circular path of radius r. Calculate the period of revolution and its speed also.

11 20 positively charged particles are kept fixed on the X-axis at points $x = 1m, 2m, 3m, \ldots, 20m$. The first particle has a charge $1.0\mu C$, the second $8.0\ \mu C$, the third $27\mu C$ and so on. Find the magnitude of the electric force acting on a $+10\mu C$ charge placed at the origin.

12 Four identical point charges each of magnitude q are placed at the corners of a square of side a. Find the net electrostatic force on any of the charges.

13 The distance between two fixed point charges $+4q$ and $+q$ is l. What charge Q is needed to achieve equilibrium for the entire system and where should it be placed?

14 A charge q_0 is placed at the centre of a regular pentagon having equal charges q at its corners. Assume that distance from centre to corner for pentagon = a

(i) Find the force on q_0.

(ii) If charge of one of the corners is removed, then find out magnitude of force acting on q_0.

(iii) If two adjacent charges are removed from corners, then find the magnitude of force acting on q_0.

15 Two spheres, each having mass m are suspended from a common point by two insulating strings each of length l. The spheres are identically charged and the separation between the spheres is d. Find the charge on each sphere.

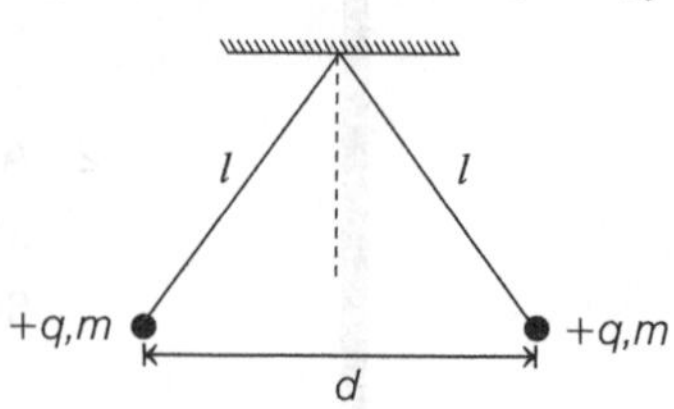

16 Find the force between a uniformly charged rod of length L and charge $+Q$ and a point charge $+q$ placed at a distance d from the centre of the rod at its axial position.

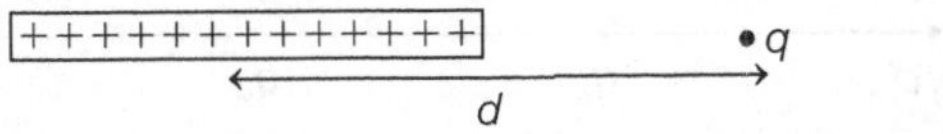

17 Two equal charges $+q$ are kept fixed at $x = -a$ and $x = +a$ along the X-axis. A particle of mass m and charge $+Q$ is brought to the origin and given a small displacement along the

(i) X-axis and (ii) Y-axis.

Describe the motion in the two cases.

18 (i) A particle A having a charge of 4μC and a mass of 600 g is fixed at the bottom of a smooth inclined plane of inclination 37°. Where should another particle B, having same charge and mass, be placed on the inclined plane, so that it may remain in equilibrium?

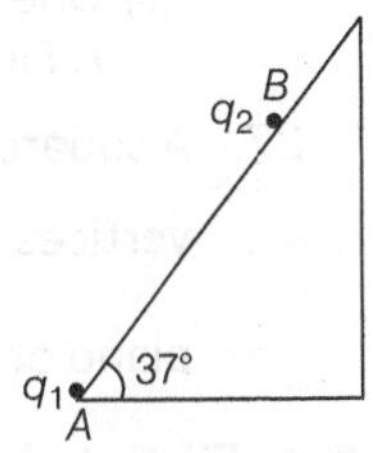

(ii) If B is replaced by a negative point charge of same mass and same amount of charge, then what will be its instantaneous acceleration? (Take, $g = 10\ \text{m/s}^2$)

19 A particle having a charge of 4μC is held fixed on a horizontal table. A second charged particle of same charge and mass 300 g stays in equilibrium on the table at a distance of d metre from the first charge. The coefficient of friction between the table and this second particle is $\mu = 0.3$. Find the range of d for which the second particle lies at rest. (Take, $g = 10\ \text{m/s}^2$)

1.3 Electric Field due to Point Charges

20 A point charge produces an electric field of magnitude 9.0 N/C at a distance of 20 cm from it. What is the magnitude of the charge?

21 An electric charge $Q = 10^{-10}$ C is placed at the point (1, 2, 3) m. Find the electric field intensity due to it at the point (2, 3, 4) m.

22 Find the magnitude of electric field intensity at point C in the following figures.

(i)

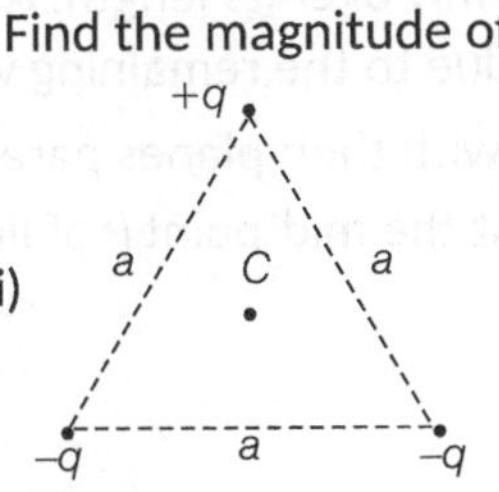

C is the centroid of equilateral triangle of side a

(ii)

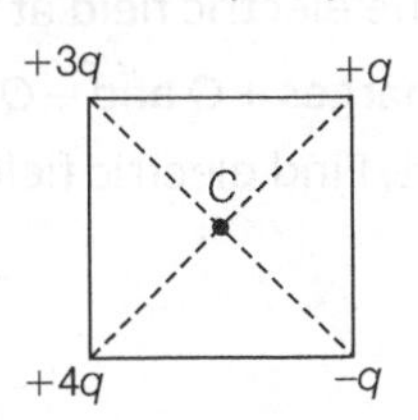

C is the centre of square of side a

(iii)

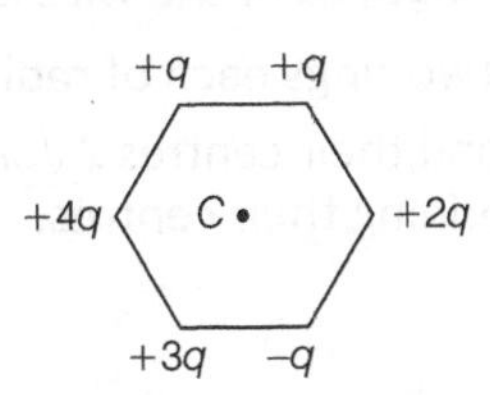

C is the centroid of regular hexagon of side a

23 Two point charges +20μC and −20μC are fixed at (5,0) and (−5,0), respectively. Find electric field intensity at the following points.

(i) (0, 0)
(ii) (10, 0)
(iii) (0, 5)

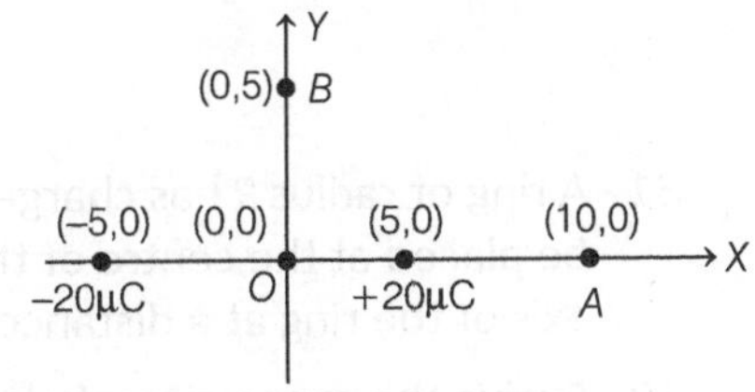

24 Two point charges A and B having charges of +4μC, and −64μC respectively are held at a separation of 90 cm. Locate the point(s) on the line AB or on its extension, where the electric field is zero.

25 Three charges $q_1 = 4nC$, $q_2 = -2nC$ and $q_3 = 1nC$ are fixed along a straight line as shown in the figure.

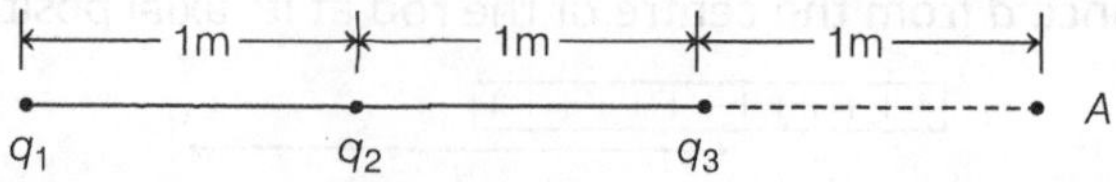

(i) Find electric field at a point A located at a distance of 1m to the right of q_3.
(ii) If a charge $Q = -2mC$ is placed at A, how much force will it experience?

26 12 equal charges q are located at the vertices of a regular 12 sided polygon.
(i) Find electric field at the centre of the polygon.
(ii) One of the charges is moved to the centre of the polygon. Its distance from all other charges is r. Find force acting on it.

27 A square of side L is placed in a horizontal plane and four equal charges $+q$ are fixed at its vertices. A point P is located vertically above the centre of the square, at a height $\frac{L}{\sqrt{2}}$ from the plane of the square. Find the electric field at point P.

1.4 Electric Field due to Continuous Charge Distributions

28 Find electric field at the centre C for each charge configuration.

(i)

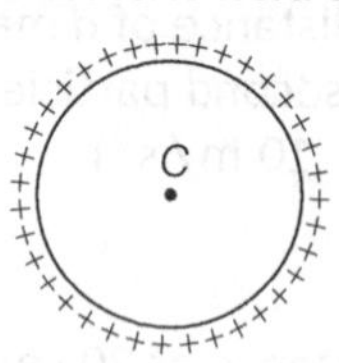

Uniformly charged ring of radius R and linear charge density λ.

(ii)

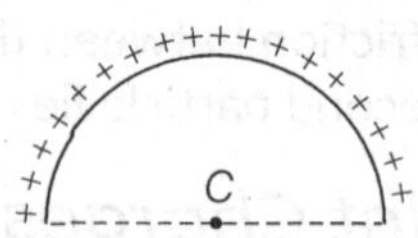

Uniformly charged half ring of radius R and linear charge density λ.

(iii)

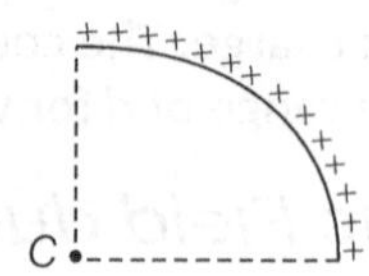

Uniformly charged quarter ring of radius R and linear charge density λ.

29 A circular wire of radius R carries a total charge Q distributed uniformly over its length. A small length dL of the wire is cut off. Find the electric field at the centre due to the remaining wire.

30 Two rings each of radius R carrying charges $+Q$ and $-Q$ are placed with their planes parallel and their centres $2\sqrt{3}R$ distance apart. Find electric field intensity at the mid-point P of line joining their centres.

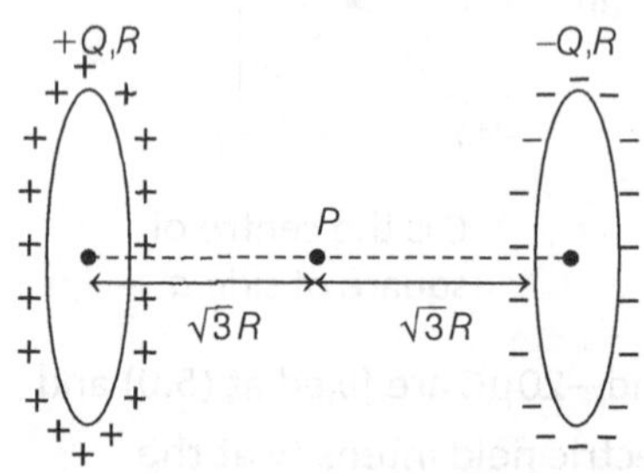

31 A ring of radius R has charge $-Q$ distributed uniformly over it. Calculate the charge that should be placed at the centre of the ring such that the electric field becomes zero at a point on the axis of the ring at a distance R from the centre of the ring.

32 Derive the expression of electric field intensity at a point P which is situated at a distance x on the axis of uniformly charged disc of radius R and surface charge density σ.

33 Find out electric field intensity due to uniformly charged thin spherical shell of radius 10 cm having charge $+20\mu C$ at the following points
(i) At the centre (ii) 5 cm from centre (iii) 20 cm from centre

34 Two concentric uniformly charged thin spherical shells of radius R and $2R$ have charges $+Q$ and $+2Q$, respectively. Find electric field as a function of r for the following positions.

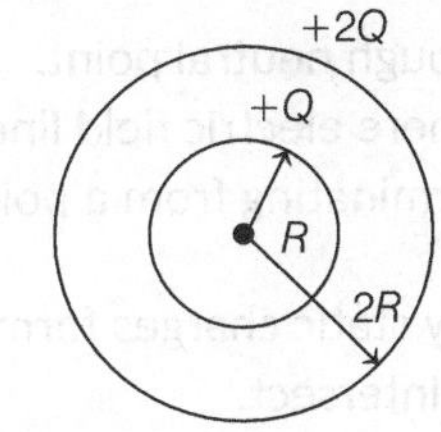

(i) $r < R$ (ii) $R \le r < 2R$ (iii) $r \ge 2R$

35 Find out electric field intensity due to uniformly charged solid non-conducting sphere of radius 10 cm having charge $+20\mu C$ at the following points.

(i) At the centre (ii) 5 cm from centre (iii) 20 cm from centre

36 A non-conducting solid sphere of radius R has volume charge density that varies as $\rho = \rho_0 r$, where ρ_0 is a constant and r is distance from centre. Find out electric field intensity at the following positions.

(i) $r < R$ (ii) $r \ge R$

37 A 10 cm long rod carries a charge of $+50\mu C$ distributed uniformly along its length. Find the magnitude of the electric field at a point 10 cm from both the ends of the rod.

38 In the following figures, find electric field intensity at point P, which is at a distance r from each line charge. λ is the linear charge density.

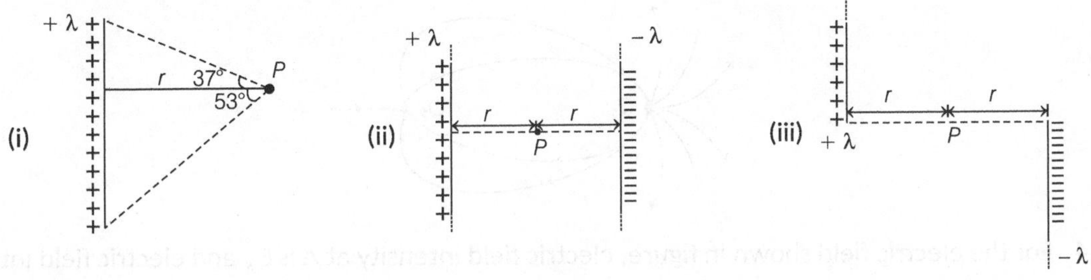

39 Find the force of interaction for the following charge configurations. λ is the linear charge density.

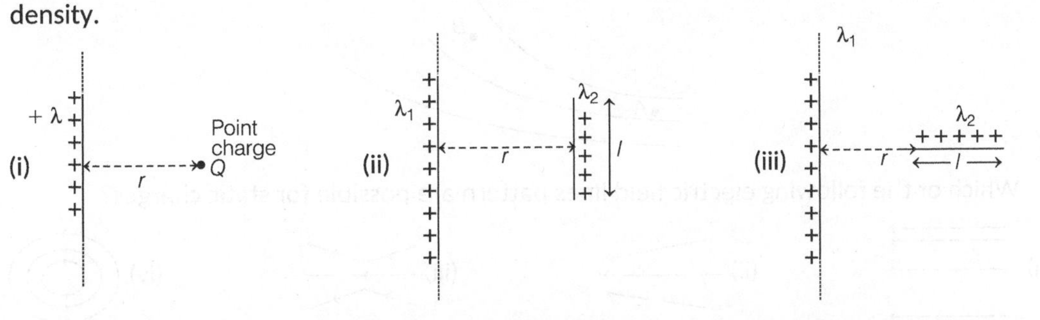

40 If three infinite charged sheets of uniform surface charge densities σ, 2σ and -4σ are placed as shown in figure, then find out electric field intensities at points A, B, C and D.

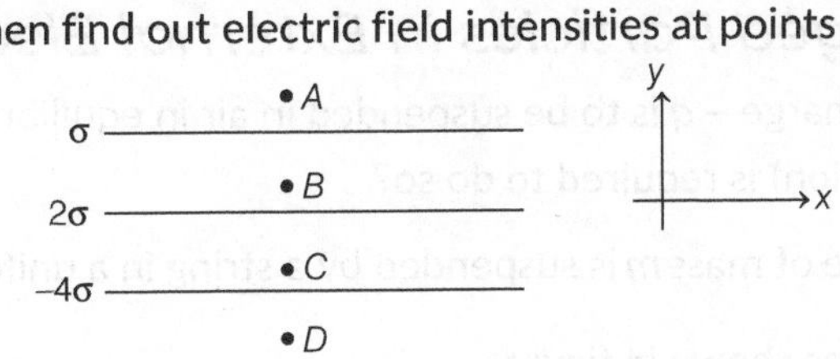

1.5 Electric Field Lines

41 State True or False.

(i) Electric line of force can pass through neutral point.

(ii) Electric field intensity is more, where electric field lines are closer.

(iii) Number of lines originating or terminating from a point charge is directly proportional to the magnitude of the charge.

(iv) Electric lines of force produced by static charges form closed loop.

(v) Two electric field lines can never intersect.

(vi) A charged particle need not always follow an electric line of force.

(vii) An electric line of force in the XY-plane is given by the equation, $x^2 + y^2 = 4$. A particle with unit positive charge, initially at rest at the point $x = 2$, $y = 0$ in the XY-plane will move along the circular line of force.

42 Draw electric field lines pattern for the following system of charges.

(i) Two point charges each $+q$ kept some distance apart.

(ii) Two point charges $+q$ and $-q$ kept some distance apart.

(iii) Three point charges each $+q$ lying on the vertices of an equilateral triangle.

(iv) Uniformly charged thin spherical shell.

(v) Uniformly charged solid non-conducting sphere.

(vi) A uniformly charged infinite thin non-conducting sheet.

43 From the field pattern shown in figure, find $\frac{Q_1}{Q_2}$.

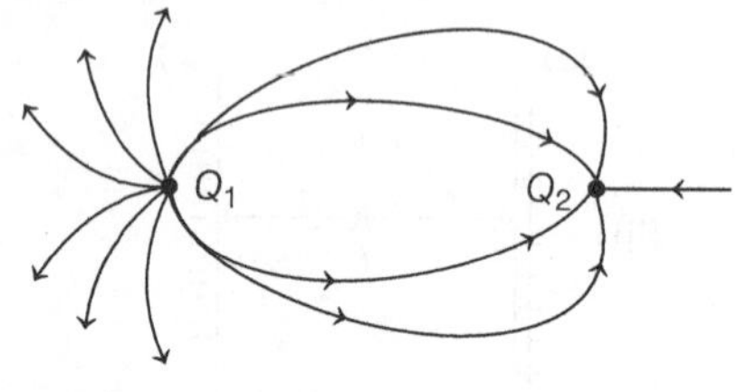

44 For the electric field shown in figure, electric field intensity at A is E_A and electric field intensity at B is E_B. Which is greater E_A or E_B?

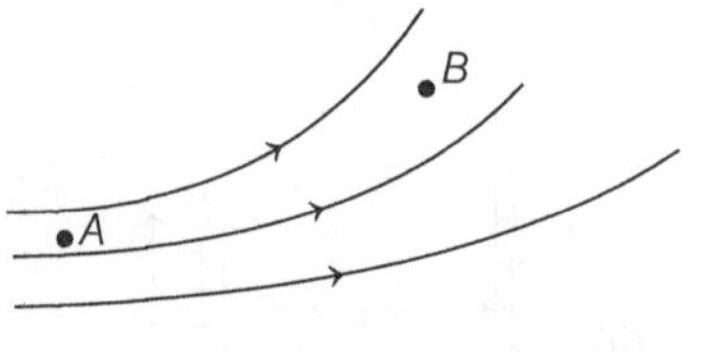

45 Which of the following electric field lines pattern are possible for static charges?

(i) 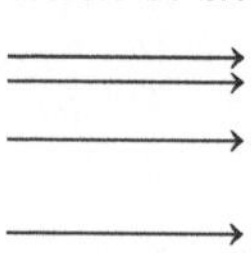 (ii) 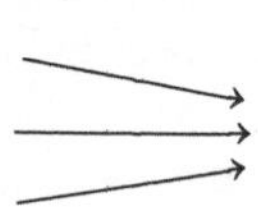 (iii) 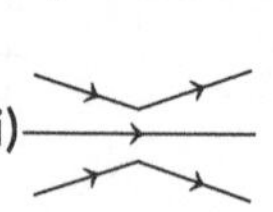 (iv) 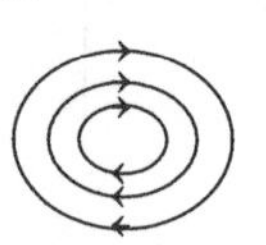

1.6 Motion of Charged Particles in External Electric Field

46 A ball of mass m and charge $-q$ is to be suspended in air in equilibrium. What electric field (magnitude and direction) is required to do so?

47 A small charged sphere of mass m is suspended by a string in a uniform field $E = \frac{mg}{q}$ at an angle of 37° with horizontal as shown in figure.

The sphere is in equilibrium. Find

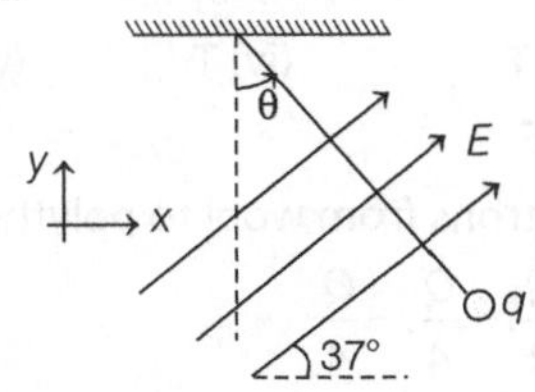

(i) the angle θ made by the string with vertical.

(ii) the tension in the string.

48 A simple pendulum has a bob of mass m and charge q connected to a string of length l, whose other end is fixed to the ceiling. Find time period of small oscillations of the simple pendulum, if

(i) a uniform vertical electric field of strength E pointing downward is switched ON.

(ii) a uniform vertical electric field of strength E pointing upward is switched ON.

(iii) a uniform horizontal electric field of strength E is switched ON.

49 A particle of mass m and charge q is thrown at a speed u against a uniform electric field E. How much distance will it travel before coming to momentarily rest? Also, find the time required till that moment. (Neglect gravity)

50 A uniform electric exists between two parallel charged plates as shown in figure. An electron-proton pair is released from rest from the middle of the gap between the plates. The electron hits the positive plate in time t_1 and the proton hits the negative plate in time t_2. Find $\frac{t_1}{t_2}$.

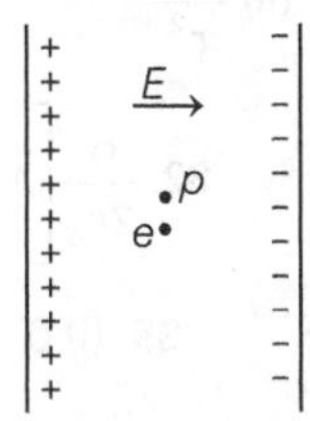

51 A uniform electric field $E = 91 \times 10^{-6}$ V/m is created between two parallel, charged plates as shown in figure. An electron enters the field symmetrically between the plates with a speed $v_0 = 4 \times 10^3$ m/s. The length of each plate is $l = 1$m. Find the angle of deviation of the path of the electron as it comes out of the field.

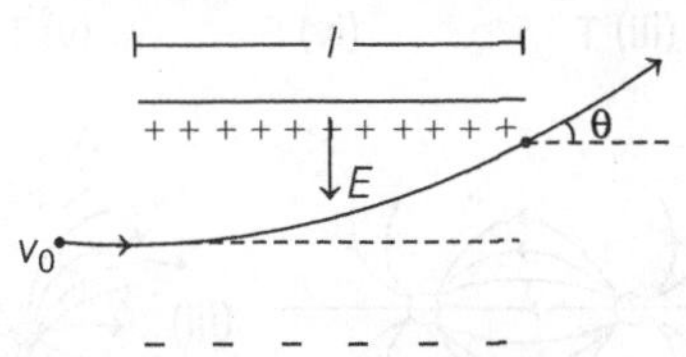

52 A particle of charge $-q$ and mass m moves in a circle of radius r around an infinitely long line charge of linear charge density $+\lambda$. Find its time period.

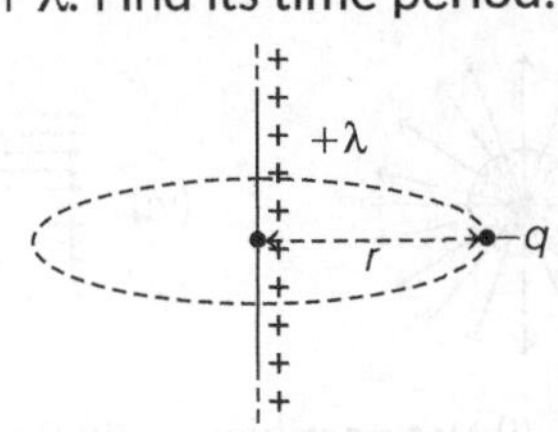

53 A proton is released from rest 10 cm from a large non-conducting thin sheet carrying a surface charge density of -2.21×10^{-9} C/m^2. Find the approximate time after which it shall strike the sheet. (Take, $\varepsilon_0 = 8.84 \times 10^{-12}$ C^2/N-m^2)

Answers

1. (i) T (ii) F (iii) T (iv) T (v) T (vi) F (vii) F
(viii) T (ix) F (x) F

2. 2.16×10^{-11} 3. (i) 1.875×10^{12} electrons from wool to polythene (ii) Yes

4. (ii), (iii), (iv) and (v) 5. $\frac{Q}{2}, -\frac{Q}{4}, -\frac{Q}{4}$

6. 1.25×10^{13} 7. 20 N 8. (i) F (ii) F/80 (iii) 79 F/80

9. $0.036(4\hat{i} - 3\hat{k})$ 10. $T = 4\pi r\sqrt{\frac{\pi\varepsilon_0 mr}{q_1 q_2}}, v = \sqrt{\frac{q_1 q_2}{4\pi\varepsilon_0 mr}}$ 11. 18.9 N 12. $\left(\frac{1}{2} + \sqrt{2}\right)\frac{q^2}{4\pi\varepsilon_0 a^2}$

13. $\frac{2l}{3}$ from charge $+4q, Q = -\frac{4q}{9}$ 14. (i) 0, (ii) $\frac{Kq_0 q}{a^2}$, (iii) $\frac{2Kq_0 q}{a^2}\cos 36°$ 15. $\left(\frac{4\pi\varepsilon_0 mgd^3}{\sqrt{4l^2 - d^2}}\right)^{1/2}$

16. $\frac{4KqQ}{4d^2 - L^2}$ 17. (i) SHM with $T = 2\pi\sqrt{\frac{\pi\varepsilon_0 ma^3}{Qq}}$ (ii) Accelerated translationally

18. (i) 20 cm from the bottom (ii) $12\ \text{m/s}^2$ 19. $d \geq 0.4$ m

20. 4×10^{-11} C 21. $\mathbf{E} = \frac{\sqrt{3}}{10}(\hat{i} + \hat{j} + \hat{k})$ N/C 22. (i) $\frac{6Kq}{a^2}$ (ii) $\frac{10Kq}{a^2}$ (iii) $\frac{4Kq}{a^2}$

23. (i) $14.4(-\hat{i})$ kN/C (ii) $6.4(\hat{i})$ kN/C (iii) $3.6\sqrt{2}(-\hat{i})$ kN/C 24. 30 cm from A along BA

25. (i) 8.5 N/C, right (ii) 0.017 N, left 26. (i) 0 (ii) $\frac{kq^2}{r^2}$ 27. $\frac{q}{\sqrt{2\pi\varepsilon_0 L^2}}$ 28. (i) 0 (ii) $\frac{2K\lambda}{R}$ (iii) $\frac{K\lambda}{R}$

29. $\frac{QdL}{8\pi^2\varepsilon_0 R^3}$ 30. $\frac{\sqrt{3}KQ}{4R^2}$ 31. $\frac{Q}{2\sqrt{2}}$ 32. $\frac{\sigma}{2\varepsilon_0}\left[1 - \frac{x}{\sqrt{R^2 + x^2}}\right]$ 33. (i) 0 (ii) 0 (iii) 4.5×10^6 N/C

34. (i) 0 (ii) $\frac{KQ}{r^2}\hat{r}$ (iii) $\frac{3KQ}{r^2}\hat{r}$ 35. (i) 0 (ii) 9×10^6 N/C (iii) 4.5×10^6 N/C

36. (i) $\frac{\rho_0 r^2}{4\varepsilon_0}\hat{r}$ (ii) $\frac{\rho_0 R^4}{4\varepsilon_0 r^2}\hat{r}$ 37. 5.2×10^7 N/C 38. (i) $\sqrt{2}\frac{K\lambda}{r}$ (ii) $\frac{4K\lambda}{r}$ (iii) $\frac{2\sqrt{2}K\lambda}{r}$

39. (i) $\frac{\lambda Q}{2\pi\varepsilon_0 r}$ (ii) $\frac{\lambda_1\lambda_2 l}{2\pi\varepsilon_0 r}$ (iii) $\frac{\lambda_1\lambda_2}{2\pi\varepsilon_0}\ln\left(1 + \frac{l}{r}\right)$ 40. $\mathbf{E}_A = \frac{-\sigma}{2\varepsilon_0}\hat{j}, \mathbf{E}_B = \frac{-3\sigma}{2\varepsilon_0}\hat{j}, \mathbf{E}_C = \frac{-7\sigma}{r}\hat{j}, \mathbf{E}_D = \frac{\sigma}{2\varepsilon_0}\hat{j}$

41. (i) F (ii) T (iii) T (iv) F (v) T (vi) T (vii) F

42.

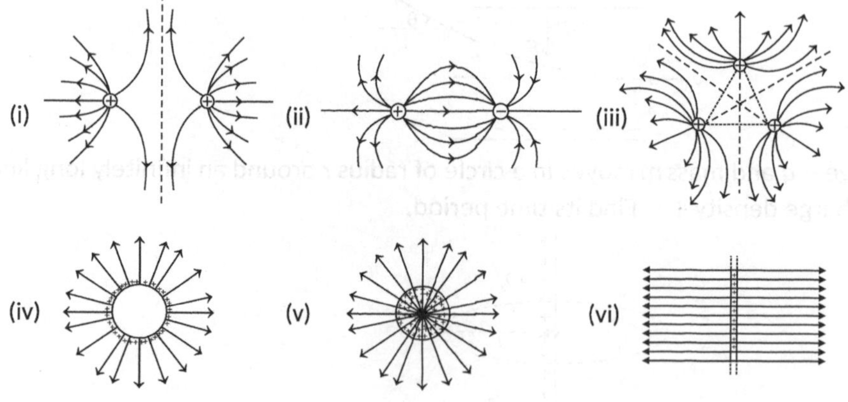

43. −2 44. $E_A > E_B$ 45. (i) Not possible (ii) Possible (iii) Not possible (iv) Not possible

46. $\frac{mg}{q}$, vertically down 47. (i) $\theta = \tan^{-1}(2)$ (ii) $T = \frac{2mg}{\sqrt{5}}$

48. (i) $2\pi\sqrt{\dfrac{l}{g+\dfrac{qE}{m}}}$ (ii) $2\pi\sqrt{\dfrac{l}{g-\dfrac{qE}{m}}}$ (iii) $2\pi\sqrt{\dfrac{l}{\sqrt{g^2+\left(\dfrac{qE}{m}\right)^2}}}$ **49.** $\dfrac{mu^2}{2qE}, \dfrac{mu}{qE}$

50. 0.0233 **51.** $\theta = \tan^{-1}\dfrac{eEl}{mv_0^2} = 45°$

52. $2\pi r\sqrt{\dfrac{2\pi\varepsilon_0 m}{\lambda q}}$ **53.** $4\,\mu s$

For Detailed Solutions
Scan the code

LEVEL 02 \ JEE Main Level

Single Choice Correct Questions

1 Two copper balls, each weighing 10 g are kept in air 10 cm apart. If one electron from every 10^6 atoms is transferred from one ball to the other, then find the Coulomb force between them. (Atomic weight of copper = 63.5 g)
(a) 2.0×10^{10} N (b) 2.0×10^{8} N (c) 2.0×10^{6} N (d) 2.0×10^{4} N

2 A charged particle q_1 is at position (2, – 1, 3). The electrostatic force on another charged particle q_2 at (0, 0, 0) is
(a) $\dfrac{q_1 q_2 (2\hat{i} - \hat{j} + 3\hat{k})}{56\,\pi\varepsilon_0}$ (b) $\dfrac{q_1 q_2 (2\hat{i} - \hat{j} + 3\hat{k})}{56\sqrt{14}\,\pi\varepsilon_0}$ (c) $\dfrac{q_1 q_2 (\hat{j} - 2\hat{i} - 3\hat{k})}{56\,\pi\varepsilon_0}$ (d) $\dfrac{q_1 q_2 (\hat{j} - 2\hat{i} - 3\hat{k})}{56\sqrt{14}\,\pi\varepsilon_0}$

3 Five balls, numbered 1 to 5, are suspended using separated threads. Pairs (1, 2), (2, 4), (4, 1) show electrostatic attraction, while pairs (2, 3) and (4, 5) show repulsion, therefore ball 1 must be
(a) positively charged (b) negatively charged (c) neutral (d) Cannot say

4 Three charges $+q$, $+2q$ and $+4q$ are connected by strings as shown in the figure. What is ratio of tensions in the strings AB and BC.

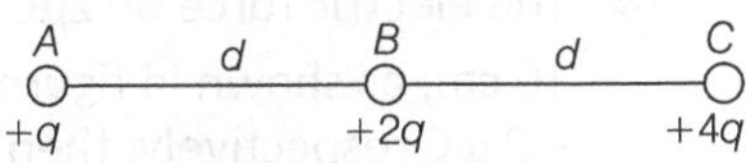

(a) 1 : 2 (b) 1 : 3 (c) 2 : 1 (d) 3 : 1

5 Two similar very small conducting spheres having charges 40 μC and – 20 μC are some distance apart. Now, they are touched and kept at same distance. The ratio of the initial to the final force between them is
(a) 8 : 1 (b) 4 : 1 (c) 1 : 2 (d) 1 : 1

6 Two point charges placed at a distance r in air exert a force F on each other. The value of distance R at which they experience force $4F$ when placed in a medium of dielectric constant $K = 16$ is
(a) r (b) $r/4$ (c) $r/8$ (d) $2r$

7 If a charge q is placed at the centre of the line joining two equal charges Q, such that the system is in equilibrium, then the value of q is
(a) $Q/2$ (b) $-Q/2$ (c) $Q/4$ (d) $-Q/4$

8 ABC is a right-angled triangle. Point charges q_1, q_2 and q_3 are at A, B and C, respectively. The force on q_3 is F as shown. Then,

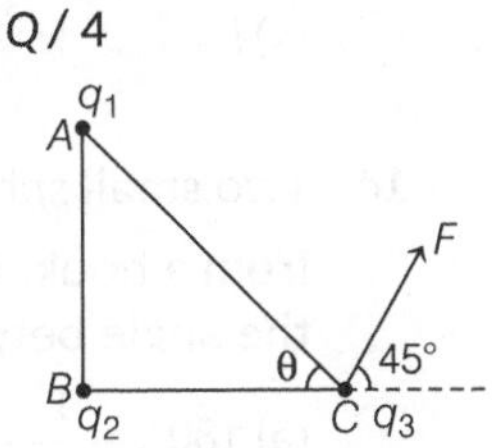

(a) $\left|\dfrac{q_1}{q_2}\right| = \dfrac{1+\tan^2\theta}{\sin\theta + \cos\theta}$ (b) $\left|\dfrac{q_1}{q_2}\right| = \dfrac{1+\cot^2\theta}{\sin\theta + \cos\theta}$
(c) $|q_1| = |q_2|$ (d) None of these

9 Two identical small conducting spheres, having charges of opposite sign, attract each other with a force of 0.108 N, when separated by 0.5 m. The spheres are connected by a conducting wire, which is then removed, and thereafter, they repel each other with a force of 0.036 N. The initial charges on the spheres are

(a) $\pm 5 \times 10^{-6}$ C and $\mp 15 \times 10^{-6}$ C (b) $\pm 1.0 \times 10^{-6}$ C and $\mp 3.0 \times 10^{-6}$ C
(c) $\pm 2.0 \times 10^{-6}$ C and $\mp 6.0 \times 10^{-6}$ C (d) $\pm 0.5 \times 10^{-6}$ C and $\mp 1.5 \times 10^{-6}$ C

10 Two spherical conductors B and C having equal radii and carrying equal charges repel each other with a force F, when kept apart at some distance. A third spherical conductor A having same radius as that of B but uncharged, is brought in contact with B, then brought in contact with C and finally removed away from both. The new force of repulsion between B and C is

(a) $\frac{F}{4}$ (b) $\frac{3F}{4}$ (c) $\frac{F}{8}$ (d) $\frac{3F}{8}$

11 A charge Q is placed at each of opposite corners of a square. A charge q is placed at each of the other two corners. If the net electrical force on Q is zero, then Q/q equals

(a) $-\frac{1}{\sqrt{2}}$ (b) $-2\sqrt{2}$ (c) -1 (d) 1

12 Two charges, each equal to q, are kept at $x = -a$ and $x = +a$ on the X-axis. A particle of mass m and charge $q_0 = \frac{q}{2}$ is placed at the origin. If charge q_0 is given a small ($y << a$) displacement along the Y-axis, the net force acting on the particle is proportional to

(a) y (b) $-y$ (c) $\frac{1}{y^2}$ (d) $-\frac{1}{y^2}$

13 Four positive charges $(2\sqrt{2}-1)Q$ each are arranged at corners of a square. Another charge q is placed at the centre of the square. The system is in equilibrium, if q is

(a) $-7Q/4$ (b) $-4Q/7$ (c) $-Q$ (d) None of these

14 The electric force on 2μC charge placed at the centre O of two equilateral triangles each of side 10 cm, as shown in figure is P. If charge A, B, C, D, E and F are $2\ \mu$C, $2\ \mu$C, $2\ \mu$C, $-2\ \mu$C, $-2\ \mu$C, $-2\ \mu$C respectively, then P is

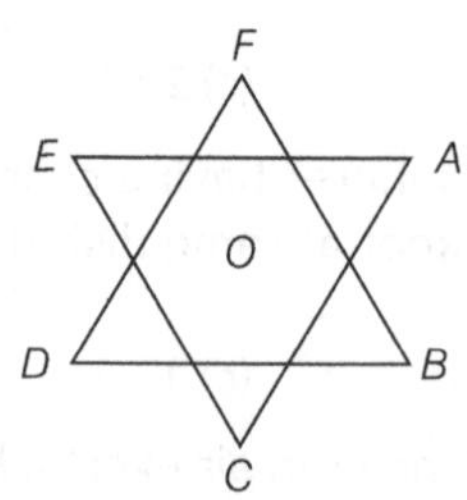

(a) 21.6 N (b) 64.8 N (c) 0 (d) 43.2 N

15 Two particles of equal mass m and charge q are placed at a distance of 16 cm. They do not experience any force. The value of $\frac{q}{m}$ is

(a) l (b) $\sqrt{\frac{\pi\varepsilon_0}{G}}$ (c) $\sqrt{\frac{G}{4\pi\varepsilon_0}}$ (d) $\sqrt{4\pi\varepsilon_0 G}$

16 Two small spheres each having the charge $+Q$ are suspended by insulating threads of length L from a hook. This arrangement is taken in space, where there is no gravitational effect, then the angle between the two suspensions and the tension in each will be

(a) $180°, \frac{1}{4\pi\varepsilon_0}\frac{Q^2}{(2L)^2}$ (b) $90°, \frac{1}{4\pi\varepsilon_0}\frac{Q^2}{L^2}$ (c) $180°, \frac{1}{4\pi\varepsilon_0}\frac{Q^2}{2L^2}$ (d) $180°, \frac{1}{4\pi\varepsilon_0}\frac{Q^2}{L^2}$

17 Two small spherical balls each carrying a charge $Q = 10\ \mu C$ are suspended by two insulating threads of equal lengths 1 m each, from a point fixed in the ceiling. It is found that in equilibrium threads are separated by an angle 60° between them, as shown in the figure. What is the tension in the threads? (Given, $\frac{1}{4\pi\varepsilon_0} = 9 \times 10^9$ N-m/C^2)

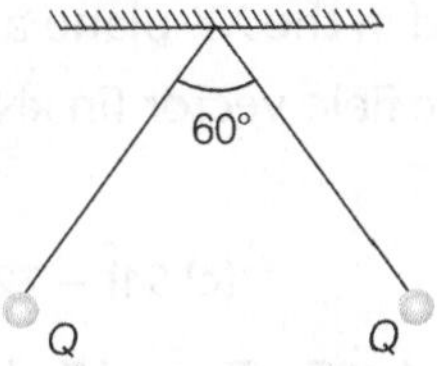

(a) 18 N (b) 1.8 N (c) 0.18 N (d) None of these

18 Two identical charged spheres are suspended by strings of equal lengths. The strings make an angle of 30° with each other. When suspended in a liquid of density 2.7 g cm^{-3}, the angle remains the same. If density of the material of the sphere is 3.6 g cm^{-3}, the dielectric constant of the liquid is

(a) 1 (b) 4 (c) 3 (d) 2

19 Two pith balls with mass m are suspended from insulating threads. When the pith balls are given equal positive charge Q, they hang in equilibrium as shown. We now increase the charge on the left pith ball from Q to 2Q, while leaving its mass essentially unchanged. Which of the following diagrams best represents the new equilibrium configuration?

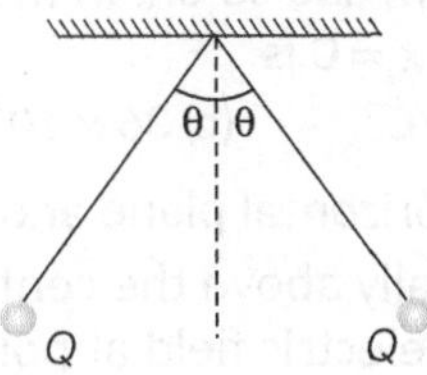

(a) 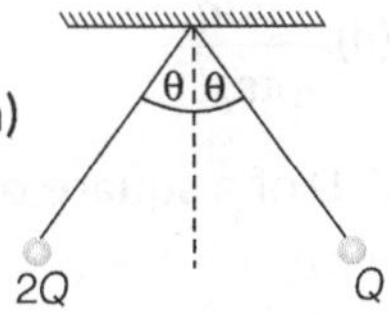

(b) 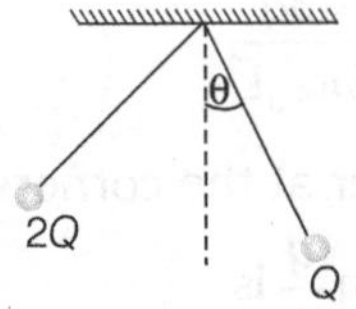

(c) 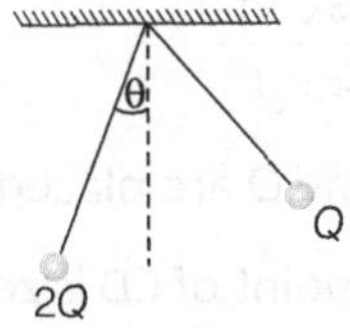

(d) 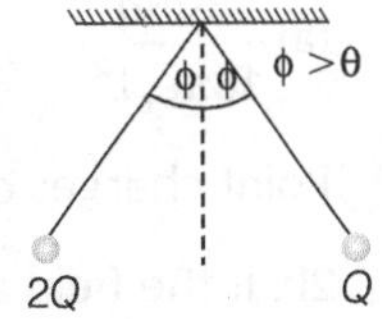

20 A point charge +Q is placed at the centroid of an equilateral triangle. When a second charge +Q is placed at a vertex of the triangle, the magnitude of the electrostatic force on the central charge is 8N. The magnitude of the net force on the central charge, when a third charge +Q is placed at another vertex of the triangle is

(a) zero (b) 4 N (c) $4\sqrt{2}$ N (d) 8 N

21 The number of electrons that should be removed from a coin of mass 1.6 g, so that it may float in electric field intensity 10^9 N/C directed upwards is

(a) 9.8×10^7 (b) 9.8×10^5 (c) 9.8×10^3 (d) 9.8×10^1

22 A particle of mass m and charge $+q$ is placed at rest in a uniform electric field as shown and released. The kinetic energy it attains after moving a distance y is (Neglect gravity)

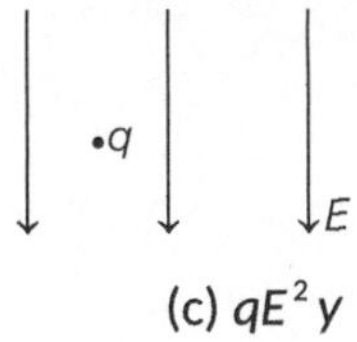

(a) $\frac{1}{2}qEy$ (b) qEy (c) qE^2y (d) $\frac{1}{2}m(qEy)$

23 A proton and an electron are placed in a uniform electric field.
(a) The electric forces acting on them will be equal
(b) The magnitudes of the forces will be equal
(c) Their acceleration will be equal
(d) The magnitudes of acceleration will be equal

24 A point charge $q = 100\ \mu C$ is located in the XY-plane at the point with position vector $\mathbf{r}_0 = 2\hat{\mathbf{i}} + 3\hat{\mathbf{j}} + \hat{\mathbf{k}}$. What is the electric field vector (in kN/C) at the point with position vector $\mathbf{r} = 8\hat{\mathbf{i}} - 5\hat{\mathbf{j}} + \hat{\mathbf{k}}$?
(a) $4.2\hat{\mathbf{i}} - 5.6\hat{\mathbf{j}}$ (b) $6\hat{\mathbf{i}} - 8\hat{\mathbf{j}}$ (c) $54\hat{\mathbf{i}} - 72\hat{\mathbf{j}}$ (d) $5.4\hat{\mathbf{i}} - 7.2\hat{\mathbf{j}}$

25 A point charge q is placed at origin. Let $\mathbf{E}_A$, $\mathbf{E}_B$ and $\mathbf{E}_C$ be the electric field at three points A (1, 2, 3), B(1, 1, – 1) and C(2, 2, 2) due to charge q. Then,
(i) $\mathbf{E}_A \perp \mathbf{E}_B$ (ii) $|\mathbf{E}_B| = 4|\mathbf{E}_C|$
Select the correct alternative.
(a) Only (i) is correct (b) Only (ii) is correct
(c) Both (i) and (ii) are correct (d) Both (i) and (ii) are incorrect

26 Two charges $+5\ \mu C$ and $-10\ \mu C$ are placed 20 cm apart. The net electric field at the mid-point is
(a) 13.5×10^6 N/C directed towards $-10\ \mu C$ (b) 4.5×10^6 N/C directed towards $-10\ \mu C$
(c) 13.5×10^6 N/C directed towards $+5\ \mu C$ (d) 4.5×10^6 N/C directed towards $+5\ \mu C$

27 An infinite number of electric charges each equal to 5nC (magnitude) are placed along X-axis at $x = 1$ cm, $x = 2$ cm, $x = 4$ cm, $x = 8$ cm and so on. In the set up, if the successive charges have opposite sign, then electric field at $x = 0$ is
(a) 12×10^4 N/C (b) 24×10^4 N/C (c) 36×10^4 N/C (d) 48×10^4 N/C

28 A hexagon of side L is placed in a horizontal plane and six equal charges q are fixed at its vertices. A point P is located vertically above the centre of the hexagon, at a height $L\sqrt{3}$ from the plane of the hexagon. Find the electric field at point P.
(a) $\dfrac{3\sqrt{3}q}{16\pi\varepsilon_0 L^2}$ (b) $\dfrac{3\sqrt{3}q}{4\pi\varepsilon_0 L^2}$ (c) $\dfrac{\sqrt{3}q}{16\pi\varepsilon_0 L^2}$ (d) $\dfrac{\sqrt{3}q}{4\pi\varepsilon_0 L^2}$

29 Point charges q, $-q$, $2Q$ and Q are placed in order at the corners A, B, C, D of a square of side $2b$. If the field at the mid-point of CD is zero, then $\dfrac{q}{Q}$ is
(a) 1 (b) 2 (c) $\dfrac{2\sqrt{2}}{5}$ (d) $\dfrac{5\sqrt{5}}{2}$

30 Two charges q_1 and q_2 are kept on X-axis and electric field at different points an X-axis is plotted against x. Choose correct statement about nature and magnitude of q_1 and q_2.

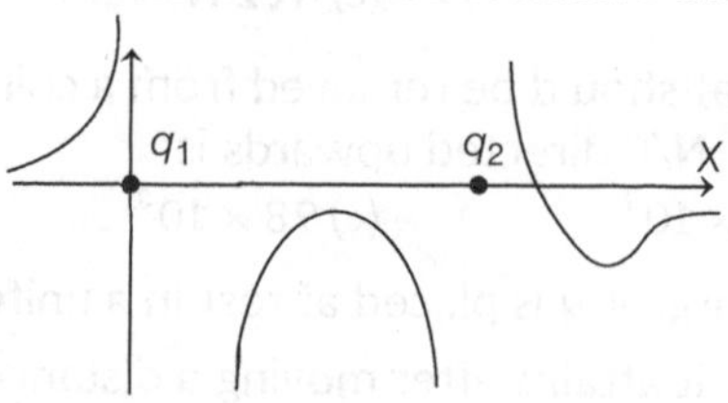

(a) q_1 +ve, q_2 – ve; $|q_1| > |q_2|$
(b) q_1 +ve, q_2 – ve; $|q_1| < |q_2|$
(c) q_1 – ve, q_2 +ve; $|q_1| > |q_2|$
(d) q_1 – ve, q_2 +ve; $|q_1| < |q_2|$

31 The figure shows some of the electric field lines corresponding to an electric field. The figure suggests

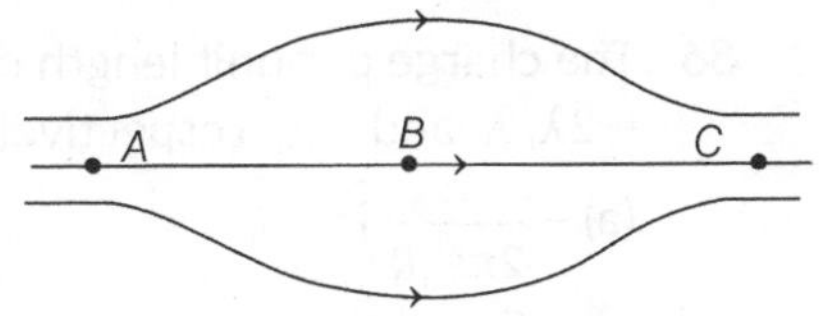

(a) $E_A > E_B > E_C$ (b) $E_A = E_B = E_C$

(c) $E_A = E_C > E_B$ (d) $E_A = E_C < E_B$

32 In an inkjet printer, an ink droplet of mass m is given a negative charge q by a computer controlled charging unit and then enters at speed v into the region midway between two deflecting plates of length L separated by distance d as shown in the figure. The electric field E is uniform in the regions between the plates. Neglecting the gravitational force on the droplet, the maximum charge that it can have, so that it does not hit a plate is most closely approximated by

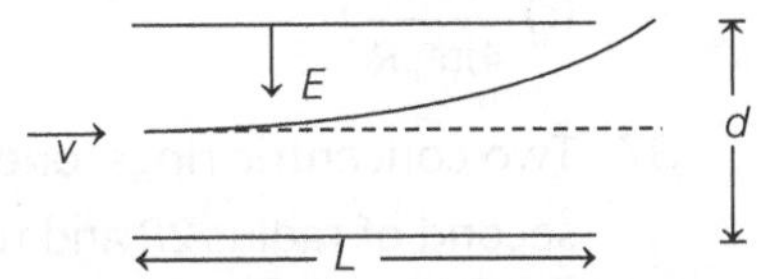

(a) $\dfrac{md}{E(vL)^2}$ (b) $\dfrac{mv^2}{dL^2}E$

(c) $\dfrac{m(vL)^2}{Ed}$ (d) $\dfrac{mv^2}{EL^2}d$

33 Two small sphere each of mass m are suspended by light string l metre in length as shown in the figure. Spheres have charges of q and $-q$ and are placed in the horizontal electric field. Electric field that enables the spheres to be in equilibrium at θ is

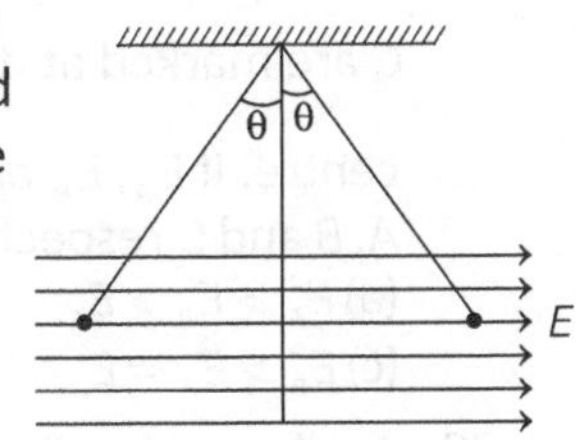

(a) $\dfrac{q}{4\pi\varepsilon_0}\dfrac{1}{(2l\sin\theta)^2}+\dfrac{mg\tan\theta}{q}$ (b) $\dfrac{mg\tan\theta}{q}-\dfrac{q}{4\pi\varepsilon_0}\dfrac{1}{(2l\sin\theta)^2}$

(c) $\dfrac{q}{4\pi\varepsilon_0}\dfrac{1}{(l\sin\theta)^2}+\dfrac{mg\tan\theta}{q}$ (d) $\dfrac{mg\tan\theta}{q}-\dfrac{q}{4\pi\varepsilon_0}\dfrac{1}{(l\sin\theta)^2}$

34 Two very long line charges of uniform charge density $+\lambda$ and $-\lambda$ are placed along same line with the separation between the nearest ends being $2a$, as shown in figure. The electric field intensity at point O is

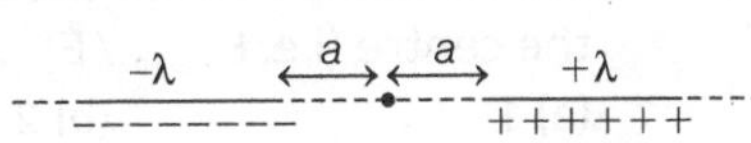

(a) $\dfrac{\lambda}{2\pi\varepsilon_0 a}$ (b) 0

(c) $\dfrac{\lambda}{\pi\varepsilon_0 a}$ (d) $\dfrac{\lambda}{4\pi\varepsilon_0 a}$

35 Which of the following graphs shows the correct variation of electric field as a function of x along the axis of a uniformly and positively charged ring of radius r and charge Q.

(a)
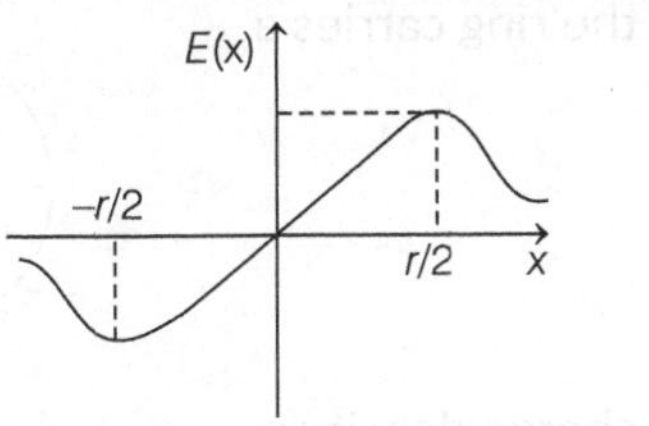

(b)
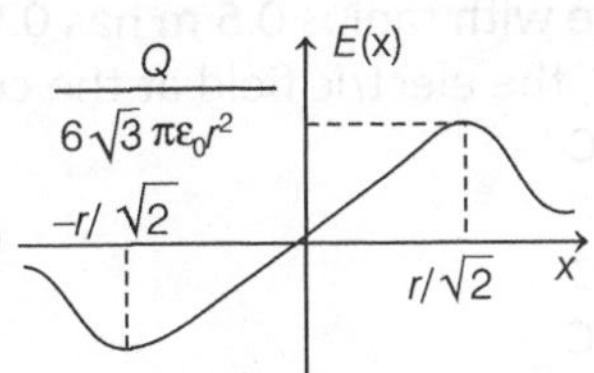

(c)
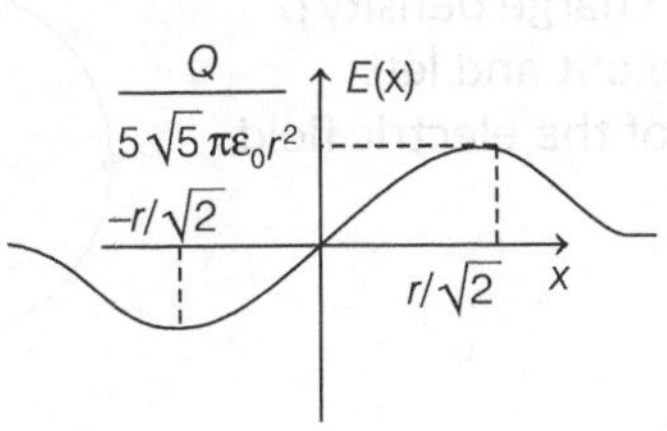

(d)
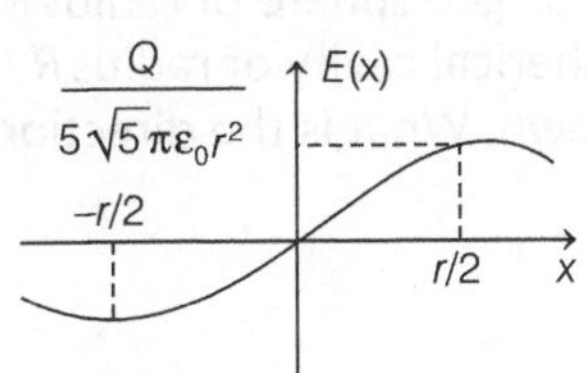

36 The charge per unit length of the four quadrant of the ring is 2λ, -2λ, λ, and $-\lambda$, respectively. The electric field at the centre is

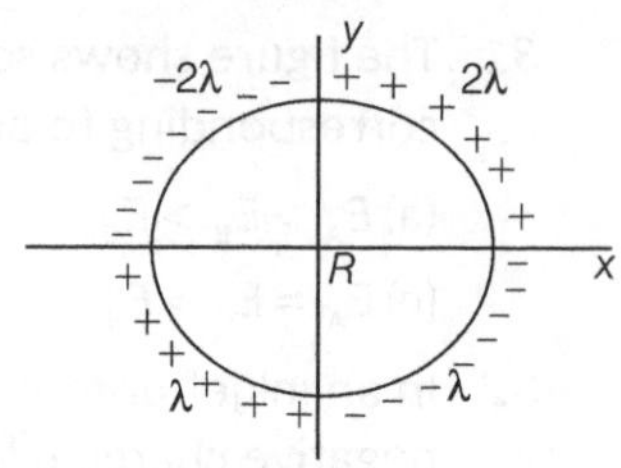

(a) $-\dfrac{\lambda}{2\pi\varepsilon_0 R}\hat{\mathbf{i}}$ (b) $\dfrac{\lambda}{2\pi\varepsilon_0 R}\hat{\mathbf{j}}$

(c) $\dfrac{\sqrt{2}\,\lambda}{4\pi\varepsilon_0 R}\hat{\mathbf{i}}$ (d) None of these

37 Two concentric rings, one of radius R and total charge $+Q$ and the second of radius $2R$ and total charge $-\sqrt{8}\,Q$, lie in XY-plane (i.e., $Z=0$ plane). The common centre of rings lies at origin and the common axis coincides with Z-axis. The charge is uniformly distributed on both rings. At what distance from origin is the net electric field on Z-axis zero?

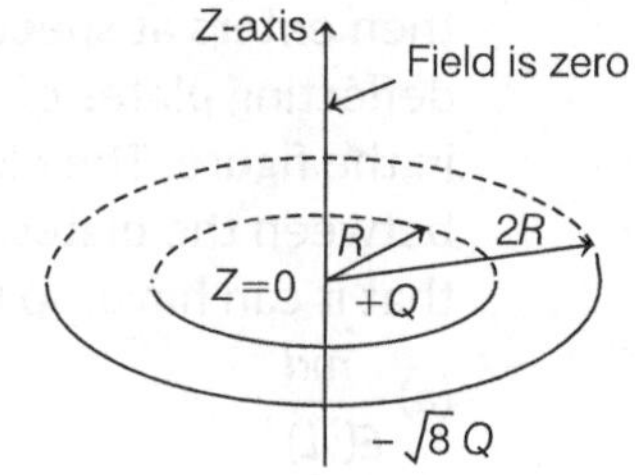

(a) $\sqrt{2}R$ (b) $2\sqrt{2}R$

(c) $2R$ (d) $\sqrt{3}R$

38 Consider two thin uniformly charged concentric shells of radii r and $2r$ having charges Q and $-Q$ respectively, as shown. Three points A, B and C are marked at distances $\dfrac{r}{2}$, $\dfrac{3r}{2}$ and $\dfrac{5r}{2}$ respectively from their common centre. If E_A, E_B and E_C are magnitudes of the electric fields at points A, B and C respectively, then

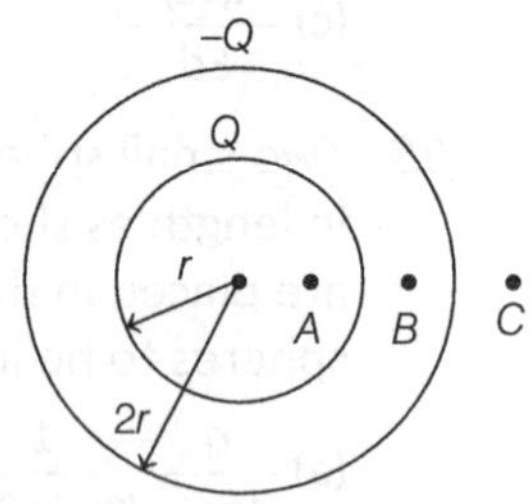

(a) $E_A > E_B > E_C$ (b) $E_C > E_B > E_A$

(c) $E_B > E_A = E_C$ (d) $E_B > E_A > E_C$

39 A sphere of radius R carries charge such that its volume charge density is proportional to the square of the distance from the centre. What is the ratio of the magnitude of the electric field at a distance $2R$ from the centre to the magnitude of the electric field at a distance of $R/2$ from the centre (i.e. $E_{r=2R}/E_{r=R/2}$)?

(a) 1 (b) 2 (c) 4 (d) 8

40 A charge is uniformly distributed on a wire, such that linear charge density is λ. Shape of the wire is shown in the figure, straight section is semi infinitely large. C is the centre of the semicircular portion of radius R. Considering all physical quantities in SI units, magnitude of electric field at C is $\left(K = \dfrac{1}{4\pi\varepsilon_0}\right)$

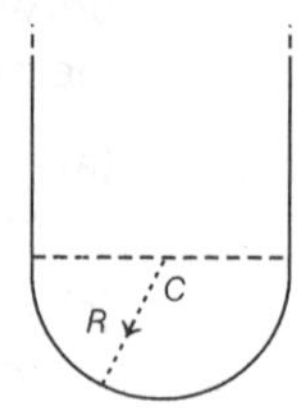

(a) $\dfrac{2K\lambda}{R}$ (b) $\dfrac{\sqrt{2}K\lambda}{R}$ (c) $\dfrac{K\lambda}{R}$ (d) zero

41 A ring of charge with radius 0.5 m has $0.002\,\pi$ m gap. If the ring carries a charge of $+1$ C, the electric field at the centre is

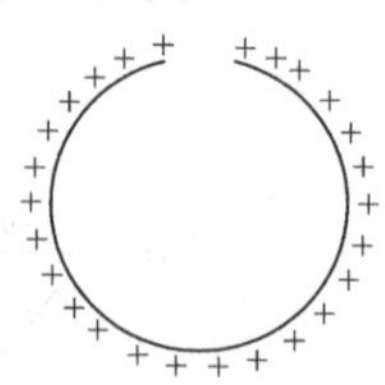

(a) 7.5×10^7 N/C
(b) 7.2×10^7 N/C
(c) 6.2×10^7 N/C
(d) 6.5×10^7 N/C

42 A positively charged sphere of radius R carries a volume charge density ρ (Figure). A spherical cavity of radius $R/2$ is then scooped out and left empty, as shown. What is the direction and magnitude of the electric field at point B?

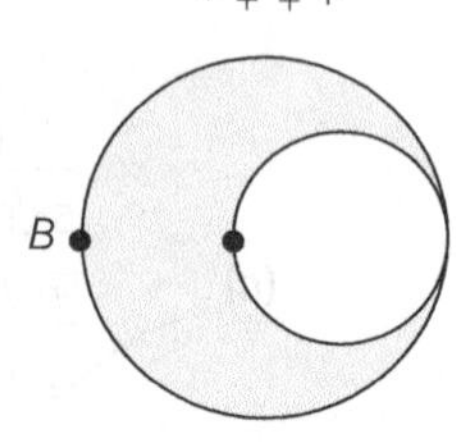

(a) $\dfrac{17\rho R}{54\varepsilon_0}$, left (b) $\dfrac{\rho R}{6\varepsilon_0}$, left

(c) $\dfrac{17\rho R}{54\varepsilon_0}$, right (d) $\dfrac{\rho R}{6\varepsilon_0}$, right

43 Consider a finite uniformly charged rod. Electric field at point P (shown) makes an angle θ with horizontal dotted line, then angle θ is

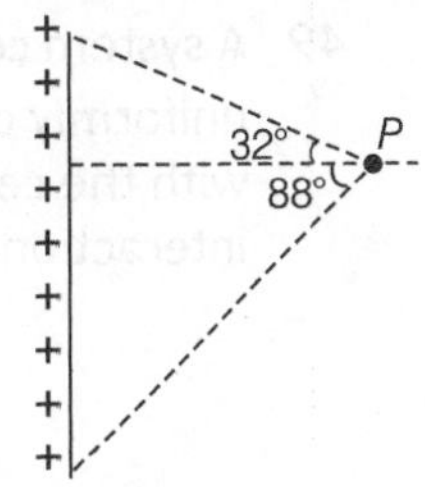

(a) 60°

(b) 28°

(c) 44°

(d) Information insufficient

44 Three charged rods of length L each are made of insulating material and form an equilateral triangle. They have linear charge densities of λ, λ and $-\lambda$ as shown in figure. Find the magnitude of electric field intensity at the centroid O of the triangle.

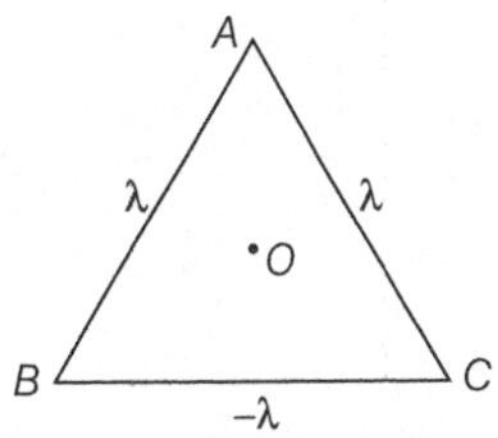

(a) $\dfrac{3\lambda}{2\pi\varepsilon_0 L}$ (b) $\dfrac{\sqrt{3}\lambda}{\pi\varepsilon_0 L}$

(c) $\dfrac{3\sqrt{3}\lambda}{2\pi\varepsilon_0 L}$ (d) $\dfrac{3\lambda}{\pi\varepsilon_0 L}$

45 Two point charges of charge $+q$ each are placed $2a$ distance apart and a ring of radius a having charge Cq distributed uniformly over its circumference is placed midways between them as shown in figure.If none of them accelerated when released, the value of C will be

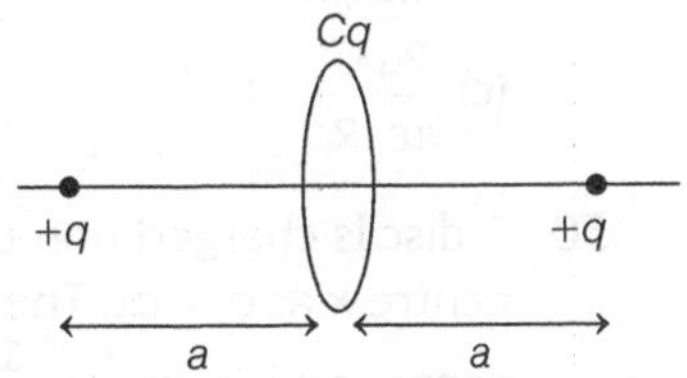

(a) $-\dfrac{1}{\sqrt{2}}$ (b) $-\sqrt{2}$

(c) $-\dfrac{1}{2\sqrt{2}}$ (d) $-2\sqrt{2}$

46 Consider a uniformly charged hemispherical shell shown alongside. Indicate the directions (not magnitude) of the electric field at the central point P_1 and an off-centre point P_2 on the drumhead of the shell.

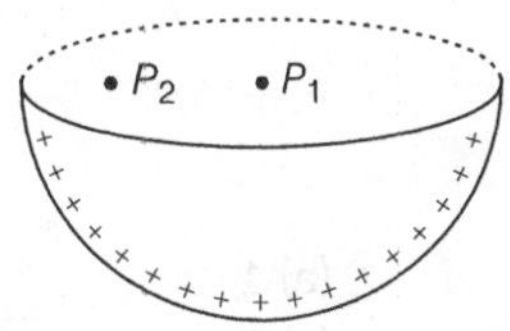

(a)  (b) 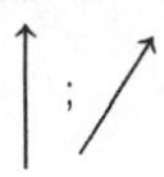

(c) 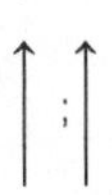 (d) 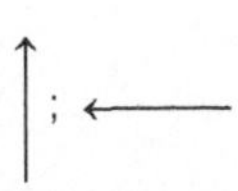

47 Using Thomson's model of the atom, consider an atom consisting of two electrons, each of charge $-e$, embedded in a sphere of charge $+2e$ and radius R. In equilibrium, each electron is at distance d from the centre of the atom. What is equilibrium separation between electrons?

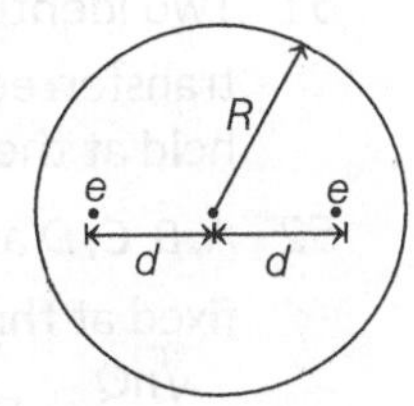

(a) R (b) R / 2

(c) R / 3 (d) R / 4

48 A particle of mass m, charge $-Q$ is constrained to move along the axis of a ring of radius a. The ring carries a uniform charge density $+\lambda$ along its circumference. Initially, the particle lies in the plane of the ring at a point, where no net force acts on it. The period of oscillation of the particle when it is displaced slightly from its equilibrium position is

(a) $4\pi\sqrt{\dfrac{\varepsilon_0 ma^2}{\lambda Q}}$ (b) $2\pi\sqrt{\dfrac{2\varepsilon_0 ma^2}{\lambda Q}}$

(c) $2\pi\sqrt{\dfrac{4\varepsilon_0 ma^2}{\lambda Q}}$ (d) $2\pi\sqrt{\dfrac{\varepsilon_0 ma^2}{2\lambda Q}}$

49 A system consists of a uniformly charged ring of radius R carrying charge Q and a very long uniformly charged thread oriented along the axis of the ring, with one of its ends coinciding with the centre of the ring. The charge of the thread (per unit length) is equal to λ. The interaction force between the ring and the thread is

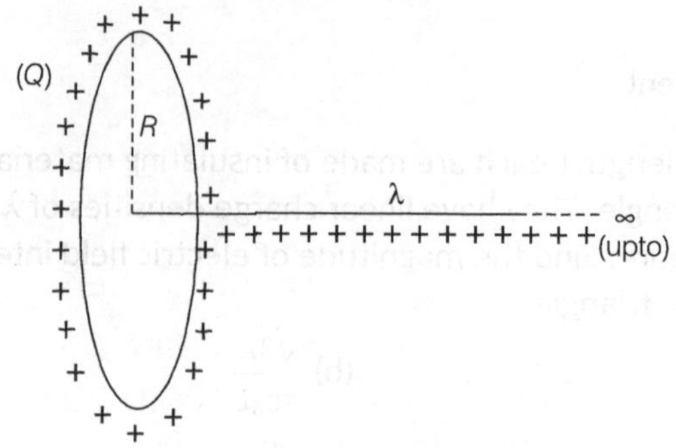

(a) $\frac{q\lambda}{4\pi\varepsilon_0 R}$

(b) $\frac{q\lambda}{2\pi\varepsilon_0 R}$

(c) $\frac{2q\lambda}{\pi\varepsilon_0 R}$

(d) zero

50 A disc is charged non-uniformly with charge per unit area depending on distance from the centre x as $\sigma = cx$. The radius of disc is 3 m and the electric field at a distance of 4 m from centre on it's axis is $\frac{1}{5\varepsilon_0}$ in SI units. Find c (in SI units).

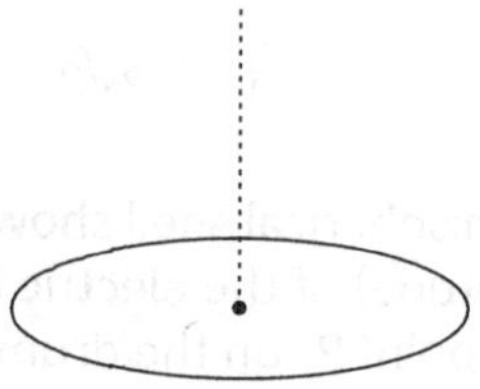

(a) 1

(b) 2

(c) 5

(d) 10

Numerical Value Questions

51 Two identical charge, 5μC each, are certain distance apart in the air. The largest charge to be transferred from one charge to another, so that their interaction force decreases 5 times when held at the same separation in a liquid of dielectric constant 2 is $\sqrt{n}$ μC. Find n.

52 A, B, C, D are the corners of a regular tetrahedron of side a. Three identical charges each Q are fixed at three corners B, C and D. The magnitude of electric field intensity at the corner A is $\frac{\sqrt{n}Q}{4\pi\varepsilon_0 a^2}$. Find n.

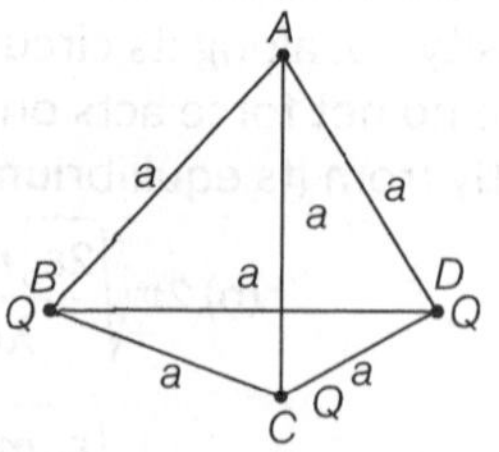

53 The curved wire, which is a part of a circle of radius R has a charge per unit length of $\lambda = \lambda_0 \sin 2\theta$, where θ is the angle shown in the figure. The electric field intensity at point O, the centre of the circle is $\frac{\sqrt{2}\lambda}{n\pi\varepsilon_0 R}$. Find n.

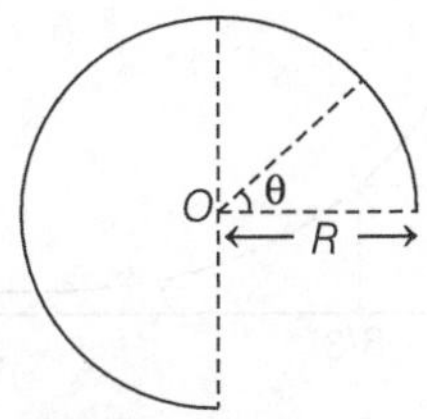

54 A disc of radius 4 m is uniformly charged with the charge of 1 μC. On the periphery of the disc, we stick a thin ring which has a uniform charge of – 1 μC. What is the electric field (in N/C) due to the combination at a point 3 m away from the centre of combination on the common axis?

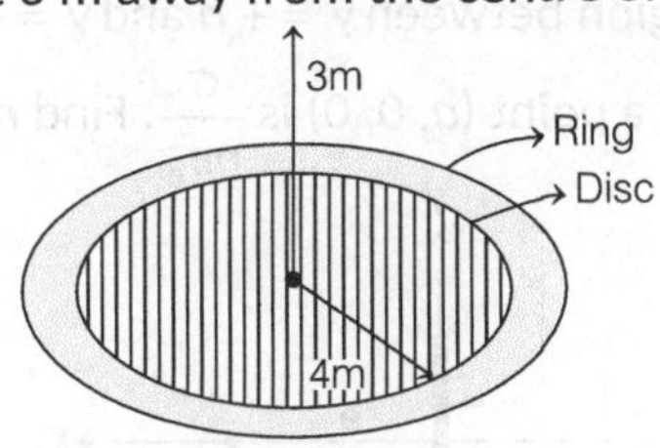

55 Three charged small spheres are suspended from point O with help of strings of length l as shown. All three charges are in equilibrium. Tension in middle string is $\frac{x}{10}$ N. Find x.

($m = 20$ g ; $q = 10$ μC; $l = 3$ m)

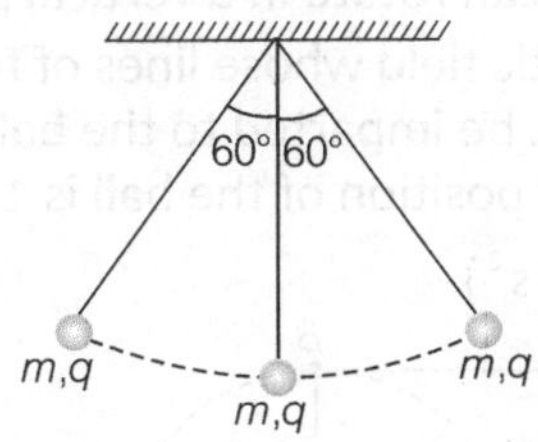

56 A point charge – Q is placed at the centre of a uniformly charged ring of charge Q. Electric field at a point on the axis of ring at a distance x ($x >> R$) varies as $E \propto \frac{1}{x^n}$. Find n.

57 In a 2-D region of xy-space, a uniform electric field exists, given by: $E = (2 \text{ V/m}) (3\hat{\mathbf{i}} + 4\hat{\mathbf{j}})$. A charged particle of mass 10^{-2} g and charge 10^{-5} C is fired from the origin with initial velocity 20 m/s, directed along the negative y-direction. Find the minimum speed (in m/s) of the particle during its subsequent motion. Neglect gravity.

58 A rod of mass m (uniformly distributed) and length l is hinged about point O in a uniform vertical electric field as shown. Linear charge density of rod varies with x as $\lambda = \lambda_0 x$. Electric field for which rod remain horizontal is $\frac{nmg}{2\lambda_0 l^2}$. Find value of n.

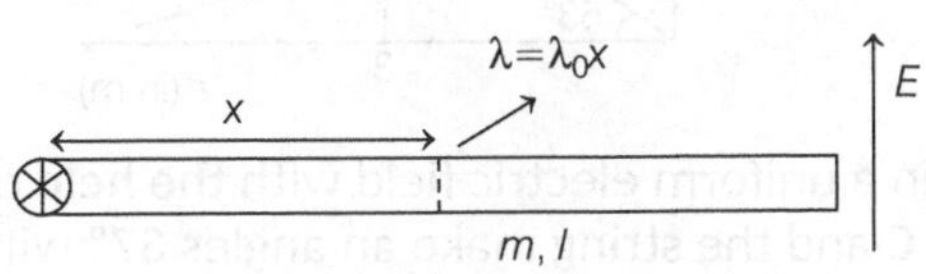

Problems in Physics

59 A uniformly charged disc lies in XY-plane with its centre at the origin. The strength of electric field (E) changes along the Z-axis according to the graph shown. Find the radius (in cm) of the disc.

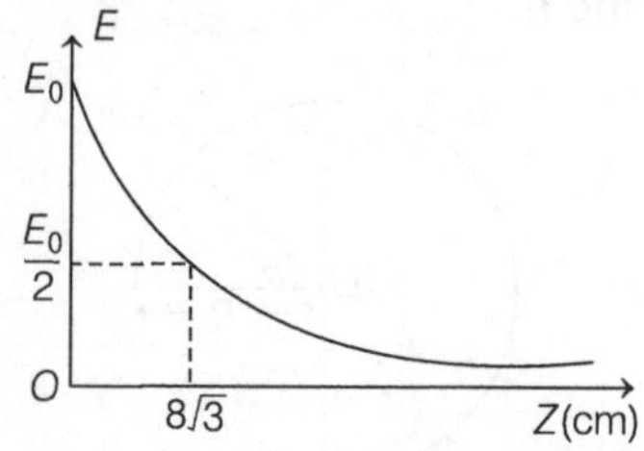

60 Electric field intensity at the centre of a uniformly charged thin hemispherical shell with surface charge density σ is $\frac{\sigma}{n\varepsilon_0}$. Find n.

61 In the YZ-plane, the entire region between $y = +a$ and $y = -a$ has a uniform surface charge density σ. The electric field at a point $(a, 0, 0)$ is $\frac{\sigma}{n\varepsilon_0}$. Find n.

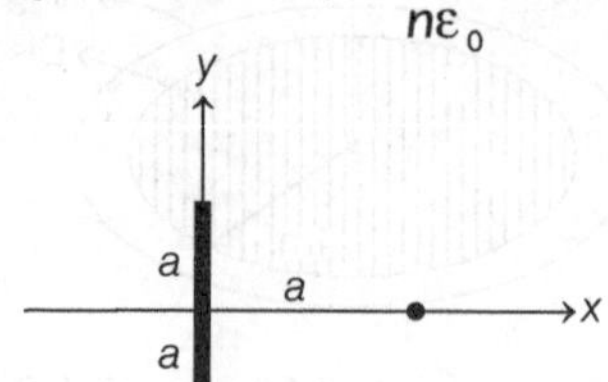

62 A ball of mass m with a charge q can rotate in a vertical plane at the end of a string of length $l = 10$ cm in a uniform electrostatic field whose lines of force are directed upwards. What horizontal velocity (in m/s) must be imparted to the ball in the upper position, so that the tension in the string in the lower position of the ball is 15 times the weight of the ball? (Given, $qE = 3\,mg$ and $g = 10\text{ m/s}^2$)

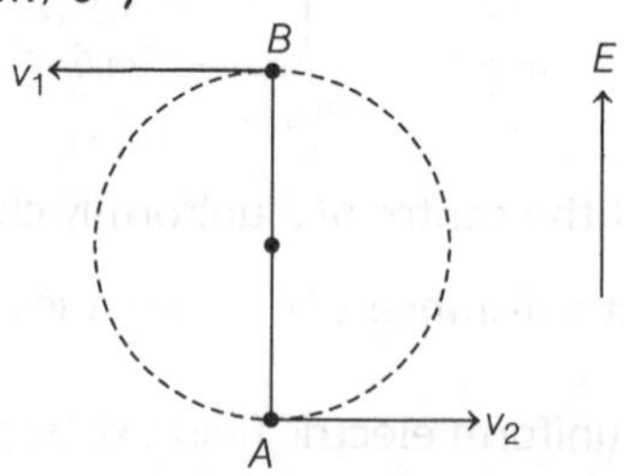

63 We have a solid non-conducting uniformly charged sphere. Its electric field variation with distance from centre is shown in graph, then the net charge on the sphere (in SI units) is $n\pi\varepsilon_0$. Find n.

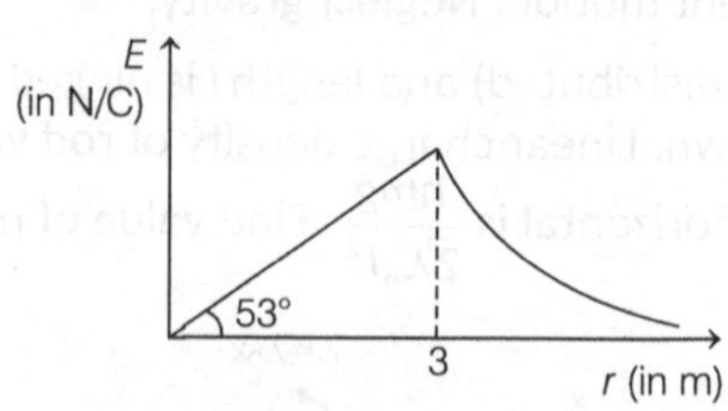

64 A 1kg ball is suspended in a uniform electric field with the help of a string fixed to a point. The ball is given a charge $\sqrt{5}$ C and the string make an angles 37° with the vertical in equilibrium position. In the equilibrium position, the tension is double the weight of the ball. Find the magnitude of the electric field in N/C.

65 An inclined plane making an angle of 37° with horizontal is placed in a uniform horizontal electric field E of 100 V/m. A block of mass 5 kg and charge 0.25 C is allowed to slide down from rest from a height of 6.75 m. If the coefficient of friction is 0.1, find the time(in seconds) taken by the block to reach the bottom. (Take, $g = 10\ \text{m/s}^2$)

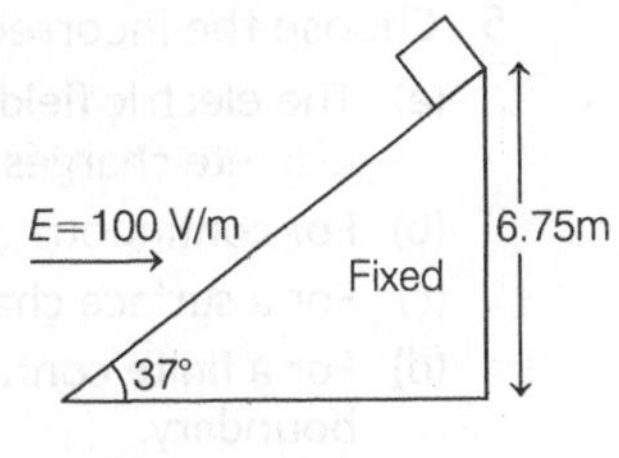

Answers

1. (b)	2. (d)	3. (c)	4. (b)	5. (a)	6. (c)	7. (d)	8. (a)	9. (b)	10. (d)
11. (b)	12. (a)	13. (a)	14. (d)	15. (d)	16. (a)	17. (b)	18. (b)	19. (d)	20. (d)
21. (a)	22. (b)	23. (b)	24. (d)	25. (c)	26. (a)	27. (c)	28. (a)	29. (d)	30. (c)
31. (c)	32. (d)	33. (a)	34. (a)	35. (b)	36. (a)	37. (a)	38. (c)	39. (b)	40. (d)
41. (b)	42. (a)	43. (b)	44. (d)	45. (a)	46. (c)	47. (a)	48. (b)	49. (a)	50. (a)

51. (15)	52. (6)	53. (6)	54. (234)	55. (3)
56. (4)	57. (12)	58. (3)	59. (24)	60. (4)
61. (4)	62. (5)	63. (144)	64. (6)	65. (5)

For Detailed Solutions
Scan the code

LEVEL 03 \ JEE Advanced Level

Single Choice Correct Questions

1 Identical charges of $-48\left(\frac{5}{4}-\frac{1}{\sqrt{3}}\right)$ μC are located at the vertices of a regular hexagon. What charge must be placed at the centre of the hexagon to set the whole system of charges in equilibrium?

(a) 29 μC (b) 45 μC (c) 59 μC (d) 75 μC

2 Six point charges are placed at six corners of a cube as shown. Unit vector along direction of net electric field at centre of cube O is equal to

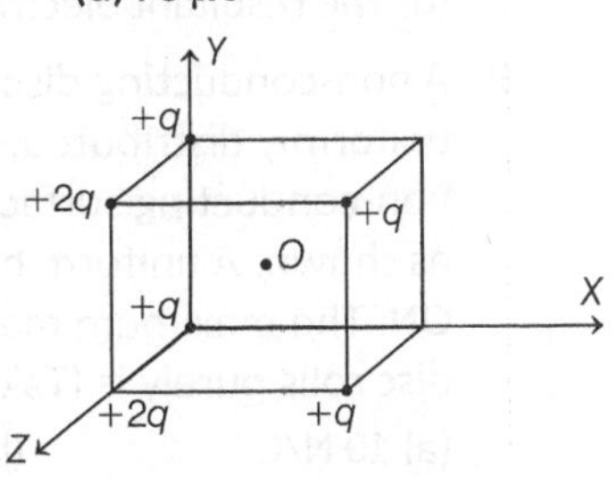

(a) $\frac{\hat{i}-\hat{k}}{\sqrt{2}}$ (b) $\frac{\hat{i}+\hat{k}}{\sqrt{2}}$

(c) $\frac{\hat{i}+\hat{j}}{\sqrt{2}}$ (d) $\frac{\hat{i}-\hat{j}}{\sqrt{2}}$

3 n identical point charges are kept symmetrically on the periphery of the circle $x^2+y^2=R^2$ in XY-plane. The resultant electric field at (0, 0, R) is E_1 and at (0, 0, 2R) is E_2, the ratio of E_1/E_2 is

(a) $\frac{5\sqrt{5}}{4\sqrt{2}}$ (b) $\frac{5}{2}$ (c) 2 (d) $\frac{5\sqrt{5}}{2\sqrt{2}}$

4 A particle is uncharged and is thrown vertically upward from ground level with a speed of $5\sqrt{5}$ m/s in a region of space having uniform electric field. As a result, it attains a maximum height h. The particle is then given a positive charge $+q$ and reaches the same maximum height h when thrown vertically upward with a speed of 13 m/s. Finally, the particle is given a negative charge $-q$. Ignoring air resistance, determine the speed with which the negatively charged particle must be thrown vertically upward, so that it attains exactly the same maximum height h.

(a) 6 m/s (b) 8 m/s (c) 9 m/s (d) 10 m/s

5 Choose the incorrect statement.

(a) The electric field due to a discrete charge configuration is not defined at the locations of the discrete charges.

(b) For continuous volume charge distribution, it is defined at any point in the distribution.

(c) For a surface charge distribution, electric field is discontinuous across the surface.

(d) For a finite continuous volume charge distribution, electric field is discontinuous at the boundary.

6 In a horizontal uniform electric field, a small charged block of charge q is gently released on the top of a fixed sperical dome. The block slides down the dome without friction and breaks away from the surface of the dome at the angular position $\theta = \sin^{-1}\left(\frac{3}{5}\right)$ from vertical. The ratio of the force of gravity acting on the block to force of its interaction with the field is

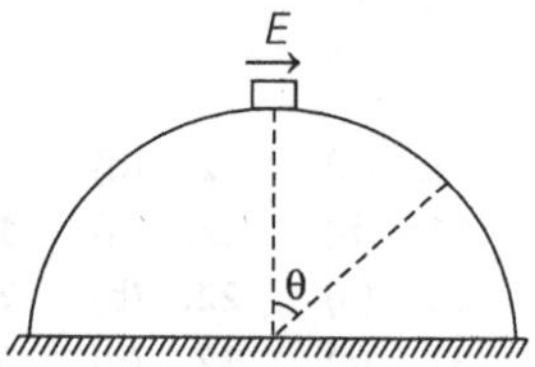

(a) 2.5 (b) 4 (c) 4.5 (d) 5

7 Two charged particles A and B are having charge -2 mC and 4 mC, and mass 2 kg and 1 kg, respectively. Both charges are released from rest in uniform electric field of strength 10 kN/C. Initially, A is at origin and B is at point (100, 100). At $t = 3$ sec, if B at a point (50, 70), then find the location of the point, where A would be at this time. (Assuming there is no gravity)

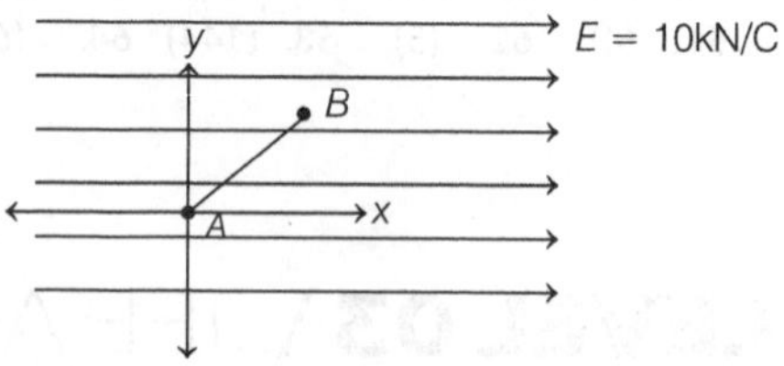

(a) (160, 30) (b) (90, 30) (c) (160, 15) (d) (70, 15)

8 A ring carries a uniform linear charge density on one half and the linear charge density of same magnitude but opposite sign on the other half.

(a) The component of electric field along the axis is zero at all points on the axis

(b) The component of electric field along the axis is zero only at the centre

(c) The resultant electric field at the centre is zero

(d) The resultant electric field at all points on the axis is zero except the centre

9 A non-conducting disc of mass = 2 kg, total charge = + 1C uniformly distributed, is placed on a rough horizontal non-conducting surface with its cross-section in vertical plane as shown. A uniform horizontal electric field E is now switched ON. The maximum magnitude of electric field E, so that the disc rolls purely is (Take, $g = 10 \text{ ms}^{-2}$)

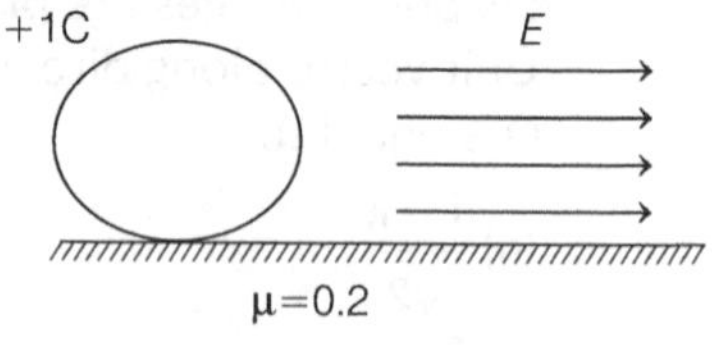

(a) 10 N/C (b) 12 N/C (c) 15 N/C (d) 20 N/C

10 A thin disc of radius $b = 2a$ has a concentric hole of radius a in it (see figure). It carries uniform surface charge σ on it. The electric field on its axis at height h ($h << a$) from its centre is given as

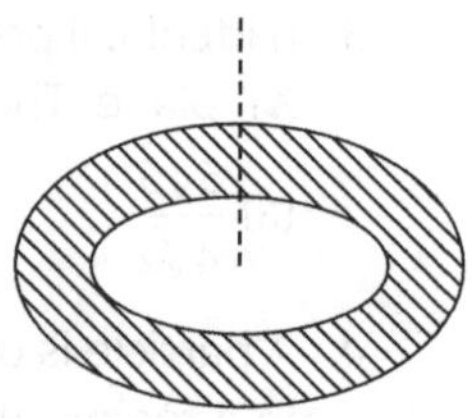

(a) $\frac{\sigma h}{4a\varepsilon_0}$ (b) $\frac{\sigma h}{2a\varepsilon_0}$

(c) $\frac{\sigma h}{8a\varepsilon_0}$ (d) $\frac{\sigma h}{a\varepsilon_0}$

11 A thin long strip whose cross-section is a semicircle carries a uniform surface charge of density σ on its inner surface. Find the electric field at a point O located midway on its axis.

(a) $\frac{\sigma}{2\pi\varepsilon_0}$ (b) $\frac{\sigma}{\pi\varepsilon_0}$ (c) $\frac{\sigma}{2\varepsilon_0}$ (d) $\frac{\sigma}{\varepsilon_0}$

12 Two small equally charged spheres, each of mass m, are suspended from the same point by silk threads of length l. The distance between the spheres $x << l$. Find the rate $\frac{dq}{dt}$ with which the charge leaks off each sphere, if they approach each other with velocity varying as $v = \frac{a}{\sqrt{x}}$, where a is a constant.

(a) $\frac{3a}{2}\sqrt{\frac{2\pi\varepsilon_0 mg}{l}}$ (b) $\frac{2a}{3}\sqrt{\frac{2\pi\varepsilon_0 mg}{l}}$

(c) $\frac{3a}{2}\sqrt{\frac{4\pi\varepsilon_0 mg}{l}}$ (d) $\frac{2a}{3}\sqrt{\frac{4\pi\varepsilon_0 mg}{l}}$

13 Two small equally charged identical conducting balls are suspended from long threads secured at one point. The charges and masses of the balls are such that they are in equilibrium when the distance between them is 10 cm (the length of the threads > > 10 cm.) One of the balls is then discharged. What will be the distance between the balls when equilibrium is restored?

(a) $10(1/2)^{1/3}$ cm (b) $10(1/2)^{3/2}$ cm

(c) $10(1/4)^{1/3}$ cm (d) $10(1/4)^{2/3}$ cm

14 A fixed non-conducting smooth track is in the shape of a quarter circle of radius R in vertical plane. A small metal ball A is fixed at the bottom of the track. Another identical ball B, which is free to move, is placed in contact with ball A. A charge Q is given to ball A which gets equally shared by the two balls. Ball B gets repelled and ultimately comes to rest in its equilibrium position, where its radius vector makes an angle $\theta = 60°$ with vertical. Mass of ball B is m. Find charge Q that was given to the balls.

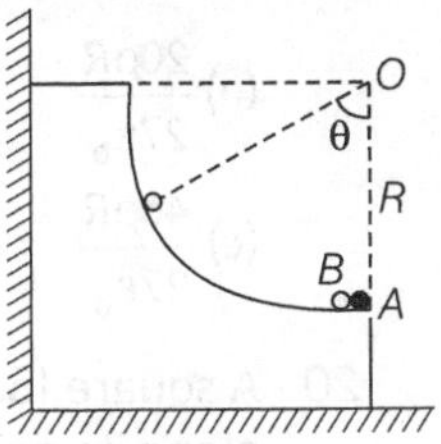

(a) $R\sqrt{\pi\varepsilon_0 mg}$ (b) $2R\sqrt{\pi\varepsilon_0 mg}$

(c) $3R\sqrt{\pi\varepsilon_0 mg}$ (d) $4R\sqrt{\pi\varepsilon_0 mg}$

15 A charged particle enters at point A and comes out from B. Its velocity makes angle α and β with electric field at these two points. Ratio of kinetic energy of charged particle at these two points will be

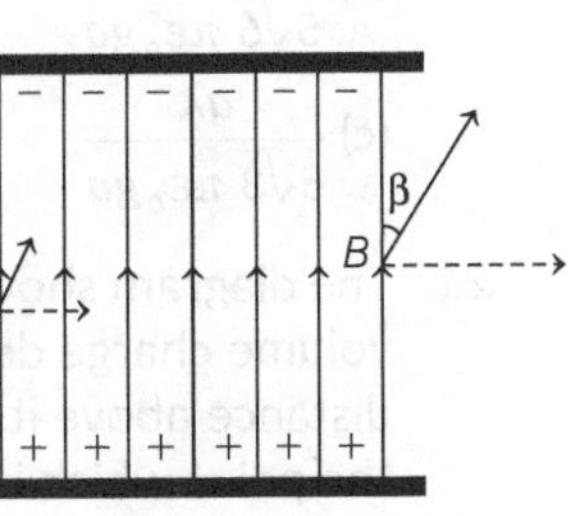

(a) $\frac{\sin^2\alpha}{\sin^2\beta}$ (b) $\frac{\sin^2\beta}{\sin^2\alpha}$

(c) $\frac{\cos^2\alpha}{\cos^2\beta}$ (d) $\frac{\cos^2\beta}{\cos^2\alpha}$

16 Two non-conducting infinite wires have shapes as shown in the figure. The linear charge densities of wire (1) and wire (2) are

$$\begin{cases} x, & \text{if } x \le -a \text{ and } x \ge a \\ 0, & \text{if } -a < x < a \end{cases}$$

$$\begin{cases} y, & \text{if } y \le -a \text{ and } y \ge a \\ 0, & \text{if } -a < y < a \end{cases}$$

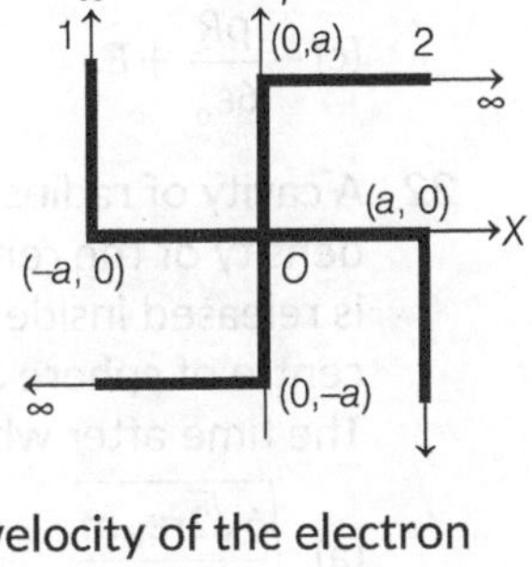

respectively, where (x, y) are coordinates of point on the wire. An electron is released from origin, the unit vector along the direction of velocity of the electron just after its release will be

(a) $\cos 0°\hat{i} + \sin 0°\hat{j}$ (b) $\cos 45°\hat{i} + \sin 45°\hat{j}$

(c) $-\cos 0°\hat{i} + \sin 0°\hat{j}$ (d) $\cos 45°\hat{i} - \sin 45°\hat{j}$

17 Find the force experienced by the semicircular wire charged with a charge q, placed as shown in figure. Radius of the wire is R and the long line of charge with linear charge density λ is passing through its centre and perpendicular to the plane of wire.

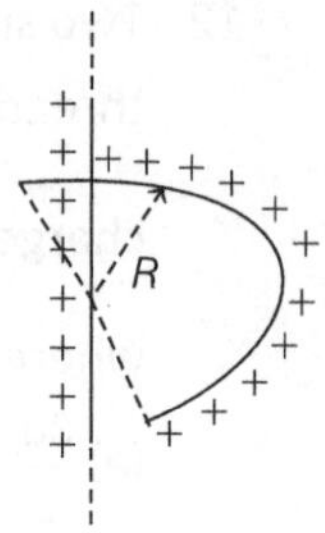

(a) $\frac{\lambda q}{2\pi^2 \varepsilon_0 R}$ (b) $\frac{\lambda q}{\pi^2 \varepsilon_0 R}$

(c) $\frac{\lambda q}{4\pi^2 \varepsilon_0 R}$ (d) $\frac{\lambda q}{4\pi \varepsilon_0 R}$

18 A thin conducting ring of radius R has an electric charge $+Q$. If a point charge q is placed at the centre of the ring, the increment in tension in the ring will be

(a) $\frac{Qq}{8\pi\varepsilon_0 R^2}$ (b) $\frac{Qq}{4\pi\varepsilon_0 R^2}$

(c) $\frac{Qq}{8\pi^2 \varepsilon_0 R^2}$ (d) $\frac{Qq}{4\pi^2 \varepsilon_0 R^2}$

19 A spherical region of radius $3R$ have three different volume charge density $+\rho$, -2ρ and $+3\rho$ as shown. Electric field at point A is

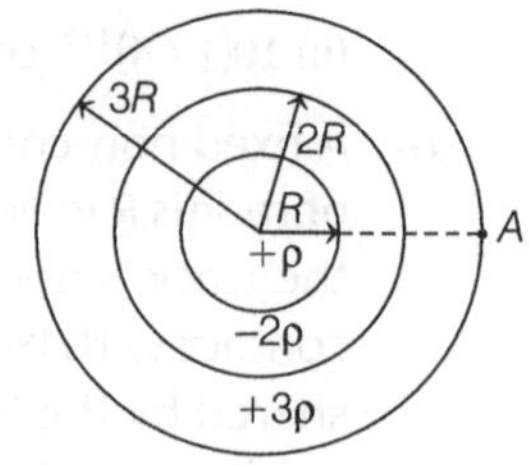

(a) $\frac{20\rho R}{27\varepsilon_0}$ (b) $\frac{25\rho R}{27\varepsilon_0}$

(c) $\frac{40\rho R}{27\varepsilon_0}$ (d) $\frac{44\rho R}{27\varepsilon_0}$

20 A square loop of side length a has a uniform linear charge density λ along its periphery. The loop lies in a horizontal plane. A particle having charge q is placed exactly above the centre of the square at a height a. The particle stays in equilibrium. Its mass is

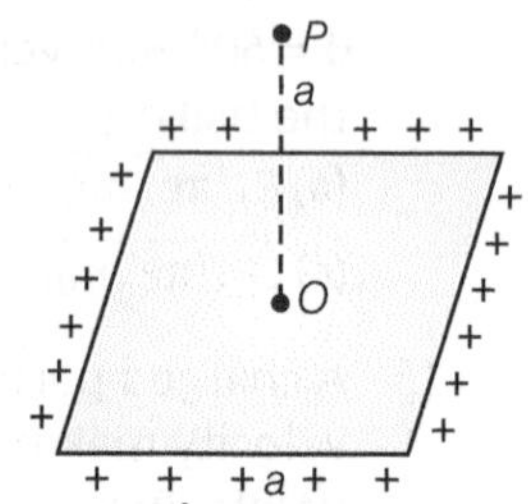

(a) $\frac{8q\lambda}{5\sqrt{6}\,\pi\varepsilon_0 ga}$ (b) $\frac{2q\lambda}{5\sqrt{3}\,\pi\varepsilon_0 ga}$

(c) $\frac{q\lambda}{5\sqrt{3}\,\pi\varepsilon_0 ga}$ (d) $\frac{2q\lambda}{5\sqrt{6}\,\pi\varepsilon_0 ga}$

21 The diagram shows a uniformly charged hemisphere of radius R. It has volume charge density ρ. If the magnitude of electric field at a point $2R$ distance above its centre is E, then what is the magnitude of electric field at the point which is $2R$ below its centre?

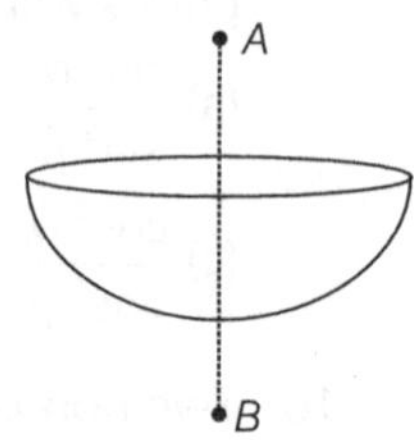

(a) $\frac{\rho R}{6\varepsilon_0} + E$ (b) $\frac{\rho R}{12\varepsilon_0} - E$

(c) $-\frac{\rho R}{6\varepsilon_0} + E$ (d) $\frac{\rho R}{24\varepsilon_0} + E$

22 A cavity of radius r is made inside a solid sphere. The volume charge density of the remaining sphere is ρ. An electron (charge e, mass m) is released inside the cavity from point P as shown in figure. The centre of sphere and centre of cavity are separated by a distance a. The time after which the electron again touches the sphere is

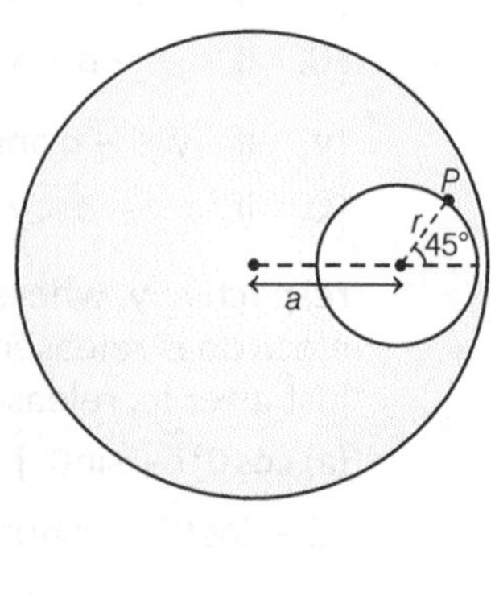

(a) $\sqrt{\frac{6\sqrt{2}r\varepsilon_0 m}{e\rho a}}$ (b) $\sqrt{\frac{\sqrt{2}r\varepsilon_0 m}{e\rho a}}$

(c) $\sqrt{\frac{6r\varepsilon_0 m}{e\rho a}}$ (d) $\sqrt{\frac{r\varepsilon_0 m}{e\rho a}}$

23 Three quarter circles (each of radius R) made of insulating material are uniformly charged with linear charge densities $-\lambda$, $+\lambda$ and $-\lambda$ and arranged in space as shown. Then, net electric field at centre is

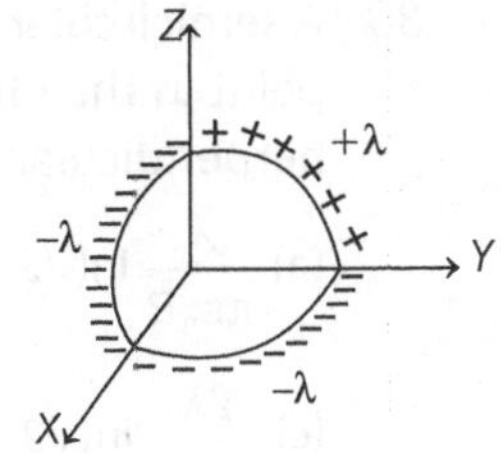

(a) $\frac{\lambda}{2\pi\varepsilon_0 R}\hat{\mathbf{i}}$
(b) $\frac{\lambda}{2\pi\varepsilon_0 R}\hat{\mathbf{j}}$
(c) $\frac{\lambda}{2\pi\varepsilon_0 R}\hat{\mathbf{k}}$
(d) $\frac{\lambda}{2\pi\varepsilon_0 R}(\hat{\mathbf{j}}+\hat{\mathbf{k}})$

24 A hemispherical bowl of inner radii R and outer radii $2R$ is charged with volume charge density $\rho = \rho_0 \frac{r}{R}$, where r is the distance from the centre. The electric field at point O is

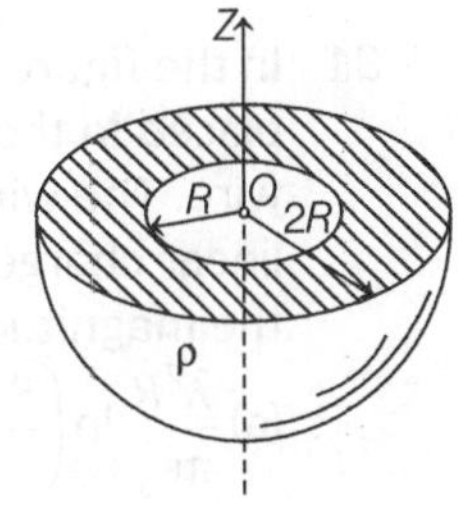

(a) $\frac{2\rho_0 R}{3\varepsilon_0}$
(b) $\frac{3\rho_0 R}{4\varepsilon_0}$
(c) $\frac{\rho_0 R}{2\varepsilon_0}$
(d) $\frac{3\rho_0 R}{8\varepsilon_0}$

25 A square of side b centred at the origin with sides parallel to axes of X and Y has surface charge density $\sigma(x, y) = \sigma_0\, xy$ within its boundaries. Total charge on the square is
(a) zero (b) $\sigma_0 b^2$ (c) $2\sigma_0 b^2$ (d) $4\sigma_0 b^2$

26 A charged particle having some mass is resting in equilibrium at a height H above the centre of a uniformly charged non-conducting horizontal ring of radius R. The force of gravity acts downwards. The equilibrium of the particle will be stable
(a) for all values of H (b) only if $H > \frac{R}{\sqrt{2}}$ (c) only if $H < \frac{R}{\sqrt{2}}$ (d) only if $H = \frac{R}{\sqrt{2}}$

27 The electric field intensity at the centre of a uniformly charged hemispherical shell is E_0. Now, two portions of the hemisphere are cut from either side and remaining portion is shown in figure. If $\alpha = \beta = \frac{\pi}{3}$, then electric field intensity at centre due to remaining portion is

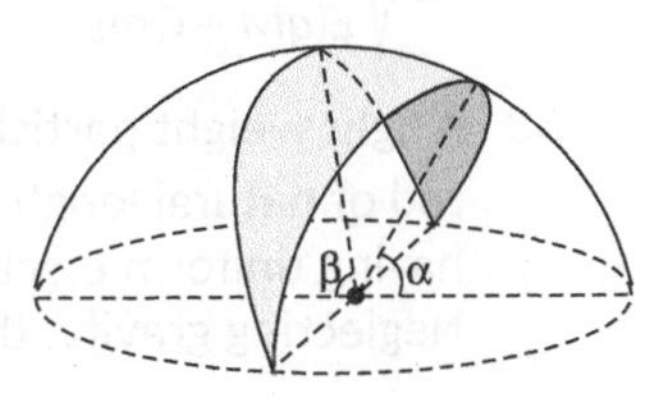

(a) $\frac{E_0}{3}$
(b) $\frac{E_0}{6}$
(c) $\frac{E_0}{2}$
(d) Information insufficient

28 Two infinitely long wires having linear charge densities λ_1 and λ_2 are placed perpendicular to one another as shown in figure. Wire 1 is in the plane of the paper and wire 2 is perpendicular to this plane. The symbol ⊙ indicates that the wire is going into the plane of the paper. Separation between the two wires is r. Find the force of interaction between them.

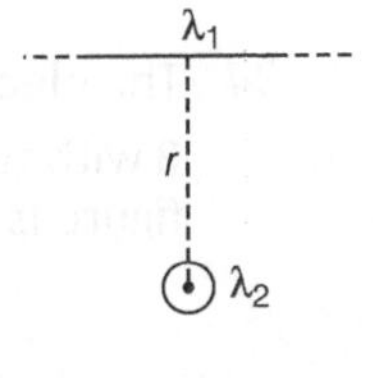

(a) $\frac{\lambda_1\lambda_2}{\varepsilon_0}$
(b) $\frac{\lambda_1\lambda_2}{2\varepsilon_0}$
(c) $\frac{\lambda_1\lambda_2}{2\pi\varepsilon_0}$
(d) $\frac{\lambda_1\lambda_2}{4\pi\varepsilon_0}$

29 A charged particle q of mass m is in equilibrium at a height h from a horizontal infinite line charge with uniform linear charge density λ. The charge lies in the vertical plane containing the line charge. The time period of small oscillations along vertical about its equilibrium position is
(a) $2\pi\sqrt{\frac{mh^2}{2Kq\lambda}}$ (b) $2\pi\sqrt{\frac{mh^2}{Kq\lambda}}$ (c) $2\pi\sqrt{\frac{2mh^2}{Kq\lambda}}$ (d) $\pi\sqrt{\frac{mh^2}{Kq\lambda}}$

30 A semicircular ring of radius R carries a uniform linear charge of λ. P is a point in the plane of the ring at a distance R from centre O. OP is perpendicular to AB. Find electric field intensity at point P.

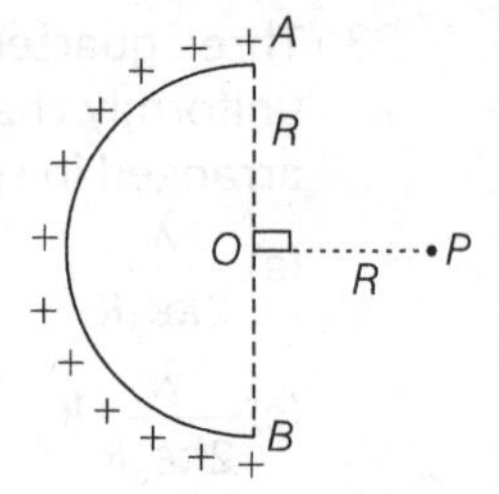

(a) $\dfrac{\lambda}{\pi\varepsilon_0 R}\ln(\sqrt{2}+1)$ (b) $\dfrac{\lambda}{2\pi\varepsilon_0 R}\ln(\sqrt{2}+1)$

(c) $\dfrac{2\lambda}{\pi\varepsilon_0 R}\ln(\sqrt{2}+1)$ (d) $\dfrac{\lambda}{4\pi\varepsilon_0 R}\ln(\sqrt{2}+1)$

31 In the figure shown, a very long wire and a semicircular ring of radius R are placed in the same plane. The centre of the ring is at a distance r from the wire. The wire has uniformly distributed line charge density λ and the ring has linear charge density $+\lambda$ on one half and $-\lambda$ on the other half as shown. Find the magnitude of net torque on the ring due to the wire.

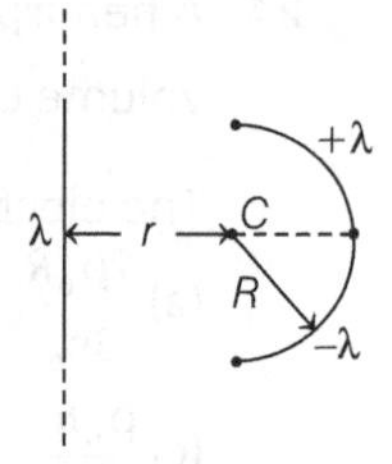

(a) $\dfrac{\lambda^2 R}{\pi\varepsilon_0}\ln\left(\dfrac{R+r}{r}\right)$ (b) $\dfrac{\lambda^2 R}{2\pi\varepsilon_0}\ln\left(\dfrac{R+r}{r}\right)$

(c) $\dfrac{\lambda^2 R}{\pi\varepsilon_0}\ln\left(\dfrac{R}{r}\right)$ (d) $\dfrac{\lambda^2 R}{2\pi\varepsilon_0}\ln\left(\dfrac{R}{r}\right)$

32 The charge and mass of two particles are $+Q$, M and $-q$, m, respectively. The particles separated by a distance L, are released from rest in a uniform electric field E. The electric field is parallel to line joining both the charges and is directed from negative to positive charge. For the separation between particles to remain constant, find the value of L. $\left(K=\dfrac{1}{4\pi\varepsilon_0}\right)$

(a) $\sqrt{\dfrac{(M+m)KQq}{E(qM+Qm)}}$ (b) $\sqrt{\dfrac{(M-m)KQq}{E(qM+Qm)}}$ (c) $\sqrt{\dfrac{(M+m)KQq^2}{E(qM-Qm)}}$ (d) $\sqrt{\dfrac{(M-m)KQ^2q}{E^2(qM-Qm)}}$

33 A light weight particle of charge Q is fixed at one end of an electrically insulated uniform elastic rod of natural length L, cross-sectional area A and Young's modulus Y. The rod is placed in space having uniform electric field of magnitude E and directed parallel to length of the rod as shown. Neglecting gravity, the magnitude of extension in this rod is

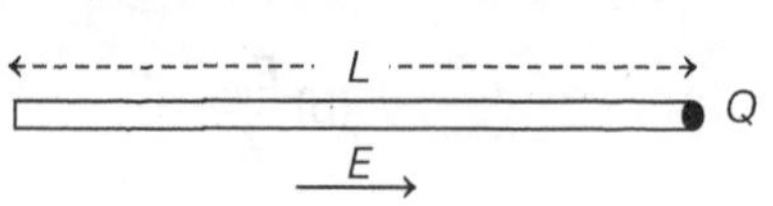

(a) 0 (b) $\dfrac{2QEL}{YA}$ (c) $\dfrac{QEL}{YA}$ (d) $\dfrac{QEL}{2YA}$

34 The electric field vector at point $P(a, a, a)$ due to three uniformly charged infinite wires 1, 2 and 3 with uniform charge density λ each kept along the x, y and z-axes respectively, as shown in figure is

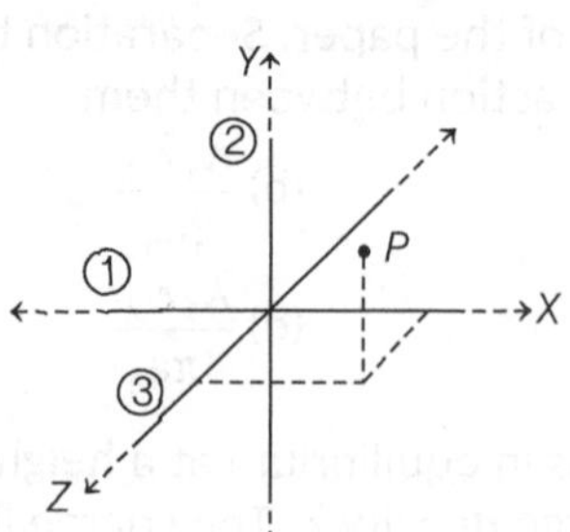

(a) $\dfrac{\lambda}{3\pi\varepsilon_0 a}(\hat{i}+\hat{j}+\hat{k})$ (b) $\dfrac{\lambda}{2\pi\varepsilon_0 a}(\hat{i}+\hat{j}+\hat{k})$ (c) $\dfrac{\lambda}{2\sqrt{2}\pi\varepsilon_0 a}(\hat{i}+\hat{j}+\hat{k})$ (d) $\dfrac{\sqrt{2}\lambda}{\pi\varepsilon_0 a}(\hat{i}+\hat{j}+\hat{k})$

35 A non-conducting infinite rod is placed along the Z-axis; the upper half of the rod (lying along $Z > 0$) is charged positively with a uniform linear charge density $+\lambda$, while the lower half ($Z < 0$) is charged negatively with a uniform linear charge density $-\lambda$. The origin is located at the junction of the positive and negative halves of the rod. A uniformly charged annular disc (surface charge density $= \sigma_0$) of inner radius R and outer radius $2R$ is placed in the XY-plane with its centre at the origin. The force on the rod due to the disc is

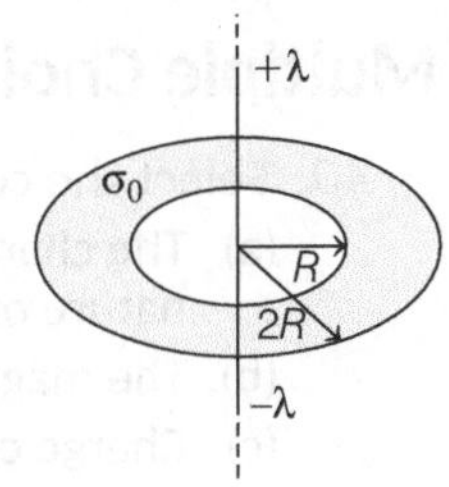

(a) $\dfrac{2\sigma_0\lambda R}{\varepsilon_0}$ (b) $\dfrac{\sigma_0\lambda R}{2\varepsilon_0}$ (c) $\dfrac{\sigma_0\lambda R}{\varepsilon_0}$ (d) $\dfrac{\sigma_0\lambda R}{3\varepsilon_0}$

36 A cylinder is uniformly charged on its curved surface with surface charge density σ. Radius of the cylinder is R. Calculate the electric field at point P on the axis of the cylinder, as shown in the figure.

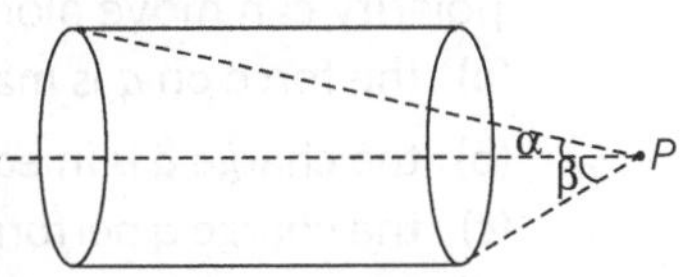

(a) $\dfrac{\sigma}{\varepsilon_0}(\sin\beta + \sin\alpha)$ (b) $\dfrac{\sigma}{\varepsilon_0}(\sin\beta - \sin\alpha)$

(c) $\dfrac{\sigma}{2\varepsilon_0}(\sin\beta + \sin\alpha)$ (d) $\dfrac{\sigma}{2\varepsilon_0}(\sin\beta - \sin\alpha)$

37 Three infinitely long, straight thin rods each carrying charge density $+\lambda$ and mass per unit length m are equally spaced as shown in figure. The middle rod is displaced through a very small distance x in the plane of figure while the other two rods held fixed. Find the frequency of small oscillations.

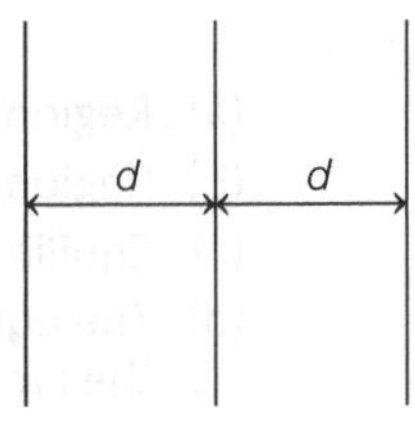

(a) $\dfrac{1}{2\pi}\sqrt{\dfrac{\lambda^2}{\pi\varepsilon_0 md^2}}$ (b) $\dfrac{1}{2\pi}\sqrt{\dfrac{\lambda^2}{2\pi\varepsilon_0 md^2}}$

(c) $\dfrac{1}{2\pi}\sqrt{\dfrac{2\lambda^2}{\pi\varepsilon_0 md^2}}$ (d) $\dfrac{1}{2\pi}\sqrt{\dfrac{\lambda^2}{4\pi\varepsilon_0 md^2}}$

38 Two identical charged particles of charge q are placed symmetrically on both the sides, on the axis of a uniformly charged ring of radius a with linear charge density λ as shown in the figure. The distance of the charge particle from centre of ring is l. Calculate tension in the ring due to presence to the charged particles.

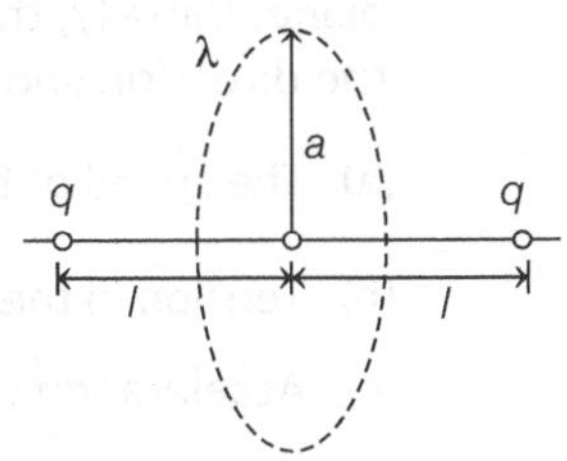

(a) $\dfrac{\lambda qa^2}{\pi\varepsilon_0(a^2+1^2)^{3/2}}$ (b) $\dfrac{\lambda qa^2}{2\pi\varepsilon_0(a^2+l^2)^{3/2}}$

(c) $\dfrac{2\lambda qa^2}{\pi\varepsilon_0(a^2+1^2)^{3/2}}$ (d) $\dfrac{\lambda qa^2}{4\pi\varepsilon_0(a^2+1^2)^{3/2}}$

39 In a free space, a thin rod carrying uniformly distributed negative charge $-q$ is placed symmetrically along the axis of a thin ring of radius R carrying uniformly distributed charge Q. The mass of the rod is m and length is $l = 2R$. The ring is fixed and the rod is free to move. The rod is displaced slightly along the axis of the ring and then released. Time period T of the small amplitude oscillations of the rod is

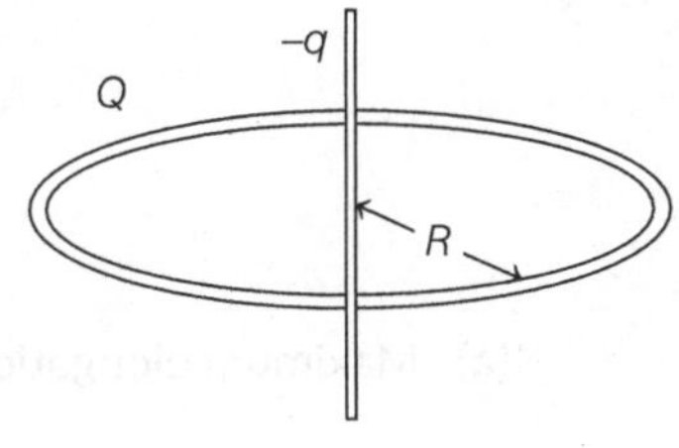

(a) $4\pi\left(\dfrac{\sqrt{2}\pi\varepsilon_0 R^3 m}{Qq}\right)^{1/2}$ (b) $4\pi\left(\dfrac{2\pi\varepsilon_0 R^3 m}{Qq}\right)^{1/2}$

(c) $4\pi\left(\dfrac{2\sqrt{2}\pi\varepsilon_0 R^3 m}{Qq}\right)^{1/2}$ (d) $4\pi\left(\dfrac{\pi\varepsilon_0 R^3 m}{Qq}\right)^{1/2}$

Multiple Choice(s) Correct Questions

40 Select the correct alternatives.

(a) The charge gained by the uncharged body from a charged body due to conduction is of same nature of charge initially present on charged body.

(b) The magnitude of charge increases with the increase in velocity of charge.

(c) Charge cannot exist without matter although matter can exist without charge.

(d) If three charges are in electrostatic equilibrium, then all three charges may be of same sign.

41 Two point charges (Q each) are placed at $(0, y)$ and $(0, -y)$. A point charge q of the same polarity can move along X-axis. Then,

(a) the force on q is maximum at $x = \pm y/\sqrt{2}$

(b) the charge q is in equilibrium at the origin

(c) the charge q performs an oscillatory motion about the origin

(d) for any position of q other than origin, the force is directed away from origin

42 The figure shows two point charges $q_1 = 2Q (> 0)$ and $q_2 = -Q$. The charges divide the line joining them in three parts I, II and III.

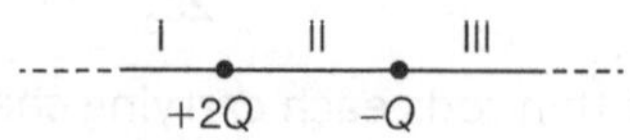

(a) Region III has a local maxima of electric field

(b) Region I has a local minima of electric field

(c) Equilibrium position for a test charge lies in region II

(d) The equilibrium (if there is any in region I,II or III) for constrained motion along the line joining the two charges is stable for a negative test charge

43 A particle of mass m and charge q is fastened to other end of a string fixed at point O. The whole system lies on a frictionless horizontal plane. Initially, the mass was at rest at A. A uniform electric field E in the direction shown is then switched ON.

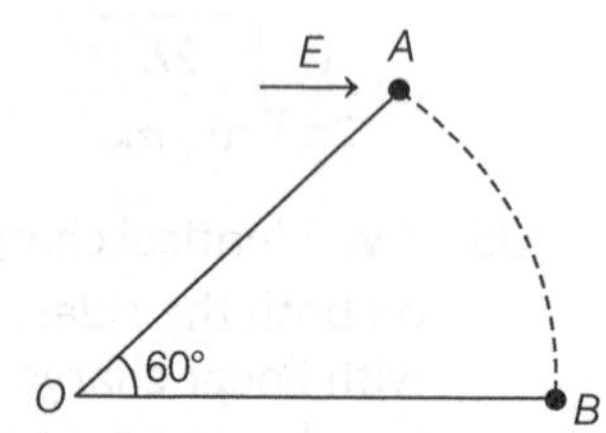

(a) The speed of the particle when it reaches point B is $\sqrt{\frac{qEl}{m}}$

(b) Tension in the string when the particle reaches point B is $2qE$

(c) Acceleration of the particle at point B is $\frac{qE}{m}$

(d) Angular momentum of particle about point O is conserved

44 In the given figure, initially an ideal sufficiently long spring is at its natural length when a constant electric field is switched ON. Choose correct option/options.

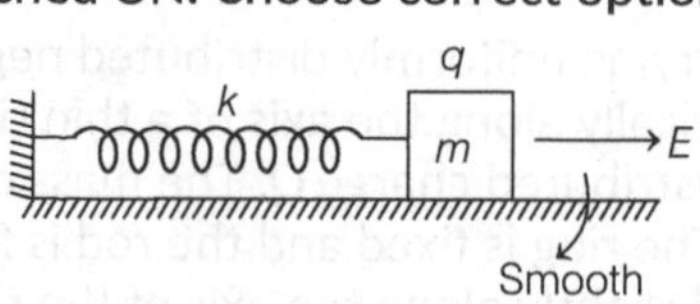

(a) Maximum elongation of spring is $\frac{2qE}{k}$

(b) Maximum compression of spring is $\frac{qE}{k}$

(c) Time period of SHM is $2\pi\sqrt{\frac{m}{k}}$

(d) It will perform SHM if and only if magnitude of field is small

45 A sphere of radius r carries a surface charge of density $\sigma = \mathbf{a} \cdot \mathbf{r}$, where $\mathbf{a}$ is a constant vector and $\mathbf{r}$ is the radius vector of a point on the sphere relative to its centre. Then,

(a) total charge on the sphere is zero

(b) if we cut the sphere in two part symmetrically by plane which is perpendicular to **a**, then each sphere will have charge of magnitude $\pi r^3 a$

(c) field at centre of the sphere $= \frac{r}{3\varepsilon_0}(-\mathbf{a})$

(d) field at centre of the sphere $= \frac{r}{2\varepsilon_0}(-\mathbf{a})$

46 An electron is placed just in the middle between two long fixed line charges of charge density $+\lambda$ each. The wires are in the XY-plane (Do not consider gravity),

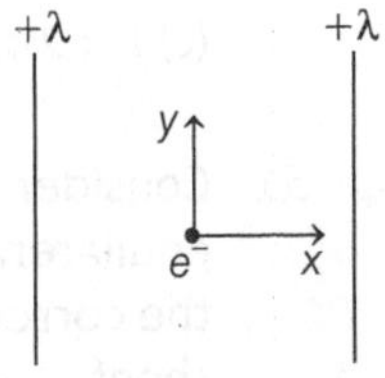

(a) the equilibrium of the electron will be unstable along x-direction

(b) the equilibrium of the electron will be neutral along y-direction

(c) the equilibrium of the electron will be stable along z-direction

(d) the equilibrium of the electron will be stable along y-direction

47 A particle of mass 2 kg and charge 1 mC is projected vertically with a velocity 10 ms^{-1}. There is a uniform horizontal electric field of 10^4 N/C,

(a) the horizontal range of the particle is 10 m

(b) the time of flight of the particle is 2 s

(c) the maximum height reached is 5 m

(d) the horizontal range of the particle is 0

48 An oil drop has a charge -9.6×10^{-19} C and has a mass 1.6×10^{-15} g. When allowed to fall, due to air resistance force, it attains a constant velocity. Then if a uniform electric field is to be applied vertically to make the oil drop ascend up with the same constant speed, which of the following are correct? ($g = 10 \text{ ms}^{-2}$)

(Assume that the magnitude of resistance force is same in both the cases)

(a) The electric field is directed upward

(b) The electric field is directed downward

(c) The intensity of electric field is $\frac{1}{3} \times 10^2 \text{ NC}^{-1}$

(d) The intensity of electric field is $\frac{1}{6} \times 10^5 \text{ NC}^{-1}$

49 There is two concentric and coplanar ring of radius a and $3a$ with charge $-Q$ and $8Q$ uniformly distributed, fixed in YZ-plane with centre at origin. A particle of small positive charge q and mass m is at origin. [Neglect gravity]

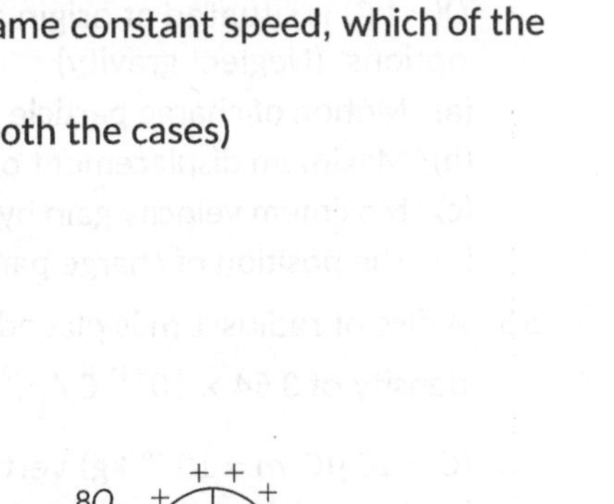

(a) If the particle is given a small push along X-axis, then it perform oscillatory motion of time period

$$T = 12\sqrt{\frac{3\pi^3 a^3 m\varepsilon_0}{19Qq}}$$

(b) If the particle is given a small push along X-axis, then it perform oscillatory motion of time period $T = 12\sqrt{\frac{7\pi^3 a^3 m\varepsilon_0}{19Qq}}$

(c) If the particle is given a velocity along X-axis and it performs oscillatory motion, then amplitude of oscillation must be less than $\sqrt{\frac{7}{3}}\, a$

(d) If the particle is given a velocity along X-axis and it performs oscillatory motion, then amplitude of oscillation must be less than $\sqrt{\frac{5}{3}}a$

50 A simple pendulum of length l has a bob of mass m, with a charge q on it. A vertical sheet of charge, with charge σ per unit area, passes through the point of suspension of the pendulum. At equilibrium, the string makes an angle θ with the vertical. Its time period of oscillation is T in this position

(a) $\tan\theta = \frac{\sigma q}{2\varepsilon_0 mg}$

(b) $\tan\theta = \frac{\sigma q}{\varepsilon_0 mg}$

(c) $T < 2\pi\sqrt{\frac{l}{g}}$

(d) $T > 2\pi\sqrt{\frac{l}{g}}$

51 Consider a uniformly charged sheet $ABCD$, which is a part of an equilateral triangular sheet of a side a as shown in the figure. Choose the correct options regarding the electric field E at point O due to this sheet.

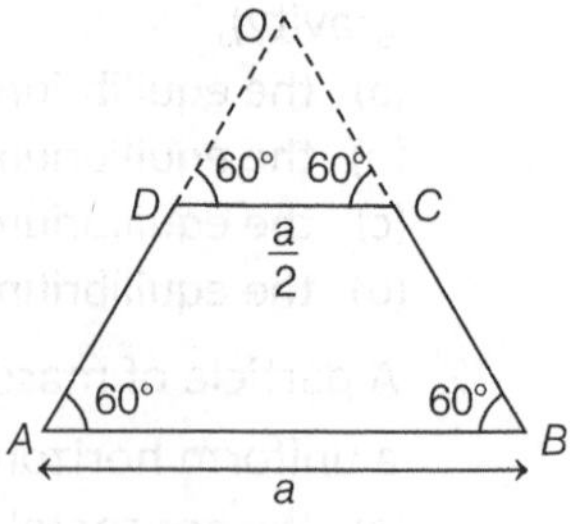

(a) Magnitude of **E** increases with the increase in a (keeping charge density same)

(b) Magnitude of **E** decreases with increase in a (keeping total charge same)

(c) If charge density is σ and $a = 1$ m, magnitude of E is equal to $\frac{7\sigma}{44\varepsilon_0}\ln\sqrt{2}$

(d) If charge density is σ and $a = 2$ m, magnitude of E is equal to $\frac{7\sigma}{22\varepsilon_0}\ln\sqrt{2}$

52 Electric field in a region is given as $\mathbf{E} = (10 - 5x)\hat{\mathbf{i}}$. A charge particle of mass 5 kg and charge $Q(= 1\,\text{C})$ is situated at origin and free to move in given electric field. Then, choose the correct options. (Neglect gravity)

(a) Motion of charge particle is oscillatory

(b) Maximum displacement of charge particle from origin is 4 SI units

(c) Maximum velocity gain by charge particle is 2 SI units

(d) The position of charge particle when velocity gained by particle is maximum, is 2 SI units

53 A disc of radius 1 m is placed on the horizontal ground. It is uniformly charged with charge density of $3.54 \times 10^{-11}\,\text{C/m}^2$. At a distance of $\frac{4}{3}$ m from the centre, we place a charge ($Q = 10\,\mu\text{C}$, $m = 10^{-6}$ kg) vertically above the centre. When we release it,

(a) it's initial acceleration is $3\,\text{m/s}^2$ upwards

(b) it's initial acceleration is $6\,\text{m/s}^2$ downwards

(c) it's in equilibrium at a height of $\frac{5}{\sqrt{3}}$ m from the centre

(d) it's in equilibrium at a height of $\frac{1}{\sqrt{3}}$ m from the centre

54 A particle of mass m and charge q is projected in a region, where an electric field is existing and given by $\mathbf{E} = E_0\hat{\mathbf{i}}$, with a velocity $v_0\hat{\mathbf{j}}$ from the origin at time $t = 0$, then choose the correct statements (assuming $m^2v_0^2 = 2qE_0mx_0$). [Neglect gravitation force]

(a) Radius of curvature of the particle when its x-coordinate becomes x_0 is $2x_0$.

(b) Radius of curvature of the particle when its x-coordinate becomes x_0 is $4\sqrt{2}\,x_0$

(c) Speed of the particle when its x-coordinate becomes x_0 is $\sqrt{2}v_0$.

(d) Speed of the particle when its x-coordinate becomes x_0 is v_0.

55 A charged particle of mass $m = 1$ kg and charge $Q = 1$ C, is projected from (0, 10) with velocity **v**. Uniform electric field $(-10\hat{\mathbf{i}} + 5\hat{\mathbf{j}})$ N/C exists in the region. Gravity acts along negative y- direction. (Take, $g = 10\ \text{m/s}^2$). For $\mathbf{v} = 10\hat{\mathbf{i}}$,

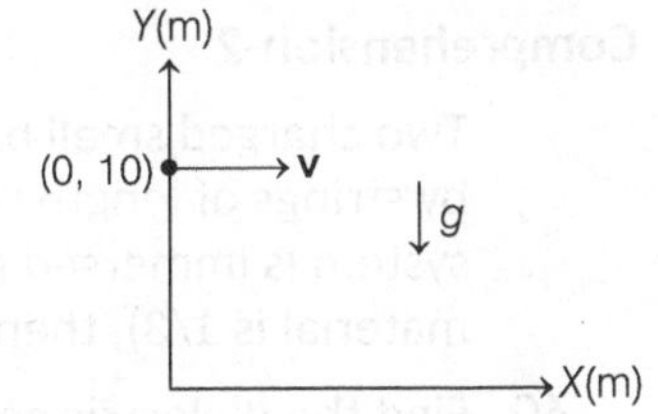

(a) particle passes through origin at some time
(b) time spent by charged particle in the first quadrant is 2 s
(c) while crossing origin, the trajectory makes equal angles with both X and Y-axes
(d) kinetic energy of particle at (0, 10) will be half the kinetic energy at origin

56 In a uniformly charged dielectric sphere, a very thin tunnel has been made along the diameter as shown in the figure below. A charge particle $-q$ having mass m is released from rest at one end of tunnel. For the situation described, mark out the correct statement (s). [Neglect gravity]

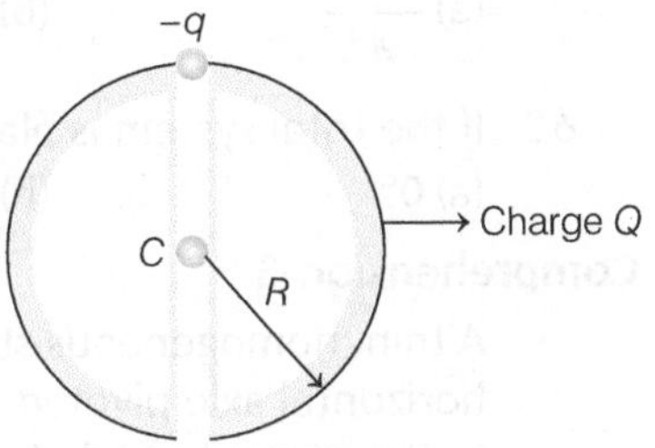

(a) Charged particle will perform SHM about centre of the sphere as mean position
(b) Time period of the particle is $2\pi\sqrt{\dfrac{4\pi\varepsilon_0 mR^3}{qQ}}$
(c) Particle will perform oscillation but not SHM
(d) Speed of the particle while crossing-mean position is $\sqrt{\dfrac{Qq}{4\pi\varepsilon_0 mR}}$

Comprehension Based Questions

Comprehension-1

Two point charges Q_a and Q_b are positioned at points A and B. The field strength to the right of charge Q_b on line that passes through two charges varies according to a law that is represented schematically in the figure. The field strength is assumed to be positive, if its direction coincides with the positive direction of X-axis.

The distance between the charges is $l = 21$ cm.

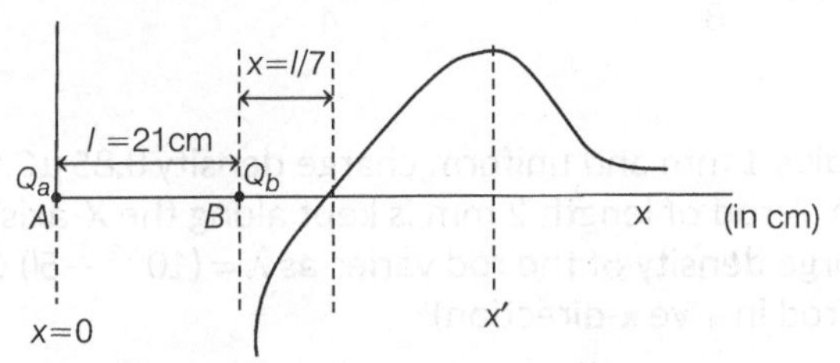

57 The signs of charges Q_a and Q_b are

(a) both positive
(b) both negative
(c) positive and negative, respectively
(d) negative and positive, respectively

58 $\dfrac{|Q_a|}{|Q_b|}$ is

(a) 8
(b) 16
(c) 32
(d) 64

59 Find the coordinate x of the point, where the field strength is maximum.

(a) 25
(b) 28
(c) 30
(d) 36

Comprehension-2

Two charged small balls each of charge q and mass m when suspended from a common point by strings of length l in air, then these strings form an angle of 120° with each other. When the system is immersed in a liquid of dielectric constant K (relative density of liquid w.r.t. ball's material is 1/3), then the angle between the strings remains same.

60 Find the dielectric constant of the liquid.

(a) 3/2 (b) $\sqrt{3/2}$ (c) $\sqrt{3}$ (d) 3

61 Find the electrostatic force exerted by the liquid on one of the balls.

(a) $\frac{4Kq^2}{9l^2}$ (b) $\frac{Kq^2}{2l^2}$ (c) $\frac{Kq^2}{3l^2}$ (d) $\frac{Kq^2}{9l^2}$

62 If the total system is placed in a gravitational free space, then angle between the strings will be

(a) 0° (b) 90° (c) 180° (d) remain unchanged

Comprehension-3

A thin, homogeneous stick of mass m and length L may rotate in the vertical plane around a horizontal axle pivoted at one end of the stick. A small ball of mass m and charge Q is attached to the opposite end of this stick. The whole system is positioned in a constant horizontal electric field of magnitude $E = \frac{mg}{2Q}$. The stick is held horizontally at the beginning.

63 What is the acceleration of the small ball at the instance of releasing the stick?

(a) $\frac{3g}{2}$ (b) $\frac{3g}{4}$ (c) $\frac{9g}{8}$ (d) None of these

64 What is the speed of ball when rod becomes vertical?

(a) $\sqrt{\frac{3gl}{2}}$ (b) $\sqrt{2gl}$ (c) $\sqrt{\frac{3gl}{4}}$ (d) None of these

65 Magnitude of acceleration of the end of the stick, when it swings through the vertical position, is

(a) $\frac{3\sqrt{5}g}{2}$ (b) $\frac{3\sqrt{17}g}{8}$ (c) $\frac{3\sqrt{2}g}{4}$ (d) None of these

Comprehension-4

Consider a disc of radius 1 mm and uniform charge density 8.85 $\mu C/m^2$ kept at origin with its axis along x-direction. A rod of length 2 mm is kept along the X-axis at a distance 3m from origin. The linear charge density of the rod varies as $\lambda = (10^{-2} - 5l)$ C / m (l measured (in m) from one end of the rod in +ve x-direction)

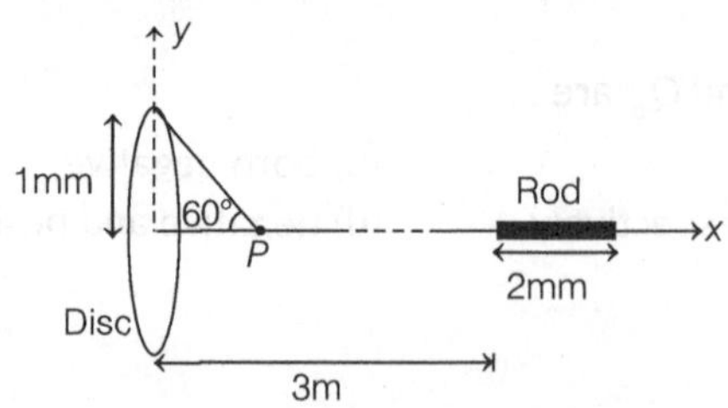

66 The magnitude of electric field at point P is

(a) 25×10^4 N/C (b) 10^4 N/C (c) 24×10^4 N/C (d) 10^5 N/C

67 Find the force experienced by rod due to the disc.

(a) $\frac{25}{9} \times 10^{-7}$ N (b) $\frac{9}{25} \times 10^{-7}$ N (c) $\frac{9}{16} \times 10^{-7}$ N (d) $\frac{16}{9} \times 10^{-7}$ N

Comprehension-5

There is an insulator rod of length L and of negligible mass with two small balls of mass m and electric charge Q attached to its ends. The rod can rotate in the horizontal plane around a vertical axis crossing it at a $L/4$ distance from one of its ends.

68 At first, the rod is in unstable equilibrium in a horizontal uniform electric field of field strength E. Then, we gently displace it from this position. Determine the maximum velocity attained by the ball which is closer to the axis in the subsequent motion.

(a) $\sqrt{\frac{2QEL}{m}}$ (b) $\sqrt{\frac{2QEL}{5m}}$

(c) $\sqrt{\frac{QEL}{5m}}$ (d) $\sqrt{\frac{4QEL}{5m}}$

69 In what position is the rod to be set so that, if displaced a little from that position it begins a harmonic oscillation about the axis A?

(a) 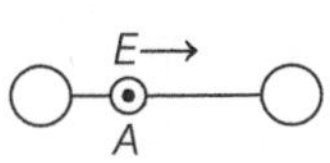

(b) 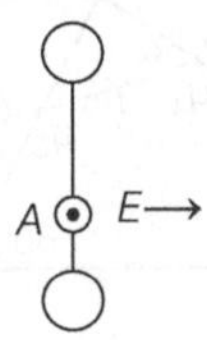

(c) 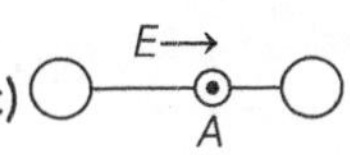

(d) 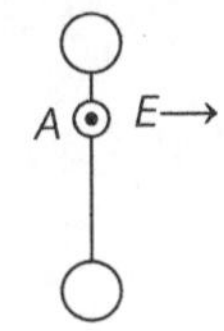

70 What is the time period of the SHM as mentioned in above question?

(a) $2\pi\sqrt{\frac{mL}{QE}}$ (b) $2\pi\sqrt{\frac{2mL}{3QE}}$ (c) $2\pi\sqrt{\frac{5mL}{QE}}$ (d) $2\pi\sqrt{\frac{5mL}{4QE}}$

Comprehension-6

You may wonder why the notion of electric field has been introduced. After all, for any system of charges, the measurable quantity is the force on a charge which can be directly determined using Coulomb's law and the superposition principle. Why then introduce this intermediate quantity called the electric field?

The true physical significance of the concept of electric field, however emerges only, when we go beyond electrostatics. Suppose we consider the force between two distant charges q_1 and q_2 in accelerated motion. Now, the greatest speed with which a signal or information can go from one point to another is c, the speed of light. Thus, the effect of any motion of q_1 on q_2 cannot arise instantaneously.

There will be some time delay between the effect (force on q_2) and the cause (motion of q_1). It is precisely here that the notion of electric field (strictly, electromagnetic field) is natural and very useful.

The field picture is this: the accelerated motion of charge q_1 produces electromagnetic waves, which then propagate with the speed c, reach q_2 and cause a force on q_2. The notion of field elegantly accounts for the time delay. Thus, even though electric and magnetic fields can be detected only by their effects (forces) on charges, they are regarded as physical entities, not merely mathematical entities.

71 There are four charge particles A, B, C and D having charge $1\mu C$, $2\mu C$, $-5\mu C$ and $8\mu C$, respectively are shown below kept at rest. Now, A is moved with velocity $v (<< c)$, then choose the correct statement.

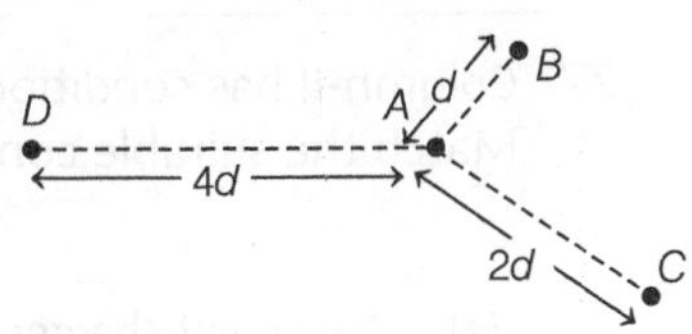

(a) Effect of motion of A is observed by B before C and D
(b) Effect of motion of A is observed by C before B and D
(c) Effect of motion of A is observed by D before B and C
(d) Effect of motion of A is observed by all the charge particles simultaneously

72 Two free particles having equal charge and mass m_1 and m_2 are released at $t = 0$. Initially $v_1 = 0$ and $v_2 = 2$ m/s, then

(a) After few instants, magnitude of $(m_1 v_1 + m_2 v_2)$ is more than $2m_2$
(b) After few instants, magnitude of $(m_1 v_1 + m_2 v_2)$ is less than $2m_2$
(c) After few instants, magnitude of $(m_1 v_1 + m_2 v_2)$ is equal to $2m_2$
(d) After few instants, magnitude of $(m_1 v_1 + m_2 v_2)$ is equal to zero

Comprehension-7

A rigid insulated wire frame in the form of a right angled triangle *ABC*, is set in a vertical plane as shown. Two beads of equal masses *m* each and carrying charges q_1 and q_2 are connected by a cord of length *l* and slide without friction on the wires. Consider the case, when the beads are stationary.

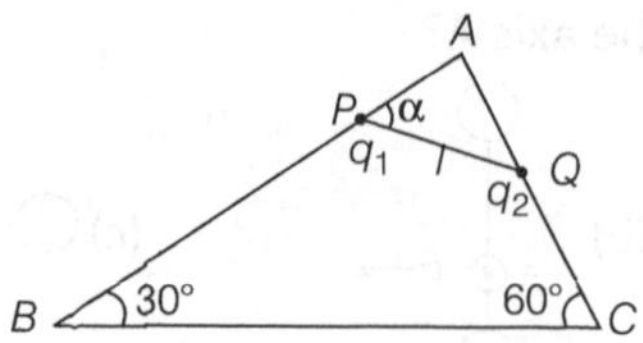

73 The angle α is
(a) 30° (b) 45° (c) 60° (d) 75°

74 The tension in the cord is
(a) $mg + \dfrac{kq_1 q_2}{l^2}$ (b) $mg - \dfrac{kq_1 q_2}{l^2}$ (c) $2mg$ (d) $\dfrac{2kq_1 q_2}{l^2}$

75 The normal reaction on bead 1 is
(a) mg (b) $\sqrt{3}\, mg$ (c) $2mg$ (d) $2\sqrt{3}\, mg$

Match the Columns Questions

76 **Match the following:** (Assume only electrostatic interaction forces) :

	Column-I		Column-II
(a)	Three charges are kept along a straight line	(p)	The system may be in equilibrium with proper choice of the value of charges.
(b)	Three charges are kept at the vertices of an equilateral triangle	(q)	The system will be in equilibrium for any value of the charges.
(c)	Three charges are kept at the three vertices of a square, and a fourth charge is kept at the point of intersection of the diagonals	(r)	The system will not be in equilibrium for any choice of the value of charges.
(d)	Three charges are kept at the vertices of an equilateral triangle with the fourth charge at the centroid	(s)	The equilibrium is unstable.

77 Column-II has conditions on electric field and Column-I has certain charge configurations. Match the suitable condition(s) with each charge configurations.

	Column-I		Column-II
(a)	Two point charges q and $-q$ are placed along *X*-axis and *C* is a point nearer to $+q$ 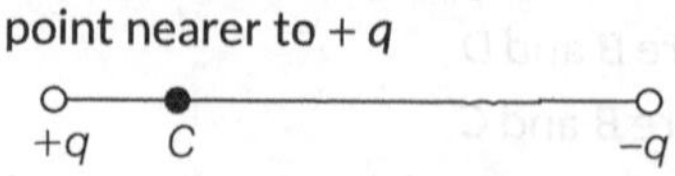	(p)	Electric field at the point *C* is towards right (positive *x*-direction)

	Column-I		Column-II
(b)	A uniformly positively charged sphere with a spherical cavity inside such that centre C and O lie on X-axis 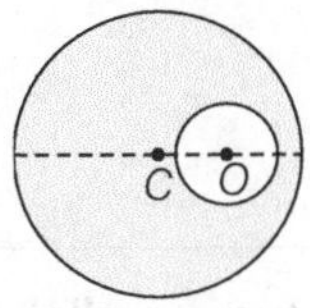	(q)	Electric field at the point C is zero
(c)	A regular hexagon with 5 charges as shown. Dotted line shows X-axis 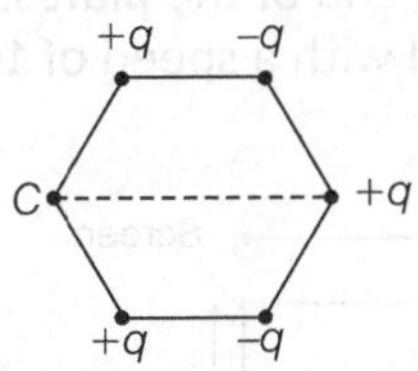	(r)	Electric field at the point C is towards left (negative x-direction)
(d)	A uniformly negatively charged sphere with a spherical cavity inside such that centre of cavity coincides with centre of sphere C 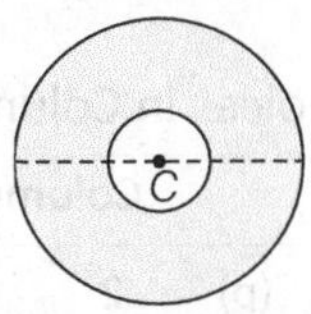	(s)	Electric field at the point C has two component one along negative x and other along positive y

78 Variation of electric field along the line joining two charges is given in Column-I. Relation between charges is given in Column-II, Match the proper variation. Take, electric field along right as positive

	Column-I		Column-II
(a)	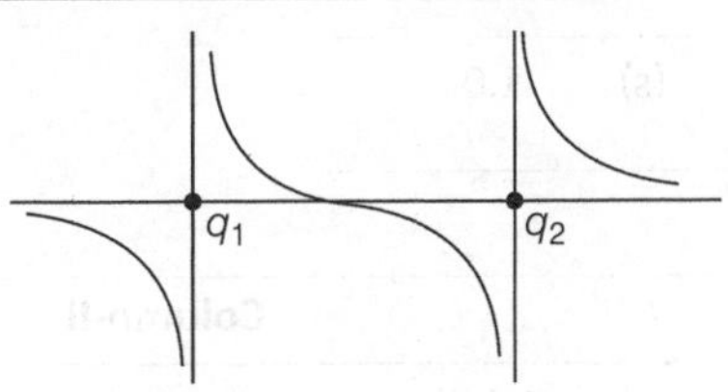	(p)	$\lvert q_1 \rvert > \lvert q_2 \rvert$ and $q_1 < 0$; $q_2 > 0$
(b)	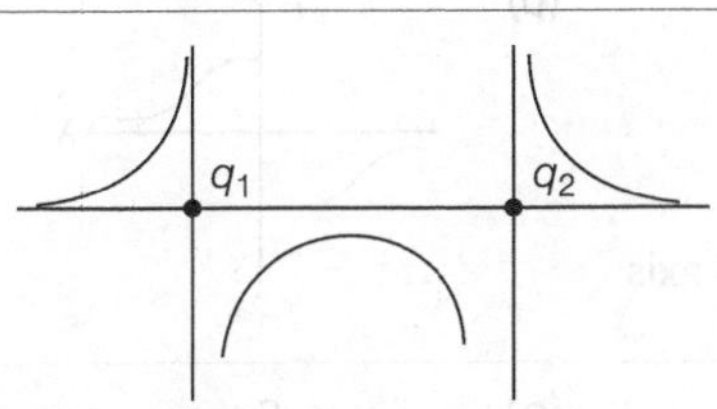	(q)	$\lvert q_1 \rvert = \lvert q_2 \rvert$ and $q_1 > 0$; $q_2 > 0$
(c)	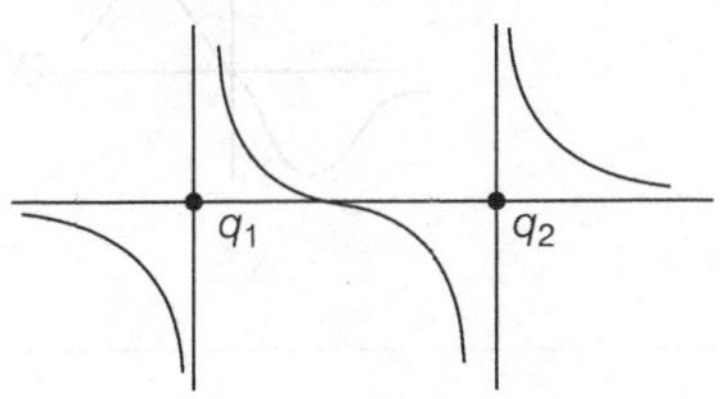	(r)	$\lvert q_1 \rvert = \lvert q_2 \rvert$ and $q_1 < 0$; $q_2 > 0$

	Column-I		Column-II
(d)	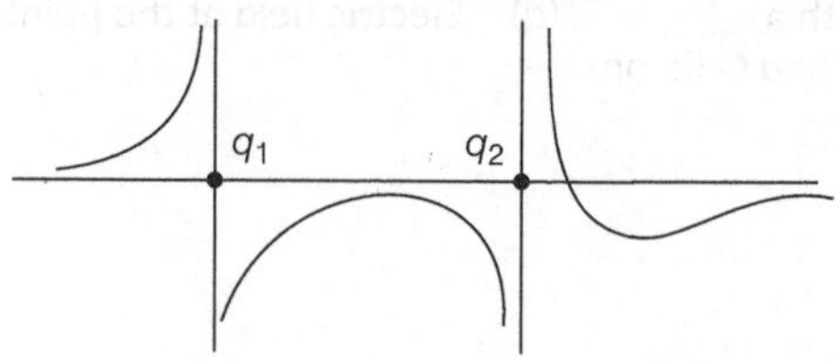	(s)	$\lvert q_2\rvert > \lvert q_1\rvert$ and $q_1 > 0$; $q_2 > 0$

79 An electron enters a uniform electric field between two parallel charged plates as shown in the diagram below. The electric field between the plates is assumed to be uniform.

The fluorescent screen is placed very close to the end of the plates. The electric field across the plates is 11.5 N/C and the electron enter the field with a speed of 10^6 m/s, parallel to the plates. It is given that 1 eV $= 1.6 \times 10^{-19}$ J.

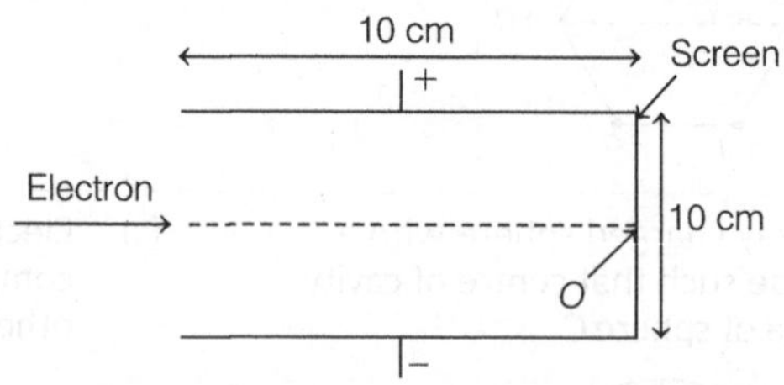

Match approximately Column-I with appropriate choices in Column-II.

	Column-I		Column-II
(a)	At what distance from O will the electron hit the screen (in cm)?	(p)	2
(b)	The speed of electron in vertical direction when it hits the plate (in 10^5 m/s)	(q)	0.2
(c)	The direction of electron's velocity with horizontal (in radians) at which it strikes the screen.	(r)	3
(d)	The kinetic energy possessed by electron (in eV) at the instant when it hits the screen.	(s)	1.0

80 Match the Column-I with Column-II.

	Column-I		Column-II
(a)	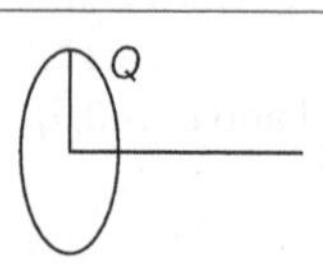 Uniformly positively charged ring in YZ-plane with axis along X-axis	(p)	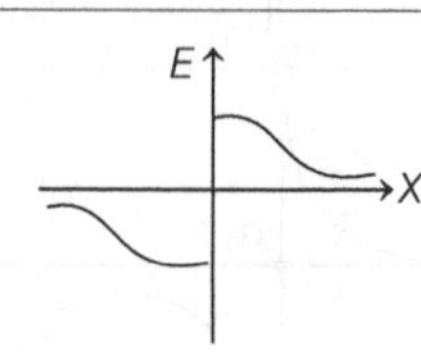
(b)	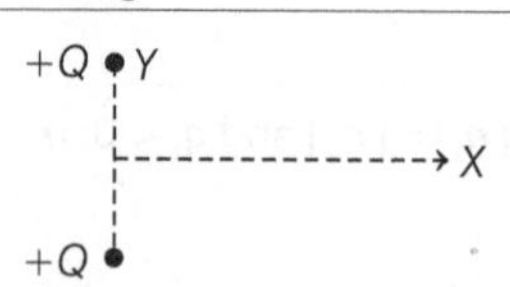 Two identical positively charged particle placed on Y-axis symmetrical to X-axis	(q)	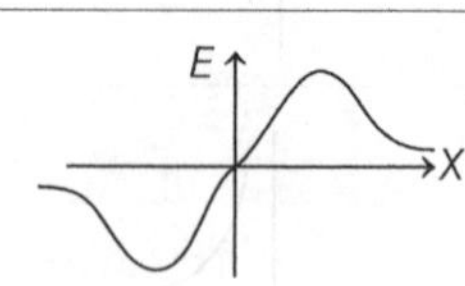

Column-I	Column-II
(c) 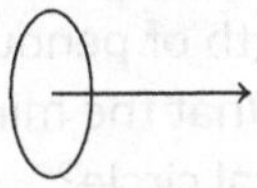 Uniformly positively charged non-conducting disc in YZ-plane with axis along X-axis	(r) 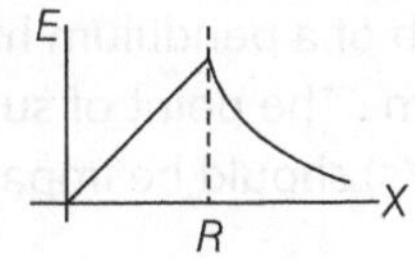
(d) 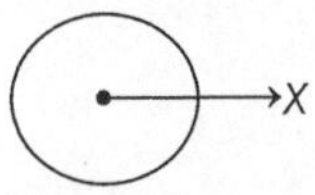 Uniformly positively charged (charge distributed throughout its volume) non-conducting sphere with diameter axis along X-axis.	(s) 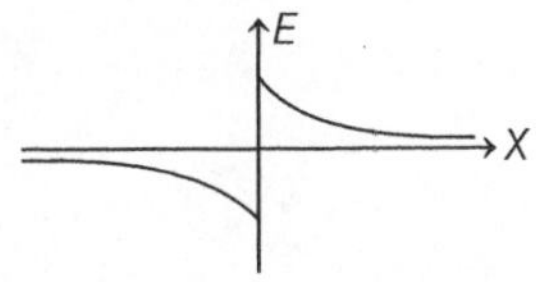

81 Two small positively charged spheres are suspended from a common point at the ceiling by the insulating light strings. The first sphere has mass m_1 and charge q_1 while the second one has mass m_2 and charge q_2. The masses are in equilibrium. Match Column-I with best possible entry in Column-II.

Column-I	Column-II
(a) 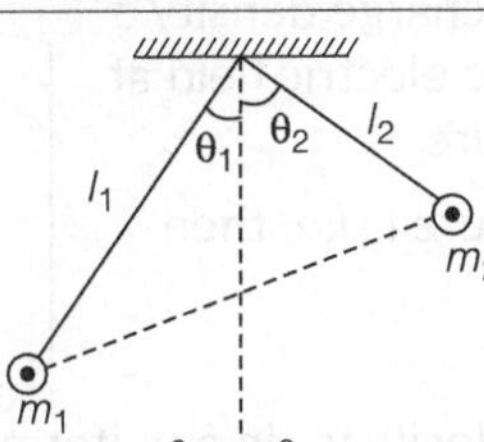 $\theta_1 = \theta_2$ $l_1 > l_2$	(p) $m_1 < m_2$
(b) 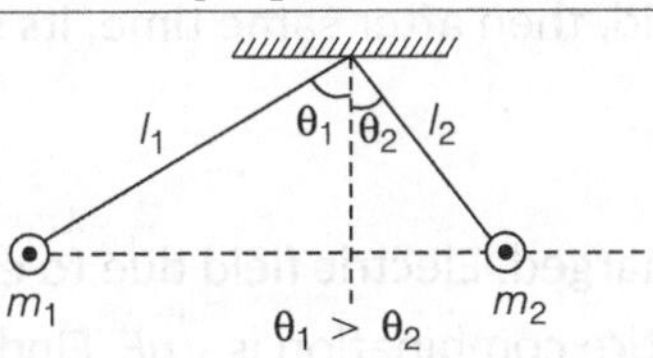 $\theta_1 > \theta_2$ $l_1 > l_2$	(q) $m_1 > m_2$
(c) 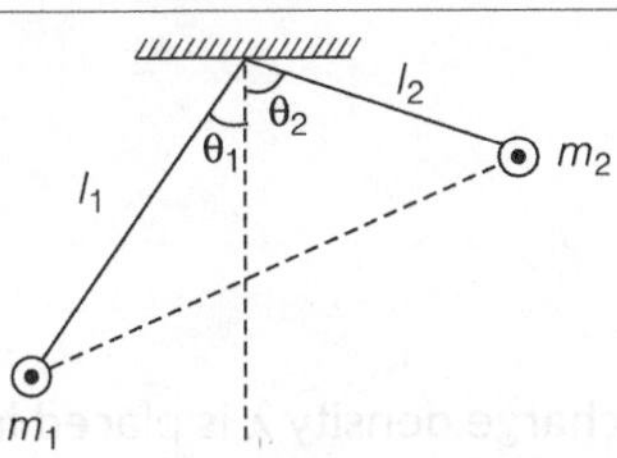 $\theta_1 < \theta_2$ $l_1 = l_2$	(r) $m_1 = m_2$
(d) 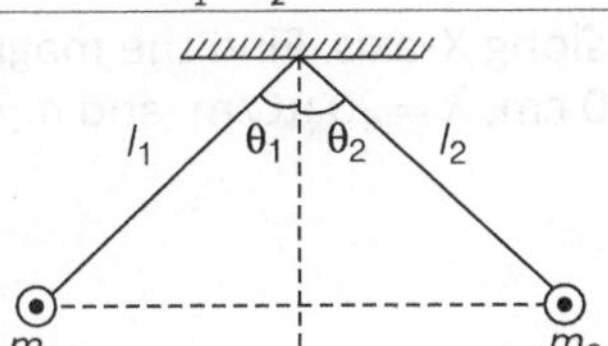 $\theta_1 = \theta_2$ $l_1 = l_2$	(s) $m_1 \neq m_2$

Numerical Value Questions

82 The bob of a pendulum has mass $m = 1$ kg and charge $q = +40\ \mu C$. Length of pendulum is $l = 0.9$ m. The point of suspension also has the same charge $+40\ \mu C$. What the minimum speed u (in m/s) should be imparted to the bob, so that it can complete vertical circle?

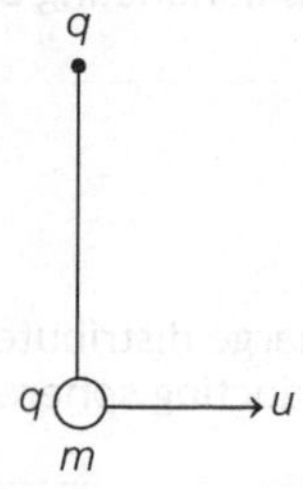

83 The charge density inside a nucleus is given by $\rho = \rho_0 \left(1 - \frac{r^2}{a^2}\right)$, where $r \leq a$, a = radius of nucleus = 3 fm and $\rho_0 = 5 \times 10^{25}\ C/m^3$. The electric field is maximum at $r = r_0$. Find r_0 (in fm) upto 2 decimal places.

84 An infinitely large non- conducting plane of uniform surface charge density σ has circular aperture of certain radius curved out from it. The electric field at a point which is at a distance a from the centre of the aperture (perpendicular to the plane) is $\frac{\sigma}{2\sqrt{2}\,\varepsilon_0}$. If the radius of aperture is ka, then find the value of k.

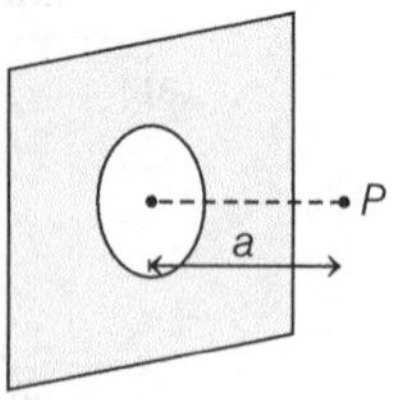

85 A positive charged particle q_0, when launched with some velocity $\mathbf{v}_0$ in a uniform field is found to deviate by 60° in a certain time, such that its speed is halved. If –ve charged particle $-2q_0$, is launched with the same velocity $\mathbf{v}_0$ in the same field, then after same time, its speed is v_2. Mass of both particles is same, then find value of $\frac{v_2^2}{v_0^2}$.

86 7 rods, AB, BC, CD, DE, EF, FG and GH are uniformly charged. Electric field due to each one is E at the centre of cube. Net electric field due to the entire combination is $\sqrt{n}E$. Find n.

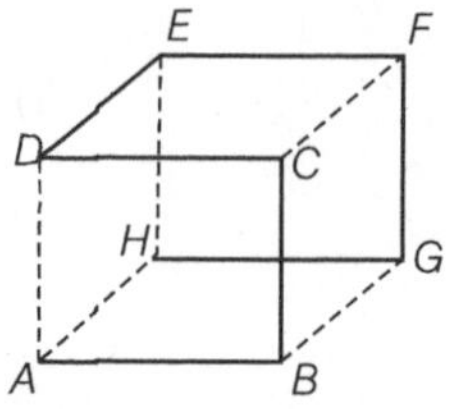

87 A square loop of side l each side having uniform linear charge density λ is placed in XY-plane as shown in the figure. There exists a non-uniform electric field $\mathbf{E} = \frac{a}{l}(x + l)\,\hat{\mathbf{i}}$, where a and l are constants and x is the position of the point from origin along X-axis. Find the magnitude of resultant electric force on the loop (in newtons), if $l = 10$ cm, $\lambda = 20\ \mu C/m$ and $a = 5 \times 10^5$ N/C.

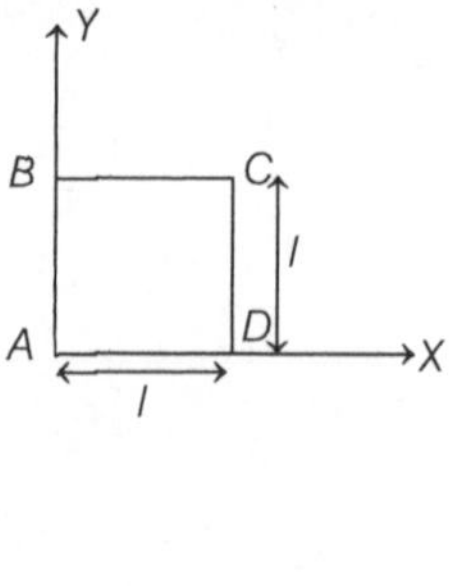

88 Find the magnitude of uniform electric field E in N/C (direction shown in figure), if an electron entering with velocity 100 m/s making 30° comes out making 60° (see figure), after a time numerically equal to $\frac{m}{e}$ of electron.

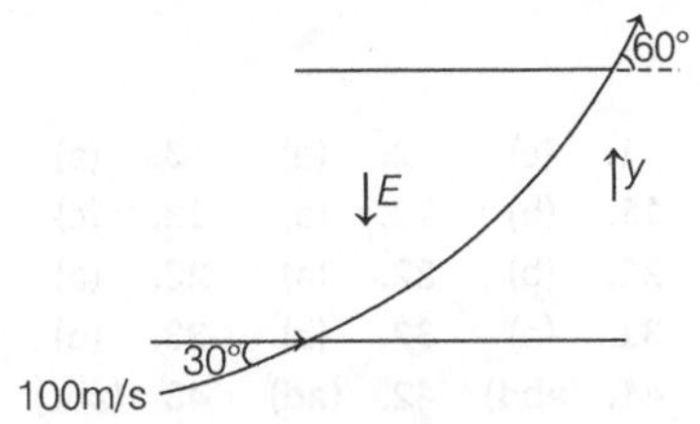

89 Two small spheres each of mass m and with a charge q lie inside a non-conducting smooth hemispherical bowl of radius R. The separation between the two charges in equilibrium is R. If $q = \frac{10^{-4}\,\text{C}}{n}$, then find n. (Take, $m = \sqrt{3}$ kg, $R = 1$ m)

90 Four point charges each of $+1\mu$C are fixed at the vertices of a square $ABCD$. The diagonals of the square intersect at O. Length of diagonal is 2m. A particle of mass 2g having a charge $+2\mu$C is released from a point 1 cm from O and 99cm from A. The motion of the particle is SHM with a period of $\frac{2\pi}{\sqrt{n}}$ s. Find the value of n neglecting gravity.

91 A small block of mass m is kept on the smooth inclined plane of angle 37°, placed in an elevator going upward with acceleration $a = \frac{g}{3}$, as shown in the figure. A horizontal electric field E perpendicular to the left and right vertical wall of the elevator exists. The charge on the block is $+q$. Find the time (in s), when the block collides with the vertical wall.

(Take, $g = 10\,\text{ms}^{-2}$ and $\frac{qE}{m} = 25\,\text{ms}^{-2}$)

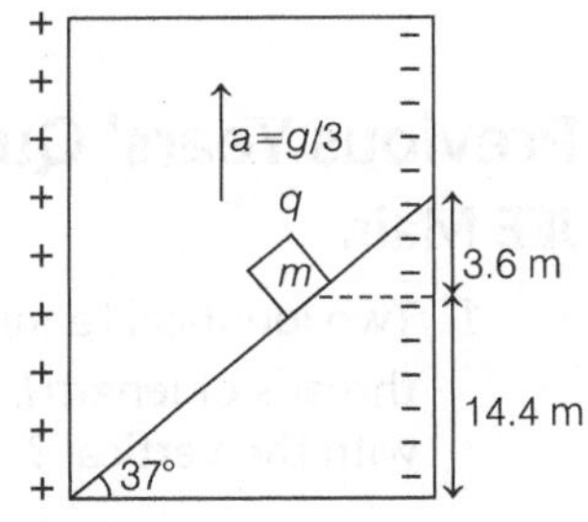

92 An infinitely long tube of semicircular cross-section is given some electric charge. One half of the tube is given a surface charge density $+\sigma$ and the other half of $-\sigma$. If $|\sigma| = 2\pi\varepsilon_0\,\text{C/m}^2$, find the magnitude of electric field(in N/C) at point O on the axis.
[The point O is some where near the centre on the axial line]

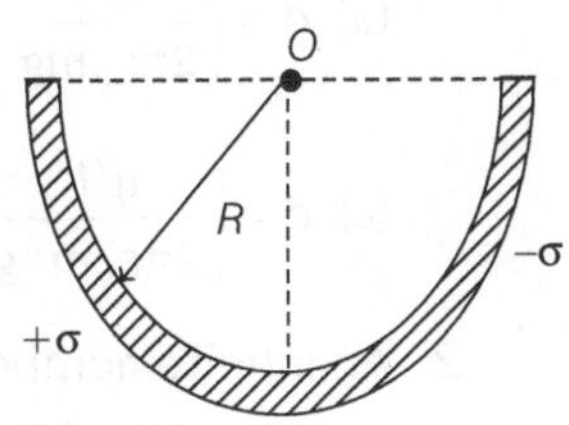

93 A thin wire is bent in the form of a semicircle of radius 1 m as shown in the figure. If carries a total charge of $16\sqrt{2}\pi^2\varepsilon_0$ coulombs uniformly distributed over its entire length. P is a point on the axis of the semicircular ring at a distance of 1 m from O. Find the component of electric field in (N/C) along Z-axis at point P.

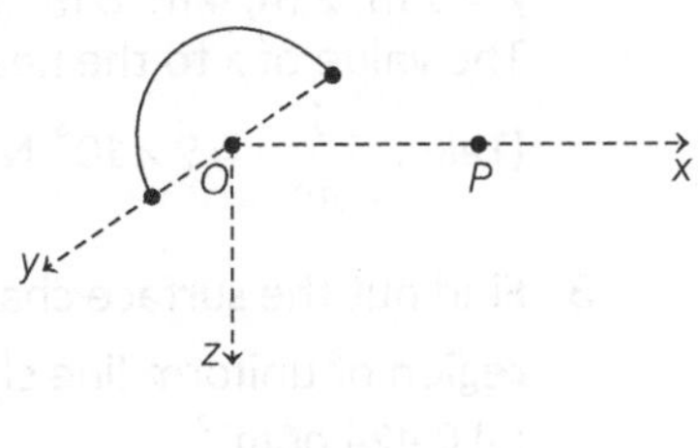

94 A point charge $Q_1 = -1\mu$C is fixed at the centre of an insulated disc of mass = 1 kg. The disc is resting on a rough horizontal surface. Another charge $Q_2 = 1\mu$C is fixed vertically above the centre of disc at a height $h = 1$ m. The disc is displaced slightly in horizontal direction by a small displacement x. Friction is sufficient to prevent slipping. Let radius of disc be R, and the corresponding angular displacement of disc is $\theta = \frac{x}{R}$. If angular acceleration is $\alpha = (y \times 10^z)\theta\,\text{rad/s}^2$, then find $y - z$.

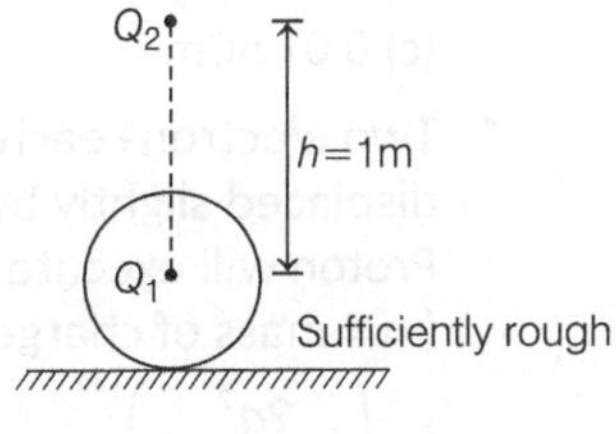

Answers

1. (c)	2. (a)	3. (a)	4. (c)	5. (d)	6. (c)	7. (d)	8. (a)	9. (b)	10. (a)
11. (b)	12. (a)	13. (c)	14. (d)	15. (b)	16. (a)	17. (b)	18. (c)	19. (d)	20. (a)
21. (b)	22. (a)	23. (a)	24. (d)	25. (a)	26. (b)	27. (c)	28. (b)	29. (a)	30. (d)
31. (a)	32. (a)	33. (d)	34. (b)	35. (c)	36. (d)	37. (a)	38. (b)	39. (c)	40. (ac)
41. (abd)	42. (ad)	43. (abc)	44. (ac)	45. (abc)	46. (abc)	47. (abc)	48. (bc)	49. (ad)	50. (ac)
51. (bc)	52. (abcd)	53. (bd)	54. (bc)	55. (abcd)	56. (abd)	57. (c)	58. (d)	59. (b)	60. (a)
61. (d)	62. (c)	63. (c)	64. (a)	65. (b)	66. (c)	67. (a)	68. (c)	69. (a)	70. (d)
71. (a)	72. (b)	73. (c)	74. (a)	75. (b)	76. (a) → p, s; (b) → r; (c) → r; (d) → p, s				

77. (a) → p; (b) → p; (c) → r; (d) → q

78. (a) → s; (b) → r; (c) → q; (d) → p

79. (a) → s; (b) → p; (c) → q; (d) → r

80. (a) → q; (b) → q; (c) → s; (d) → r

81. (a) → ps; (b) → ps; (c) → q,s; (d) → r

82. (6)	83. (2.24)	84. (1)	85. (7)	86. (1)
87. (6)	88. (100)	89. (3)	90. (18)	91. (1)
92. (2)	93. (4)	94. (9)		

For Detailed Solutions
Scan the code

Previous Years' Questions

JEE Main

1 Two identical tennis balls each having mass m and charge q are suspended from a fixed point by threads of length l. What is the equilibrium separation when each thread makes a small angle θ with the vertical ? **[JEE Main 2021]**

(a) $d = \left(\dfrac{q^2 l}{2\pi\varepsilon_0 mg}\right)^{\frac{1}{2}}$

(b) $d = \left(\dfrac{q^2 l}{2\pi\varepsilon_0 mg}\right)^{\frac{1}{3}}$

(c) $d = \left(\dfrac{q^2 l^2}{2\pi\varepsilon_0 m^2 g}\right)^{\frac{1}{3}}$

(d) $d = \left(\dfrac{q^2 l^2}{2\pi\varepsilon_0 m^2 g^2}\right)^{\frac{1}{3}}$

2 An infinite number of point charges, each carrying 1 μC charge, are placed along the Y-axis at $y = 1$ m, 2 m, 4 m, 8 m. The total force on a 1 C point charge, placed at the origin, is $x \times 10^3$ N. The value of x to the nearest integer, is ....

(Take, $\dfrac{1}{4\pi\varepsilon_0} = 9 \times 10^9$ N-m^2/C^2) **[JEE Main 2021]**

3 Find out the surface charge density at the intersection of point $X = 3$ m plane and X-axis, in the region of uniform line charge of 8 nC/m lying along the Z-axis in free space. **[JEE Main 2021]**

(a) 0.424 nCm^{-2}

(b) 47.88 nCm^{-2}

(c) 0.07 nCm^{-2}

(d) 4.0 nCm^{-2}

4 Two electrons each are fixed at a distance $2d$. A third charge proton placed at the mid-point is displaced slightly by a distance $x (x << d)$ perpendicular to the line joining the two fixed charges. Proton will execute simple harmonic motion having angular frequency?
(m = mass of charged particle) **[JEE Main 2021]**

(a) $\left(\dfrac{2q^2}{\pi\varepsilon_0 md^3}\right)^{1/2}$

(b) $\left(\dfrac{\pi\varepsilon_0 md^3}{2q^2}\right)^{1/2}$

(c) $\left(\dfrac{q^2}{2\pi\varepsilon_0 md^3}\right)^{1/2}$

(d) $\left(\dfrac{2\pi\varepsilon_0 md^3}{q^2}\right)^{1/2}$

5 What will be the magnitude of electric field at point O as shown in figure? Each side of the figure is l and perpendicular to each other. **[JEE Main 2021]**

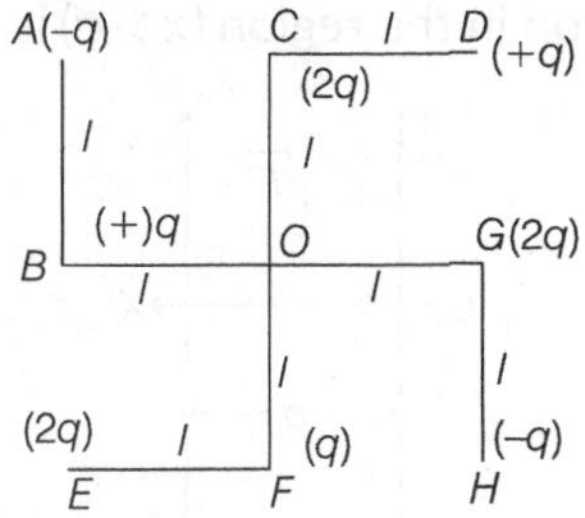

(a) $\frac{1}{4\pi\varepsilon_0}\frac{q}{l^2}$ (b) $\frac{1}{4\pi\varepsilon_0}\frac{q}{(2l^2)}(2\sqrt{2}-1)$

(c) $\frac{q}{4\pi\varepsilon_0 (2l)^2}$ (d) $\frac{1}{4\pi\varepsilon_0}\frac{2q}{2l^2}(\sqrt{2})$

6 A body having specific charge 8 μC/g is resting on a frictionless plane at a distance 10 cm from the wall (as shown in the figure). It starts moving towards the wall, when a uniform electric field of 100 V/m is applied horizontally towards the wall. If the collision of the body with the wall is perfectly elastic, then the time period of the motion will be ...... s. **[JEE Main 2021]**

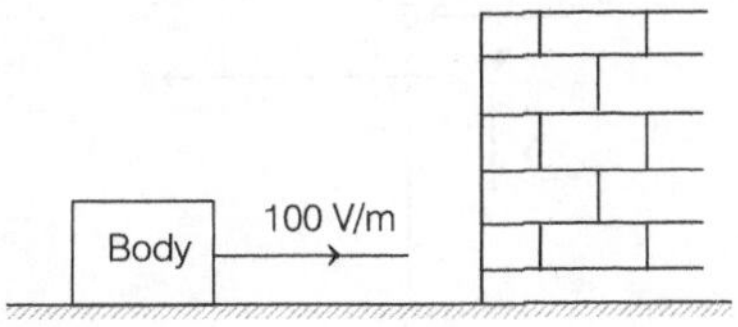

7 Find the electric field at point P (as shown in figure) on the perpendicular bisector of a uniformly charged thin wire of length L carrying a charge Q. The distance of the point P from the centre of the rod is $a = \frac{\sqrt{3}}{2}L$. **[JEE Main 2021]**

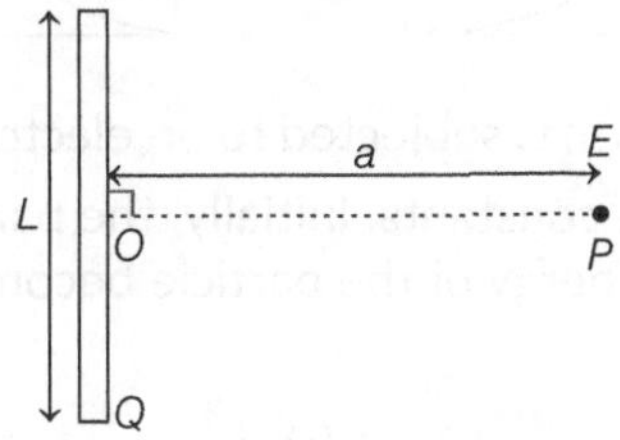

(a) $\frac{\sqrt{3}Q}{4\pi\varepsilon_0 L^2}$ (b) $\frac{Q}{3\pi\varepsilon_0 L^2}$

(c) $\frac{Q}{2\sqrt{3}\pi\,\varepsilon_0 L^2}$ (d) $\frac{Q}{4\pi\varepsilon_0 L^2}$

8 A cube of side a has point charges $+Q$ located at each of its vertices except at the origin, where the charge is $-Q$. The electric field at the centre of cube is **[JEE Main 2021]**

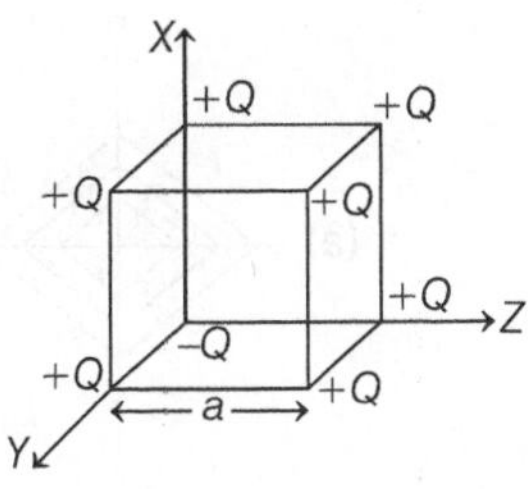

(a) $\frac{-Q}{3\sqrt{3}\pi\varepsilon_0 a^2}(\hat{x}+\hat{y}+\hat{z})$ (b) $\frac{Q}{3\sqrt{3}\pi\varepsilon_0 a^2}(\hat{x}+\hat{y}+\hat{z})$

(c) $\frac{-2Q}{3\sqrt{3}\pi\varepsilon_0 a^2}(\hat{x}+\hat{y}+\hat{z})$ (d) $\frac{2Q}{3\sqrt{3}\,\pi\varepsilon_0 a^2}(\hat{x}+\hat{y}+\hat{z})$

9 A charged particle (mass m and charge q) moves along X-axis with velocity v_0. When it passes through the origin it enters a region having uniform electric field $\mathbf{E} = -E\hat{\mathbf{j}}$ which extends upto $x = d$. Equation of path of electron in the region $(x > d)$ is **[JEE Main 2020]**

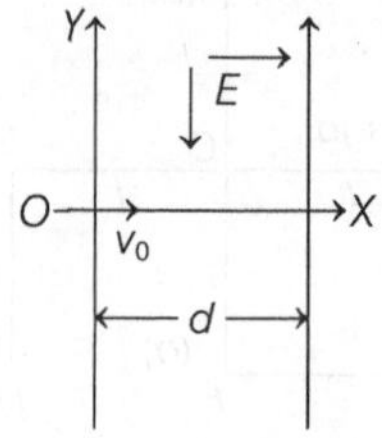

(a) $y = \dfrac{qEd}{mv_0^2}x$

(b) $y = \dfrac{qEd}{mv_0^2}(x - d)$

(c) $y = \dfrac{qEd}{mv_0^2}\left(\dfrac{d}{2} - x\right)$

(d) $y = \dfrac{qEd^2}{mv_0^2}x$

10 A small point mass carrying some positive charge on it, is released from the edge of a table. There is an uniform electric field in this region in the horizontal direction. Which of the following options then correctly describe the trajectory of the mass? (Curves are drawn schematically and are not to scale) **[JEE Main 2020]**

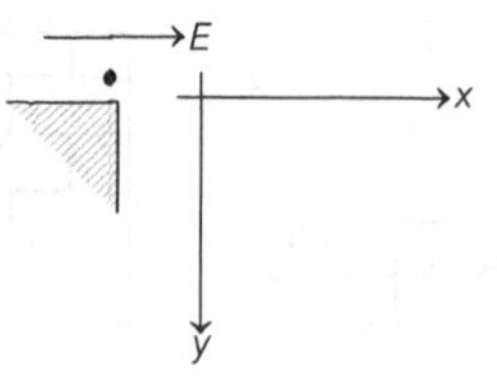

(a) 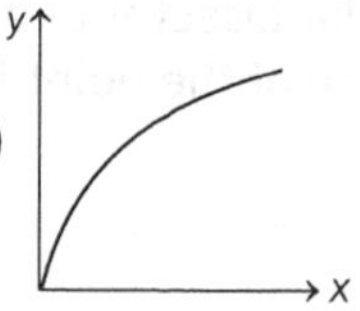

(b) 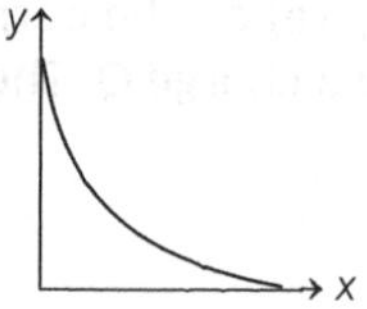

(c) 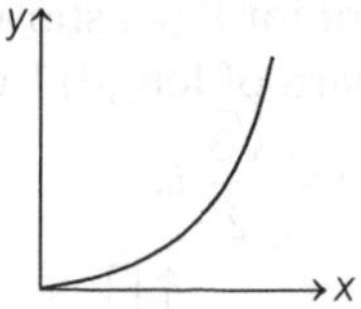

(d) 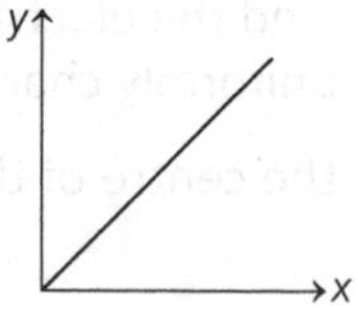

11 A particle of charge q and mass m is subjected to an electric field $E = E_0(1 - ax^2)$ in the x-direction, where a and E_0 are constants. Initially, the particle was at rest at $x = 0$. Other than the initial position, the kinetic energy of the particle becomes zero when the distance of the particle from the origin is **[JEE Main 2020]**

(a) $\sqrt{\dfrac{1}{a}}$

(b) a

(c) $\sqrt{\dfrac{3}{a}}$

(d) $\sqrt{\dfrac{2}{a}}$

12 Two charged thin infinite plane sheets of uniform surface charge densities σ_+ and σ_-, where $|\sigma_+| > |\sigma_-|$, intersect at right angle. Which of the following best represents the electric field lines for this system? **[JEE Main 2020]**

(a) 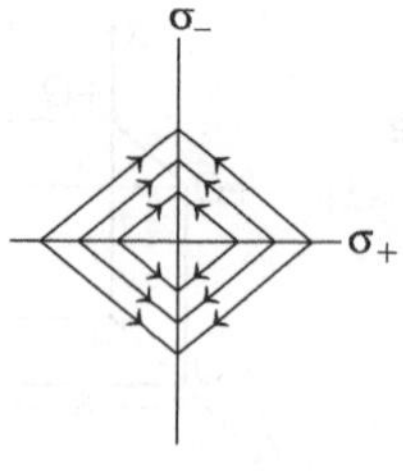

(b) 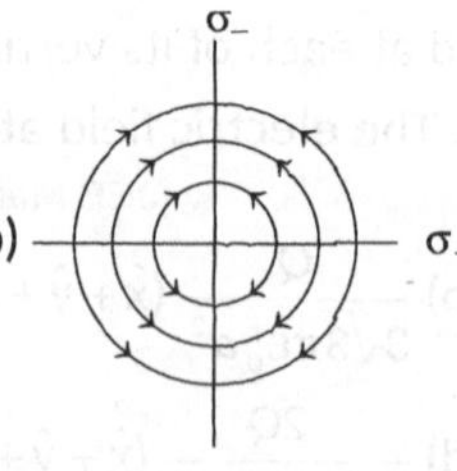

(c) 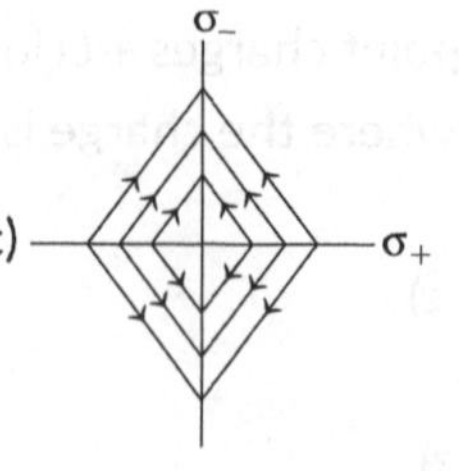

(d) 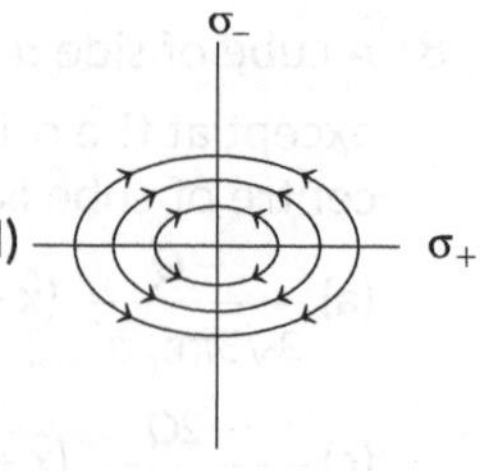

13 Charges Q_1 and Q_2 are at points A and B of a right angle triangle OAB (see figure). The resultant electric field at point O is perpendicular to the hypotenuse, then Q_1/Q_2 is proportional to **[JEE Main 2020]**

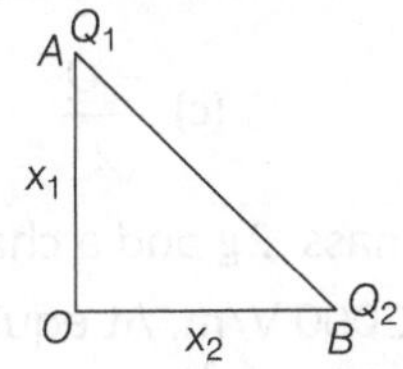

(a) $\frac{x_1^3}{x_2^3}$ (b) $\frac{x_2}{x_1}$ (c) $\frac{x_1}{x_2}$ (d) $\frac{x_2^2}{x_1^2}$

14 Three charged particles A, B and C with charges $-4q$, $2q$ and $-2q$ are present on the circumference of a circle of radius d. The charged particles A,C and centre O of the circle formed an equilateral triangle as shown in figure. Electric field at O along x-direction is **[JEE Main 2020]**

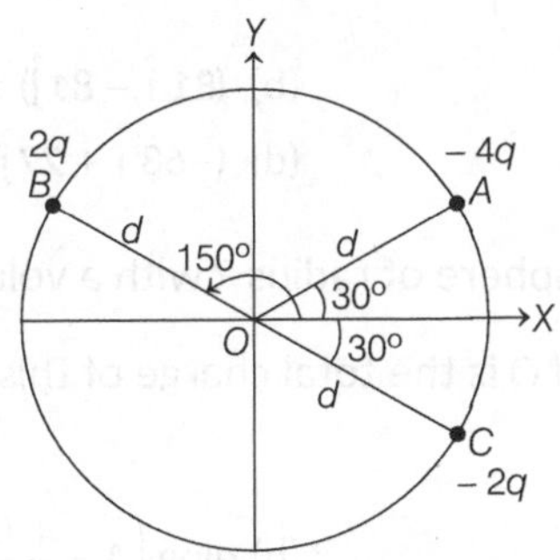

(a) $\frac{2\sqrt{3}q}{\pi\varepsilon_0 d^2}$ (b) $\frac{\sqrt{3}q}{4\pi\varepsilon_0 d^2}$ (c) $\frac{3\sqrt{3}q}{4\pi\varepsilon_0 d^2}$ (d) $\frac{\sqrt{3}q}{\pi\varepsilon_0 d^2}$

15 A particle of mass m and charge q is released from rest in a uniform electric field. If there is no other force on the particle, the dependence of its speed v on the distance x travelled by it is correctly given by (graphs are schematic and not drawn to scale) **[JEE Main 2020]**

(a) 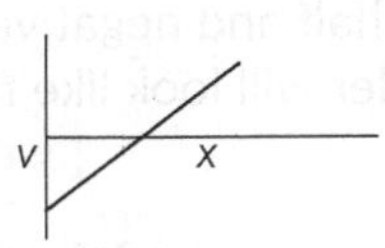

(b) 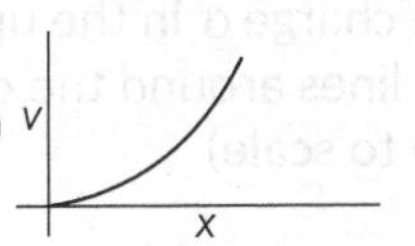

(c) 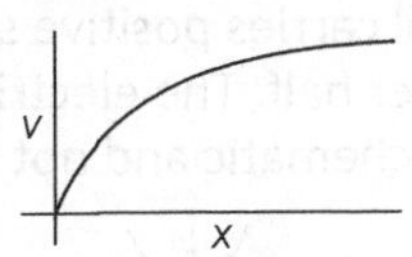

(d) 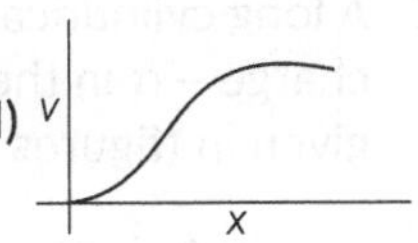

16 Two infinite planes each with uniform surface charged density $+\sigma$ are kept in such a way that the angle between them is 30°. The electric field in the region shown between them is given by **[JEE Main 2020]**

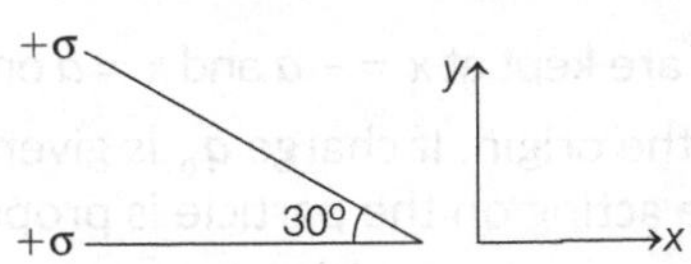

(a) $\frac{\sigma}{2\varepsilon_0}\left[(1+\sqrt{3})\hat{y}-\frac{\hat{x}}{2}\right]$ (b) $\frac{\sigma}{2\varepsilon_0}\left[\left(1-\frac{\sqrt{3}}{2}\right)\hat{y}-\frac{\hat{x}}{2}\right]$

(c) $\frac{\sigma}{2\varepsilon_0}\left[(1+\sqrt{3})\hat{y}+\frac{\hat{x}}{2}\right]$ (d) $\frac{\sigma}{\varepsilon_0}\left[\left(1+\frac{\sqrt{3}}{2}\right)\hat{y}+\frac{\hat{x}}{2}\right]$

17 Three charges $+Q$, q, $+Q$ are placed, respectively at distance 0, $d/2$ and d from the origin on the X-axis. If the net force experienced by $+Q$ placed at $x = 0$ is zero, then value of q is **[JEE Main 2019]**

(a) $\frac{+Q}{2}$ (b) $\frac{+Q}{4}$ (c) $\frac{-Q}{2}$ (d) $\frac{-Q}{4}$

18 The bob of a simple pendulum has mass 2 g and a charge of 5.0 μC. It is at rest in a uniform horizontal electric field of intensity 2000 V/m. At equilibrium, the angle that the pendulum makes with the vertical is (take, $g = 10\ \text{m/s}^2$) **[JEE Main 2019]**

(a) $\tan^{-1}(2.0)$ (b) $\tan^{-1}(0.2)$
(c) $\tan^{-1}(5.0)$ (d) $\tan^{-1}(0.5)$

19 Two point charges q_1 ($\sqrt{10}\ \mu$C) and q_2 ($-25\ \mu$C) are placed on the X-axis at $x = 1$ m and $x = 4$ m, respectively. The electric field (in V/m) at a point $y = 3$ m on Y-axis is
$\left(\text{Take, } \frac{1}{4\pi\varepsilon_0} = 9 \times 10^9\ \text{N-m}^2\text{C}^{-2}\right)$ **[JEE Main 2019]**

(a) $(63\,\hat{i} - 27\,\hat{j}) \times 10^2$ (b) $(81\,\hat{i} - 81\,\hat{j}) \times 10^2$
(c) $(-81\,\hat{i} + 81\,\hat{j}) \times 10^2$ (d) $(-63\,\hat{i} + 27\,\hat{j}) \times 10^2$

20 Charge is distributed within a sphere of radius R with a volume charge density $\rho(r) = \frac{A}{r^2} e^{\frac{-2r}{a}}$, where A and a are constants. If Q is the total charge of this charge distribution, the radius R is **[JEE Main 2019]**

(a) $a \log \left(\frac{1}{1 - \frac{Q}{2\pi a A}} \right)$ (b) $a \log \left(1 - \frac{Q}{2\pi a A} \right)$

(c) $\frac{a}{2} \log \left(1 - \frac{Q}{2\pi a A} \right)$ (d) $\frac{a}{2} \log \left(\frac{1}{1 - \frac{Q}{2\pi a A}} \right)$

21 A long cylindrical shell carries positive surface charge σ in the upper half and negative surface charge $-\sigma$ in the lower half. The electric field lines around the cylinder will look like figure given in (figures are schematic and not drawn to scale) **[JEE Main 2015]**

(a) 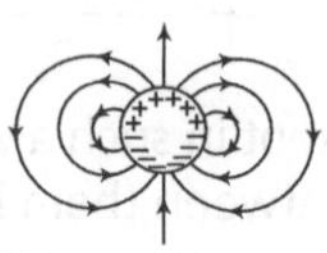 (b) 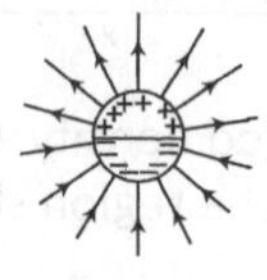 (c) 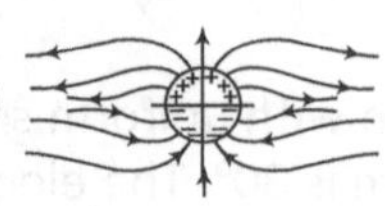 (d) 

22 Two charges each equal to q, are kept at $x = -a$ and $x = a$ on the X-axis. A particle of mass m and charge $q_0 = q/2$ is placed at the origin. If charge q_0 is given, a small displacement ($y << a$) along the Y-axis, the net force acting on the particle is proportional to **[JEE Main 2013]**

(a) y (b) $-y$ (c) $1/y$ (d) $-1/y$

JEE Advanced

23 A particle of mass 10^{-3} kg and charge 1.0 C is initially at rest. At time $t = 0$, the particle comes under the influence of an electric field $\mathbf{E}(t) = E_0 \sin \omega t\,\hat{\mathbf{i}}$, where $E_0 = 1.0\ \text{NC}^{-1}$ and $\omega = 10^3\ \text{rad s}^{-1}$. Consider the effect of only the electrical force on the particle. Then, the maximum speed (in ms^{-1}), attained by the particle at subsequent times is .... **[JEE Advanced 2019]**

24. The figures alongside depict two situations in which two infinitely long static line charges of constant positive line charge density λ are kept parallel to each other. In their resulting electric field, point charges q and $-q$ are kept in equilibrium between them. The point charges are confined to move in the x direction only. If they are given a small displacement about their equilibrium positions, then the correct statements is/are

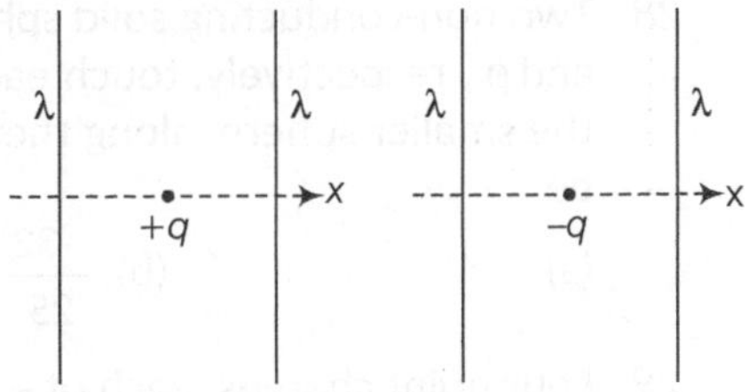

[JEE Advanced 2015]

(a) both charges execute simple harmonic motion.
(b) both charges will continue moving in the direction of their displacement.
(c) charge $+q$ executes simple harmonic motion while charge $-q$ continues moving in the direction of its displacement.
(d) charge $-q$ executes simple harmonic motion while charge $+q$ continues moving in the direction of its displacement.

25 Charges Q, 2Q and 4Q are uniformly distributed in three dielectric solid spheres 1, 2 and 3 of radii $\frac{R}{2}$, R and $2R$ respectively, as shown in figure. If magnitudes of the electric fields at point P at a distance R from the centre of spheres 1, 2 and 3 are E_1, E_2 and E_3 respectively, then **[JEE Advanced 2014]**

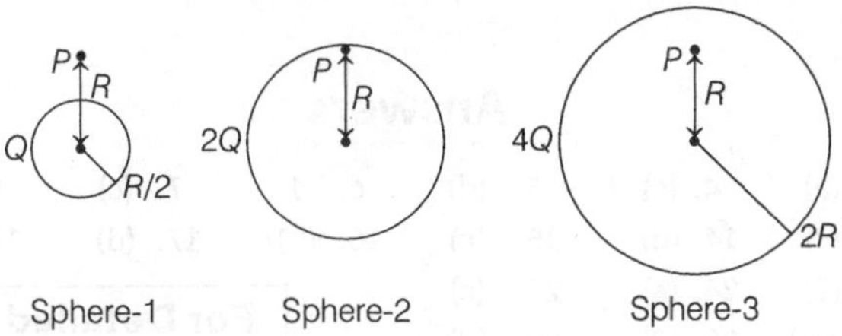

(a) $E_1 > E_2 > E_3$ (b) $E_3 > E_1 > E_2$ (c) $E_2 > E_1 > E_3$ (d) $E_3 > E_2 > E_1$

26 Let $E_1(r)$, $E_2(r)$ and $E_3(r)$ be the respective electric fields at a distance r from a point charge Q, an infinitely long wire with constant linear charge density λ, and an infinite plane with uniform surface charge density σ. If $E_1(r_0) = E_2(r_0) = E_3(r_0)$ at a given distance r_0, then **[JEE Advanced 2014]**

(a) $Q = 4\sigma\pi r_0^2$ (b) $r_0 = \dfrac{\lambda}{2\pi\sigma}$ (c) $E_1(r_0/2) = 2E_2(r_0/2)$ (d) $E_2(r_0/2) = 4E_3(r_0/2)$

27 Four charges Q_1, Q_2, Q_3 and Q_4 of same magnitude are fixed along the X-axis at $X = -2a, -a, +a$ and $+2a$, respectively. A positive charge q is placed on the positive Y-axis at a distance $b > 0$. Four options of the signs of these charges are given in Column-I. The direction of the forces on the charge q is given in Column-II. Match Column I with Column II and select the correct answer using the code given below the lists. **[JEE Advanced 2014]**

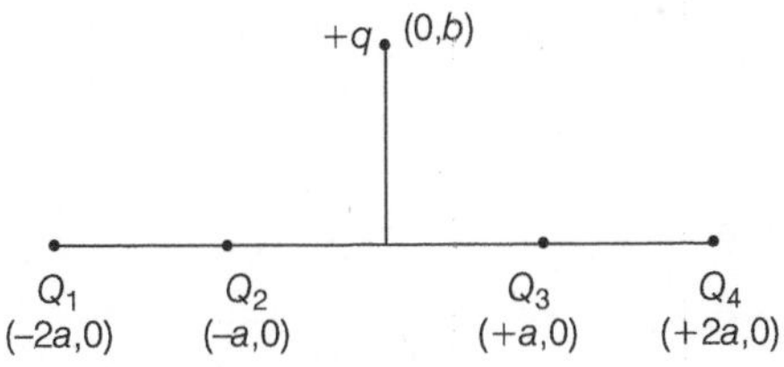

Column-I		Column-II	
A.	Q_1, Q_2, Q_3, Q_4 all positive	p.	$+X$
B.	Q_1, Q_2 positive; Q_3, Q_4 negative	q.	$-X$
C.	Q_1, Q_4 positive; Q_2, Q_3 negative	r.	$+Y$
D.	Q_1, Q_3 positive; Q_2, Q_4 negative	s.	$-Y$

Codes

(a) A-r, B-p, C-s, D-q
(b) A-s, B-q, C-r, D-p
(c) A-r, B-p, C-q, D-s
(d) A-s, B-q, C-p, D-r

28 Two non-conducting solid spheres of radii R and $2R$, having uniform volume charge densities ρ_1 and ρ_2 respectively, touch each other. The net electric field at a distance $2R$ from the centre of the smaller sphere, along the line joining the centre of the spheres, is zero. The ratio ρ_1/ρ_2 can be **[JEE Advanced 2013]**

(a) -4 (b) $\frac{-32}{25}$ (c) $\frac{32}{25}$ (d) 4

29 Four point charges, each of $+q$, are rigidly fixed at the four corners of a square planar soap film of side a. The surface tension of the soap film is γ. The system of charges and planar film are in equilibrium, and $a = k\left[\frac{q^2}{\gamma}\right]^{1/N}$, where k is a constant. Then N is .... **[JEE Advanced 2011]**

30 A tiny spherical oil drop carrying a net charge q is balanced in still air with a vertical uniform electric field of strength $\frac{81\pi}{7} \times 10^5$ Vm^{-1}. When the field is switched off, the drop is observed to fall with terminal velocity 2×10^{-3} ms^{-1}. Given $g = 9.8$ ms^{-2}, viscosity of the air $= 1.8 \times 10^{-5}$ Ns-m^{-2} and the density of oil $= 900$ kg m^{-3}, the magnitude of q is **[JEE Advanced 2010]**

(a) 1.6×10^{-19} C (b) 3.2×10^{-19} C
(c) 4.8×10^{-19} C (d) 8.0×10^{-19} C

Answers

1. (b)	2. 12	3. (a)	4. (c)	5. (d)	6. 1	7. (c)	8. (c)	9. (c)	10. (d)
11. (c)	12. (c)	13. (c)	14. (d)	15. (c)	16. (b)	17. (d)	18. (d)	19. (a)	20. (d)
21. (a)	22. (b)	23. (2)	24. (c)	25. (c)					
26. (c)	27. (a)	28. (b,d)	29. (3)	30. (d)					

For Detailed Solutions
Scan the code

CHAPTER

02

ELECTRIC POTENTIAL

LEVEL 01 \ Starter Level

1.1 Electric Potential

1 A point charge + 2 C is taken from infinity to a point in an electric field, without changing its speed. If work done against electrostatic forces is – 40 J, then find the potential at that point.

2 A point charge + 20 μC is shifted slowly from infinity to a point *P* in an electric field. The potential of the point *P* is 1000 V.

(i) What is the work done by external agent against electric field?

(ii) What is the work done by electric field?

(iii) If a point charge + 30 μC is released from rest at point *P*, then find its kinetic energy at a large distance.

3 A point charge + 20 μC is shifted from infinity to a point *P* in an electric field. The potential of the point *P* is 1000 V. If the kinetic energy of the charged particle is found to increase by 10 mJ, when it is brought from infinity to point *P*, then

(i) what is the work done by electric field and

(ii) what is the total work done by external agent?

4 An electron is shifted slowly from infinity to a point *P* in an electric field. The potential of the point *P* is – 10 V.

(i) Find out work done by external agent against electric field.

(ii) What is the work done by electric field?

1.2 Electric Potential due to Point Charge(s)

5 Some point charges are kept at the vertices of the following shapes as shown. Find electric potential at point C for each case.

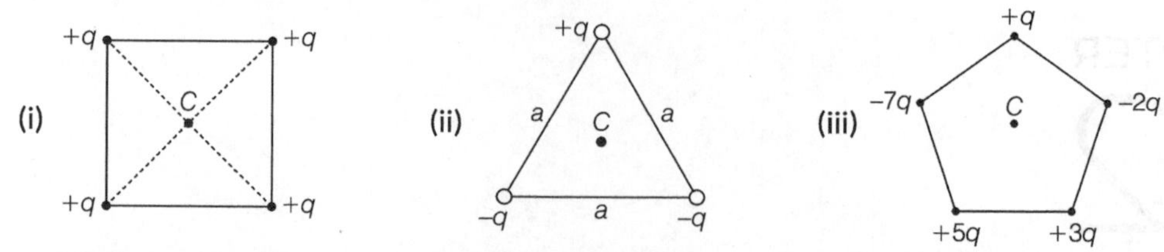

C is the centre of the square of side a

C is the centroid of the equilateral triangle of side a

C is the centroid of regular pentagon of side a

6 Two point charges $2\mu C$ and $-4\mu C$ are situated at points $(-2,0)$ and $(2,0)$, respectively. Find out potential at points $A(4,0)$ and $B(0,\sqrt{5})$.

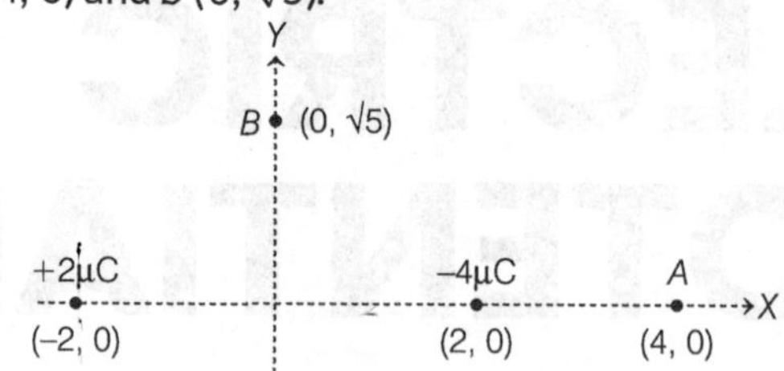

7 An infinite number of charges, each equal to q are placed along the X-axis at $x=1$, $x=2$, $x=4$, $x=8$, ... and so on. Find the potential at the point $x=0$ due to this set of charges. What will be the potential in the above set, if the consecutive charges have opposite signs?

8 Two particles A and B having charges of $+4\mu C$ and $-8\mu C$, respectively are held fixed at a separation of 60 cm. Locate the point(s) on the line AB, where the electric potential is zero.

9 Two point charges $+q$ and $-3q$ are placed at $(0,0)$ and $(8,0)$, respectively. Find the locus of the zero potential points in the XY-plane.

10 A regular hexagon of side L is placed in a horizontal plane and six equal charges $+q$ are fixed at its vertices. A point P is located vertically above the centre of the hexagon, at a height $\sqrt{3}L$ from the plane of the hexagon. Find the electric potential at point P.

1.3 Electric Potential due to Continuous Charge Configurations

11 Find electric potential at the centre C for each charge configuration.

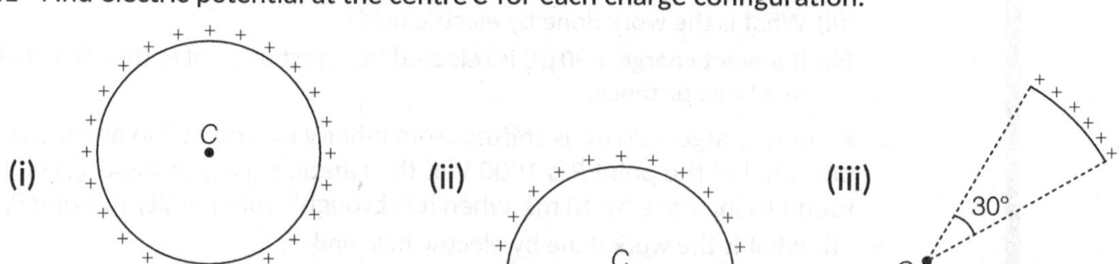

A non-uniformly charged ring of radius R carrying charge Q

A half ring of radius R carrying charge Q

An arc of radius R subtending angle 30° at centre and carrying charge Q

12 A charge $Q = 10$ nC is distributed non-uniformly over the circumference of a ring of radius 3 m placed on XY-plane with its centre at origin. Find the electric potential at point $P(0, 0, 4\text{m})$.

13 Find the electric potential at a point P, located on the axis of a uniformly charged thin circular disc of radius R and surface charge density σ at a distance x from the plane of the disc.

14 Find electric potential due to uniformly charged thin spherical shell of radius 10 cm carrying charge + 1.6 μC at the following points.
(i) At the centre (ii) 4 cm from centre (iii) 12 cm from centre

15 Two concentric spherical shells of radius R_1 and R_2 ($R_2 > R_1$) are having uniformly distributed charges Q_1 and Q_2, respectively. Find out potential at points A, B and C. Distances of A, B and C from the common centre O can be taken as OA, OB and OC respectively, if required.

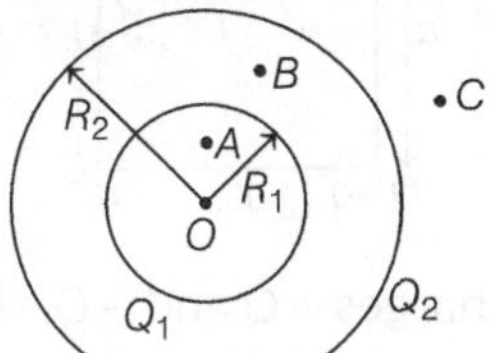

16 Find electric potential due to uniformly charged solid non-conducting sphere of radius 10 cm carrying charge + 1.6 μC at the following points.
(i) At the centre (ii) 5 cm from centre (iii) 12 cm from centre

17 A solid sphere of radius R is uniformly charged with charge density ρ in its volume.
A spherical cavity of radius $\frac{R}{2}$ is made in the sphere as shown in the figure.
Find the electric potential at the centre of the sphere.

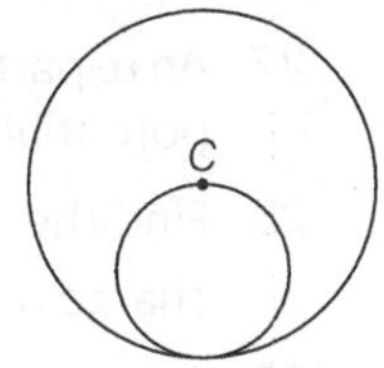

18 A charge Q is uniformly distributed over a thin rod AB of length L as shown in the figure. Find the electric potential at point P.

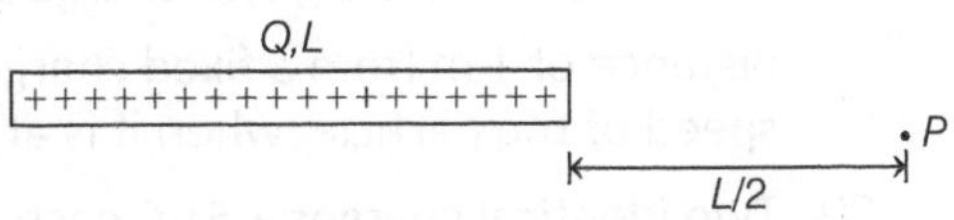

1.4 Potential Difference

19 20 J of work has to be done against an existing electric field to take a charge of 0.05 C from A to B. How much is the potential difference $V_B - V_A$?

20 An electron is accelerated through a potential difference of 45.5 V from rest. Find its final kinetic energy and speed.

21 The kinetic energy of a charged particle decreases by 20 J as it moves from a point at potential 50 V to a point at potential 100 V. Find the charge on the particle .

22 A particle of charge $+3 \times 10^{-9}$ C is in a uniform field directed to the left. It is released from rest and moves a distance of 5 cm, after which its kinetic energy is found to be 4.5×10^{-5} J.
(i) What work was done by the electrical force?
(ii) What is the magnitude of the electrical field?
(iii) What is the potential of the starting point with respect to the end point?

23 In the previous problem, suppose that another force in addition to the electrical force acts on the particle, so that when it is released from rest, it moves to the right. After, it has moved 5 cm, the additional force has done 9×10^{-5} J of work and the particle has 4.5×10^{-5} J of kinetic energy.
(i) What work was done by the electrical force?
(ii) What is the magnitude of the electric field?
(iii) What is the potential of the starting point with respect to the end point?

24 A charge of 8 mC is located at the origin. Find the work done by external agent in slowly taking a small charge of -2×10^{-9} C from a point A(0,0,0.03 m) to a point B(0,0.04m,0) *via* a point C(0,0.06m, 0.09m).

25 Four charges $+q, +q, -q, -q$ are fixed respectively at the corners *A*, *B*, *C* and *D* of a square of side *a* arranged in the given order. If *E* and *F* are the mid-points of sides *BC*, *CD* respectively, what will be the work done by external agent in carrying a charge Q slowly from *E* to *F*?

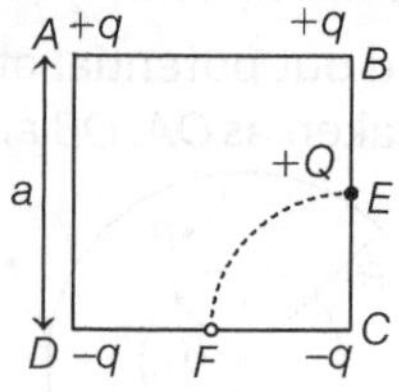

26 Two rings, each of radius *R* carry charges $+Q$ and $-Q$, respectively and they are held fixed with their planes parallel, such that their axes coincide and their centres are $\sqrt{3}R$ distance apart. Find the work done by external agent in slowly shifting a test charge $+q$ from the centre of the second ring to the centre of the first ring.

1.5 Potential Energy of a Point Charge

27 An α-particle is placed in an electric field at a point having electric potential 5 V. Find its potential energy.

28 Find the potential energy of a charge q_0 placed at the centre of regular hexagon of side *a*, if charge *q* is placed at each vertex of regular hexagon.

29 A particle of mass 2 g and charge 1 μC is held at rest on a frictionless horizontal surface at a distance of 1 m from a fixed charge 1 mC. If the particle is released, it will be repelled. Find the speed of the particle, when it is at distance of 10 m from the fixed charge.

30 Two identical charges $+5\,\mu$C each are fixed at a distance 8 cm and a charged particle of mass 9×10^{-6} kg and charge $-10\,\mu$C is placed at a distance 5 cm from each of them and is released from rest. Find the speed of the particle, when it is nearest to the two charges.

31 A particle of mass *m*, charge $q>0$ and initial kinetic energy *K* is projected from infinity towards a heavy nucleus of charge Q assumed to have a fixed position. If the aim is perfect, find the distance of the closest approach.

32 A small point charge ($+q=1\,\mu$C, $m=3$ kg) is thrown with speed $v_0=2$ m/s from a very long distance towards a fixed point charge ($+Q=1$ mC). The initial line of motion of *q* is at a perpendicular distance of $d=1$ m from the line passing from the centre of Q (as shown in figure). Find the distance of the closest approach between two point charges.

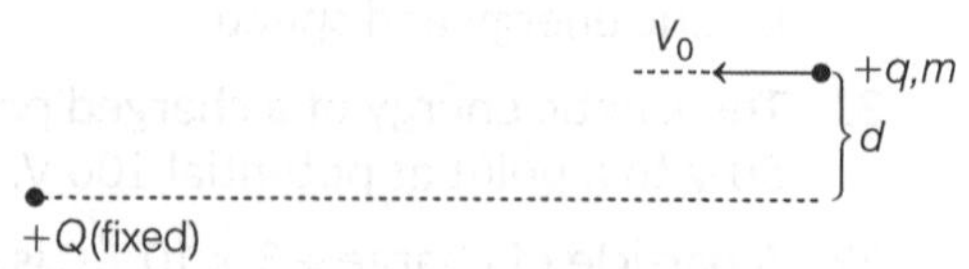

33 A particle having charge $-q$ and mass *m* is released from rest on the axis of a uniformly charged fixed ring of total charge Q and radius *R* from a distance $\sqrt{3}R$. Find out its speed, when it reaches the centre of ring.

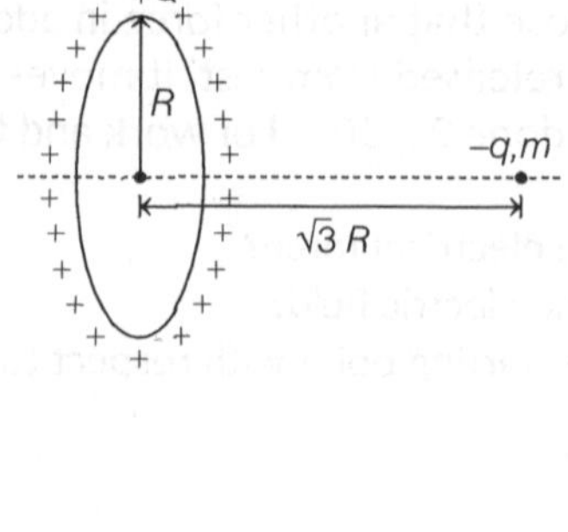

34 A solid uniformly charged fixed non-conducting sphere of total charge Q and radius R contains a tunnel of negligible diameter. If a point charge q of mass m is projected from point P as shown in figure, then find the minimum speed of projection u_{min}, so that the point charge comes out of the tunnel from the other end.

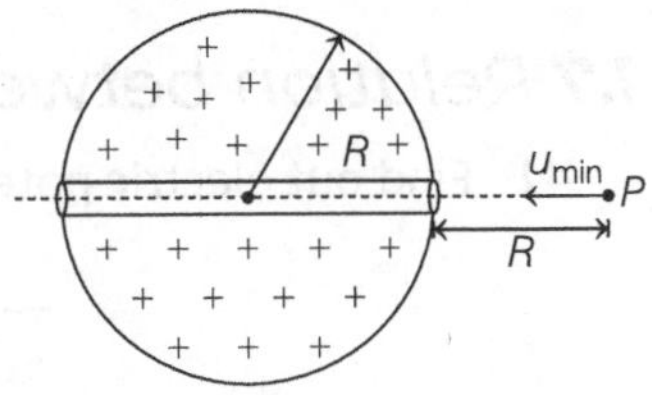

1.6 Potential Energy of a System of Point Charges

35 Calculate the electrostatic potential energy of the following systems of point charges.

(i)
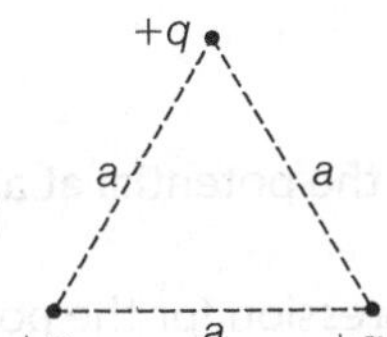

(ii)
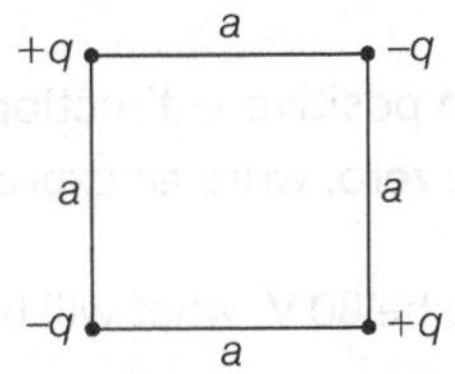

(iii)
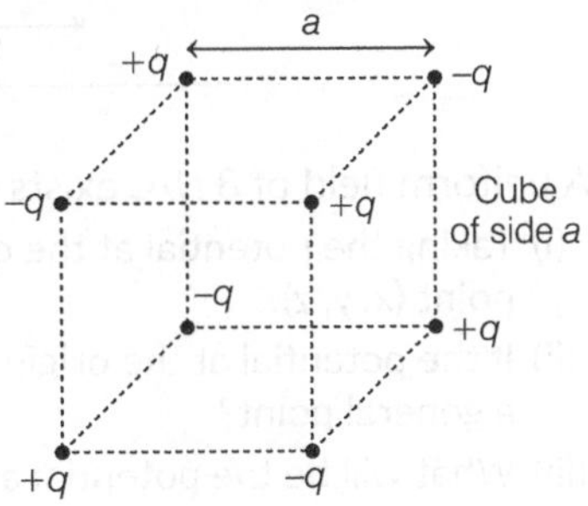

36 Two point charges + 15 μC and + 10 μC are 30 cm apart. Calculate the work done in bringing them closer to each other by 15 cm.

37 Four point charges $+q, +q, +q$ and $-q$ are placed at the vertices of a square of side a as shown in Fig. (i). The arrangement is changed and the charges are positioned at the vertices of a rhombus of side a as shown in Fig.(ii). Find the work done by external agent in changing the configuration.

(i)
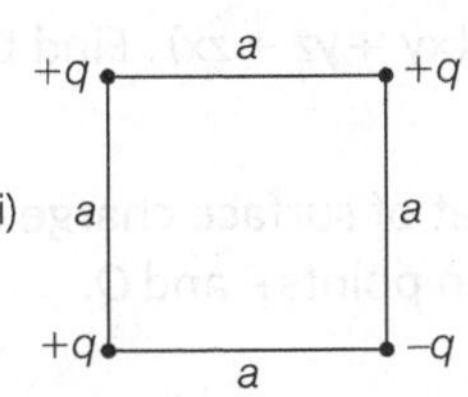

(ii)
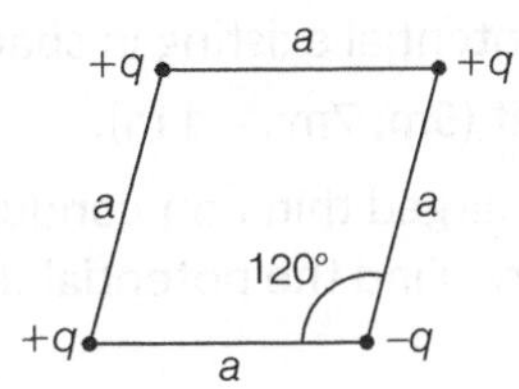

38 Two identical particles of mass m carry a charge Q each. Initially, one is at rest on a smooth horizontal plane and the other is projected along the plane directly towards the first particle from a large distance with speed u. Find the distance of the closest approach of them.

39 A particle having charge $+q$ is fixed at a point O and a second particle of mass m and having charge $-q_0$ moves with constant speed in a circle of radius r about the charge $+q$. Find energy required to be supplied to the moving charge to increase radius of the path to $2r$.

40 Three point charges each of charge + 20 μC and mass 10 g are placed at the corners of an equilateral triangle of side 10 cm.

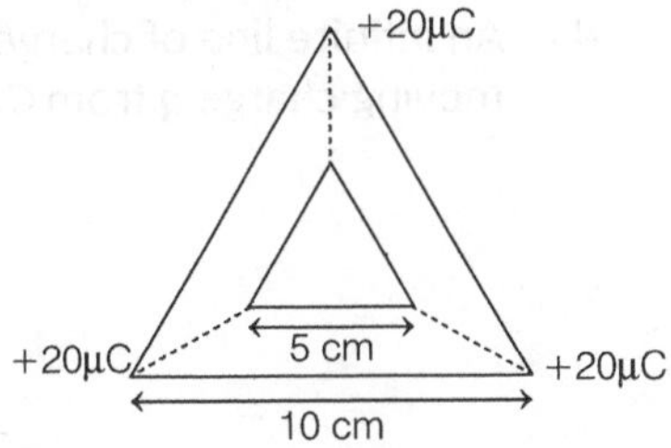

(i) Calculate the work required to move these charges to the corners of a smaller equilateral triangle of side 5 cm as shown in adjacent figure.

(ii) If all the charges are released from rest, when they are on equilateral triangle of side 10 cm, find their speeds when they lie on an equilateral triangle of side 20 cm.

(iii) If all the charges are released from rest, when they are on equilateral triangle of side 10 cm, find their speeds when they are at very large distance from each other.

(iv) If one of the charges is released from rest and the other two are held fixed when they are on equilateral triangle of side 10 cm, find its speed, when it reaches infinity.

1.7 Relation between Electric Field and Electric Potential

41 Find out electric potential for points A, C and D, if electric potential at B is 10 V.

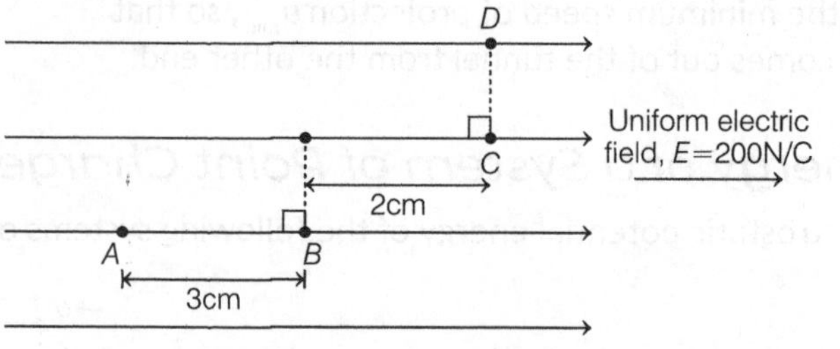

42 A uniform field of 8 N/C exists in space in positive x-direction.

(i) Taking the potential at the origin to be zero, write an expression for the potential at a general point (x, y, z).

(ii) If the potential at the origin is taken to be 80 V, what will be the expression for the potential at a general point?

(iii) What will be the potential at the origin, if the potential at infinity is taken to be zero?

43 An electric field $\mathbf{E} = (10\hat{\mathbf{i}} + 20\hat{\mathbf{j}})$ N/C exists in the space. If the potential at the origin is taken to be 100 V, find the potential at (3m, 4m).

44 If $\mathbf{E} = 6x^2\hat{\mathbf{i}} - 3y^2\hat{\mathbf{j}}$, then find $V(x, y)$.

45 If $\mathbf{E} = 2y\hat{\mathbf{i}} + 2x\hat{\mathbf{j}}$, then find $V(x, y)$, if the potential at (2,3) is 100 V.

46 If $V = x^2y + y^2z$, then find $\mathbf{E}(x, y, z)$.

47 The electric potential existing in space is $V(x, y, z) = 2(xy + yz + zx)$. Find the magnitude of the electric field at (5m, 7m, − 1m).

48 A uniformly charged thin non-conducting infinite sheet of surface charge density σ is kept as shown in figure. Find the potential difference between points P and Q.

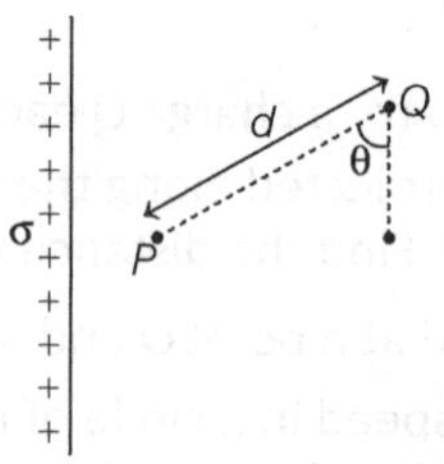

49 An infinite line of charge having charge density λ is lying along Y-axis. Find the work required in moving charge q from C to A along arc CA.

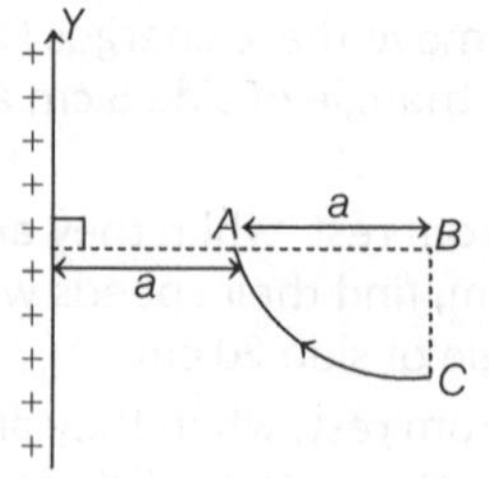

1.8 Equipotential Surfaces

50 Some equipotential surfaces are shown in the figure.

(i) Compare electric field intensities at points *A*, *B* and *C*.

(ii) Find the work done by external agent in shifting + 4 μC charge slowly from *P* to *Q* along the dotted path as shown in the figure.

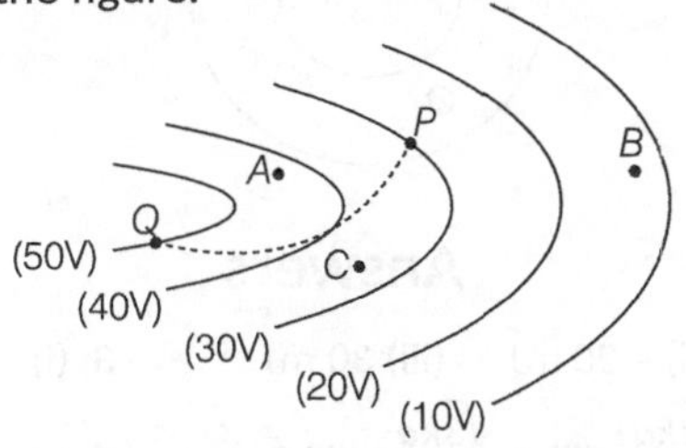

51 Some equipotential surfaces which are parallel planes are shown in figure.

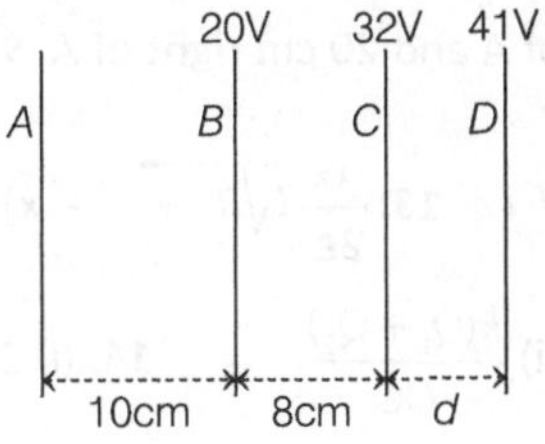

Find

(i) magnitude and direction of electric field,

(ii) separation *d* between equipotential surfaces *C* and *D* and

(iii) potential of equipotential surface *A*.

52 Some equipotential surfaces are shown in figure. Find the magnitude and direction of the electric field.

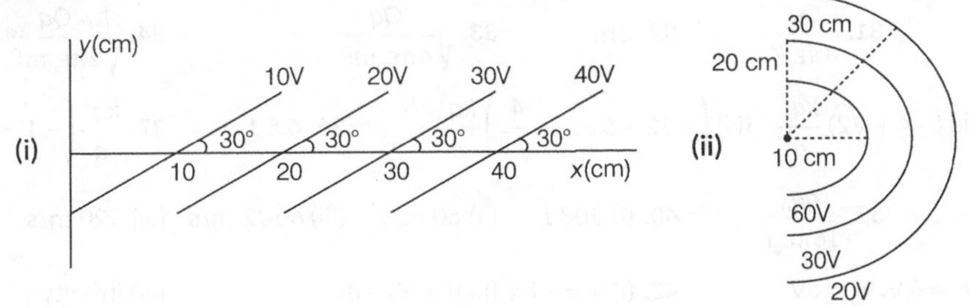

53 State True or False.

(i) Equipotential surfaces and electric field lines cut at right angle.

(ii) Two equipotential surfaces never intersect each other.

(iii) Work done by external agent in shifting a test charge from one point to another on an equipotential surface is always zero.

(iv) Electric field intensity is higher, where equipotential surfaces are closer.

1.9 Self-Energy

54 Find self-energy of the following.

(i) Uniformly charged thin spherical shell of radius *R* carrying charge *Q*.

(ii) Uniformly charged solid non-conducting sphere of radius *R* carrying charge *Q*.

55 A spherical shell of radius *R* with a uniform charge *q* has point charge q_0 at its centre. Find the work performed by the electric forces during the shell expansion slowly from radius *R* to 2*R*. Also, find out work done by external agent against electric forces.

56 Two concentric spherical shells of radius R_1 and R_2 $(R_2 > R_1)$ are having uniformly distributed charges Q_1 and Q_2, respectively. Find out total energy of the system.

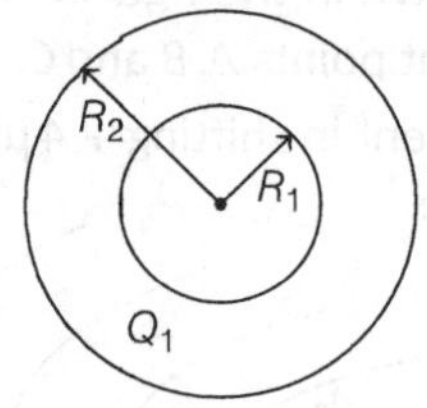

Answers

1. -20 V **2.** (i) 20 mJ (ii) -20 mJ (iii) 30 mJ **3.** (i) -20 mJ (ii) 30 mJ

4. (i) 10 eV (ii) -10 eV **5.** (i) $\frac{4\sqrt{2}kq}{a}$ (ii) $-\frac{\sqrt{3}kq}{a}$ (iii) 0 **6.** (i) $V_A = -15000$ V (ii) $V_B = -6000$ V

7. $V = \frac{q}{2\pi\varepsilon_0}, V = \frac{q}{6\pi\varepsilon_0}$ **8.** 60 cm left of A and 20 cm right of A **9.** $x^2 + y^2 + 2x - 8 = 0$ **10.** $3kq/L$

11. (i) $\frac{kQ}{R}$ (ii) $\frac{kQ}{R}$ (iii) $\frac{kQ}{R}$ **12.** 18 V **13.** $\frac{\sigma}{2\varepsilon_0}(\sqrt{R^2+x^2} - x)$ **14.** (i) 144 kV (ii) 144 kV (iii) 120 kV

15. (i) $\frac{kQ_1}{R_1} + \frac{kQ_2}{R_2}$ (ii) $\frac{kQ_1}{OB} + \frac{kQ_2}{R_2}$ (iii) $\frac{k(Q_1+Q_2)}{OC}$ **16.** (i) 216 kV, (ii) 198 kV (iii) 120 kV

17. $V = \frac{5\rho R^2}{12\varepsilon_0}$ **18.** $\frac{Q\ln 3}{4\pi\varepsilon_0 L}$ **19.** 400 V **20.** 45.5 eV, 4×10^6 m/s **21.** 0.4 C

22. (i) $+4.5 \times 10^{-5}$ J (ii) 3×10^5 N/C (iii) 1.5×10^4 V **23.** (i) -4.5×10^{-5} J (ii) 3×10^5 N/C (iii) -1.5×10^4 V

24. 1.2 J **25.** $\frac{Qq}{\pi\varepsilon_0 a}\left(\frac{1}{\sqrt{5}} - 1\right)$ **26.** $\frac{kQq}{R}$ **27.** 10 eV **28.** $\frac{6kqq_0}{a}$ **29.** 90 m/s

30. 10^3m/s **31.** $\frac{Qq}{4\pi\varepsilon_0 K}$ **32.** 2m **33.** $\sqrt{\frac{Qq}{4\pi\varepsilon_0 mR}}$ **34.** $\sqrt{\frac{Qq}{2\pi\varepsilon_0 mR}}$

35. (i) $\frac{3kq^2}{a}$ (ii) $(-4+\sqrt{2})\frac{kq^2}{a}$ (iii) $\left(-12+6\sqrt{2}-\frac{4}{\sqrt{3}}\right)\frac{kq^2}{a}$ **36.** 4.5 J **37.** $\frac{kq^2}{a}\left(-1+\frac{1}{\sqrt{3}}\right)$

38. $\frac{Q^2}{\pi\varepsilon_0 mu^2}$ **39.** $\frac{qq_0}{16\pi\varepsilon_0 r}$ **40.** (i) 108 J (ii) 60 m/s (iii) $60\sqrt{2}$ m/s (iv) 120 m/s

41. $V_A = 16$ V, $V_C = 6$ V, $V_D = 6$V **42.** (i) $V = -8x$ (ii) $V = 80 - 8x$ (iii) infinity

43. -10 V **44.** $V = -4x^3 + y^3 + C$ **45.** $V = 112 - 2xy$ **46.** $-2xy\hat{i} - (x^2+2yz)\hat{j} - y^2\hat{k}$

47. 28 V/m **48.** $V_P - V_Q = \frac{\sigma d \sin\theta}{2\varepsilon_0}$ **49.** $\frac{q\lambda \ln 2}{2\pi\varepsilon_0}$ **50.** (i) $E_B > E_C > E_A$ (ii) 80 μJ

51. (i) 150 V/m towards left (ii) 6 cm (iii) 5 V

52. (i) 200 V/m at an angle of 120° with the X-axis (ii) $(6/r^2)$ V-m radially outward,

53. (i) True (ii) True (iii) False (iv) True **54.** (i) $\frac{kQ^2}{2R}$ (ii) $\frac{3kQ^2}{5R}$

55. $\frac{q(q_0 + q/2)}{8\pi\varepsilon_0 R}, -\frac{q(q_0 + q/2)}{8\pi\varepsilon_0 R}$

56. $\frac{Q_1^2}{8\pi\varepsilon_0 R_1} + \frac{Q_2^2}{8\pi\varepsilon_0 R_2} + \frac{Q_1 Q_2}{4\pi\varepsilon_0 R_2}$

For Detailed Solutions
Scan the code

LEVEL 02 \ JEE Main Level

Single Choice Correct Questions

1 The electric field intensity at a point at a distance 2 m from a point charge q is 4 N/C. The amount of work done by the external agent in slowly bringing a charge of 2 C from infinity to this point is

(a) 4 J (b) 8 J (c) 16 J (d) 32 J

2 Two point charges $+12\ \mu C$ and $-6\ \mu C$ are placed 20 cm apart in air. There will be a point P on the line joining these charges and outside the region between them, at which the electric potential is zero. The distance of P from $-6\mu C$ is

(a) 0.10 m (b) 0.15 m (c) 0.20 m (d) 0.25 m

3 Figure represents a square carrying charges $+q, +q, -q, -q$ at its four corners as shown. A, C, P and Q are mid-points of respective sides. Then, the potential will be zero at points

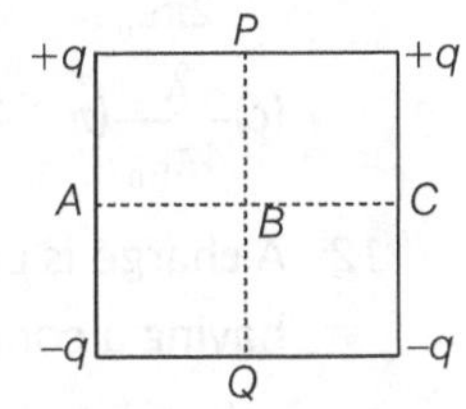

(a) A, B, C, P and Q (b) A, B and C

(c) A, P, C and Q (d) P, B and Q

4 In a regular polygon of n sides, each corner is at a distance r from the centre. Identical charges are placed at $(n-1)$ corners. At the centre, the intensity is E and the potential is V. The ratio V/E has magnitude

(a) rn (b) $r(n-1)$ (c) $(n-1)/r$ (d) $r(n-1)/n$

5 An infinite number of concentric rings carry a charge Q each alternately positive and negative. Their radii are 1m, 2m, 4m, 8m, ... , in geometric progression as shown in the figure. The potential at the centre of the rings will be

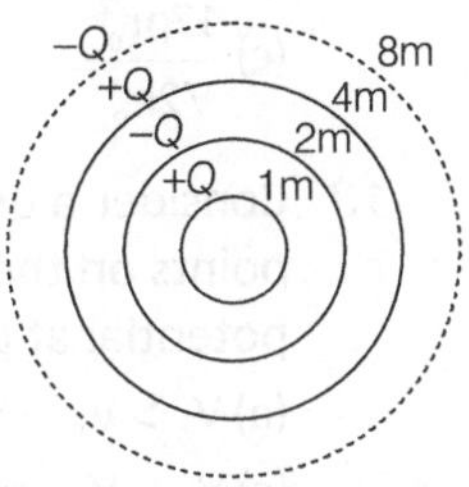

(a) zero (b) $\dfrac{Q}{12\pi\varepsilon_0}$

(c) $\dfrac{Q}{8\pi\varepsilon_0}$ (d) $\dfrac{Q}{6\pi\varepsilon_0}$

6 Figure shows three circular arcs, each of radius R and total charge as indicated. The net electric potential at the centre of curvature is

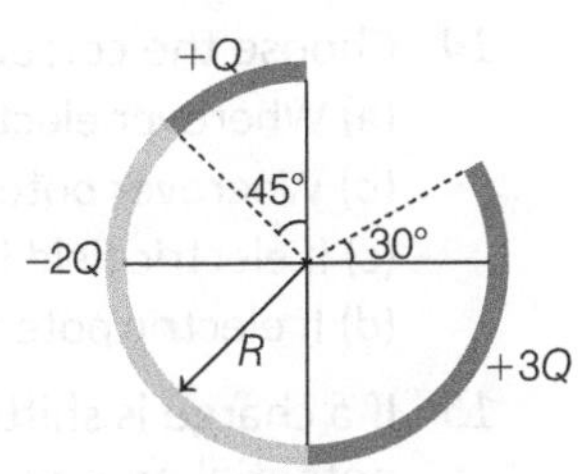

(a) $\dfrac{Q}{2\pi\varepsilon_0 R}$ (b) $\dfrac{Q}{4\pi\varepsilon_0 R}$

(c) $\dfrac{2Q}{\pi\varepsilon_0 R}$ (d) $\dfrac{Q}{\pi\varepsilon_0 R}$

7 If a uniformly charged spherical shell of radius 10 cm has a potential V at a point distant 5 cm from its centre, then the potential at a point distant 15 cm from the centre will be

(a) $\dfrac{V}{3}$ (b) $\dfrac{2V}{3}$ (c) $\dfrac{3}{2}V$ (d) 3V

8 A solid sphere of radius R is charged uniformly. At what distance from its surface is the electrostatic potential half of the potential at the centre?

(a) R (b) $R/2$ (c) $R/3$ (d) $2R$

9 Three concentric spherical metallic shells A, B and C of radii a, b and $c (c > b > a)$ have charge densities σ, $-\sigma$ and σ, respectively. The potential of shell B is

(a) $(a+b+c)\dfrac{\sigma}{\varepsilon_0}$ (b) $\left(\dfrac{a^2}{b}-b+c\right)\dfrac{\sigma}{\varepsilon_0}$ (c) $\left(\dfrac{a^2}{b}-\dfrac{b^2}{c}+c\right)\dfrac{\sigma}{\varepsilon_0}$ (d) None of these

10 A total charge Q is distributed over two concentric hollow spheres of radii a and b ($b > a$), such a way that their surface charge densities are equal. The potential at the common centre is given by

(a) $\frac{Q}{4\pi\varepsilon_0} \cdot \frac{(a+b)}{(a^2+b^2)}$ (b) $\frac{Q}{4\pi\varepsilon_0} \cdot \frac{(b-a)}{(a^2+b^2)}$

(c) $\frac{Q}{4\pi\varepsilon_0} \cdot \frac{(a+b)}{(a+b)^2}$ (d) $\frac{Q}{4\pi\varepsilon_0} \cdot \frac{(b-a)}{(a+b)^2}$

11 A wire having a uniform linear charge density λ is bent into the shape as shown in figure. Find the electric potential at point O.

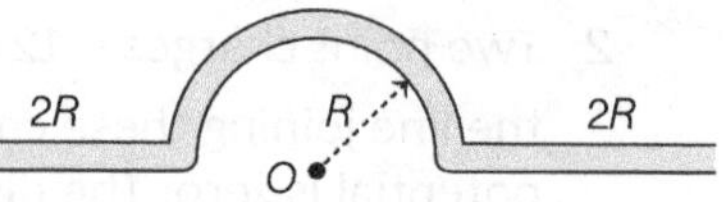

(a) $\frac{\lambda}{2\pi\varepsilon_0}(\pi + \ln 3)$ (b) $\frac{\lambda}{2\pi\varepsilon_0}(\pi + 2\ln 3)$

(c) $\frac{\lambda}{4\pi\varepsilon_0}(\pi + 2\ln 3)$ (d) $\frac{\lambda}{4\pi\varepsilon_0}(\pi + \ln 3)$

12 A charge is uniformly distributed inside a spherical body of radius $r_1 = 2r_0$ having a concentric cavity of radius $r_2 = r_0$ (ρ is charge density inside the sphere). Potential of a point P at a distance $\frac{3r_0}{2}$ from centre is

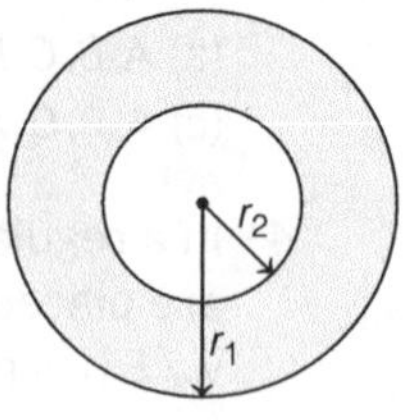

(a) $\frac{7\rho r_0^2}{6\varepsilon_0}$ (b) $\frac{101\rho r_0^2}{72\varepsilon_0}$

(c) $\frac{17\rho r_0^2}{72\varepsilon_0}$ (d) $\frac{84\rho r_0^2}{72\varepsilon_0}$

13 Consider a uniformly charged thin hemispherical shell. A, B and C are three points on the circular base of hemisphere, such that A is the centre. Let the potential at points A, B and C be V_A, V_B and V_C, respectively. Then,

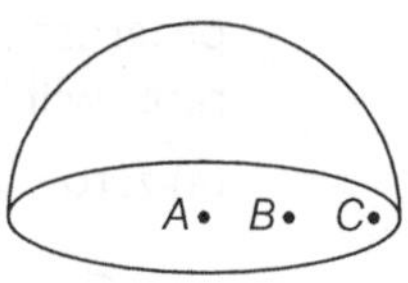

(a) $V_A > V_B > V_C$ (b) $V_C > V_B > V_A$

(c) $V_B > V_A$ and $V_B > V_C$ (d) $V_A = V_B = V_C$

14 Choose the correct statement.

(a) Wherever electric field is zero, potential has to be zero.

(b) Wherever potential is zero, electric field has to be zero.

(c) If electric field is constant in some part of space, electric potential will be zero in that space.

(d) If electric potential is constant in some part of space, electric field will be zero in that space.

15 If a charge is shifted from a low potential region to high potential region, the electrical potential energy

(a) increases (b) decreases

(c) remains constant (d) may increase or decrease

16 A ball of mass 1 g and charge 10^{-8} C moves from a point $A(V_A = 600\text{ V})$ to the point B whose potential is zero. Velocity of the ball at the point B is 20 cm/s. The velocity of the ball at the point A is

(a) 16.7 m/s (b) 16.7 cm/s (c) 2.8 m/s (d) 2.8 cm/s

17 When a negative charge is released and moves in electric field, it moves toward a position of

(a) lower electric potential and lower potential energy

(b) lower electric potential and higher potential energy

(c) higher electric potential and lower potential energy

(d) higher electric potential and higher potential energy

18 A particle of charge Q and mass m travels through a potential difference V from rest. The final linear momentum of the particle is

(a) $\frac{mV}{Q}$ (b) $2Q\sqrt{mV}$ (c) $\sqrt{2mQV}$ (d) $\sqrt{\frac{2QV}{m}}$

19 A particle A has charge $+q$ and particle B has charge $+4q$ with each of them having the same mass m. When allowed to fall from rest through same electrical potential difference, the ratio of their speed $v_A : v_B$ will be

(a) 2 : 1 (b) 1 : 2 (c) 4 : 1 (d) 1 : 4

20 An α-particle moves towards a nucleus always at rest. The kinetic energy of the α-particle at a large distance is 10 MeV. If the atomic number of the nucleus is 50, then the minimum distance of approach is

(a) 1.44×10^{-14} m (b) 2.88×10^{-14} m (c) 1.44×10^{-10} m (d) 2.88×10^{-10} m

21 Three particles of charge $+2Q$, $-Q$ and $-Q$ are placed at the corners of an equilateral triangle ABC as shown in the figure. Now, a fourth particle of charge $+Q$ is to be moved slowly from the centroid G to the mid-point D of side BC of the triangle. The work done by the external agent in order to move this fourth charge is given by

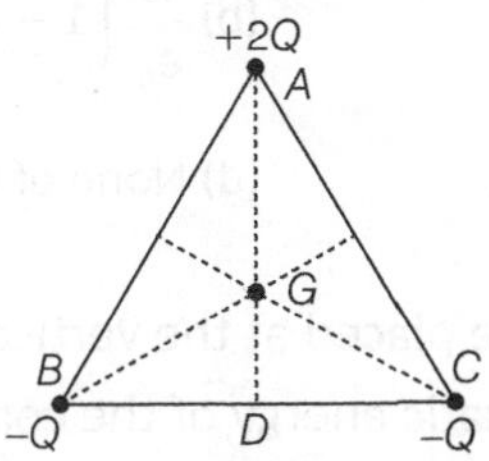

(a) $W = \frac{-Q^2}{\pi\varepsilon_0 a}$ (b) $W = \frac{-Q^2}{\pi\varepsilon_0 a}\left(1-\frac{1}{\sqrt{3}}\right)$

(c) $W = \frac{Q^2}{\pi\varepsilon_0 a}\left(1-\frac{1}{\sqrt{3}}\right)$ (d) $W = \frac{Q^2}{\pi\varepsilon_0 a}$

22 A point charge Q is fixed. A small charge q and mass m is given a velocity v_0 from infinity and perpendicular distance r_0 as shown in figure. If distance of closest approach is $r_0/2$, the value of q is (Given, $mv_0^2 = \frac{Q^2}{4\pi\varepsilon_0 r_0}$)

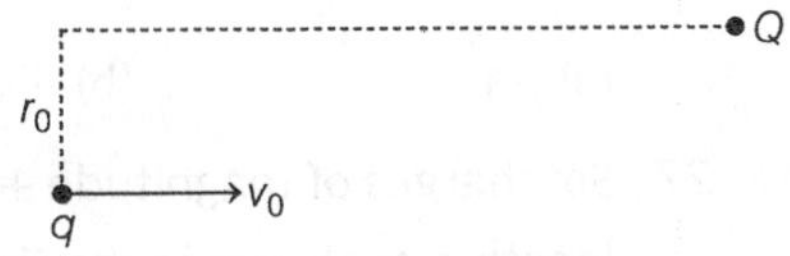

(a) $q = -\frac{Q}{4}$ (b) $q = -\frac{Q}{2}$

(c) $q = -\frac{3Q}{4}$ (d) $q = -Q$

23 A point charge q and mass m is just dropped from a height $a = 4R/3$ above a ring of charge Q and radius R. If q just reaches the centre of the ring, the value of q is

(Given, $\frac{Q^2}{4\pi\varepsilon_0 R^2} = mg$)

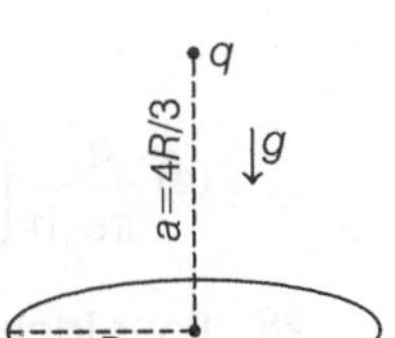

(a) $\frac{Q}{3}$ (b) 3Q

(c) $\frac{10Q}{3}$ (d) $\frac{3Q}{10}$

24 A bullet of mass m and charge q is fired towards a solid uniformly charged sphere of radius R and total charge $+q$. If it strikes the surface of sphere with speed u, find the minimum speed u, so that it can penetrate through the sphere. (Neglect all resistance forces or friction acting on bullet except electrostatic forces.)

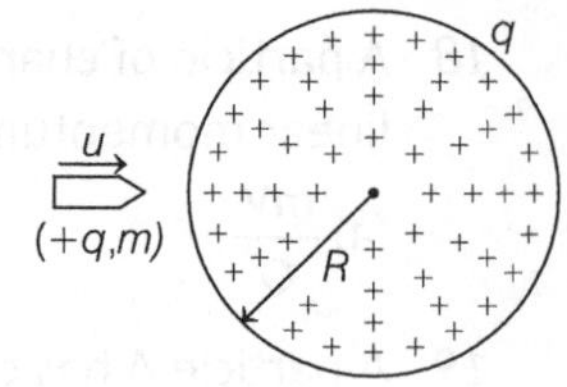

(a) $\frac{q}{\sqrt{2\pi\varepsilon_0 mR}}$ (b) $\frac{q}{\sqrt{4\pi\varepsilon_0 mR}}$

(c) $\frac{q}{\sqrt{8\pi\varepsilon_0 mR}}$ (d) $\frac{\sqrt{3}q}{\sqrt{4\pi\varepsilon_0 mR}}$

25 Two fixed insulating rings A and B carry charges with uniform linear charge density $+\lambda$ and $-\lambda$ respectively, as shown in the figure. The planes of the rings are parallel to each other and their axes are coinciding. A particle of charge q and mass m is released with zero velocity from centre P of the positively charged ring. The kinetic energy of the particle, when it reaches centre Q of the negatively charged ring will be

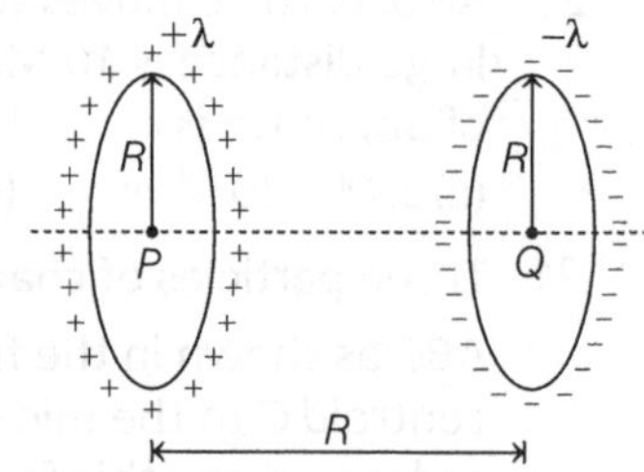

(a) $\sqrt{\frac{2\lambda q}{m\varepsilon_0}\left(1-\frac{1}{\sqrt{2}}\right)}$ (b) $\frac{\lambda}{\varepsilon_0}\left(1-\frac{1}{\sqrt{2}}\right)q$

(c) $\frac{\lambda}{2\varepsilon_0}\left(1-\frac{1}{\sqrt{2}}\right)q$ (d) None of these

26 Three point charges Q, q and q are placed at the vertices of a right angle isosceles triangle as shown in figure. The net electrostatic energy of the configuration is zero, if q/Q is equal to

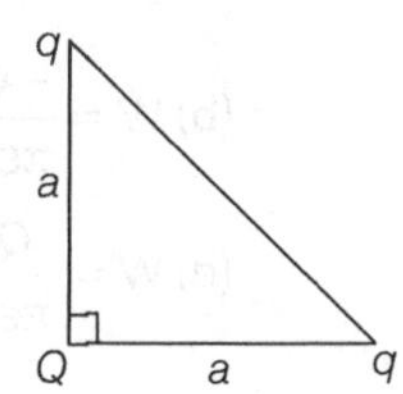

(a) -1 (b) $-\sqrt{2}$ (c) -2 (d) $-2\sqrt{2}$

27 Six charges of magnitude $+q$ and $-q$ are fixed at the corners of a regular hexagon of edge length a as shown in the figure. The electrostatic potential energy of the system of charged particles is

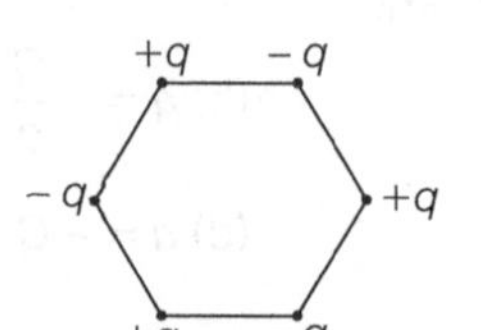

(a) $\frac{q^2}{\pi\varepsilon_0 a}\left(\frac{\sqrt{3}}{8}-\frac{15}{4}\right)$ (b) $\frac{q^2}{\pi\varepsilon_0 a}\left(\frac{\sqrt{3}}{2}-\frac{9}{4}\right)$ (c) $\frac{q^2}{\pi\varepsilon_0 a}\left(\frac{\sqrt{3}}{4}-\frac{15}{2}\right)$ (d) $\frac{q^2}{\pi\varepsilon_0 a}\left(\frac{\sqrt{3}}{2}-\frac{15}{8}\right)$

28 Four identical particles, each of charge q and mass m are released from rest simultaneously at the vertices of a square of side L. When their distance from the centre of the square doubles, the speed of each particle is

(a) $\sqrt{\frac{q^2(1+2\sqrt{2})}{16\sqrt{2}\pi\varepsilon_0 mL}}$ (b) $\sqrt{\frac{q^2(1+2\sqrt{2})}{16\pi\varepsilon_0 mL}}$ (c) $\sqrt{\frac{q^2(1+2\sqrt{2})}{8\sqrt{2}\pi\varepsilon_0 mL}}$ (d) $\sqrt{\frac{q^2(1+2\sqrt{2})}{4\sqrt{2}\pi\varepsilon_0 mL}}$

29 Three point charges q, $2q$ and $8q$ are to be placed on a straight line 9 cm long. The system possesses minimum potential energy, when

(a) $2q$ and q lie at ends with $8q$ at 3 cm from $2q$

(b) $2q$ and $8q$ lie at ends with q at 6 cm from $8q$

(c) q and $8q$ lie at ends with $2q$ at 6 cm from q

(d) $2q$ and $8q$ lie at ends with q at 3 cm from $8q$

30 A charge 3 C experiences a force 3000 N, when placed in a uniform electric field. The potential difference between two points separated by a distance of 1 cm along the field lines is

(a) 10 V (b) 90 V

(c) 1000 V (d) 9000 V

31 There is a uniform electric field in x-direction. If the work done by external agent in moving a charge of 0.2 C through a distance of 2 m slowly along the line making an angle of 60° with x-direction is 4 J, then the magnitude of E is

(a) $\sqrt{3}$N/C (b) 4 N/C

(c) 5 N/C (d) 20 N/C

32 The potential difference between points A and B in the given uniform electric field is

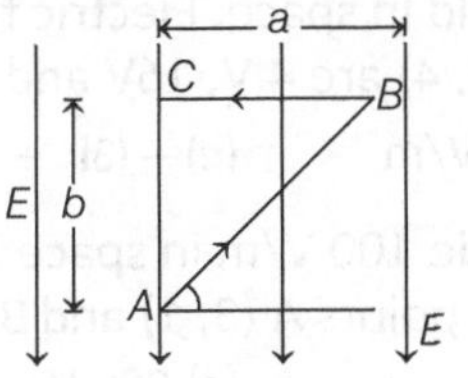

(a) Ea (b) Eb

(c) $E\sqrt{(a^2 + b^2)}$ (d) $Eb/\sqrt{2}$

33 Let V_0 be the potential at the origin in an electric field $\mathbf{E} = E_x\hat{\mathbf{i}} + E_y\hat{\mathbf{j}}$. The potential at the point (x, y) is

(a) $V_0 - xE_x - yE_y$ (b) $V_0 + xE_x + yE_y$

(c) $xE_x + yE_y - V_0$ (d) $(\sqrt{x^2 + y^2})\sqrt{E_x^2 + E_y^2} - V_0$

34 Electric potential at (x, y, z) is given as $V = -xy^2\sqrt{z}$. The electric field at (2, 1, 1) is

(a) $\hat{\mathbf{i}} + 4\hat{\mathbf{j}} + \hat{\mathbf{k}}$ (b) $-\hat{\mathbf{i}} - 4\hat{\mathbf{j}} - 2\hat{\mathbf{k}}$

(c) $\hat{\mathbf{i}} + 2\hat{\mathbf{j}} + \hat{\mathbf{k}}$ (d) $-\hat{\mathbf{i}} - 2\hat{\mathbf{j}} - \hat{\mathbf{k}}$

35 Figure shows variation of potential V with distance y from origin along Y-axis. What is electric field at $y = 2.5$ m and $y = 5.5$ m?

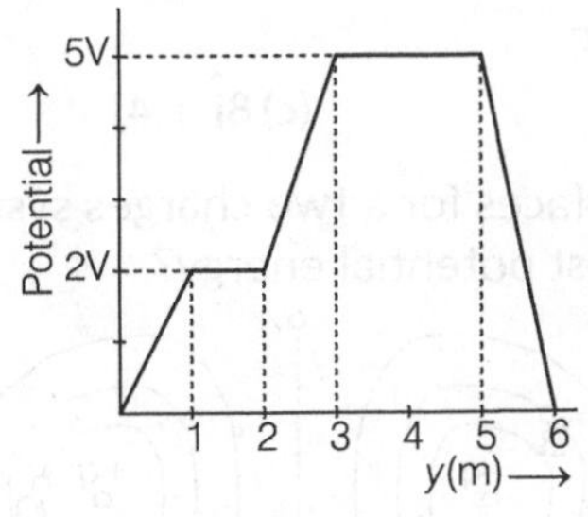

(a) $-3\hat{\mathbf{j}}, 5\hat{\mathbf{j}}$ (b) $2\hat{\mathbf{j}}, -5\hat{\mathbf{j}}$

(c) $3.5\hat{\mathbf{j}}, 2.5\hat{\mathbf{j}}$ (d) $3.5\hat{\mathbf{j}}, -2.5\hat{\mathbf{j}}$

36 In the space, electric field exists only in x-direction. The diagram shows V *versus* x graph (variation of potential along X-axis). Which of the following graphs most appropriately represent the E *versus* x graph? (Here, E = electric field and V = potential)

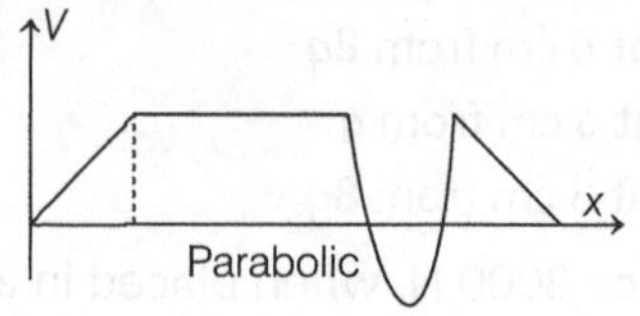

(a) 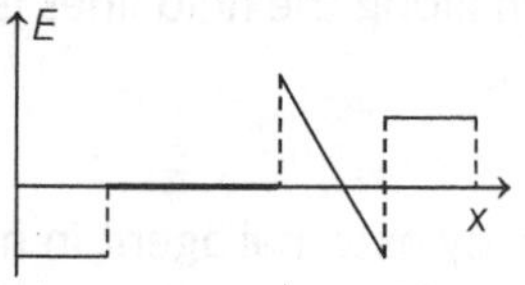

(b) 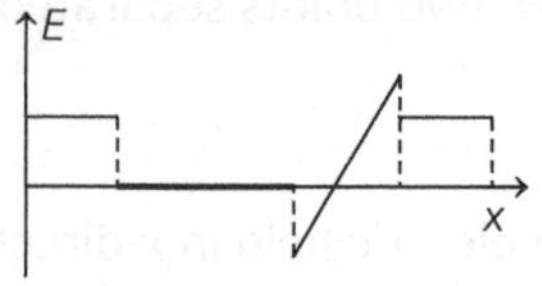

(c) 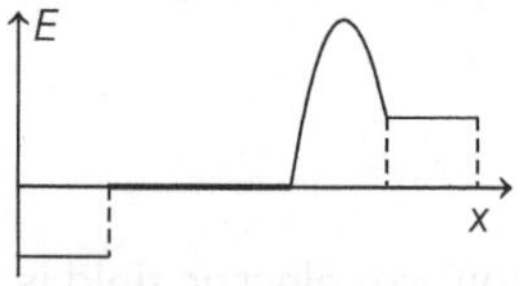

(d) 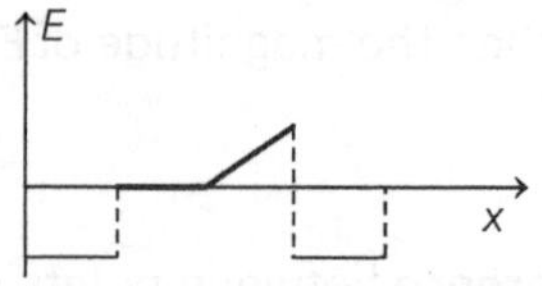

37 There exists a uniform electric field in space. Electric field is parallel to XY-plane. The potential of points $A(2, 2)$, $B(-2, 2)$ and $C(2, 4)$ are 4 V, 16V and 12V, respectively. The electric field is

(a) $(4\hat{\mathbf{i}} + 5\hat{\mathbf{j}})$ V/m (b) $(3\hat{\mathbf{i}} + 4\hat{\mathbf{j}})$ V/m (c) $-(3\hat{\mathbf{i}} + 4\hat{\mathbf{j}})$ V/m (d) $(3\hat{\mathbf{i}} - 4\hat{\mathbf{j}})$ V/m

38 Uniform electric field of magnitude 100 V/m in space is directed along the line $y = 3 + x$. Find the potential difference between points A (3, 1) and B (5, 3).

(a) 100 V (b) $200\sqrt{2}$ V (c) 200 V (d) 0

39 A, B, C, D, P and Q are points in a uniform electric field. The potentials at these points are $V(A) = 2$ V, $V(P) = V(B) = V(D) = 5$ V and $V(C) = 8$ V. The electric field at P is

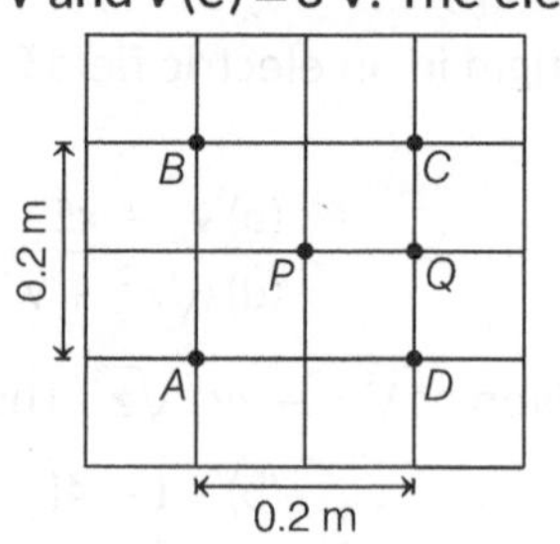

(a) 10 Vm^{-1} along PQ (b) $15\sqrt{2}$ Vm^{-1} along PA
(c) 5 Vm^{-1} along PC (d) 5 Vm^{-1} along PA

40 The equation of an equipotential line in an electric field is $y = 2x$, then the electric field strength vector at (1, 2) may be

(a) $4\hat{\mathbf{i}} + 8\hat{\mathbf{j}}$ (b) $4\hat{\mathbf{i}} - 8\hat{\mathbf{j}}$ (c) $8\hat{\mathbf{i}} + 4\hat{\mathbf{j}}$ (d) $-8\hat{\mathbf{i}} + 4\hat{\mathbf{j}}$

41 Figure shows equipotential surfaces for a two charges system. At which of the labelled points, will an electron have the highest potential energy?

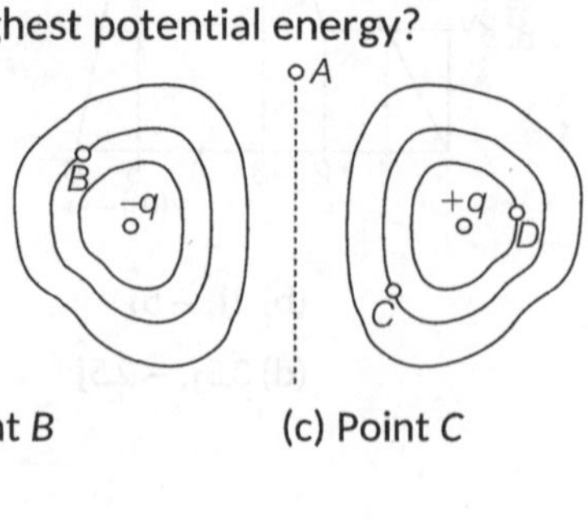

(a) Point A (b) Point B (c) Point C (d) Point D

42 An infinite non-conducting sheet of charge has a surface charge density of $10^{-7}\ C/m^2$. The separation between two equipotential surfaces near the sheet whose potential differ by 5V is

(a) 0.88 cm (b) 0.88 mm (c) 0.88 m (d) 5×10^{-7} m

43 Figure below shows two equipotential lines in *XY*-plane for an electric field. The scales are marked. Electric field in the space between these equipotential lines are respectively

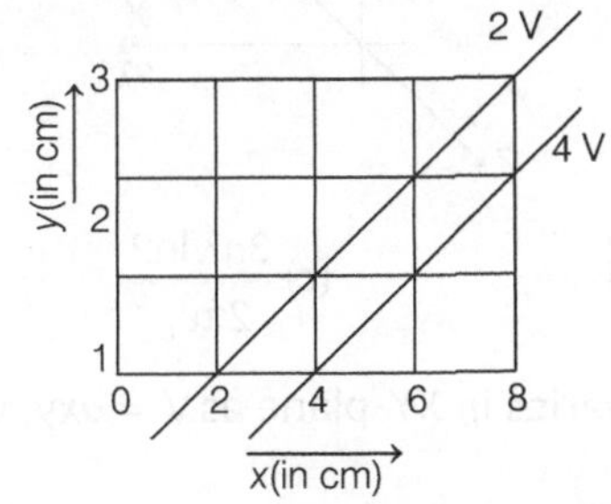

(a) $(100\hat{i} - 200\hat{j})$V/m (b) $(-100\hat{i} + 200\hat{j})$V/m

(c) $(200\hat{i} + 100\hat{j})$V/m (d) $(-200\hat{i} - 100\hat{j})$V/m

44 A non-conducting ring of radius 0.5 m carries a total charge of 1.11×10^{-10} C distributed non-uniformly on its circumference producing an electric field **E** everywhere in space. The value of the line integral $\int_{l=\infty}^{l=0} -\mathbf{E}\cdot d\mathbf{l}$ (where, $l = 0$ being centre of the ring) in volts is

(a) +2 (b) – 1

(c) – 2 (d) zero

45 The particle of mass *m* and charge *q* will touch the infinitely large plate of uniform charge density σ, if its velocity v is more than (Given that, $\sigma q > 0$)

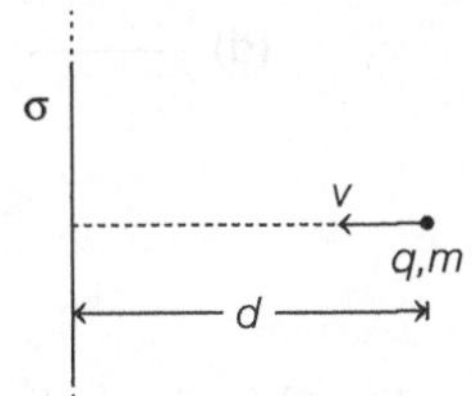

(a) zero (b) $\sqrt{\dfrac{2\sigma q d}{m\varepsilon_0}}$ (c) $\sqrt{\dfrac{\sigma q d}{m\varepsilon_0}}$ (d) None of these

46 A long line charge of charge density λ lies along the X-axis and let the surface of zero potential passes through (0, 5,12)m. The potential at a point (2, 3, – 4) m is

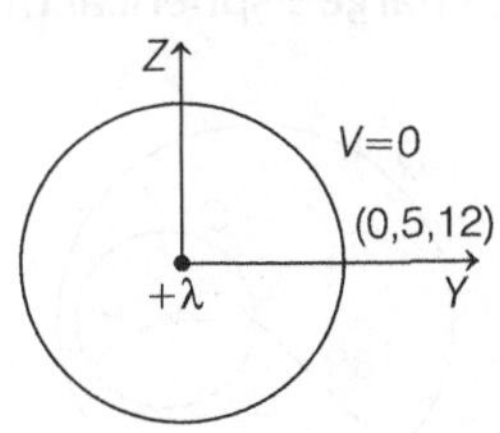

(a) $\dfrac{\lambda}{2\pi\varepsilon_0}\ln\left(\dfrac{13}{5}\right)$ (b) $\dfrac{2\lambda}{\pi\varepsilon_0}\ln\left(\dfrac{13}{5}\right)$

(c) $\dfrac{\lambda}{4\pi\varepsilon_0}\ln\left(\dfrac{13}{5}\right)$ (d) None of these

47 The diagram shows three infinitely long uniform line charges placed on the X, Y and Z-axes. The work done in moving a point charge $+q$ from (1, 1, 1) to (0, 1, 1) is equal to

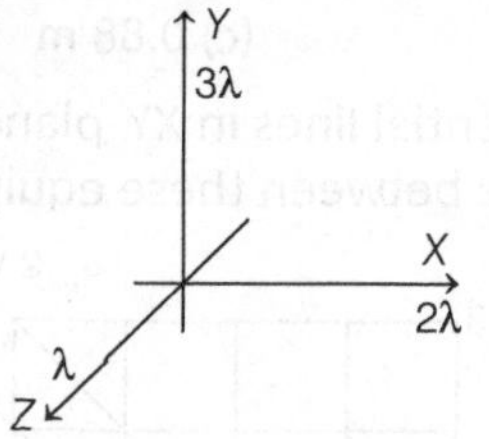

(a) $\frac{q\lambda \ln 2}{2\pi\varepsilon_0}$ (b) $\frac{q\lambda \ln 2}{\pi\varepsilon_0}$ (c) $\frac{3q\lambda \ln 2}{2\pi\varepsilon_0}$ (d) None of these

48 If the potential at any region varies in XY-plane as $V = axy$, which of the following field pattern is correct?

(a) 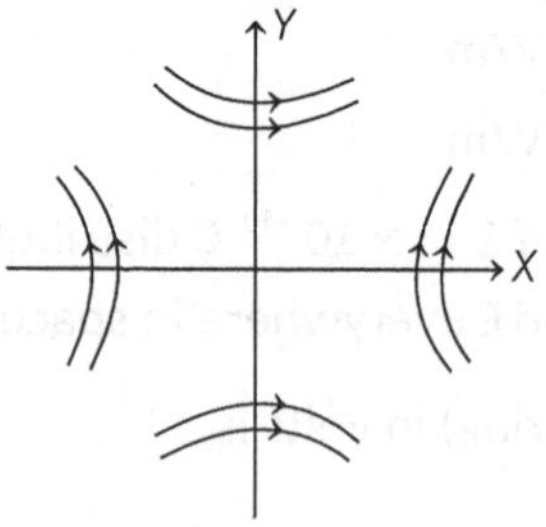

(b) 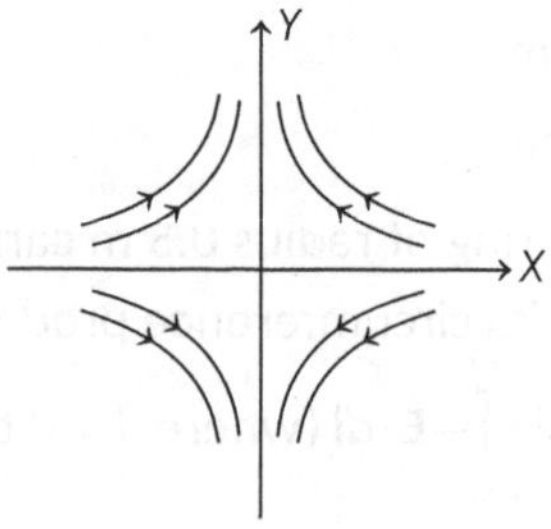

(c) 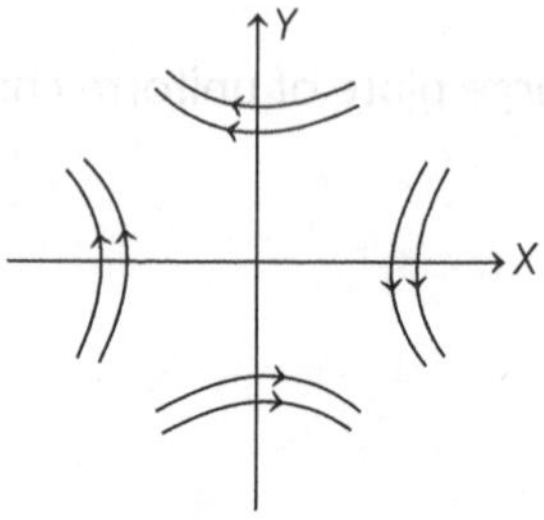

(d) 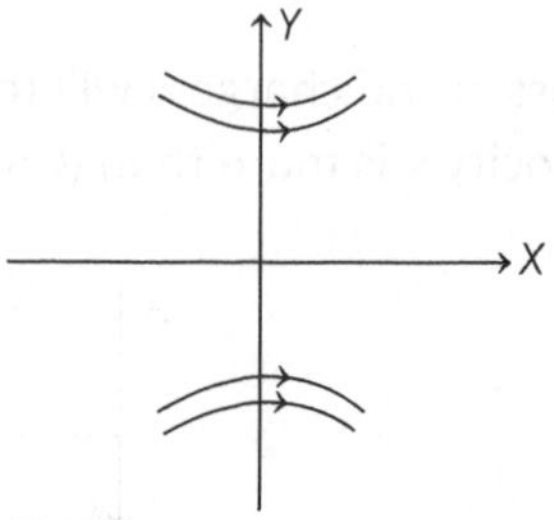

49 A soap bubble of spherical shape (radius R) is given charge Q as a result of which it expands to double its initial radius. The electrostatic potential energy stored in it will change by (as the radius expands from initial to final value)

(a) $\Delta U = \frac{-Q^2}{16\pi\varepsilon_0 R}$ (b) $\Delta U = \frac{Q^2}{8\pi\varepsilon_0 R}$ (c) $\Delta U = \frac{-Q^2}{4\pi\varepsilon_0 R}$ (d) $\Delta U = \frac{-Q^2}{32\pi\varepsilon_0 R}$

50 Shown in figure are two uniformly charged spherical thin shells. Total electric potential energy of the system will be

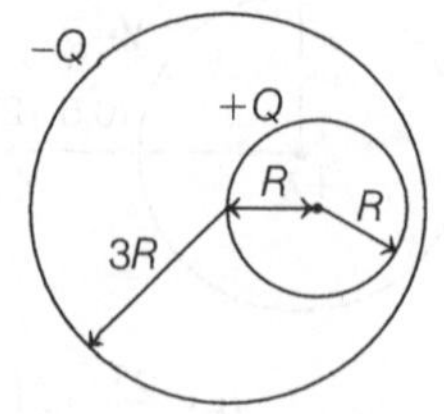

(a) $\frac{kQ^2}{12R}$ (b) $\frac{kQ^2}{6R}$ (c) $\frac{kQ^2}{2R}$ (d) $\frac{kQ^2}{3R}$

Numerical Value Questions

51 At the end points of a line segment of a length of $d = \frac{\sqrt{337}}{84}$ m, there are two identical positive electric charges q. What is the ratio of the electric field strength and the electric potential (in m^{-1}) at a point characterised by an angle $\alpha = 37°$ on the circle drawn around the line segment.

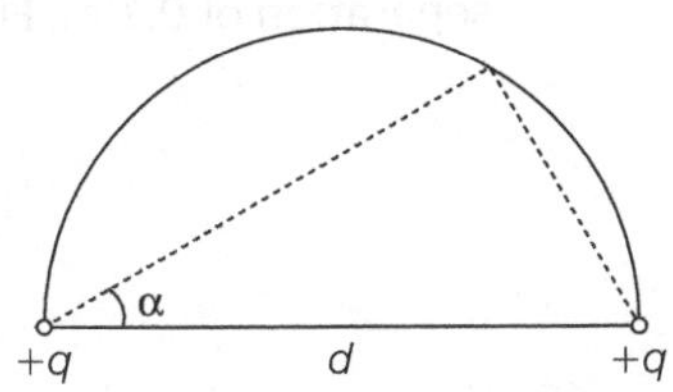

52 Two positive and two negative point charges $|Q| = 10$ nC are placed at the corners of a square insulator frame of side 5 m and the side of the frame are hinged together. The potential at the centre of the square frame is V_1. The frame is deformed into a rhombus as shown in figure. The potential at the centre of the rhombus frame is V_2. Find $V_2 - V_1$ (in volt).

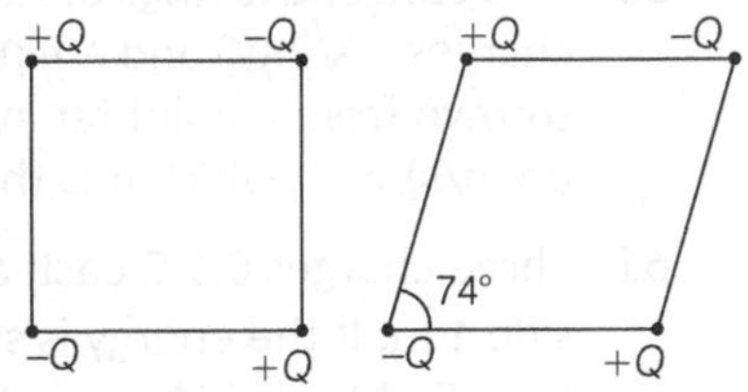

53 Two charges q and $-2q$ are placed at (−3, 0) and (3,0) in XY-plane. The locus of the point in the plane of the charges, where the field potential is zero is $(x + \alpha)^2 + y^2 = 4\beta^2$. Find the value of $(\alpha + \beta)$.

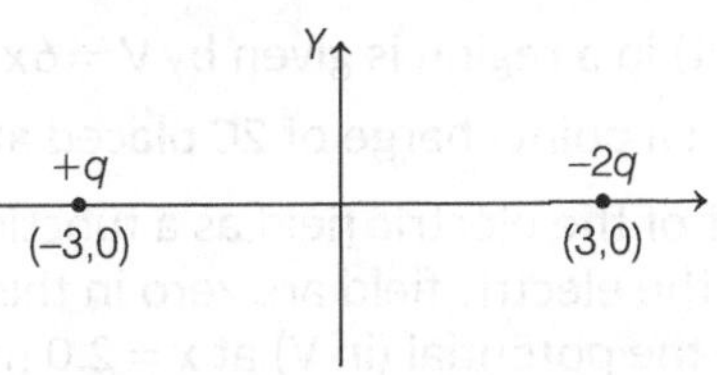

54 The charge on a ring of radius R varies according to the relation $\lambda = \lambda_0 \cos^2 \phi$, where λ_0 is a positive constant and ϕ is the azimuthal angle. The electric potential due to this ring at the centre is $\frac{\lambda_0}{n\varepsilon_0}$. Find the value of n.

55 A solid sphere of radius $\sqrt{60}$ cm is charged uniformly. Find the distance (in cm) of a point from its centre, where the electrostatic potential is 1.2 times the potential at the surface.

56 Figure shows the part of a hemisphere of radius $R = 2$ m and surface charge density $(\sigma) = 2\varepsilon_0 C/m^2$. Calculate the electric potential (in volt) at centre O.

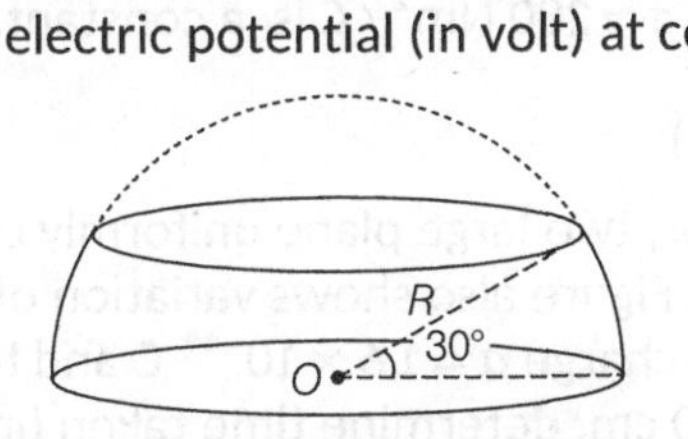

57 A positively charged particle starts at rest 25 cm from a second positively charged particle which is held stationary throughout the experiment. The first particle is released and accelerates directly away from the second particle. When the first particle has moved 25 cm, it has reached a velocity of $10\sqrt{2}$ m/s. What is the maximum velocity (in m/s) that the first particle will reach?

58 Two charge particles of opposite sign and magnitude Q each are projected directly away from each other with the same speed $\left(\frac{3Q^2}{32\pi\varepsilon_0 mr}\right)^{1/2}$, where r is the initial distance between the particles. Their masses are m and $2m$. The maximum distance between them is nr. Find the value of n.

59 Two small charged blocks of charges 5μC and 3μC are kept on a rough surface ($\mu = 0.5$) at a separation of 0.1 m. Find the separation (in cm) between the two blocks, when they come to rest.

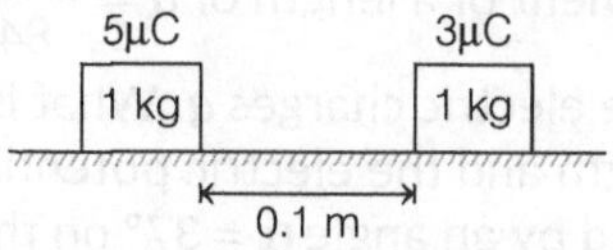

60 Two concentric rings of radii 2m and 4m are placed in YZ-plane with centre at origin and having charges $-\sqrt{2}$ μC and $+4$ μC, respectively. A particle with charge $+20\sqrt{6}$ μC and mass 10 g is thrown from a point far away from origin on X-axis towards origin. Find the minimum speed (in m/s) with which it is thrown, so that it will cross the origin.

61 Three charges 0.1 C each are placed on the corners of an equilateral triangle of side 1 m. If the energy is supplied to this system at the rate of 10 kW, how much time (in h) would be required to move one of the charges onto the mid-point of the line joining the other two?

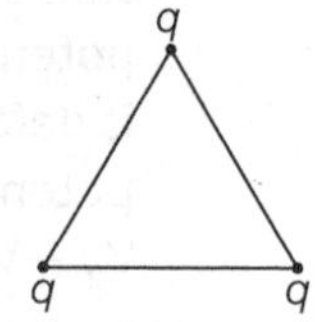

62 In a uniform electric field, the potential is 100 V at the origin of coordinates and 84 V at each of the points (1, 0, 0), (0, 2, 0) and (0, 0, 8). Find the potential (in V) at the point (3, 4, 5).

63 The electric potential (in volt) in a region is given by $V = 6x - 8xy^2 - 8y + 6yz - 4x^2$. Find electric force (in N) acting on a point charge of 2C placed at the origin.

64 A graph of the x-component of the electric field as a function of x in a region of space is shown. The y and z-components of the electric field are zero in this region. If the electric potential is 50 V at the origin, then find the potential (in V) at x = 2.0 m.

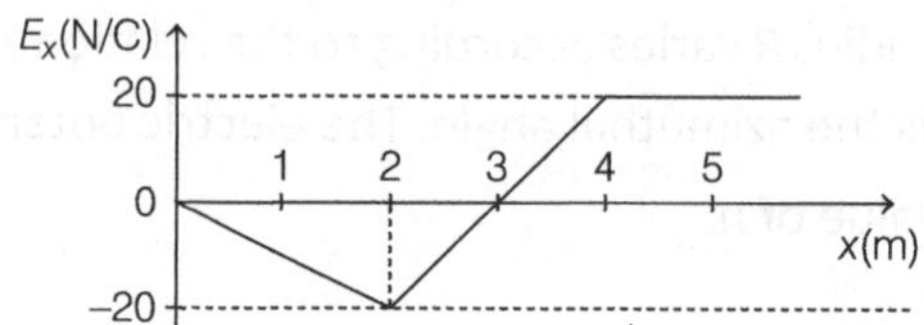

65 The electric field strength depends only on the x, y and z-coordinates according to the law $E = \dfrac{a(x\hat{i} + y\hat{j} + z\hat{k})}{(x^2 + y^2 + z^2)^{3/2}}$, where $a = 280\ \text{Nm}^2/\text{C}$ is a constant. Find the potential difference (in V) between (3, 2, 6) and (0, 3, 4).

66 In a linear particle accelerator, two large plane uniformly charged are placed parallel to each other as shown in the figure. Figure also shows variation of potential between planes. A charged particle carrying a charge $q = 1.6 \times 10^{-19}$ C and having mass $m = 16 \times 10^{-31}$ kg is released from rest at x = – 10 cm, determine time taken (in ns) by particle to strike the other plane. Neglect relativistic considerations.

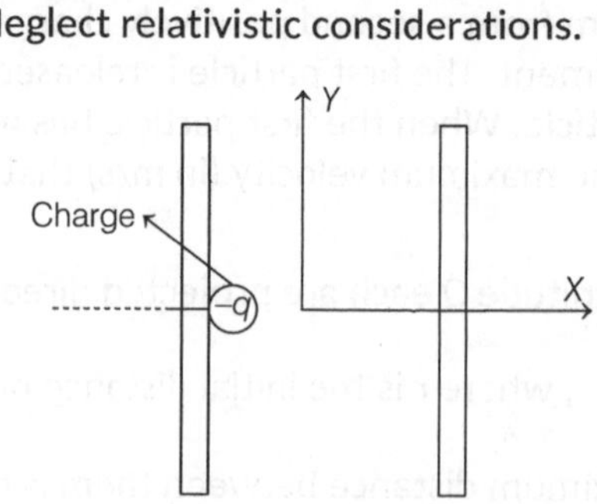

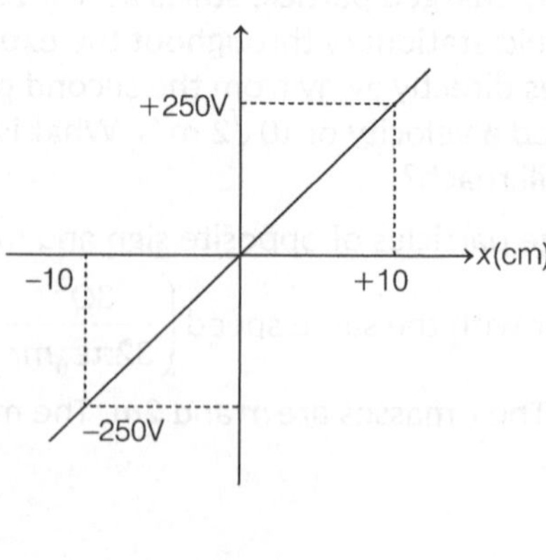

Problems in Physics

67 An infinite plane of charge with $\sigma = +1.77\ \mu C/m^2$ is tilted at a 37° angle to the vertical direction as shown below. What is the potential difference $V_A - V_B$ in volts, between points A and B at 4.50 cm distance apart?

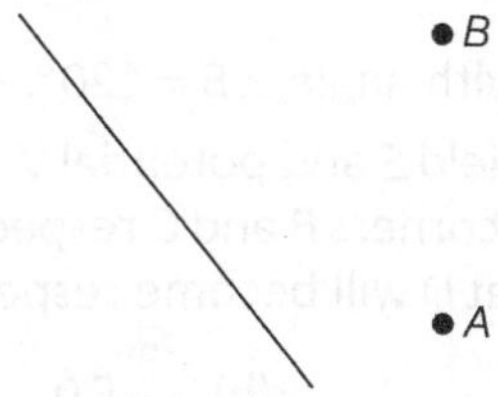

68 For a hypothetical electric field in x-direction, its potential is continuously decreasing and becoming half after every 20 cm. The rate of decreasing of potential with distance is also directly proportional to the potential itself. If the ratio of electric field and potential is $x \ln 2 (m^{-1})$. Find the value of x.

69 The charge $Q = \pi$ C is distributed on a thin semi-circular ring of radius $R = 2$m.

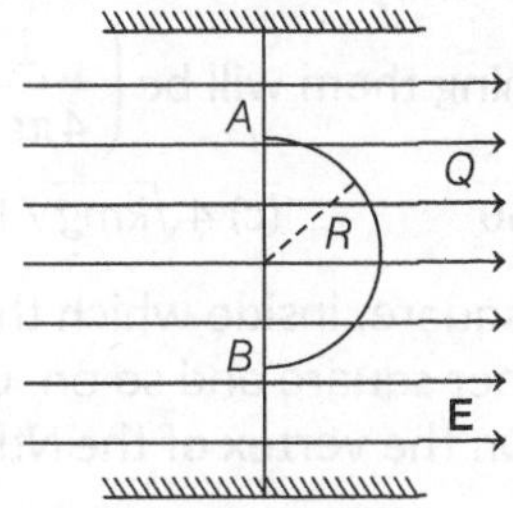

There is a uniform electrostatic field $|\mathbf{E}| = 2$N/C directed horizontally. The semi-circular ring can rotate freely about a fixed vertical axis AB. Initially, the ring is in static equilibrium as shown in figure. If we want to rotate it about the fixed axis by 90°, then minimum work required on the ring is x J. Find the value of x.

70 A particle of mass 2 g and charge $+1\ \mu C$ is projected towards a non-conducting fixed spherical shell having the charge $+3\mu C$ uniformly distributed on its surface. Find the minimum initial velocity (in m/s) of projection required, if the particle just grazes the shell.

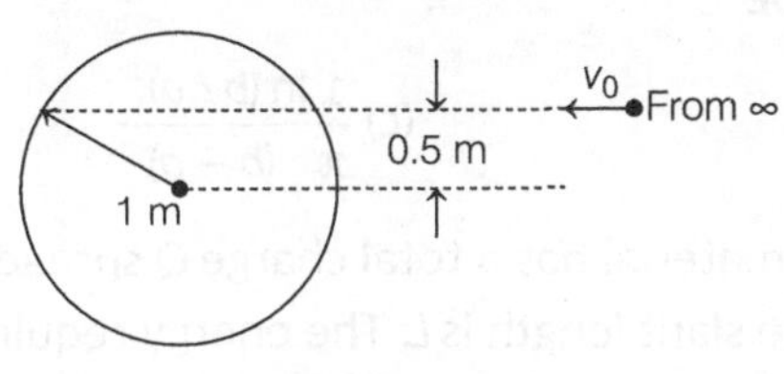

Answers

1. (c)	**2.** (c)	**3.** (b)	**4.** (b)	**5.** (d)	**6.** (a)	**7.** (b)	**8.** (c)	**9.** (b)	**10.** (a)
11. (c)	**12.** (b)	**13.** (d)	**14.** (d)	**15.** (d)	**16.** (b)	**17.** (c)	**18.** (c)	**19.** (b)	**20.** (a)
21. (b)	**22.** (c)	**23.** (c)	**24.** (b)	**25.** (b)	**26.** (d)	**27.** (d)	**28.** (c)	**29.** (b)	**30.** (a)
31. (d)	**32.** (b)	**33.** (a)	**34.** (a)	**35.** (a)	**36.** (a)	**37.** (d)	**38.** (b)	**39.** (b)	**40.** (d)
41. (b)	**42.** (b)	**43.** (b)	**44.** (a)	**45.** (c)	**46.** (a)	**47.** (b)	**48.** (c)	**49.** (a)	**50.** (d)
51. (5)	**52.** (15)	**53.** (7)	**54.** (4)	**55.** (6)					
56. (1)	**57.** (20)	**58.** (2)	**59.** (27)	**60.** (6)					
61. (5)	**62.** (10)	**63.** (20)	**64.** (70)	**65.** (16)					
66. (40)	**67.** (2700)	**68.** (5)	**69.** (8)	**70.** (6)					

For Detailed Solutions
Scan the code

LEVEL 03 \ JEE Advanced Level

Single Choice Correct Questions

1 Consider a parallelogram $ABCD$, with angle $\angle B = 120°$. A charge $+Q$ placed at the corner A produces field E and potential V at corner D. If we now add charges $-2Q$ and $+Q$ at corners B and C respectively, then magnitude of field and potential at D will become respectively

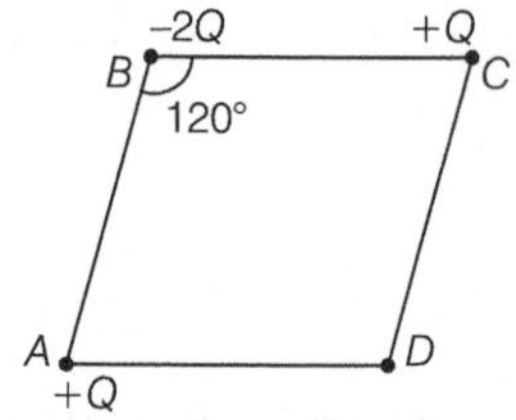

(a) $\frac{\sqrt{3}}{2}E, V$ (b) $\frac{\sqrt{3}}{2}E, 0$

(c) E, 0 (d) $\frac{E}{\sqrt{2}}, \frac{V}{\sqrt{2}}$

2 Two small equal point charges of magnitude q are suspended from a common point on the ceiling by insulating massless strings of equal lengths. They come to equilibrium with each string making angle θ from the vertical. If the mass of each charge is m, then the electrostatic potential at the centre of line joining them will be $\left(\frac{1}{4\pi\varepsilon_0} = k\right)$

(a) $2\sqrt{kmg\tan\theta}$ (b) $\sqrt{kmg\tan\theta}$ (c) $4\sqrt{kmg/\tan\theta}$ (d) $4\sqrt{kmg\tan\theta}$

3 In the diagram shown, there is a square, inside which there is another square and inside it there is another square and so on. On the vertex of each square, a charge is placed. On the vertex of the Nth square, a charge $+\frac{Q}{2^N}$ is placed. If the dimension of the outermost square is L and number of squares formed in this manner is infinite, then potential at the centre is

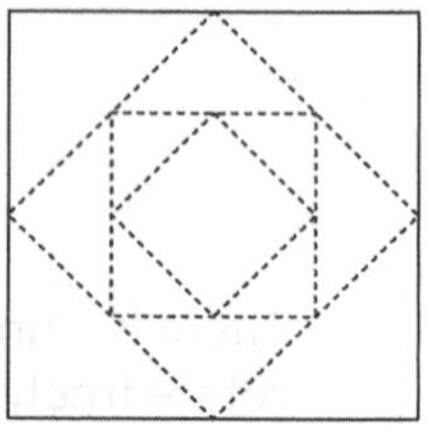

(a) $\frac{2Q(\sqrt{2}+1)}{\pi\varepsilon_0 L}$ (b) $\frac{Q(\sqrt{2}+1)}{\pi\varepsilon_0 L}$ (c) $\frac{4Q(\sqrt{2}+1)}{\pi\varepsilon_0 L}$ (d) None of these

4 A non-conducting semi-circular disc (as shown in figure) has a uniform surface charge density σ. The ratio of electric field to electric potential at the centre of the disc will be

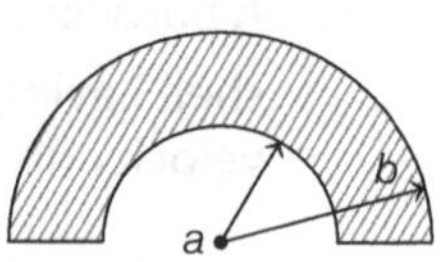

(a) $\frac{\ln b/a}{\pi(b-a)}$ (b) $\frac{2}{\pi}$ (c) $\frac{1}{\pi}\frac{\ln(b/a)^2}{(b-a)}$ (d) $\frac{\pi(b-a)}{2\ln(b/a)}$

5 A cone made of insulating material has a total charge Q spread uniformly over its sloping surface. The slant length is L. The energy required to take a test charge q from infinity to apex A of cone is

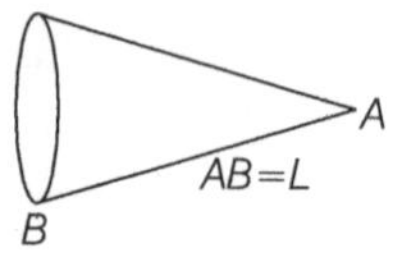

(a) $\frac{Qq}{2\pi\varepsilon_0 L}$ (b) $\frac{2Qq}{\pi\varepsilon_0 L}$ (c) $\frac{4Qq}{\pi\varepsilon_0 L}$ (d) $\frac{Qq}{\pi\varepsilon_0 L}$

6 The potential at the vertex A of a uniformly charged hemispherical surface of radius R and uniform surface charge density σ is

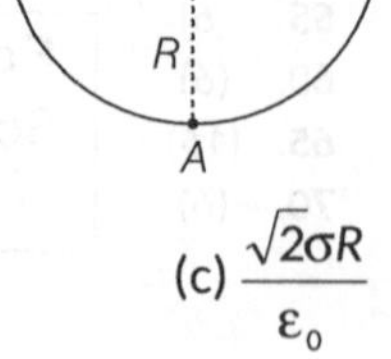

(a) $\frac{\sigma R}{2\varepsilon_0}$ (b) $\frac{2\sigma R}{\varepsilon_0}$ (c) $\frac{\sqrt{2}\sigma R}{\varepsilon_0}$ (d) $\frac{\sigma R}{\sqrt{2}\varepsilon_0}$

7 Q charge given to a uniform hemispherical charged distribution of radius R, then sum of potential at two diametrically opposite points $V_A + V_B$ is equal to

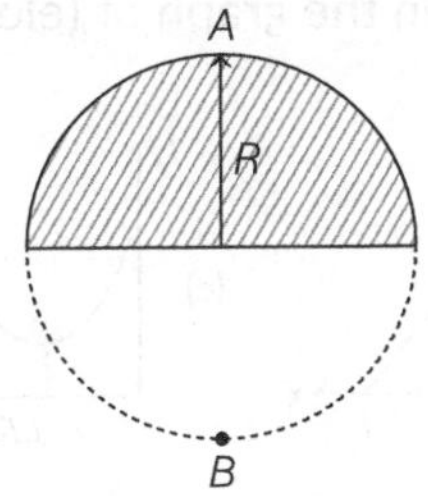

(a) $\frac{kQ}{R}$ (b) $\frac{2kQ}{R}$ (c) $\frac{kQ}{2R}$ (d) $\frac{kQ}{3R}$

8 Consider the potential at the corner of a uniformly charged cube of dimension L to be directly proportional to ρL^2, where ρ is the volume charge density. What is the value of the ratio of the potential at the centre to the potential at the corner of the cube?

(a) 4 : 1 (b) 3 : 2 (c) 8 : 1 (d) 2 : 1

9 A non-conducting sphere of radius $R = 5$ cm has its centre at origin O. It has a spherical cavity of radius $r = 1$ cm whose centre is at (0, 3 cm). Solid material of sphere has uniform positive charge density $\rho = \frac{10^{-6}}{\pi}$ C/m^3, then potential at point P (4 cm, 0) is

(a) 35.16 V (b) 34.92 V (c) 34.50 V (d) 34.20 V

10 Two charges lie on the X-axis. A rough sketch of the equipotential lines of these two charges is shown in figure. Which one of the following statement is true?

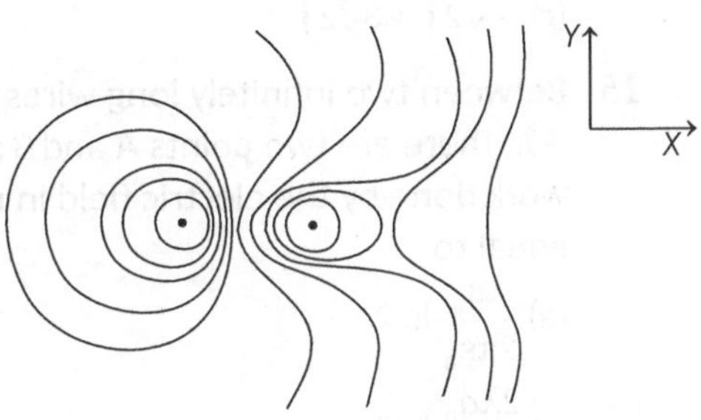

(a) The two charges have opposite signs and the charge on the left is smaller in magnitude than the charge on the right.
(b) The two charges have opposite signs and the charge on the left is larger in magnitude than the charge on the right.
(c) The two charges have the same sign and the charge on the left is smaller in magnitude than the charge on the right.
(d) The two charges have the same sign and the charge on the left is larger in magnitude than the charge on the right.

11 Two concentric spherical shells of radii r and $R(r < R)$ have surface charge densities $-\sigma$ and $+\sigma$, respectively. The variation of electric potential V with distance x from the centre O of the shells is plotted. Which of the following graphs best depict the variation qualitatively?

(a) 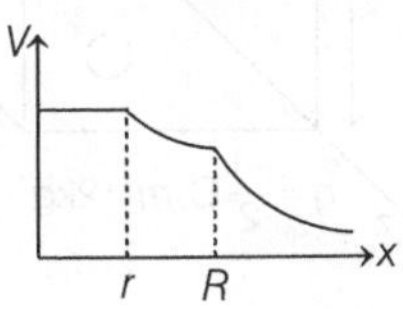

(b) 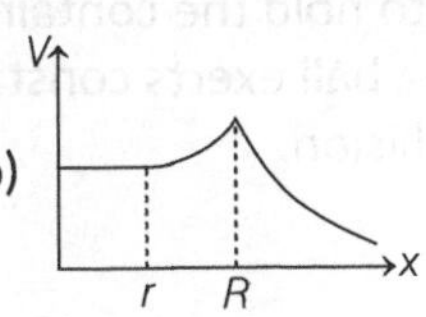

(c) 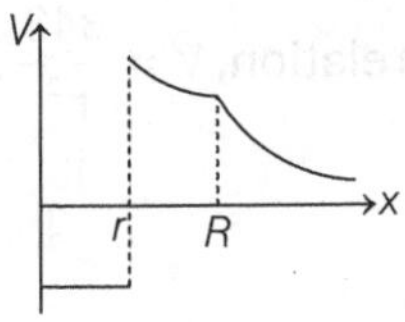

(d) 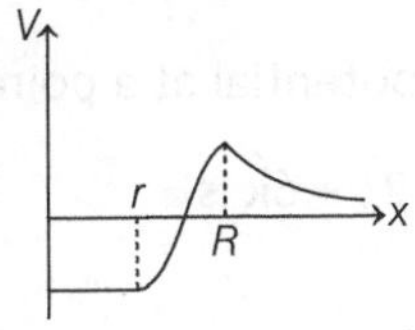

12 Two positive charges $+Q$ and $+4Q$ are placed at the ends of a straight line of length L. A negative charge $-Q$ is placed between the first two charges on the straight line at a distance x $(0 < x < L)$ from the charge $+Q$, then the graph of (electrostatic potential energy) of the system of three charges *versus* x is

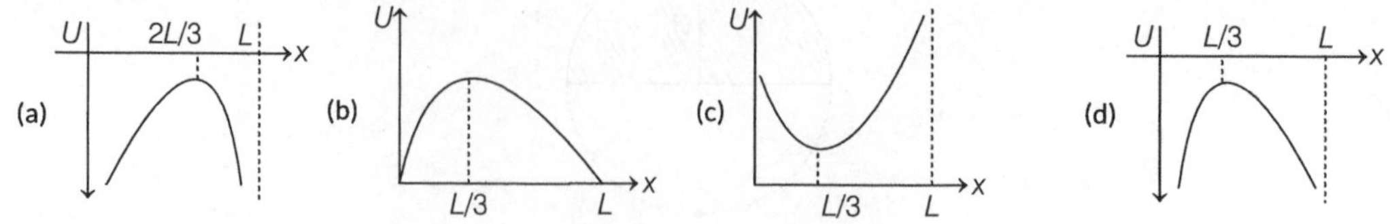

13 During lighting, a current pulse, shown in figure, flows from the cloud at a height 1.5 km to the ground. If the breakdown electric field of humid air is about 400 kVm^{-1}, the energy released during lighting would be (in unit of 10^9 J)

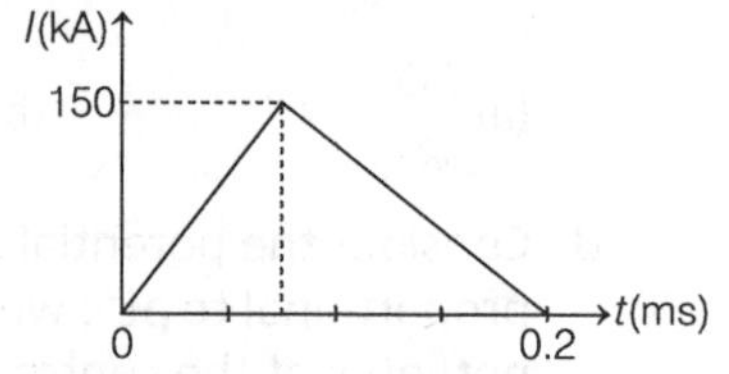

(a) 9
(b) 3.0
(c) 4.5
(d) 6.0

14 A uniform electric field exists only in XY-plane. When a student moves a distance 0.01 m along OP line, potential decreases by 0.02 V. The magnitude of electric field is $2\sqrt{5}$ V/m. When a student walks along OQ line (which is perpendicular to OP line), potential decreases. Then, electric field in vector form is

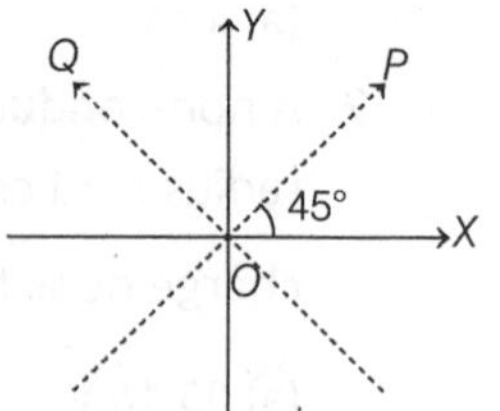

(a) $\sqrt{2}\hat{i} + 3\sqrt{2}\hat{j}$
(b) $3\sqrt{2}\hat{i} + \sqrt{2}\hat{j}$
(c) $-\sqrt{2}\hat{i} + 3\sqrt{2}\hat{j}$
(d) $-3\sqrt{2}\hat{i} + \sqrt{2}\hat{j}$

15 Between two infinitely long wires having linear charge densities $+\lambda$ and $-\lambda$, there are two points A and B as shown in the figure. The amount of work done by the electric field in moving a point charge q_0 from A to B is equal to

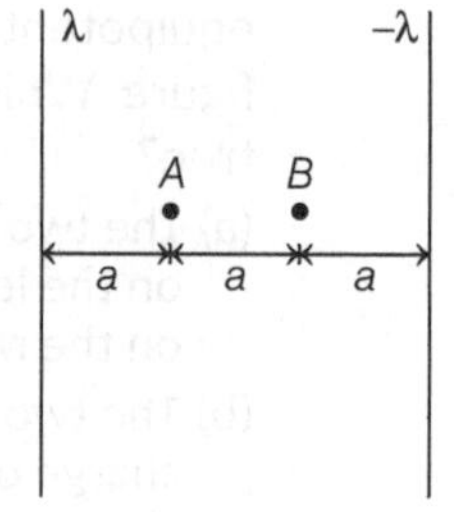

(a) $\frac{\lambda q_0}{2\pi\varepsilon_0}\ln 2$
(b) $-\frac{2\lambda q_0}{\pi\varepsilon_0}\ln 2$
(c) $\frac{2\lambda q_0}{\pi\varepsilon_0}\ln 2$
(d) $\frac{\lambda q_0}{\pi\varepsilon_0}\ln 2$

16 Consider a gravity free container as shown. System is initially at rest and electric potential in the region is $V = (y^3 + 2)$ J/C. A ball of charge q and mass m released from rest from base starts to move up due to electric field and collides with the shaded face as shown in figure. If its speed just after collision is 1.5 m/s and time for which ball is in contact with shaded face is 0.1 s, find external force required to hold the container fixed in its position during collision assuming ball exerts constant force on wall during entire span of collision.

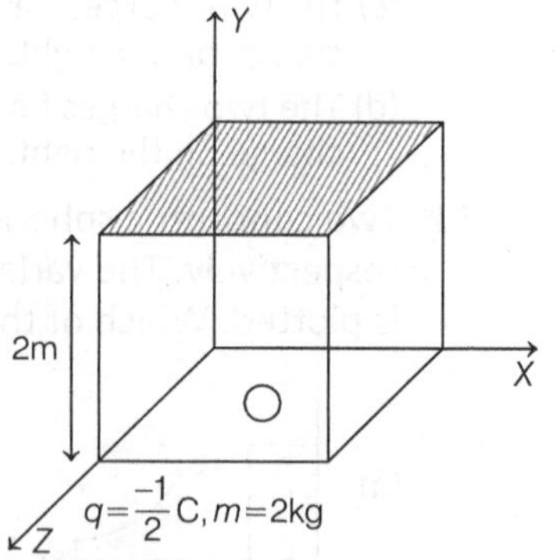

(a) 70 N
(b) 72 N
(c) 74 N
(d) 76 N

17 In an electric field, the potential at a point is given by the following relation, $V = \frac{343}{r^2}$. The electric field at $\mathbf{r} = 3\hat{i} + 2\hat{j} + 6\hat{k}$ is

(a) $\frac{2}{7}(3\hat{i} + 2\hat{j} + 6\hat{k})$
(b) $\frac{1}{7}(3\hat{i} + 2\hat{j} + 6\hat{k})$
(c) $6\hat{i} + 4\hat{j} + 12\hat{k}$
(d) $-(3\hat{i} + 2\hat{j} + 6\hat{k})$

18 A non-conducting rod of length $2l$ with its left and right half given uniform linear charge densities $-\lambda$ and $+\lambda$ respectively is placed on the X-axis as shown in figure, which of the following is correct?

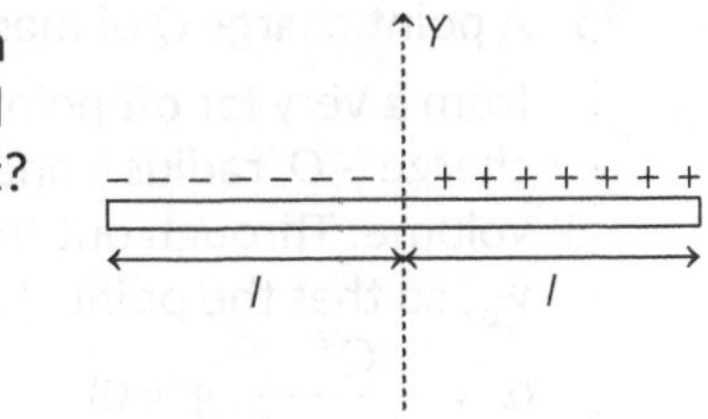

(a) The YZ-plane is an equipotential surface.

(b) The ZX-plane is an equipotential surface.

(c) The electric field at any point on the Y-axis is along the positive x-direction.

(d) The electric field at any point on the Z-axis is along the positive x-direction.

19 Two point charges of $+Q$ each have been placed at the positions $(-a/2, 0, 0)$ and $(a/2, 0, 0)$. The locus of the points in YZ-plane, where $-Q$ charge can be placed such that the total electrostatic potential energy of the system can become equal to zero, is represented by which of the following equations?

(a) $z^2 + (y-a)^2 = 2a$ (b) $z^2 + (y-a)^2 = 27a^2/4$

(c) $z^2 + y^2 = 15a^2/4$ (d) None of these

20 On a semi-circular ring of radius $= 4R$, charge $+3q$ is distributed in such a way that on one quarter $+q$ is uniformly distributed and on another quarter $+2q$ is uniformly distributed. Along its axis, a smooth non-conducting and uncharged pipe of length $6R$ is fixed axially as shown in figure. A small ball of mass m and charge $+q$ is thrown from one end of pipe with speed u. The minimum speed u_{min}, so that ball will come out from the other end of pipe is

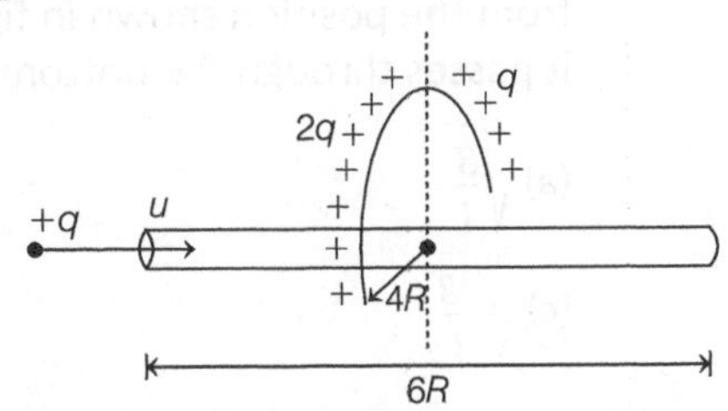

(a) $\sqrt{\dfrac{7q^2}{40\pi\varepsilon_0 mR}}$ (b) $\sqrt{\dfrac{3q^2}{40\pi\varepsilon_0 mR}}$

(c) $\sqrt{\dfrac{q^2}{40\pi\varepsilon_0 mR}}$ (d) $\sqrt{\dfrac{9q^2}{40\pi\varepsilon_0 mR}}$

21 An infinite dielectric sheet having charge density σ has a hole of radius R in it. An electron is released on the axis of the hole at a distance $\sqrt{3}R$ from the centre. The speed with which it crosses the plane of sheet is (e = charge on electron and m = mass of electron)

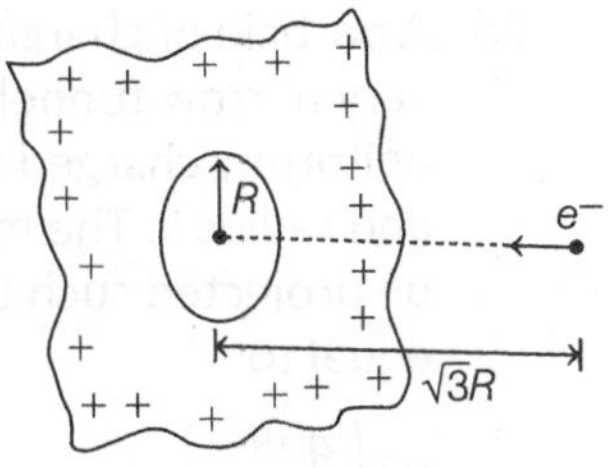

(a) $\sqrt{\dfrac{\sigma eR}{2m\varepsilon_0}}$ (b) $\sqrt{\dfrac{\sigma eR}{m\varepsilon_0}}$

(c) $\sqrt{\dfrac{2\sigma eR}{m\varepsilon_0}}$ (d) $\sqrt{\dfrac{3\sigma eR}{m\varepsilon_0}}$

22 A particle of mass m and charge $+q$ approaches from a very large distance towards a uniformly charged ring of radius R and charge $+q$, mass same as that of particle, with initial velocity v_0 along the axis of the ring as shown in the figure. The closest distance of approach between the ring and the particle is [Assume the space to be gravity free and frictionless]

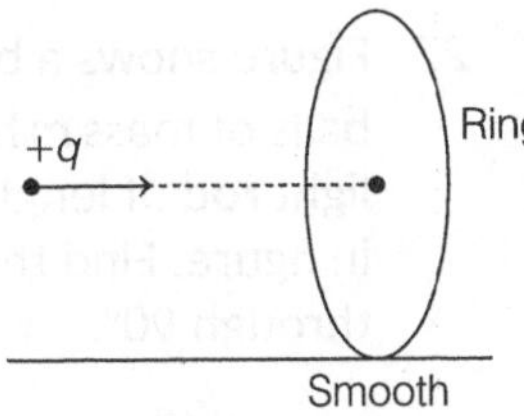

(a) $\sqrt{\dfrac{q^4}{\pi^2\varepsilon_0^2 m^2 v_0^4} - R^2}$ (b) $\sqrt{\dfrac{3q^4}{2\pi^2\varepsilon_0^2 m^2 v_0^4} + R^2}$

(c) $\sqrt{\dfrac{m^2 v_0^4}{2\pi^2 q^4 \varepsilon_0^2} - R^2}$ (d) $\sqrt{\dfrac{q^4}{4\pi^2\varepsilon_0^2 m^2 v_0^4} - R^2}$

23 A point charge Q of mass m is projected with some initial velocity v_0 from a very far off point as shown in figure. The sphere shown has a charge $-Q$, radius R and charge is uniformly distributed over its entire volume. Through out the process, this sphere stays stationary. Find, v_0, so that the point charge strikes the sphere at 45°. (Given that, $\alpha = \frac{Q^2}{4\pi\varepsilon_0 mR}$, $g = 0$)

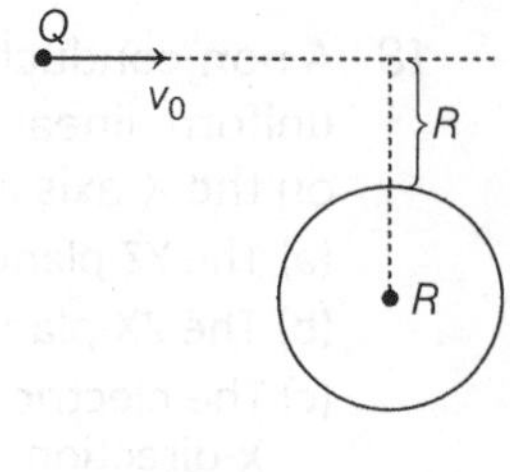

(a) $v_0 = \sqrt{\frac{2}{7}\alpha}$
(b) $v_0 = \sqrt{\frac{2}{3}\alpha}$
(c) $v_0 = \sqrt{\frac{1}{7}\alpha}$
(d) $v_0 = \sqrt{\frac{1}{3}\alpha}$

24 In space, a horizontal electric field $(E = (mg)/q)$ exists as shown in figure and a mass m is attached at the end of a light rod. If mass m is released from the position shown in figure, the angular velocity of the rod when it passes through the bottom most position is

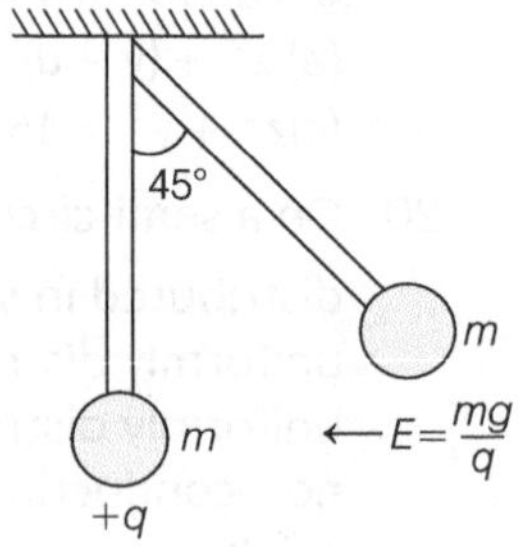

(a) $\sqrt{\frac{g}{l}}$
(b) $\sqrt{\frac{2g}{l}}$
(c) $\sqrt{\frac{3g}{l}}$
(d) $\sqrt{\frac{5g}{l}}$

25 A charged bead (charge + 4μC) is threaded on a fixed smooth circular ring of radius 2 m in gravity free space. Another charge + 4μC is fixed at 1 m from centre as shown. The bead is projected with just sufficient velocity to complete the circular path. If mass of bead is 100 g, what is initial speed v_0?

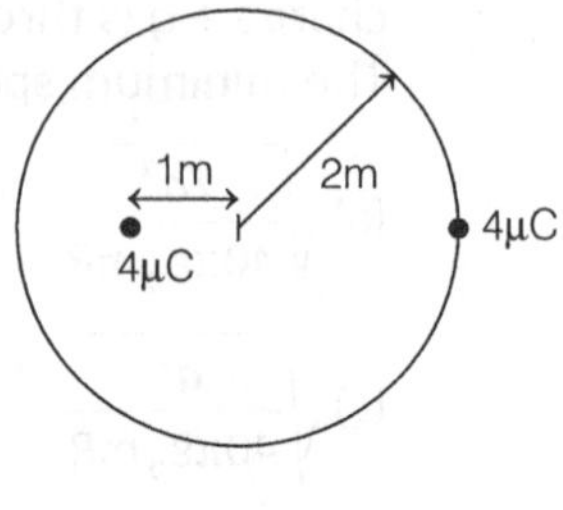

(a) $80\sqrt{3}$ cm/s
(b) $40\sqrt{2}$ cm/s
(c) 10 m/s
(d) $4\sqrt{5}$ m/s

26 A particle of charge + q and mass m is projected with a velocity v inside a very narrow tunnel as shown. The tunnel has been made inside a uniformly charged non-conducting sphere of volume charge density ρ and radius R. The minimum velocity with which the point charge should be projected such that it can it reach the opposite end of the tunnel, is equal to

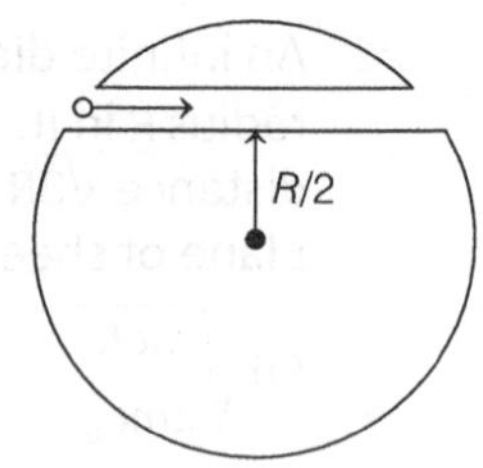

(a) $\left(\frac{q\rho R^2}{4m\varepsilon_0}\right)^{1/2}$
(b) $\left(\frac{q\rho R^2}{24m\varepsilon_0}\right)^{1/2}$
(c) $\left(\frac{q\rho R^2}{6m\varepsilon_0}\right)^{1/2}$

(d) zero because the initial and the final points are at same potential

27 Figure shows a ball having a charge q fixed at a point A. Two identical balls of mass m having charges + q and − q are attached to the end of a light rod of length $2a$. The system is released from the situation shown in figure. Find the angular velocity of the rod, when the rod turns through 90°.

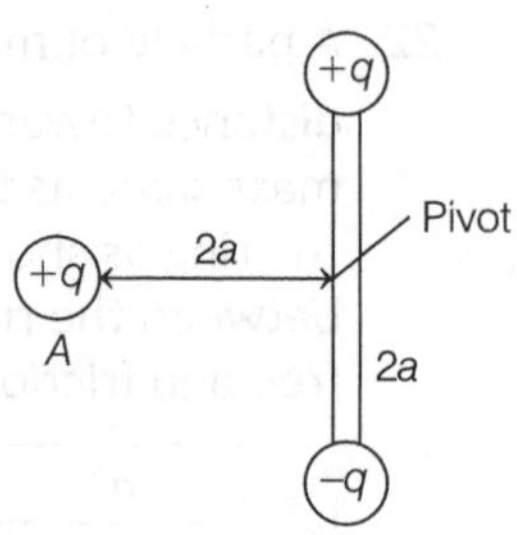

(a) $\frac{\sqrt{2}q}{3\pi\varepsilon_0 ma^3}$
(b) $\frac{q}{\sqrt{3\pi\varepsilon_0 ma^3}}$
(c) $\frac{q}{\sqrt{6\pi\varepsilon_0 ma^3}}$
(d) $\frac{\sqrt{2}q}{4\pi\varepsilon_0 ma^3}$

28 The mass m_1 and charge q_1 is released from rest from a very large distance along the axis of ring of radius R, charge q_2 and mass m_2.

Case-I Ring is kept fixed and speed of m_1, when it crosses the centre of ring is v_1.

Case-II Ring is free to move along the axis and speed of m_1, when it crosses the centre of the ring is v_2.

The value of $\left(\frac{v_1}{v_2}\right)$ is

(a) $\frac{m_1 + m_2}{m_2}$ (b) $\frac{m_2}{m_1}$ (c) $\sqrt{\frac{m_1 + m_2}{m_2}}$ (d) None of these

29 Electric field given by the vector $\mathbf{E} = \frac{E_0}{l}(x\hat{\mathbf{i}} + y\hat{\mathbf{j}})$ N/C is present in the XY-plane. A small ring of mass M carrying charge $+Q$, which can slide freely on a smooth non-conducting rod, is projected along the rod from the point $(0, l)$ such that, it can reach the other end of the rod. Assuming there is no gravity in the region, the minimum velocity which should be given to the ring is $\left(\frac{QE_0 l}{m} = 8\right)$

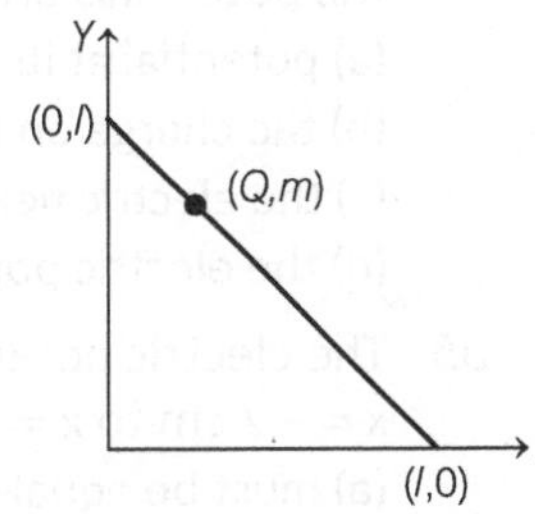

(a) 2 m/s (b) 3 m/s (c) 2.5 m/s (d) 3.5 m/s

30 Two small balls of mass m, bearing a charge q each, are connected by a non-conducting thread of length $2l$. At a certain instant, the middle of the thread starts moving at a constant velocity v perpendicular to the direction of the thread at the initial instant. The minimum distance between the balls will be

(Use, $q^2 = 4\pi\varepsilon_0 mv^2 l$)

(a) $\frac{l}{3}$ (b) $\frac{2l}{3}$ (c) l (d) $\frac{4l}{3}$

31 In a certain region, free from gravity, electric field is along negative x-direction and it is constant. A particle having mass m and charge q is projected along x-direction with speed v_0. An additional force $\mathbf{F} = \mathbf{C} \times \mathbf{v}$ is acting on the charge, where $\mathbf{v}$ is velocity vector and $\mathbf{C}$ is a constant vector. The charge comes out of region with speed $\frac{v_0}{2}$, then the magnitude of electric field will be

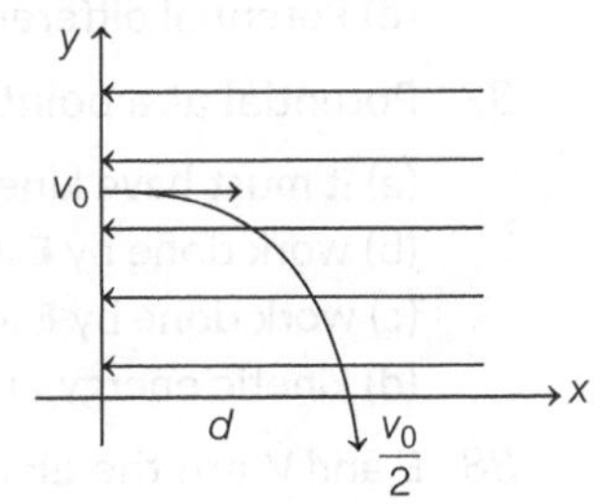

(a) $\frac{3}{4}\frac{mv_0^2}{qd}$ (b) $\frac{4}{3}\frac{mv_0^2}{qd}$

(c) $\frac{3}{8}\frac{mv_0^2}{qd}$ (d) Cannot be determined

32 Four point charges are fixed along Y-axis as shown in the figure. A particle of mass m and charge $+Q$ is projected along the $+X$-axis towards centre O from large distance. Find the minimum velocity v_{min} of particle, so that it will reach at origin O. Take $Q^2 = 12\sqrt{5}\pi\varepsilon_0 m$. Neglect gravity.

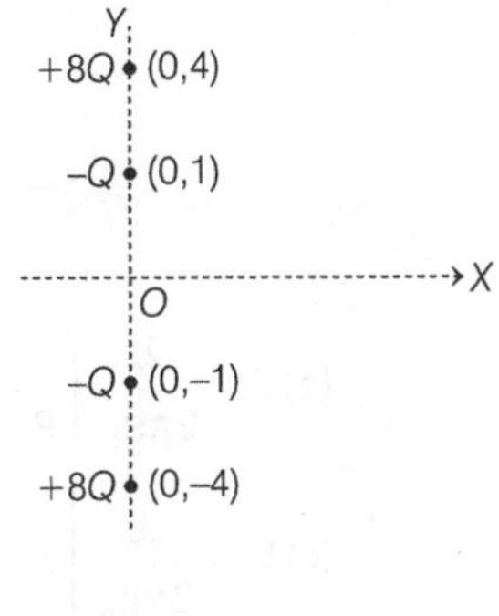

(a) 5 m/s
(b) 6 m/s
(c) 8 m/s
(d) 10 m/s

Multiple Choice(s) Correct Questions

33 A small sphere is charged uniformly and placed at point $A(x, y)$ m, so that at point $B(8, 7)$ m, electric field strength is $\mathbf{E} = (54\hat{\mathbf{i}} + 72\hat{\mathbf{j}})$ N/C and potential is + 900 V. The dielectric strength of air is 3×10^6 V/m. Choose the correct option(s).

(a) Co-ordinate of point A are (2, – 1)m

(b) Charge of the sphere is 2μC

(c) Minimum possible radius of sphere is $\sqrt{30}$ cm, so that air does not breakdown

(d) None of the above

34 At distance of 5 cm and 10 cm outwards from the surface of a uniformly charged solid sphere, the potentials are 100V and 75V, respectively. Then,

(a) potential at its surface is 150V

(b) the charge on the sphere is $(5/3) \times 10^{-9}$ C

(c) the electric field on the surface is 1500 V/m

(d) the electric potential at its centre is 225V

35 The electric potential decreases uniformly from 180 V to 20 V as one moves on the X-axis from $x = -2$ cm to $x = +2$ cm. The electric field at the origin

(a) must be equal to 40 V/cm

(b) may be equal to 40 V/cm

(c) may be greater than 40 V/cm

(d) may be less than 40 V/cm

36 Which of the following quantites depends on the choice of zero potential or zero potential energy?

(a) Potential at a particular point

(b) Change in potential energy of a two-charge system

(c) Potential energy of a two-charge system

(d) Potential difference between two points

37 Potential at a point x is 5 V and at a point y is – 5V. A proton is moving towards x from y, then

(a) it must have kinetic energy more than 10 eV to reach x from y

(b) work done by **E** in motion from x to y is + 10 eV

(c) work done by **E** in motion from y to x is zero

(d) kinetic energy at y is more than kinetic energy at x, if particle is released at x

38 E and V are the electric field and electric potential at O. If the linear charge density of the system is λ, then

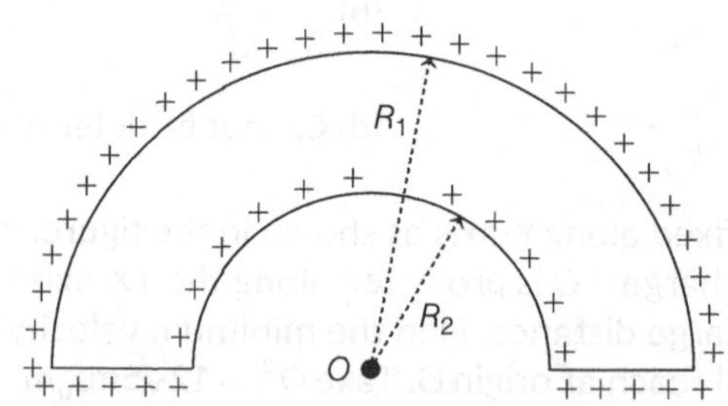

(a) $E = \dfrac{\lambda}{2\pi\varepsilon_0}\left(\dfrac{1}{R_1} + \dfrac{1}{R_2}\right)$

(b) $E = \dfrac{\lambda}{4\pi\varepsilon_0}\left(\dfrac{1}{R_1} + \dfrac{1}{R_2}\right)$

(c) $V = \dfrac{\lambda}{2\pi\varepsilon_0}\left[\pi + \ln\left(\dfrac{R_1}{R_2}\right)\right]$

(d) $V = \dfrac{\lambda}{4\pi\varepsilon_0}\left(\pi + \ln\dfrac{R_1}{R_2}\right)$

39 A wire having a uniform linear charge density λ, is bent in the form of a ring of radius R. Point A as shown in the figure, is in the plane of the ring but not at the centre. Two elements a_1 and a_2 of the ring subtend very small and equal angles at the point A. They are at distances r_1 and r_2 from the point A respectively

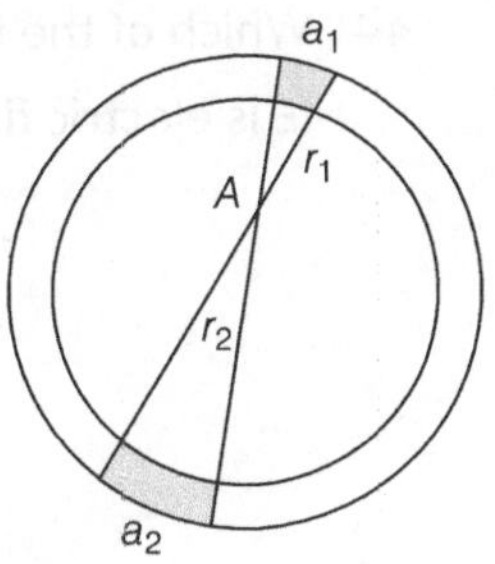

(a) The ratio of charge on elements a_1 and a_2 is r_1 / r_2.

(b) The element a_1 produces greater magnitude of electric field at A than element a_2.

(c) The elements a_1 and a_2 produce same potential at A.

(d) The element a_1 produces greater potential at A than element a_2.

40 A sphere of radius R_1 has charge density ρ uniform within its volume, except for a small spherical hollow region of radius R_2 located a distance a from centre (Take, zero potential at infinity).

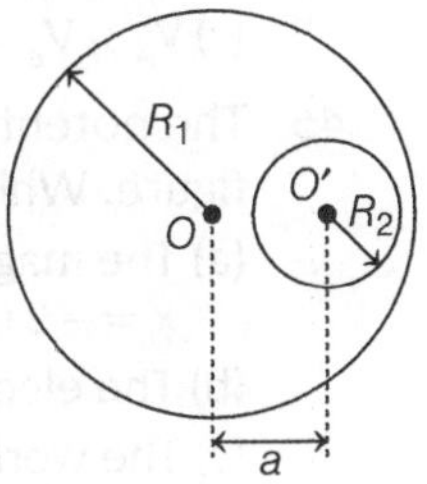

(a) Electric field at O' is $\frac{\rho a}{3\varepsilon_0}$

(b) Electric field at O' is $\frac{\rho a^3}{3\varepsilon_0}\hat{a}$

(c) Electric potential at O' is $\frac{\rho}{6\varepsilon_0}[3(R_1^2 - R_2^2) - a^2]$

(d) Electric potential at any point in hollow region is constant.

41 The electric field in XY-plane is given by $\mathbf{E} = x^2\hat{\mathbf{i}} + y^2\hat{\mathbf{j}}$. Potential at point (3, 3) is taken as zero, then

(a) work done in moving a test charge q slowly from a point (a, b) to a point (b, a) is zero.

(b) potential at any point on XY-plane can't exceed 18 unit for x and y in 1st quadrant.

(c) potential can't be defined as field is non- conservative in nature.

(d) equipotential curve are concentric circles with centre (3,3).

42 S_1 and S_2 are two equipotential surfaces on which the potentials are not equal.

(a) S_1 and S_2 cannot intersect

(b) S_1 and S_2 cannot both be plane surfaces

(c) In the region between S_1 and S_2, the field is maximum, where they are closest to each other

(d) A line of force from S_1 to S_2 must be perpendicular to both

43 Figure shows three spherical shells in separate situations, with each shell having the same uniform positive net charge. Points 1, 4 and 7 are at the same radial distances from the centre of their respective shells; so are points 2, 5 and 8 and so are points 3, 6 and 9. With the electric potential taken equals to zero at an infinite distance, choose correct statement(s).

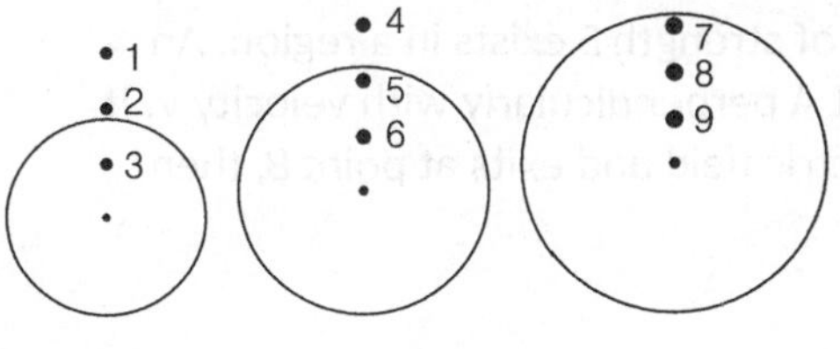

(a) Point 3 has highest potential.

(b) Points 1, 4 and 7 are at same potential.

(c) Point 1 has lowest potential.

(d) Points 5 and 8 are at different potential.

44 Which of the following is true for the figure showing electric lines of force?
(E is electric field, V is potential)

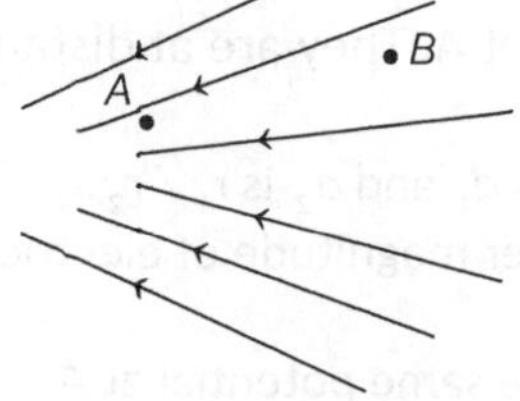

(a) $E_A > E_B$ (b) $E_B > E_A$
(c) $V_A > V_B$ (d) $V_B > V_A$

45 The potential as a function of x is plotted in figure. Which of the following is/are correct?

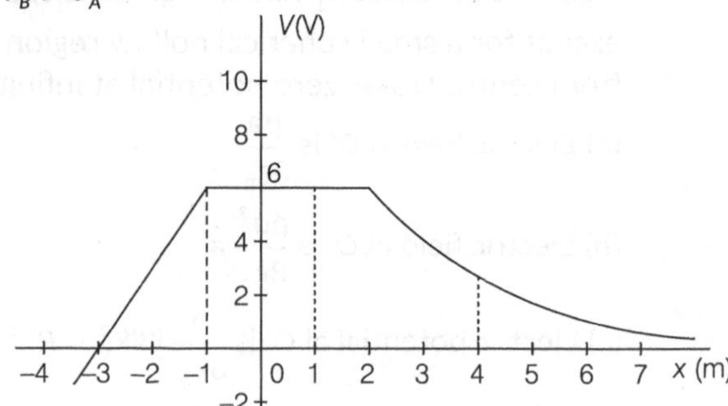

(a) The magnitude of electric field E_x at $x = -2$ m is 3 N/C
(b) The electric field E_x at $x = 1$ m is 6N/C
(c) The work done in moving a charge particle per unit charge from $x = -3$ m to $x = 1$ m is 6 J/C
(d) The work done in moving a charge particle per unit charge from $x = 2$ to $x = \infty$ is -6 J/C

46 In an electric field shown in figure, three equipotential surfaces are shown.

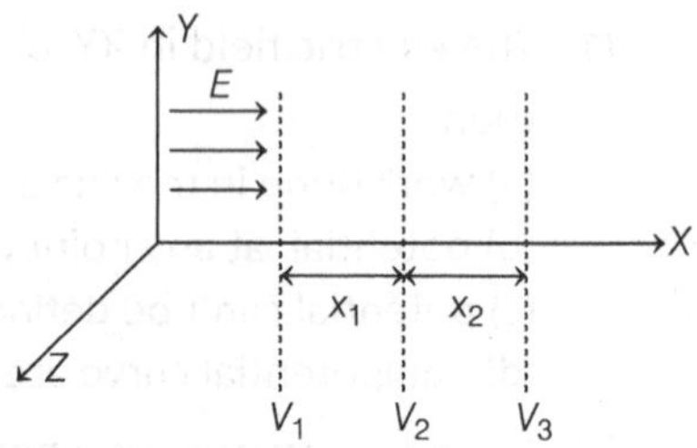

The function of electric field is $E = 2x^2$ V/m.
(a) If $V_1 - V_2 = V_2 - V_3$, then $x_1 > x_2$
(b) If $V_1 - V_2 = V_2 - V_3$, then $x_1 < x_2$
(c) $V_1 > V_2 > V_3$
(d) $V_1 < V_2 < V_3$

47 A fixed point charge Q is at origin. At $t = 0$, a charge q with mass m is at $x = a$ with leftward velocity v_0, which satisfies $\frac{kQq}{a} = 3mv_0^2$. The particle turns around and starts to move rightward at the position $b < a$.
(a) The ratio b/a is 6/7
(b) The ratio b/a is 3/7
(c) The velocity of the particle at a large distance from the origin is $\sqrt{7}\,v_0$
(d) The velocity of the particle at a large distance from the origin is zero

48 A uniform electric field of strength E exists in a region. An electron enters at point A perpendicularly with velocity v. It moves through the electric field and exits at point B, then

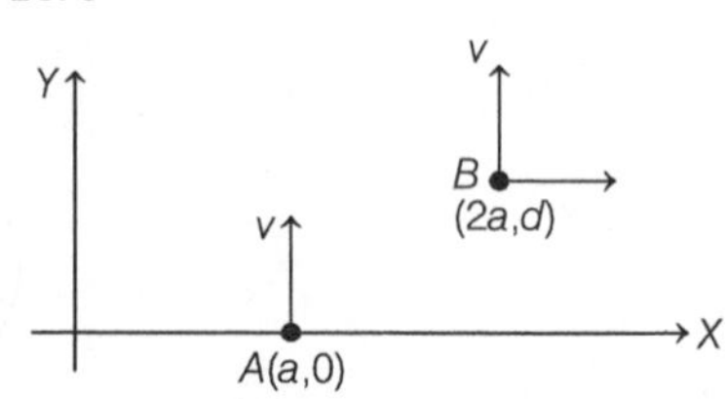

(a) $E = \frac{-2amv^2}{ed^2}\hat{i}$

(b) velocity at B is $\left(\frac{2av}{d}\right)\hat{i} + v\hat{j}$

(c) rate of work done by the electric field at B is $\frac{4ma^2v^3}{d^3}$

(d) rate of work done by the electric field at A is zero

49 Three charges are kept at the corners of a square as shown. A fourth charge q is brought slowly at the fourth corner from infinity, then choose the correct statement(s). (Take, reference potential energy at infinity to be zero and $k = \frac{1}{4\pi\varepsilon_0}$)

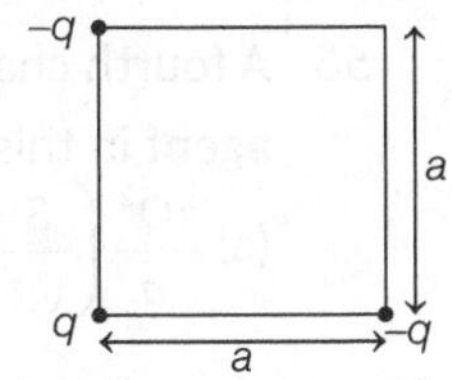

(a) Work done by electrostatic field is $\frac{kq^2}{a}\left(\frac{1-2\sqrt{2}}{\sqrt{2}}\right)$.

(b) Work done by external agent is $\frac{kq^2}{a}\left(\frac{1-2\sqrt{2}}{\sqrt{2}}\right)$

(c) Total potential energy of the system after the charge is placed at the corner is $\frac{kq^2}{a}(\sqrt{2}-4)$.

(d) For the system thus achieved centre of square will be a point of equilibrium.

50 Given figure is a graph of the electric potential in a region of gravity free space, where electric field is parallel to the X-axis. A particle having specific charge $\left(\frac{q}{m}=2\right)$ is projected from origin along positive X-axis with speed v_0, then

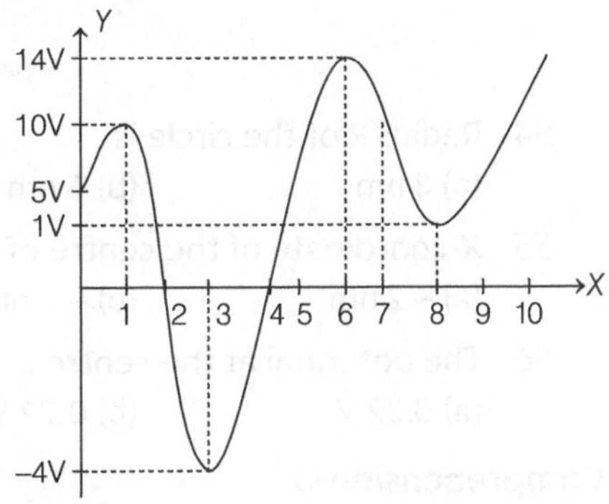

(a) minimum velocity for which charge can pass through $x = 8$ is 6m/s

(b) minimum velocity for which charge can pass through $x = 8$ is 4 m/s

(c) if $v_0 > \sqrt{20}$ m/s, particle passes through $x = 4$

(d) if $v_0 = 8$ m/s, then maximum speed of particle during its forward motion is equal to 10 m/s

Comprehension Based Questions

Comprehension-1

Three charges $+3Q$, $+2Q$ and $-Q$ respectively, are located at a distance a from the origin as shown in the figure. The point P is located at a distance a from the origin. $\left(\text{Take, } k = \frac{1}{4\pi\varepsilon_0}\right)$

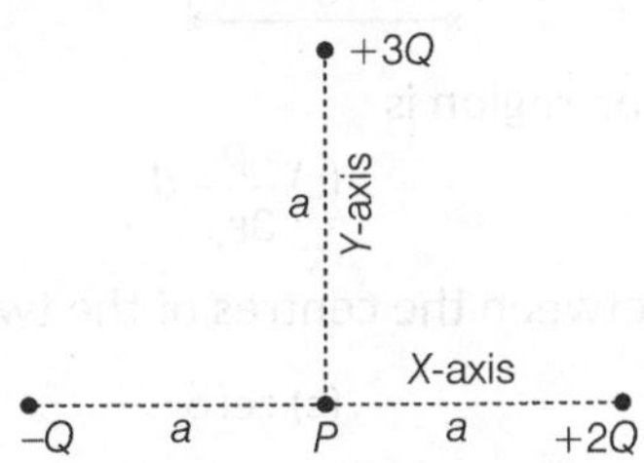

51 The magnitude of x and y-components of the electric field $\mathbf{E}$ at P are

(a) $E_x = 2, E_y = \frac{k3Q}{a^2}$ (b) $E_x = E_y = -\frac{k3Q}{a^2}$

(c) $E_x = \frac{1}{2}, E_y = \frac{k3Q}{a^2}$ (d) $E_x = \frac{kQ}{a^2}, E_y = \frac{k3Q}{a^2}$

52 The electric potential energy U for the configuration of the three charges is

(a) $\frac{kQ^2}{a}\left(1-\frac{1}{\sqrt{2}}\right)$ (b) $\frac{kQ^2}{a}\left(\frac{1}{\sqrt{2}}+1\right)$ (c) $\frac{kQ^2}{a}\left(\frac{3}{\sqrt{2}}+1\right)$ (d) $\frac{kQ^2}{a}\left(\frac{3}{\sqrt{2}}-1\right)$

53 A fourth charge $+3Q$ is slowly moved in from infinity to point P. The work done by an external agent in this process must be

(a) $\frac{kQ^2}{a}\left(\frac{3}{\sqrt{2}}-1\right)$ (b) $\frac{kQ^2}{a}\left(\frac{3}{\sqrt{2}}+1\right)$ (c) $\frac{kQ^2}{a}\frac{4}{3}$ (d) $\frac{kQ^2}{a}(12)$

Comprehension-2

A point charge $q_1 = +6e$ fixed at the origin of a coordinate system and another point charge $q_2 = -10e$ is fixed at $x = 8$ nm, $y = 0$. The locus of all points in the XY-plane for which potential $V = 0$ (other than infinity) is a circle centred on the X-axis, as shown in figure.

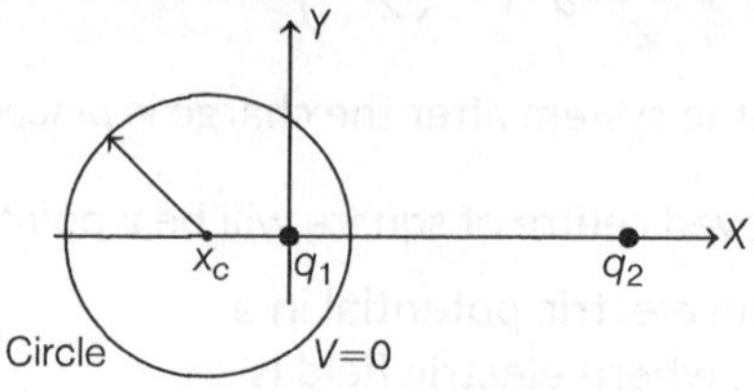

54 Radius R of the circle is

(a) 3 nm (b) 6 nm (c) 7.5 nm (d) 9 nm

55 X-coordinate of the centre of the circle is

(a) -2 nm (b) -3 nm (c) -4.5 nm (d) -7.5 nm

56 The potential at the centre of the circle is

(a) 0.32 V (b) 0.77 V (c) 1.2 V (d) -1.2 V

Comprehension-3

There are two non-conducting spheres having uniform volume charge densities ρ and $-\rho$. Both spheres have equal radius R. The spheres are now laid down such that they overlap as shown in the figure.

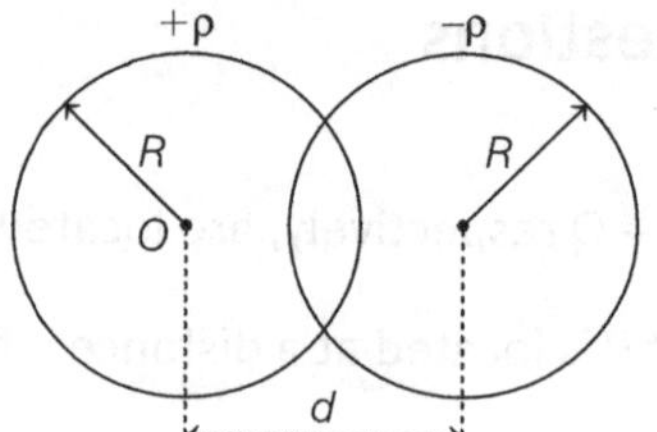

57 The electric field **E** in the overlap region is

(a) non-uniform (b) zero (c) $\frac{\rho}{3\varepsilon_0}\mathbf{d}$ (d) $\frac{\rho}{3\varepsilon_0}\mathbf{r}$

58 The potential difference ΔV between the centres of the two spheres for $d = R$, is

(a) $\frac{\rho}{3\varepsilon_0}d^2$ (b) $\frac{\rho}{\varepsilon_0}d^2$ (c) zero (d) $\frac{2\rho}{\varepsilon_0}d^2$

Comprehension-4

An uncharged ball of radius R is placed at a point in space and the region outside (from R to ∞ measured from centre of the ball) the ball is non-uniformly charged with a charge density $\rho = \frac{B}{r^3}$ C/m^3, where B is a positive constant and r is the distance of a point measured from centre of the ball.

59 Electric potential at the centre of the ball is

(a) directly proportional to R (b) directly proportional to R^2

(c) inversely proportional to R (d) inversely proportional to R^2

60 Electric potential at a distance x from centre of the ball $(x > R)$ is

(a) $\frac{B}{\varepsilon_0 x}\left[1+\ln\left(\frac{x}{R}\right)\right]$ (b) $\frac{B}{\varepsilon_0 x}\ln\left(\frac{x}{R}\right)$

(c) $\frac{B}{\varepsilon_0 x}\ln\left(\frac{x-R}{R}\right)$ (d) $\frac{B}{\varepsilon_0 x}\left[1+\ln\left(\frac{x-R}{R}\right)\right]$

61 As we move away from ball's surface, electric potential

(a) decreases (b) increases

(c) decreases then increases (d) increases then decreases

Comprehension-5

The sketch below shows cross-sections of equipotential surfaces between a solid non-conducting sphere and large non-conducting sheet that are shown in black. Some points on the equipotential surfaces near them are marked as $A, B, C,$....... The arrangement lies in air. (Take, $\varepsilon_0 = 8.85 \times 10^{-12}\ C^2/Nm^2$)

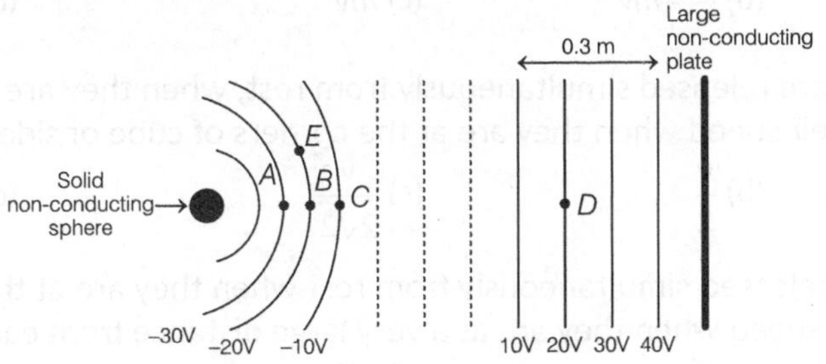

62 Surface charge density of the plate is equal to

(a) $8.85 \times 10^{-10}\ C/m^2$ (b) $-8.85 \times 10^{-10}\ C/m^2$

(c) $17.7 \times 10^{-10}\ C/m^2$ (d) $-17.7 \times 10^{-10}\ C/m^2$

63 A positive charge is placed at B. When it is released,

(a) no force will be exerted on it (b) it will move towards A

(c) it will move towards C (d) it will move towards E

64 How much work is required to slowly move a $-1\ \mu C$ charge from E to D?

(a) 2×10^{-5} J (b) -2×10^{-5} J (c) 4×10^{-5} J (d) -4×10^{-5} J

Comprehension-6

Two identical blocks A and B, each of mass $m = 600$ g, each carrying positive charge of $q = 40\mu C$ are kept stationary on a smooth horizontal floor, When the blocks are released, due to electrostatic repulsion, A moves towards left while B towards right. After moving 40 cm, A comes in contact with a massless spring of force constant $k = 6400\ Nm^{-1}$, while after moving 20 cm, B collides inelastically with a rigid wall as shown in figure.

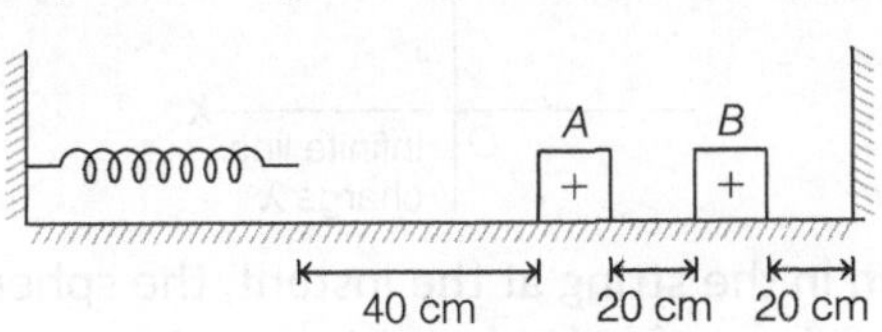

65 Velocity of A, when it comes in contact with the spring, is

(a) $10\ ms^{-1}$ (b) $15\ ms^{-1}$ (c) $20\ ms^{-1}$ (d) $30\ ms^{-1}$

66 Maximum compression of the spring is

(a) 20 cm (b) 10 cm (c) 40 cm (d) 60 cm

Comprehension-7

Eight equal point charges each of charge q and mass m are placed at eight corners of a cube of side a. If keeping all other charges fixed, charge of corner A is released, then its speed is v when it is far away from the cube.

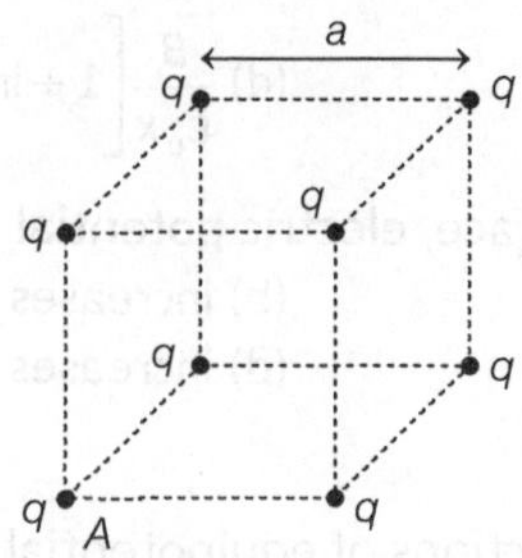

67 If all the eight charges are at the corners of cube of side a, then work done by external agent against electrostatic forces to increase all sides of cube from a to $2a$ is

(a) $\frac{1}{2}mv^2$ (b) $-\frac{1}{2}mv^2$ (c) mv^2 (d) $-mv^2$

68 If all the charges are released simultaneously from rest, when they are at the corners of a cube of side a, then their speed when they are at the corners of cube of side $2a$ is

(a) $\frac{v}{\sqrt{2}}$ (b) $\frac{v}{2}$ (c) $\frac{v}{2\sqrt{2}}$ (d) $\frac{v}{4}$

69 If all charges are released simultaneously from rest when they are at the corners of a cube of side a, then their speed when they are at a very large distance from each other is

(a) $\frac{v}{\sqrt{2}}$ (b) $\frac{v}{2}$ (c) $\frac{v}{2\sqrt{2}}$ (d) $\frac{v}{4}$

Comprehension-8

A small charged sphere (mass M, charge Q) is connected by a light, inextensible, non-conducting string of length L to a fixed point P (0, 2 L). The sphere is at rest in equilibrium in the position A. The sphere is free to slide on a non-conducting, frictionless and horizontal surface (designated as OXY-plane).

Along the X-axis lies an infinite line charge. The sphere is to be imparted a velocity $u\hat{i}$ at position A, so that it moves in a complete circle about the centre P. B, C, D are points on the circumference of the circle, BD parallel to line charge and AC normal to line charge.

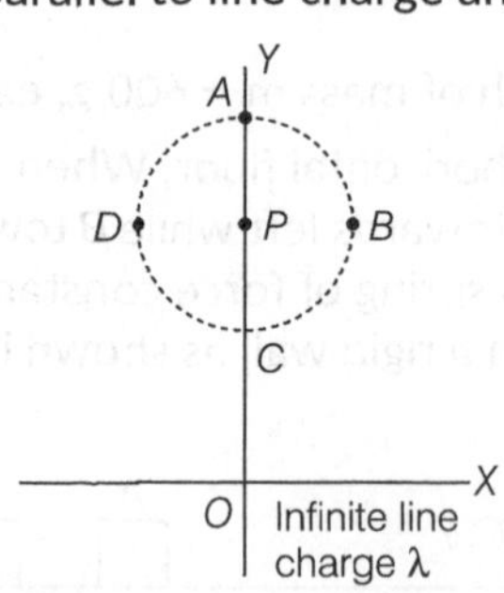

70 The instantaneous tension in the string at the instant, the sphere is given the velocity $u\hat{i}$ is

(a) $\frac{Q\lambda}{6\pi\varepsilon_0 L}$ (b) $\frac{Mu^2}{L} - \frac{Q\lambda}{6\pi\varepsilon_0 L}$ (c) $\frac{Mu^2}{L} + \frac{Q\lambda}{6\pi\varepsilon_0 L}$ (d) $\frac{Q\lambda}{6\pi\varepsilon_0 L} - \frac{Mu^2}{L}$

71 The potential difference between points A and $B(V_A - V_B)$ is

(a) $\frac{\lambda}{4\pi\varepsilon_0}\ln\left(\frac{4}{3}\right)$ (b) $\frac{\lambda}{3\pi\varepsilon_0}\ln\left(\frac{3}{5}\right)$ (c) $\frac{\lambda}{2\pi\varepsilon_0}\ln\left(\frac{3}{2}\right)$ (d) $\frac{\lambda}{2\pi\varepsilon_0}\ln\left(\frac{2}{3}\right)$

72 The tension in the string, when the sphere reaches position B is

(a) $\dfrac{Q\lambda}{4\pi\varepsilon_0 L}$

(b) $\dfrac{Mu^2}{L} - \dfrac{Q\lambda}{\pi\varepsilon_0 L}\ln\left(\dfrac{3}{2}\right)$

(c) $\dfrac{Mu^2}{L} + \dfrac{Q\lambda}{\pi\varepsilon_0 L}\ln\left(\dfrac{3}{2}\right)$

(d) None of these

Match the Columns Questions

73

Column I (Potential function)	Column II (Electric lines of force)
(a) $V = \dfrac{x}{y}$	(p) 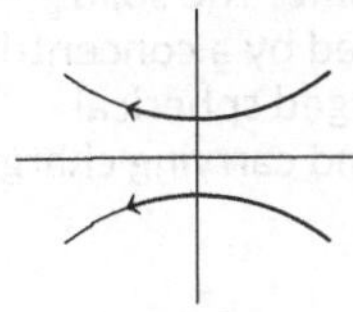
(b) $V = x^2 - y^2$	(q) 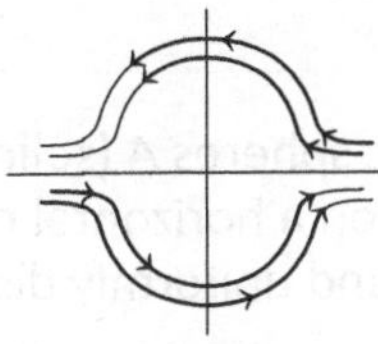
(c) $V = x^2 + y^2$	(r) 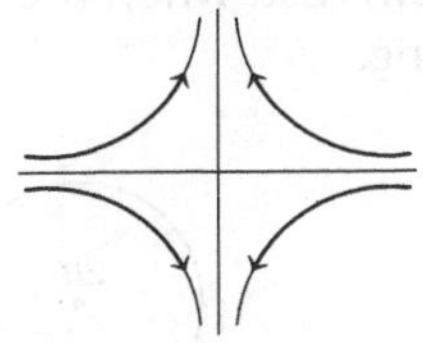
(d) $V = xy$	(s) 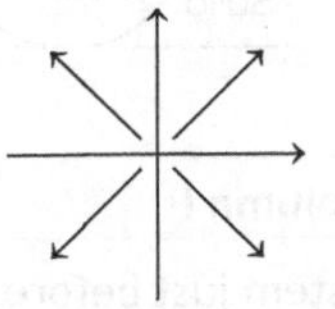

74 In each situation of Column I, some charge distributions are given with all details explained. The electrostatic potential energy and its nature is given situation in Column II. Then, match situation in Column I with the corresponding results in Column II.

Column I		Column II
(a) A thin shell of radius a and having a charge $-Q$ uniformly distributed over its surface as shown	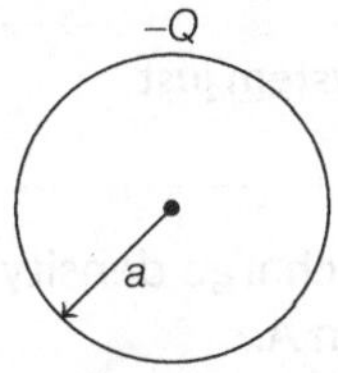	(p) $\dfrac{1}{8\pi\varepsilon_0}\dfrac{Q^2}{a}$ in magnitude
(b) A thin shell of radius $\dfrac{5a}{2}$ and having a charge $-Q$ uniformly distributed over its surface and a point charge $-Q$ placed at its centre as shown	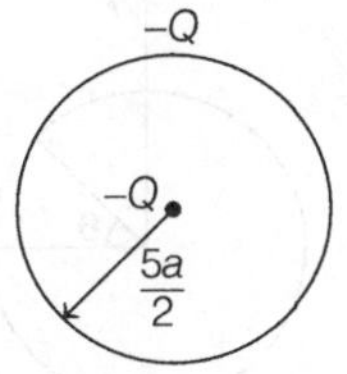	(q) $\dfrac{3}{20\pi\varepsilon_0}\dfrac{Q^2}{a}$ in magnitude

	Column I			Column II
(c)	A solid sphere of radius a and having a charge $-Q$ uniformly distributed throughout its volume as shown	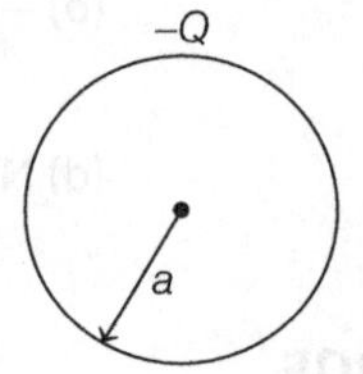	(r)	$\frac{2}{5\pi\varepsilon_0}\frac{Q^2}{a}$ in magnitude
(d)	A solid sphere of radius a and having a charge $-Q$ uniformly distributed throughout its volume. The solid sphere is surrounded by a concentric thin uniformly charged spherical shell of radius $2a$ and carrying charge $-Q$ as shown	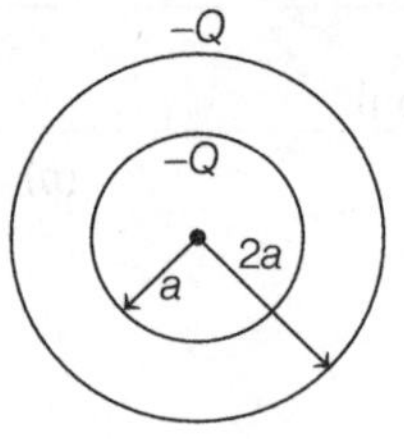	(s)	positive in sign
			(t)	negative in sign

75 Two charged dielectric spheres A (solid) and B (hollow) of same radius R and masses $2\,m$ and m respectively, are kept on a horizontal rough surface. The two spheres A and B have charges $2Q$ and $-Q$, respectively and uniformly distributed.

They are released from rest, when their centres are separated by very large distance. Assume that, there is no sliding.

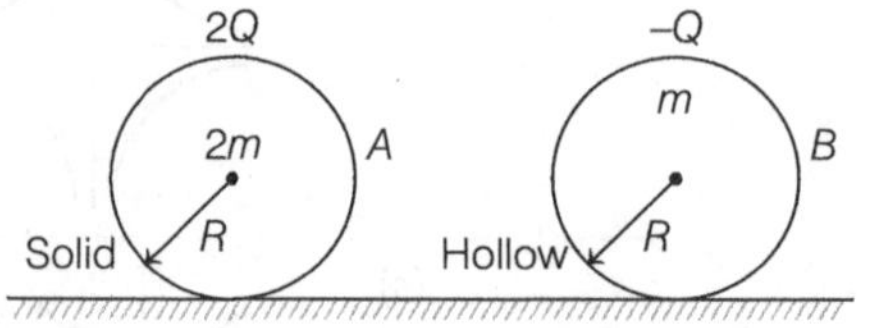

	Column I		Column II
(a)	Kinetic energy of system just before A and B meet	(p)	$\frac{838kQ^2}{2345R}$
(b)	Kinetic energy of A just before A and B meet	(q)	$\frac{kQ^2}{R}$
(c)	Kinetic energy of B just before A and B meet	(r)	$\frac{25kQ^2}{67R}$
(d)	Rotational kinetic energy of system just before A and B meet	(s)	$\frac{42kQ^2}{67R}$

76 A non-conducting ring has linear charge density λ. Match the following column regarding this ring. θ is the angle measured from A.

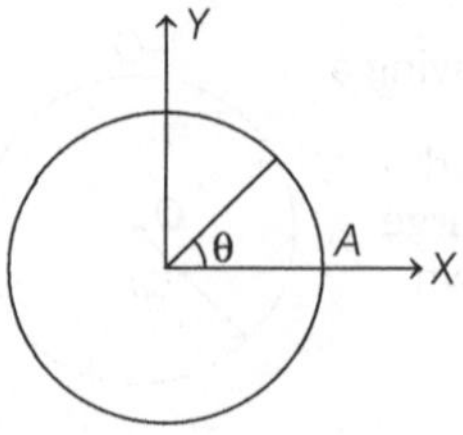

Column I		Column II	
(a)	$\lambda = \lambda_0$ C/m	(p)	Electric field at centre of the ring is in the direction to $(-\hat{j})$
(b)	$\lambda = \lambda_0 \cos\theta$ C/m, where $0 \le \theta \le 2\pi$	(q)	Electric field at centre of the ring is zero
(c)	$\lambda = \lambda_0 \sin\theta$ C/m, where $0 \le \theta \le 2\pi$	(r)	Electric potential at centre of ring is zero
(d)	$\lambda = \lambda_0(1 - \cos 2\theta)$ C/m, where $0 \le \theta \le 2\pi$	(s)	Electric field intensity at centre of the ring is direction $(-\hat{i})$
		(t)	Electrostatics potential at centre of ring is positive

Numerical Value Questions

77 A uniform electric field of 400 V/m exists in space as shown in graph. Two points A and B are also shown with their co-ordinates. If the potential of A is 20 V, then find the potential of B (in V).

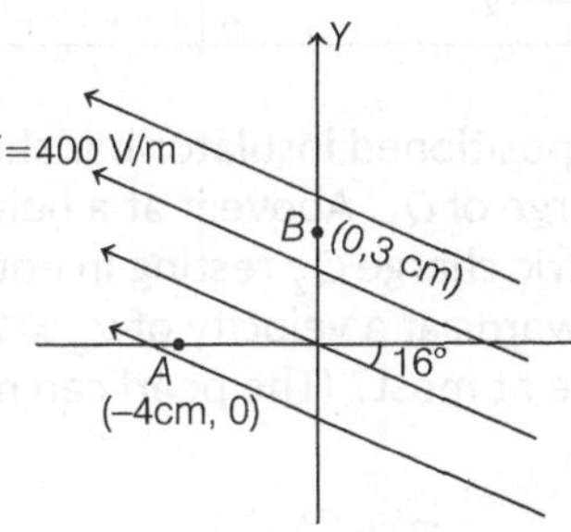

78 Consider the E versus x graph. What is the minimum velocity that should be given to a point charge $-Q$ of mass M, so that it can reach the origin?

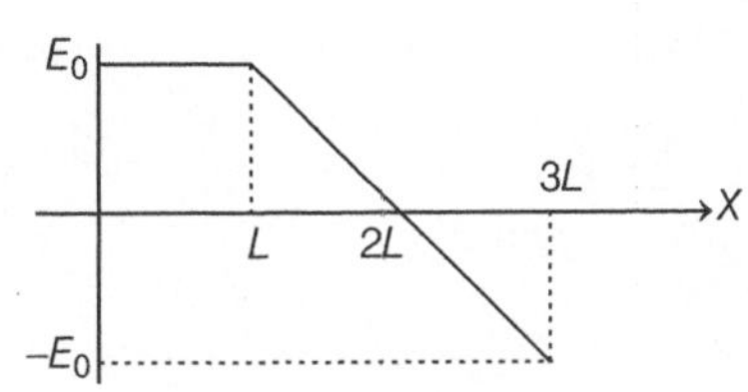

Take, $Q = 1$ mC, mass of the charge as 1 mg, $E_0 = 20$ N/C, $L = 5$ m and $\sqrt{10} = 3.162$. (Assume particle is starting from $x = 3L$ and moving towards origin along X-axis)

79 A total amount of positive charge Q is spread into a non-conducting flat circular annulus of inner radius a and outer radius $3a$. The charge is distributed, so that the charge density is given by $\sigma = \frac{k}{r^3}$, where r is the distance from the centre of the annulus to any point on it. The potential at the centre of annulus is $\frac{Q}{n\pi\varepsilon_0 a}$. Find n.

80 A uniformly charged hemisphere of radius b and charge density ρ has a hemispherical cavity of radius $a \left(a = \frac{b}{2} \right)$ cut from its centre.

If the potential at the centre of the cavity is $\frac{n\rho b^2}{16\varepsilon_0}$, then $n = ?$

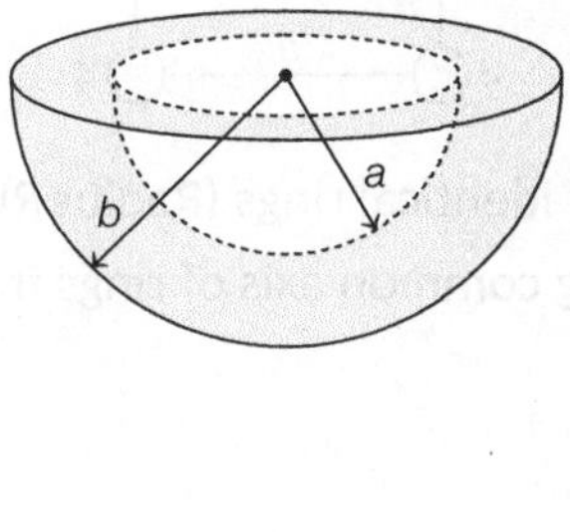

Problems in Physics

81 A non-conducting disc of radius $R = 6$ m and uniform positive surface charge density σ is placed on the ground, with its axis vertical. A particle of mass m and positive charge q is dropped, along the axis of the disc, from a height H with zero initial velocity. The particle has $\frac{q}{m} = \frac{4\varepsilon_0 g}{\sigma}$.

Find the value of H (in m), if the particle just reaches the disc.

82 Figure shows a thin rod of length L, which is uniformly charged with linear charge density λ kept on a smooth horizontal surface. Right end of rod is in contact with a vertical fixed wall. A block of mass m and charge q is projected with a velocity $v_0 = \sqrt{\frac{\lambda q}{m\pi\varepsilon_0}\ln 2}$ from a point very far from rod in the line of rod. If the distance of closest approach between the block and the left end A of the rod is $\frac{L}{n}$, then find n.

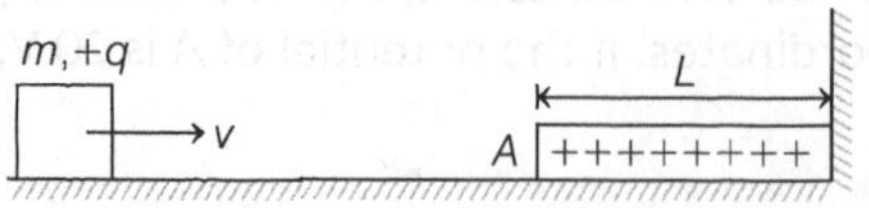

83 At the lower end of a vertically positioned insulator bar shown in the figure, there is a pearl of the mass $m = 10^{-4}$ kg and a charge of Q_1. Above it at a height $h_0 = 20$ cm, there is another pearl with the same mass m and electric charge Q_2 resting in equilibrium. At a given moment, we kick the lower pearl and it starts upwards at a velocity of $v_0 = 2$ m/s. How close (in cm) can the lower pearl get to the upper one at most? (The pearl can move along the bar without friction)

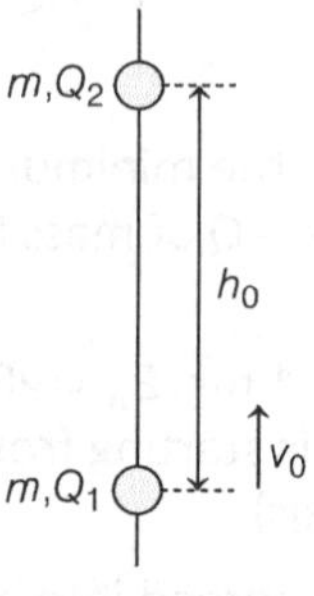

84 Four balls, each with mass m, are connected by four non-conducting strings to form a square with side a as shown in figure. The assembly is placed on a horizontal, non-conducting, frictionless surface. Balls 1 and 2 each have charge q and balls 3 and 4 are uncharged. Find the maximum speed of balls 3 and 4 after the string connected balls 1 and 2 is cut. If your answer is $\sqrt{\frac{Ykq^2}{Xam}}$ (where, $k = \frac{1}{4\pi\varepsilon_0}$), then fill value of $X + Y$, where X and Y are smallest positive integer.

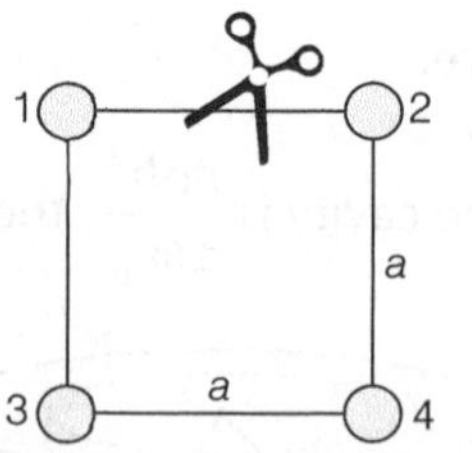

85 Two uniformly charged co-axial, identical rings (Radius R) are fixed as shown. A point charge q_0 is projected with speed v_0 along common axis of rings from large distance. Minimum value of

v_0 for which the charge particle can just pass through both rings is given as $v_0 = \sqrt{\dfrac{kqq_0}{nmR}}$.

Find n. $\left(\text{Take, } k = \dfrac{1}{4\pi\varepsilon_0}\right)$

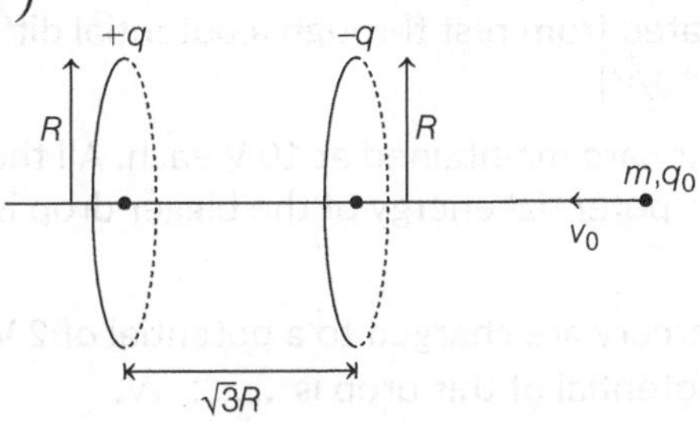

86 A bob of mass m and charge q is suspended from a point O of a vertical large thin non-conducting sheet of a surface charge density $+\sigma$. The bob is released from rest from horizontal. If $l = 1$ m, $\sigma q = 4\varepsilon_0 mg$, find velocity of the bob (in m/s), when $\theta = \sin^{-1}\left(\dfrac{3}{5}\right)$.

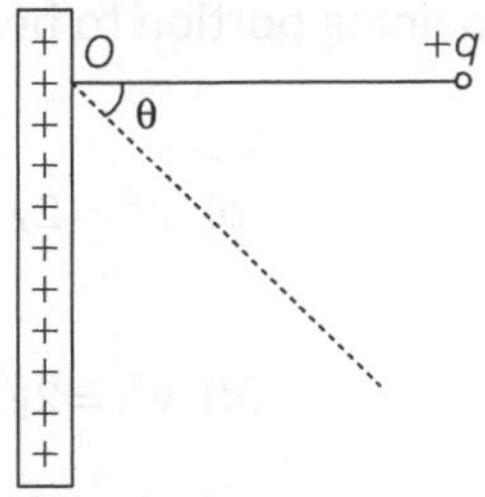

87 In a dark abandoned hall, there are a robber and a policeman standing against each other by the opposite sides of the hall. Having heard steps echoing in the darkness, the criminal points his gun horizontally on the unsuspecting cop and is about to shoot. What he doesn't know is that a potential difference $U = 144$ V was brought between the floor and the ceiling of the hall, thus creating a homogeneous field acting (vertically downwards) on the fired bullet, since the bullet acquires a charge q as it travels through the muzzle? Find the least charge-to-mass ratio q/m of the bullet, such that it hits the floor before it reaches the policeman (m denote the mass of the bullet).

The robber and the policeman are separated by a distance of $D = 100$ m, the ceiling is $d = 8$m above the floor and the bullet is fired horizontally with muzzle velocity $v_0 = 400\ \text{ms}^{-1}$. Assume that, the robber holds his gun $h = 2$m above the floor. (Take, $g = 10\ \text{ms}^{-2}$)

Answers

1. (c)	**2.** (d)	**3.** (b)	**4.** (c)	**5.** (a)	**6.** (d)	**7.** (b)	**8.** (d)	**9.** (a)	**10.** (b)
11. (b)	**12.** (d)	**13.** (a)	**14.** (c)	**15.** (d)	**16.** (d)	**17.** (a)	**18.** (a)	**19.** (c)	**20.** (b)
21. (b)	**22.** (a)	**23.** (a)	**24.** (b)	**25.** (a)	**26.** (a)	**27.** (c)	**28.** (c)	**29.** (a)	**30.** (b)
31. (c)	**32.** (b)	**33.** (ac)	**34.** (abcd)	**35.** (bc)	**36.** (ac)	**37.** (abd)	**38.** (ac)	**39.** (abc)	**40.** (ac)
41. (ab)	**42.** (acd)	**43.** (ad)	**44.** (ad)	**45.** (acd)	**46.** (ac)	**47.** (ac)	**48.** (abcd)	**49.** (bcd)	**50.** (acd)
51. (b)	**52.** (d)	**53.** (d)	**54.** (c)	**55.** (c)	**56.** (b)	**57.** (c)	**58.** (a)	**59.** (c)	**60.** (a)
61. (a)	**62.** (c)	**63.** (b)	**64.** (d)	**65.** (a)	**66.** (b)	**67.** (d)	**68.** (b)	**69.** (a)	**70.** (c)

71. (d) **72.** (b) **73.** a → q; b → r; c → s; d→p **74.** a → p, s; b → q,s; c → q,s; d → s

75. a → q; b → r, c → s, d → p **76.** a → q, t; b → s, r; c → p, r; d → q, t

77. 32 **78.** 316.20 **79.** 6 **80.** 3

81. 8 **82.** 3 **83.** 13.33 **84.** 4

85. 1 **86.** 2 **87.** 3

For Detailed Solutions
Scan the code

Previous Years' Questions

JEE Main

1 The average translational kinetic energy of N_2 gas molecules at ....°C becomes equal to the KE of an electron accelerated from rest through a potential difference of 0.1 V. [Given, $K_B = 1.38 \times 10^{-23}$ J/K] **[JEE Main 2021]**

2 27 similar drops of mercury are maintained at 10 V each. All these spherical drops combine into a single big drop. The potential energy of the bigger drop is .... times that of a smaller drop. **[JEE Main 2021]**

3 512 identical drops of mercury are charged to a potential of 2 V each. The drops are joined to form a single drop. The potential of this drop is .... V. **[JEE Main 2021]**

4 A solid sphere of radius R carries a charge $Q+q$ distributed uniformly over its volume. A very small point-like piece of it of mass m gets detached from the bottom of the sphere and falls down vertically under gravity. This piece carries charge q. If it acquires a speed v, when it has fallen through a vertical height y (see figure), then (Assume the remaining portion to be spherical.) **[JEE Main 2020]**

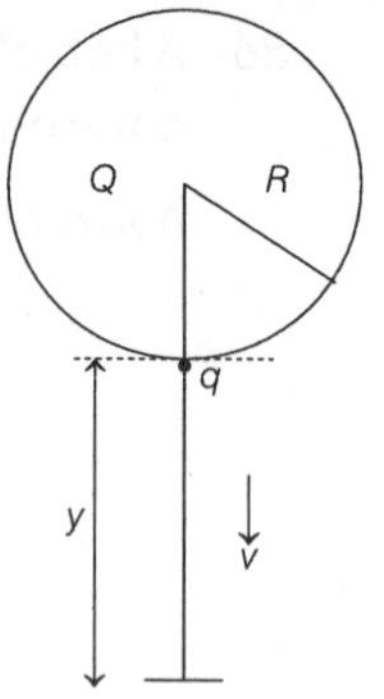

(a) $v^2 = y\left[\dfrac{qQ}{4\pi\varepsilon_0 R^2 ym} + g\right]$

(b) $v^2 = 2y\left[\dfrac{qQ}{4\pi\varepsilon_0 R(R+y)m} + g\right]$

(c) $v^2 = y\left[\dfrac{qQ}{4\pi\varepsilon_0 R(R+y)m} + g\right]$

(d) $v^2 = 2y\left[\dfrac{qQR}{4\pi\varepsilon_0 (R+y)^3 m} + g\right]$

5 A charge Q is distributed over two concentric conducting thin spherical shells radii r and R $(R > r)$. If the surface charge densities on the two shells are equal, the electric potential at the common centre is **[JEE Main 2020]**

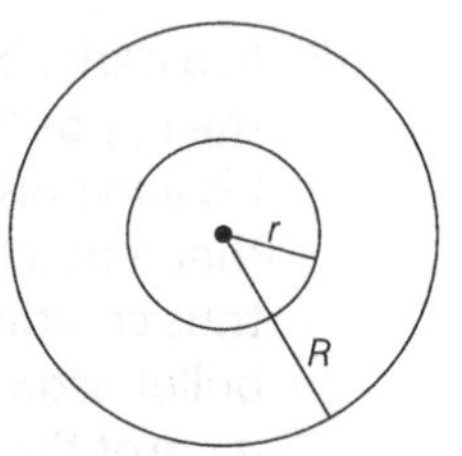

(a) $\dfrac{1}{4\pi\varepsilon_0}\dfrac{(R+r)}{2(R^2+r^2)}Q$

(b) $\dfrac{1}{4\pi\varepsilon_0}\dfrac{(2R+r)}{(R^2+r^2)}Q$

(c) $\dfrac{1}{4\pi\varepsilon_0}\dfrac{(R+2r)Q}{2(R^2+r^2)}$

(d) $\dfrac{1}{4\pi\varepsilon_0}\dfrac{(R+r)}{(R^2+r^2)}Q$

6 Concentric metallic hollow spheres of radii R and $4R$ hold charges Q_1 and Q_2, respectively. Given that, surface charge densities of the concentric spheres are equal. The potential difference $V(R) - V(4R)$ is **[JEE Main 2020]**

(a) $\dfrac{3Q_2}{4\pi\varepsilon_0 R}$

(b) $\dfrac{3Q_1}{16\pi\varepsilon_0 R}$

(c) $\dfrac{3Q_1}{4\pi\varepsilon_0 R}$

(d) $\dfrac{Q_2}{4\pi\varepsilon_0 R}$

7 Two point charges $4q$ and $-q$ are fixed on the X-axis at $x = \dfrac{-d}{2}$ and $x = \dfrac{d}{2}$, respectively. If a third point charge q is taken from the origin to $x = d$ along the semi-circle as shown in the figure, the energy of the charge will **[JEE Main 2020]**

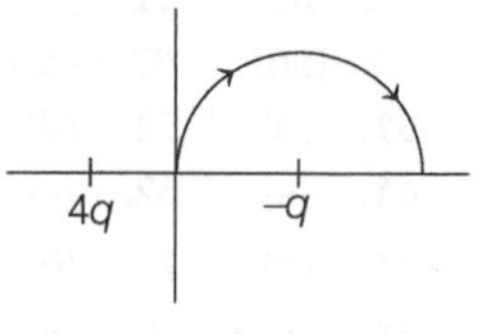

(a) increase by $\dfrac{2q^2}{3\pi\varepsilon_0 d}$

(b) decrease by $\dfrac{4q^2}{3\pi\varepsilon_0 d}$

(c) increase by $\dfrac{3q^2}{4\pi\varepsilon_0 d}$

(d) decrease by $\dfrac{q^2}{4\pi\varepsilon_0 d}$

8 Ten charges are placed on the circumference of a circle of radius R with constant angular separation between successive charges. Alternate charges 1, 3, 5, 7, 9 have charge $+q$ each, while 2, 4, 6, 8, 10 have charge $-q$ each. The potential V and the electric field E at the centre of the circle respectively, are (Take, $V = 0$ at infinity) **[JEE Main 2020]**

(a) $V = \dfrac{10q}{4\pi\varepsilon_0 R}; E = 0$
(b) $V = 0; E = \dfrac{10q}{4\pi\varepsilon_0 R^2}$
(c) $V = 0; E = 0$
(d) $V = \dfrac{10q}{4\pi\varepsilon_0 R}; E = \dfrac{10q}{4\pi\varepsilon_0 R^2}$

9 A charge Q is distributed over three concentric spherical shells of radii a, b, c $(a < b < c)$, such that their surface charge densities are equal to one another. The total potential at a point at distance r from their common centre, where $r < a$ would be **[JEE Main 2019]**

(a) $\dfrac{Q(a^2 + b^2 + c^2)}{4\pi\varepsilon_0 (a^3 + b^3 + c^3)}$
(b) $\dfrac{Q(a + b + c)}{4\pi\varepsilon_0 (a^2 + b^2 + c^2)}$
(c) $\dfrac{Q}{4\pi\varepsilon_0 (a + b + c)}$
(d) $\dfrac{Q}{12\pi\varepsilon_0} \cdot \dfrac{ab + bc + ca}{abc}$

10 The given graph shows variation (with distance r from centre) of **[JEE Main 2019]**

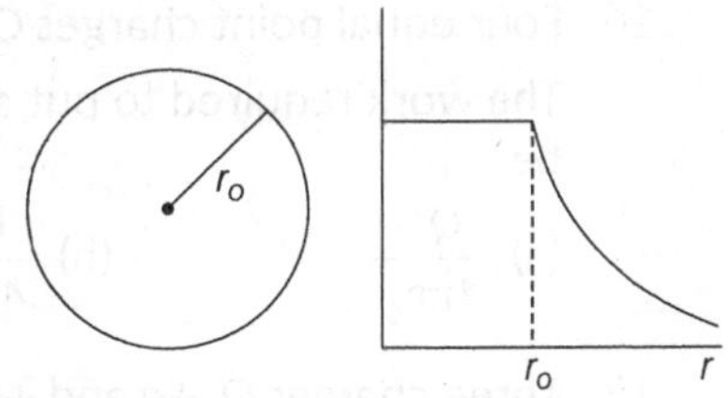

(a) electric field of a uniformly charged spherical shell
(b) potential of a uniformly charged spherical shell
(c) electric field of a uniformly charged sphere
(d) potential of a uniformly charged sphere

11 A positive point charge is released from rest at a distance r_0 from a positive line charge with uniform density. The speed (v) of the point charge, as a function of instantaneous distance r from line charge, is proportional to **[JEE Main 2019]**

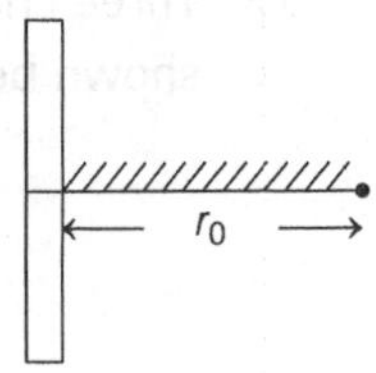

(a) $v \propto \left(\dfrac{r}{r_0}\right)$
(b) $v \propto e^{+\frac{r}{r_0}}$
(c) $v \propto \ln\left(\dfrac{r}{r_0}\right)$
(d) $v \propto \sqrt{\ln\left(\dfrac{r}{r_0}\right)}$

12 The electric field in a region is given by $\mathbf{E} = (Ax + B)\hat{\mathbf{i}}$, where E is in NC^{-1} and x is in metres. The values of constants are $A = 20$ SI unit and $B = 10$ SI unit. If the potential at $x = 1$ is V_1 and that at $x = -5$ is V_2, then $V_1 - V_2$ is **[JEE Main 2019]**

(a) -48 V
(b) -520 V
(c) 180 V
(d) 320 V

13 A system of three charges are placed as shown in the figure

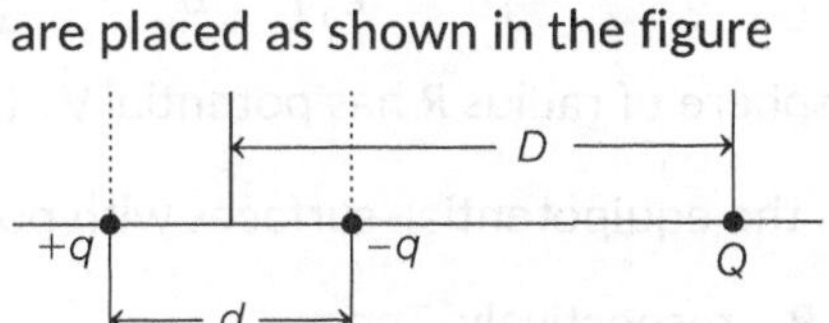

If $D >> d$, the potential energy of the system is best given by **[JEE Main 2019]**

(a) $\dfrac{1}{4\pi\varepsilon_0}\left[-\dfrac{q^2}{d} + \dfrac{2qQd}{D^2}\right]$
(b) $\dfrac{1}{4\pi\varepsilon_0}\left[+\dfrac{q^2}{d} + \dfrac{qQd}{D^2}\right]$
(c) $\dfrac{1}{4\pi\varepsilon_0}\left[-\dfrac{q^2}{d} - \dfrac{qQd}{2D^2}\right]$
(d) $\dfrac{1}{4\pi\varepsilon_0}\left[-\dfrac{q^2}{d} - \dfrac{qQd}{D^2}\right]$

14 In free space, a particle A of charge 1 μC is held fixed at a point P. Another particle B of the same charge and mass 4 μg is kept at a distance of 1 mm from P. If B is released, then its velocity at a distance of 9 mm from P is

$$\left[\text{Take,} \frac{1}{4\pi\varepsilon_0} = 9\times 10^9 \text{ N-m}^2\text{C}^{-2}\right]$$

[JEE Main 2019]

(a) 1.5×10^2 m/s (b) 3.0×10^4 m/s
(c) 1.0 m/s (d) 2.0×10^3 m/s

15 A uniformly charged ring of radius $3a$ and total charge q is placed in XY-plane centred at origin. A point charge q is moving towards the ring along the Z-axis and has speed v at $z = 4a$. The minimum value of v, such that it crosses the origin is [JEE Main 2019]

(a) $\sqrt{\frac{2}{m}}\left(\frac{1}{5}\frac{q^2}{4\pi\varepsilon_0 a}\right)^{1/2}$ (b) $\sqrt{\frac{2}{m}}\left(\frac{4}{15}\frac{q^2}{4\pi\varepsilon_0 a}\right)^{1/2}$

(c) $\sqrt{\frac{2}{m}}\left(\frac{1}{15}\frac{q^2}{4\pi\varepsilon_0 a}\right)^{1/2}$ (d) $\sqrt{\frac{2}{m}}\left(\frac{2}{15}\frac{q^2}{4\pi\varepsilon_0 a}\right)^{1/2}$

16 Four equal point charges Q each are placed in the XY-plane at (0, 2), (4, 2), (4, −2) and (0, −2). The work required to put a fifth charge Q at the origin of the coordinate system (in joule) will be [JEE Main 2019]

(a) $\frac{Q^2}{4\pi\varepsilon_0}$ (b) $\frac{Q^2}{4\pi\varepsilon_0}\left(1+\frac{1}{\sqrt{3}}\right)$ (c) $\frac{Q^2}{2\sqrt{2}\pi\varepsilon_0}$ (d) $\frac{Q^2}{4\pi\varepsilon_0}\left(1+\frac{1}{\sqrt{5}}\right)$

17 Three charges Q, $+q$ and $+q$ are placed at the vertices of a right angle isosceles triangle as shown below. The net electrostatic energy of the configuration is zero, if the value of Q is [JEE Main 2019]

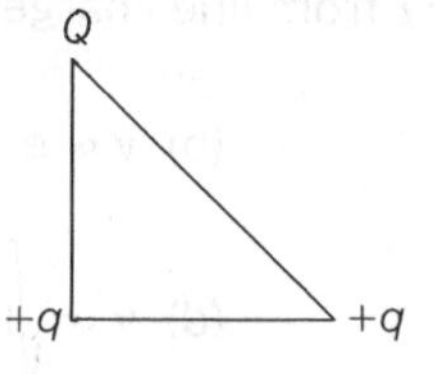

(a) $-2q$ (b) $\frac{-q}{1+\sqrt{2}}$ (c) $+q$ (d) $\frac{-\sqrt{2}q}{\sqrt{2}+1}$

18 Three concentric metal shells A, B and C of respective radii a, b and c ($a < b < c$) have surface charge densities $+\sigma$, $-\sigma$ and $+\sigma$, respectively. The potential of shell B is [JEE Main 2018]

(a) $\frac{\sigma}{\varepsilon_0}\left[\frac{a^2-b^2}{a}+c\right]$ (b) $\frac{\sigma}{\varepsilon_0}\left[\frac{a^2-b^2}{b}+c\right]$ (c) $\frac{\sigma}{\varepsilon_0}\left[\frac{b^2-c^2}{b}+a\right]$ (d) $\frac{\sigma}{\varepsilon_0}\left[\frac{b^2-c^2}{c}+a\right]$

19 A uniformly charged solid sphere of radius R has potential V_0 (measured with respect to ∞) on its surface. For this sphere, the equipotential surfaces with potentials $\frac{3V_0}{2}, \frac{5V_0}{4}, \frac{3V_0}{4}$ and $\frac{V_0}{4}$ have radius R_1, R_2, R_3, and R_4, respectively. Then, [JEE Main 2015]

(a) $R_1 = 0$ and $R_2 > (R_4 - R_3)$ (b) $R_1 \neq 0$ and $(R_2 - R_1) > (R_4 - R_3)$
(c) $R_1 = 0$ and $R_2 < (R_4 - R_3)$ (d) $2R < R_4$

20 Assume that an electric field $\mathbf{E} = 30x^2\ \hat{\mathbf{i}}$ exists in space. Then, the potential difference $V_A - V_O$, where V_O is the potential at the origin and V_A is the potential at $x = 2$ m, is [JEE Main 2014]

(a) 120 J (b) −120 J
(c) −80 J (d) 80 J

21 A charge Q is uniformly distributed over a long rod AB of length L as shown in the figure. The electric potential at the point O lying at distance L from the end A is **[JEE Main 2013]**

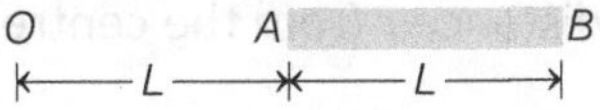

(a) $\frac{Q}{8\pi\varepsilon_0 L}$ (b) $\frac{3Q}{4\pi\varepsilon_0 L}$ (c) $\frac{Q}{4\pi\varepsilon_0 L \ln 2}$ (d) $\frac{Q\ln 2}{4\pi\varepsilon_0 L}$

JEE Advanced

Passage *(1-2) There are two questions based on the following passage.*

Consider an evacuated cylindrical chamber of height *h* having rigid conducting plates at the ends and an insulating curved surface as shown in the figure. A number of spherical balls made of a light weight and soft material and coated with a conducting material are placed on the bottom plate. The balls have a radius $r << h$. Now, a high voltage source (HV) connected across the conducting plates, such that the bottom plate is at $+V_0$ and the top plate at $-V_0$. Due to their conducting surface, the balls will get charge, will become equipotential with the plate and are repelled by it. The balls will eventually collide with the top plate, where the coefficient of restitution can be taken to be zero due to te soft nature of the material of the balls. The electric field in the chamber can be considered to be that of a parallel plate capacitor. Assume that there are no collisions between the balls and the interaction between them is negligible. (Ignore gravity)

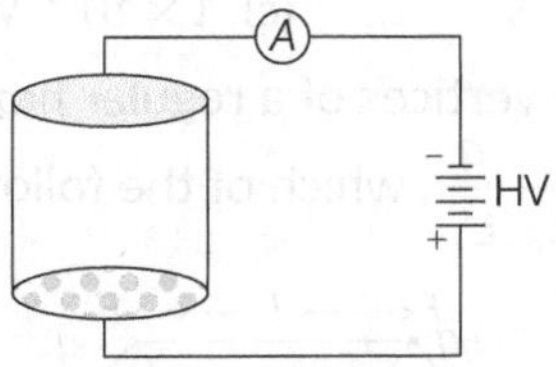

22 Which one of the following statement is correct? **[JEE Advanced 2016]**
(a) The balls will execute simple harmonic motion between the two plates.
(b) The balls will bounce back to the bottom plate carrying the same charge they went up with.
(c) The balls will stick to the top plate and remain there.
(d) The balls will bounce back to the bottom plate carrying the opposite charge they went up with.

23 The average current in the steady state registered by the ammeter in the circuit will be **[JEE Advanced 2016]**
(a) proportional to V_0^2. (b) proportional to the potential V_0.
(c) zero (d) proportions to $V_0^{1/2}$

24 Two non-conducting spheres of radii R_1 and R_2 and carrying uniform volume charge densities $+\rho$ and $-\rho$, respectively, are placed such that they partially overlap, as shown in the figure. At all points in the overlapping region **[JEE Advanced 2013]**

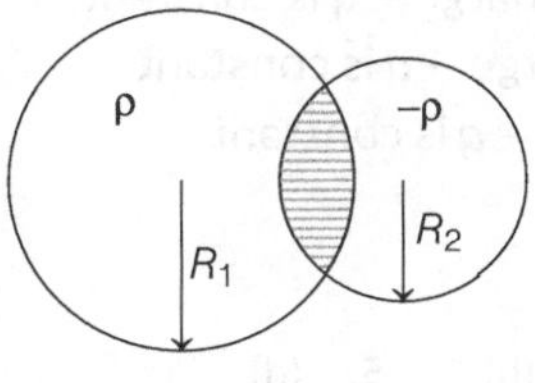

(a) the electrostatic field is zero.
(b) the electrostatic potential is constant.
(c) the electrostatic field is constant in magnitude.
(d) the electrostatic field has same direction.

25 Consider a thin spherical shell of radius R with its centre at the origin, carrying uniform positive surface charge density. The variation of the magnitude of the electric field $|\mathbf{E}(r)|$ and the electric potential $V(r)$ with the distance r from the centre, is best represented by which graph? **[JEE Advanced 2012]**

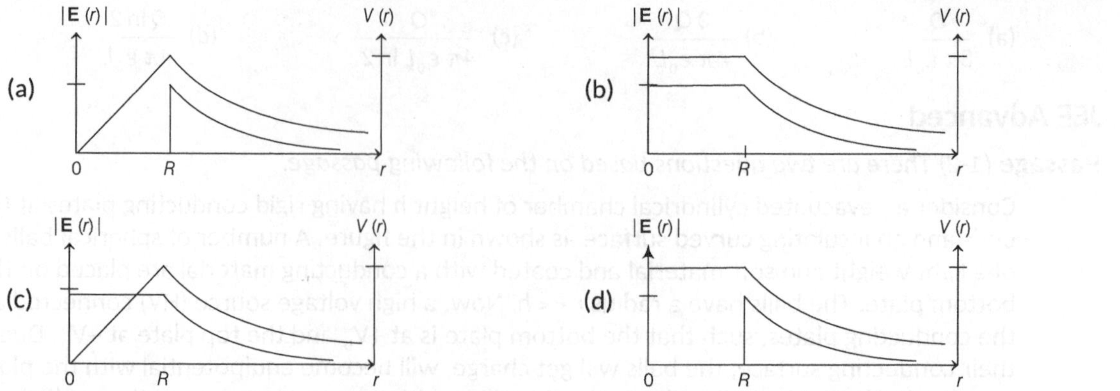

26 Two large vertical and parallel metal plates having a separation of 1 cm are connected to a DC voltage source of potential difference X. A proton is released at rest midway between the two plates. It is found to move at 45° to the vertical just after release. Then X is nearly **[JEE Advanced 2012]**

(a) 1×10^{-5} V (b) 1×10^{-7} V (c) 1×10^{-9} V (d) 1×10^{-10} V

27 Six point charges are kept at the vertices of a regular hexagon of side L and centre O as shown in the figure. Given that $K=\frac{1}{4\pi\varepsilon_0}\frac{q}{L^2}$, which of the following statements(s) is/are correct. **[JEE Advanced 2012]**

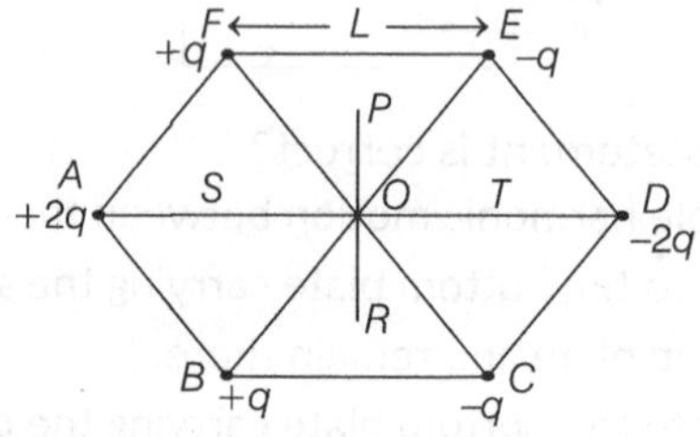

(a) The electric field at O is $6K$ along OD
(b) The potential at O is zero
(c) The potential at all points on the line PR is same
(d) The potential at all points on the line ST is same

28 Under the influence of the coulomb field of charge $+Q$, a charge $-q$ is moving around it in an elliptical orbit. Find out the correct statement(s). **[JEE Advanced 2009]**

(a) The angular momentum of the charge $-q$ is constant
(b) The linear momentum of the charge $-q$ is constant
(c) The angular velocity of the charge $-q$ is constant
(d) The linear speed of the charge $-q$ is constant

Answers

1. 500	2. 243	3. 128	4. (b)	5. (d)
6. (b)	7. (b)	8. (c)	9. (b)	10. (b)
11. (d)	12. (c)	13. (d)	14. (d)	15. (d)
16. (d)	17. (d)	18. (b)	19. (cd)	20. (c)
21. (d)	22. (d)	23. (a)	24. (cd)	25. (d)
26. (c)	27. (abc)	28. (a)		

For Detailed Solutions
Scan the code

CHAPTER

03

ELECTRIC DIPOLE

LEVEL 01 \ Starter Level

1.1 Electric Dipole Moment

1 An electron and a proton are at a distance of 1 Å. Find the electric dipole moment.

2 Find electric dipole moment of the following charge configurations.

(a) Three point charges are placed as shown.

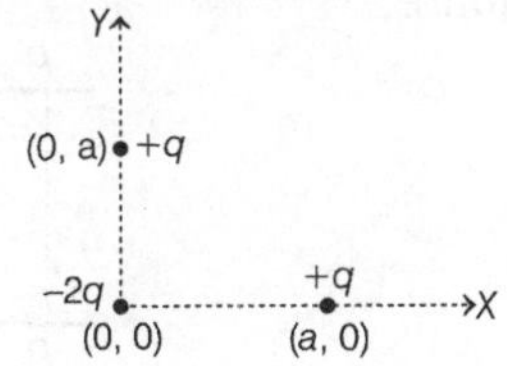

(b) Four point charges kept on a circle of radius R as shown.

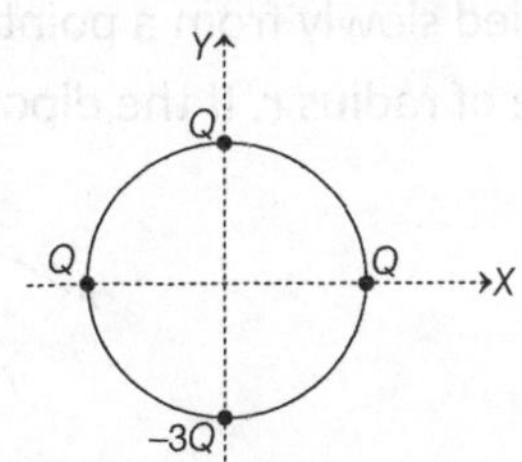

(c) Charge $+q$ is uniformly distributed on half-ring and a point charge $-q$ is kept at its centre.

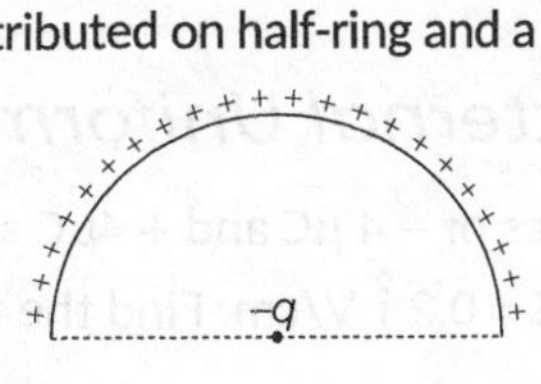

1.2 Electric Field due to Dipole

3 A dipole is placed at origin of coordinate system as shown in figure, find the electric field at point $P(0, y)$.

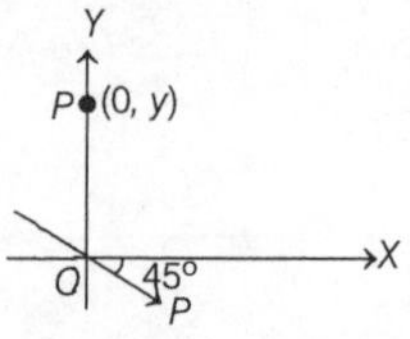

4 Find out the magnitude of electric field intensity at point (2 m, 0, 0) due to a dipole of dipole moment, $\mathbf{p} = (\hat{\mathbf{i}} + \sqrt{3}\hat{\mathbf{j}})$ C-m kept at origin.

5 Find the magnitude of the electric field at the point P in the configuration shown in figure for $d >> a$.

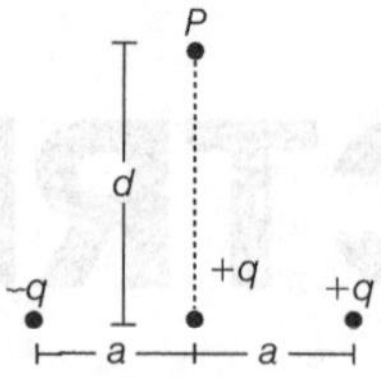

6 Two small electric dipoles of moment $p\hat{\mathbf{i}}$ and $-64\, p\, \hat{\mathbf{i}}$ are placed at (0,0) and (30,0) respectively, where will electric field intensity be zero?

1.3 Electric Potential due to Dipole

7 A dipole of dipole moment $\mathbf{p} = (2\hat{\mathbf{i}} - 3\hat{\mathbf{j}} + 4\hat{\mathbf{k}})$ μC-m is placed at point $A(2, -3, 1)$. Find the electric potential due to this dipole at the point $B(4, -1, 0)$.

8 Four short dipoles, each of dipole moment p are placed at the vertices of a square of side a. The direction of the dipole moments are shown in the figure. Find the electric potential at the centre O of the square.

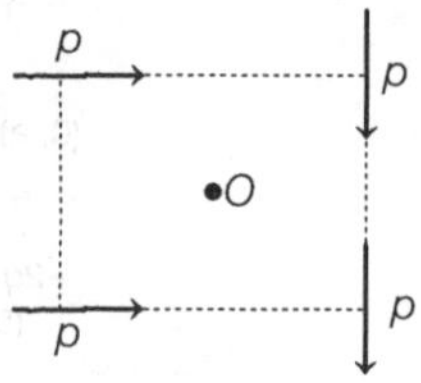

9 A charge q is carried slowly from a point $A(r, 135°)$ to point $B(r, 45°)$ following a path which is a quadrant of circle of radius r. If the dipole moment is p, then find out the work done by external agent.

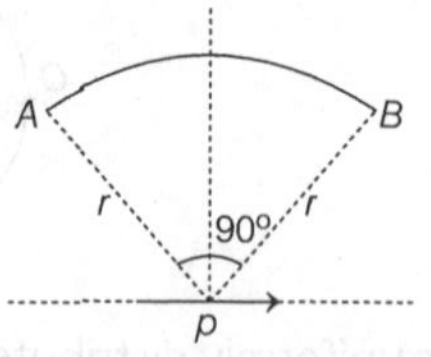

1.4 Dipole in External Uniform Electric Field

10 Two point charges of -4 μC and $+4$μC are placed at points A (1,0,4) and $B(2, -1, 5)$ located in an electric field $\mathbf{E} = 0.2\,\hat{\mathbf{i}}$ V/cm. Find the force and torque acting on the dipole.

11 Figure shows an electric dipole formed by two particles fixed at the ends of a light rod of length l. The mass of each particle is m and the charges are $-q$ and $+q$. The system is placed in such a way that, the dipole axis is parallel to a uniform electric field E that exists in the region. The dipole is slightly rotated about its centre and released. Show that for small angular displacements, the motion is angular simple harmonic and find its time period.

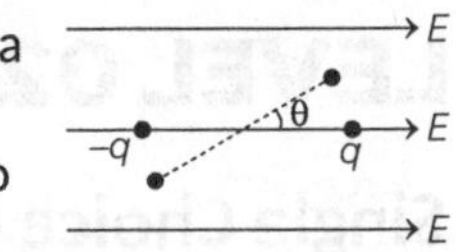

12 An electric dipole with dipole moment $\mathbf{p} = (2\hat{\mathbf{i}} + 3\hat{\mathbf{j}} - \hat{\mathbf{k}})\,\mu$C-m is placed in a uniform field $\mathbf{E} = (3\hat{\mathbf{i}} + 2\hat{\mathbf{k}}) \times 10^5$ N/C.

(i) What is the potential energy of the dipole?

(ii) If the dipole were to be rotated in the field **E**, what is the maximum potential energy of the dipole?

13 What is the work done by the external agent in slowly rotating a dipole of dipole moment of p in a uniform electric field E from stable equilibrium orientation to unstable equilibrium orientation?

1.5 Dipole in External Non-Uniform Electric Field

14 Electric dipole of moment $\mathbf{p} = p\hat{\mathbf{i}}$ is kept at a point (x, y) in an electric field $\mathbf{E} = 4xy^2\,\hat{\mathbf{i}} + 4x^2y\,\hat{\mathbf{j}}$. Find the magnitude of force acting on the dipole.

15 Find out force experienced by short dipole of dipole moment p due to a point charge Q in following different arrangements as shown in figures.

(i) 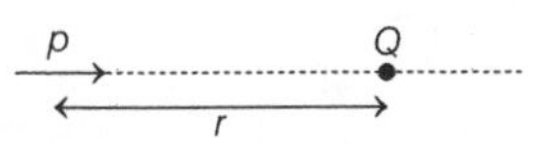

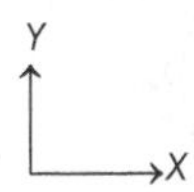

(ii) 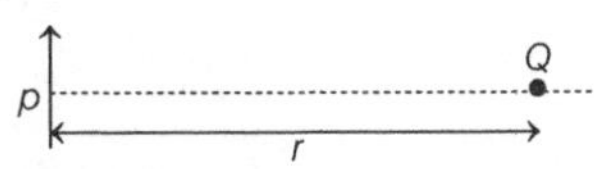

16 Find out force experienced by short dipole p_2 in the following arrangements as shown in figures.

(i) 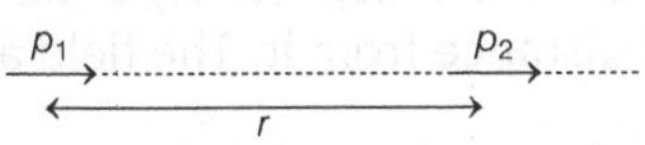

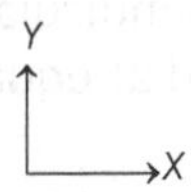

(ii) 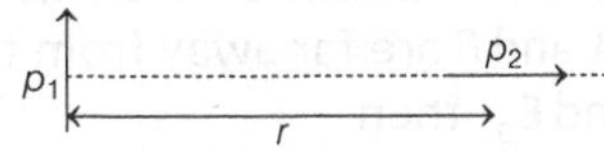

17 Find electrostatic potential energy of the following arrangements as shown in figures. Neglect self-energy of dipole.

(i) 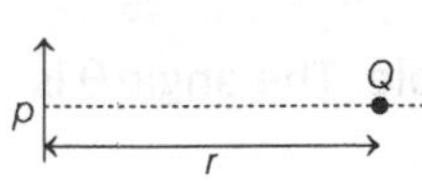

(ii) 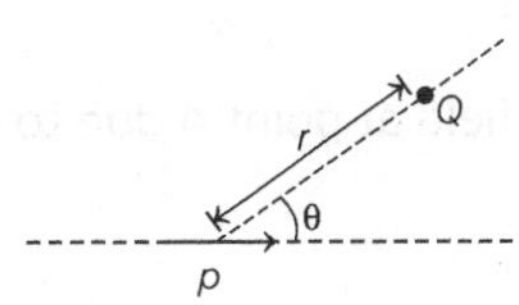

Answers

1. 1.6×10^{-29} C-m, directed from electron to proton
2. (a) $qa(\hat{\mathbf{i}} + \hat{\mathbf{j}})$ (b) $4QR\hat{\mathbf{j}}$ (c) $\frac{2qR}{\pi}\hat{\mathbf{j}}$
3. $\frac{kp}{\sqrt{2}y^3}(-\hat{\mathbf{i}} - 2\hat{\mathbf{j}})$
4. $\frac{9\sqrt{7}}{8} \times 10^9$ V/m
5. $\frac{q\sqrt{d^2 + 4a^2}}{4\pi\varepsilon_0 d^3}$
6. $(-10, 0)$ and $(6, 0)$
7. -2000 V
8. $\frac{p}{\sqrt{2}\pi\varepsilon_0 a^2}$
9. $\frac{\sqrt{2}\,qp}{4\pi\varepsilon_0 r^2}$
10. $F = 0$, $\tau = 8 \times 10^{-5}(\hat{\mathbf{j}} + \hat{\mathbf{k}})$ N-m
11. $T = 2\pi\sqrt{\frac{ml}{2qE}}$
12. (i) -0.4 J (ii) 1.35 J
13. $2pE$
14. $4py\sqrt{y^2 + 4x^2}$
15. (i) $\frac{2kpQ}{r^3}(-\hat{\mathbf{i}})$ (ii) $\frac{kpQ}{r^3}\hat{\mathbf{j}}$
16. (i) $\frac{6kp_1p_2}{r^4}(-\hat{\mathbf{i}})$ (ii) $\frac{3kp_1p_2}{r^4}(\hat{\mathbf{j}})$
17. (i) 0 (ii) $\frac{kpQ\cos\theta}{r^2}$

For Detailed Solutions
Scan the code

LEVEL 02 \ JEE Main Level

Single Choice Correct Questions

1 Four charges are placed each at a distance a from origin as shown in figure. The dipole moment of configuration is

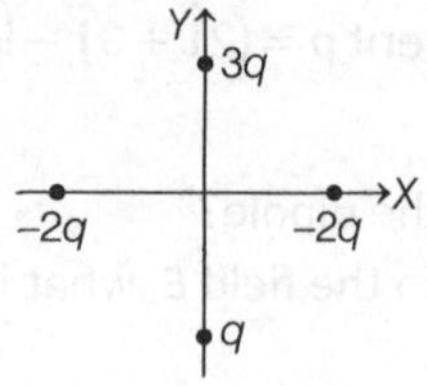

(a) $2qa\hat{j}$ (b) $3qa\hat{j}$ (c) $2aq(\hat{i}+\hat{j})$ (d) None of these

2 The dipole moment of a system of charge $+q$ distributed uniformly on an arc of radius R subtending an angle $\pi/2$, at its centre, where another charge $-q$ is placed is

(a) $\frac{2\sqrt{2}qR}{\pi}$ (b) $\frac{\sqrt{2}qR}{\pi}$ (c) $\frac{qR}{\pi}$ (d) $\frac{2qR}{\pi}$

3 A thin non-conducting ring of radius R has a linear charge density $\lambda = \lambda_0 \cos\theta$, where λ_0 is a constant. Find net electric dipole moment for this charge distribution.

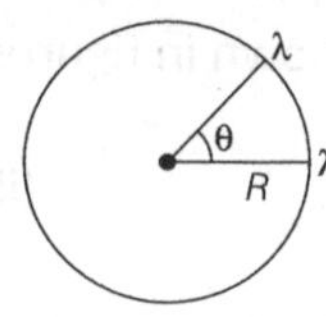

(a) $\pi R^2 \lambda_0$ (b) $2\pi R^2 \lambda_0$ (c) $3\pi R^2 \lambda_0$ (d) $4\pi R^2 \lambda_0$

4 A and B are two points on the axis and the perpendicular bisector, respectively of an electric dipole. A and B are far away from the dipole and at equal distance from it. The field at A and B are $\mathbf{E}_A$ and $\mathbf{E}_B$, then

(a) $\mathbf{E}_A = \mathbf{E}_B$ (b) $\mathbf{E}_A = 2\mathbf{E}_B$

(c) $\mathbf{E}_A = -2\mathbf{E}_B$ (d) $|\mathbf{E}_B| = \frac{1}{2}|\mathbf{E}_A|$ and $\mathbf{E}_B$ is perpendicular to $\mathbf{E}_A$

5 The electric field at point A due to dipole p is perpendicular to dipole. The angle θ is

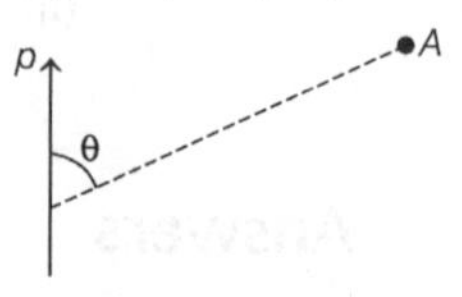

(a) 0° (b) 90° (c) $\tan^{-1}(2)$ (d) $\tan^{-1}\sqrt{2}$

6 Two equal charges Q are at opposite corners of a square of side a and an electric dipole of moment p is at a third corner, pointing towards one of the charges. If $p = 2\sqrt{2}\,aQ$, the field strength at the fourth corner of the square is

(a) $\frac{\sqrt{17/2}}{4\pi\varepsilon_0 a^2}Q$ (b) $\frac{\sqrt{17/4}}{4\pi\varepsilon_0 a^2}Q$ (c) $\frac{Q}{4\sqrt{2}\pi\varepsilon_0 a^2}$ (d) $\frac{\sqrt{3}Q}{4\pi\varepsilon_0 a^2}$

7 Two short dipoles $p(\hat{i}+\hat{k})$ and $4p\hat{k}$ are located at (0, 0, 0) and (1m, 0, 2m), respectively. The resultant electric field due to the two dipoles at the point (1 m, 0, 0) is

(a) $\frac{p}{\pi\varepsilon_0}$ (b) $\frac{p}{2\pi\varepsilon_0}$ (c) $\frac{p}{\sqrt{2}\pi\varepsilon_0}$ (d) $\frac{\sqrt{2}p}{\pi\varepsilon_0}$

8 Figure shows an electric dipole of dipole moment $p = q(2a)$. For $r >> a$, what is potential of dipole at a point P with coordinates (x, y, z)?

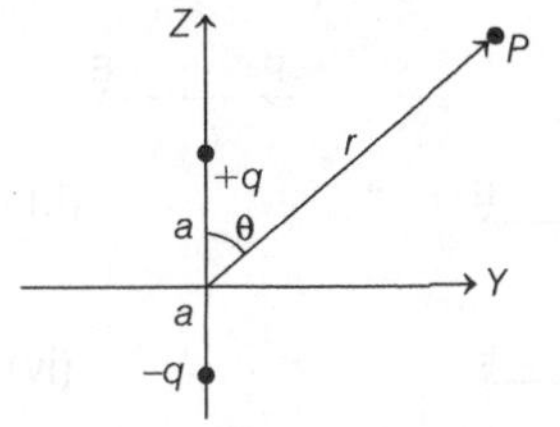

(a) $\dfrac{p}{4\pi\varepsilon_0} \dfrac{x}{(x^2 + y^2 + z^2)^{3/2}}$ (b) $\dfrac{p}{4\pi\varepsilon_0} \dfrac{y}{(x^2 + y^2 + z^2)^{3/2}}$

(c) $\dfrac{p}{4\pi\varepsilon_0} \dfrac{z}{(x^2 + y^2 + z^2)^{3/2}}$ (d) $\dfrac{p}{4\pi\varepsilon_0} \dfrac{xyz}{(x^2 + y^2 + z^2)^{3/2}}$

9 Find out work done by electric field in shifting a point charge $\dfrac{4\sqrt{2}}{27}$ μC from point P to S, which are shown in the figure. At point O, a dipole of dipole moment 2×10^{-6} C-m is kept

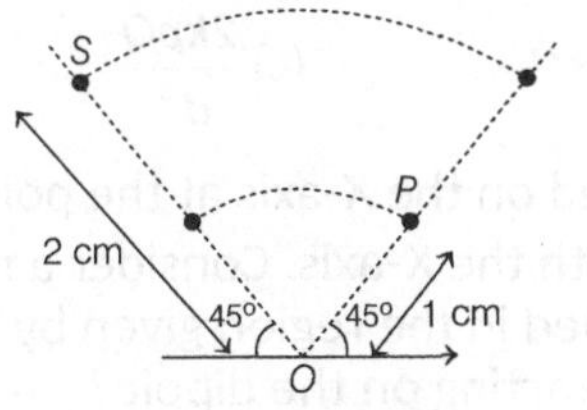

(a) (100/3) J (b) (200/3) J (c) 100 J (d) 200 J

10 Which of the following represents the equipotential lines of a dipole (two equal and opposite charges placed at small separation)?

(a) 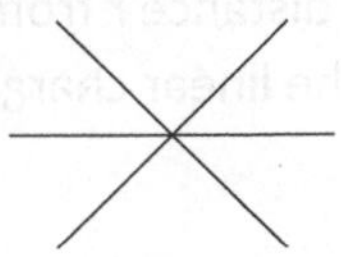 (b)  (c) 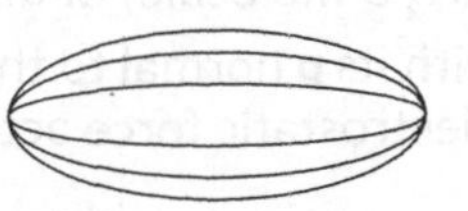 (d) 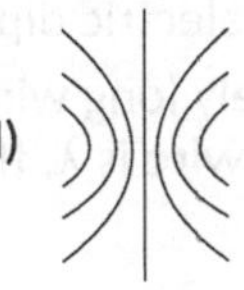

11 An electric dipole consists of two opposite charges each of magnitude 1.0 μC separated by a distance of 2.0 cm. The dipole is placed in an external field of 1.0×10^5 N/C. The maximum torque on the dipole is

(a) 0.2×10^{-3} N-m (b) 1.0×10^{-3} N-m (c) 2.0×10^{-3} N-m (d) 4.0×10^{-3} N-m

12 An electric dipole with dipole moment $\mathbf{p} = (3\hat{\mathbf{i}} + 4\hat{\mathbf{j}}) \times 10^{-30}$ C-m is placed in an electric field $\mathbf{E} = 4000\,\hat{\mathbf{i}}$ (N/C). An external agent turns the dipole slowly until its electric dipole moment becomes $(-4\hat{\mathbf{i}} + 3\hat{\mathbf{j}}) \times 10^{-30}$ C-m. The work done by the external agent is equal to

(a) 4×10^{-28} J (b) -4×10^{-28} J (c) 2.8×10^{-26} J (d) -2.8×10^{-26} J

13 An electric dipole is placed (not at infinity) in an electric field generated by a point charge,

(a) the net force on the dipole must be zero
(b) the net force on the dipole may be zero
(c) torque on the dipole due to the field must be zero
(d) torque on the dipole due to the field may be zero

14 An electric dipole of length 2 cm is placed with its axis making an angle of 30° to a uniform electric field of 10^5 N/C. If it experiences a torque of $10\sqrt{3}$ N-m, then potential energy of dipole is

(a) -10 J (b) -20 J (c) -30 J (d) -40 J

15 An electric dipole of dipole moment **p** is oriented parallel to a uniform electric field **E**, as shown. It is rotated to one of the four orientations shown below. Rank the final orientations according to the change in the potential energy of the dipole-field system, most negative to most positive.

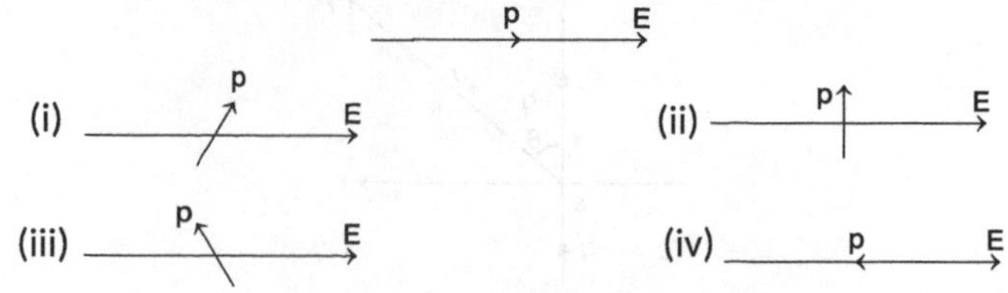

(a) (i), (ii), (iv), (iii)
(b) (iv), (iii), (ii), (i)
(c) (i), (ii), (iii), (iv)
(d) (iii), (ii) and (iv) tie, then (i)

16 Two point charges each Q are placed at a distance *d* from equatorial line of dipole as shown in figure. Find net force acting on the dipole (where, $d >> $ size of dipole).

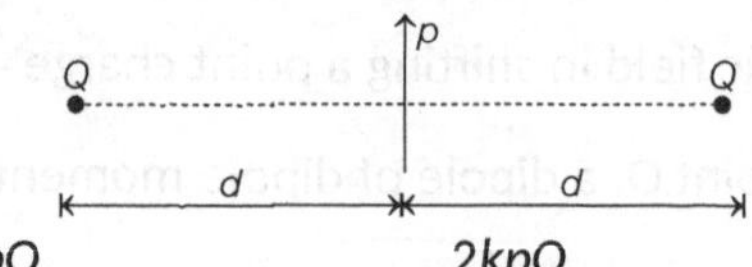

(a) 0
(b) $\frac{kpQ}{d^3}$
(c) $\frac{2kpQ}{d^3}$
(d) $\frac{4kpQ}{d^3}$

17 A small electric dipole **p** is placed on the *X*-axis at the point (1,0). The dipole vector forms an angle of 30° with the *X*-axis. Consider a non-uniform electric field to have been applied in the region given by the vector $\mathbf{E} = x^2\hat{\mathbf{i}} + y^2\hat{\mathbf{j}}$. What is the force acting on the dipole?

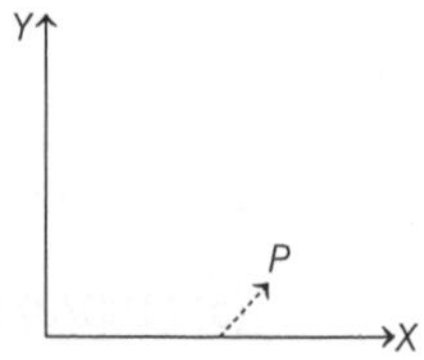

(a) $2\mathbf{p}\cos 30° (\hat{\mathbf{i}} + 2\hat{\mathbf{j}})$
(b) $2\mathbf{p}\cos 30° (\hat{\mathbf{i}})$
(c) $2\mathbf{p}\cos 30° (2\hat{\mathbf{j}})$
(d) None of these

18 A tiny electric dipole (H_2O molecule) of dipole moment **p** is placed at a distance *r* from an infinitely long wire, with its **p** normal to the wire in the same plane. If the linear charge density of the wire is λ, the electrostatic force acting on the dipole is equal to

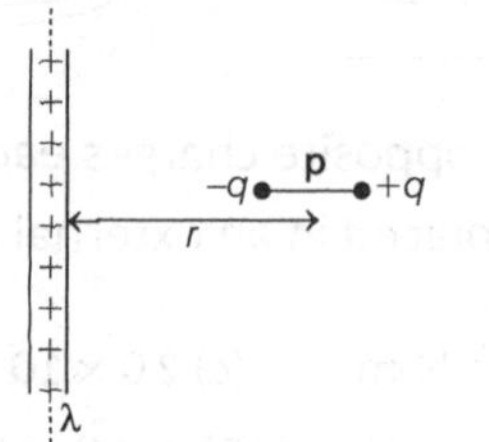

(a) $\frac{\lambda p}{4\pi\varepsilon_0 r}$
(b) $\frac{\lambda p}{\pi\varepsilon_0 r}$
(c) $\frac{2\lambda p}{\pi\varepsilon_0 r}$
(d) $\frac{\lambda p}{2\pi\varepsilon_0 r^2}$

19 Two short electric dipoles are placed as shown in figure. The energy of electric interaction between these dipoles will be

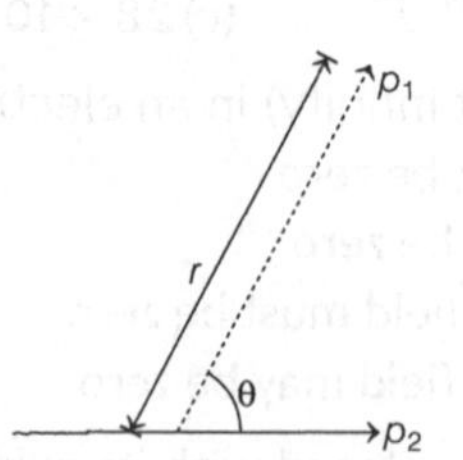

(a) $\frac{2kp_1p_2\cos\theta}{r^3}$
(b) $\frac{-2kp_1p_2\cos\theta}{r^3}$
(c) $\frac{-2kp_1p_2\sin\theta}{r^3}$
(d) $\frac{-4kp_1p_2\cos\theta}{r^3}$

Numerical Type Questions

20 An electric dipole is placed along X-axis at the origin as shown in the figure. Electric field at the point P is parallel to the Y-axis. If X-coordinate of P is $\sqrt{2}$ m, what is its Y-coordinate in metres?

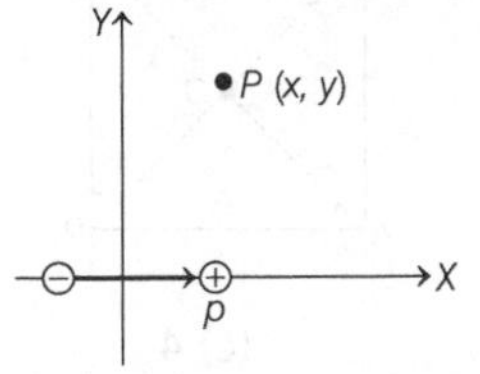

21 $ABCDEF$ is a regular hexagon of side a. At each vertex A, B and C, a point charge $+q$ is fixed and D, E, F, a point charge $-q$ is fixed. Intensity of electric field at a point lying on the line joining FC at a distance x away from the centre of hexagon($x >> a$) is found to be $\frac{\sqrt{\alpha}aq}{2\pi\varepsilon_0 x^3}$, where α is an integer. Find the value of α.

22 Three small dipoles of dipole moment p, each are placed at the three vertices of an equilateral triangle of side a as shown in figure. If total potential at the centroid G is given by $V = \frac{\sqrt{nkp}}{2a^2}$, Find the value of n. $\left(\text{Take, } k = \frac{1}{4\pi\varepsilon_0}\right)$

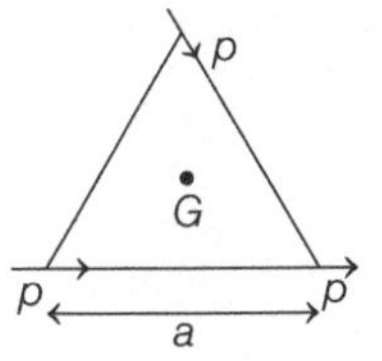

23 A non-conducting ring of mass m and radius R is charged as shown in figure. The charged density, i.e. charged per unit length is λ. It is then placed on a rough non-conducting horizontal surface plane. At time $t = 0$, a uniform electric field $\mathbf{E} = E_0\hat{\mathbf{i}}$ is switched on and the ring starts rolling without sliding. The friction force acting on the ring, when it starts moving is $n\lambda RE_0\hat{\mathbf{i}}$. Find the value of n.

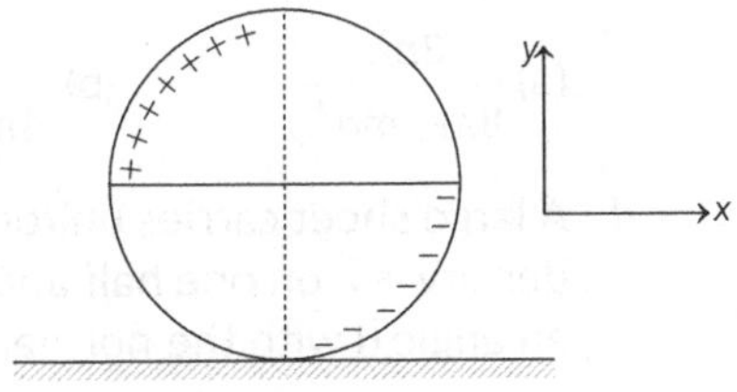

Answers

1. (a)	2. (a)	3. (a)	4. (c)	5. (d)
6. (a)	7. (b)	8. (c)	9. (a)	10. (d)
11. (c)	12. (c)	13. (d)	14. (c)	15. (c)
16. (c)	17. (b)	18. (d)	19. (b)	20. (2)
21. (7)	22. (27)	23 (1)		

For Detailed Solutions
Scan the code

LEVEL 03 \ JEE Advanced Level

Single Choice Correct Questions

1 Three point charges $+2q$, $+q$ and $-q$ are located, respectively at $(0, a, a)$, $(0, -a, a)$ and $(0, 0, -a)$ as shown in figure. The dipole moment of this distribution is

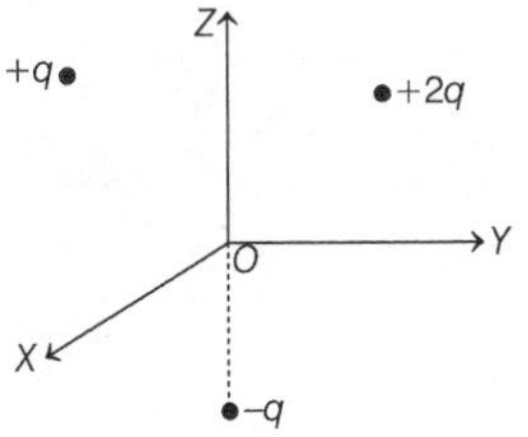

(a) $2qa$ in the YOZ plane at $\tan^{-1}\left(\frac{1}{4}\right)$ with Z-axis

(b) $\sqrt{17}qa$ in the YOZ plane at $\tan^{-1}\left(\frac{1}{4}\right)$ with Z-axis

(c) $\sqrt{5}\ qa$ in the XOY plane at $\tan^{-1}(4)$ with Y-axis

(d) $4qa$ in the XOY plane at $\tan^{-1}(4)$ with Y-axis

2 Two short dipoles of dipole moment p are placed at two corners of a square as shown in figure. What is the ratio of magnitudes of electric field intensity at two points O and A i.e. $E_{\text{at } O} : E_{\text{at } A}$?

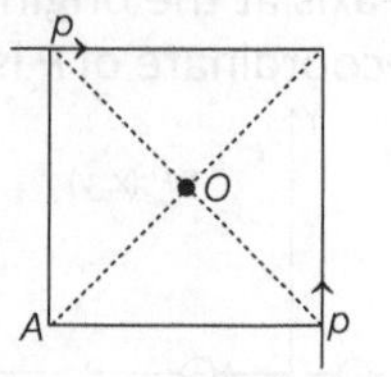

(a) 2 (b) $2\sqrt{2}$ (c) 4 (d) $\sqrt{2}$

3 A dipole of dipole moment p is placed at large distance a from a uniformly charged wire of mass m, length a and charge per unit length λ as shown in figure. The wire is free to rotate in the vertical plane about a horizontal axis passing through its lower end. Initial angular acceleration of the wire is

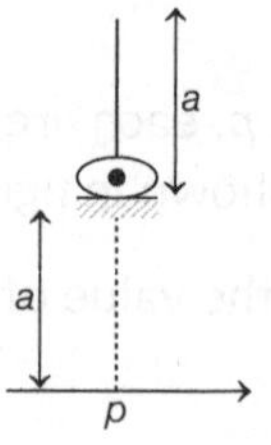

(a) $\dfrac{3p\lambda}{8\pi\varepsilon_0 ma^3}$ (b) $\dfrac{3p\lambda}{4\pi\varepsilon_0 ma^3}$ (c) $\dfrac{3p\lambda}{2\pi\varepsilon_0 ma^3}$ (d) $\dfrac{3p\lambda}{\pi\varepsilon_0 ma^3}$

4 A large sheet carries uniform surface charge density σ. A rod of length $2l$ has a linear charge density $+\lambda$ on one half and $-\lambda$ on the second half. The rod is hinged at mid-point O and makes an angle θ with the normal to the sheet. The torque experienced by the rod is

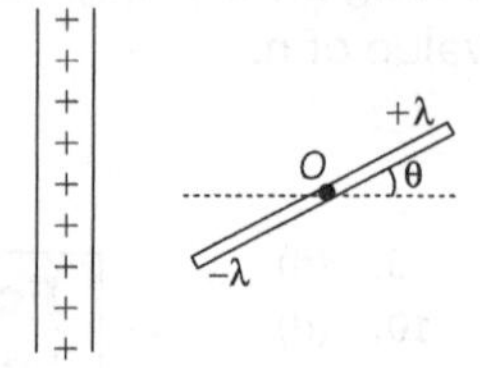

(a) 0 (b) $\dfrac{\sigma\lambda l^2}{2\varepsilon_0}\sin\theta$ (c) $\dfrac{\sigma\lambda l^2}{\varepsilon_0}\sin\theta$ (d) $\dfrac{\sigma\lambda l}{2\varepsilon_0}$

5 A particle A of mass $2m$ and carrying charge q is connected by a light rigid rod of length L to another particle B of mass m and carrying charge $-q$ as shown in figure. The system is placed in an electric field $\mathbf{E}$. After the system settles into equilibrium, one particle is given a small push in the transverse direction, so that the rod makes a small angle θ_0 with the electric field. Find the maximum tension in the rod.

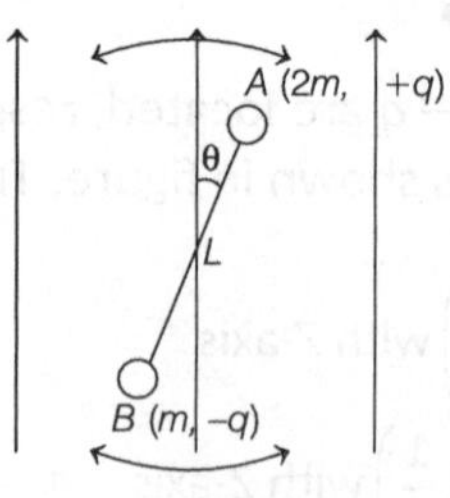

(a) $qE(1+\theta_0^2)$ (b) $qE\left(1+\dfrac{\theta_0^2}{2}\right)$ (c) $qE\left(1+\dfrac{\theta_0^2}{3}\right)$ (d) $qE\left(1-\dfrac{\theta_0^2}{2}\right)$

6 An electric dipole is kept on the axis of a uniformly charged ring at a distance $\frac{R}{\sqrt{2}}$ from the centre of the ring. The direction of the dipole moment is along the axis. If the dipole moment is p, charge of the ring is Q and radius of the ring is R, then the force on the dipole is

(a) $\frac{4kpQ}{3\sqrt{3}R^2}$ (b) $\frac{4kpQ}{3\sqrt{3}R^3}$ (c) $\frac{2kpQ}{3\sqrt{3}R^3}$ (d) zero

7 The equation of equipotential surfaces in XY-plane for a dipole kept at the origin and oriented along X-axis is

(a) $\frac{x}{(x^2+y^2)^{3/2}}=c$ (b) $\frac{y}{(y^2+x^2)^{3/2}}=c$ (c) $\frac{y^2}{(x^2+y^2)^{3/2}}=c$ (d) $\frac{x^2}{(x^2+y^2)^{3/2}}=c$

8 The electrostatic potential due to the charge configuration at point P as shown in figure for $b << a$.

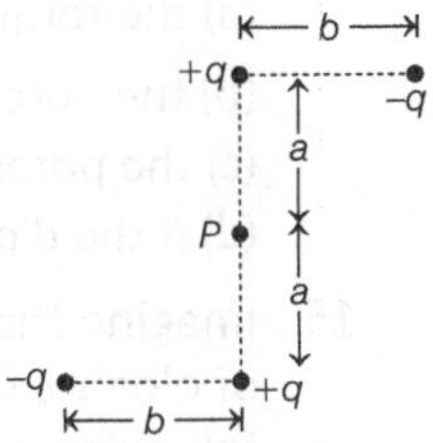

(a) $\frac{2q}{4\pi\varepsilon_0 a}$ (b) $\frac{2qb^2}{4\pi\varepsilon_0 a^3}$

(c) $\frac{qb^2}{4\pi\varepsilon_0 a^3}$ (d) zero

9 A point negative charge $-Q$ is placed at a distance r from a dipole with dipole moment p as shown in figure. The x-component of force acting on the charge $-Q$ is

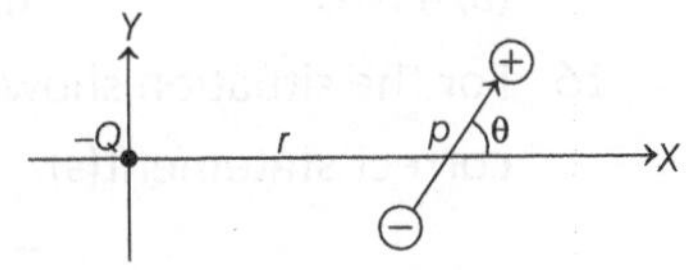

(a) $-\frac{pkQ}{r}\cos\theta\hat{i}$ (b) $\frac{pkQ}{r}\cos\theta\hat{i}$

(c) $-\frac{2pkQ}{r^3}\cos\theta\hat{i}$ (d) $\frac{2pkQ}{r^3}\cos\theta\hat{i}$

10 A wheel having mass m has charges $+q$ and $-q$ on diametrically opposite points. It remains in equilibrium on a rough inclined plane in the presence of uniform vertical electric field E. The value of E is

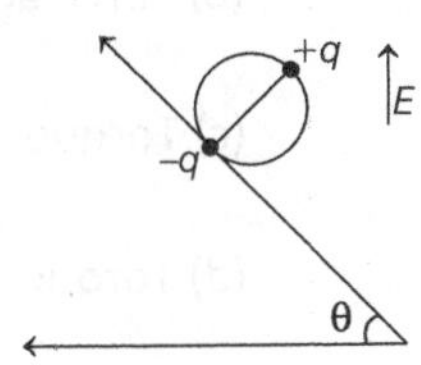

(a) $\frac{mg}{q}$ (b) $\frac{mg}{2q}$

(c) $\frac{mg\tan\theta}{2q}$ (d) None of these

11 The electrostatic potential energy due to interaction of a system of three point charges each $+q$ and a short dipole of dipole moment $p = qa/8$ as shown in figure is (neglect self-energy of dipole)

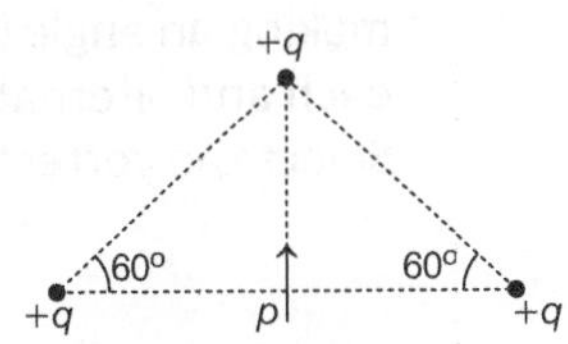

(a) $\frac{19kq^2}{6a}$ (b) $\frac{17kq^2}{6a}$

(c) $\frac{10kq^2}{3a}$ (d) $\frac{8kq^2}{3a}$

12 Three identical small electric dipoles are arranged parallel to each other at equal separation a as shown in the figure and there total interaction energy is U. Now, one of the end dipole is gradually reversed, how much work is done by the electric forces?

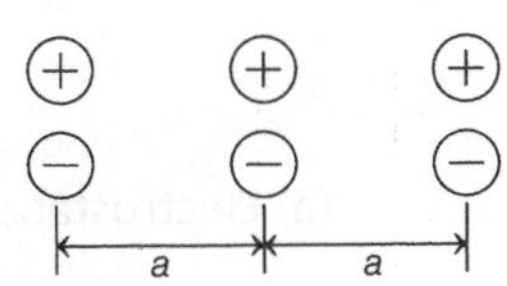

(a) $\frac{17U}{8}$ (b) $\frac{16U}{17}$

(c) $\frac{16U}{8}$ (d) $\frac{18U}{17}$

Multiple Choice(s) Correct Questions

13 Three points charges are placed at the corners of an equilateral triangle of side L as shown in the figure. Then,

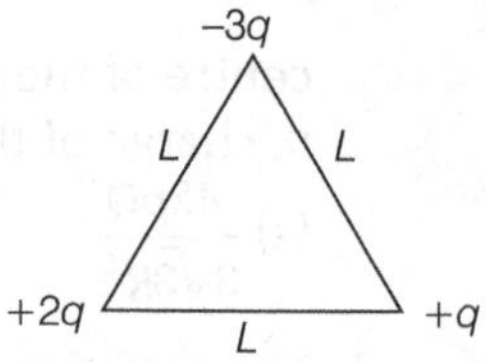

(a) the potential at the centroid of the triangle is zero

(b) the electric field at the centroid of the triangle is zero

(c) the dipole moment of the system is $\sqrt{7}\, qL$

(d) the dipole moment of the system is $\sqrt{3}\, qL$

14 An electric dipole moment $\mathbf{p} = (2.0\,\hat{\mathbf{i}} + 3.0\,\hat{\mathbf{j}})\ \mu$C-m is placed in an uniform electric field $\mathbf{E} = (3.0\,\hat{\mathbf{i}} + 2.0\,\hat{\mathbf{k}}) \times 10^5\ \text{NC}^{-1}$,

(a) the torque that $\mathbf{E}$ exerts on $\mathbf{p}$ is $(0.6\hat{\mathbf{i}} - 0.4\hat{\mathbf{j}} - 0.9\hat{\mathbf{k}})$ N-m

(b) the potential energy of the dipole is − 0.6 J

(c) the potential energy of the dipole is 0.6 J

(d) If the dipole is rotated in the electric field, the maximum potential energy of the dipole is 1.3 J

15 Imagine that, a dipole is at the centre of a spherical non-conducting surface. If the magnitude of electric field at a certain point on the surface of sphere is 10 N/C, then which of the following cannot be the magnitude of electric field anywhere on the surface of sphere?

(a) 4 N/C (b) 8 N/C (c) 16 N/C (d) 32 N/C

16 For the situation shown in the figure below (assume $r >>$ length of dipole), mark out the correct statement(s).

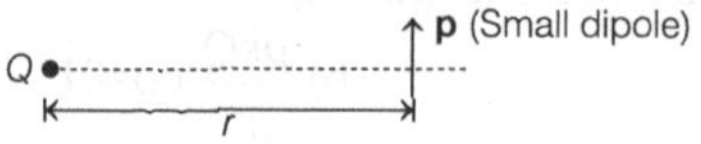

(a) Force acting on the dipole is zero.

(b) Force acting on the dipole is approximately $\dfrac{pQ}{4\pi\varepsilon_0 r^3}$ and is acting upward.

(c) Torque acting on the dipole is $\dfrac{pQ}{4\pi\varepsilon_0 r^2}$ in clockwise direction.

(d) Torque acting on the dipole is $\dfrac{pQ}{4\pi\varepsilon_0 r^2}$ in anti-clockwise direction.

17 Given a square frame of diagonal length $2r$ made of insulating wires. There is a short dipole, having dipole moment p, fixed in the plane of the figure lying at the centre of the square, making an angle θ as shown in figure. Four identical particles having charges of magnitude q each and alternatively positive and negative signs are placed at the four corners of the square. Select the correct alternative(s).

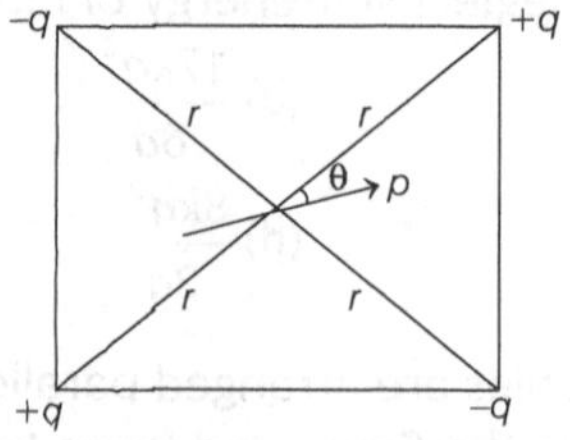

(a) Electrostatic force on the system of four charges due to dipole is $\dfrac{6kpq}{r^3}$

(b) Electrostatic force on the system of four charges due to dipole is $\dfrac{6kpq}{r^3}\cos\theta$

(c) Net torque on the system of four charges about the centre of the square due to dipole is zero

(d) Net torque on the system of four charges about the centre of the square due to dipole is $\dfrac{3kpq}{r^2}$

18 For a system of two dipoles $\mathbf{p}_1$ and $\mathbf{p}_2$ as shown in the figure, both dipoles are at origin and perpendicular to each other along X and Y-axes respectively, then

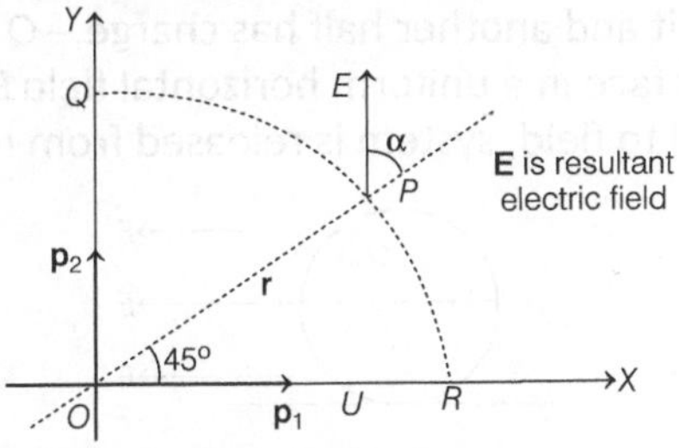

(a) work done in taking electron from P to R on QPR is zero

(b) $\tan\alpha = \frac{(p_1 + p_2)}{2(p_1 - p_2)}$

(c) $\tan\alpha = \frac{(p_1 - p_2)}{2(p_1 + p_2)}$

(d) $-\int_{\infty}^{r} \mathbf{E} \cdot d\mathbf{r} = \frac{p_1 + p_2}{4\sqrt{2}\pi\varepsilon_0 r^2}$

Comprehension Based Questions

Comprehension-1

A thin ring of radius R metres is placed in XY-plane, such that its centre lies on origin. The half ring in region $x < 0$ carries uniform linear charge density $+\lambda$ C/m and the remaining half-ring in region $x > 0$ carries uniform linear charge density $-\lambda$ C/m.

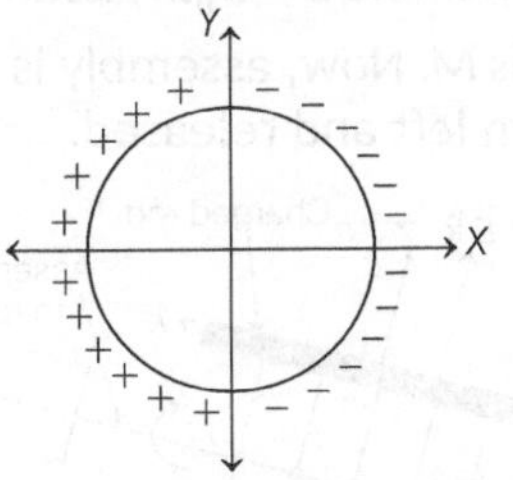

19 The dipole moment of the ring in C-m is

(a) $-(2\pi R^2\lambda)\,\hat{\mathbf{i}}$
(b) $(2\pi R^2\lambda)\,\hat{\mathbf{i}}$
(c) $-(4R^2\lambda)\,\hat{\mathbf{i}}$
(d) $(4R^2\lambda)\,\hat{\mathbf{i}}$

20 The electric potential (in V) at point P, whose coordinates are $\left(0\text{ m}, +\frac{R}{2}\text{m}\right)$ is

(a) $\frac{1}{4\pi\varepsilon_0}\frac{\lambda}{2}$
(b) 0
(c) $\frac{1}{4\pi\varepsilon_0}\frac{\lambda}{4}$
(d) Cannot be determined

21 The direction of electric field at point P, whose coordinates are $\left(0\text{ m}, +\frac{R}{2}\text{ m}\right)$ is

(a) along positive x-direction
(b) along negative x-direction
(c) along negative y-direction
(d) None of these

Comprehension-2

A uniform disc of mass m and radius R is made up of two halves, one half has charge $+Q$ uniformly distributed over it and another half has charge $-Q$ uniformly distributed over it. This system is kept on rough surface in a uniform horizontal field E in a way, so that line of separation become parallel to field, system is released from rest, at $t = 0$ as shown in figure.

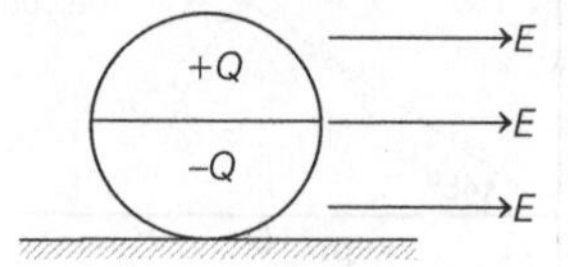

22 Torque of the electrostatic force about centre of mass at $t = 0$ is

(a) $\frac{8QER}{3\pi}$ (b) $\frac{QER}{9}$ (c) $\frac{16QER}{9\pi}$ (d) QER

23 Acceleration of centre of the disc at $t = 0$, if friction is sufficient to prevent slipping is

(a) $\frac{4QE}{9\pi m}$ (b) $\frac{8QE}{9\pi m}$ (c) $\frac{16QE}{9\pi m}$ (d) $\frac{2QE}{3m}$

24 Minimum value of coefficient of friction to prevent slipping at $t = 0$, is

(a) $\frac{4QE}{9\pi mg}$ (b) $\frac{8QE}{9\pi mg}$ (c) $\frac{16QE}{9\pi mg}$ (d) $\frac{2QE}{3mg}$

Comprehension-3

Figure shows two non-conducting charged rods separated by massless plastic spacers of length L. Entire assembly lies in an electric field $E = E_0\hat{\mathbf{k}}$. Assembly can rotate about Y-axis and initially is in YZ-plane. Mass of each rod is M. Now, assembly is turned through an angle $\theta = 60°$ in anti-clockwise sense as seen from left and released.

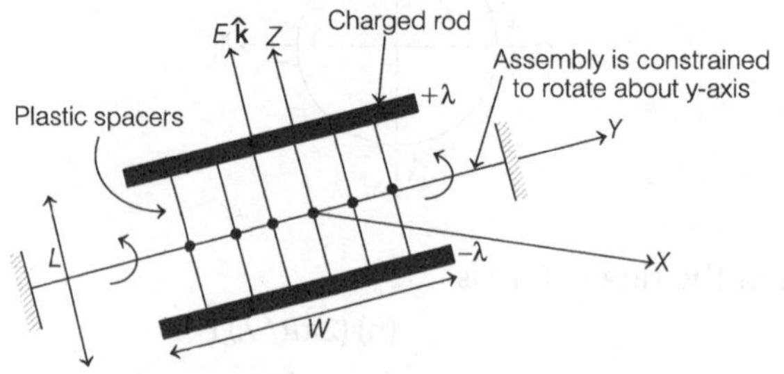

25 What is total mechanical energy of the assembly, taking reference position for potential energy, when it lies in XY-plane?

(a) $+\lambda WLE$ (b) $\frac{\lambda WLE}{2}$ (c) $-\frac{\lambda WLE}{2}$ (d) $-\lambda WLE$

26 What is angular velocity of assembly, when it comes back to YZ-plane?

(a) $\sqrt{\frac{4\lambda WE}{ML}}$ (b) $\sqrt{\frac{\lambda WE}{2ML}}$ (c) $\sqrt{\frac{\sqrt{3}\lambda WE}{2ML}}$ (d) $\sqrt{\frac{2\lambda WE}{ML}}$

27 If assembly is turned through a small angle from its stable equilibrium position, what is minimum time in which it will attain its equilibrium position?

(a) $2\pi\sqrt{\frac{ML}{\lambda WE}}$ (b) $\pi\sqrt{\frac{ML}{2\lambda WE}}$ (c) $3\pi\sqrt{\frac{ML}{\lambda WE}}$ (d) $\pi\sqrt{\frac{ML}{8\lambda WE}}$

Match the Columns Questions

28 Column I gives a situation in which two dipoles of dipole moment $p\hat{i}$ and $\sqrt{3}\,p\hat{j}$ are placed at origin. A circle of radius R with centre at origin is drawn as shown in figure. Column II gives co-ordinates of certain positions on the circle. Match the statements in Column I with the statements in Column II.

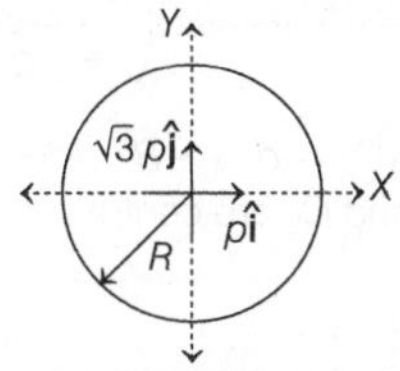

	Column I		Column II
(a)	The coordinate(s) of point on circle, where potential is maximum.	(p)	$\left(\frac{R}{2}, \frac{\sqrt{3}R}{2}\right)$
(b)	The coordinate(s) of point on circle, where potential is zero.	(q)	$\left(-\frac{R}{2}, -\frac{\sqrt{3}R}{2}\right)$
(c)	The coordinate(s) of point on circle, where magnitude of electric field intensity is $\frac{1}{4\pi\varepsilon_0}\frac{4p}{R^3}$.	(r)	$\left(-\frac{\sqrt{3}R}{2}, \frac{R}{2}\right)$
(d)	The coordinate(s) of point on circle, where magnitude of electric field intensity is $\frac{1}{4\pi\varepsilon_0}\frac{2p}{R^3}$.	(s)	$\left(\frac{\sqrt{3}R}{2}, -\frac{R}{2}\right)$

29

	Column I (Dipole **p** in an electric field **E**)		Column II (F = force, τ = torque and U = energy of dipole)
(a)	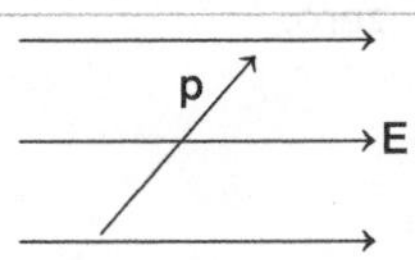 Uniform electric field	(p)	$F \neq 0, \tau = 0$
(b)	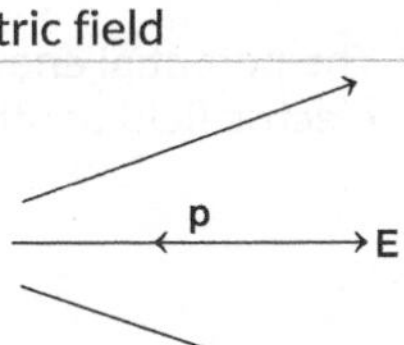 Non-uniform electric field	(q)	$U = 0, F = 0$
(c)	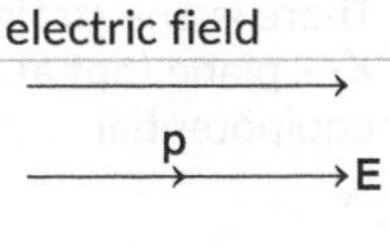 Uniform electric field	(r)	$\tau \neq 0, F = 0$ and U = negative
(d)	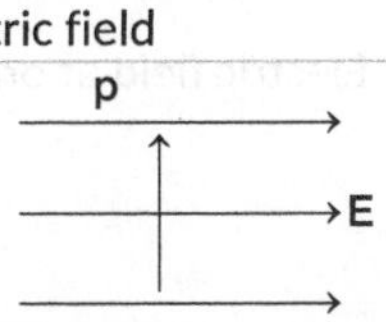 Uniform electric field	(s)	U is minimum and $\tau = 0$

30 Two identical charged rings lie in XY-plane both having linear charge density λ varying as per $\lambda = \lambda_0 \cos\theta$ (λ_0 = constant), where θ is measured from + X-axis. Radius for both the rings is R.

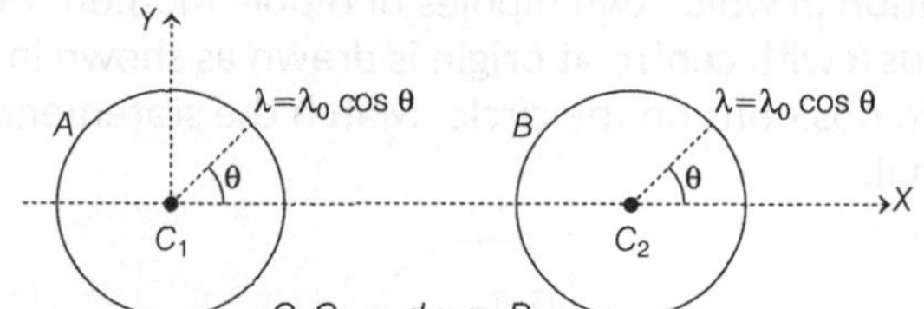

$C_1C_2 = d >> R$
C_1 and C_2 are centres

	Column I	Column II
(a)	Dipole moment of each ring is $x\lambda_0\pi R^2$, then x is	(p) 0
(b)	Magnitude of work done by electric field in rotating ring B by 90° about X-axis is $\frac{xk\lambda_0^2\pi^2R^4}{2d^3}$, then x is	(q) 1
(c)	Magnitude of work done by electric field in rotating ring B by 90° about Z-axis is $\frac{xk\lambda_0^2\pi^2R^4}{2d^3}$, then x is	(r) 6
(d)	Electric force between the two rings is $\frac{xk\lambda_0^2\pi^2R^4}{d^4}$, then x is	(s) 4

31 In each situation of Column I, two short electric dipoles having dipole moments $\mathbf{p}_1$ and $\mathbf{p}_2$ of same magnitude (i.e. $p_1 = p_2$) are placed on X-axis symmetrically about origin in different orientations as shown in figure. In Column II, certain inferences are shown for these dipoles. Match the different orientations of dipole in Column I with the corresponding results in Column II.

	Column I	Column II
(a)	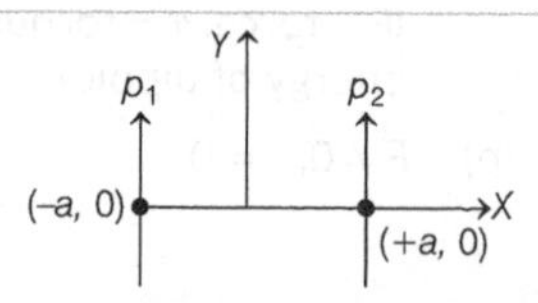 p_1 and p_2 are perpendicular to X-axis	(p) The torque on one dipole due to other is zero.
(b)	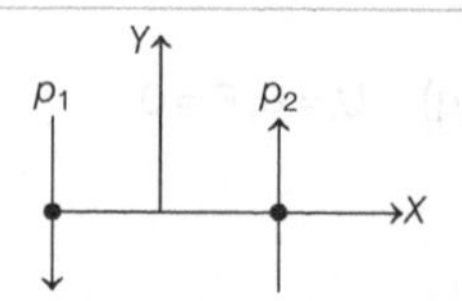 p_1 and p_2 are perpendicular to X-axis	(q) The potential energy of one dipole in electric field of other dipole is negative.
(c)	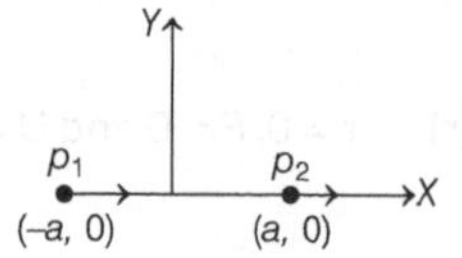 p_1 and p_2 are parallel to X-axis	(r) There is one straight line in XY- plane (not at infinity) which is equipotential.
(d)	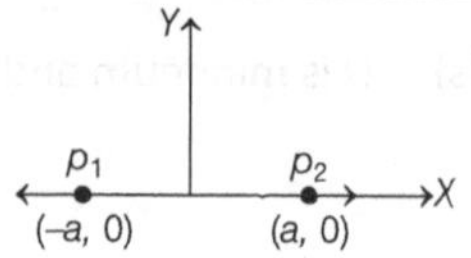	(s) Electric field at origin is zero.
		(t) Repulsive force will act between dipoles.

Numerical Type Questions

32 Electric field intensity at point P due to below configuration for $r >> a$ is given by $n \times \frac{kpa}{r^m}$, where $p = qa$. Find the value of $m + n$.

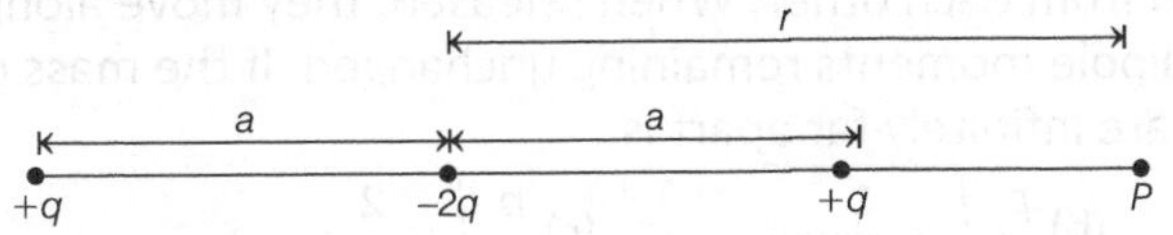

33 An infinite non-conducting plane with uniform charge density σ is kept parallel to YZ-plane and at a distance d from a dipole $\mathbf{p}$, which itself is located at the origin. An equipotential surface for this system is spherical, centred at origin, having radius $R(< d)$. Given that, $R = \left(\frac{p}{n\pi\sigma}\right)^{1/3}$, find the integer n.

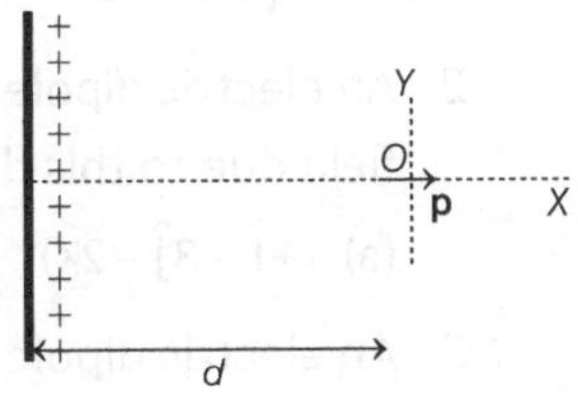

34 A uniform ring of mass m and radius R is placed freely in a gravity free region. Half of the ring positively charged with uniform linear charge density $+\lambda$ and other half is negatively charged with uniform charged density $-\lambda$. A uniform electric field E is switched on along the normal to the plane of the ring. If the angular speed of the ring, when it turns through an angle 90° is $n\left(\sqrt{\frac{\lambda E}{m}}\right)$, find the value of n.

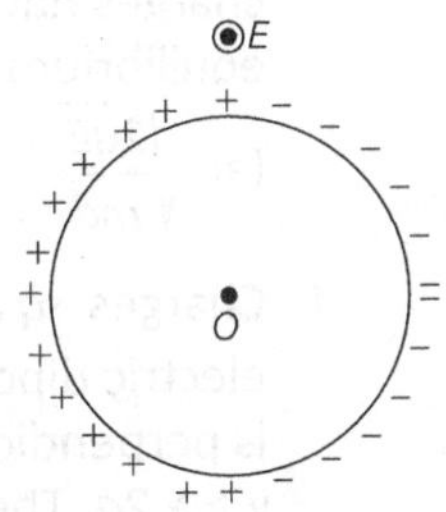

35 A non-conducting solid sphere of mass m and radius R having uniform charge densities $+\rho$ and $-\rho$ as shown in figure. It is then placed on a rough non-conducting horizontal plane. At $t = 0$, a uniform electric field $\mathbf{E} = E_0 \hat{\mathbf{i}}$ N/C is switched on and the solid sphere starts rolling without sliding. The magnitude of frictional force at $t = 0$ is $\frac{2\rho\pi R^3 E_0}{K}$, find the value of K.

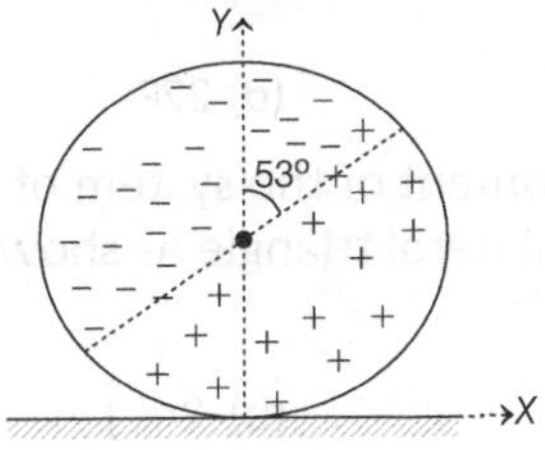

Answers

1. (b)	**2.** (b)	**3.** (a)	**4.** (b)	**5.** (a)	**6.** (d)	**7.** (a)	**8.** (c)	**9.** (c)	**10.** (b)
11. (a)	**12.** (d)	**13.** (ac)	**14.** (abc)	**15.** (ad)	**16.** (bc)	**17.** (ac)	**18.** (cd)	**19.** (c)	**20.** (b)
21. (a)	**22.** (a)	**23.** (c)	**24.** (c)	**25.** (c)	**26.** (d)	**27.** (d)	**28.** a → p; b → rs; c → pq; d → rs		

29. a → r; b → p; c → s; d → q

30. a → q; b → p; c → s; d → r

31. a → prt; b → pqrs; c → pqr; d → pst

32. 10 **33.** 2 **34.** 4 **35.** 7

For Detailed Solutions
Scan the code

Previous Years' Questions

JEE Main

1 Two identical electric point dipoles have dipole moments $\mathbf{p}_1 = p\hat{\mathbf{i}}$ and $\mathbf{p}_2 = -p\hat{\mathbf{i}}$ are held on the X-axis at distance a from each other. When released, they move along the X-axis with the direction of their dipole moments remaining unchanged. If the mass of each dipole is m, their speed, when they are infinitely far apart is **[JEE Main 2020]**

(a) $\frac{p}{a}\sqrt{\frac{1}{\pi\varepsilon_0 ma}}$ (b) $\frac{p}{a}\sqrt{\frac{1}{2\pi\varepsilon_0 ma}}$ (c) $\frac{p}{a}\sqrt{\frac{2}{\pi\varepsilon_0 ma}}$ (d) $\frac{p}{a}\sqrt{\frac{3}{2\pi\varepsilon_0 ma}}$

2 An electric dipole of moment $\mathbf{p} = (-\hat{\mathbf{i}} - 3\hat{\mathbf{j}} + 2\hat{\mathbf{k}}) \times 10^{-29}$ C-m is at the origin (0, 0, 0). The electric field due to this dipole at $\mathbf{r} = +\hat{\mathbf{i}} + 3\hat{\mathbf{j}} + 5\hat{\mathbf{k}}$ (note that $\mathbf{r} \cdot \mathbf{p} = 0$) is parallel to **[JEE Main 2020]**

(a) $(+\hat{\mathbf{i}} - 3\hat{\mathbf{j}} - 2\hat{\mathbf{k}})$ (b) $(-\hat{\mathbf{i}} - 3\hat{\mathbf{j}} + 2\hat{\mathbf{k}})$ (c) $(+\hat{\mathbf{i}} + 3\hat{\mathbf{j}} - 2\hat{\mathbf{k}})$ (d) $(-\hat{\mathbf{i}} + 3\hat{\mathbf{j}} - 2\hat{\mathbf{k}})$

3 An electric dipole is formed by two equal and opposite charges q with separation d. The charges have same mass m. It is kept in a uniform electric field E. If it is slightly rotated from its equilibrium orientation, then its angular frequency ω is **[JEE Main 2019]**

(a) $\sqrt{\frac{2qE}{md}}$ (b) $2\sqrt{\frac{qE}{md}}$ (c) $\sqrt{\frac{qE}{md}}$ (d) $\sqrt{\frac{qE}{2md}}$

4 Charges $-q$ and $+q$ located at A and B, respectively, constitute an electric dipole. Distance $AB = 2a$, O is the mid-point of the dipole and OP is perpendicular to AB. A charge Q is placed at P, where $OP = y$ and $y >> 2a$. The charge Q experiences an electrostatic force F.

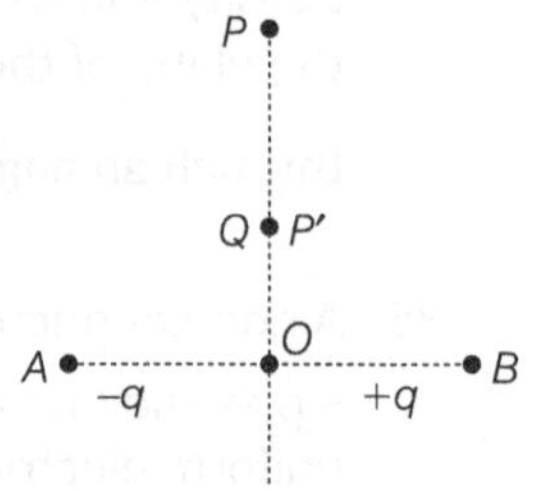

If Q is now moved along the equatorial line to P', such that $OP' = \left(\frac{y}{3}\right)$, the force on Q will be close to $\left(\frac{y}{3} >> 2a\right)$ **[JEE Main 2019]**

(a) $\frac{F}{3}$ (b) $3F$

(c) $9F$ (d) $27F$

5 Determine the electric dipole moment of the system of three charges, placed on the vertices of an equilateral triangle as shown in the figure. **[JEE Main 2019]**

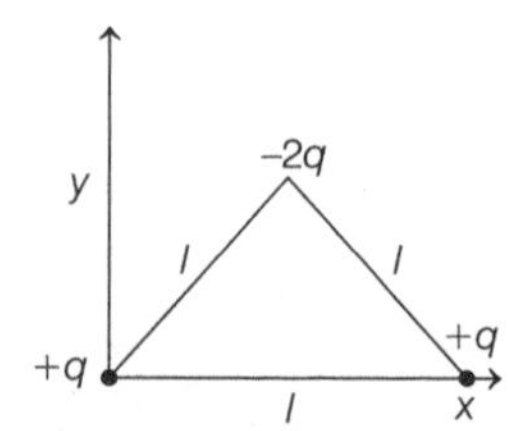

(a) $\sqrt{3}\, ql \frac{\hat{\mathbf{j}} - \hat{\mathbf{i}}}{\sqrt{2}}$ (b) $2ql\,\hat{\mathbf{j}}$

(c) $-\sqrt{3}\, ql\,\hat{\mathbf{j}}$ (d) $(ql)\frac{\hat{\mathbf{i}} + \hat{\mathbf{j}}}{\sqrt{2}}$

6 A point dipole $\mathbf{p} = -p_0\,\hat{\mathbf{x}}$ is kept at the origin. The potential and electric field due to this dipole on the Y-axis at a distance d are, respectively. [Take, $V = 0$ at infinity] **[JEE Main 2019]**

(a) $\frac{|\mathbf{p}|}{4\pi\varepsilon_0 d^2}, \frac{\mathbf{p}}{4\pi\varepsilon_0 d^3}$ (b) $0, \frac{-\mathbf{p}}{4\pi\varepsilon_0 d^3}$

(c) $0, \frac{\mathbf{p}}{4\pi\varepsilon_0 d^3}$ (d) $\frac{|\mathbf{p}|}{4\pi\varepsilon_0 d^2}, \frac{-\mathbf{p}}{4\pi\varepsilon_0 d^3}$

7 An electric field of 1000 V/m is applied to an electric dipole at angle of 45°. The value of electric dipole moment is 10^{-29} C-m. What is the potential energy of the electric dipole? **[JEE Main 2019]**

(a) -9×10^{-20} J (b) -10×10^{-29} J (c) -20×10^{-18} J (d) -7×10^{-27} J

8 An electric dipole has a fixed dipole moment **p**, which makes angle θ with respect to X-axis. When subjected to an electric field $\mathbf{E}_1 = E\hat{\mathbf{i}}$, it experiences a torque $\mathbf{T}_1 = \tau\hat{\mathbf{k}}$. When subjected to another electric field $\mathbf{E}_2 = \sqrt{3}E_1\hat{\mathbf{j}}$, it experiences a torque $\mathbf{T}_2 = -\mathbf{T}_1$. The angle θ is [JEE Main 2017]

(a) 45° (b) 60° (c) 90° (d) 30°

JEE Advanced

9 The electric field E is measured at a point $P(0, 0, d)$ generated due to various charge distributions and the dependence of E on d is found to be different for different charge distributions. Column I contains different relations between E and d. Column II describes different electric charge distributions, along with their locations. Match the functions in Column I with the related charge distributions in Column II.

	Column I		Column II
A.	E is independent of d	p.	A point charge Q at the origin
B.	$E \propto \frac{1}{d}$	q.	A small dipole with point charges Q at $(0, 0, l)$ and $-Q$ at $(0, 0, -l)$. (Take, $2l << d$)
C.	$E \propto \frac{1}{d^2}$	r.	An infinite line charge coincident with the X-axis, with uniform linear charge density λ.
D.	$E \propto \frac{1}{d^3}$	s.	Two infinite wires carrying a uniform linear charge density parallel to the X- axis. The one along $(y = 0, z = l)$ has a charge density $+\lambda$ and the one along $(y = 0, z = -l)$ has a charge density $-\lambda$. (Take, $2l << d$).
		t.	Infinite plane charge coincident with the XY-plane with uniform surface charge density.

(a) A $\to$ t; B $\to$ r,s; C $\to$ p; D $\to$ q
(b) A $\to$ t; B $\to$ r; C $\to$ p, s; D $\to$ q [JEE Advanced 2018]
(c) A $\to$ t; B $\to$ r; C $\to$ p, q; D $\to$ s
(d) A $\to$ s; B $\to$ q, r; C $\to$ p; D $\to$ t

Answers

1. (b)	2. (c)	3. (a)	4. (d)	5. (c)
6. (b)	7. (d)	8. (b)	9. (b)	

For Detailed Solutions
Scan the code

CHAPTER

04

ELECTRIC FLUX

LEVEL 01 \ Starter Level

1.1 *Electric Flux*

1 Find out the electric flux through an area $10\,m^2$ lying in *XY*-plane due to an electric field $\mathbf{E} = (3\hat{\mathbf{i}} - 4\hat{\mathbf{j}} + 10\hat{\mathbf{k}})\,Vm^{-1}$.

2 Find the electric flux through the ring of radius *R* in each of the following cases.

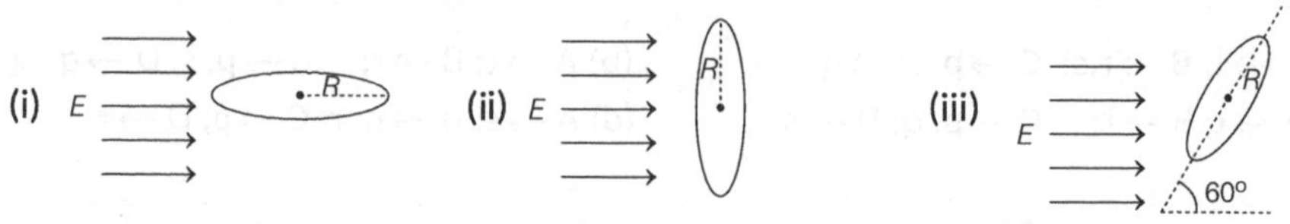

(i) Uniform electric field *E* is parallel to the plane of the ring.

(ii) Uniform electric field *E* is normal to the plane of the ring.

(iii) Uniform electric field *E* makes an angle of 60° with the plane of the ring.

3 Find the electric flux through the curved surface of the hemisphere in each of the following cases.

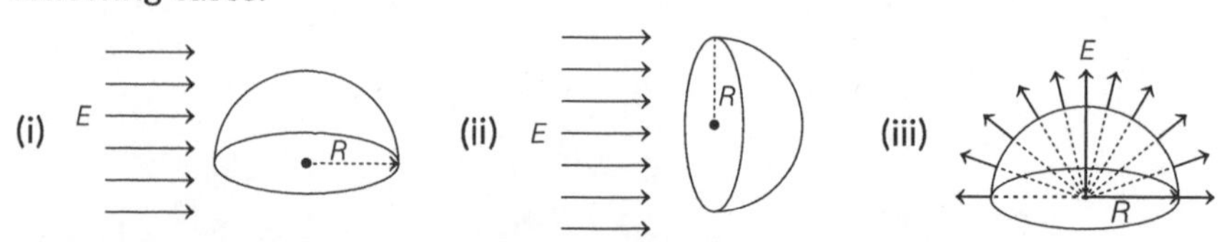

(i) A uniform horizontal electric field *E* is parallel to the base of the hemisphere.

(ii) A uniform horizontal electric field *E* is perpendicular to the base of the hemisphere.

(iii) Electric field is radial and magnitude of electric field at any point on the curved surface of the hemisphere is *E*.

4 A cube of side 2 m has one corner at the origin of coordinates and extends along the positive *X*, *Y* and *Z*-axis. An electric field of magnitude 20 N/C exists in *XY*-plane and it makes an angle of 37° with the *X*-axis. Find the electric flux passing through each face of the cube.

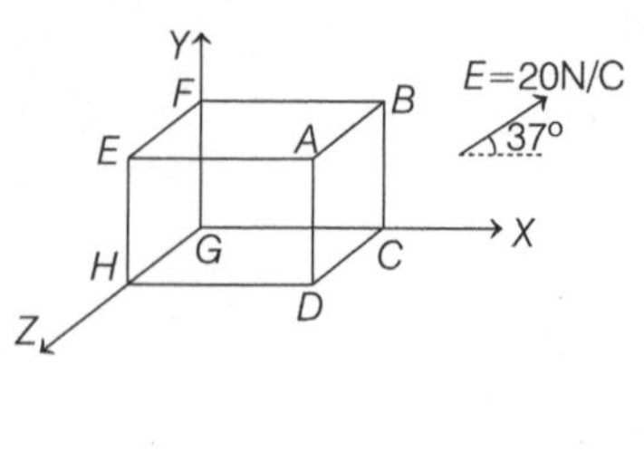

5 A uniform electric field $\mathbf{E} = E_0 \hat{\mathbf{i}}$ exists in the region. Find the electric flux through each of the five faces of the triangular prism.

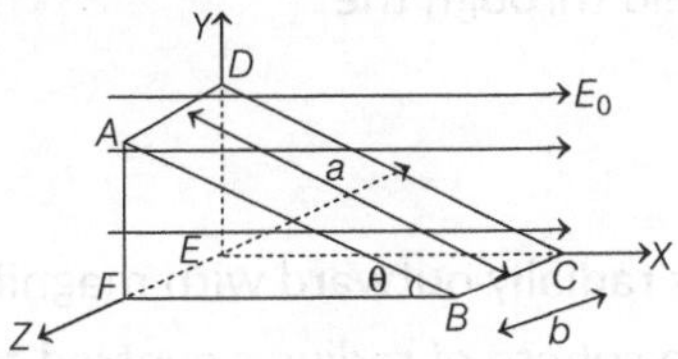

6 Find the electric flux entering the cone in each of the following cases.

(i) 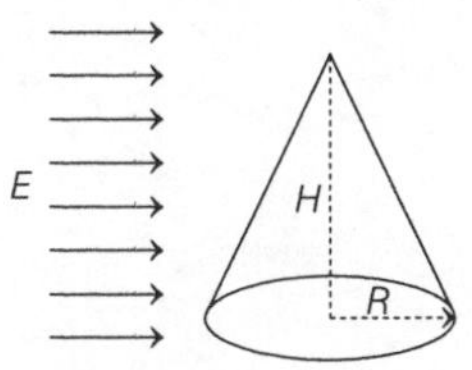

A uniform horizontal electric field E is present and axis of the cone is vertical.

(ii) 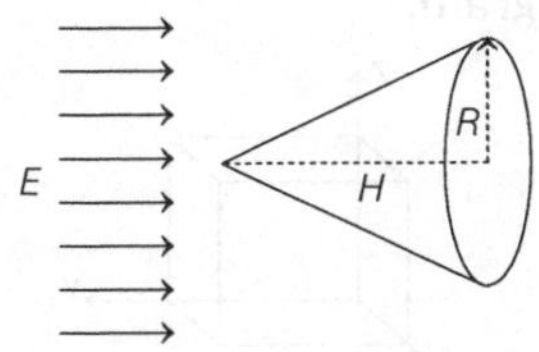

A uniform electric field E directed along the axis of cone exists in the region.

(iii) 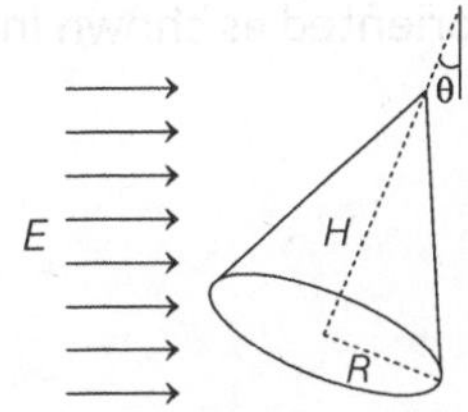

A uniform horizontal electric field E is present and axis of cone is tilted by an angle θ from vertical.

1.2 Gauss Law

7 A spherical surface encloses a point charge q. How the flux through the surface changes in each of the following cases?

(i) Radius of the spherical surface is doubled.
(ii) Charge q is shifted within the spherical surface.
(iii) Spherical surface is replaced with a cubical surface.
(iv) Another point charge q' is placed outside the surface.

8 The flux of the electric field, through the closed spherical surface S', is found to be four times that through the closed spherical surface S. Find the magnitude of the charge Q.

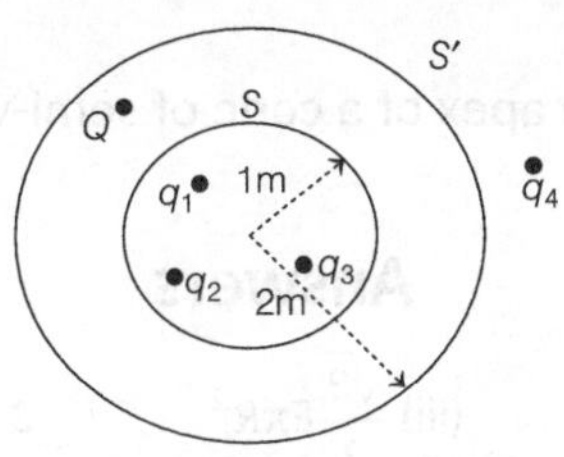

[Given, $q_1 = +20\,\mu C$, $q_2 = -25\,\mu C$, $q_3 = +9.75\,\mu C$ and $q_4 = +30\,\mu C$.]

9 A point charge q is placed at various positions as shown in the following cases. Find the electric flux due to this charge through the hemispherical surface for each case.

(i) 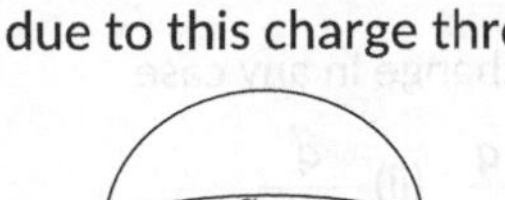

(ii) 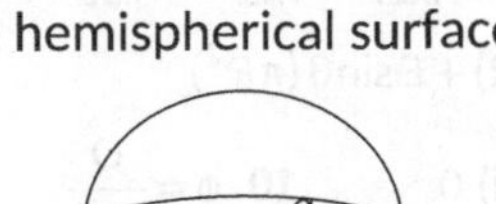

(iii) 

10 A point charge Q is placed at a distance $a/2$ above the centre of a horizontal, square surface of edge a as shown in figure. Find the flux of the electric field through the square surface.

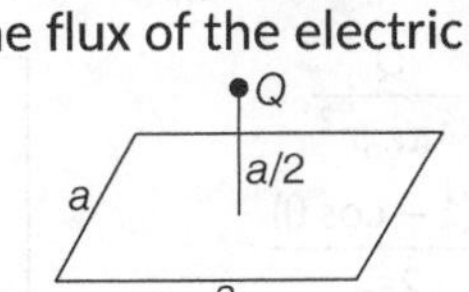

Problems in Physics

11 A point charge Q is placed at a corner of a cube as shown. Find the flux of the electric field through the

(i) cube

(ii) face $ABCD$

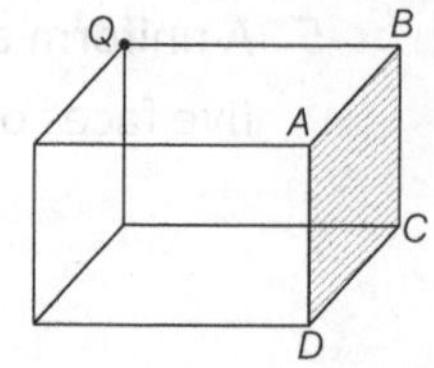

12 The electric field in a region is radially outward with magnitude $E = Ba$, where B is a constant. Find the charge contained in a sphere of radius a centred at the origin.

13 Electric field in a region is given by $\mathbf{E} = -4x\hat{\mathbf{i}} + 6y\hat{\mathbf{j}}$. Find the charge enclosed in the cube of side 1 m oriented as shown in the diagram.

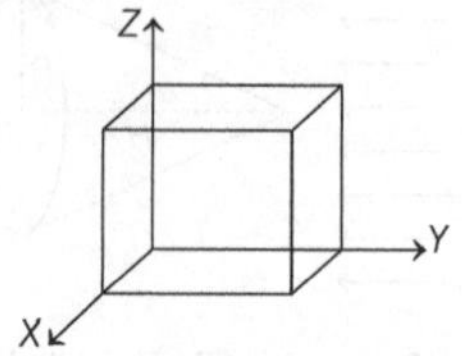

14 A non-conducting solid sphere of radius R has volume charge density that varies as $\rho = \rho_0 r$, where ρ_0 is a constant and r is distance from centre. Find out electric field intensities for the following positions.

(i) $r < R$ (ii) $r \geq R$

15 A charge Q is uniformly distributed throughout the volume of non-conducting spherical shell of inner radius a and outer radius b. Find the electric field intensity for the following points.

(i) $r < a$ (ii) $a \leq r < b$

(iii) $r \geq b$, where r is the distance of the point from the centre.

16 A long cylindrical volume (of radius R) contains a uniformly distributed charge of density ρ. Find the electric field at a point P inside the cylindrical volume at a distance x from its axis as shown in the figure for the following cases.

(i) $x < R$

(ii) $x \geq R$

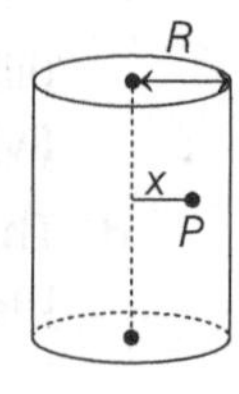

17 A point charge q is placed at the apex of a cone of semi-vertex angle θ. Find the electric flux through the base of the cone.

Answers

1. 100 V-m **2.** (i) 0 (ii) $E\pi R^2$ (iii) $\frac{\sqrt{3}}{2}E\pi R^2$ **3.** (i) 0 (ii) $E\pi R^2$ (iii) $2E\pi R^2$

4. $\phi_{ABCD} = 64$ V - m, $\phi_{EFGH} = -64$ V - m, $\phi_{ABEF} = 48$ V - m, $\phi_{CDGH} = -48$ V - m, $\phi_{ADEH} = \phi_{BCFG} = 0$

5. $\phi_{ADEF} = -E_0\, ab \sin\theta$, $\phi_{ABCD} = E_0\, ab \sin\theta$, $\phi_{BCEF} = \phi_{ABF} = \phi_{CDE} = 0$

6. (i) ERH (ii) $E\pi R^2$ (iii) $E \cos\theta\,(HR) + E \sin\theta\,(\pi R^2)$ **7.** No change in any case

8. 14.25 μC **9.** (i) $\frac{q}{2\varepsilon_0}$ (ii) $\frac{q}{2\varepsilon_0}$ (iii) 0 **10.** $\phi = \frac{Q}{6\varepsilon_0}$ **11.** (i) $\frac{q}{8\varepsilon_0}$ (ii) $\frac{q}{24\varepsilon_0}$

12. $4\pi\varepsilon_0 Ba^3$ **13.** $q = 2\varepsilon_0$ **14.** (i) $\frac{\rho_0 r^2}{4\varepsilon_0}\hat{\mathbf{r}}$ (ii) $\frac{\rho_0 R^4}{4\varepsilon_0 r^2}\hat{\mathbf{r}}$

15. (i) 0 (ii) $\frac{Q(r^3 - a^3)}{4\pi\varepsilon_0 r^2 (b^3 - a^3)}$ (iii) $\frac{Q}{4\pi\varepsilon_0 r^2}$

16. (i) $\frac{\rho x}{2\varepsilon_0}$ (ii) $\frac{\rho R^2}{2\varepsilon_0 x}$ **17.** $\frac{q(1 - \cos\theta)}{2\varepsilon_0}$

For Detailed Solutions
Scan the code

LEVEL 02 \ JEE Main Level

Single Choice Correct Questions

1 A square of side a is lying in XY-plane, such that two of its sides are lying on the axes. If an electric field $\mathbf{E} = E_0 x\hat{\mathbf{k}}$ is applied on the square, the flux passing through the square is

(a) $E_0 a^3$ (b) $\frac{E_0 a^3}{2}$ (c) $\frac{E_0 a^3}{3}$ (d) $\frac{E_0 a^2}{2}$

2 A $+q_1$ charge is at centre of an imaginary spherical Gaussian surface S and $-q_1$ charge is placed nearby this $+q_1$ charge inside S. A charge $+q_2$ is located outside this Gaussian surface. Then, electric field on Gaussian surface will be

(a) due to $-q_1$ and q_2 (b) uniform (c) due to all charges (d) zero

3 Electric charges are distributed in a small volume. The flux of the electric field through a spherical surface of radius 20 cm surrounding the total charge is 50 V-m. The flux over a concentric sphere of radius 40 cm will be

(a) 50 V-m (b) 75 V-m (c) 100 V-m (d) 200 V-m

4 If the electric flux entering and leaving a closed surface are respectively, of magnitude ϕ_1 and ϕ_2, then the electric charge inside the surface will be

(a) $\frac{\phi_2 - \phi_1}{\varepsilon_0}$ (b) $(\phi_1 - \phi_2)\varepsilon_0$ (c) $\varepsilon_0(\phi_2 - \phi_1)$ (d) $\varepsilon_0(\phi_2 + \phi_1)$

5 An electric dipole is placed at the centre of a sphere. Mark the correct option.

(a) The electric field is zero at every point of the sphere.
(b) The flux of the electric field through the sphere is non-zero.
(c) The electric field is zero on a circle on the sphere.
(d) The electric field is not zero anywhere on the sphere.

6 Which of the following is false for a closed Gaussian surface?

(a) If the electric field is zero everywhere on the surface, then there can be no net charge enclosed by the surface.
(b) If the total electric flux through the surface is zero, then the total charge enclosed by the surface is zero.
(c) If the electric field is zero everywhere on the surface, then the total electric flux through the surface is zero.
(d) If the total electric flux through the surface is zero, then the electric field must be zero everywhere on the surface.

7 A point charge $+Q$ is positioned at the centre of the base of a square pyramid as shown. The flux through one of the four identical upper faces of the pyramid is

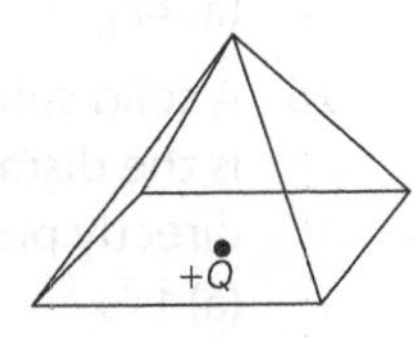

(a) $\frac{Q}{16\varepsilon_0}$ (b) $\frac{Q}{4\varepsilon_0}$ (c) $\frac{Q}{8\varepsilon_0}$ (d) None of these

8 A point charge q is placed at the corner of an equilateral prism. Then, the electric flux through the prism is

(a) $\frac{q}{8\varepsilon_0}$ (b) $\frac{q}{6\varepsilon_0}$ (c) $\frac{q}{12\varepsilon_0}$ (d) $\frac{q}{24\varepsilon_0}$

9 Four point charges q, $2q$, $3q$ and $4q$ are placed at the centre of the cube, at the mid-point of the edge, at the centre of the top face and at the corner of the cube, respectively as shown in figure. The total flux passing through the cube is

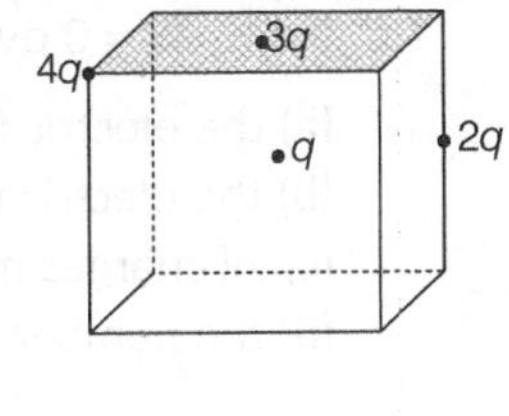

(a) $\frac{3q}{2\varepsilon_0}$ (b) $\frac{7q}{2\varepsilon_0}$ (c) $\frac{11q}{4\varepsilon_0}$ (d) $\frac{13q}{4\varepsilon_0}$

10 Eight point charges (can be assumed as uniformly charged small spheres and their centres at the corner of the cube) having value q each are fixed at vertices of a cube. The electric flux through square surface $ABCD$ of the cube is

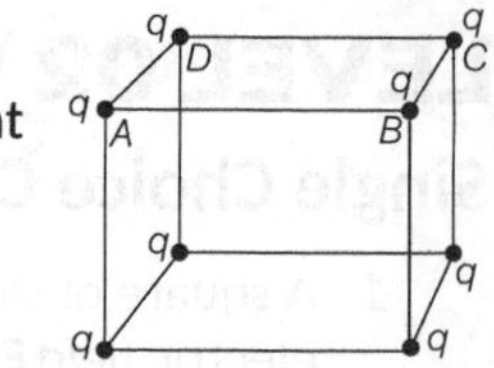

(a) $\frac{q}{24\varepsilon_0}$ (b) $\frac{q}{12\varepsilon_0}$ (c) $\frac{q}{6\varepsilon_0}$ (d) $\frac{q}{8\varepsilon_0}$

11 A long wire with a charge of λ per unit length passes through an imaginary cube of edge a. The maximum flux of the electric field through the cube will be

(a) $\frac{\lambda a}{\varepsilon_0}$ (b) $\frac{\sqrt{2}\lambda a}{\varepsilon_0}$ (c) $\frac{6\lambda a^2}{\varepsilon_0}$ (d) $\frac{\sqrt{3}\lambda a}{\varepsilon_0}$

12 Three uniformly charged wires with linear charge density λ are placed along X, Y and Z-axes, respectively. What is flux of electric field through Gaussian surface given by $x^2 + y^2 + z^2 = 1;\ x > 0;\ y > 0;\ z > 0$?

(a) $\frac{3\lambda}{2\,\varepsilon_0}$ (b) $\frac{3\lambda}{\varepsilon_0}$ (c) $\frac{6\lambda}{\varepsilon_0}$ (d) $\frac{3\lambda}{4\,\varepsilon_0}$

13 Figure (i) shows an imaginary cube of edge length L. An uniformly charged rod of length $2L$ moves towards left at a small but constant speed v. At $t = 0$, the left end of the rod just touches the centre of the face of the cube opposite to it. Which of the graphs shown in Fig. (ii) represents the flux of the electric field through the cube as the rod goes through it?

(i)

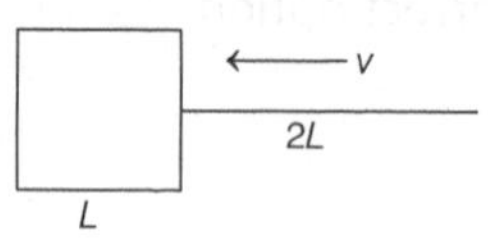

(ii)

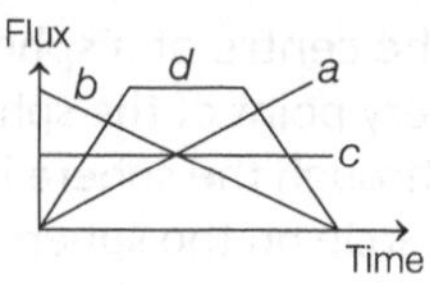

(a) a (b) b (c) c (d) d

14 The electric field in a region is given by $\mathbf{E} = 200\hat{\mathbf{i}}$ N/C for $x > 0$ and $-200\,\hat{\mathbf{i}}$ N/C for $x < 0$. A closed cylinder of length 2 m and cross-section area $10^2\ m^2$ is kept in such a way that the axis of cylinder is along X-axis and its centre coincides with origin. The total charge inside the cylinder is

(a) zero (b) 1.86×10^{-5} C (c) 1.77×10^{-11} C (d) 35.4×10^{-8} C

15 An electric field given by $\mathbf{E} = 4\hat{\mathbf{i}} - 3(y^2 + 2)\hat{\mathbf{j}}$ passes through Gaussian cube of side 1 m placed at origin, such that its three sides represent X, Y and Z-axes. The magnitude of net charge enclosed within the cube is

(a) $4\varepsilon_0$ (b) $3\varepsilon_0$ (c) $5\varepsilon_0$ (d) zero

16 A solid sphere of radius R has a volume charge density $\rho = \rho_0 r^2$ (where, ρ_0 is a constant and r is the distance from centre). At a distance x from its centre (for $x < R$), the electric field is directly proportional to

(a) $1/x^2$ (b) $1/x$ (c) x^3 (d) x^2

17 1 μC charge is uniformly distributed on a spherical shell given by equation $x^2 + y^2 + z^2 = 25$. What will be intensity of electric field at a point (1, 1, 2)?

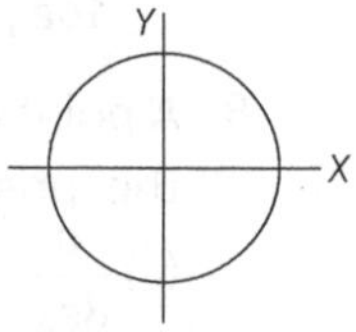

(a) 5 N/C (b) 45 N/C (c) $\frac{5\sqrt{3}}{2}$ N/C (d) Zero

18 If $\oint \mathbf{E} \cdot d\mathbf{S} = 0$ over a surface, then

(a) the electric field inside the surface and on it must be zero

(b) the electric field inside the surface is necessarily uniform

(c) all charges may be inside the surface

(d) the number of flux lines entering the surface must be equal to the number of flux lines leaving it

19 Figure shows four charges q_1, q_2, q_3 and q_4 fixed in space, then the total flux of electric field through a closed surface S, due to all charges q_1, q_2, q_3 and q_4 is

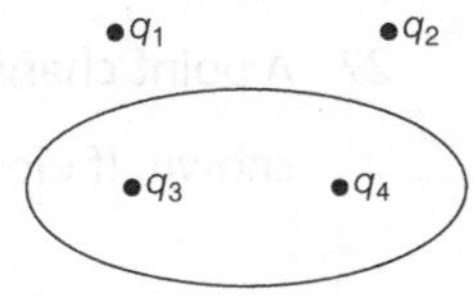

(a) not equal to the total flux through S due to charges q_3 and q_4
(b) equal to the total flux through S due to charges q_3 and q_4
(c) zero, if $q_1 + q_2 = q_3 + q_4$
(d) twice the total flux through S due to charges q_3 and q_4, if $q_1 + q_2 = q_3 + q_4$

20 In a certain region, **E** increases radially as $\mathbf{E} = 90r(-\hat{\mathbf{r}})$. The electric charge contained within a sphere of radius 2 m centred at the origin is
(a) 80 nC (b) – 80 nC
(c) 160 nC (d) – 160 nC

21 An infinite, uniformly charged sheet with surface charge density σ cuts through a spherical Gaussian surface of radius R at a distance x from its centre, as shown in the figure. The electric flux ϕ through the Gaussian surface is

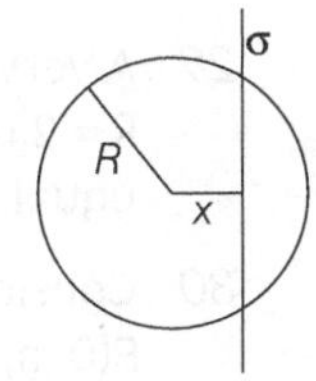

(a) $\frac{\pi R^2 \sigma}{\varepsilon_0}$ (b) $\frac{2\pi(R^2 - x^2)\sigma}{\varepsilon_0}$
(c) $\frac{\pi(R - x)^2 \sigma}{\varepsilon_0}$ (d) $\frac{\pi(R^2 - x^2)\sigma}{\varepsilon_0}$

22 A point charge q is placed at the centre of left circular end of a cylinder of length $l = 4$ cm and radius $R = 3$ cm as shown. Then, the electric flux through the curved surface of the cylinder is

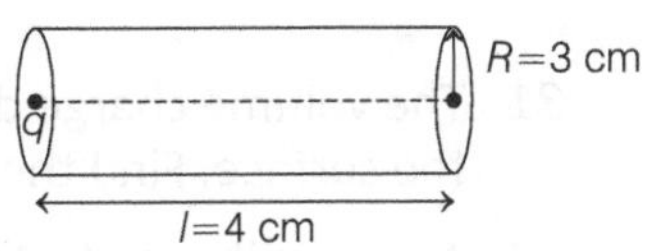

(a) $\frac{q}{5\varepsilon_0}$ (b) $\frac{2q}{5\varepsilon_0}$ (c) $\frac{3q}{5\varepsilon_0}$ (d) $\frac{4q}{5\varepsilon_0}$

23 Inside a uniformly charged infinitely long cylinder of radius R and volume charge density ρ there is a spherical cavity of radius $R/2$. A point P is located at a distance $2R$ from the axis of the cylinder as shown. Then, the electric field strength at the point P is

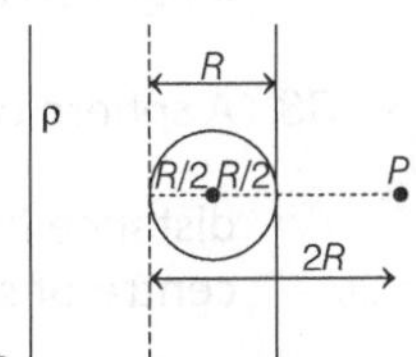

(a) $\frac{23\rho R}{54\varepsilon_0}$ (b) $\frac{23\rho R}{108\,\varepsilon_0}$ (c) $\frac{25\rho R}{54\varepsilon_0}$ (d) $\frac{25\rho R}{108\varepsilon_0}$

24 A point charge placed on the axis of an uniformly charged disc experiences a force F due to the disc. If the charge density on the disc is σ, the electric flux through the disc, due to the point charge will be
(a) $\frac{2\pi F}{\sigma}$ (b) $\frac{F}{2\pi\sigma}$ (c) $\frac{F^2}{\sigma}$ (d) $\frac{F}{\sigma}$

25 For a spherically symmetrical charge distribution, electric field at a distance r from the centre of sphere is $\mathbf{E} = kr^7\hat{\mathbf{r}}$, where k is a constant. What will be the volume charge density at a distance r from the centre of sphere?
(a) $\rho = 9k\varepsilon_0 r^6$ (b) $\rho = 5k\varepsilon_0 r^3$ (c) $\rho = 3k\varepsilon_0 r^4$ (d) $\rho = 9k\varepsilon_0 r^0$

Numerical Value Questions

26 A uniform electric field $E = 500$ N/C passes through a hemispherical surface of radius $R = 1.2$ m as shown in figure. If the net electric flux (in SI units) through the hemispherical surface only is $N\pi$, then find the value of N.

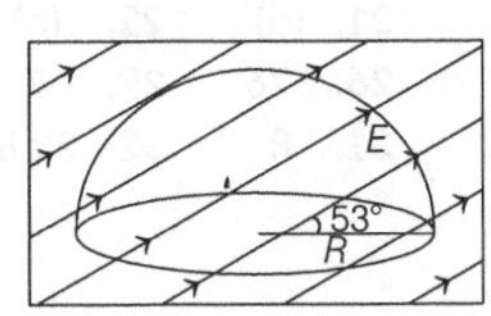

27 A point charge $+q$ is placed on the axis of a closed cylinder of radius R and height $25R/12$ as shown. If electric flux coming out from the curved surface of cylinder is $\frac{xq}{10\varepsilon_0}$, then calculate x.

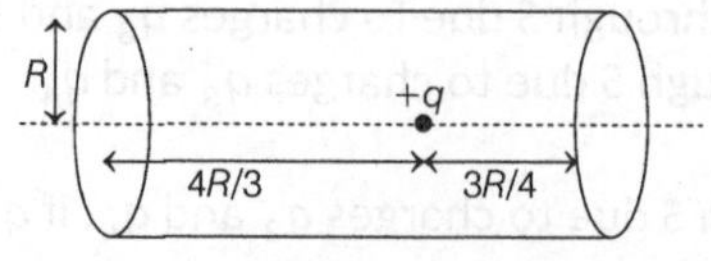

28 The length of each side of a cubical closed surface is l. A charge q is situated on one of the vertices of the cube as shown in figure. The flux passing through shaded faces of the cube is $\frac{q}{n\varepsilon_0}$. Find n.

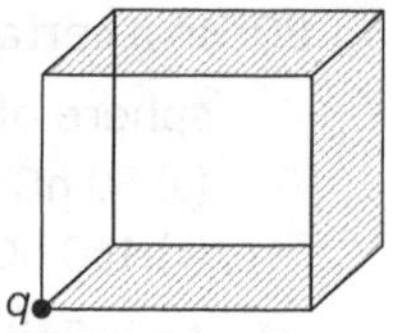

29 A very long uniformly charged thread oriented along the axis of a ring of radius $R = 2$ m rests on its centre with one of the ends. The charge on the thread per unit length is equal to $\lambda = 8.85 \times 10^{-10}$ Cm^{-1}. Find the flux (in SI unit) of the electric field through the ring.

30 Consider a triangular surface whose vertices are three points having coordinate $A(2a, 0, 0)$, $B(0, a, 0)$, $C(0, 0, a)$. If there is an uniform electric field $E_0\hat{\mathbf{i}} + 2E_0\hat{\mathbf{j}} + 3E_0\hat{\mathbf{k}}$, then flux linked to triangular surface ABC is $\frac{E_0a^2}{2} \times N$. Find the value of N.

31 The volume charge density of a solid sphere of radius 24 m is proportional to the depth below the surface. Find the depth (in m) at which the electric intensity is maximum.

32 A non-uniform electric field in x-direction is increasing uniformly from 2 N/C at $x = 1$ m to 8 N/C at $x = 7$ m. The centre of cube is at $x = 2$ m. If the charge enclosed inside a small cube of side length 10 cm is $\beta \times 10^{-\alpha}$ C, then find the value of $(\alpha + \beta)$.

33 A sphere of radius R has charge density given by $\rho = \rho_0\left(1 - \frac{nr}{3R}\right)$, where ρ_0 is a constant, r is distance from centre of sphere. For a spherical Gaussian surface of radius R centred at the centre of sphere, the flux is zero. Find n.

34 A charge q is placed a distance above a square sheet of size $a \times a$ as shown in figure. If the flux through the square is $\frac{q}{n\varepsilon_0}$, then find n.

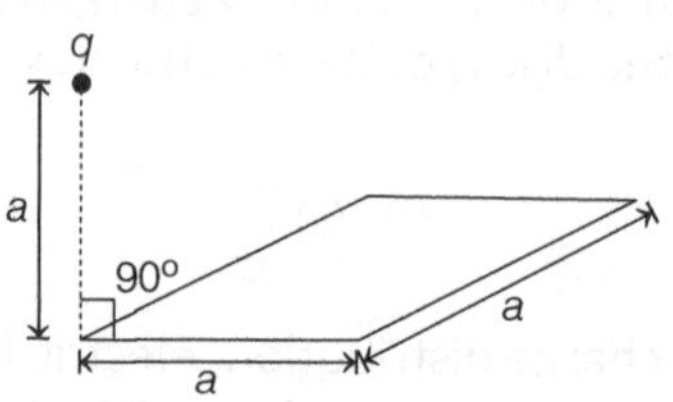

Answers

1. (b)	**2.** (c)	**3.** (a)	**4.** (c)	**5.** (d)	**6.** (d)	**7.** (c)	**8.** (c)	**9.** (b)	**10.** (c)
11. (d)	**12.** (d)	**13.** (d)	**14.** (d)	**15.** (b)	**16.** (c)	**17.** (d)	**18.** (d)	**19.** (b)	**20.** (b)
21. (d)	**22.** (b)	**23.** (d)	**24.** (d)	**25.** (a)					
26. 576	**27.** 7	**28.** 12	**29.** 100	**30.** 11					
31. 8	**32.** 23.85	**33.** 4	**34.** 24						

For Detailed Solutions
Scan the code

LEVEL 03 \ JEE Advanced Level

Single Choice Correct Questions

1 Electric field at point P is given by $\mathbf{E} = \mathbf{r}E_0$. The total flux through the given cylinder of radius R and height h is

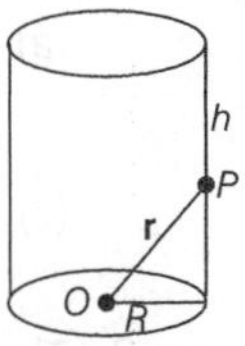

(a) $E_0\pi R^2 h$ (b) $2E_0\pi R^2 h$

(c) $3E_0\pi R^2 h$ (d) $4E_0\pi R^2 h$

2 An infinitely long metal cylinder of radius R and surface charge density σ is placed symmetrically with an imaginary surface of the shape of a prism. The length of prism is L and its 3 sides are all equal to $3R$. The flux through the prism is

(a) $\frac{\pi}{3}\frac{RL\sigma}{\varepsilon_0}$ (b) $\frac{\pi}{2}\frac{RL\sigma}{\varepsilon_0}$ (c) $\frac{\pi RL\sigma}{\varepsilon_0}$ (d) $\frac{2\pi RL\sigma}{\varepsilon_0}$

3 A ring of radius R having a linear charge density λ moves towards a solid imaginary sphere of radius $\frac{R}{2}$, so that the centre of ring passes through the centre of sphere. The axis of the ring is perpendicular to the line joining the centres of the ring and the sphere. The maximum flux through the sphere in this process is

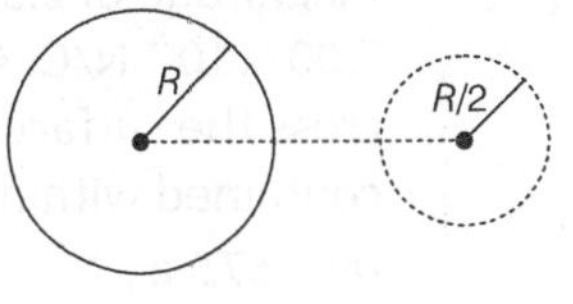

(a) $\frac{\lambda R}{\varepsilon_0}$ (b) $\frac{\lambda R}{2\varepsilon_0}$ (c) $\frac{\lambda \pi R}{4\varepsilon_0}$ (d) $\frac{\lambda \pi R}{3\varepsilon_0}$

4 Consider a hollow sphere of radius R, its quarter part is cut and kept, such that geometrical centre lies at origin as shown in figure. XY-axis lies in plane of paper and Z-axis is perpendicular to the plane of the paper. Sphere is kept in a manner, such that plane portions of sphere lies in ZX and YZ-plane. Uniform electric field $\mathbf{E} = E_0\hat{\mathbf{i}} + 2E_0\hat{\mathbf{j}} + 2E_0\hat{\mathbf{k}}$ (N/C) lies in the complete space. Electric flux coming out of the curved surface of sphere is

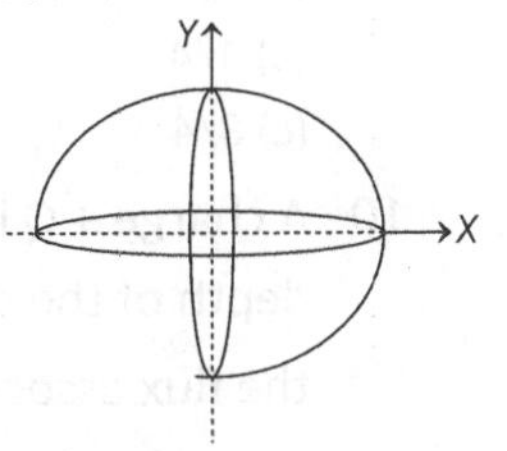

(a) $\frac{3}{2}\pi R^2 E_0$ (b) $\frac{5}{2}\pi R^2 E_0$ (c) $2\pi R^2 E_0$ (d) zero

5 Consider an imaginary hemispherical surface. A semi-infinite wire of charge density λ is kept with one of its ends coinciding with centre of hemisphere and wire is kept along the symmetric axis of the hemisphere as shown in the figure. The electric flux passing through the spherical surface of the hemisphere is

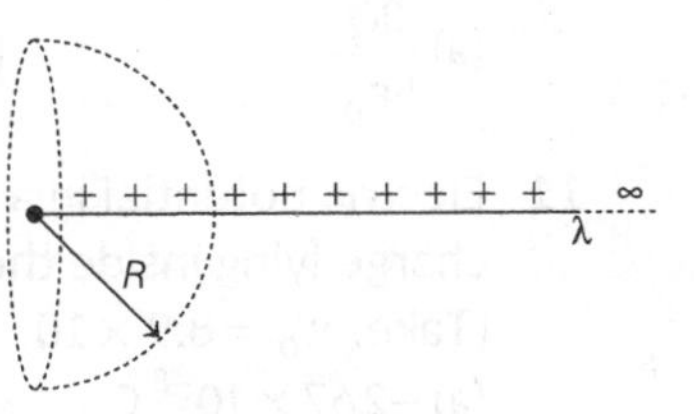

(a) $\frac{\lambda R}{3\varepsilon_0}$ (b) $\frac{\lambda R}{4\varepsilon_0}$ (c) $\frac{3}{4}\frac{\lambda R}{\varepsilon_0}$ (d) $\frac{\lambda R}{2\varepsilon_0}$

6 A charge Q is uniformly distributed over a square plate of side a. Another point charge q is placed at a height $\frac{a}{2}$ from the centre of square plate as shown in diagram. Force acting on the point charge q is

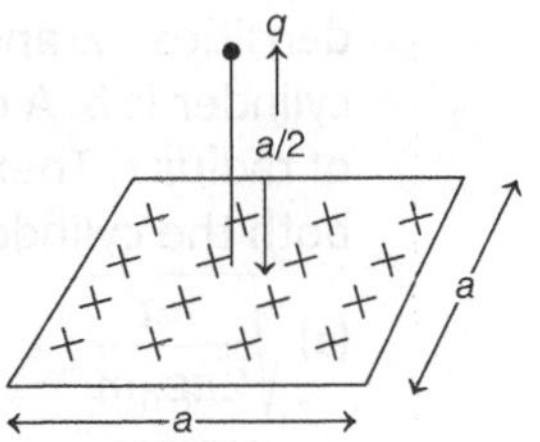

(a) $\frac{Qq}{4\pi\varepsilon_0 a^2}$ (b) $\frac{Qq}{6\pi\varepsilon_0 a^2}$

(c) $\frac{Qq}{6\varepsilon_0 a^2}$ (d) $\frac{Qq}{4\varepsilon_0 a^2}$

7 Two point charges $+q$ and $-q$ are separated by the distance $2l$ as shown in figure. The flux of electric field passing through a circle of radius R will be

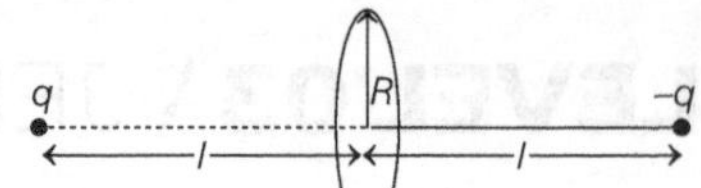

(a) $\frac{q}{\varepsilon_0}\left(1+\frac{l}{\sqrt{l^2+R^2}}\right)$ (b) $\frac{q}{\varepsilon_0}\left(1-\frac{l}{\sqrt{l^2+R^2}}\right)$

(c) $\frac{q}{\varepsilon_0}$ (d) zero

8 The electric field $\mathbf{E}_1$ at one face of a parallelopiped is uniform over the entire face and is directed out of the face. At the opposite face, the electric field $\mathbf{E}_2$ is also uniform over the entire face and is directed into that face (figure). The two faces in question are inclined at 30.0° from the horizontal, while $\mathbf{E}_1$ and $\mathbf{E}_2$ are both horizontal; $\mathbf{E}_1$ has a magnitude of 2.50×10^4 N/C, and $\mathbf{E}_2$ has a magnitude of 7.00×10^4 N/C. Assuming that, no other electric field lines cross the surfaces of the parallelopiped, the net charge contained within the parallelopiped is

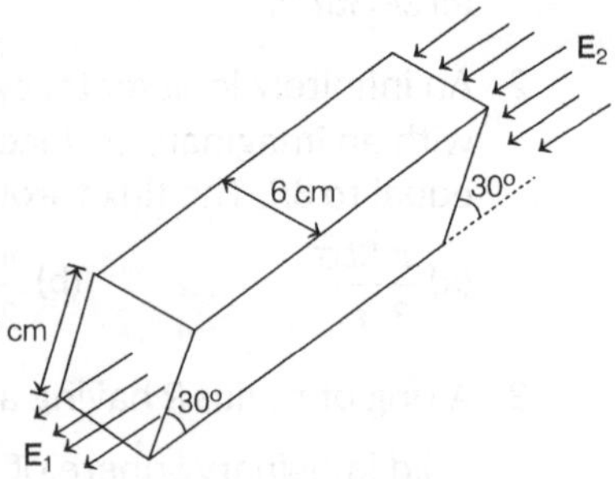

(a) $-67.5\,\varepsilon_0$ (b) $67.5\,\varepsilon_0$ (c) $105\,\varepsilon_0$ (d) $-105\,\varepsilon_0$

9 The volume charge density as a function of distance x from one face inside a unit cube is varying as shown in the figure. Then, the total flux (in SI units) through the cube is ($\rho_0=8.85\times10^{-12}$ C/m^3)

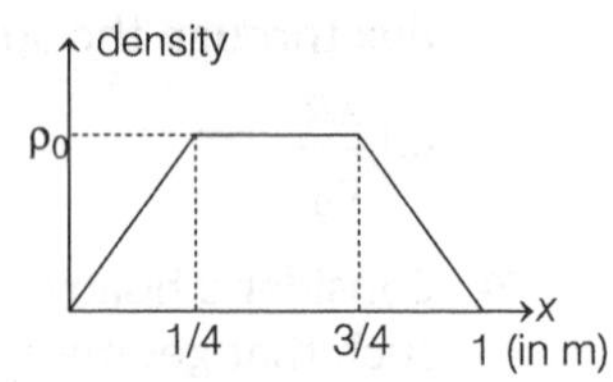

(a) 1/4 (b) 1/2

(c) 3/4 (d) 1

10 A charge $+Q$ is located somewhere inside a vertical cone, such that the depth of the charge from the free surface of the cone is H. It is found that the flux associated with the cone with the curved surface is $\frac{3Q}{5\varepsilon_0}$. If the charge is raised vertically through a height $2H$, then the flux through the curved surface is

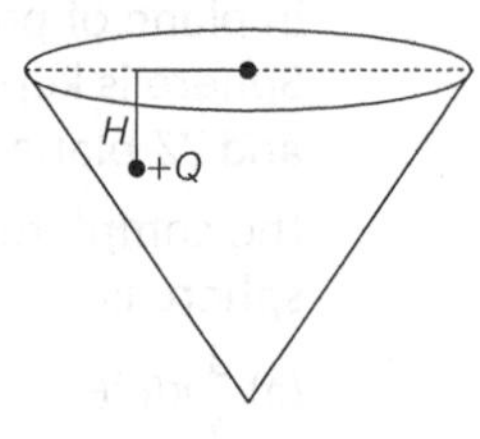

(a) $\frac{3Q}{5\varepsilon_0}$ (b) $\frac{2Q}{5\varepsilon_0}$ (c) $\frac{4Q}{5\varepsilon_0}$ (d) zero

11 Electric potential in a region varies as $V=(1500\ \text{kV/m}^2)x^2$. Total electric charge lying inside the cube of side 10 cm shown in figure is (Take, $\varepsilon_0=8.9\times10^{-12}\ \text{C}^2/\text{N-m}^2$)

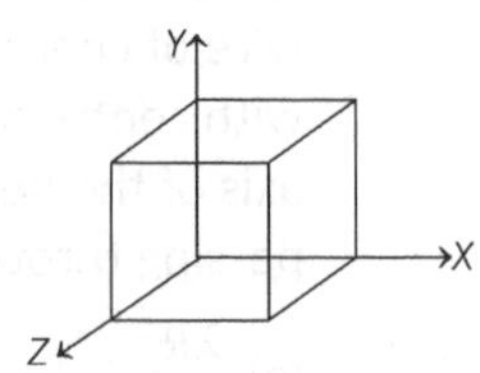

(a) -2.67×10^{-8} C (b) $+2.67\times10^{-8}$ C

(c) -5.24×10^{-8} C (d) $+5.24\times10^{-8}$ C

12 Figure shows two large cylindrical shells having uniform linear charge densities $+\lambda$ and $-\lambda$. Radius of inner cylinder is a and that of outer cylinder is b. A charged particle of mass m, charge q revolves in a circle of radius r. Then, its speed v is (Neglect gravity and assume the radii of both the cylinders to be very small in comparison to their lengths.)

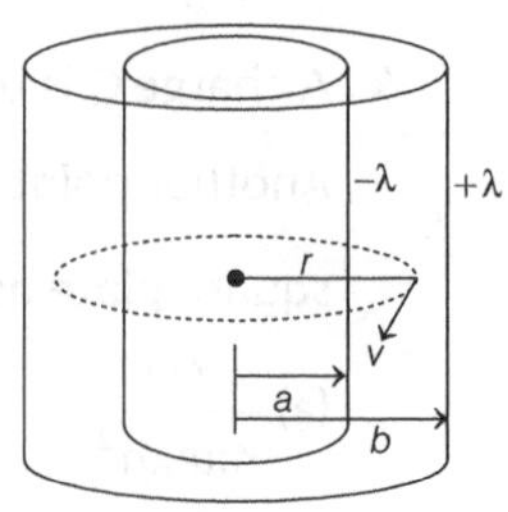

(a) $\sqrt{\frac{\lambda q}{2\pi\varepsilon_0 m}}$ (b) $\sqrt{\frac{2\lambda q}{\pi\varepsilon_0 m}}$

(c) $\sqrt{\frac{\lambda q}{\pi\varepsilon_0 m}}$ (d) $\sqrt{\frac{\lambda q}{4\pi\varepsilon_0 m}}$

13 Consider a solid insulating sphere of radius R with charge density varying as $\rho = \rho_0 r^2$ (ρ_0 is a constant and r is measured from centre). Consider two points A and B at distance x and y, respectively ($x < R$, $y > R$) from the centre. If magnitudes of electric fields at points A and B are equal, then

(a) $x^2 y = R^3$ (b) $x^3 y^2 = R^5$ (c) $x^2 y^3 = R^5$ (d) $\frac{x^4}{y} = R^5$

14 A system consists of a uniformly charged sphere of radius R and a surrounding medium filled by a charge with the volume density $\rho = \frac{\alpha}{r}$, where α is a positive constant and r is the distance from the centre of the sphere ($r > R$). The charge of the sphere for which the electric field intensity E outside the sphere is independent of R is

(a) $\frac{\alpha}{2\varepsilon_0}$ (b) $\frac{2}{\alpha\varepsilon_0}$ (c) $2\pi\alpha R^2$ (d) None of these

15 A large insulating thick sheet of thickness $2d$ carries a uniform charge per unit volume ρ. A particle of mass m, carrying a charge q having a sign opposite to that of the sheet, is released from the surface of the sheet. The sheet does not offer any mechanical resistance to the motion of the particle. The oscillation frequency ν of the particle inside the sheet is

(a) $\nu = \frac{1}{4\pi}\sqrt{\frac{2q\rho}{m\varepsilon_0}}$ (b) $\nu = \frac{1}{2\pi}\sqrt{\frac{2q\rho}{m\varepsilon_0}}$ (c) $\nu = \frac{1}{4\pi}\sqrt{\frac{q\rho}{m\varepsilon_0}}$ (d) $\nu = \frac{1}{2\pi}\sqrt{\frac{q\rho}{m\varepsilon_0}}$

16 An infinite long cylinder of radius R has a cylindrical cavity of radius $R/2$ just touching its surface. The region inside the cylinder is filled with uniform charge density ρ except in the cavity. A charge particle of charge q and mass m is thrown from point A in the cavity. AB is perpendicular to the diameter of the cavity passing through the axis of the cylinder. The minimum velocity u required at A for it to reach point B is

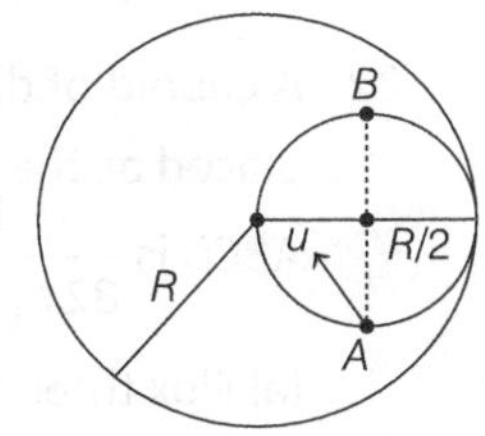

(a) $\sqrt{\frac{q\rho R^2}{4\varepsilon_0 m}}$ (b) $\sqrt{\frac{q\rho R^2}{2\varepsilon_0 m}}$ (c) $\sqrt{\frac{2q\rho R^2}{\varepsilon_0 m}}$ (d) $\sqrt{\frac{4q\rho R^2}{\varepsilon_0 m}}$

17 Two charges $+q_1$ and $-q_2$ are placed at A and B, respectively. A line of force emanates from q_1 at angle α with line AB and it terminates at an angle β as shown, then β is

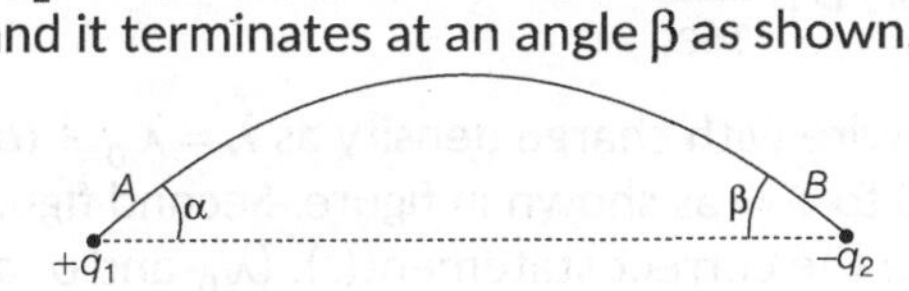

(a) $\sin^{-1}\left(\sqrt{\frac{q_1}{q_2}}\sin\frac{\alpha}{2}\right)$ (b) $2\sin^{-1}\left(\sqrt{\frac{q_1}{q_2}}\sin\alpha\right)$ (c) $\sin^{-1}\left(\sqrt{\frac{q_2}{q_1}}\sin\frac{\alpha}{2}\right)$ (d) $2\sin^{-1}\left(\sqrt{\frac{q_1}{q_2}}\sin\frac{\alpha}{2}\right)$

18 A cylindrical portion of radius r is removed from a solid sphere of radius R and uniform volume charge density ρ in such a way that the axis of the hollow cylinder coincides with one of the diameters of the sphere. (r is negligible compared to R). Then, the electric field intensity at point A is

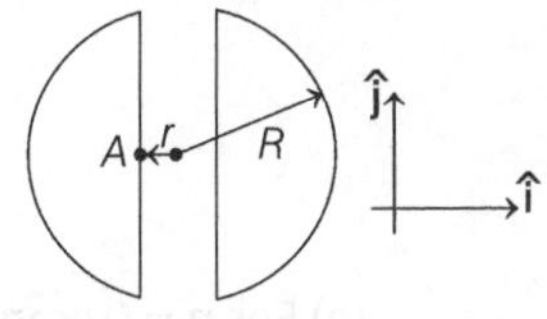

(a) $\frac{\rho r}{3\varepsilon_0}\hat{i}$ (b) $-\frac{\rho r}{3\varepsilon_0}\hat{i}$ (c) $\frac{\rho r}{6\varepsilon_0}\hat{i}$ (d) $-\frac{\rho r}{6\varepsilon_0}\hat{i}$

19 A long solid non-conducting cylinder of radius R has a uniform volume charge density ρ. The potential difference between the axis and the surface of the cylinder is

(a) $\frac{\rho R^2}{\varepsilon_0}$ (b) $\frac{\rho R^2}{2\varepsilon_0}$ (c) $\frac{\rho R^2}{4\varepsilon_0}$ (d) $\frac{2\rho R^2}{\varepsilon_0}$

Multiple Choice(s) Correct Questions

20 In the given hypothetical electric field $\mathbf{E} = [(d + x)\hat{\mathbf{i}} - E_0\hat{\mathbf{j}}]$ N/C, a hypothetical closed surface is taken as shown in figure.

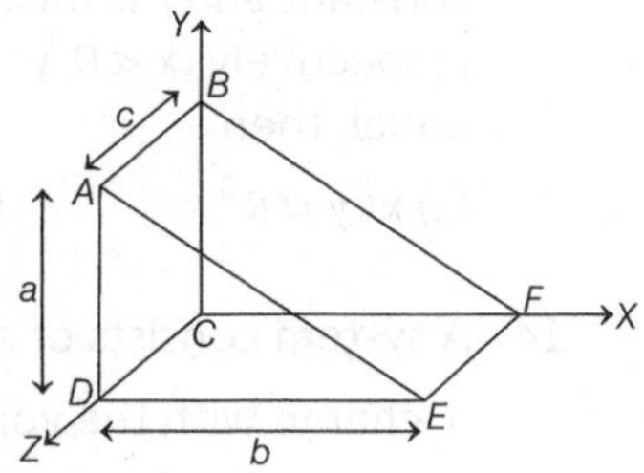

(a) Net flux through plane $ABCD$ is $-acd$ unit.
(b) Net flux through plane $CDEF$ is bcE_0 unit.
(c) Net flux through plane $ABEF$ is $\left[-bcE_0 + acd + \frac{acb}{2}\right]$ unit.
(d) Net charge enclosed by the closed surface is $abc\varepsilon_0$ unit.

21 A particle with charge Q is located immediately above the centre of the flat face of a hemisphere of radius R as shown in figure. Choose the correct statement(s).

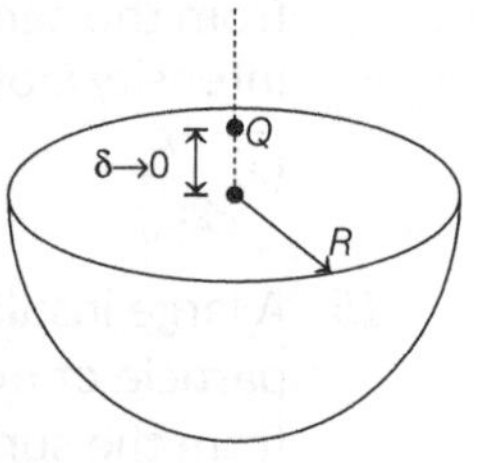

(a) Flux through the flat surface is $-\frac{Q}{2\varepsilon_0}$.
(b) Flux through the curved surface is $\frac{Q}{2\varepsilon_0}$.
(c) If charge Q is displaced up, flux through flat surface is less than $\frac{Q}{2\varepsilon_0}$ in magnitude.
(d) If charge Q is displaced up, flux through curved surface is less than $\frac{Q}{2\varepsilon_0}$ in magnitude.

22 A cuboid of dimension $[a \times a \times b]$ is shown in figure. Charge q is placed at the centre of edge having length b. If flux through face $ABCD$ is $\frac{q}{32\,\varepsilon_0}$, then select the correct statement(s).

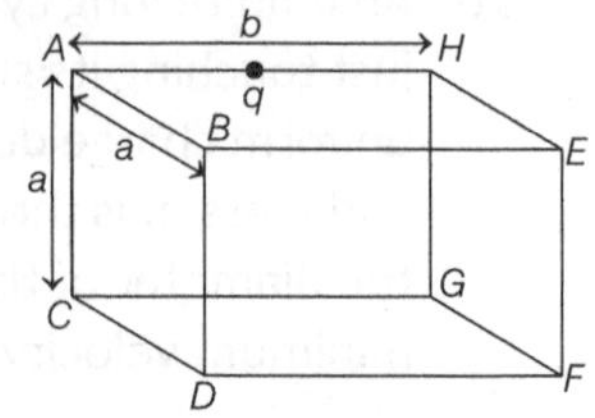

(a) Flux through the entire cuboid is $\frac{q}{4\varepsilon_0}$.
(b) Flux through the face $ABEH$ is zero.
(c) Flux through the face '$BEFD$' is $\frac{3q}{32\varepsilon_0}$.
(d) Flux through the face $BEFD$ is $\frac{q}{16\varepsilon_0}$.

23 Consider an infinite long wire with charge density as $\lambda = \lambda_0 + \alpha x^n$, kept along X-axis. A cube of side a is moved from 0 to $+\infty$ as shown in figure. Second figure shows the variation of flux through the cube. Choose the correct statement(s). (λ_0 and α are positive constants.)

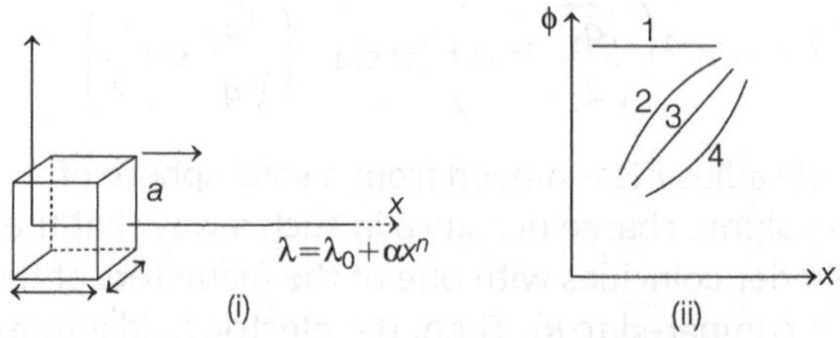

(i) (ii)

(a) For $n = 0$ graph is (1)
(b) For $n = 1$ graph is (3)
(c) For $n = 1$ graph is (4)
(d) For $n = 0$ graph is (2)

24 A Gaussian surface
(a) must not pass through any discrete charge
(b) must pass through any material medium, whether conducting or non-conducting
(c) may pass through a continuous charge distribution
(d) may pass through a region, where no material medium is present

25 In the statement of Gauss's law, $\oint \mathbf{E} \cdot d\mathbf{s} = \frac{q_{\text{enclosed}}}{\varepsilon_0}$. Select the correct alternative(s).

(a) The value of $\oint \mathbf{E} \cdot d\mathbf{s}$ does not depend on the shape of gaussian surface, if it encloses same charge.

(b) The electric field **E** also includes the contribution of charges outside the gaussian surface.

(c) If charge is moved inside the gaussian surface, $\mathbf{E} \cdot d\mathbf{s}$ at any point remains constant.

(d) If charge is moved inside the gaussian surface, electric field at every point on surface may change but $\oint \mathbf{E} \cdot d\mathbf{s}$ remains unchanged.

26 Consider a gaussian spherical surface, covering a dipole of charge q and $-q$, then

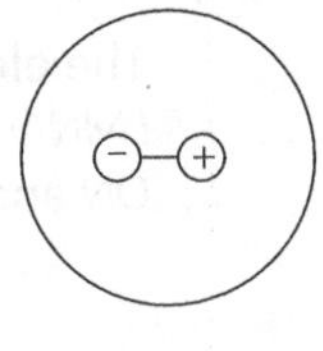

(a) $q_{\text{enclosed}} = 0$ (Net charge enclosed by the spherical surface)

(b) $\phi_{\text{net}} = 0$ (Net flux coming out the spherical surface)

(c) $E = 0$ at all points on the spherical surface

(d) $\oint \mathbf{E} \cdot d\mathbf{s} = 0$ (Surface integral of **E** over the spherical surface)

27 A spherical volume centred at the origin of coordinates with radius a, has a volume charge density $\rho(r) = +\frac{Q}{a}\left(\frac{1}{r^2}\right)$. The electric field

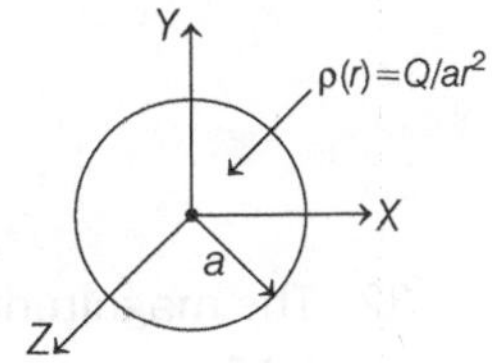

(a) at $r = \frac{a}{2}, E = \frac{2Q}{a^2\varepsilon_0}$ (b) at $r = a, E = \frac{Q}{a^2\varepsilon_0}$

(c) at $r = 2a, E = \frac{Q}{2a^2\varepsilon_0}$ (d) at $r = 4a, E = \frac{Q}{16a^2\varepsilon_0}$

28 A non-conducting rod AB of length $\sqrt{3}R$ has uniformly distributed charge of linear charge density λ. A non-conducting ring has uniformly distributed charge Q. Point A is the centre of ring and line AB is the axis of ring, perpendicular to plane of ring. Choose the correct statement(s).

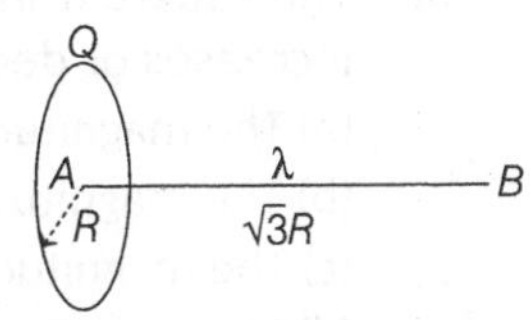

(a) The net electrostatic force on the ring due to charge on the rod is $\frac{Q\lambda}{8\pi\varepsilon_0 R}$.

(b) The net electrostatic force on the ring due to charge on the rod is $\frac{Q\lambda}{4\pi\varepsilon_0 R}$.

(c) The electrostatic flux through ring is $\frac{\lambda R}{\varepsilon_0}(\sqrt{3}-1)$.

(d) The electrostatic flux through ring is $\frac{\lambda R}{2\varepsilon_0}(\sqrt{3}-1)$.

Comprehension Based Questions

Comprehension-1

A cube has side of length, $L = 0.3$ m. It is placed with one corner at origin as shown in figure. The electric field is not uniform but is given by $\mathbf{E} = (-5\text{NC}^{-1})x\hat{\mathbf{i}} + (3\text{NC}^{-1})z\hat{\mathbf{k}}$

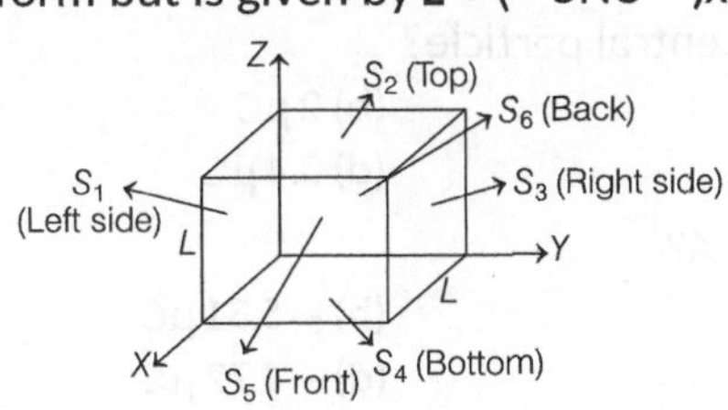

29 Which of the surfaces have zero flux?

(a) S_2, S_4 and S_5 (b) S_1, S_3, S_4 and S_6 (c) S_1, S_2 and S_3 (d) S_2, S_3 and S_4

30 The flux passing through the surface S_5 will be

(a) $-0.135\ N\text{-}m^2C^{-1}$
(b) $-0.054\ N\text{-}m^2C^{-1}$
(c) $0.81\ N\text{-}m^2C^{-1}$
(d) $0.054\ N\text{-}m^2C^{-1}$

31 The total electric charge inside the cube is

(a) $-0.054\varepsilon_0$
(b) $0.081\varepsilon_0$
(c) $0.135\varepsilon_0$
(d) $0.054\varepsilon_0$

Comprehension-2

The electric field intensity at all points in space is given by $\mathbf{E} = \sqrt{3}\hat{\mathbf{i}} - \hat{\mathbf{j}}$ N/C. A square frame *LMNO* of side 1 m is shown in figure. The point *N* lies in *XY*-plane. The initial angle between line *ON* and *X*-axis is $\theta = 60°$.

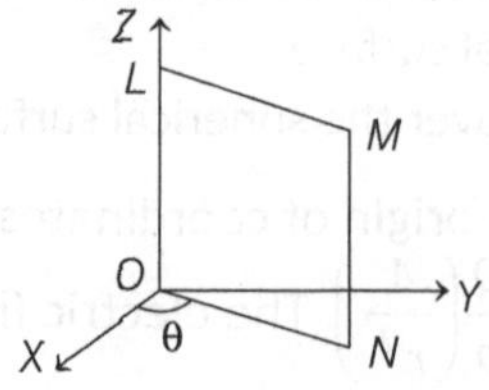

32 The magnitude of electric flux through area enclosed in square frame *LMNO* is

(a) 0 V-m
(b) 1 V-m
(c) 2 V-m
(d) 4 V-m

33 The square frame *LMNO* is now rotated about *Z*-axis by an angle 30°, such that θ either increases or decreases. Then, pick up the correct statement.

(a) The magnitude of electric flux increases from initial value as θ is increased.
(b) The magnitude of electric flux increases from initial value as θ is decreased.
(c) The magnitude of electric flux may increase or decrease from initial value as θ is changed.
(d) The magnitude of electric flux will decrease from initial value as θ is changed.

Comprehension-3

A charged particle is suspended at the centre of two thin concentric spherical charged shells, made of non-conducting material. Figure (i) shows cross-section of the arrangement. Figure (ii) gives the net flux ϕ through a Gaussian sphere centred on the particle, as a function of the radius *r* of the sphere.

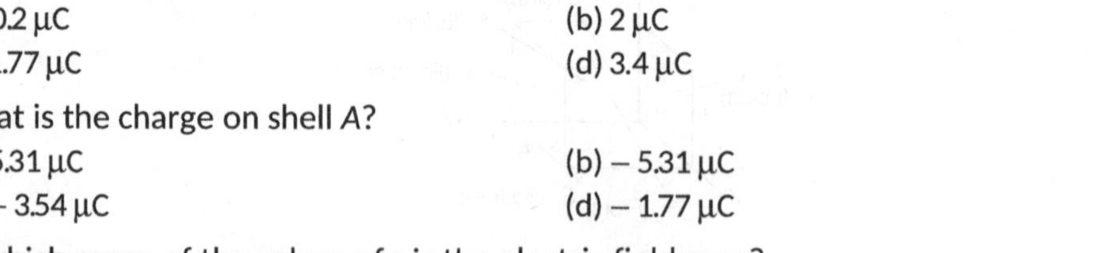

34 What is the charge on the central particle?

(a) 0.2 μC
(b) 2 μC
(c) 1.77 μC
(d) 3.4 μC

35 What is the charge on shell *A*?

(a) 5.31 μC
(b) – 5.31 μC
(c) – 3.54 μC
(d) – 1.77 μC

36 In which range of the values of *r* is the electric field zero?

(a) 0 to r_A
(b) r_A to r_B
(c) For $r > r_B$
(d) For no range of *r*, electric field is zero

Comprehension-4

Two non-conducting plates A and B of radii $2R$ and $4R$ respectively, are kept at distances x and $2x$ from the point charge q. A surface cutout of a non-conducting shell C is kept, such that its centre coincides with the point charge. Each plate and spherical surface carries surface charge density σ.

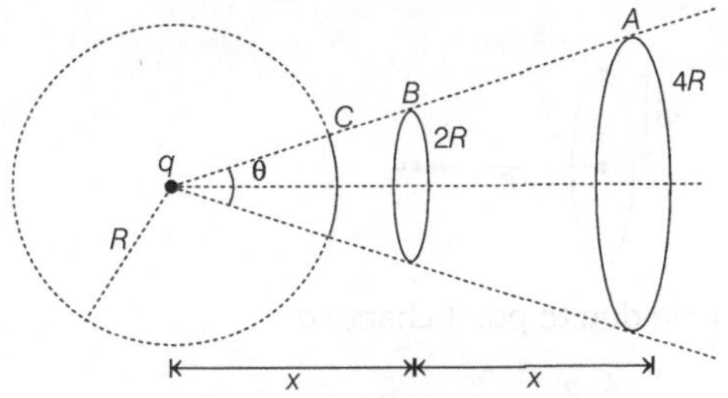

37 If F_A, F_B and F_C be the forces on plate A, plate B and spherical surface C due to charge q respectively, then

(a) $F_A = F_B = F_C$
(b) $F_C < F_B < F_A$
(c) $F_C > F_B > F_A$
(d) $F_A = F_B > F_C$

38 If ϕ_1 is flux through surface of B due to electric field of A and ϕ_2 be the flux through A due to electric field of B, then

(a) $\phi_1 = \phi_2$
(b) $\phi_1 > \phi_2$
(d) $\phi_1 < \phi_2$
(d) it depends on x and R

39 If distance x and radius R are doubled, so that F_A becomes F_A', F_B becomes F_B', then the correct option is

(a) $F_B > F_B'$
(b) $F_B' > F_B$
(c) $F_A' > F_B'$
(d) None of these

Comprehension-5

Electric field in a region is given by following equation $\mathbf{E} = \dfrac{K(\mathbf{r} - \mathbf{r}_0)}{|\mathbf{r} - \mathbf{r}_0|^2}$, where $\mathbf{r} = x\hat{\mathbf{i}} + y\hat{\mathbf{j}}$ and $\mathbf{r}_0 = \hat{\mathbf{i}} + \hat{\mathbf{j}}$ and an imaginary cube is made as shown in the diagram.

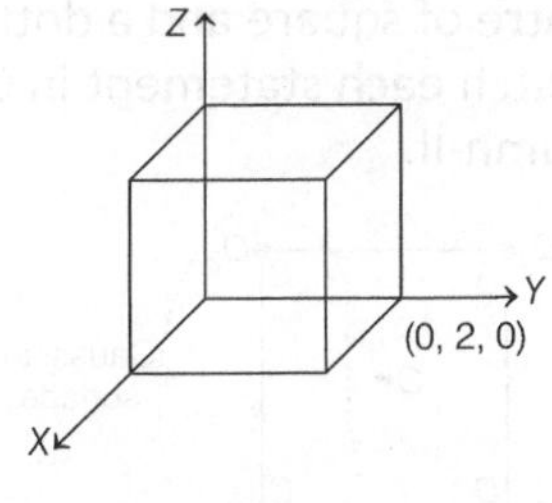

40 Total flux through this cube due to the given electric field is

(a) $4\pi K$
(b) $6\pi K$
(c) $16K$
(d) $8K$

41 Total charge encloses in this imaginary shown cube is

(a) $4K\pi\varepsilon_0$
(b) $8K\pi\varepsilon_0$
(c) $6K\pi\varepsilon_0$
(d) $2K\pi\varepsilon_0$

42 If $\mathbf{r}_0 = (2\hat{\mathbf{i}} + 2\hat{\mathbf{j}})$, then the value of total flux through the walls of the shown cube is

(a) $2K\pi$
(b) $K\pi$
(c) $4K\pi$
(d) $4K$

Match the Columns Questions

43 Match the Column I with Column II.

	Column-I		Column-II
(a)	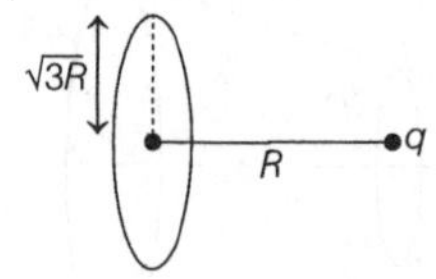 Flux through given circle due to point charge q	(p)	0
(b)	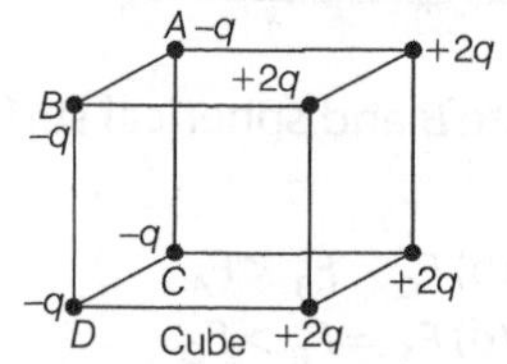 Net flux through surface $ABCD$	(q)	$\dfrac{q}{2\varepsilon_0}$
(c)	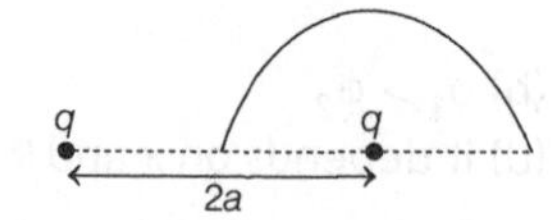 Net flux through given hemispherical surface	(r)	$\dfrac{q}{3\varepsilon_0}$
(d)	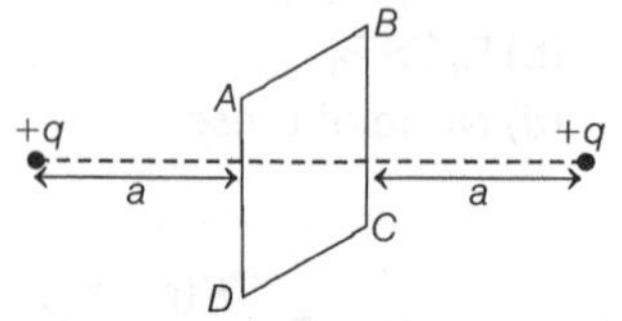 Net flux through surface $ABCD$	(s)	$\dfrac{q}{4\varepsilon_0}$

44 Four particles having charges Q_1, Q_2, Q_3 and Q_4 are fixed at vertices of a square A, B, C and D as shown in figure. Let O be the centre of square and a dotted gaussian (closed) surface encloses charges Q_2 and Q_3 as shown. Match each statement in Column-I with the corresponding correct condition(s) given in Column-II.

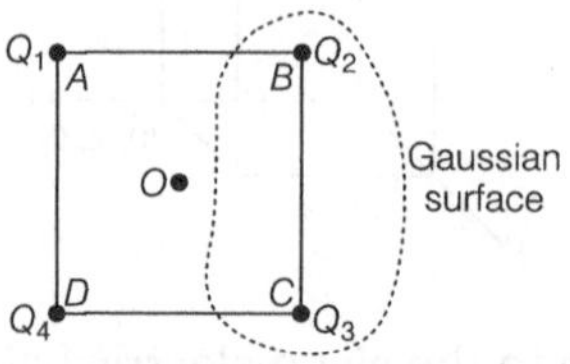

Column-I	Column-II
(a) The electric field at O is zero, if	(p) $Q_1 = Q_3 = +10\,\mu C$ and $Q_2 = Q_4 = -10\,\mu C$
(b) The potential at O is zero, if	(q) $Q_1 = Q_2 = +10\,\mu C$ and $Q_3 = Q_4 = -10\,\mu C$
(c) The flux through shown Gaussian surface is zero, if	(r) $Q_1 = Q_2 = +20\,\mu C$ and $Q_3 = Q_4 = -10\,\mu C$
(d) The electric field at O is parallel to **BC**, if	(s) $Q_1 = +10\,\mu C, Q_2 = +40\,\mu C, Q_3 = -40\,\mu C$ and $Q_4 = -10\,\mu C$
	(t) $Q_1 = Q_3 = +10\,\mu C$ and $Q_2 = Q_4 = -5\,\mu C$

45 Graph between electric potential and the X-coordinate is given and the electric field is along the X-axis. Then, find total charge in various cubes placed with one edge along positive X-axis.

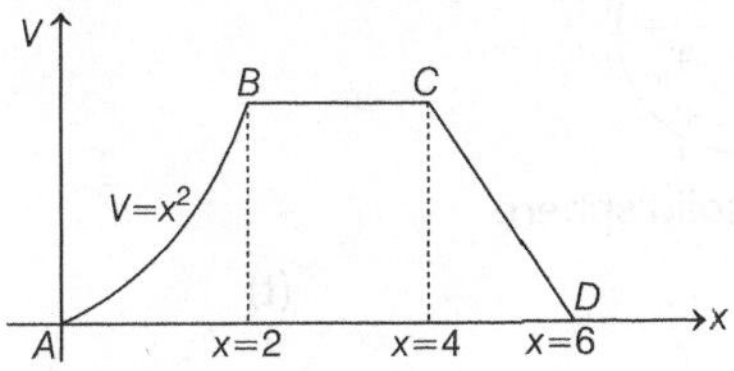

	Column-I		Column-II
(a)	Cube of 1 m edge with one corner at origin.	(p)	Positive
(b)	Cube of 1 m edge which lies between X = 3 to X = 4.	(q)	Negative
(c)	Cube of 2 m edge placed between the coordinate X = 4 to X = 6.	(r)	Zero
(d)	Cube of 6 m edge placed between the coordinate X = 0 to X = 6.	(s)	Magnitude of charge is $2\varepsilon_0$ units.

46 Column-I contains few charge distributions. Column-II contains graphs of electric field *E* vs *r* or electric potential *V* vs *r* for the charge distributions. Here, *r* is the distance from the geometrical mid-plane or geometrical centre of given body.
(Take, field towards right as positive and left as negative. Take, value of potential zero at $r = 0$.)

	Column-I		Column-II
(a)	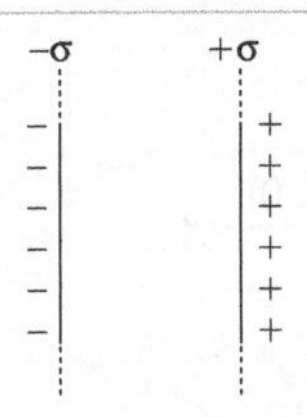 Two infinite charged sheets. Surface charge density σ is uniform.	(p)	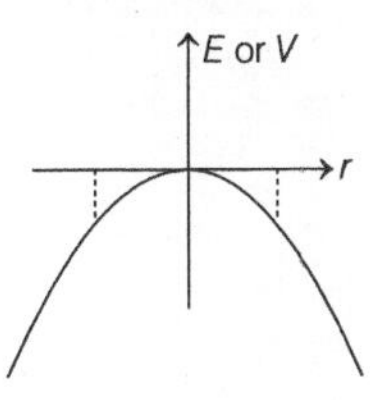
(b)	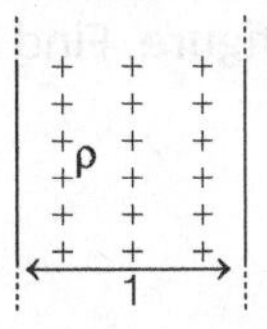 Large plate having some thickness. Volume charge density ρ is uniform.	(q)	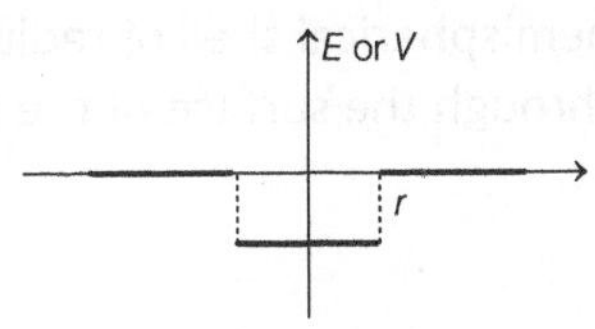
(c)	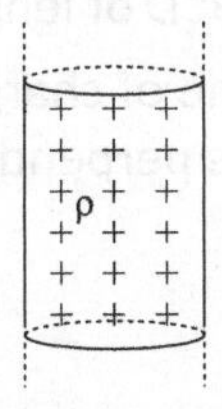 Long solid cylinder having uniform volume charge density ρ.	(r)	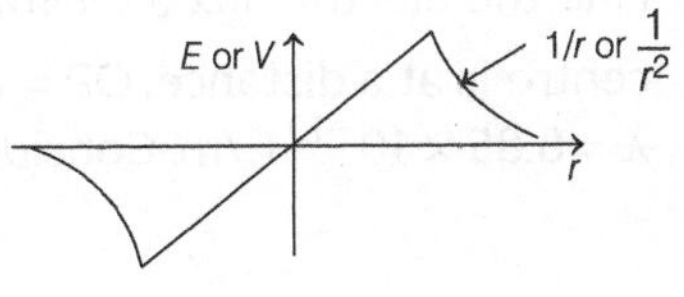

	Column-I		Column-II
(d)	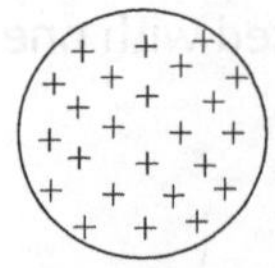 Uniformly charged solid sphere	(s)	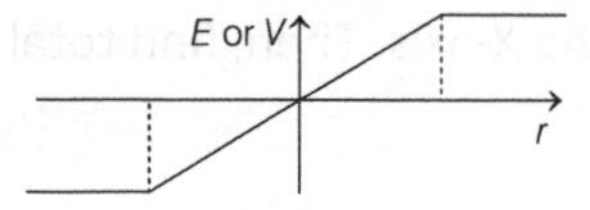
		(t)	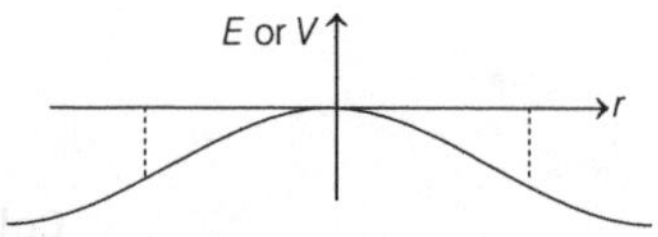

Numerical Value Questions

47 Find the electrostatic flux (in V-m) through a rectangular area of $\frac{7}{47}$m^2 lying in a plane $2x+3y+6z=10$ due to an electric field, $\mathbf{E}=(8\hat{\mathbf{i}}+6\hat{\mathbf{j}}+10\hat{\mathbf{k}})$ V/m.

48 A point charge is kept on the axis of circular non-conducting disc. The distance is, such that 1/10th of the total electric flux coming from the charge passes through the disc. By what distance (in cm) should the charge be moved along the axis, so that 1/5th of the total electric flux coming from the charge passes through the disc? The radius of the disc is 30 cm.

49 A point charge $q=8.85\times10^{-9}$ C is placed at (0, d, 0), where $d=1$ m. An insulating surface $ABCA$ is on the XZ-plane, which is bounded by a quarter circle AB and two equal length straight lines BC and CA as shown in the figure. Electric flux through the surface $ABCA$ in Nm^2/C is (nearest integer).

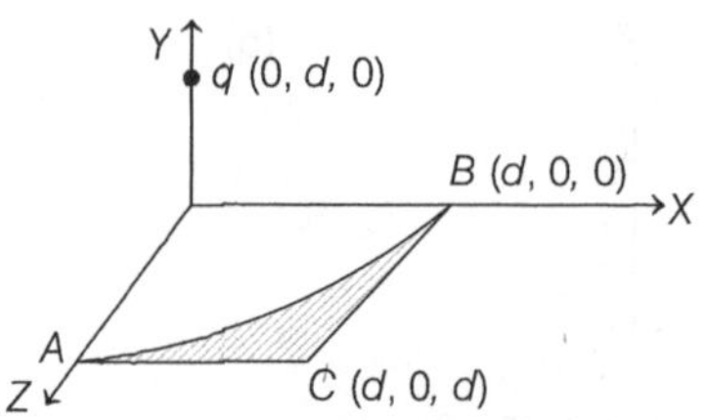

50 A dipole of dipole moment $p=17.7$nC-m , is kept at the centre, along the axis of the hemispherical shell of radius $R=50$ m as shown in the figure. Find the electric flux (in V-m) through the surface of the shell.

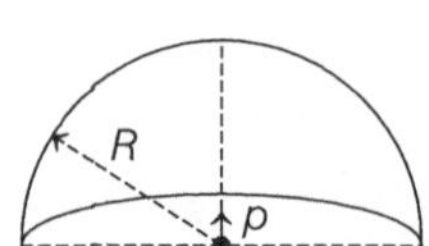

51 Find the electric flux (in V-m) crossing the wire frame $ABCD$ of length $l=1$ m, width b and whose centre is at a distance, $OP=d\,(=b/2)$ from an infinite line of charge with linear charge density, $\lambda=8.85\times10^{-10}$ C/m. Consider that the plane of frame is perpendicular to line OP.

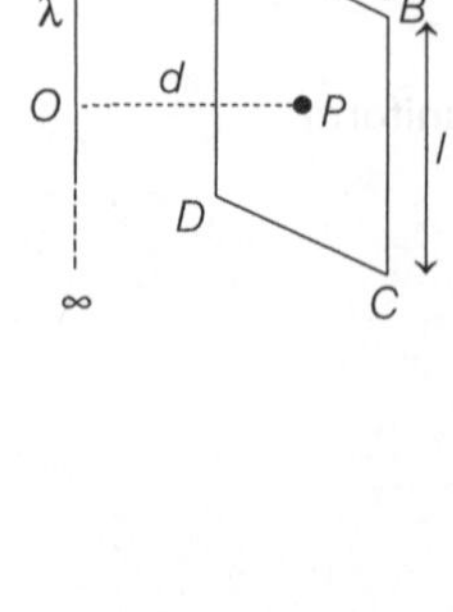

52 An infinitely long solid cylinder of radius R has a uniform charge density ρ. It has a spherical cavity of radius $R/2$ with its centre on the axis of the cylinder(as shown in figure). The magnitude of the electric field at point P, which is at a distance $2R$ from the axis of the cylinder, is given by $\frac{23\rho R}{6K\varepsilon_0}$.

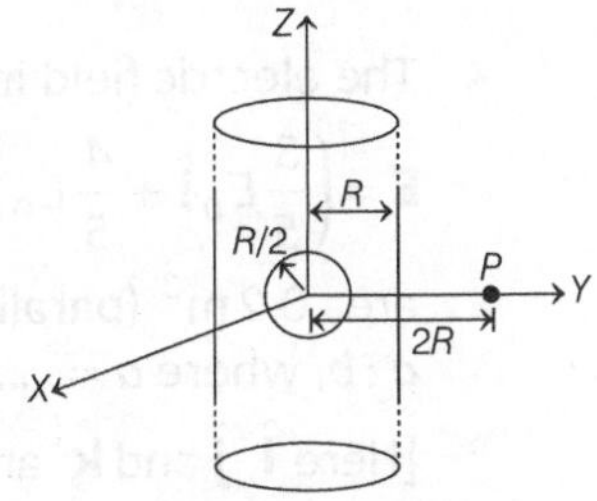

The value of K is ....

53 The visible portion of a lightning strike is preceded by an invisible stage in which a column of electrons extends downward from a cloud to the ground. Assume the, linear charge density along the column is 1.00×10^{-3} C/m. Treat the column of charge as, if it were straight and infinitely long. The dielectric strength for air is 3.00×10^{6} V/m? What will be the radius (in m) of a visible lightning bolt?

Answers

1. (c)	2. (c)	3. (d)	4. (a)	5. (d)	6. (c)	7. (b)	8. (a)	9. (c)	10. (b)
11. (a)	12. (a)	13. (b)	14. (c)	15. (d)	16. (a)	17. (d)	18. (c)	19. (c)	20. (abc)

21. (abcd) 22. (abc) 23. (ab) 24. (acd) 25. (abd) 26. (abd) 27. (abd) 28. (ad)

29. (b) 30. (a) 31. (a) 32. (c) 33. (d) 34. (c) 35. (b) 36. (d) 37. (d) 38. (a)

39. (d) 40. (a) 41. (a) 42. (b) 43. a → s; b → r; c → q; d → p

44. a → p,t; b → p,q,s; c → p,q,s; d → q,r,s 45. a → q,s; b → r; c → r; d → p

46. a → q,s; b → p,s; c → p,r; d → r,t

47. 2 48. 17.5 49. 5 50. 20 V-m

51. 25 52. 16 53. 6.00

For Detailed Solutions
Scan the code

Previous Years' Questions

JEE Main

1 The total charge enclosed in an incremental volume of 2×10^{-9} m^3 located at the origin is ...... nC, if electric flux density of its field is found as $D = e^{-x} \sin y \hat{\mathbf{i}} - e^{-x} \cos y \hat{\mathbf{j}} + 2z\hat{\mathbf{k}}$ C/m^2.
[JEE Main 2021]

2 The electric field in a region is given by $\mathbf{E} = \frac{2}{5}E_0\,\hat{\mathbf{i}} + \frac{3}{5}E_0\,\hat{\mathbf{j}}$ with $E_0 = 4.0 \times 10^3$ N/C. The flux of this field through a rectangular surface area 0.4 m^2 parallel to the YZ-plane is .... N-m^2 C^{-1}.
[JEE Main 2021]

3 A charge q is placed at one corner of a cube as shown in figure. The flux of electrostatic field **E** through the shaded area is
[JEE Main 2021]

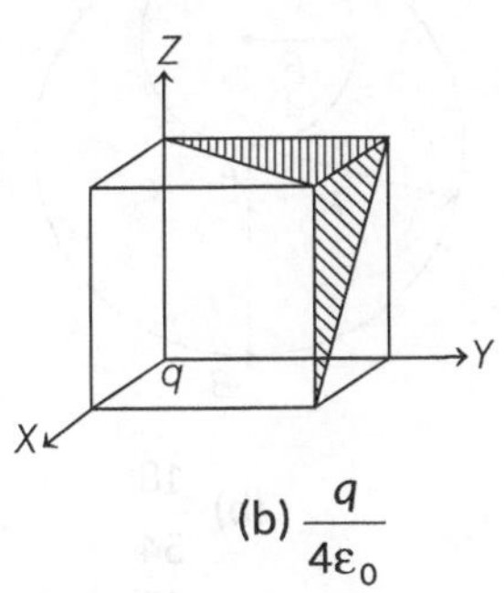

(a) $\frac{q}{48\varepsilon_0}$
(b) $\frac{q}{4\varepsilon_0}$
(c) $\frac{q}{8\varepsilon_0}$
(d) $\frac{q}{24\varepsilon_0}$

Problems in Physics

4 The electric field in a region is given $\mathbf{E}=\left(\frac{3}{5}E_0\hat{\mathbf{i}}+\frac{4}{5}E_0\hat{\mathbf{j}}\right)\frac{N}{C}$. The ratio of flux of reported field through the rectangular surface of area $0.2\,m^2$ (parallel to YZ-plane) to that of the surface of area $0.3\,m^2$ (parallel to XZ- plane) is $a:b$, where $a=$......... .

[Here $\hat{\mathbf{i}}$, $\hat{\mathbf{j}}$ and $\hat{\mathbf{k}}$ are unit vectors along X, Y and Z-axes, respectively] **[JEE Main 2021]**

5 A point charge of $+12\,\mu C$ is at a distance 6 cm vertically above the centre of a square of side 12 cm as shown in figure. The magnitude of the electric flux through the square will be .... $\times 10^3$ N-m^2/C. **[JEE Main 2021]**

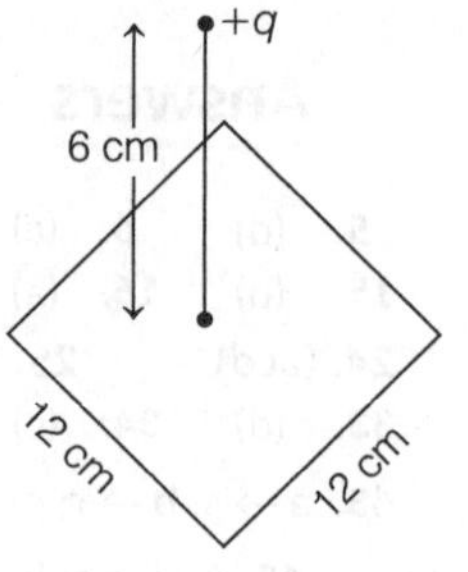

6 In finding the electric field using Gauss law, the formula $|\mathbf{E}|=\frac{q_{enc}}{\varepsilon_0|A|}$ is applicable. In the formula, ε_0 is permittivity of free space, A is the area of Gaussian surface and q_{enc} is charge enclosed by the Gaussian surface. This equation can be used in which of the following situation? **[JEE Main 2020]**

(a) Only when the Gaussian surface is an equipotential surface and $|E|$ is constant on the surface.
(b) Only when the Gaussian surface is an equipotential surface.
(c) For any choice of Gaussian surface.
(d) Only when $|E|=$ constant on the surface.

7 Consider a sphere of radius R which carries a uniform charge density ρ. If a sphere of radius $\frac{R}{2}$ is carved out of it, as shown in the figure the ratio $\frac{|\mathbf{E}_A|}{|\mathbf{E}_B|}$ of magnitude of electric field $\mathbf{E}_A$ and $\mathbf{E}_B$ respectively, at points A and B due to the remaining portion is **[JEE Main 2020]**

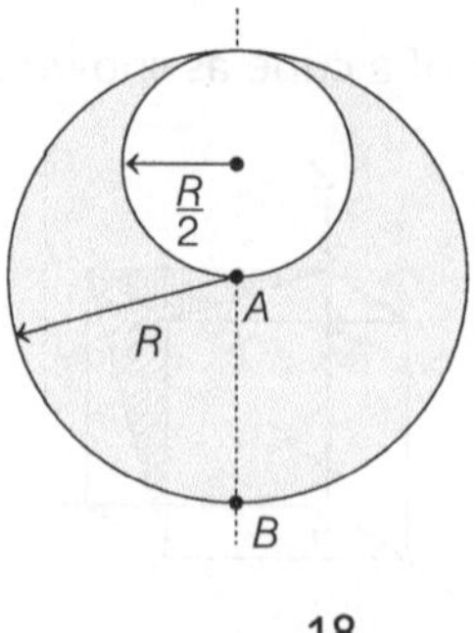

(a) $\frac{21}{34}$
(b) $\frac{18}{54}$
(c) $\frac{17}{54}$
(d) $\frac{18}{34}$

8 An electric field $\mathbf{E} = 4x\hat{\mathbf{i}} - (y^2 + 1)\hat{\mathbf{j}}$ N/C passes through the box shown in figure. The flux of the electric field through surfaces *ABCD* and *BCGF* are marked as ϕ_1 and ϕ_2, respectively. The difference between $(\phi_1 - \phi_2)$ is (in N-m^2/C) .... **[JEE Main 2020]**

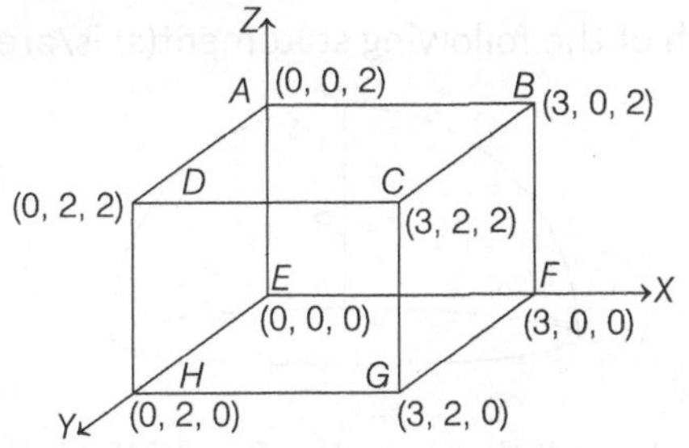

9 Let a total charge 2*Q* be distributed in a sphere of radius *R*, with the charge density given by $\rho(r) = kr$, where *r* is the distance from the centre. Two charges *A* and *B*, of $-Q$ each, are placed on diametrically opposite points, at equal distance *a*, from the centre. If *A* and *B* do not experience any force, then **[JEE Main 2019]**

(a) $a = 8^{-1/4}R$ (b) $a = \dfrac{3R}{2^{1/4}}$ (c) $a = 2^{-1/4}R$ (d) $a = R/\sqrt{3}$

10 The region between two concentric spheres of radii *a* and *b*, respectively (see the figure), has volume charge density $\rho = \dfrac{A}{r}$, where *A* is a constant and *r* is the distance from the centre. At the centre of the spheres is a point charge *Q*. The value of *A*, such that the electric field in the region between the spheres will be constant is **[JEE Main 2016]**

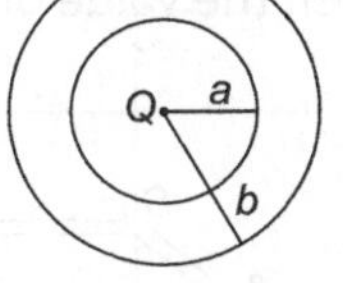

(a) $\dfrac{Q}{2\pi a^2}$ (b) $\dfrac{Q}{2\pi(b^2 - a^2)}$ (c) $\dfrac{2Q}{\pi(a^2 - b^2)}$ (d) $\dfrac{2Q}{\pi a^2}$

JEE Advanced

11 An infinitely long thin non-conducting wire is parallel to the *Z*-axis and carries a uniform line charge density λ. It pierces a thin non-conducting spherical shell of radius *R* in such a way that the arc *PQ* subtends an angle 120° at the centre *O* of the spherical shell, as shown in the figure. The permittivity of free space is ε_0. Which of the following statement(s) is/are true? **[JEE Advanced 2018]**

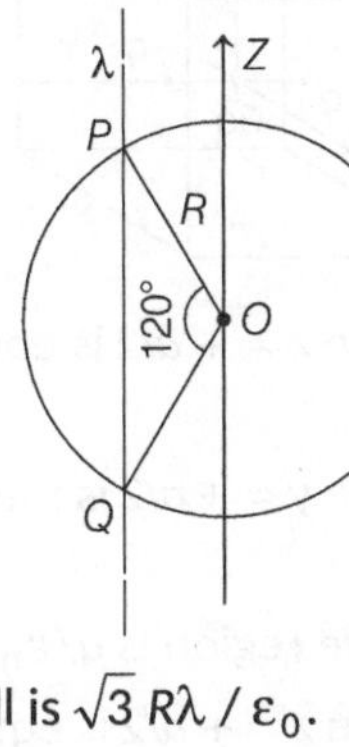

(a) The electric flux through the shell is $\sqrt{3}\,R\lambda/\varepsilon_0$.

(b) The *z*-component of the electric field is zero at all the points on the surface of the shell.

(c) The electric flux through the shell is $\sqrt{2}\,R\lambda / \varepsilon_0$.

(d) The electric field is normal to the surface of the shell at all points.

12 A point charge $+Q$ is placed just outside an imaginary hemispherical surface of radius R as shown in the figure. Which of the following statement(s) is/are correct? **[JEE Advanced 2017]**

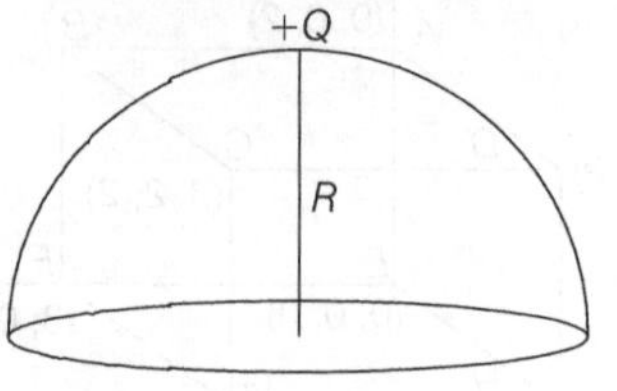

(a) The electric flux passing through the curved surface of the hemisphere is $-\dfrac{Q}{2\varepsilon_0}\left(1-\dfrac{1}{\sqrt{2}}\right)$.

(b) The component of the electric field normal to the flat surface is constant over the surface.

(c) Total flux through the curved and the flat surfaces, is $\dfrac{Q}{\varepsilon_0}$.

(d) The circumference of the flat surface is an equipotential.

13 An infinitely long uniform line charge distribution of charge per unit length λ lies parallel to the y-axis in the yz-plane at $z = \dfrac{\sqrt{3}}{2}a$ (see figure). If the magnitude of the flux of the electric field through the rectangular surface $ABCD$ lying in the xy-plane with its centre at the origin is $\dfrac{\lambda L}{n\varepsilon_0}$ (ε_0 = permittivity of free space), then the value of n is .... **[JEE Advanced 2015]**

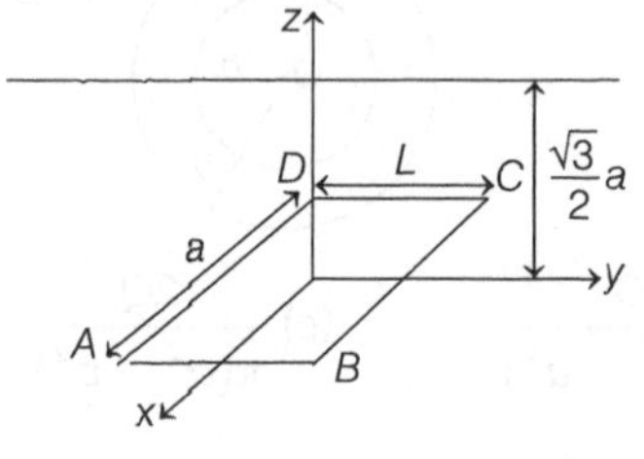

14 A cubical region of side a has its centre at the origin. It encloses three fixed point charges, $-q$ at $(0, -a/4, 0)$, $+3q$ at $(0, 0, 0)$ and $-q$ at $(0, +a/4, 0)$. Choose the correct option(s). **[JEE Advanced 2012]**

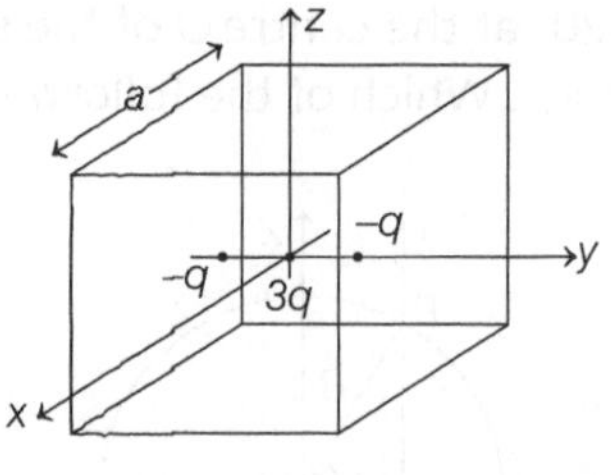

(a) The net electric flux crossing the plane $x = +a/2$ is equal to the net electric flux crossing the plane $x = -a/2$

(b) The net electric flux crossing the plane $y = +a/2$ is more than the net electric flux crossing the plane $y = -a/2$

(c) The net electric flux crossing the entire region is q/ε_0

(d) The net electric flux crossing the plane $z = +a/2$ is equal to the net electric flux crossing the plane $x = +a/2$

15 An infinitely long solid cylinder of radius R has a uniform volume charge density ρ. It has a spherical cavity of radius $\frac{R}{2}$ with its centre on the axis of the cylinder as shown in the figure. The magnitude of the electric field at the point P, which is at a distance $2R$ from the axis of the cylinder is given by the expression $\frac{23\rho R}{16k\varepsilon_0}$. The value of k is ....

[JEE Advanced 2012]

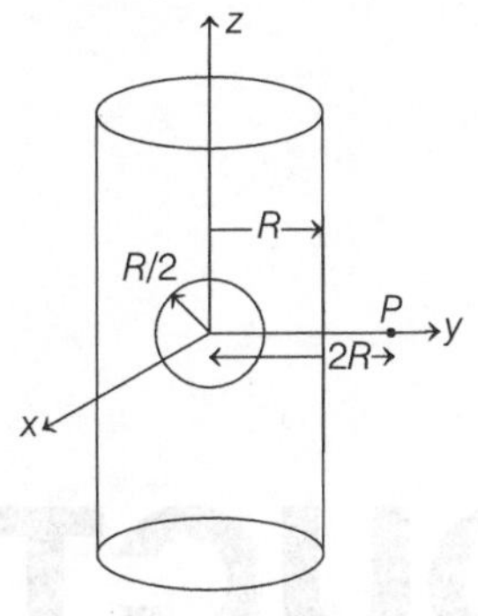

16 Consider an electric field $\mathbf{E} = E_0\,\hat{\mathbf{x}}$, where E_0 is a constant. The flux through the shaded area (as shown in the figure) due to this field is

[JEE Advanced 2011]

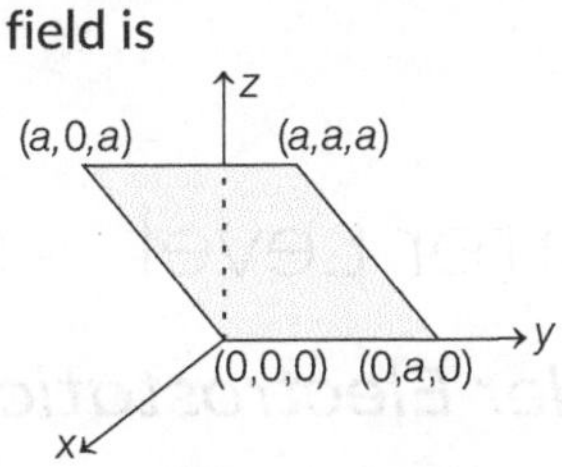

(a) $2E_0a^2$ (b) $\sqrt{2}E_0a^2$ (c) E_0a^2 (d) $\frac{E_0a^2}{\sqrt{2}}$

17 A disc of radius $\frac{a}{4}$ having a uniformly distributed charge $6C$ is placed in the xy-plane with its centre at $\left(\frac{-a}{2}, 0, 0\right)$. A rod of length a carrying a uniformly distributed charge $8C$ is placed on the x-axis from $x = \frac{a}{4}$ to $x = \frac{5a}{4}$. Two point charges $-7C$ and $3C$ are placed at $\left(\frac{a}{4}, \frac{-a}{4}, 0\right)$ and $\left(\frac{-3a}{4}, \frac{3a}{4}, 0\right)$, respectively. Consider a cubical surface formed by six surfaces $x = \pm\frac{a}{2}$, $y = \pm\frac{a}{2}$, $z = \pm\frac{a}{2}$. The electric flux through this cubical surface is

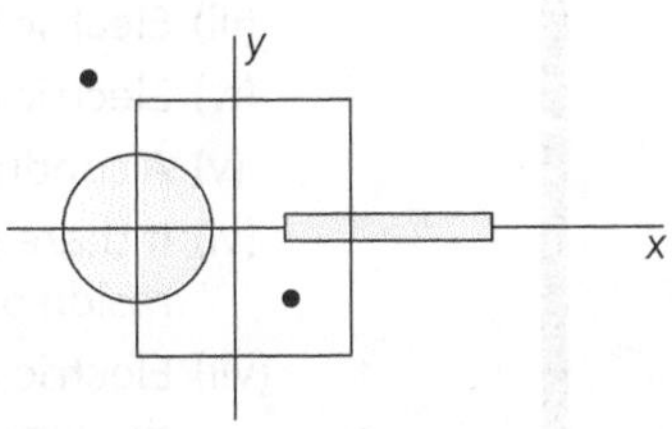

[JEE Advanced 2009]

(a) $\frac{-2C}{\varepsilon_0}$ (b) $\frac{2C}{\varepsilon_0}$ (c) $\frac{10\,C}{\varepsilon_0}$ (d) $\frac{12C}{\varepsilon_0}$

Answers

1. 4	**2.** 640	**3.** (d)	**4.** 1	**5.** 226
6. (a)	**7.** (d)	**8.** – 48	**9.** (a)	**10.** (a)
11. (a,b)	**12.** (a,d)	**13.** 6	**14.** (a,c)	**15.** 6
16. (c)	**17.** (a)			

For Detailed Solutions
Scan the code

CHAPTER

05

CONDUCTORS

LEVEL 01 \\ Starter Level

1.1 Conductors Under Electrostatic Conditions

1 State true or false.

(i) Any excess charge given to a conductor resides on its surface.

(ii) Electric field inside a conductor is zero.

(iii) Electric lines of force can enter into conductors.

(iv) Electric field lines are always perpendicular to surface of a conductor.

(v) A conductor is an equipotential body.

(vi) If there is a cavity inside the conductor having no charge, then charge will always reside only on the outer surface of conductor.

(vii) Electric field due to the charges inside the cavity and due to the induced charges on the inner surface of the cavity is zero at any point outside the cavity.

(viii) Electric field due to the charges outside the conductor and due to the induced charges on the outer surface of the conductor is zero at any point inside the conductor.

(ix) When any conductor is grounded, then its charge becomes zero.

(x) When two conductors are connected, there will be charge flow till their charges become equal.

2 Find induced charges on inner and outer surfaces of conductor in the following cases.

(i)

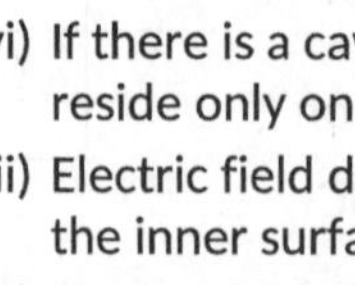

A neutral conductor with a cavity is given charge Q.

(ii)

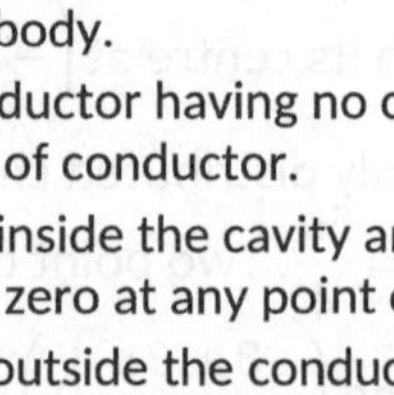

A neutral spherical conductor with a concentric spherical cavity has a point charge q at its centre C.

(iii)

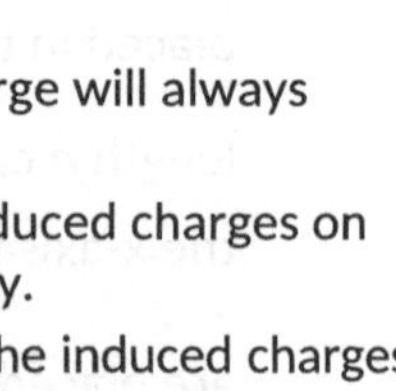

A spherical conductor with a concentric spherical cavity carries a charge Q. A point charge q is kept in the cavity OFF centre.

3 Find induced charges on all surfaces in the following cases. Surfaces are marked as 1, 2, 3, 4, 5, 6.

(i)

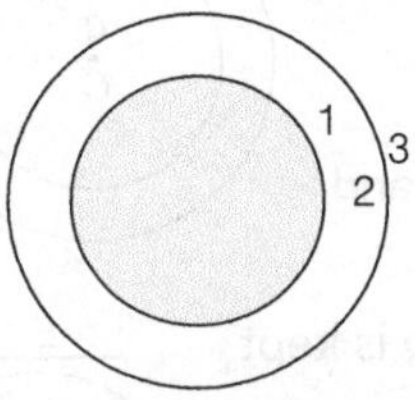

A solid metal sphere of charge $+20\,\mu C$ is surrounded by a thin metallic shell of charge $+10\,\mu C$.

(ii)

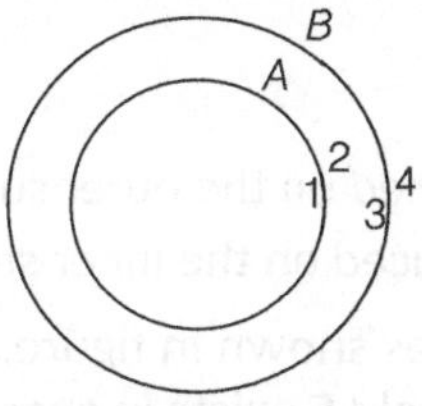

Two concentric thin metallic spherical shells A and B are given charges $+10\,\mu C$ and $-20\,\mu C$ respectively.

(iii)

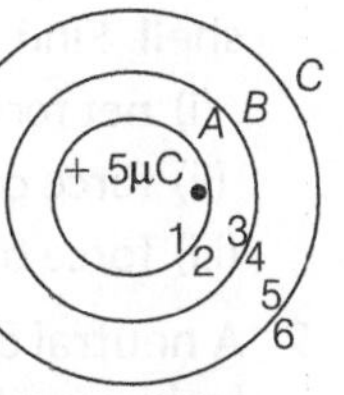

Three concentric thin metallic spherical shells A, B and C are given charges $+10\,\mu C$, 0 and $-20\,\mu C$, respectively. A point charge $+5\,\mu C$ is kept off centre as shown.

4 Find E and V at points A, B and C for each case.

(i)

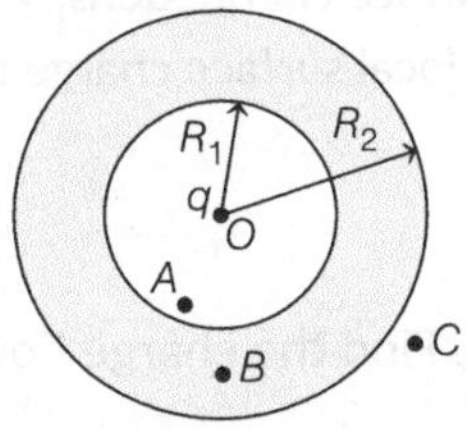

A neutral spherical conductor of inner radius R_1 and outer radius R_2 contains a point charge q placed at centre O.

(ii)

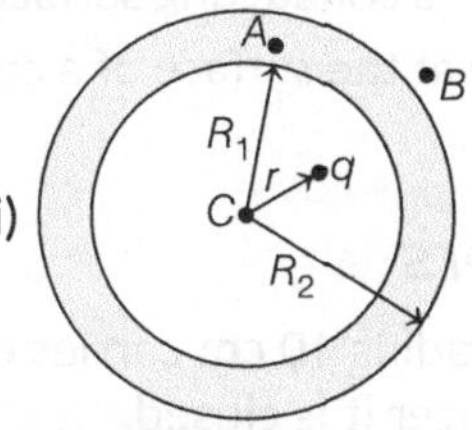

A neutral spherical conductor of radius R_2 with a concentric cavity of radius R_1 contains a point charge q placed at a distance r from its centre C.

(iii)

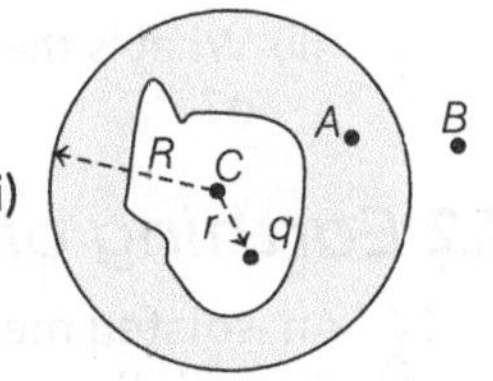

A neutral conductor of radius R with an irregular shaped cavity has a point charge q kept at a distance r from its centre C.

5 A spherical conductor of charge Q has an irregular shaped cavity as shown in figure. A point charge q is kept at the centre O of the conductor and another point charge q_0 is kept outside the conductor at point P. Find

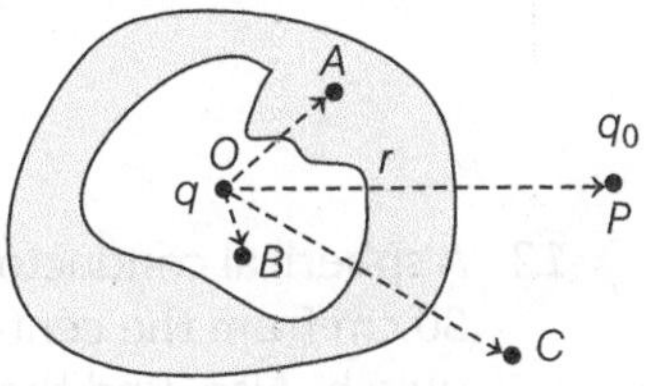

(i) induced charges on the inner and outer surfaces of the conductor.

(ii) electric field at points A, B and C due to the induced charges on the inner surface of cavity.

(iii) electric field at points A, B and C due to induced charges on the outer surface of the conductor.

6 An uncharged spherical conductor of radius R has a cubical shaped cavity as shown in figure. There is a point charge q outside the conductor at point P. Find

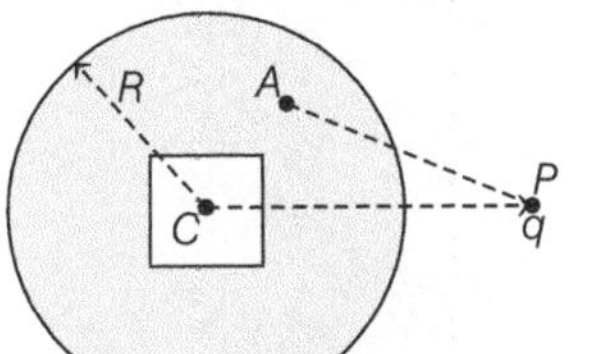

(i) electric potential at C,

(ii) electric potential at A and

(iii) electric potential at A due to induced charges on the conductor.

7 A neutral solid cubical metallic box has an irregular shaped cavity as shown in figure. A point charge q is kept at point A in the cavity. Find

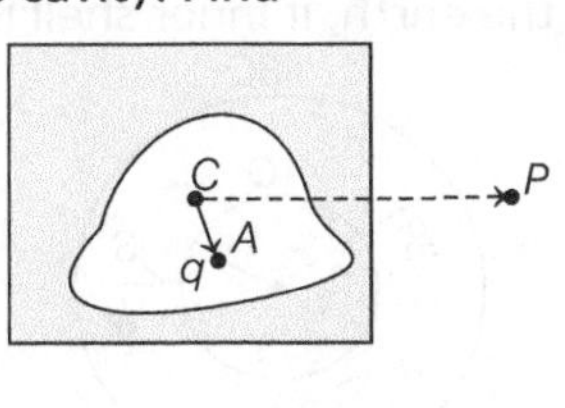

(i) potential at point P and

(ii) potential at point P due to induced charges on the cavity.

8 A point charge q is at the centre of a conducting spherical shell and another point charge q' is kept at distance r from centre outside the shell. Find

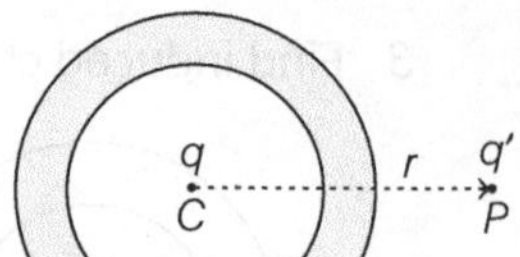

(i) net force acting on q,
(ii) force on q due to charges induced on the outer surface of the shell and
(iii) force on q' due to charges induced on the inner surface of the shell.

9 A neutral conductor has a cavity as shown in figure. A point charge q is kept inside cavity. A uniform electric field E exists in space. Find

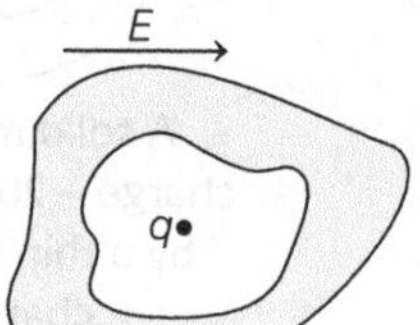

(i) net force acting on q,
(ii) net force acting on the conductor,
(iii) force on the conductor due to electric field,
(iv) net force on the induced charges on the outer surface of the conductor and
(v) net force on the induced charges on the inner surface of the cavity.

10 (i) What is the electric field near a conducting surface, if the local surface charge density is σ?
(ii) What is the electric pressure at the surface of a conductor, if the local surface charge density is σ?

1.2 Earthing of Conductors

11 An isolated metallic sphere of radius 10 cm carries charge $+50\ \mu C$. Find the charge flown through the switch to ground after it is closed.

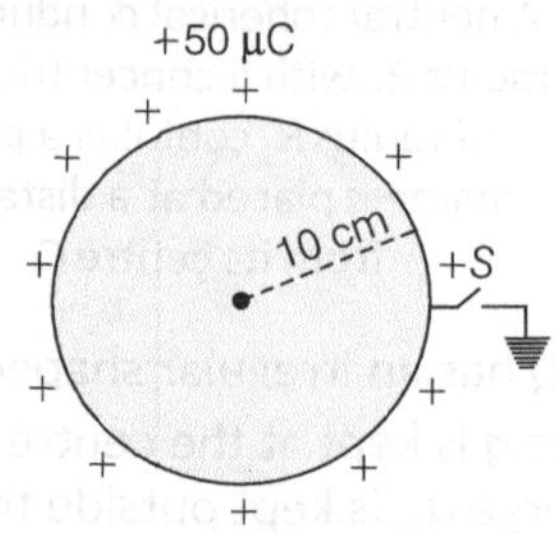

12 A spherical conductor of radius 10 cm carries charge $+50\ \mu C$. A point charge $+60\ \mu C$ is placed 30 cm from the centre of the conductor. Find the charge on the conductor after closing the switch. Also, find the charge flown to the ground.

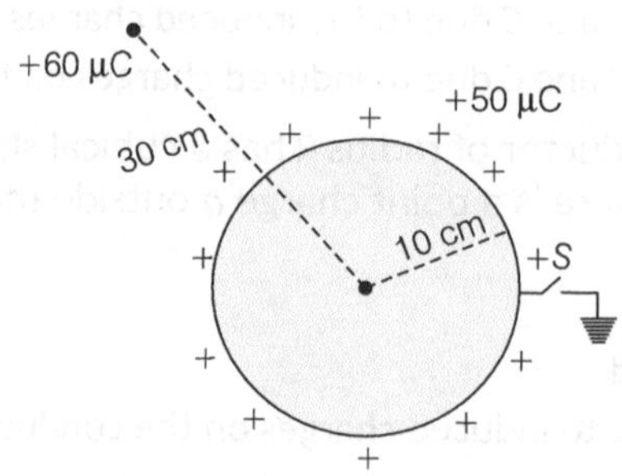

13 Two conducting thin spherical shells of radii R and $2R$ carry charges $-Q$ and $3Q$, respectively. How much charge will flow into the earth, if inner shell is grounded ?

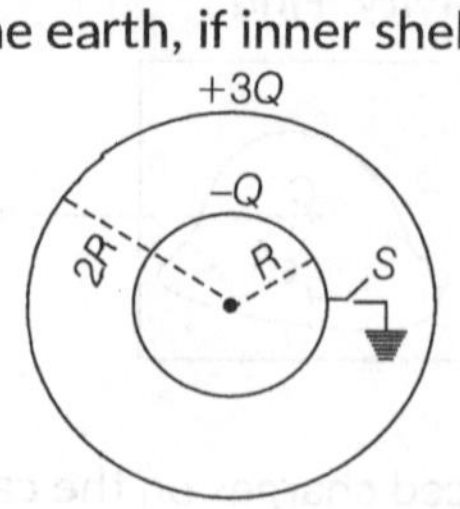

14 There are three concentric metallic spherical shells A, B and C of radius R, $2R$ and $3R$ with charges Q, $-3Q$, $4Q$, respectively as shown in figure. If switch is closed (i.e. its middle shell is earthed), then how much charge flows through switch to earth?

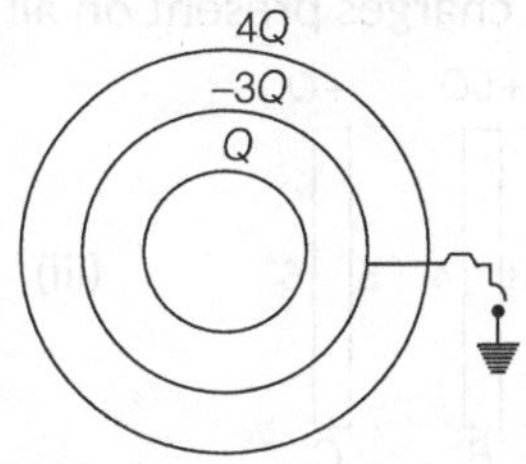

15 There are three concentric metallic thin spherical shells A, B and C with charges $+10\ \mu C$, $-20\ \mu C$, $+25\ \mu C$, respectively as shown in figure. Now, the outermost shell is earthed. How much charge flows through the switch to earth?

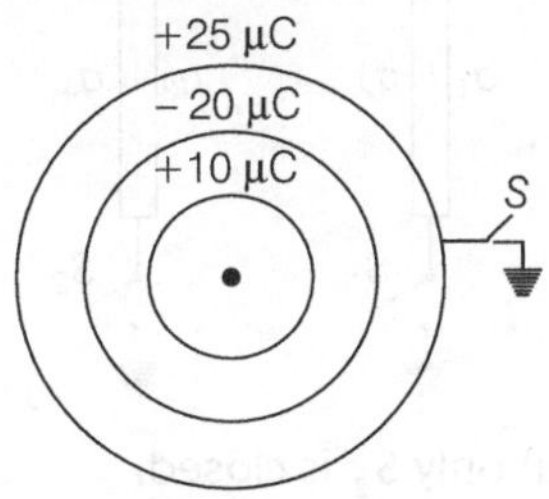

1.3 Connection of Conductors

16 The two conducting thin spherical shells are joined by a conducting wire and cut after sometime when charges stop flowing. Find out the charge on each sphere R after that. Also, find the heat produced in this process.

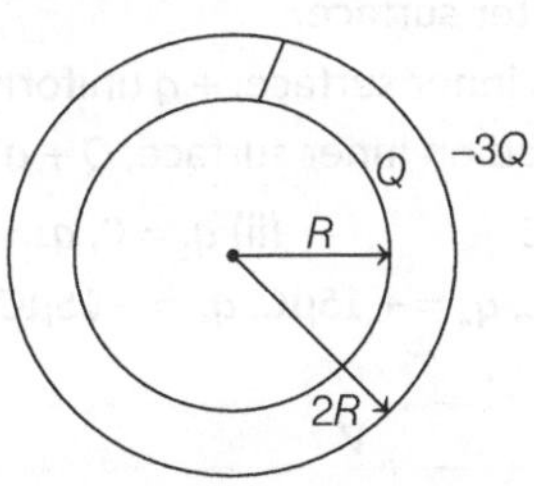

17 There are three concentric metallic thin spherical shells of radii R, $2R$ and $3R$ with charges Q, $-2Q$, $5Q$, respectively as shown in figure. Find charge on each spherical shell after joining the inner most shell and outer most shell by a conducting wire.

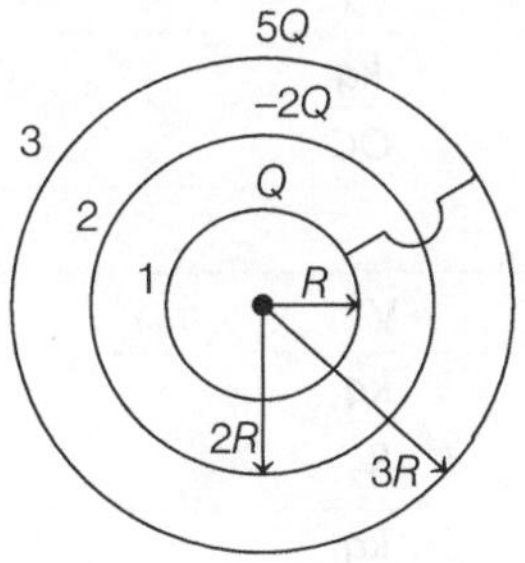

1.4 Parallel Conducting Plates

18 In each of the following cases, the parallel metallic plates kept at close separation are given charges as shown in figures. Find charges present on all surfaces marked as 1,2,3 for each case.

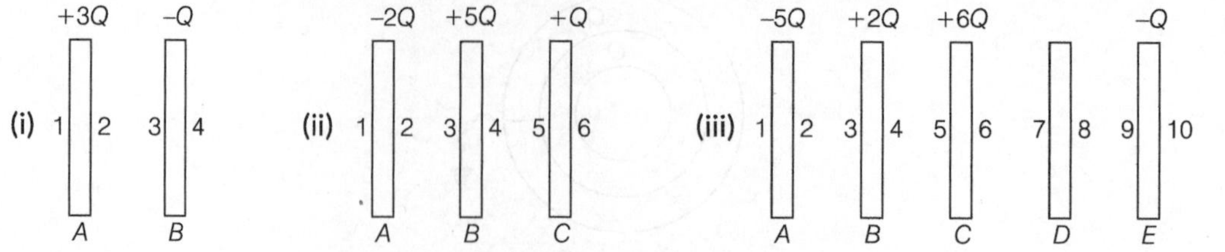

19 In figure, plate A has charge 100 μC and plate B has charge 60 μC. Find q_1, q_2, q_3 and q_4 when

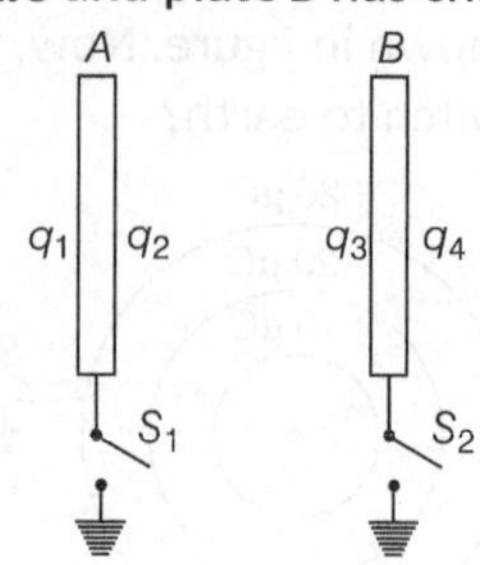

(i) only S_1 is closed, (ii) only S_2 is closed, (iii) both S_1 and S_2 are closed.

Answers

1. (i) T (ii) T (iii) F (iv) T (v) T (vi) T (vii) T
(viii) T (ix) F (x) F

2. (i) No charge on inner surface, Q on outer surface.
(ii) – q charge uniformly distributed on inner surface, + q uniformly distributed on outer surface.
(iii) – q charge non-uniformly distributed on inner surface, $Q+q$ uniformly distributed on outer surface.

3. (i) $q_1 = +20\mu C$, $q_2 = -20\mu C$, $q_3 = +30\mu C$ (ii) $q_1 = 0$, $q_2 = +10\mu C$, $q_3 = -10\mu C$, $q_4 = -10\mu C$
(iii) $q_1 = -5\mu C$, $q_2 = +15\mu C$, $q_3 = -15\mu C$, $q_4 = +15\mu C$, $q_5 = -15\mu C$, $q_6 = -5\mu C$

4. (i)

	E	V
A	$\frac{kq}{(OA)^3}$ OA	$\frac{kq}{OA} + \frac{k(-q)}{R_1} + \frac{kq}{R_2}$
B	0	$\frac{kq}{R_2}$
C	$\frac{kq}{(OC)^3}$ OC	$\frac{kq}{OC}$

(ii)

	E	V
A	0	$\frac{kq}{R_2}$
B	$\frac{kq}{(CB)^3}$ CB	$\frac{kq}{CB}$
C	Can't find using the knowledge within JEE syllabus	$\frac{kq}{r} + \frac{k(-q)}{R_1} + \frac{kq}{R_2}$

(iii)

	E	V
A	0	$\frac{kq}{R}$
B	$\frac{kq}{(CB)^3}$ **CB**	$\frac{kq}{CB}$
C	Can't find	Can't find

5. (i) $-q$ on the inner surface, $Q+q$ on the outer surface.

(ii) $\mathbf{E}_A = \frac{kq}{(OA)^2}$ towards q, $\mathbf{E}_B$ cannot be found, $\mathbf{E}_C = \frac{kq}{(OC)^2}$ towards q

(iii) $\mathbf{E}_A = \frac{kq_0}{(PA)^2}$ towards q_0, $\mathbf{E}_B = \frac{kq_0}{(PB)^2}$ towards q_0, $\mathbf{E}_C$ cannot be found

6. (i) $\frac{kq}{CP}$ (ii) $\frac{kq}{CP}$ (iii) $\frac{kq}{CP} - \frac{kq}{AP}$ **7.** (i) Can't find (ii) $-\frac{kq}{AP}$

8. (i) 0 (ii) $\frac{kqq'}{r^2}$, rightwards (iii) $\frac{kqq'}{r^2}$, leftwards

9. (i) Can't find (ii) Can't find (iii) 0 (iv) qE, rightwards (v) Cannot find

10. (i) $\frac{\sigma}{\varepsilon_0}$ (ii) $\frac{\sigma^2}{2\varepsilon_0}$ **11.** $+50\mu C$ **12.** $-20\mu C, 70\mu C$ **13.** $Q/2$ **14.** $2Q/3$

15. $+15\mu C$ **16.** Charge on the inner shell = 0 and outer shell = $-2Q$, heat produced = $\frac{kQ^2}{4R}$

17. Charge on innermost shell = $\frac{Q}{2}$, middle shell = $-2Q$, outermost shell = $\frac{5Q}{2}$

18. (i) $q_1 = Q, q_2 = 2Q, q_3 = -2Q, q_4 = Q$ (ii) $q_1 = 2Q, q_2 = -4Q, q_3 = 4Q, q_4 = Q, q_5 = -Q, q_6 = 2Q$

(iii) $q_1 = Q, q_2 = -6Q, q_3 = 6Q, q_4 = -4Q, q_5 = 4Q, q_6 = 2Q, q_7 = -2Q, q_8 = 2Q, q_9 = -2Q, q_{10} = Q$

19. (i) $q_1 = 0, q_2 = -60\,\mu C, q_3 = 60\,\mu C, q_4 = 0$

(ii) $q_1 = 0, q_2 = 100\,\mu C, q_3 = -100\,\mu C, q_4 = 0$

(iii) $q_1 = q_2 = q_3 = q_4 = 0$

For Detailed Solutions
Scan the code

LEVEL 02 \ JEE Main Level

Single Choice Correct Questions

1 Which of the following statements about solid conductors in electrostatics is/are true?

I. The electric field inside the conductor is always zero.

II. The electric potential inside the conductor is always zero.

III. Any net charge is on the surface.

(a) Only I (b) Only III (c) Both II and III (d) Both I and III

2 For a spherical shell,

(a) if potential inside it is zero, then it is necessarily electrically neutral

(b) electric field in a charged conducting spherical shell can be zero only when the charge is uniformly distributed

(c) electric potential due to induced charges at a point inside it will always be zero

(d) None of the above

3 Figure shows a closed surface which intersects a conducting sphere. If a positive charge is placed at the point P, then the flux of the electric field through the closed surface will

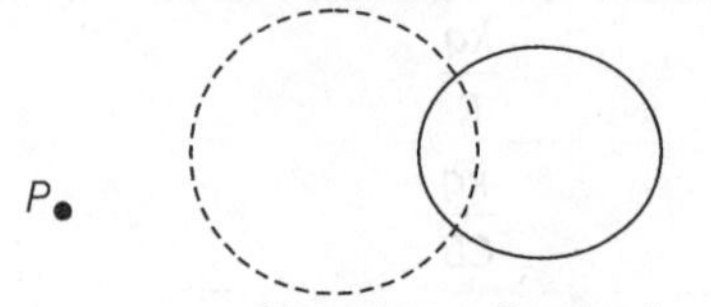

(a) remain zero (b) become positive (c) become negative (d) become undefined

4 An isolated positively charged spherical conducting shell has a small hole at its surface. The field at the centre of the hole is (σ is surface charge density)

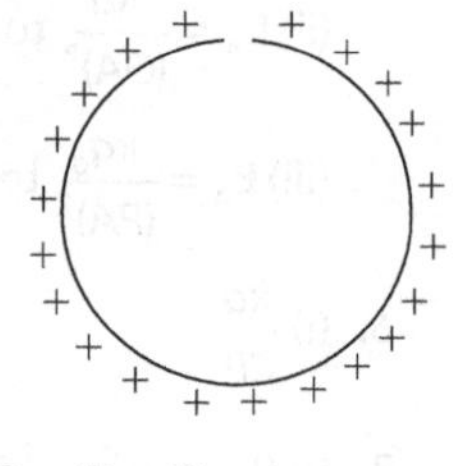

(a) $\frac{\sigma}{4\varepsilon_0}$ (b) $\frac{\sigma}{2\varepsilon_0}$

(c) $\frac{\sigma}{\varepsilon_0}$ (d) zero

5 Three concentric metallic spherical shells of radii R, $2R$, $3R$ are given charges Q_1, Q_2, Q_3, respectively. It is found that the surface charge densities on the outer surfaces of the shells are equal, then the ratio of the charges given to the shells Q_1, Q_2, Q_3 is

(a) 1 : 2 : 3 (b) 1 : 3 : 5 (c) 1 : 4 : 9 (d) 1 : 8 : 18

6 A solid metallic sphere has a charge $+3Q$. Concentric with this sphere is a conducting thin spherical shell having charge $-Q$. The radius of the sphere is a and that of the spherical shell is $b(>a)$. What is the electric field at a distance $r(a<r<b)$ from the centre?

(a) $\frac{Q}{2\pi\varepsilon_0 r^2}$ (b) $\frac{2Q}{4\pi\varepsilon_0 r^2}$ (c) $\frac{3Q}{4\pi\varepsilon_0 r^2}$ (d) None of these

7 A charge $+q$ is placed at a distance d from the centre of the uncharged metallic cube of side a. The electric field at the centre of the cube due to induced charges on the cube will be

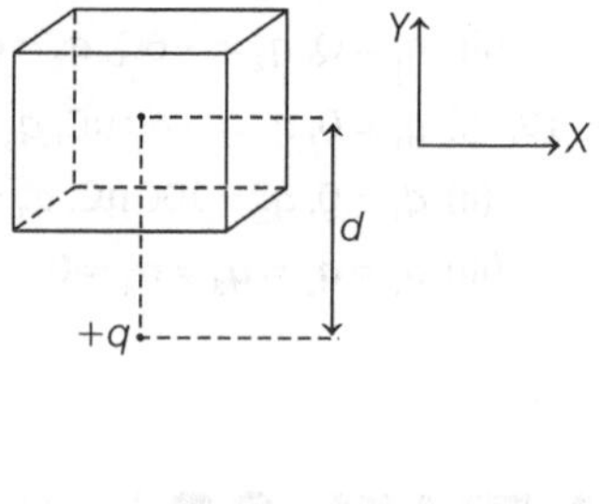

(a) zero (b) $\frac{q}{4\pi\varepsilon_0 d^2}(-\hat{\mathbf{j}})$

(c) $\frac{q}{4\pi\varepsilon_0 d^2}(\hat{\mathbf{j}})$ (d) $\frac{q}{4\pi\varepsilon_0\left(d-\frac{a}{2}\right)^2}(-\hat{\mathbf{j}})$

8 A spherical cavity of radius r is made in a conducting sphere of radius $2r$. A charge q is kept at the centre of cavity as shown in the figure. Find the magnitude of the total electric field at $(4r, 0, 0)$.

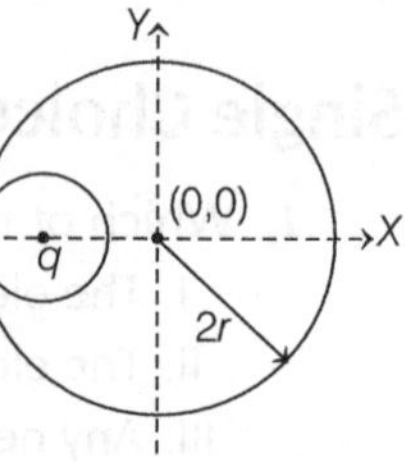

(a) $\frac{kq}{16r^2}$ (b) $\frac{4kq}{r^2}$

(c) $\frac{kq}{25r^2}$ (d) $\frac{4kq}{9r^2}$

9 A hollow spherical conductor has inner radius a and outer radius b. A positive point charge $+q$ is placed at the centre of the sphere. The conductor is uncharged. The potential at a distance r from centre $(a<r<b)$ is

(a) $V = kq\left[\frac{1}{r}+\frac{(b-a)}{ab}\right]$ (b) $V = \frac{kq}{r}$

(c) zero (d) $V = \frac{kq}{b}$

10 A point charge q is placed inside a neutral conducting spherical shell of inner radius $2R$ and outer radius $3R$ at a distance of R from the centre of the shell. The electric potential at the centre of shell will be $\frac{1}{4\pi\varepsilon_0}$ times, is

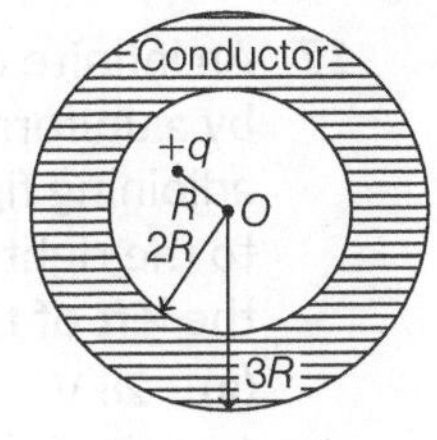

(a) $\frac{q}{2R}$
(b) $\frac{4q}{3R}$
(c) $\frac{5q}{6R}$
(d) $\frac{2q}{3R}$

11 Consider two concentric conducting spheres as shown in figure. The outer sphere is hollow and has a total charge of $+5\,\mu C$ charge on it; its inner radius is $R_1 = 9$ cm and its outer radius is $R_2 = 10$ cm. The inner sphere has a radius of 1 cm, is solid and has a charge $-3\mu C$ on it. What is the potential of the inner surface of the spherical shell?

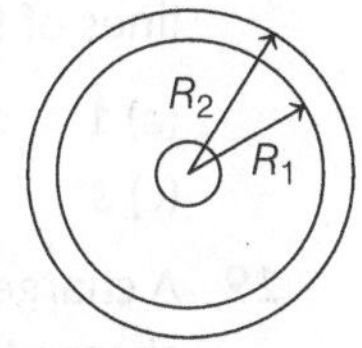

Assume that, the potential at infinite distance is zero.
(a) 180 kV (b) 300 kV (c) 720 kV (d) None of these

12 A point charge q is placed at distance a from the centre of an uncharged thin spherical conducting shell of radius $R = 2a$. A point P is located at a distance $4a$ from the centre of the conducting shell as shown in figure. The electric potential due to induced charge on the inner surface of the conducting shell at point P is

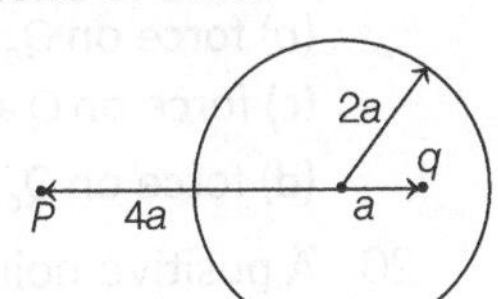

(a) $\frac{kq}{5a}$ (b) $-\frac{kq}{5a}$ (c) $\frac{kq}{4a}$ (d) $-\frac{kq}{4a}$

13 A small spherical conductor A of radius a is initially charged to a potential V and then placed inside an uncharged spherical conducting shell B of radius $6a$ as shown in figure. The potential of the spherical conductor A after closing the switch S is

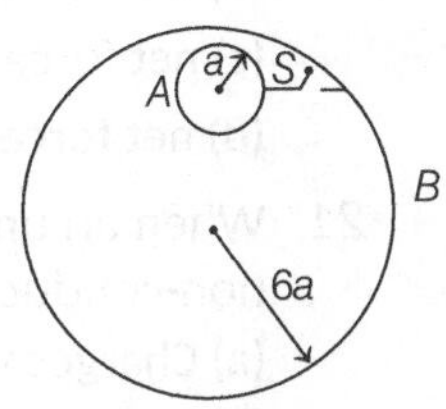

(a) V
(b) $V/3$
(c) $V/2$
(d) $V/6$

14 S is a solid neutral conducting sphere. A point charge $q = 1\times10^{-6}$ C is placed at point A. C is the centre of sphere and AB is a tangent $BC = 3$ m and $AB = 4$ m. The electric potential at B due to induced charge on the sphere is

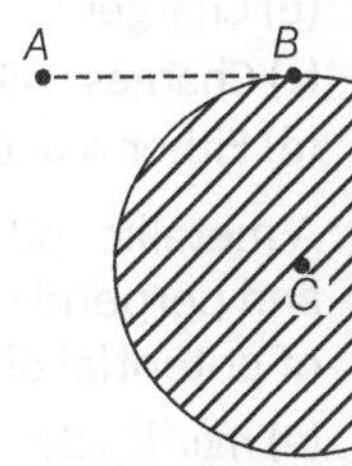

(a) 1.2 kV
(b) – 1.2 kV
(c) – 0.45 kV
(d) 0.45 kV

15 Two conducting, concentric, hollow spheres A and B have radii a and b respectively, with A inside B. They have the same potential V. A is now given some charge such that its potential becomes zero. The potential of B will now be
(a) 0
(b) $V(1 - a/b)$
(c) Va/b
(d) $V(b - a)/(b + a)$

16 A point charge q is placed at a distance $3R$ from the centre of a hollow thin earthed conducting thin shell of radius R as shown in the figure. What is the electric potential due to the charge on spherical shell at a point A at a distance $\frac{R}{2}$ from centre of shell ?

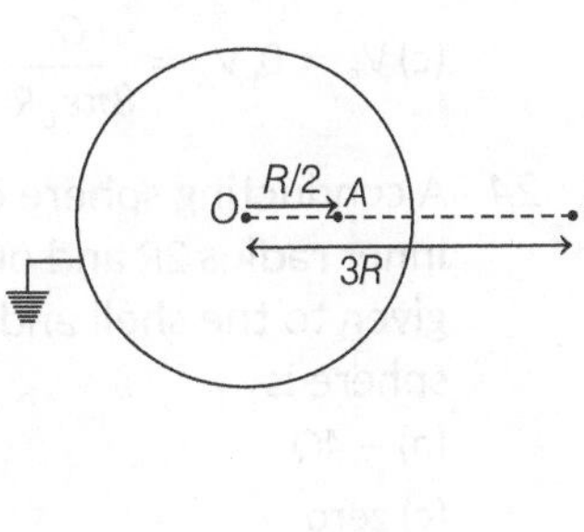

(a) $-\frac{1}{4\pi\varepsilon_0}\frac{q}{3R}$
(b) $-\frac{1}{4\pi\varepsilon_0}\frac{q}{2.5R}$
(c) Zero
(d) None of these

17 An infinite conducting plate of thickness 0.020 m is surrounded by a uniform field $E = 400$ V/m directed left to right. Refer adjoining figure. Let the potential $V_0 = 0$ at a distance 0.020 m to the right of the plate. What is V_3, if the potential 0.030 m to the left of the plate?

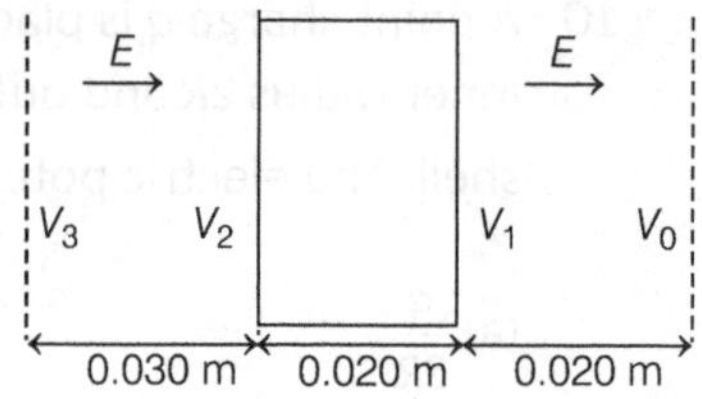

(a) – 28 V (b) – 20 V
(c) + 28 V (d) + 20 V

18 A metallic solid sphere is placed in a uniform electric field. The lines of force follow the path(s) shown in figure as

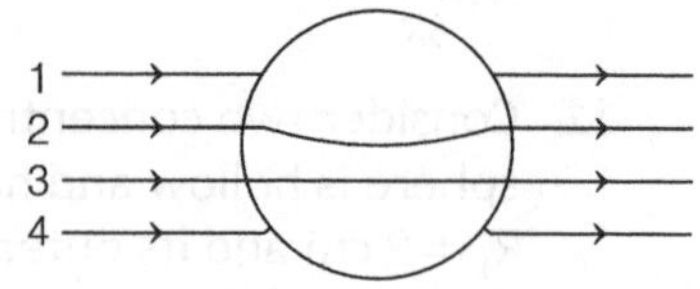

(a) 1 (b) 2
(c) 3 (d) 4

19 A charge Q_0 is placed at centre of spherical cavity of a conductor and another charge Q is placed outside the conductor. The correct statement is

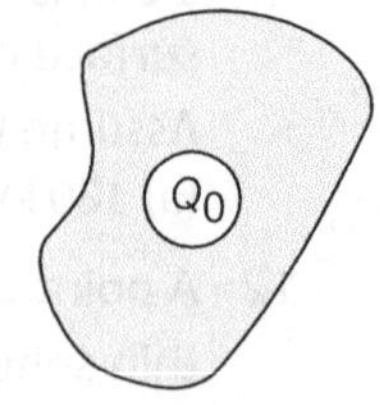

(a) force on Q_0 and Q are equal and opposite
(b) force on Q_0 is zero and force on Q is non-zero
(c) force on Q and Q_0 is zero
(d) force on Q_0 is non-zero and force on Q is zero

20 A positive point charge Q is kept (as shown in figure) inside a neutral conducting shell, whose centre is at C. An external uniform electric field E is applied, then

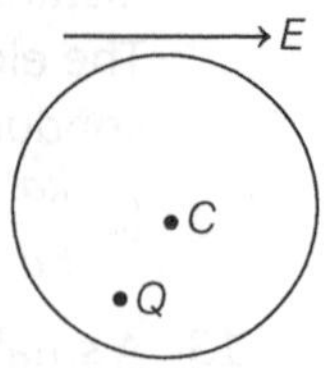

(a) force on Q due to E is zero
(b) net force on Q is zero
(c) net force acting on Q and conducting shell considered as a system is zero
(d) net force acting on the shell due to E is zero

21 When an uncharged cubical metal block is brought near an infinite uniformly charged non-conducting plane, choose the correct option.
(a) Charges will be induced on block and it will be attracted towards the plane
(b) Charges will be induced on block and it will be repelled from the plane
(c) Charges will be induced on block but it won't experience any force
(d) Either A or B depending on nature of metal

22 A metallic rod of length l rotates at angular velocity ω about an axis passing through one end and perpendicular to the rod . If mass of electron is m and its charge is $-e$, then the magnitude of potential difference between its two ends is
(a) $m\omega^2 l^2 /2e$ (b) $m\omega^2 l^2 / e$ (c) $m\omega^2 l / e$ (d) None of these

23 A and B are two concentric spheres of radii R and $2R$ respectively. A is given a charge $+Q$ and B is earthed. Which of the following is correct?

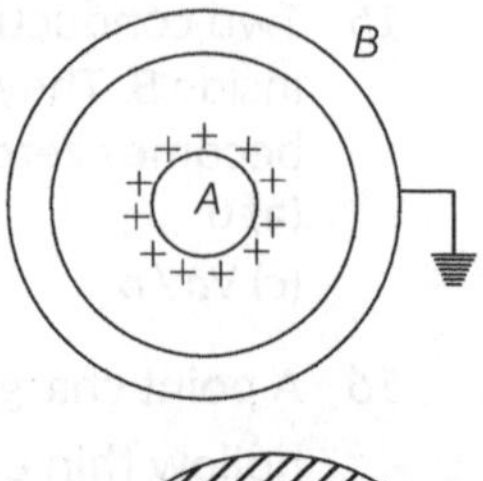

(a) $V_B = 0, V_A = \dfrac{Q}{4\pi\varepsilon_0 R}$ (b) $V_A = V_B = 0$

(c) $V_B = 0, V_A = \dfrac{Q}{8\pi\varepsilon_0 R}$ (d) $V_A = 0, V_B = \dfrac{Q}{2\pi\varepsilon_0 R}$

24 A conducting sphere of radius R and a concentric thick spherical shell of inner radius $2R$ and outer radius $3R$ is shown in figure. A charge $+10\,Q$ is given to the shell and inner sphere is earthed, then charge on inner sphere is

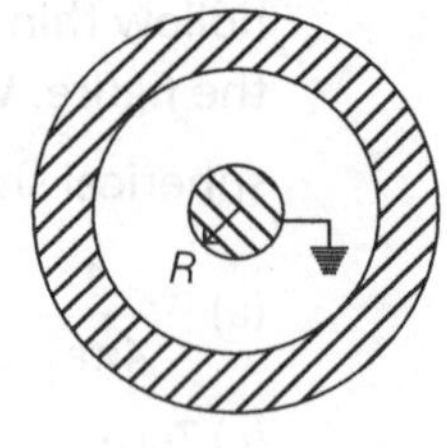

(a) – 4Q (b) – 10Q
(c) zero (d) None of these

25 There are two uncharged identical metallic spheres 1 and 2 of radius r separated by a distance $d(d >> r)$. A charged metallic sphere of same radius having charge q is touched with one of the spheres. After sometime, it is moved away from the system. Now, the uncharged sphere is earthed, charge on the earthed sphere is

(a) q (b) $-q$

(c) $-qr/2d$ (d) zero

26 A system of three thin, concentric spherical metallic shells have radii a, b, c, where $a:b:c = 1:2:5$. Initially, charges $Q_1 = 24\ \mu C$ and $Q_2 = 30\ \mu C$ are given to the innermost and outermost. How much charge passes through the switch S, when it is closed?

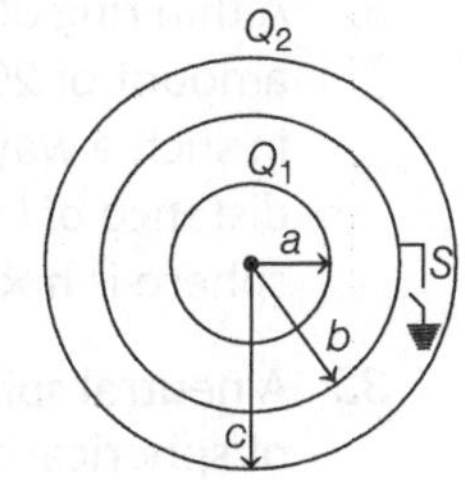

(a) 120 μC

(b) 60 μC

(c) 36 μC

(d) zero

27 Consider a thin conducting shell of radius R carrying total charge Q. Two point charges Q and $2Q$ are placed on points A and B, which are at a distances of $R/2$ and $2R$ from the centre of the cell respectively as shown in the figure. If the cell is earthed, how much charge will flow to the earth?

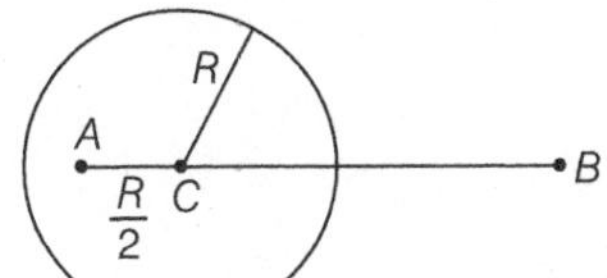

(a) $3Q$ (b) Q

(c) $2Q$ (d) $Q/2$

28 Three concentric spherical conducting shells A, B and C of radii a, $2a$ and $4a$ are initially given charges $+Q$, $-2Q$ and $+Q$, respectively. The charge on the middle spherical shell B after switches S_1 and S_2 are simultaneously closed, will be

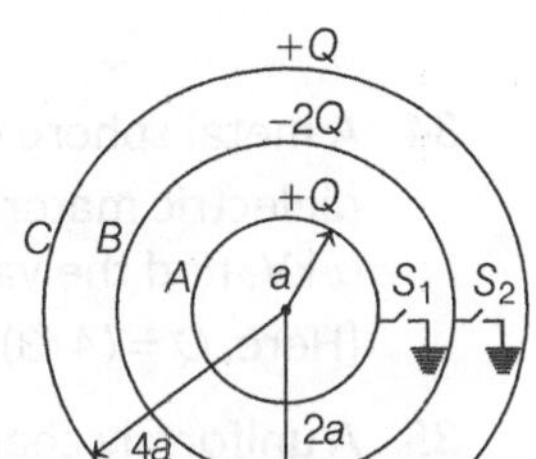

(a) $-Q$ (b) $-\frac{Q}{2}$

(c) $+Q$ (d) $+\frac{3Q}{2}$

29 In the figure shown below, the electric potential energy of the system is (q is at the centre of the conducting neutral spherical shell of inner radius a and outer radius b and ignore self-energy of point charge)

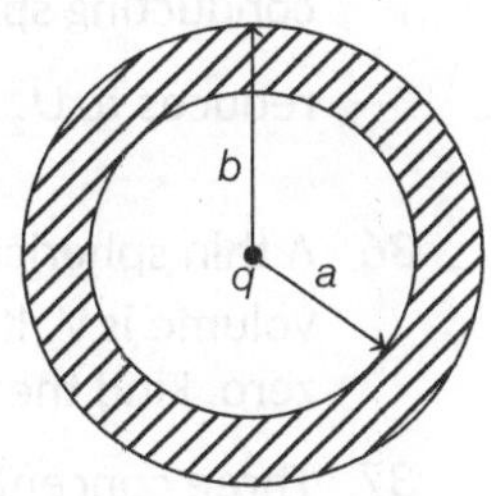

(a) zero (b) $\frac{kq^2}{2b}$

(c) $\frac{kq^2}{2b} - \frac{kq^2}{2a}$ (d) $\frac{kq^2}{2a} - \frac{kq^2}{2b}$

30 Two concentric spherical conducting shells of radii a and $2a$ are initially given charges $+2Q$ and $+Q$ respectively as shown in figure. The total heat dissipated after switch S has been closed is (Take, $k = \frac{1}{4\pi\varepsilon_0}$)

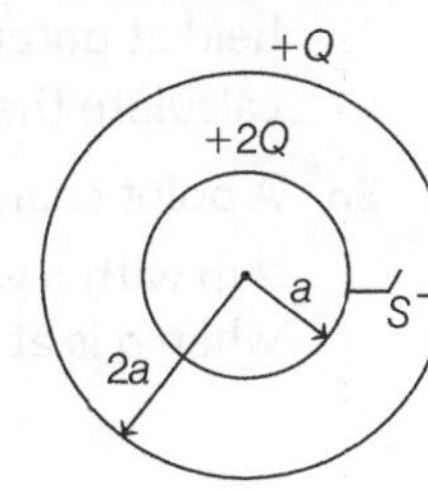

(a) $\frac{5kQ^2}{4a}$ (b) $\frac{kQ^2}{a}$

(c) $\frac{kQ^2}{2a}$ (d) $\frac{kQ^2}{4a}$

Numerical Value Questions

31 Mercury from a metal vessel maintained at a potential of 3V falls in spherical drops of 2 mm diameter through a small hole into a thin walled metal sphere of diameter 8 cm, isolated and placed in air until the sphere is filled with mercury. Calculate the final potential of the metal sphere (in volt).

32 A thin ring of radius $R = 3$m has been uniformly charged with an amount of 20 μC and placed near to an uncharged conducting sphere in such a way that the centre of the sphere O, lies on the rings axis at a distance of $l = 4$ m from the plane of the ring. The potential of the sphere is $n \times 10^4$ volt, find the value of n.

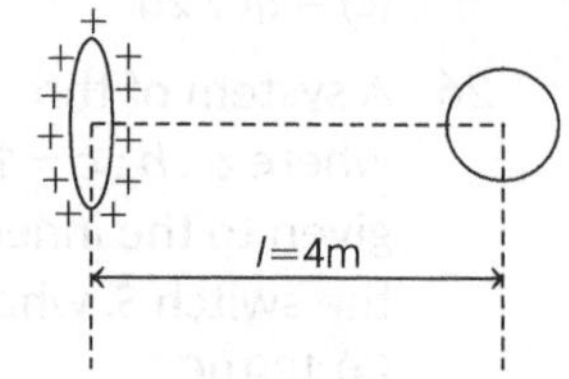

33 A neutral spherical conductor having cavity A and B of radius 1cm and 2cm, respectively. Radius of spherical conductor is 10 cm. If cavity A contains charge 2μC and B contains charge of -1 μC, electric potential at the centre of conductor is $k \times 10^4$ volt, then what is the value of k?

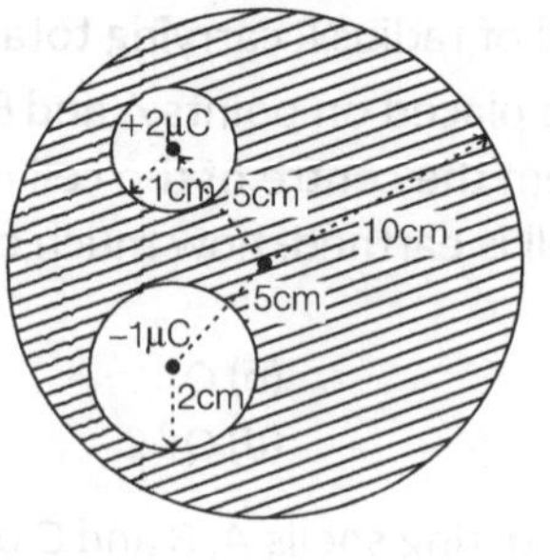

34 A metal sphere of radius a carries a charge Q. It is surrounded, out to radius b, by linear dielectric material of permittivity ε. Then, potential at the centre (relative to infinity) is given by α kV, find the value of α.
[Here, $Q = (4/3)$ μC, $a = 1$m, $b = 2$m and $\varepsilon = 17.7 \times 10^{-12}$ SI unit]

35 A uniformly charged non-conducting solid sphere has a radius R. Total energy of system (inside + outside) is U_1. If this sphere is now covered by a conducting spherical layer of thickness $4R$, the total energy of system reduces to U_2. Find the ratio of $\frac{U_1}{U_2}$.

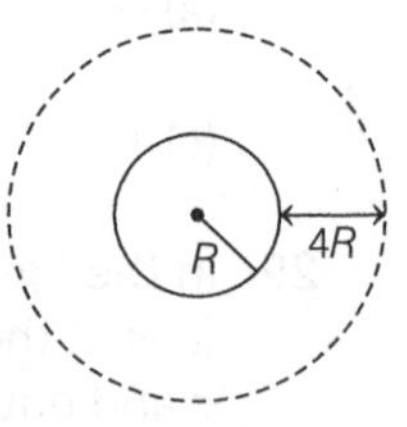

36 A thin spherical soap bubble has surface tension S. Its surface is charged with charge Q and its volume is V. It is found that when $Q^2 = n\pi\varepsilon_0 SV$, the excess pressure inside the bubble becomes zero. Find the value of n.

37 Three concentric, thin, spherical, metallic shells have radii 1 cm, 2 cm and 4 cm and they are held at potentials 10 V, 0 V and 40 V, respectively. Taking the origin at the common centre, calculate the potential (in volt) at $r = 1.25$ cm.

38 A point charge $q = 2$mC is moving towards the centre of an earthed conducting sphere of radius 2 m with a velocity 5 m/s. Find the current (in μA) flowing in the ammeter as shown in figure, when q is at a distance 5 m from the centre of sphere.

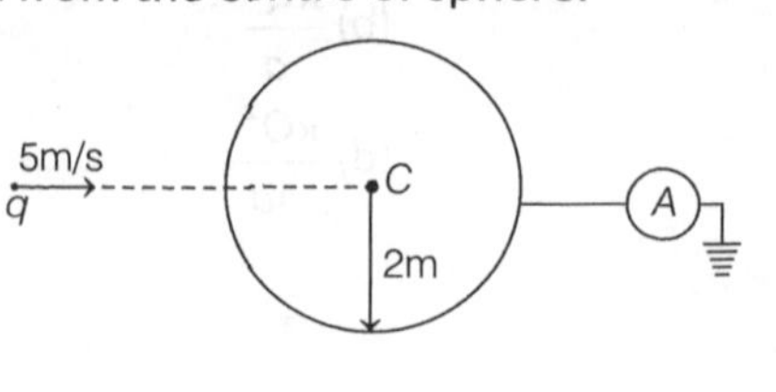

39 In the figure shown, a point charge 2Q is placed at the centre. The two concentric thin conducting spherical shells of radii R and $2R$ shown have charges $+Q$ and $-2Q$, respectively. Find the amount of heat generated (in joule) after the shell of radius R is earthed. (Take, $R=1$ cm and $Q=1\,\mu C$)

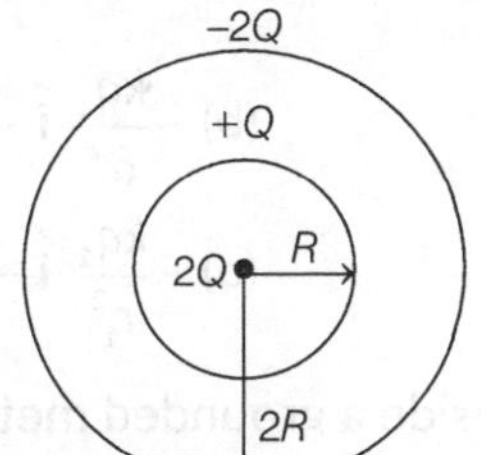

40 Three conducting concentric spherical shells of radii R, $2R$ and $3R$ carry equal charge densities. The innermost and the outermost spheres are now connected by a conducting wire. The ratio of potentials at the centre C before and after the connection is β. Find the value of 14β.

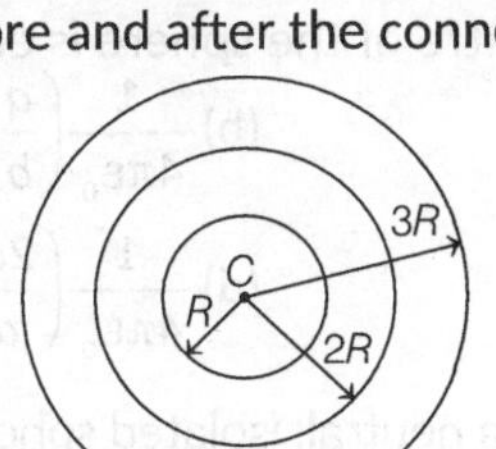

Answers

1. (d)	2. (d)	3. (c)	4. (b)	5. (b)	6. (c)	7. (b)	8. (a)	9. (d)	10. (c)
11. (a)	12. (b)	13. (d)	14. (c)	15. (b)	16. (b)	17. (d)	18. (d)	19. (b)	20. (d)
21. (c)	22. (a)	23. (c)	24. (a)	25. (c)	26. (c)				
27. (a)	28. (b)	29. (c)	30. (b)	31. 4800	32. 3.6				
33. 9	34. 9	35. 3	36. 96	37. 6	38. 800				
39. 1.8	40. 18								

For Detailed Solutions
Scan the code

LEVEL 03 \ JEE Advanced Level

Single Choice Correct Questions

1 Two concentric conducting spherical shells of radii a_1 and $a_2 (a_2 > a_1)$ are charged to potentials V_1 and V_2, respectively. The charge on the inner shell is

(a) $4\pi\varepsilon_0 \left(\frac{V_1 - V_2}{a_2 - a_1}\right) a_1 a_2$

(b) $4\pi\varepsilon_0 \left(\frac{V_1 - V_2}{a_2 + a_1}\right) a_1 a_2$

(c) $4\pi\varepsilon_0 \left(\frac{V_1 + V_2}{a_2 - a_1}\right) a_1 a_2$

(d) $4\pi\varepsilon_0 \left(\frac{V_1 + V_2}{a_2 + a_1}\right) a_1 a_2$

2 An ideal dipole of dipole moment **p** is placed infront of an uncharged conducting sphere of radius R as shown in figure. Potential at A due to the induced charges on the sphere is

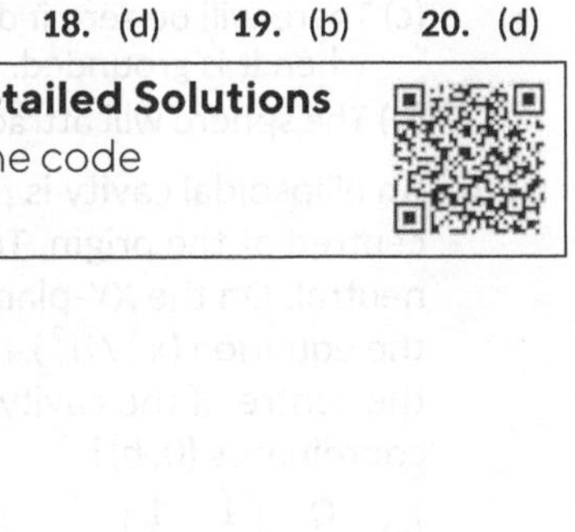

(a) $\frac{kp}{(r-R)^2}$

(b) $\frac{kp}{r^2} + \frac{kp}{(r-R)^2}$

(c) $\frac{kp}{r^2} - \frac{kp}{(r-R)^2}$

(d) $\frac{kp}{r^2}$

3 Two charges q_1 and q_2 are placed inside spherical cavities made inside uncharged conducting sphere of radius R as shown in figure. Electric field at P due to all the induced charges on the sphere is

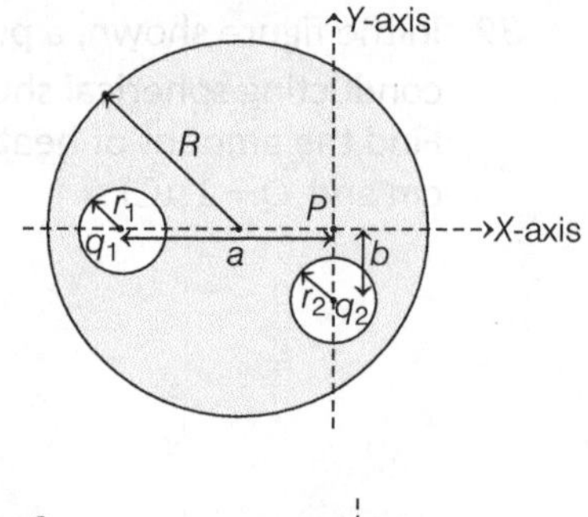

(a) $\frac{kq_1}{a^2}\hat{i}+\frac{kq_2}{b^2}\hat{j}$ (b) $-\frac{kq_1}{a^2}\hat{i}-\frac{kq_2}{b^2}\hat{j}$

(c) $\frac{kq_1}{r_1^2}\hat{i}+\frac{kq_2}{r_2^2}\hat{j}$ (d) $-\frac{kq_1}{r_1^2}\hat{i}-\frac{kq_2}{r_2^2}\hat{j}$

4 A charged particle $+q$ is placed inside a grounded metallic shell of inner and outer radii a and b respectively as shown in the figure. The particle is at a distance of $\left(\frac{a}{2}\right)$ from the centre of the shell, then the magnitude of potential at the centre of the sphere is equal to

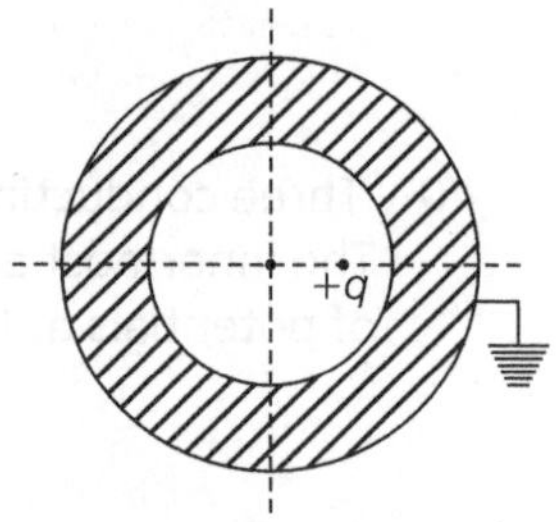

(a) zero (b) $\frac{1}{4\pi\varepsilon_0}\left(\frac{q}{b}+\frac{q}{a}\right)$

(c) $\frac{1}{4\pi\varepsilon_0}\left(\frac{q}{a}\right)$ (d) $\frac{1}{4\pi\varepsilon_0}\left(\frac{2q}{a}+\frac{q}{b}\right)$

5 A charged particle is brought near a neutral, isolated spherical conductor. Now, the switch S connecting the conductor to ground is closed. Which of the following statement(s) is/are incorrect?

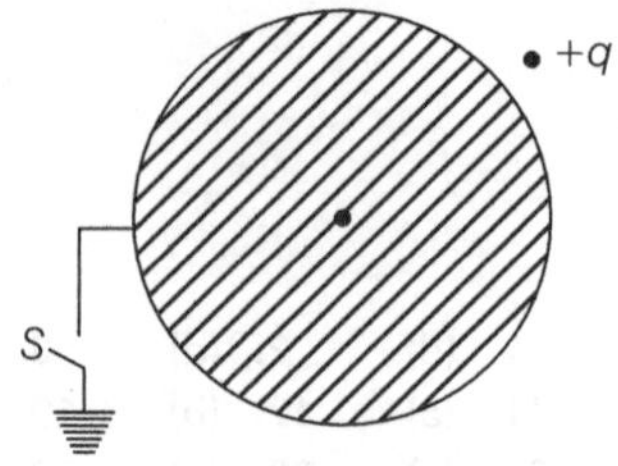

(a) The sphere draws negative charge, when grounded.
(b) The sphere's potential reduces to zero, when it is grounded.
(c) There will be zero induced charges on the surface of the sphere, when it is grounded.
(d) The sphere will attract the charged particle.

6 An ellipsoidal cavity is present inside a solid conducting sphere centred at the origin. The radius of the sphere is R and it is electrically neutral. On the XY-plane, the equation of the cavity can be given by the equation $(x^2/a^2)+(y^2/b^2)=1$. If a point charge $+Q$ is placed at the centre of the cavity, then what is the potential at point A, having coordinates $(0, b)$?

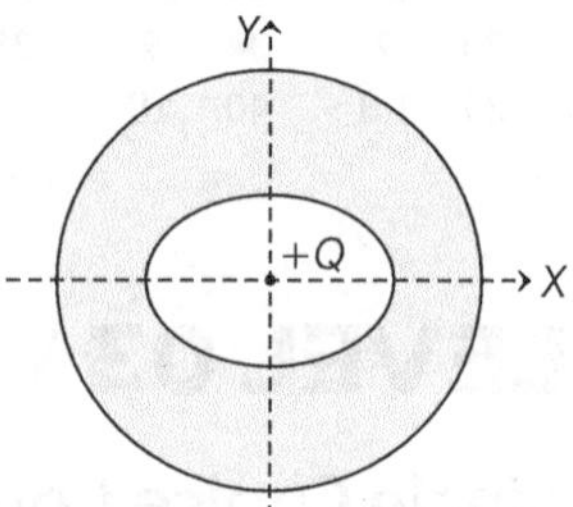

(a) $\frac{Q}{4\pi\varepsilon_0}\left(\frac{1}{b}-\frac{1}{R}\right)$ (b) $\frac{Q}{4\pi\varepsilon_0}\left(\frac{1}{b}+\frac{1}{R}\right)$

(c) $\frac{Q}{4\pi\varepsilon_0}\left(\frac{1}{a+b}-\frac{1}{R}\right)$ (d) $\frac{Q}{4\pi\varepsilon_0 R}$

7 A non-conducting sphere with radius a is concentric with and surrounded by a conducting spherical shell with inner radius b and outer radius c. The inner sphere has a negative charge uniformly distributed throughout its volume, while the spherical shell has no net charge. The potential $V(r)$ as a function of distance from the centre is given by

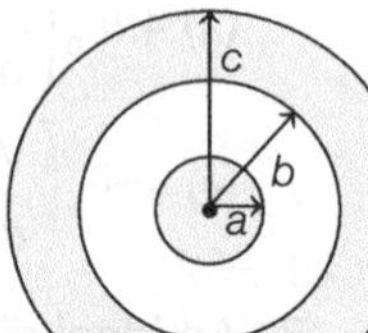

(a) 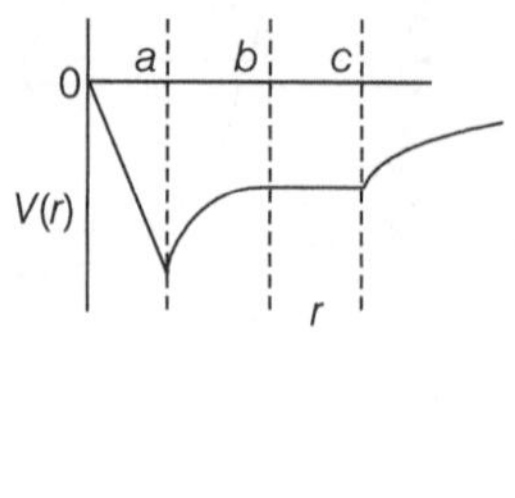

(b) 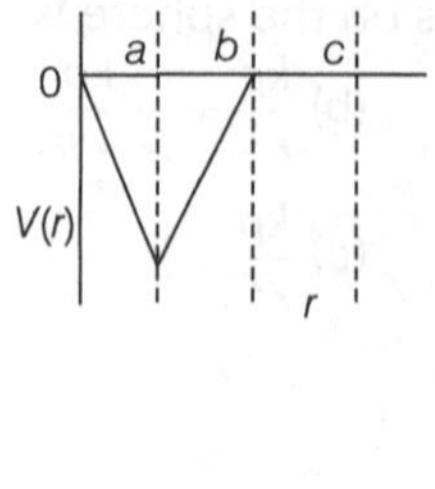

(c) 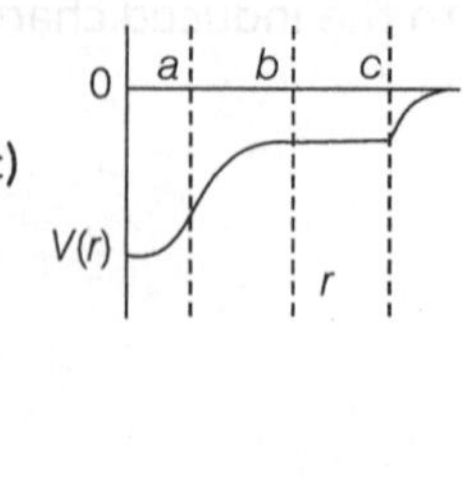

(d) 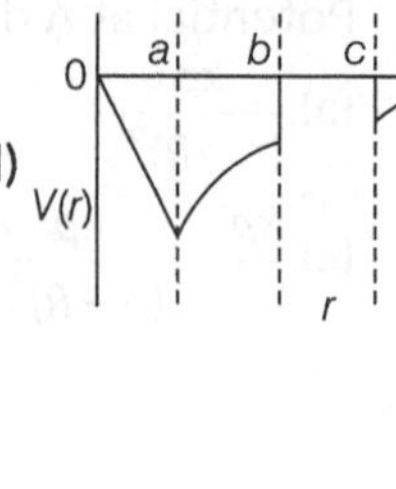

8 A charge q_1 is placed at O, inside a hollow conducting sphere having inner and outer radii as 10m and 11m as shown in figure. The force experienced by charge q_2 placed at point P is F_1. Instead of O, if q_1 is placed at O, force experience by q_2 is F_2 ($OO' = 5$ m, $OP = 12$ m). Then, F_1/F_2 is

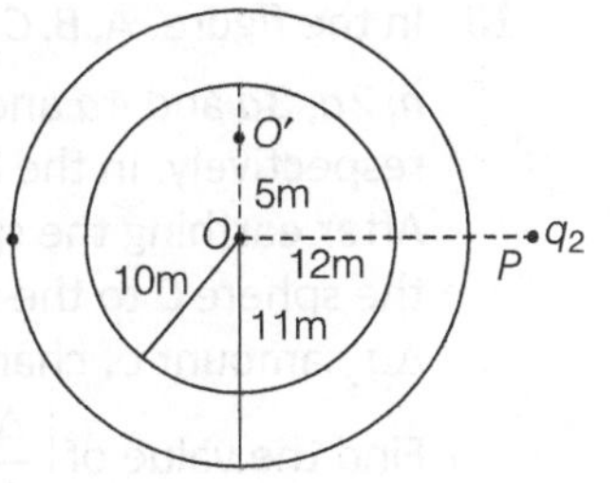

(a) 1
(b) $\left(\frac{12}{13}\right)^2$
(c) $\left(\frac{13}{12}\right)^2$
(d) $\left(\frac{4}{3}\right)^2$

9 There are two conducting concentric spherical thin shells of radius $\frac{3R}{4}$ and R. Thin shell of radius R is given a charge $4Q$. A point charge $-Q$ is also placed in the cavity at a distance $R/2$ from the centre as shown in figure. The electric potential at a distance $2R$ from the centre of concentric spherical shells along the line joining point charge is

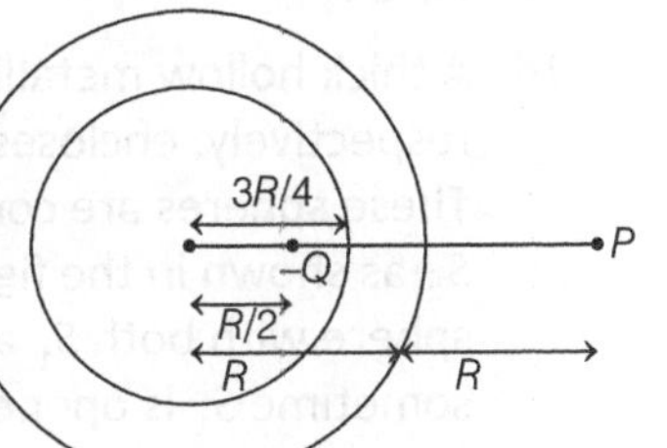

(a) $\frac{1}{8\pi\varepsilon_0}\frac{5Q}{R}$
(b) $-\frac{1}{8\pi\varepsilon_0}\frac{5Q}{R}$
(c) $\frac{1}{8\pi\varepsilon_0}\frac{3Q}{R}$
(d) $\frac{1}{2\pi\varepsilon_0}\frac{Q}{3R}$

10 Figure shows a large conducting sheet having charge density σ on its both surfaces. The surface potential of the sheet is V_0 with respect to earth. A conducting sphere of radius r is placed at a separation $l (l >> r)$ from the sheet as shown in figure. Find the charge that will appear on the sphere after it is connected to the earth.

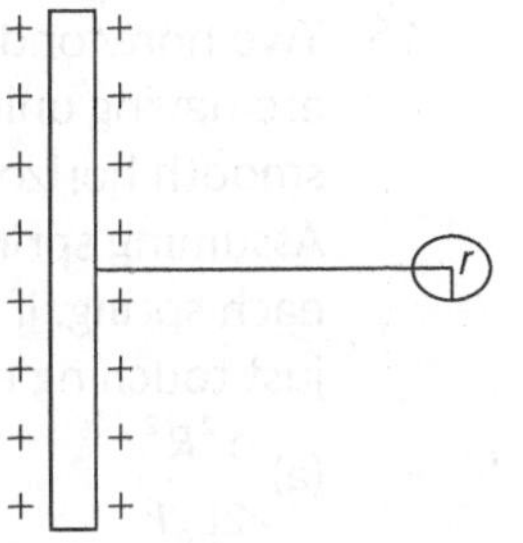

(a) $4\pi r(\sigma l - V_0\varepsilon_0)$
(b) $4\pi r(\sigma l + V_0\varepsilon_0)$
(c) $\pi r(\sigma l - V_0\varepsilon_0)$
(d) Zero

11 A conducting sphere of radius R and charge Q is placed near a uniformly charged non-conducting infinitely large thin plate having surface charge density σ. Then, find the potential at point A (on the surface of sphere) due to charge on sphere.

(Take, $k = \frac{1}{4\pi\varepsilon_0}$ and $\theta_0 = \frac{\pi}{3}$)

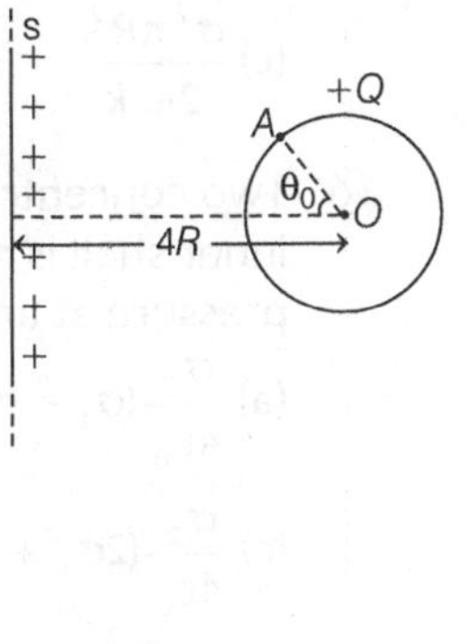

(a) $k\frac{Q}{R} - \frac{\sigma}{4\varepsilon_0}R$
(b) $k\frac{Q}{R} - \frac{\sigma R}{\varepsilon_0}$
(c) $k\frac{Q}{R}$
(d) None of these

12 A straight chain consisting of n identical metal balls is at rest in a region of free space as shown in figure. In the chain, each ball is connected with adjacent balls by identical conducting wires. Length l of a connecting wire is much larger than the radius r of a ball. A uniform electric field E pointing along the chain is switched ON in the region. Find magnitude of induced charges on one of the end ball.

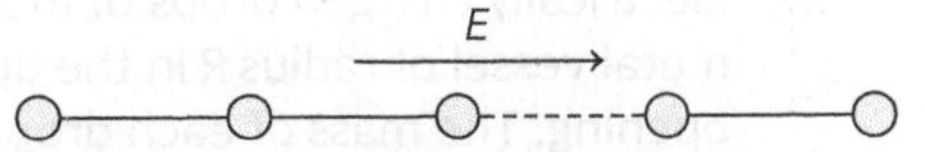

(a) $2\pi\varepsilon_0 rnlE$
(b) $2\pi\varepsilon_0 r(n-1)lE$
(c) $4\pi\varepsilon_0 r(n-1)lE$
(d) $4\pi\varepsilon_0 (n-1)^2 l^2 E$

13 In the figure, A, B, C and D are four concentric spheres of radii a, $2a$, $3a$ and $4a$ and contain charges $+3q$, $+6q$, $-9q$ and $+12q$, respectively. In the beginning, switches S_1 and S_2 are open. After earthing the sphere C, Δq_1 amount of charge flows from the sphere C to the earth. If the sphere B is also earthed, then Δq_2 amount of charge flows from the sphere B to the earth. Find the value of $\left|\dfrac{\Delta q_1}{\Delta q_2}\right|$.

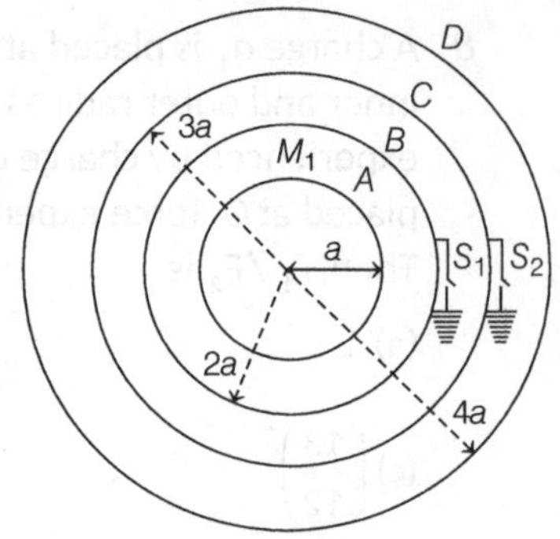

(a) 1 (b) 3/5

(c) 2/3 (d) 1/2

14 A thick hollow metallic sphere of inner and outer radii $2a$ and $4a$ respectively, encloses a solid concentric metallic sphere of radius a. These spheres are connected to the earth through switches S_1 and S_2 as shown in the figure. Initially, charge Q is given to the outer sphere with both S_1 and S_2 open. First, switch S_1 is closed and after sometime S_1 is opened. Then, S_2 is closed and after sometime S_2 is opened. This completes one cycle. The charge on the inner sphere after n cycles of earthing the inner and outer spheres is

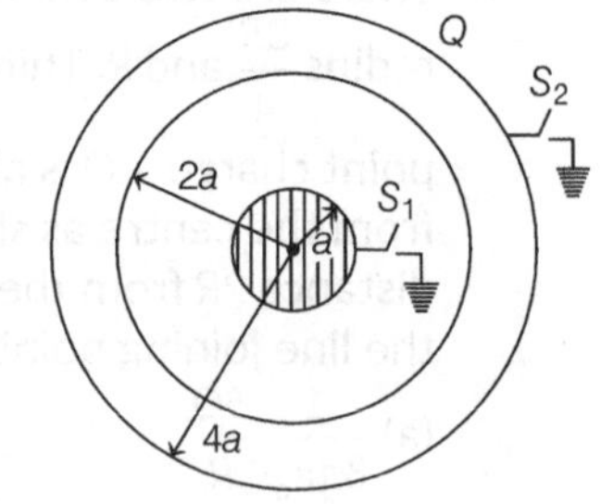

(a) $-Q/3$ (b) $-Q/3^n$

(c) $Q/3^n$ (d) None of these

15 Two non-conducting hemispherical surfaces, which are having uniform charge density σ are placed on smooth horizontal surface as shown in figure. Assuming springs are ideal, calculate compression in each spring, if both the hemispherical surfaces are just touching each other.

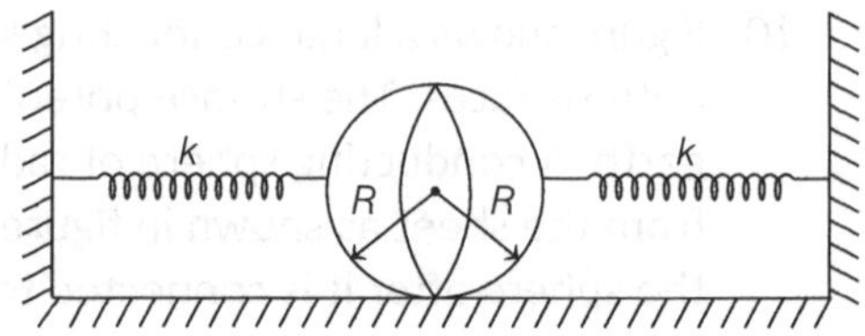

(a) $\dfrac{\sigma^2 R^2}{2\varepsilon_0 k}$ (b) R

(c) $\dfrac{\sigma^2 \pi R^2}{2\varepsilon_0 k}$ (d) None of these

16 Two concentric thin metal shells are shown in figure. Charge density on inner shell is σ_1, while that on outer shell is σ_2. Find the electrostatic pressure at any point of outer shell.

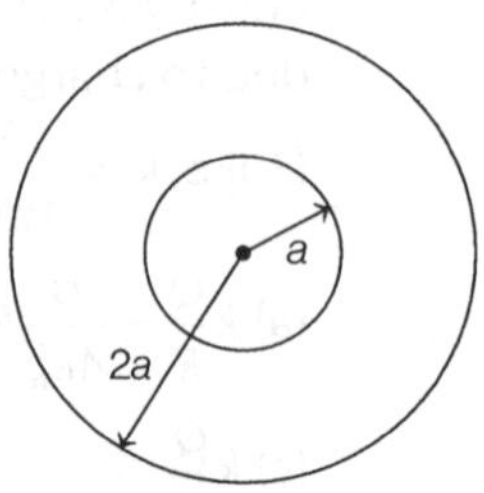

(a) $\dfrac{\sigma_2}{4\varepsilon_0}(\sigma_1 + \sigma_2)$ (b) $\dfrac{\sigma_2}{4\varepsilon_0}(\sigma_1 + 2\sigma_2)$

(c) $\dfrac{\sigma_2}{4\varepsilon_0}(2\sigma_1 + \sigma_2)$ (d) $\dfrac{\sigma_2}{2\varepsilon_0}(\sigma_1 + \sigma_2)$

17 Electrically charged drops of mercury fall from altitude h into a spherical metal vessel of radius R in the upper part of which there is a small opening. The mass of each drop is m and charge is Q. What is the number n of last drop that can still enter the sphere?

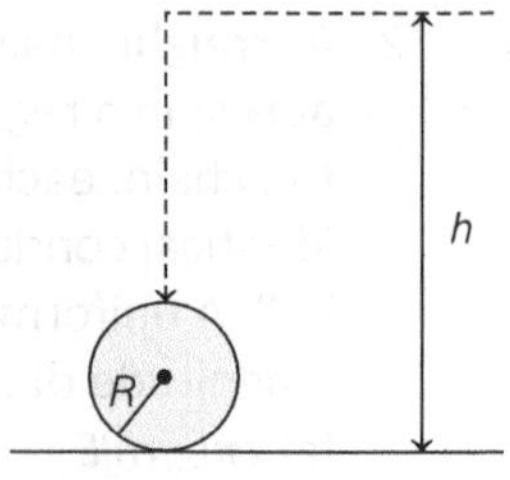

(a) $\dfrac{mg(h-R)R}{kQ^2}$ (b) $\dfrac{mg(h-2R)R}{kQ^2}$

(c) $\dfrac{mghR}{kQ^2}$ (d) $\dfrac{2mg(h-R)R}{kQ^2}$

18 The diagram shows three concentric conducting spherical shells having radii R, $2R$ and $3R$. The initial potential of each shell is as mentioned in the figure. If the inner most shell is earthed, then the charge present on its outer surface would be equal to

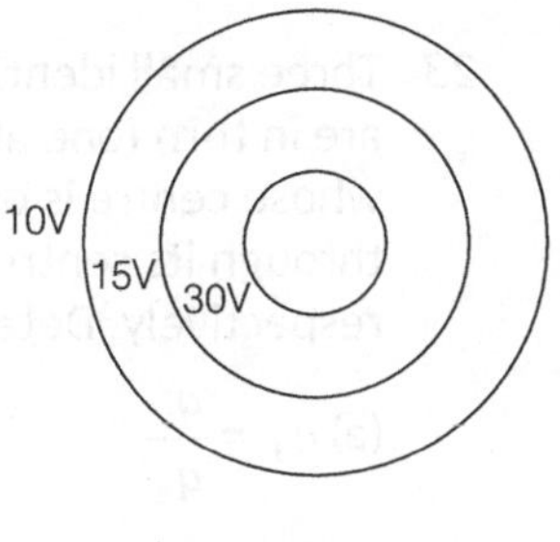

(a) zero
(b) $40\pi\varepsilon_0 R$
(c) $4\pi\varepsilon_0 R$
(d) $-\pi\varepsilon_0 R$

19 Three identical metallic uncharged spheres A, B and C of radius a are kept on the corners of an equilateral triangle of side d. The fourth sphere of radius a $(d >> a)$ which has a charge q touches A and is then removed to a position far away. B is earthed and then the earth connection is removed. C is then earthed, the charge on C is

(a) $\frac{qa}{2d}\left(\frac{2d-a}{2d}\right)$ (b) $\frac{qa}{2d}\left(\frac{2d-a}{d}\right)$

(c) $-\frac{qa}{2d}\left(\frac{d-a}{d}\right)$ (d) $\frac{2qa}{d}\left(\frac{d-a}{2d}\right)$

20 A point charge q is located at the centre O of a spherical uncharged conducting layer provided with small orifice. The inside and outside radii of the layer are equal to a and b, respectively. The amount of work that has to be performed to slowly transfer the charge q from the point O through the orifice and into infinity is

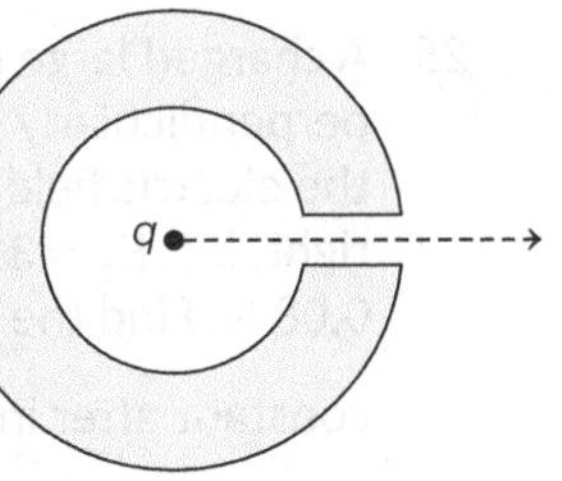

(a) $\frac{q^2}{4\pi\varepsilon_0}\left(\frac{1}{a}-\frac{1}{b}\right)$ (b) $\frac{q^2}{8\pi\varepsilon_0}\left(\frac{1}{a}-\frac{1}{b}\right)$

(c) $\frac{3q^2}{4\pi\varepsilon_0}\left(\frac{1}{a}-\frac{1}{b}\right)$ (d) $\frac{3q^2}{8\pi\varepsilon_0}\left(\frac{1}{a}-\frac{1}{b}\right)$

21 The density of charge at point P on the conductor is σ C/m^2. The resultant electric field near P (just outside) will be

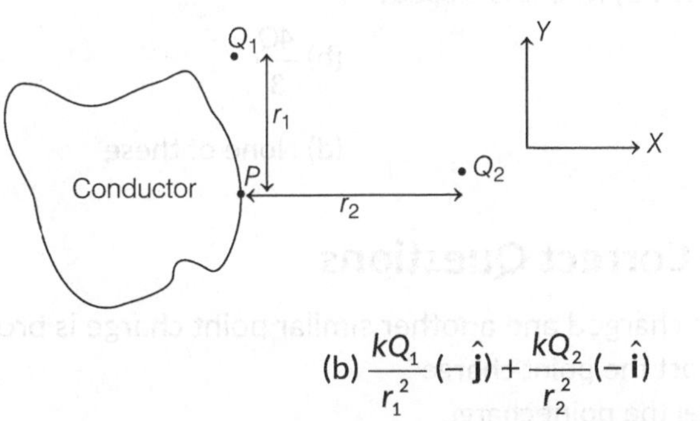

(a) $\frac{\sigma}{\varepsilon_0}\hat{i}$ (b) $\frac{kQ_1}{r_1^2}(-\hat{j})+\frac{kQ_2}{r_2^2}(-\hat{i})$

(c) $\frac{kQ_1}{r_1^2}(-\hat{j})+\frac{kQ_2}{r_2^2}(-\hat{i})+\frac{\sigma}{\varepsilon_0}(\hat{i})$ (d) $\frac{kQ_1}{r_1^2}(\hat{j})+\frac{kQ_2}{r_2^2}(\hat{i})+\frac{\sigma}{\varepsilon_0}(-\hat{i})$

22 An insulated soap bubble has radius r. It is then charged uniformly by total charge q. Neglecting surface tension, find the small increase in radius of bubble due to charging. Assume atmospheric pressure is p_0, temperature remains constant.

(a) $\frac{q^2}{32\pi^2 r^3\varepsilon_0 p_0}$ (b) $\frac{q^2}{48\pi^2 r^3\varepsilon_0 p_0}$

(c) $\frac{q^2}{96\pi^2 r^3\varepsilon_0 p_0}$ (d) None of these

23 Three small identical neutral metal balls are at the vertices of an equilateral triangle. The balls are in turn (one after another is removed) connected to an isolated large conducting sphere, whose centre is on the perpendicular erected from the plane of the triangle and passing through its centre. As a result, the first and second balls have acquired charges q_1 and q_2, respectively. Determine the charge q_3 of the third ball.

(a) $q_3 = \frac{q_1^2}{q_2}$ (b) $q_3 = \frac{q_2^2}{q_1}$ (c) $q_3 = q_1$ (d) $q_3 = q_2$

24 Charge Q is distributed on a small hollow conducting sphere (radius r) as shown inside a spherical cavity of radius r_1 made inside a solid conducting sphere of radius r_2 as shown in the figure. The total energy of the system is

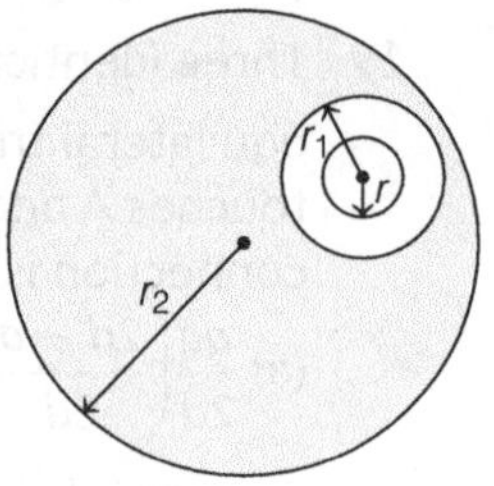

(a) $kQ^2\left(\frac{1}{2r_2} - \frac{1}{2r_1} + \frac{1}{2r}\right)$ (b) $kQ^2\left(\frac{1}{r_1} + \frac{1}{r_2} + \frac{1}{2r}\right)$

(c) $kQ^2\left(\frac{1}{2r_1} + \frac{1}{4r_2} + \frac{1}{2r}\right)$ (d) $kQ^2\left(\frac{1}{4r_1} + \frac{1}{2r_2} + \frac{1}{2r}\right)$

25 A charged large metal sheet is placed into uniform electric field, perpendicularly to the electric field lines. After placing the sheet into the field, the electric field on the left side of the sheet is $E_1 = 5 \times 10^5$ V/m and on the right, it is $E_2 = 3 \times 10^5$ V/m. The sheet experiences a net electric force of 0.08 N. Find the area of one face of the sheet. Assume, external field to remain constant after introducing the large sheet. $\left(\text{Take, } \frac{1}{4\pi\varepsilon_0} = 9 \times 10^9 \text{ N-m}^2/\text{C}^2\right)$

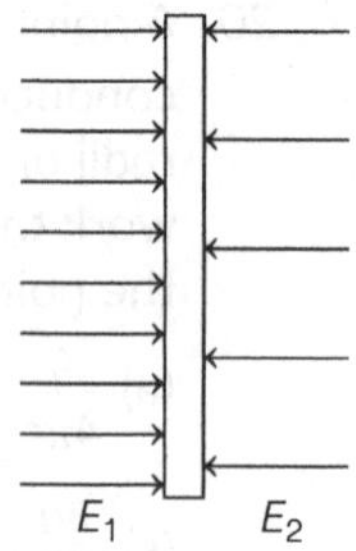

(a) $3.6\,\pi \times 10^{-2}\,\text{m}^2$ (b) $0.9\,\pi \times 10^{-2}\,\text{m}^2$
(c) $1.8\,\pi \times 10^{-2}\,\text{m}^2$ (d) None of these

26 Three large plates are arranged as shown in figure. How much charge will flow through the key K, if it is closed?

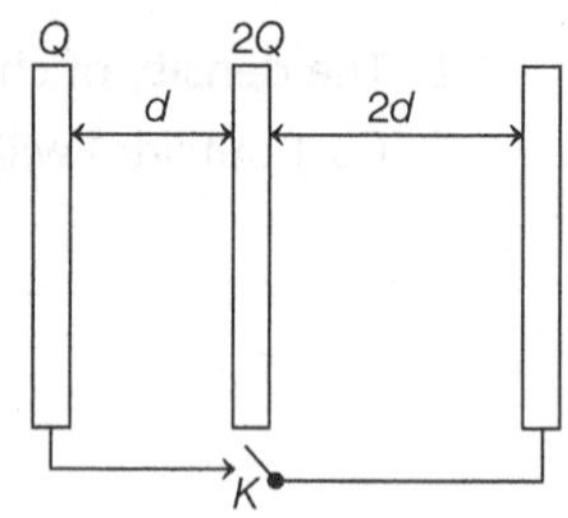

(a) $\frac{5Q}{6}$ (b) $\frac{4Q}{3}$

(c) $\frac{3Q}{2}$ (d) None of these

Multiple Choice(s) Correct Questions

27 A conducting ball is charged and another similar point charge is brought closer to the ball,
(a) the ball may attract the point charge
(b) the ball may repel the point charge
(c) the electric field inside the ball due to ball's charges is non- zero
(d) the net electric field inside the ball is zero

28 A conductor having total charge $+Q$ with a cavity, is placed inside a non-uniform electric field and there is no charge inside cavity. Then, in electrostatic conditions, the electric field
(a) inside the conductor is zero but non-zero in the cavity
(b) just outside the conductor is $\frac{\sigma}{\varepsilon_0}$, where σ is the local surface charge density
(c) inside the conductor is non-zero
(d) inside the conductor as well as cavity is zero

29 Figure shows a neutral metallic sphere with a point charge $+Q$ placed near its surface. Electrostatic equilibrium conditions exist on metallic sphere, mark the correct statements.

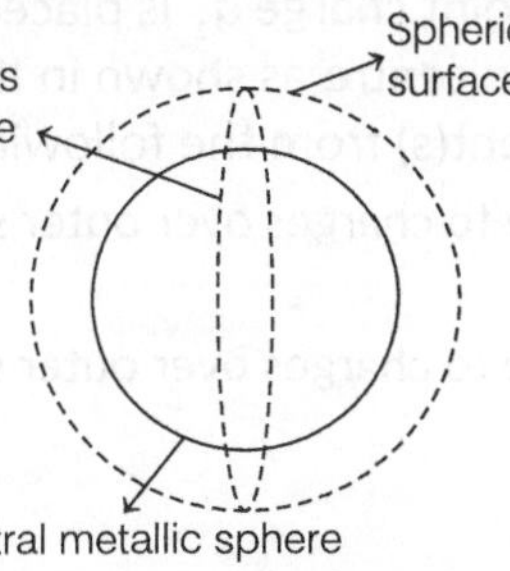

(a) Net flux through Gaussian surface due to charge Q is zero.
(b) Net flux through Gaussian surface due to charges appearing on the outer surface of metallic sphere must be zero.
(c) If point charge Q is displaced towards metallic sphere, magnitude of net flux through right hemispherical closed Gaussian surface increases.
(d) If point charge Q is displaced towards metallic sphere, charge distribution on outer surface of sphere will change.

30 There is a conductor with cavity of radius R as shown in figure. A point charge Q is placed at $R/2$ distance from centre of cavity. A point P is at r distance from centre of conductor, then

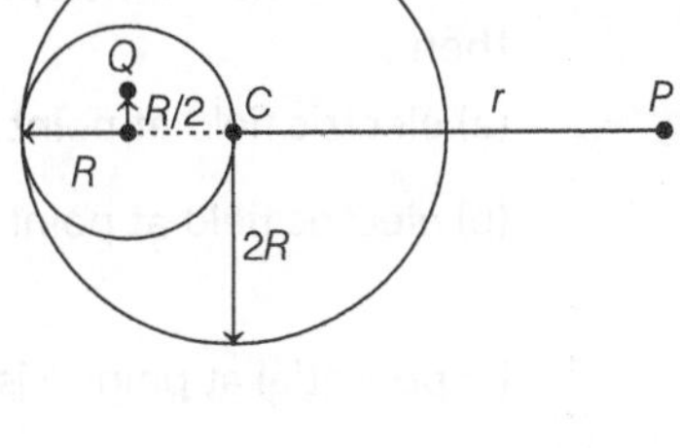

(a) charge density at inner surface of cavity will be non-uniform
(b) charge density at surface of conductor will be uniform
(c) electric field at point P is $\frac{Q}{4\pi\varepsilon_0 r^2}$
(d) force on charge Q is zero

31 For the situation shown in the figure below, mark out the correct statement(s).

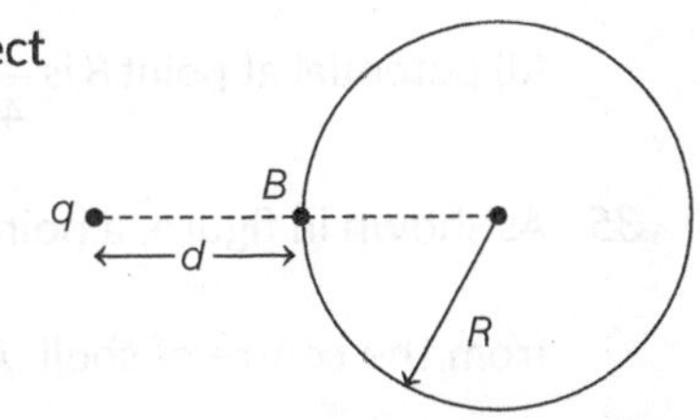

Hollow neutral conductor

(a) Potential of the conductor is $\frac{q}{4\pi\varepsilon_0 (d+R)}$.
(b) Potential of the conductor is $\frac{q}{4\pi\varepsilon_0 d}$.
(c) Potential of the conductor cannot be determined as nature of distribution of induced charges is not known.
(d) Potential at point B due to induced charges is $\frac{-qR}{4\pi\varepsilon_0 (d+R)d}$.

32 An earthed solid metallic cuboid with a cavity has a charged particle $+Q$ placed inside it and A and B are two points on the inner surface of the cavity as shown in figure. If σ, E and V represent the induced surface charge density, magnitude of electric field and electric potential, then which of the following statement(s) is/are correct?

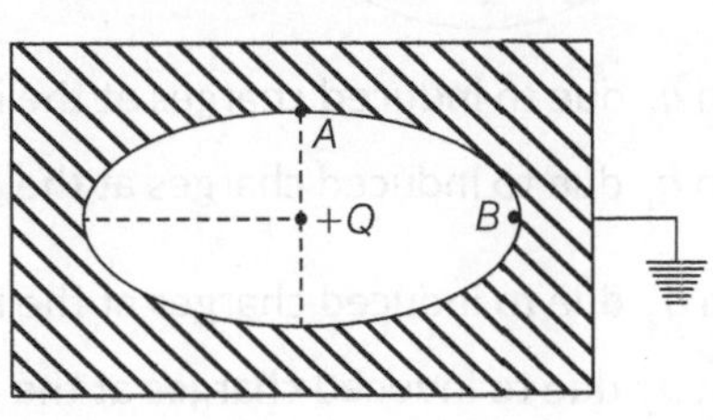

(a) $\sigma_A = \sigma_B$ (b) $V_A = V_B$ (c) $E_A = E_B$
(d) Total electric flux through the inner surface of the cavity is Q/ε_0.

33 A thin metallic spherical shell contains a charge Q over it. A point charge $+q$ is placed inside the shell at point T separated from the centre by distance a. Another point charge q_1 is placed outside the shell at a distance b from the centre as shown in the figure. Now, select the correct statement(s) from the following.

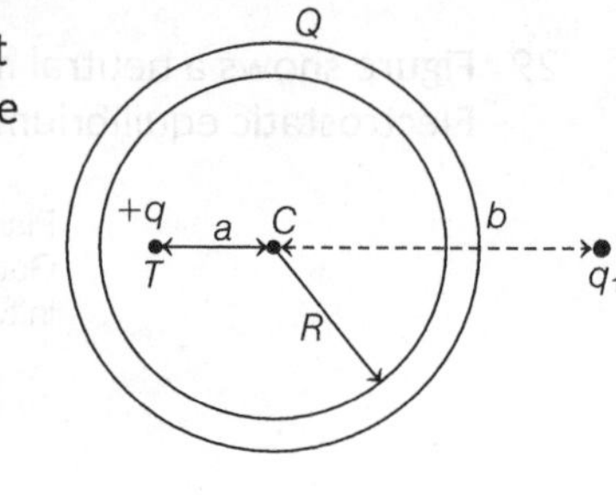

(a) Electric field at the centre due to charges over outer surface of the shell is zero.

(b) Electric field at the centre due to charges over outer surface of the shell is $\frac{q_1}{4\pi\varepsilon_0 b^2}$.

(c) Electric potential at the centre due to all charges in space is $\frac{1}{4\pi\varepsilon_0}\left(\frac{q}{a}-\frac{q}{R}+\frac{Q}{R}+\frac{q_1}{b}\right)$.

(d) Electric potential at the centre due to all charges in space is $\frac{1}{4\pi\varepsilon_0}\left(\frac{q}{a}+\frac{Q}{R}+\frac{q_1}{b}\right)$.

34 A cavity is made inside a solid conducting sphere and a charge q is placed inside the cavity at the centre. A charge q_1 is placed outside the sphere as shown in the figure. Point A is inside the sphere and point B is inside the cavity, then

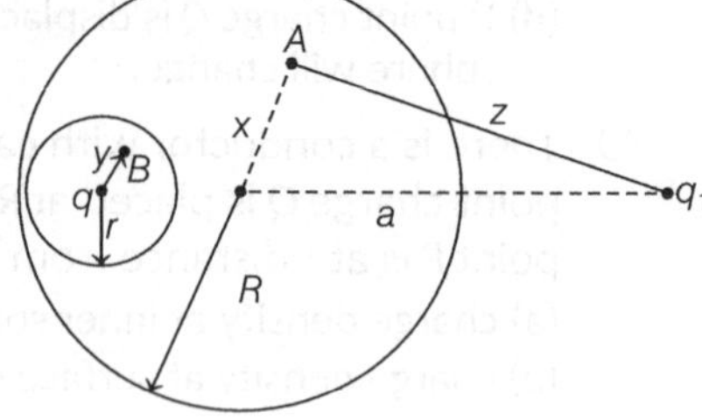

(a) electric field at point A is zero

(b) electric field at point B is $\frac{1}{4\pi\varepsilon_0}\frac{q}{y^2}$

(c) potential at point A is $\frac{1}{4\pi\varepsilon_0}\left(\frac{q_1}{a}+\frac{q}{R}\right)$

(d) potential at point B is $\frac{1}{4\pi\varepsilon_0}\left(\frac{q_1}{a}+\frac{q}{R}+\frac{q}{y}-\frac{q}{r}\right)$

35 As shown in figure, a point charge q_1 is placed inside a neutral conducting shell at a distance $\frac{R}{2}$ from the centre of shell. Another point charge q_2 is placed at a distance $\frac{\sqrt{7}}{2}R$ from the centre of shell. Choose the correct statement(s).

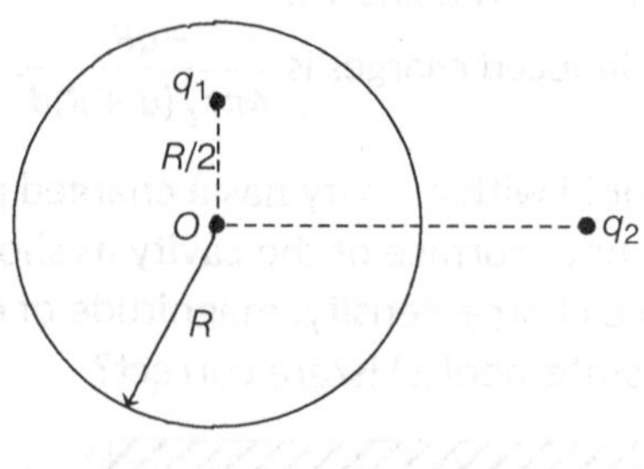

(a) The magnitude of force on q_1 due to induced charges at the inner surface of the shell is zero.

(b) The magnitude of force on q_1 due to induced charges at the outer surface of the shell is $\frac{q_1 q_2}{8\pi\varepsilon_0 R^2}$.

(c) The magnitude of force on q_2 due to induced charges at the inner surface of the shell is zero.

(d) The magnitude of force on q_2 due to induced charges at the inner surface of the shell is $\frac{q_1 q_2}{8\pi\varepsilon_0 R^2}$.

36 A thin conducting rod AB is introduced in between the two charges $+q_1$ and $-q_2$ as shown in figure. For this situation, mark the correct statement(s).

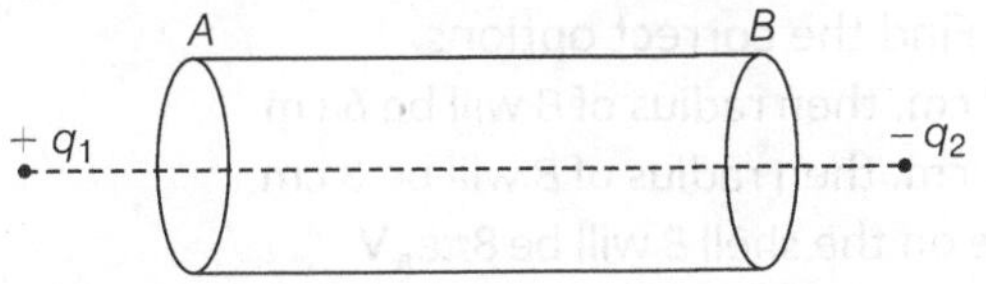

(a) The total force experienced by q_1 is vector sum of electric force experienced by $+q_1$ due to $-q_2$ and due to induced charges on rod.
(b) The end A will become negatively charged.
(c) The total magnitude of force acting on $+q_1$ will be greater than as compared to the case without rod.
(d) The total magnitude of force acting on $-q_2$ will be greater than as compared to the case without rod.

37 Two concentric shells have radii R and $2R$ charges q_A and q_B and potentials $2V$ and $(3/2)V$ respectively, if potential at infinity is zero. Now, shell B is earthed and let charges on them become q'_A and q'_B. Then,

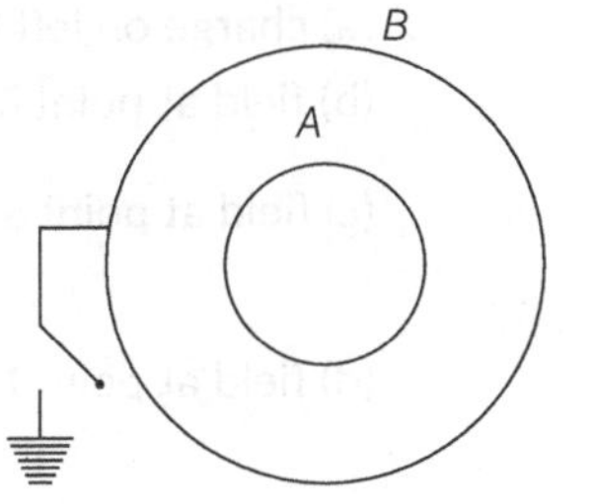

(a) $q_A / q_B = 1/2$
(b) $q'_A / q'_B = 1/2$
(c) potential of A after earthing becomes $(3/2)$ V
(d) potential difference between A and B after earthing becomes $V/2$

38 A solid conducting sphere of radius R is initially uncharged and Q charge is brought from infinity to point A, which is a distance $2R$ from centre of the sphere. B and D are points on sphere as shown in figure. Switch S is initially open, then

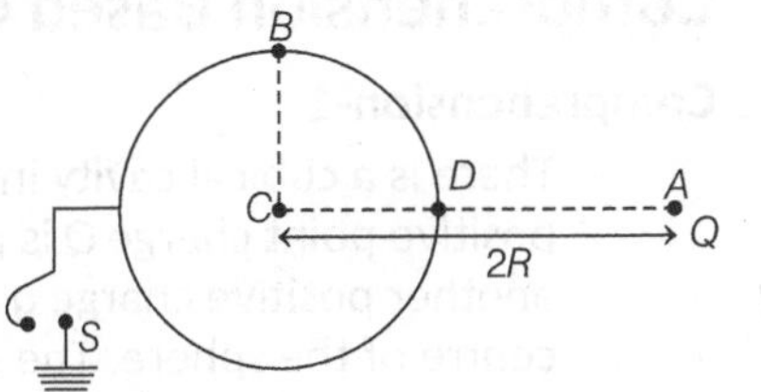

(a) $\frac{V_D}{V_B} = \frac{\sqrt{5}}{2-\sqrt{5}}$, where V_B and V_D are potential at B and D respectively due to induced charge on the sphere
(b) $V_B - V_D = \frac{Q}{4\pi\varepsilon_0 R}\left(\frac{1}{2} - \frac{1}{\sqrt{5}}\right)$, where V_B are V_D potential at B and D respectively due to induced charge on the sphere
(c) If switch is closed, then $Q/2$ flows from sphere to ground
(d) If switch is closed, then Q flows from sphere to ground

39 A spherical shell of radius R is given charge $3q$ on its surface and a point charge q is placed at distance $R/2$ from its centre C. Also, there is a charge $2q$ placed outside the shell at a distance of $2R$ as shown in figure. Then,

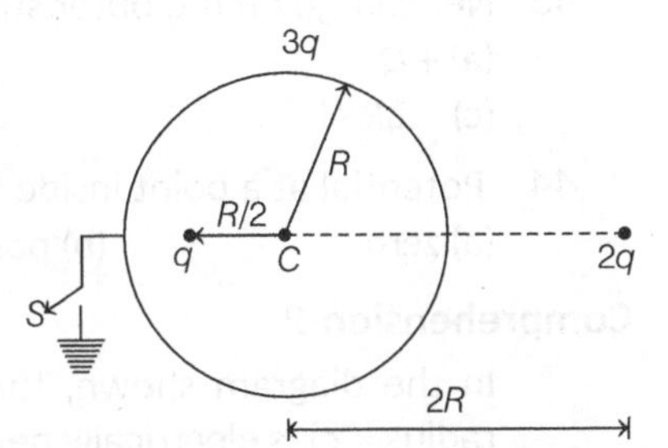

(a) the magnitude of electric field at the centre C due to charges on the outer surface of shell is $\frac{kq}{2R^2}$ before closing the switch S
(b) the electric potential at the centre C due to charges on the outer surface of shell is $\left(\frac{-kq}{R}\right)$ before closing the switch S
(c) the electric potential at the centre C due to charges on the outer surface of shell is $\left(\frac{-kq}{R}\right)$ after closing the switch S
(d) charge flown through the switch into earth after closing the switch S is $5q$

40 When isolated, the spherical shell A has a potential V. When connected by a conducting wire with initially uncharged spherical shell B and inside it, its potential is (2/3) V. Find the correct options.

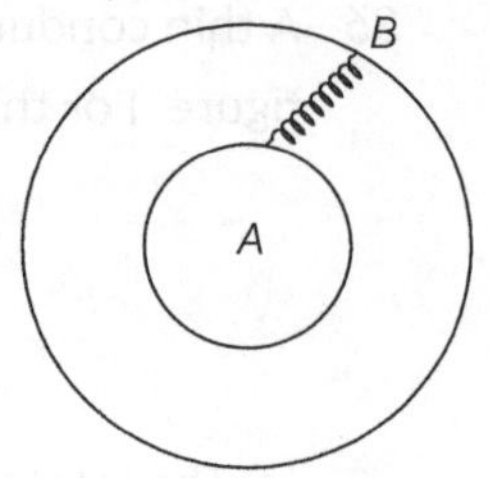

(a) If the radius of A is 2 cm, then radius of B will be 6 cm
(b) If the radius of A is 2 cm, then radius of B will be 3 cm
(c) The minimum charge on the shell B will be $8\pi\varepsilon_0 V$
(d) The minimum charge on the shell B will be $4\pi\varepsilon_0 aV$, where a is the radius of sphere A

41 Two large thin conducting plates with small gap in between are placed in a uniform electric field E (perpendicular to the plates). Area of each plate is A and charges $+Q$ and $-Q$ are given to these plates as shown in the figure. If points R, S and T as shown in the figure are three points in space, then the

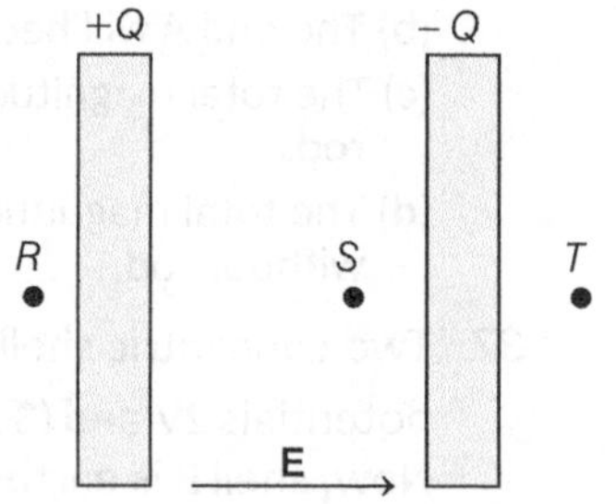

(a) charge on left face of left plate is $-\varepsilon_0 AE$
(b) field at point R is E
(c) field at point S is $\left(E + \dfrac{Q}{\varepsilon_0 A}\right)$
(d) field at point T is $\left(E + \dfrac{Q}{A\varepsilon_0}\right)$

Comprehension Based Questions

Comprehension-1

There is a cubical cavity inside a conducting sphere of radius R. A positive point charge Q is placed at the centre of the cube and another positive charge q is placed at a distance $l(> R)$ from the centre of the sphere. The sphere is earthed.

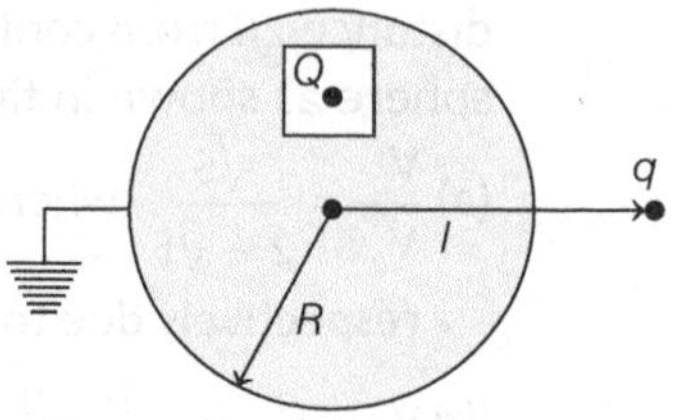

42 Charge induced on the inner surface of cavity is
(a) $-Q$, uniformly distributed
(b) $-Q$, non-uniformly distributed
(c) $-(Q + q)$ non-uniformly distributed
(d) None of the above

43 Net charge on the outer surface of conducting sphere is
(a) $+Q$
(b) $Q - qR/l$
(c) $-qR/l$
(d) None of these

44 Potential at a point inside the cavity is
(a) zero
(b) positive
(c) negative
(d) Cannot be determined

Comprehension-2

In the diagram shown, the outer conducting spherical shell (inner radius $= b$ and the outer radius $= c$) is electrically neutral and the inner conducting sphere (radius $= a$) carries charge $+Q$.

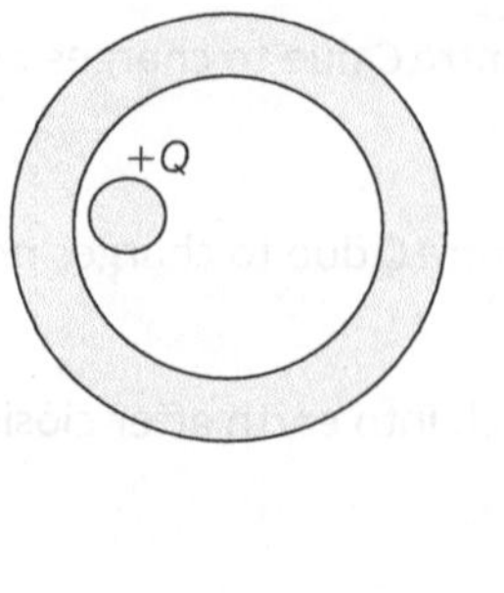

45 What is the potential of the outer spherical shell?

(a) $\left(\frac{1}{4\pi\varepsilon_0}\right)\frac{Q}{c}$ (b) $\left(\frac{Q}{4\pi\varepsilon_0}\right)\left(\frac{1}{c}-\frac{1}{a}\right)$ (c) $\left(\frac{Q}{4\pi\varepsilon_0}\right)\left(\frac{1}{b}-\frac{1}{c}\right)$ (d) None of these

46 If V_s is potential of inner sphere and V_c is the potential at the centre of spherical shell, then

(a) $V_c > V_s$ (b) $V_c < V_s$ (c) $V_c = V_s$

(d) None of the conclusions can be drawn from the given information

Comprehension-3

A conducting sphere of radius *R* has two spherical cavities of radius *a* and *b*. The cavities have charges q_a and q_b respectively at their centres, find

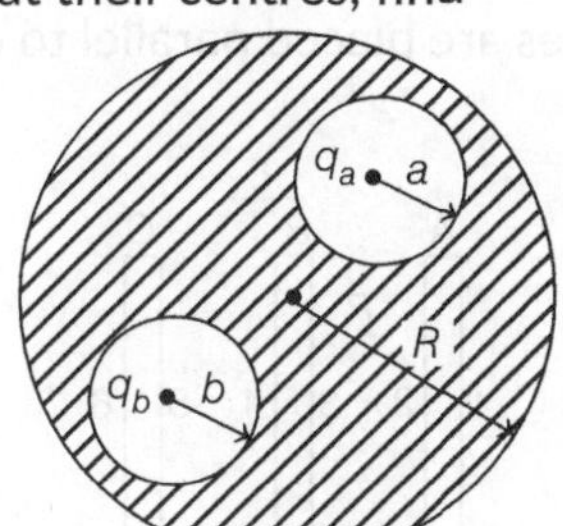

47 The electric field at a point outside conductor at a distance *r* from its centre is

(a) $\frac{q_a+q_b}{4\pi\varepsilon_0 r^2}$ (b) $\frac{q_a+q_b}{4\pi\varepsilon_0 R^2}$ (c) zero (d) Cannot find

48 The electric potential at a point inside cavity of radius *b* at a distance *r* from its centre is

(a) $\frac{q_b}{4\pi\varepsilon_0 r}$ (b) $-\frac{q_b}{4\pi\varepsilon_0 r}$ (c) $\frac{q_a+q_b}{4\pi\varepsilon_0 R}$ (d) $\frac{1}{4\pi\varepsilon_0}\left(\frac{q_a+q_b}{R}+\frac{q_b}{r}-\frac{q_b}{b}\right)$

49 Force on q_a is

(a) zero (b) $\frac{kq_a^2}{a^2}$ (c) $\frac{kq_a(q_a+q_b)}{R^2}$ (d) Cannot find

Comprehension-4

Four concentric hollow metallic spheres of radii *R*, 2*R*, 3*R* and 4*R* have charges as shown in the figure, then the spheres 1 and 3, 2 and 4 are connected by conducting wires (both the connections are made at the same time).

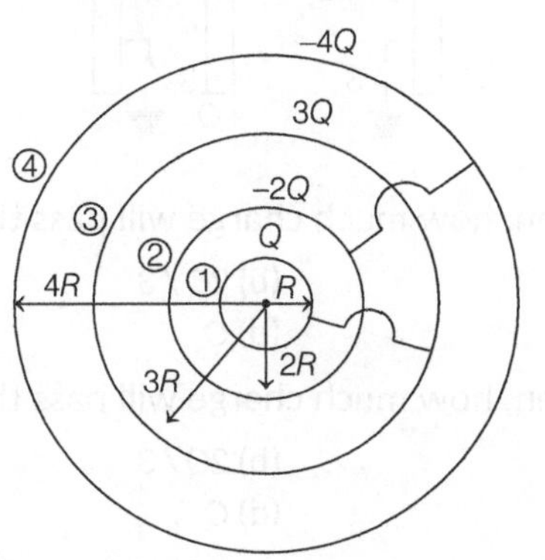

50 The charge on inner surface of third conductor is

(a) $-\frac{6Q}{5}$ (b) $+\frac{6Q}{5}$ (c) – 2Q (d) + 2Q

51 The charge on fourth conductor is

(a) $+\frac{22Q}{5}$ (b) $+\frac{11Q}{3}$ (c) $-\frac{11Q}{3}$ (d) $-\frac{22Q}{5}$

52 The potential of conductor 1 is

(a) $\frac{3Q}{40\pi\varepsilon_0 R}$ (b) $-\frac{19Q}{40\pi\varepsilon_0 R}$ (c) $-\frac{3Q}{40\pi\varepsilon_0 R}$ (d) $\frac{19Q}{40\pi\varepsilon_0 R}$

53 The potential of conductor 2 is

(a) $-\frac{Q}{8\pi\varepsilon_0 R}$ (b) $\frac{Q}{8\pi\varepsilon_0 R}$ (c) $\frac{Q}{32\pi\varepsilon_0 R}$ (d) $-\frac{Q}{32\pi\varepsilon_0 R}$

Comprehension-5

Three neutral metallic large plates are placed parallel to each other in a uniform electric field E as shown in the figure.

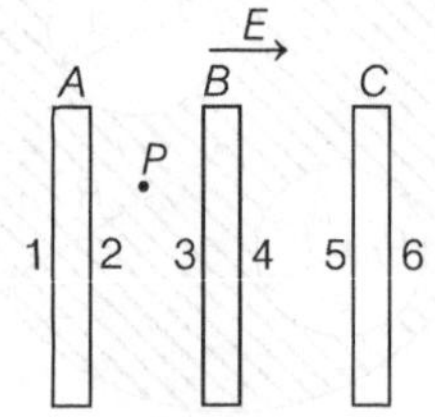

54 The induced charge density on surface 1 is

(a) $\varepsilon_0 E$ (b) $-\varepsilon_0 E$ (c) zero (d) Cannot find

55 The induced charge density on surface 4 is

(a) $\varepsilon_0 E$ (b) $-\varepsilon_0 E$ (c) zero (d) Cannot find

56 The electric field at point P in the region between plates A and B is

(a) E towards right (b) E towards left

(c) zero (d) Cannot find

Comprehension-6

Three identical parallel conducting plates A, B and C are placed as shown in figure. Switches S_1 and S_2 are open, and can connect A and C to earth when closed. $+Q$ charge is now given to B.

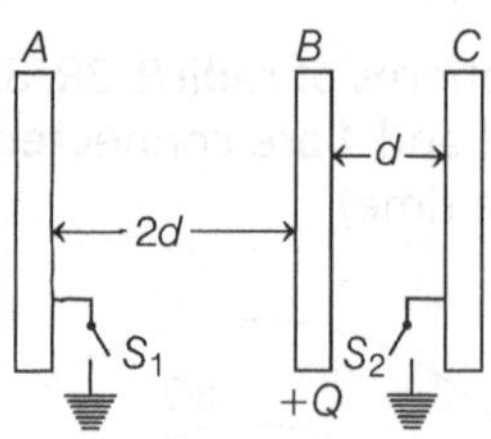

57 If S_1 is closed while S_2 is open, how much charge will pass through S_1?

(a) $Q/2$ (b) $2Q/3$

(c) $3Q/4$ (d) Q

58 If S_2 is closed while S_1 is open, how much charge will pass through S_2?

(a) $Q/2$ (b) $2Q/3$

(c) $3Q/4$ (d) Q

59 If S_1 and S_2 are closed simultaneously, how much charge will pass through S_2?

(a) $Q/2$ (b) $2Q/3$

(c) $3Q/4$ (d) Q

Match the Columns Questions

60 Figure shows three concentric thin spherical shells A, B and C of radii a, b and c, respectively. The shells A and C are given charges q and $-q$ respectively and the shell B is earthed. Find the charges appearing on the surfaces of B and C.

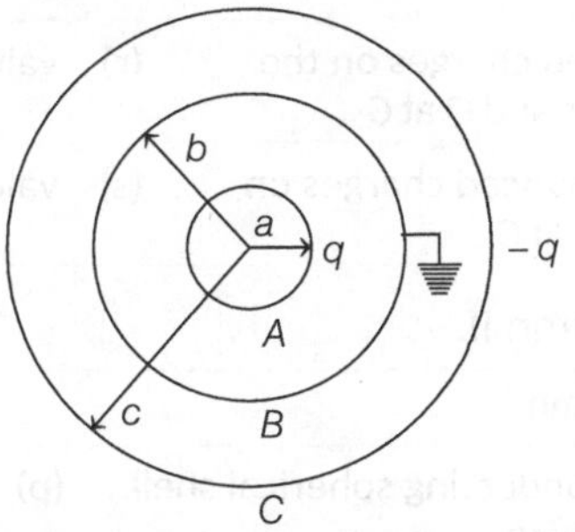

Column I	Column II
(a) Charge on outer surface of spherical shell C	(p) $q/2$
(b) Charge on inner surface of spherical shell C	(q) $-q$
(c) Charge on outer surface of spherical shell B	(r) bq/c
(d) Charge on inner surface of spherical shell B	(s) $-bq/c$
	(t) $q(b/c-1)$

61 For the situation shown in the figure below, match the entries of Column I with entries of Column II

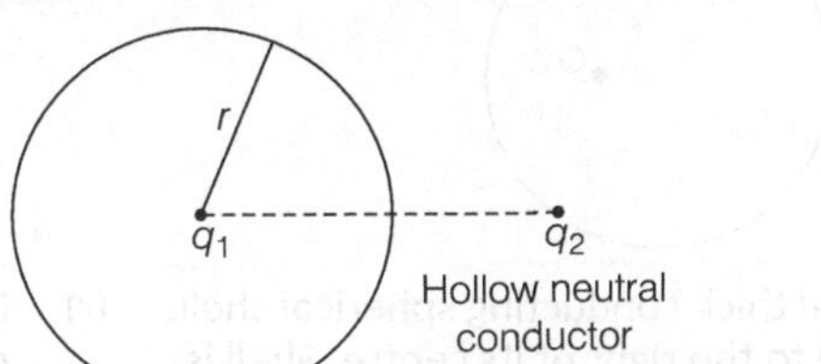

Column I	Column II
(a) In the situation shown,	(p) distribution of charge on inner surface of conductor is uniform
(b) If outside charge is not present,	(q) distribution of charge on inner surface of conductor is non-uniform
(c) If we displace the outside charge while the inside charge remains at centre,	(r) distribution of charge on outer surface of conductor is uniform
(d) If the inside charge is displaced by small amount from centre,	(s) distribution of charge on outer surface of conductor is non-uniform

62 In the shown figure, the conductor is uncharged and a charge q is placed inside a spherical cavity at a distance a from its centre C. Point P and charge $+Q$ are as shown in figure. a, b, c, d are known.

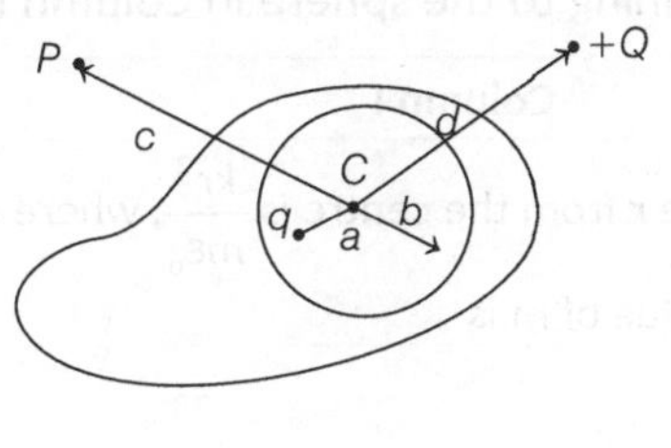

	Column I		Column II
(a)	Electric field due to induced charges on the inner surface of cavity at point P	(p)	zero
(b)	Electric potential due to charges on the inner surface of cavity and q at P	(q)	non-zero
(c)	Electric field due to induced charges on the outer surface of conductor and Q at C	(r)	value can be stated with the given data
(d)	Electric potential due to induced charges on the inner surface of cavity at C	(s)	value cannot be stated from the given data

63 Match the Column I with Column II.

	Column I		Column II
(a)	Electrically neutral thick conducting spherical shell, with point charge at its centre. 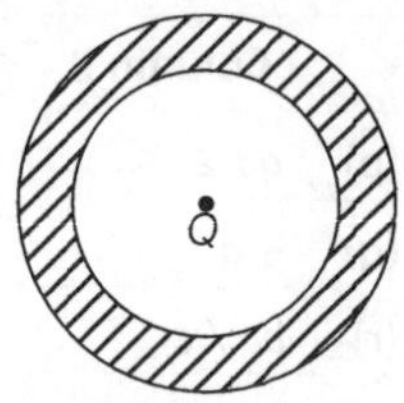	(p)	Electric field everywhere inside the cavity due to charges induced on the inner surface of conductor is zero.
(b)	Electrically neutral thin conducting spherical shell, with point charge to the right of its centre 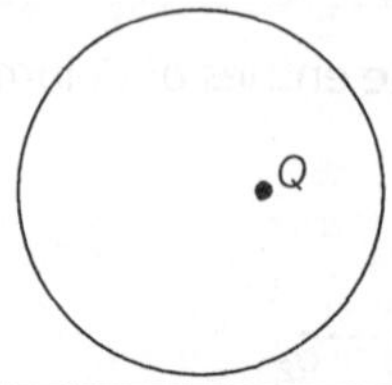	(q)	Electric field everywhere inside the cavity due to charges induced on the outer surface of conductor is zero.
(c)	Electrically neutral thick conducting spherical shell, with point charge to the right of its centre. Shell is earthed. 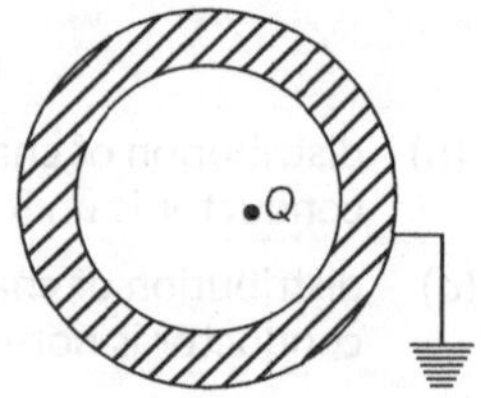	(r)	Electric potential at the centre of the cavity due to charges induced on inner and outer surfaces of conductor is zero.
		(s)	Electric potential everywhere inside the cavity due to charges induced on the inner and outer surfaces of conductor is zero.

64 A non-conducting sphere of radius R is given a volume charge density $\rho = kr$, where r is the distance of a point in the spherical volume from its centre and k is a positive constant. Match the physical quantities pertaining to the sphere in Column I with the values given in Column II.

	Column I		Column II
(a)	Electric field at a distance r from the centre is $\frac{kr^2}{m\varepsilon_0}$, where m is a positive constant. The value of m is	(p)	7

	Column I	Column II
(b)	If electrostatics pressure at distance r from the centre is given by $p(r) = a(b - r^n)$, where a, b and n are constants, then $n =$	(q) 4
(c)	Electrostatic force of interaction between the left and right hemispheres is $\dfrac{k^2 \pi R^6}{4l\varepsilon_0}$, where l is a positive constant. The value of l is	(r) 5
(d)	Self-energy (Electrostatic) of the sphere is given by dR^e, where d and e are constants, then value of e is	(s) 6

Numerical Value Questions

65 The radius and surface tension of a spherical soap bubble be r and T, respectively. A charge is to be given to the bubble, so that its radius will become $2r$. Here, atmospheric pressure is p_0, temperature of air inside bubble remains constant and charge is assumed to uniformly distributed on the surface of bubble.
The final surface charge density is $\alpha \times 10^{-4}$ C/m^2, find the value of α.
(Take, $p_0 = 10^4$ Pa, $T = 0.2$ N/m, $r = 0.12$ mm, $\varepsilon_0 = 9 \times 10^{-12}$ C^2/N-m^2)

66 Two metal balls of radii R_1 and R_2, such that $R_2 = 2R_1$ are connected by a long thin conducting wire and has total charge $Q = 21\,\mu$C. Then, the ball of radius R_1 is placed inside a grounded metal sphere of radius $R = 3R_1$ as shown in figure. What charge (in μC) will flow through the wire in this process?

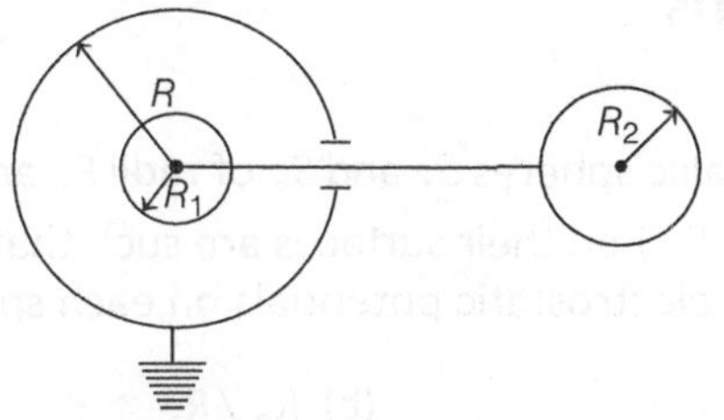

67 The diagram shows a conducting sphere A of radius a which is surrounded by a neutral conducting spherical shell B of radius $b\,(> a)$. Initially switches S_1, S_2 and S_3 are open and sphere A carries a charge Q. First, the switch S_1 is closed to connect the shell B with the ground and then opened. Now, the switch S_2 is closed, so that the sphere A is grounded and then S_2 is opened. Finally, the switch S_3 is closed to connect the spheres together. Find the heat (in joule) which is produced after closing the switch S_3.
(Take, $b = 4$ cm, $a = 2$ cm and $Q = 8\,\mu$C)

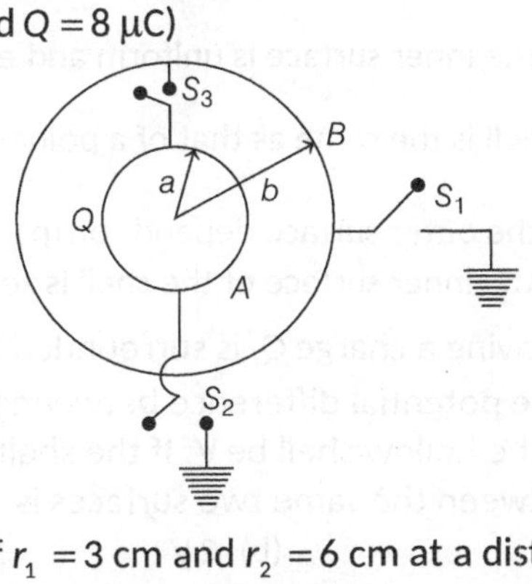

68 Two metal balls with radii of $r_1 = 3$ cm and $r_2 = 6$ cm at a distance of $R = 100$ cm from each other are connected to a battery with an electromotive force of $E = 100$ V. If the electric force between the balls is $\left(\dfrac{x}{9 \times 10^9}\right)$N ,then find the value of x.

69 A uniformly charged non-conducting shell of radius 1m and thickness $t = 1$ cm has a positive charge $Q = 10$C and breaking stress $\frac{100}{\varepsilon_0}$ N/m^2. Determine the minimum magnitude of a charge q (in C) located at the centre of shell for which the shell will tear apart. (Take, $\pi^2 = 10$)

Answers

1. (a)	**2.** (c)	**3.** (b)	**4.** (c)	**5.** (c)	**6.** (d)	**7.** (c)	**8.** (a)	**9.** (c)	**10.** (a)
11. (a)	**12.** (b)	**13.** (a)	**14.** (b)	**15.** (c)	**16.** (b)	**17.** (b)	**18.** (a)	**19.** (c)	**20.** (b)
21. (a)	**22.** (c)	**23.** (b)	**24.** (a)	**25.** (a)	**26.** (a)	**27.** (abcd)	**28.** (bd)	**29.** (abcd)	**30.** (abc)
31. (ad)	**32.** (bd)	**33.** (bd)	**34.** (abcd)	**35.** (bd)	**36.** (abcd)	**37.** (ad)	**38.** (ac)	**39.** (acd)	**40.** (bd)
41. (abc)	**42.** (b)	**43.** (c)	**44.** (b)	**45.** (a)	**46.** (d)	**47.** (a)	**48.** (d)	**49.** (a)	**50.** (b)
51. (d)	**52.** (c)	**53.** (a)	**54.** (b)	**55.** (a)	**56.** (a)	**57.** (d)	**58.** (d)	**59.** (b)	

60. (a) $\rightarrow$ t, (b) $\rightarrow$ s, (c) $\rightarrow$ r, (d) $\rightarrow$ q

61. (a) $\rightarrow$ p, s; (b) $\rightarrow$ p, r; (c) $\rightarrow$ p, s; (d) $\rightarrow$ q, s

62. (a) $\rightarrow$ q, s; (b) $\rightarrow$ p, r; (c) $\rightarrow$ p,r; (d) $\rightarrow$ q, r

63. (a) $\rightarrow$ p, q; (b) $\rightarrow$ q, r; (c) $\rightarrow$ q

64. (a) $\rightarrow$ q, (b) $\rightarrow$ q, (c) $\rightarrow$ s, (d) $\rightarrow$ p

65. (4.5) **66.** (2) **67.** (1.8) **68.** (4) **69.** (27)

For Detailed Solutions
Scan the code

Previous Years' Questions

JEE Main

1 Consider two charged metallic spheres S_1 and S_2 of radii R_1 and R_2, respectively. The electric fields E_1 (on S_1) and E_2 (on S_2) on their surfaces are such that $E_1/E_2 = R_1/R_2$. Then the ratio V_1(on S_1)/V_2 (on S_2) of the electrostatic potentials on each sphere is **[JEE Main 2020]**

(a) $\left(\frac{R_1}{R_2}\right)^3$ (b) R_2/R_1

(c) R_1/R_2 (d) $(R_1/R_2)^2$

2 Shown in the figure is a shell made of a conductor. It has inner radius a and outer radius b and carries charge Q. At its centre is a dipole **p** as shown. In this case, **[JEE Main 2019]**

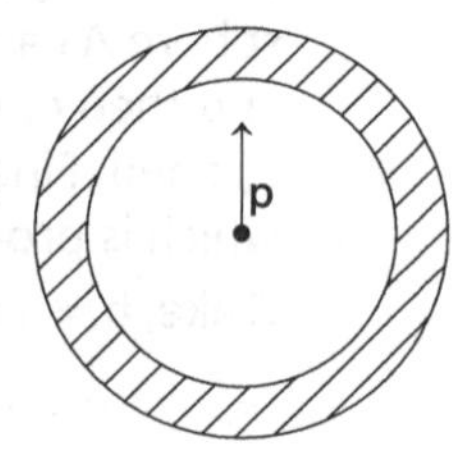

(a) surface charge density on the inner surface is uniform and equal to $\frac{\left(\frac{Q}{2}\right)}{4\pi a^2}$

(b) electric field outside the shell is the same as that of a point charge at the centre of the shell

(c) surface charge density on the outer surface depends on $|\mathbf{p}|$

(d) surface charge density on the inner surface of the shell is zero everywhere

3 A solid conducting sphere, having a charge Q, is surrounded by an uncharged conducting hollow spherical shell. Let the potential difference between the surface of the solid sphere and that of the outer surface of the hollow shell be V. If the shell is now given a charge of $-4Q$, the new potential difference between the same two surfaces is **[JEE Main 2019]**

(a) $-2V$ (b) $2V$

(c) $4V$ (d) V

JEE Advanced

4 A spherical metal shell A of radius R_A and a solid metal sphere B of radius $R_B (< R_A)$ are kept far apart and each is given charge $+Q$. Now they are connected by a thin metal wire. Then **[JEE Advanced 2011]**

(a) $(E)_{(A)\text{ inside}} = 0$ (b) $Q_A > Q_B$ (c) $\frac{\sigma_A}{\sigma_B} = \frac{R_B}{R_A}$ (d) $(E)_{(A)\text{ on surface}} < (E)_{(B)\text{ on surface}}$

5 A uniformly charged thin spherical shell of radius R carries uniform surface charge density of σ per unit area. It is made of two hemispherical shells, held together by pressing them with force F (see figure). F is proportional to **[JEE Advanced 2010]**

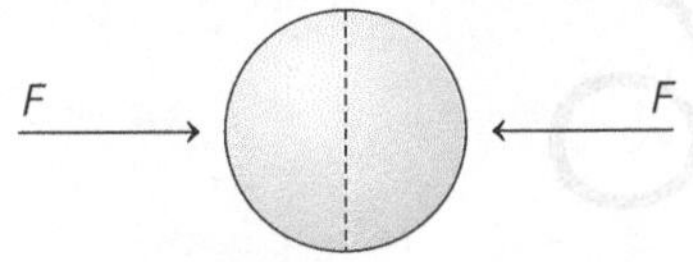

(a) $\frac{1}{\varepsilon_0}\sigma^2 R^2$ (b) $\frac{1}{\varepsilon_0}\sigma^2 R$ (c) $\frac{1}{\varepsilon_0}\frac{\sigma^2}{R}$ (d) $\frac{1}{\varepsilon_0}\frac{\sigma^2}{R^2}$

6 Three concentric metallic spherical shells of radii R, $2R$ and $3R$ are given charges Q_1, Q_2 and Q_3, respectively. It is found that the surface charge densities on the outer surfaces of the shells are equal. Then, the ratio of the charges given to the shells, $Q_1 : Q_2 : Q_3$, is **[JEE Advanced 2009]**

(a) 1 : 2 : 3 (b) 1 : 3 : 5 (c) 1 : 4 : 9 (d) 1 : 8 : 18

Answers

1. (d)	**2.** (b)	**3.** (d)	**4.** (a,b,c,d)	**5.** (a)	**6.** (b)

For Detailed Solutions
Scan the code

CHAPTER

06

CURRENT ELECTRICITY

LEVEL 01 \ Starter Level

1.1 Electric Current, Current Density and Drift Velocity

1 The electric charge flowing through a cross-section of a wire is given by $q = (5t^2 - 2t + 3)$ C. Find

(i) average current over time interval $t = 0$ to $t = 2$ s

(ii) current at $t = 2$ s.

2 The current through a wire depends on time as $i = i_0 + \alpha t$, where $i_0 = 10$ A and $\alpha = 4$ A/s. Find the charge crossed through a section of the wire in 10 s and average current for that interval.

3 The current passing through a wire decreases linearly from 10 A to 0 A in 8 s. Find total charge flown through the wire.

4 A conductor wire have 10^{23} free electrons/cm^3 carries a current of 20A. If the cross-section of the wire is 1 mm^2, then find the drift velocity of electrons.

5 An insulating pipe of cross-section area A contains an electrolyte which has two types of ions, their charges being $-e$ and $+2e$. A potential difference applied between the ends of the pipe result in the drifting of the two types of ions, having drift speed $= v$ ($-$ ve ion) and $v/4$ (+ve ion). Both ions have the same number per unit volume $= n$. Find the current flowing through the pipe.

6 The current density across a cylindrical wire of radius R varies according to the equation $J = J_0\left(1 - \frac{r}{R}\right)$, where J_0 is a constant and r is distance from the axis. Calculate the current flowing through the wire.

7 Calculate the electric field in a copper wire of cross-sectional area 2.0 mm^2 carrying a current of 10 A. The resistivity of copper $= 1.7 \times 10^{-8}$ Ω-m.

1.2 Resistance

8 What length of a copper wire of cross-sectional area 0.01 mm^2 will be needed to prepare a resistance of 340Ω? Resistivity of copper $= 1.7 \times 10^{-8}$ Ω-m.

9 If a wire of resistance *R* is stretched to double its length, find the new resistance.

10 A wire of resistance *R* is stretched, so that radius of its cross-section becomes one-third of its original radius. Find the new resistance.

11 Find the percentage change in resistance of a wire, if it is stretched to increase its length by
(i) 10% (ii) 0.3%

12 The dimensions of a conductor of specific resistance ρ are shown below. Find the resistance of the conductor across *AB*, *CD* and *EF*.

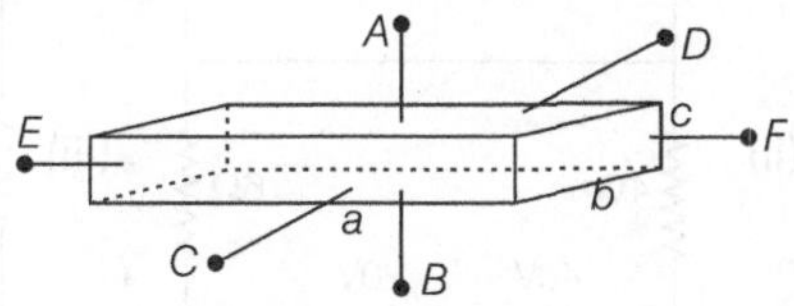

13 The temperature coefficient of resistivity of a wire is 1.25×10^{-3} per degree celsius. At 27°C, its resistance is 50 Ω. Find the temperature at which resistance of the wire will be 60Ω ?

14 Consider a conductor of length 40 cm, where a potential difference of 10V is maintained between the ends of the conductor. Find out the mobility of the electrons provided that the drift velocity of electrons is 5×10^{-6} m/s.

15 A resistor has colour code of brown, violet, yellow and gold as shown in the figure. What is its resistance?

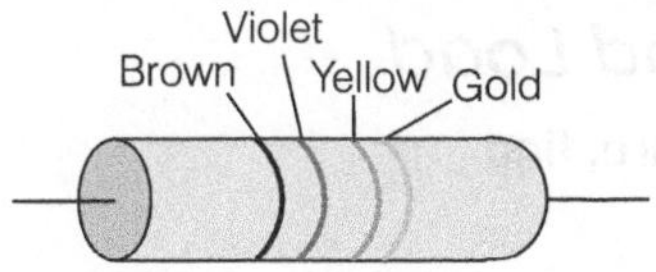

1.3 Kirchhoff's Laws

16 Find potential difference between points A and B.

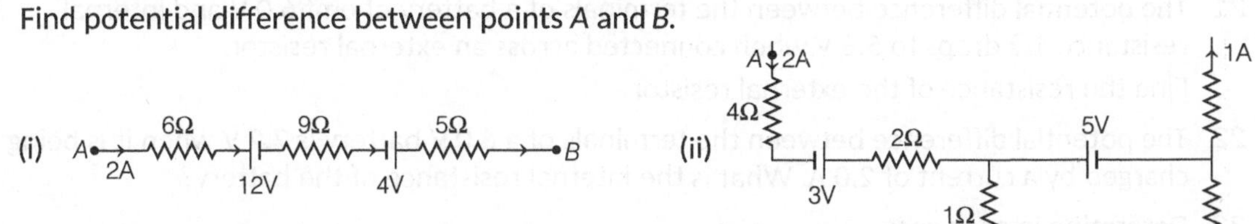

17 Using reference point, find current *i*.

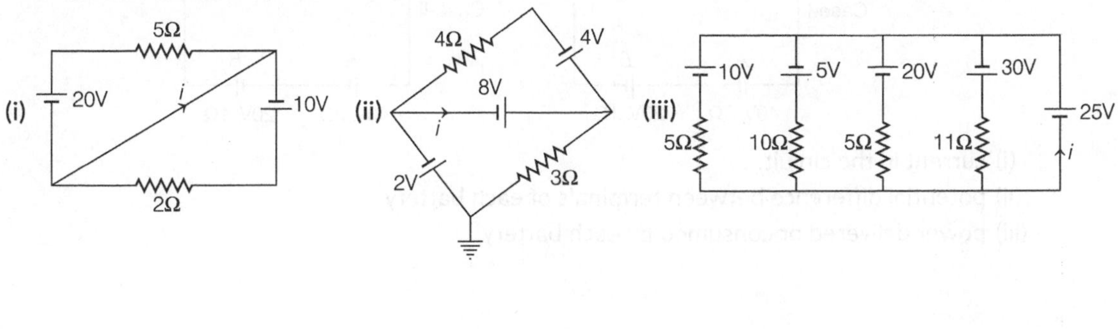

18 Using Kirchhoff's current law, find current through each resistor in the following circuits.

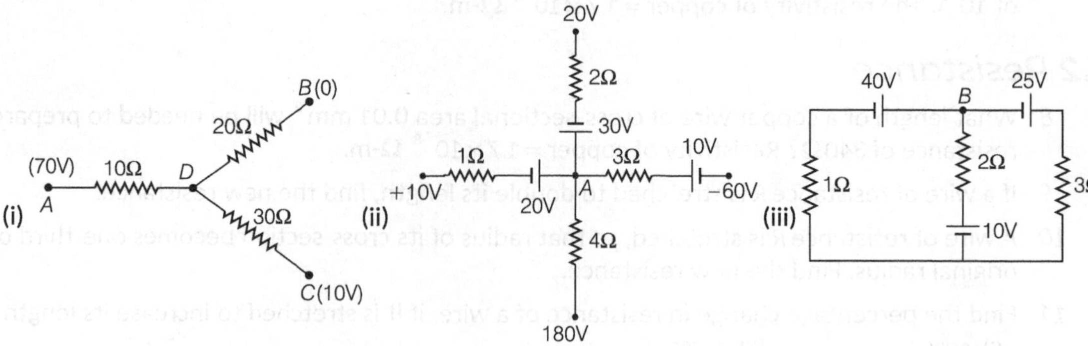

19 Using Kirchhoff's voltage law, find current in all the branches of the following circuits.

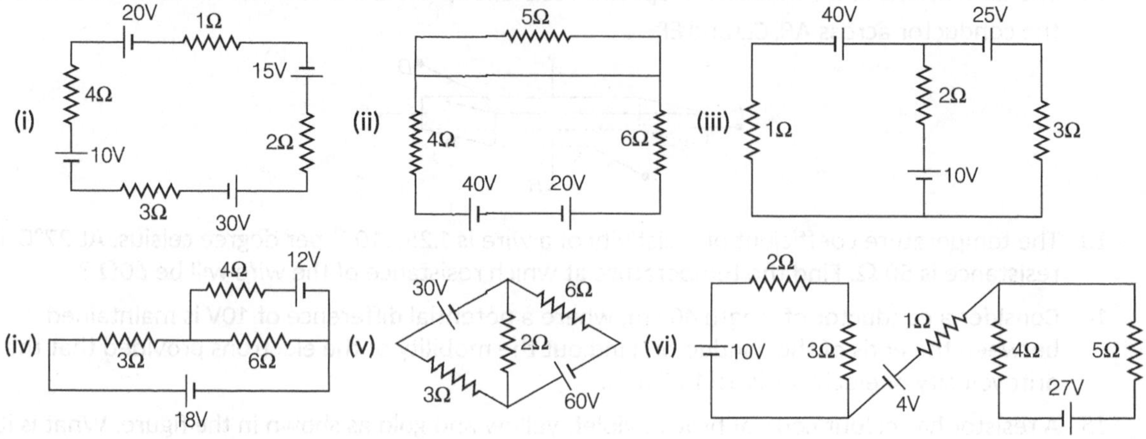

1.4 Battery as Source and Load

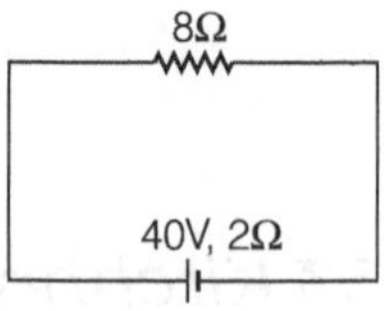

20 In the circuit shown in the figure, find

(i) the current in the circuit,

(ii) the potential difference across the terminals of the battery,

(iii) net power supplied by the battery.

21 The potential difference between the terminals of a battery of emf 6.0 V and internal resistance 1Ω drops to 5.8 V when connected across an external resistor. Find the resistance of the external resistor.

22 The potential difference between the terminals of a 6.0 V battery is 7.2 V when it is being charged by a current of 2.0 A. What is the internal resistance of the battery?

23 Determine in each case,

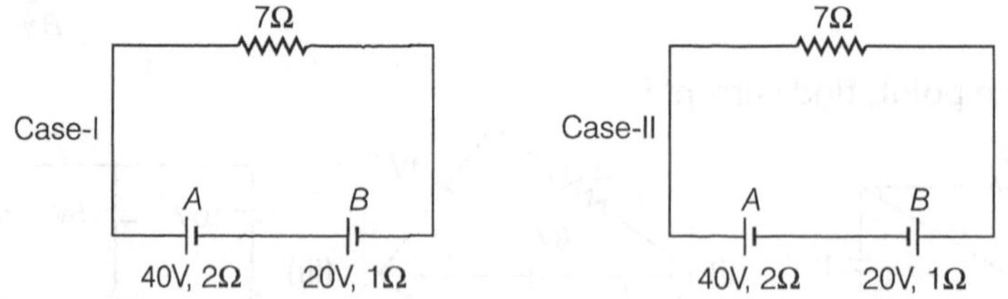

(i) current in the circuit,

(ii) potential difference between terminals of each battery,

(iii) power delivered or consumed by each battery.

1.5 Series and Parallel Combination Circuits

24 In the given circuit, determine
  (i) current,
  (ii) potential difference across 4 Ω resistance,
  (iii) power dissipated in 3Ω resistance.

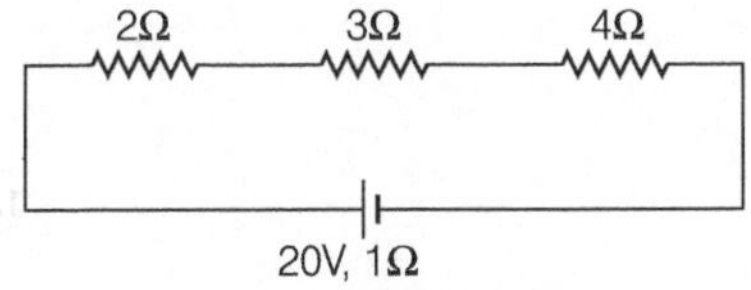

25 In the given circuit, determine
  (i) current through the battery and 6Ω resistor,
  (ii) potential difference across 12Ω resistor,
  (iii) power dissipated in 4Ω resistor.

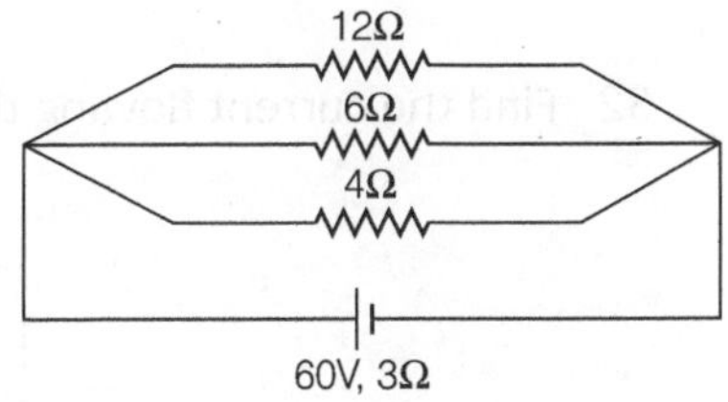

26 Find the current through the 3Ω resistor.

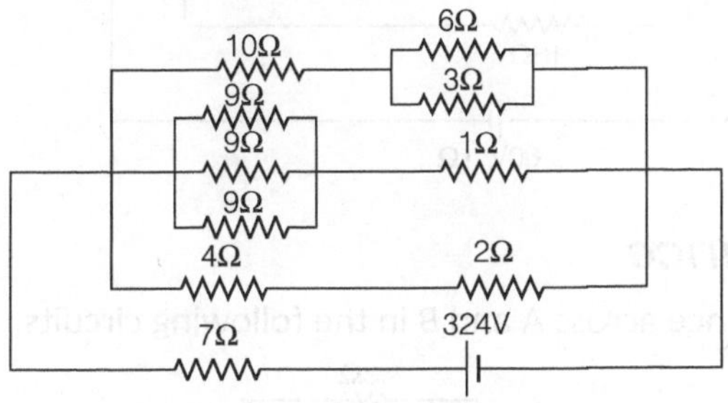

27 Find the current in 2Ω resistance.

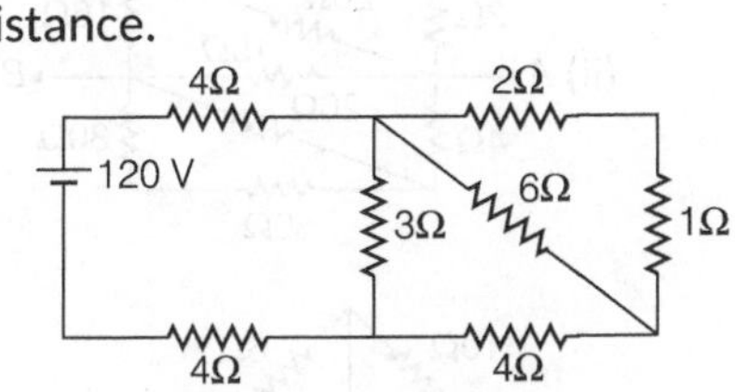

28 In the given circuit, current through 6Ω resistor is 4A. Find emf of the battery.

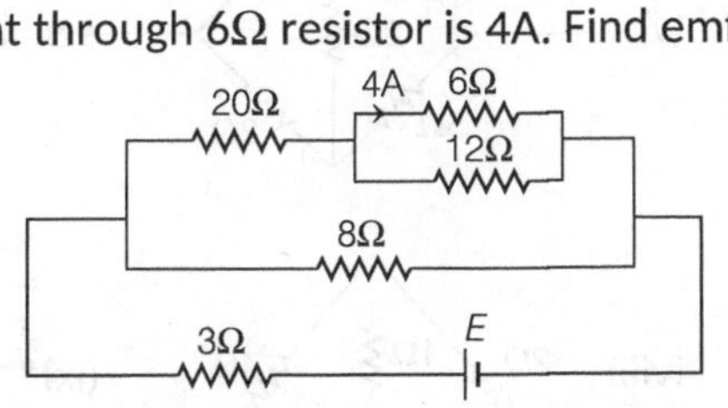

29 Find the current flowing through the segment *AB* of the circuit shown in figure.

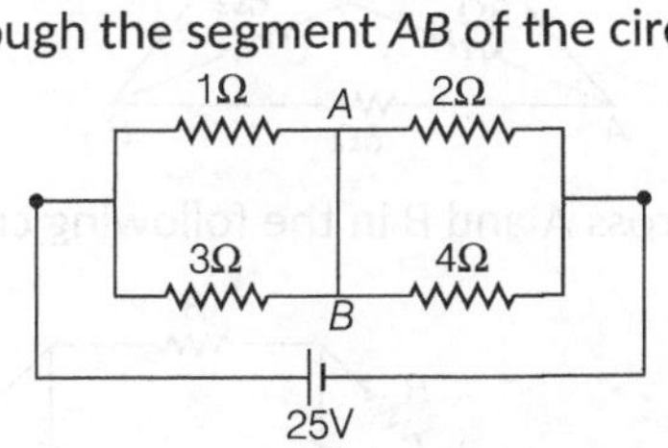

30 A current of 10A flows in a system of conductors as shown in the following figure. Find the potential difference $(V_A - V_B)$.

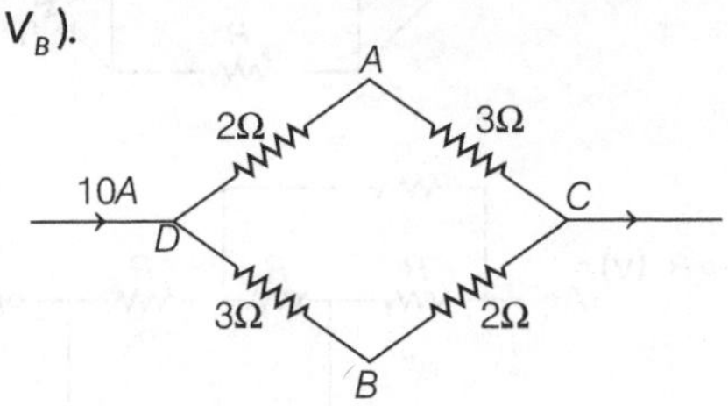

31 In the given circuit, if potential difference across R is 100 V, then find the value of R.

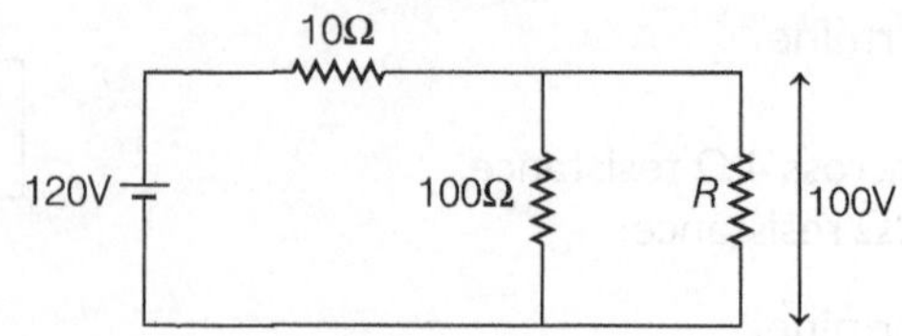

32 Find the current flowing through the battery.

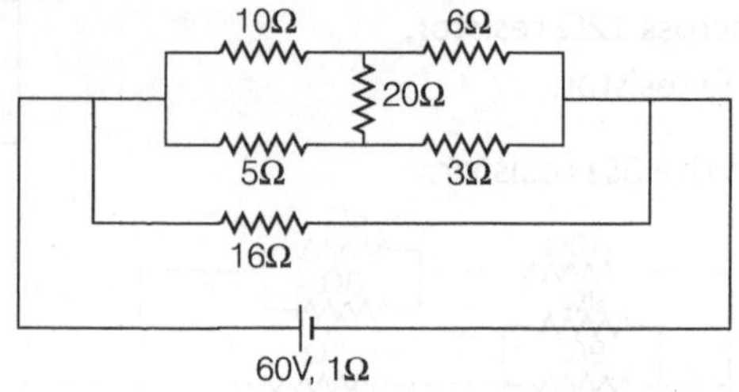

1.6 Equivalent Resistance

33 Find the equivalent resistance across A and B in the following circuits.

(i) 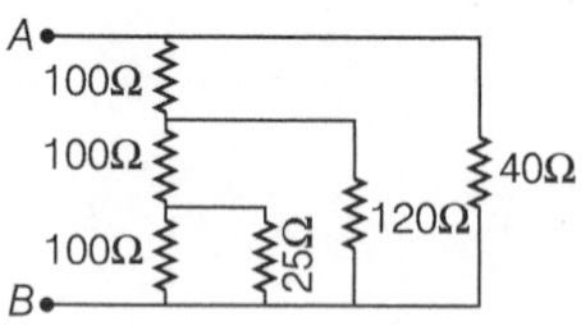

(ii) 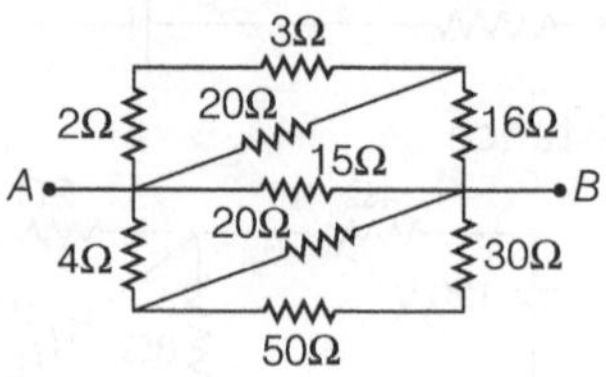

(iii) 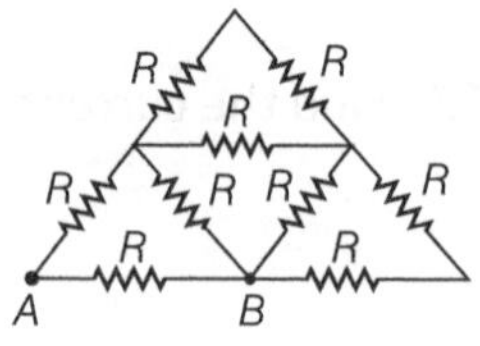

(iv) 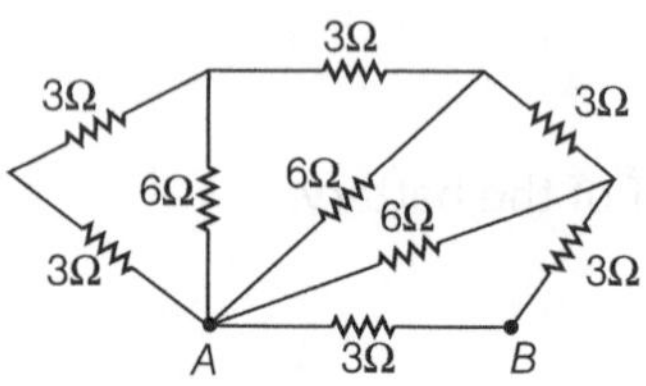

(v) 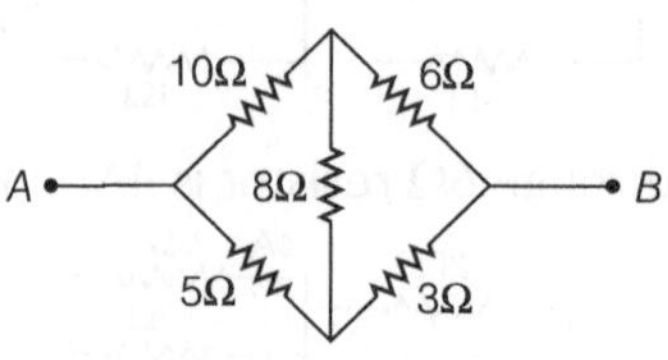

(vi) 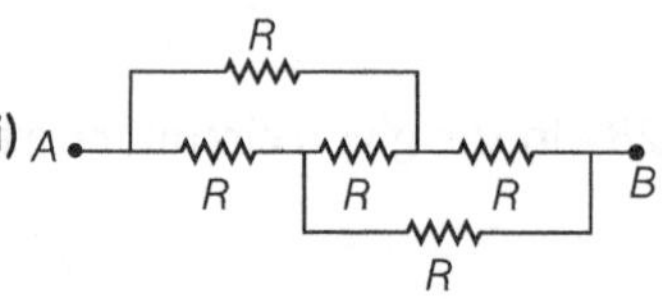

(vii) 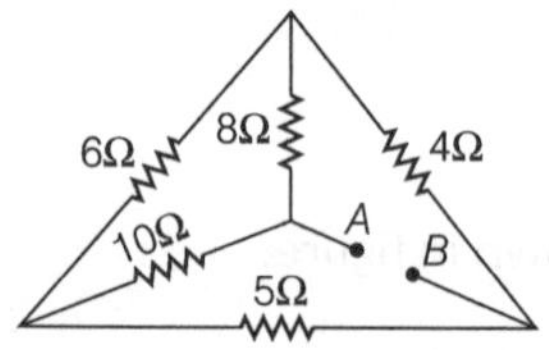

(viii) 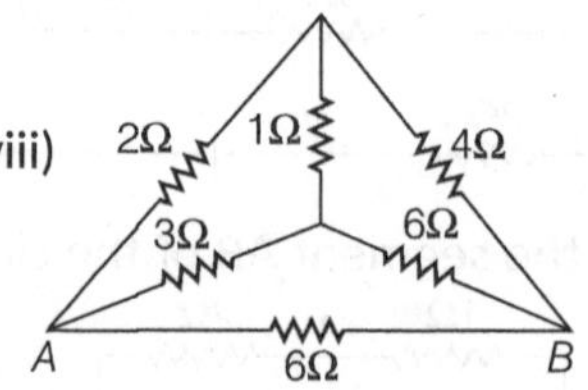

(ix) 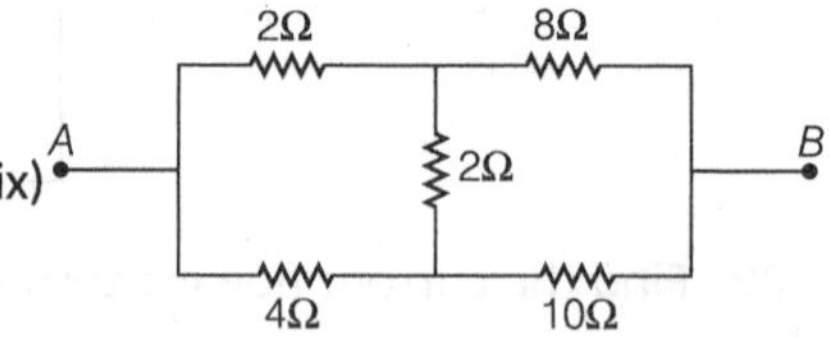

34 Find equivalent resistance across A and B in the following circuits.

(i) 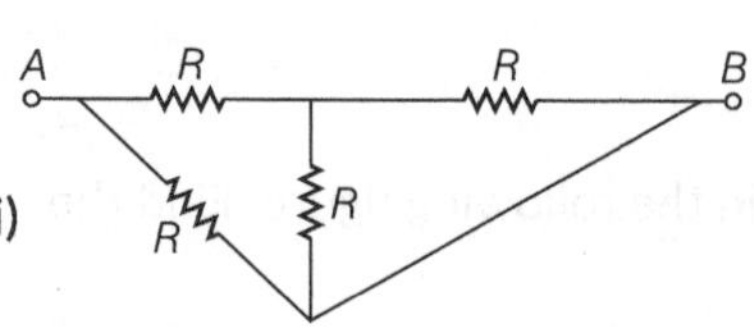

(ii) 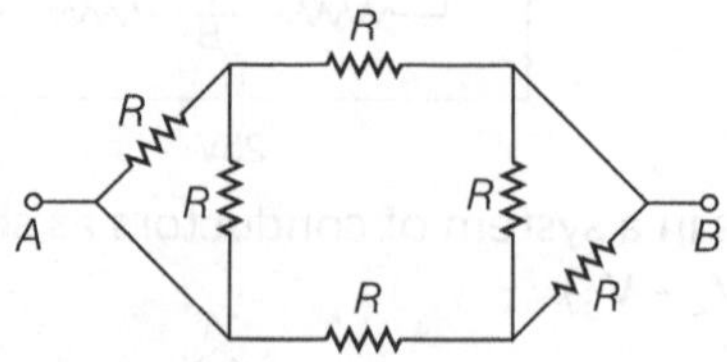

(iii) 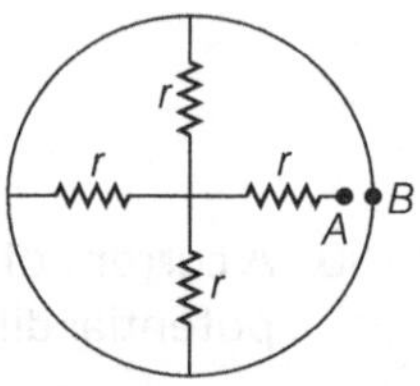

(iv) 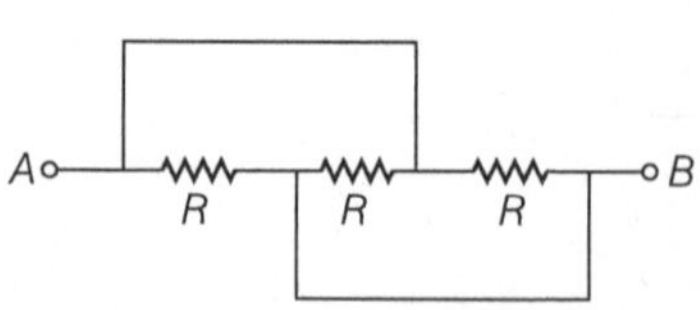

(v) 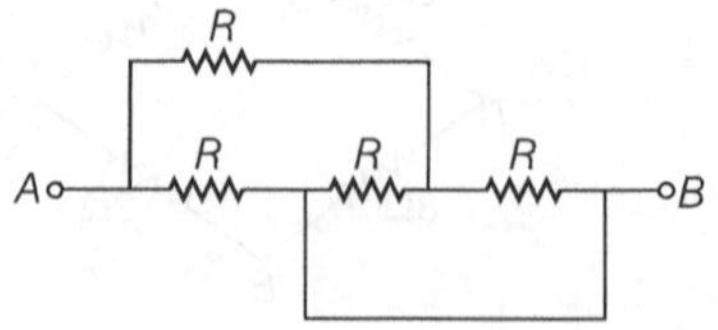

(vi) 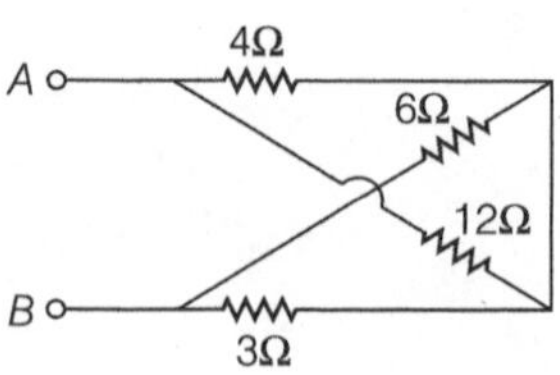

(vii) 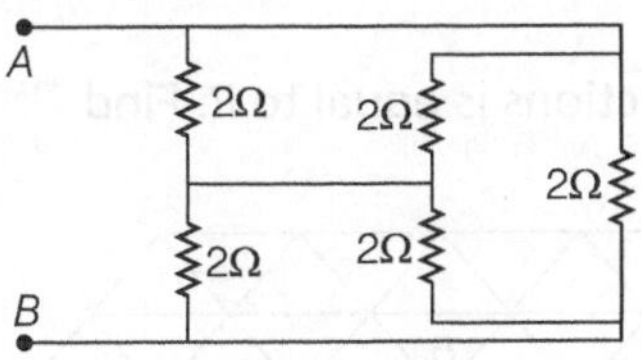

(viii) 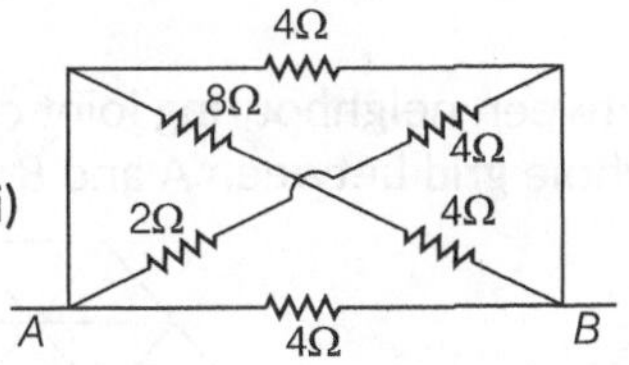

(ix) 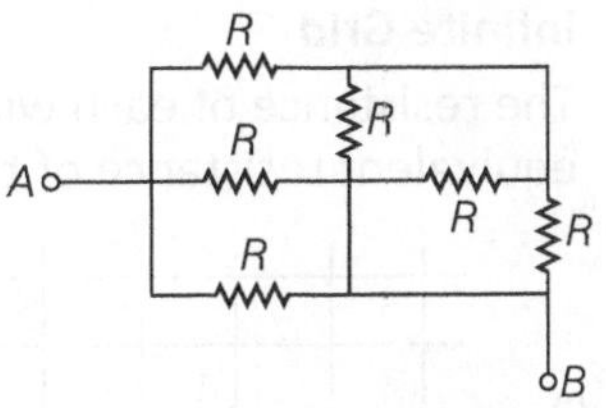

(x) 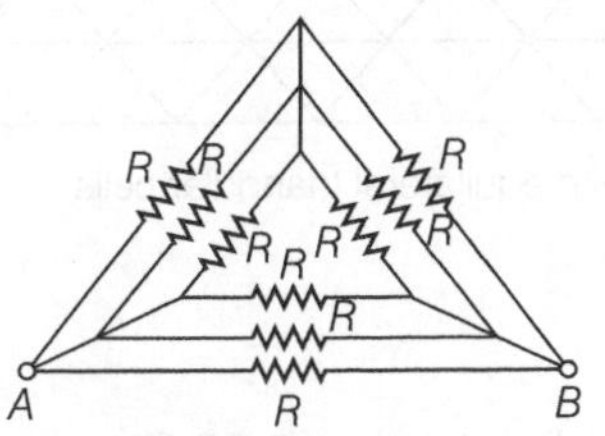

(xi) 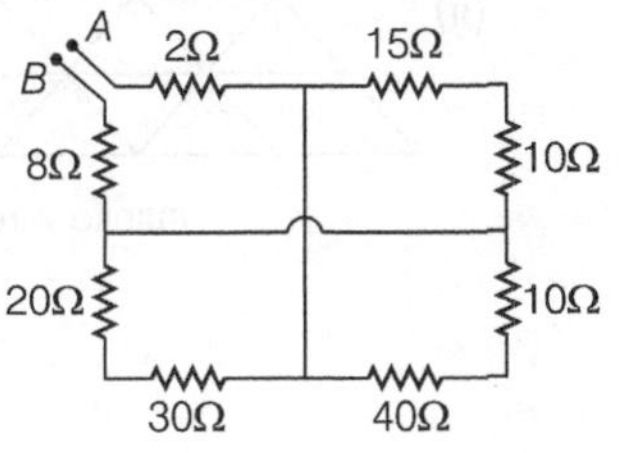

(xii) 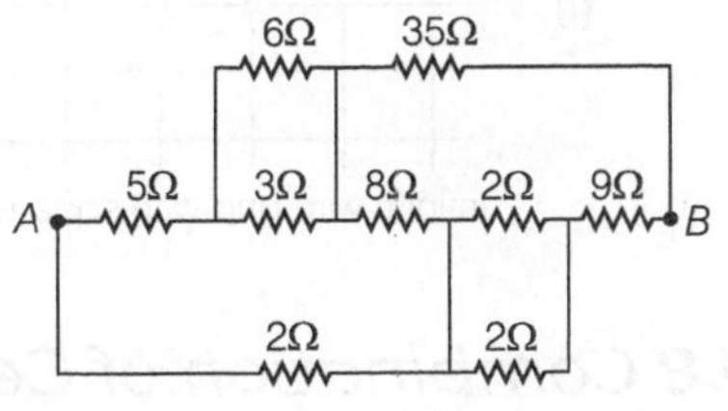

1.7 *Symmetric Circuits*

35 Find equivalent resistance across *A* and *B* in the following circuits.

(i) 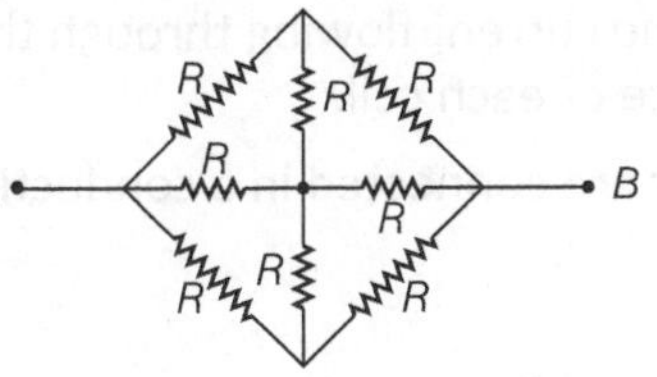

(ii) 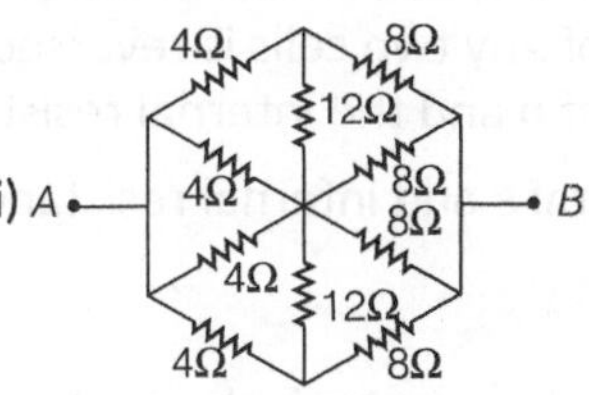

(iii) 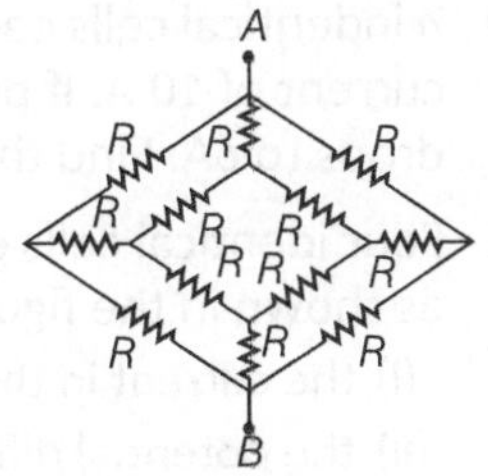

(iv) 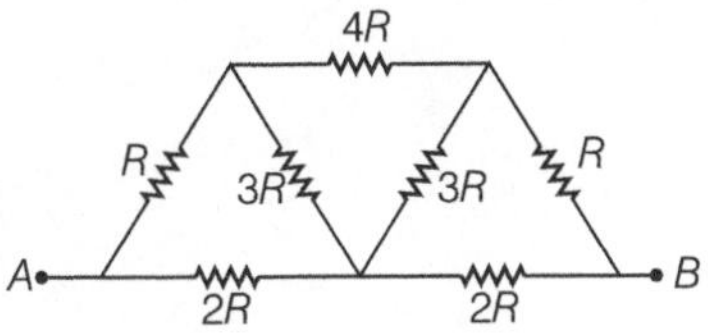

(v) 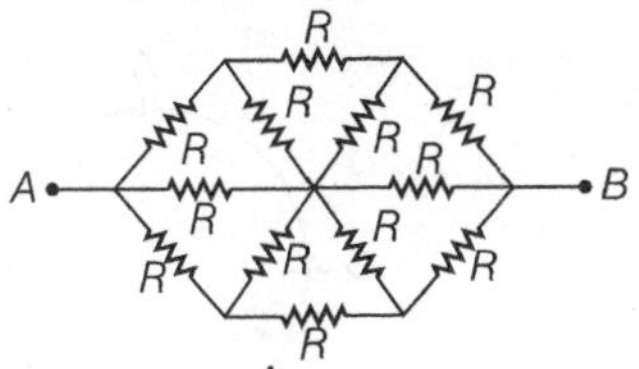

(vi) 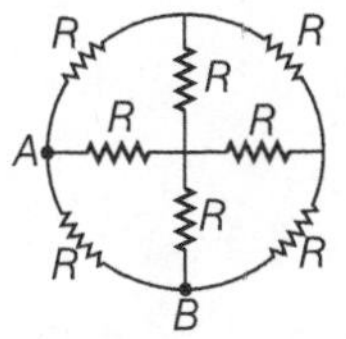

(vii) 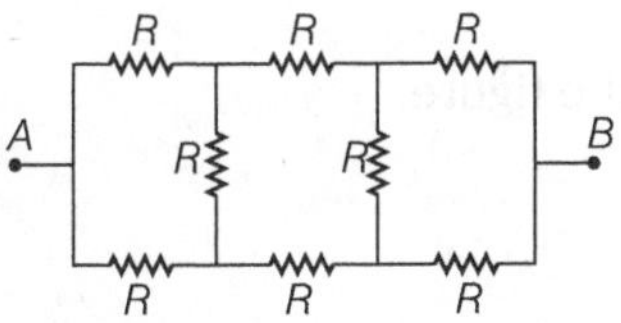

(viii) 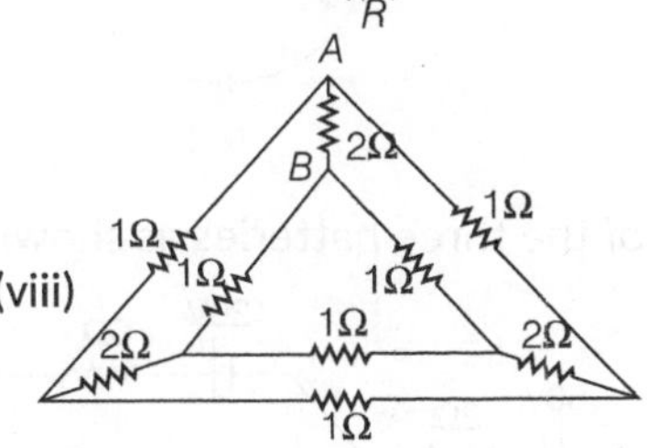

(ix) 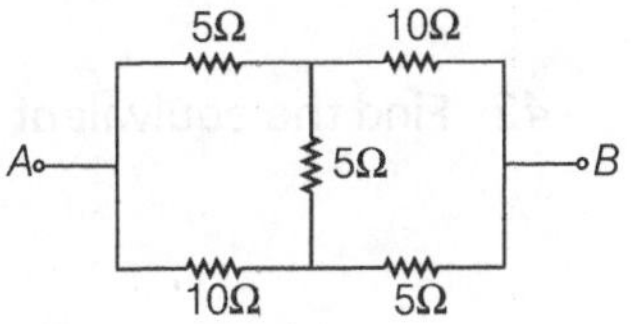

36 **Cube Problem**

Find the resistance of a wire frame shaped as a cube (figure) when measured between points

(i) 1-7

(ii) 1-3

(iii) 1-2

The resistance of each edge of the frame is *R*.

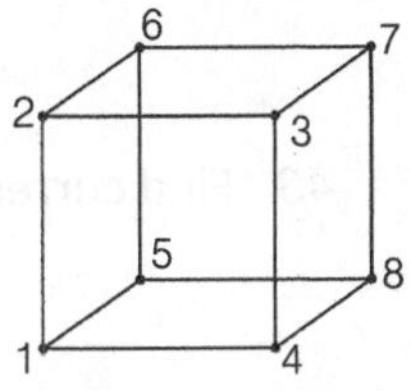

37 **Infinite Ladder Circuits**

Find the equivalent resistance across *A* and *B* in the following circuits.

(i) 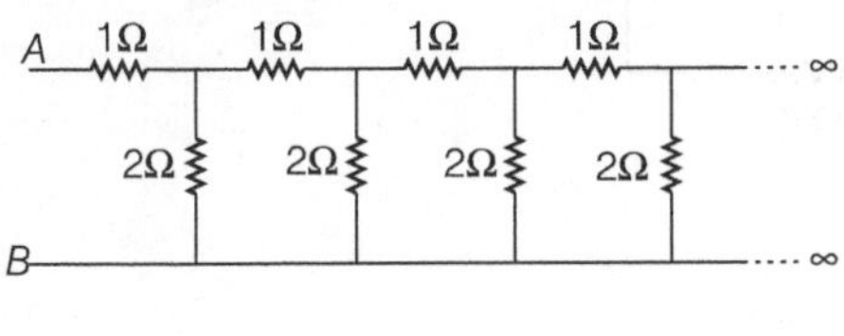

(ii) 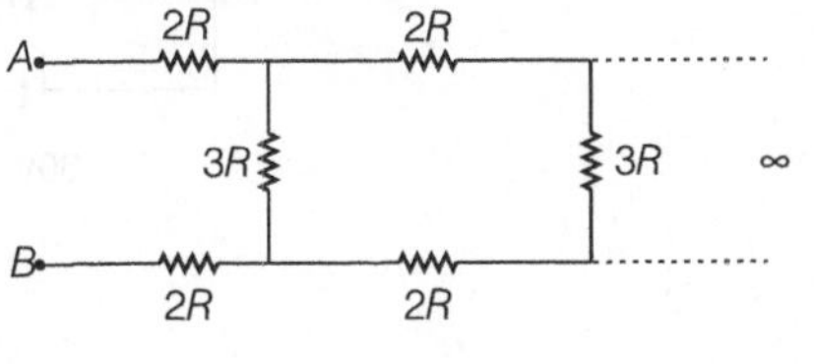

Problems in Physics

38 Infinite Grid

The resistance of each wire between neighbouring joint connections is equal to R. Find equivalent resistance of the whole grid between A and B.

(i) 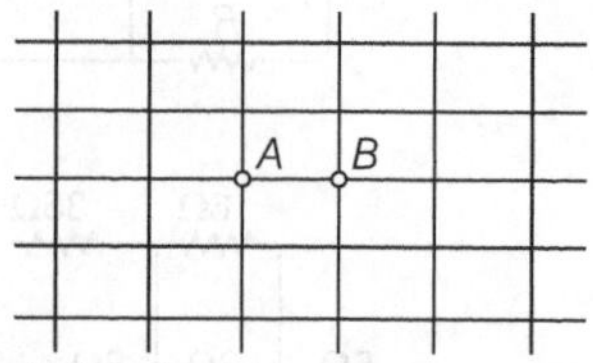

Infinite wire grid with square cells

(ii) 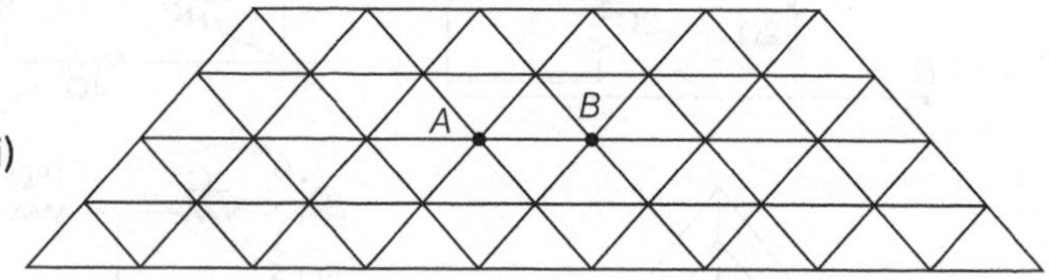

Infinite wire grid with equilateral triangular cells

1.8 Combination of Cells

39 10 lead-acid type of secondary cells, each of emf 3.0 V and internal resistance 0.05 Ω, are joined in series to provide a supply to a resistance of 4.5 Ω. Determine

(i) the current drawn from the supply and (ii) its terminal voltage.

40 n indentical cells each of emf 6 V, connected in series with an external resistor of 5Ω, carry a current of 10 A. If polarties of any two cells is reversed, the current flowing through the cells drops to 6A. Find the value of n and the internal resistance of each cell.

41 Four identical cells each of emf ε and internal resistance r are connected in a conducting loop as shown in the figure. Find

(i) the current in the loop

(ii) the potential difference between terminals of cell A and cell B.

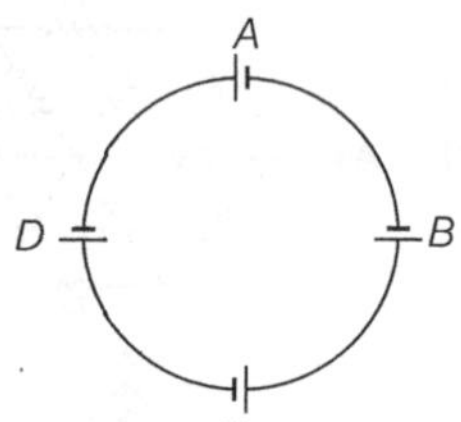

42 Find the equivalent emf of the three batteries as shown in the figure.

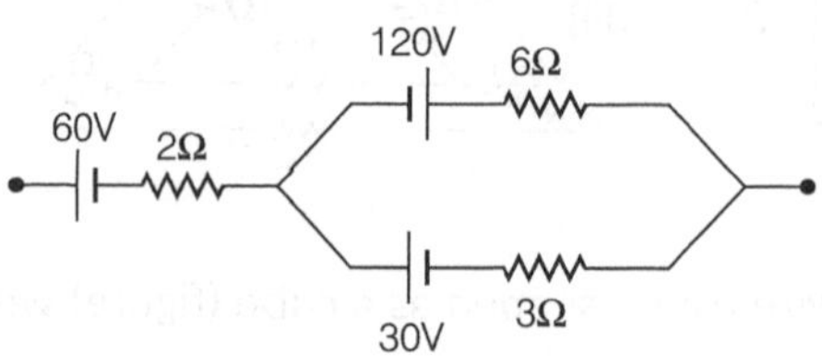

43 Find current flowing in 4Ω resistor.

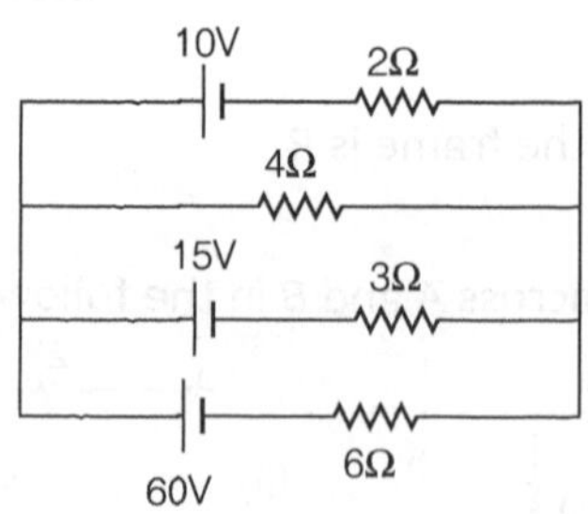

44 All batteries are having emf 10 V and internal resistance negligible. All resistors are in ohms. Calculate the current in the right most 2Ω resistor.

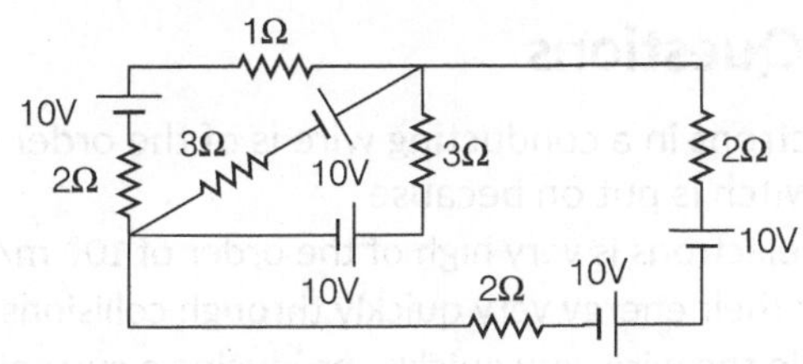

45 In a mixed grouping of identical cells, 8 rows are connected in parallel and each row contains 20 cells. This combination is connected across a 20 Ω external resistor. If the emf and internal resistance of each cell are 1.5 V and 2Ω respectively, find the current flowing through the external resistor.

46 *n* rows each containing *m* cells in series, are joined in parallel Maximum current is taken from this combination in a 3 Ω resistance. If the total number of cells used is 24 and internal resistance of each cell is 0.5 Ω, find the value of *m* and *n*.

Answers

1. (i) 8A (ii) 18 A **2.** 300 C, 30 A **3.** 40 C **4.** 1.25×10^{-3} ms^{-1} **5.** 1.5 *nevA* **6.** $\frac{J_0 \pi R^2}{3}$

7. 8.5×10^{-2} V/m **8.** 200 m **9.** 4*R* **10.** 81*R* **11.** (i) 21% (ii) 0.6%

12. $R_{AB} = \frac{\rho c}{ab}, R_{CD} = \frac{\rho b}{ac}, R_{EF} = \frac{\rho a}{bc}$ **13.** 187°C **14.** 2×10^{-7} m^2/V-s **15.** $(17 \times 10^4 \pm 5\%)$ Ω

16. (i) 48 V (ii) 7V **17.** (i) 1A (ii) 3A (iii) 12 A **18.** (i) $i_{10\Omega} = 3A, i_{20\Omega} = 2A, i_{30\Omega} = 1A$
(ii) $i_{1\Omega} = 10A, i_{2\Omega} = 5A, i_{3\Omega} = 30A, i_{4\Omega} = 35A$ (iii) $i_{1\Omega} = 20A, i_{2\Omega} = 5A, i_{3\Omega} = 15A$

19. (i) 2.5 A (ii) $i_{4\Omega} = 2A, i_{5\Omega} = 0, i_{6\Omega} = 2A$ (iii) $i_{1\Omega} = 20A, i_{2\Omega} = 5A, i_{3\Omega} = 15A$
(iv) $i_{3\Omega} = 2A$, $i_{6\Omega} = 2A, i_{4\Omega} = 0$ (v) $i_{2\Omega} = 10A$, $i_{3\Omega} = 10/3A$, $i_{6\Omega} = 20/3A$
(vi) $i_{1\Omega} = 0$, $i_{2\Omega} = i_{3\Omega} = 2A$, $i_{4\Omega} = i_{5\Omega} = 3A$

20. (i) 4A (ii) 32 V (iii) 128 W **21.** 29 Ω **22.** 0.6 Ω

23. **Case I** (i) 6 A (ii) 28 V and 14 V (iii) 168W (delivered) and 84W (delivered)
Case II (i) 2A (ii) 36 V and 22 V (iii) 72 W (delivered) and 44 W (consumed)

24. (i) 2A (ii) 8V (iii) 12 W **25.** (i) 12 A and 4 A (ii) 24 V (iii) 144 W

26. 4 A **27.** 8/3 A **28.** 216 V **29.** 1A **30.** 5 V **31.** 100 Ω **32.** 12 A

33. (i) 32 Ω (ii) 6Ω (iii) 11*R*/18 (iv) 2 Ω (v) 16/3 Ω (vi) *R* (vii) 20/3 Ω
(viii) 9/4Ω (ix) 226/39 Ω

34. (i) 3*R*/5 (ii) 3*R*/4 (iii) 4*r*/3 (iv) *R*/3 (v) 3*R*/5 (vi) 5Ω (vii) 1Ω
(viii) 4/3 Ω (ix) 4*R*/11 (x) 2*R*/9 (xi) 22.5 Ω (xii) 28/3 Ω

35. (i) 2*R*/3 (ii) 3Ω (iii) 3*R*/4 (iv) 44*R*/21 (v) 4*R*/5 (vi) 8*R*/15
(vii) 3*R*/2 (viii) 1Ω (ix) 7Ω **36.** (i) $\frac{5}{6}R$ (ii) $\frac{3}{4}R$ (iii) $\frac{7}{12}R$

37. (i) 2Ω (ii) 6*R* **38.** (i) *R*/2 (ii) *R*/3

39. (i) 6A (ii) 27 V **40.** $n = 10, r = 0.1$ Ω

41. (i) $\frac{\varepsilon}{2r}$ (ii) $\frac{\varepsilon}{2}$ and $\frac{3\varepsilon}{2}$ **42.** 40 V **43.** 4A

44. 2A **45.** 1.2 A **46.** $n = 2$ and $m = 12$

For Detailed Solutions
Scan the code

LEVEL 02 \ JEE Main Level

Single Choice Correct Questions

1 The drift velocity of electrons in a conducting wire is of the order of 1 mm/s, yet the bulb glows very quickly after the switch is put on because

(a) the random speed of electrons is very high of the order of 10^6 m/s

(b) the electrons transfer their energy very quickly through collisions

(c) electric field is set up in the wire very quickly, producing a current through each cross-section, almost instantaneously

(d) All of the above

2 Two cylindrical rods of uniform cross-section radii r and $2r$, having free electrons per unit volume $2n$ and n respectively are joined in series. A current I flows through them in steady state. Then, the ratio of drift velocity of free electron in left rod to drift velocity of electron in the right rod is $\left(\frac{V_L}{V_R}\right)$ is

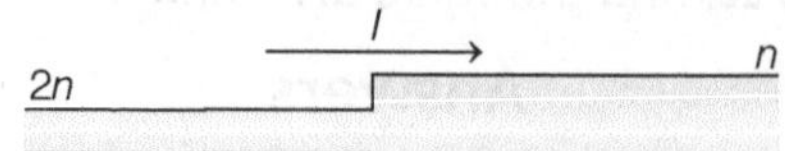

(a) 0.5 (b) 1 (c) 2 (d) 4

3 Find the average drift speed of conduction electrons in a copper wire of cross-sectional area 1.0×10^{-7} m^2 carrying a current of 1.5 A. Assume that each copper atom contributes roughly one conduction electron. The density of copper is 9.0×10^3 kg m^{-3} and its atomic mass is 63.5 u.

(a) 1.1 mm/s (b) 1.5 mm/s (c) 2.2 mm/s (d) 2.5 mm/s

4 A current I flows through a cylindrical rod of uniform cross-section area A and resistivity ρ. The electric flux through the shaded cross-section of rod as shown in figure, is

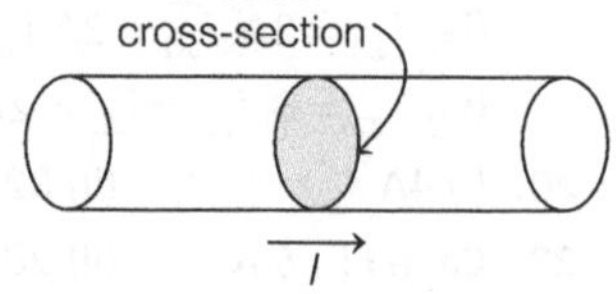

(a) $\frac{I}{\rho}$ (b) ρI

(c) $\frac{\rho I}{A}$ (d) $\frac{\rho A}{I}$

5 In a straight conductor of uniform cross-section charge q is flowing for time t. Let s be the specific charge of an electron. The momentum of all the free electrons per unit length of the conductor, due to their drift velocity only, is

(a) $\frac{q}{ts}$ (b) $\left(\frac{q}{ts}\right)^2$ (c) $\sqrt{\frac{q}{ts}}$ (d) qts

6 The current in a metallic conductor is plotted against voltage at two different temperatures T_1 and T_2. Which is correct?

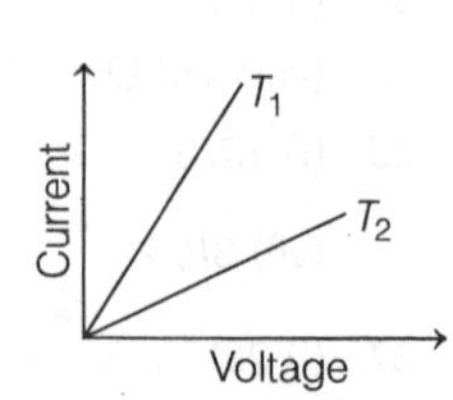

(a) $T_1 > T_2$ (b) $T_1 < T_2$

(c) $T_1 = T_2$ (d) None of these

7 A resistance wire has a resistance R. Half of this wire is stretched to double its length and half is twisted to double its thickness, then its new resistance becomes

(a) $\frac{17R}{8}$ (b) $\frac{17R}{16}$

(c) $\frac{65R}{32}$ (d) $\frac{65R}{16}$

8 The resistance of a wire at 20° C is 20Ω and at 500° C is 60Ω. At what temperature, resistance will be 25 Ω?

(a) 50°C (b) 60°C (c) 70°C (d) 80°C

9 Two resistors of resistances 400Ω and 800Ω at 0°C, having thermal coefficient of resistance α and 4α respectively, are connected in parallel. The equivalent thermal coefficient of resistance is

(a) 5α (b) $\frac{3\alpha}{2}$ (c) 2α (d) $\frac{4\alpha}{5}$

10 A non-conducting ring of radius R has a charge Q distributed unevenly over it. If it rotates with an angular velocity ω, the equivalent current will be

(a) 0 (b) $Q\omega$ (c) $\frac{Q\omega}{2\pi}$ (d) $\frac{Q\omega}{2R}$

11 Six resistors are connected between A and B as shown. The equivalent resistance between A and B of this network will be independent of the value of R, if x is equal to

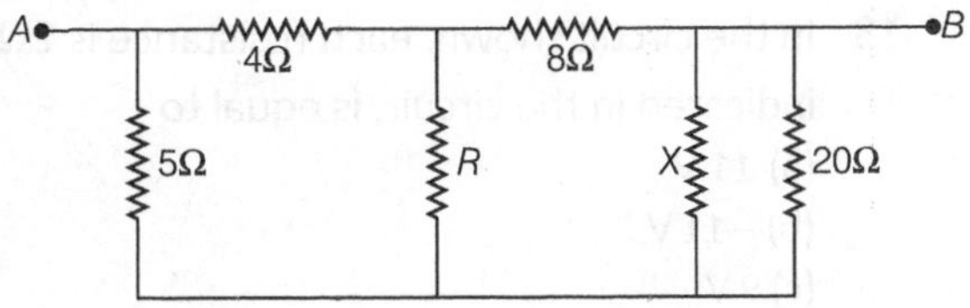

(a) 5 Ω (b) 10 Ω
(c) 15 Ω (d) 20 Ω

12 The equivalent resistance between the points A and B is

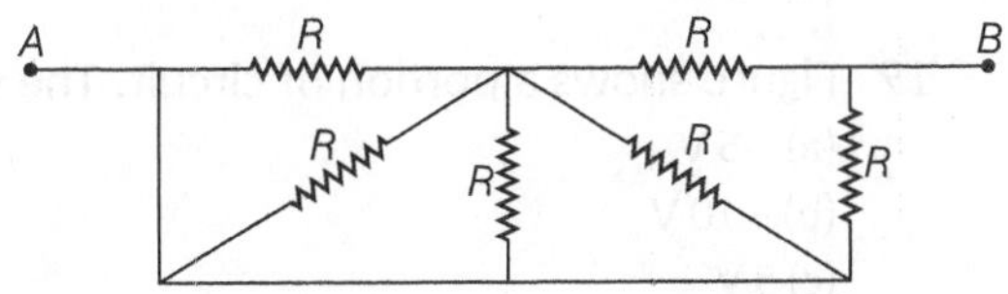

(a) 5R/9
(b) 2R/3
(c) 5R/6
(d) 3R/5

13 Sixteen resistors each of resistance R are connected in the circuit as shown in figure. The equivalent resistance across A and B is

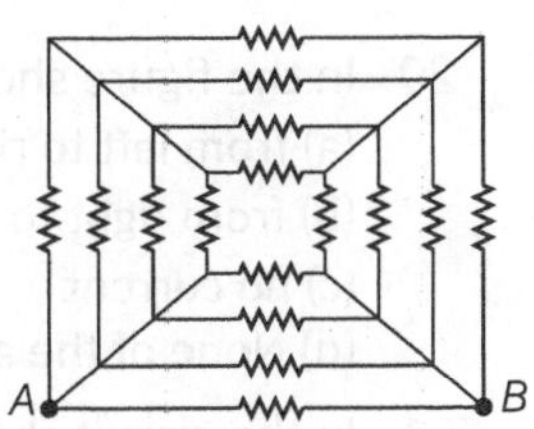

(a) R / 16
(b) R / 8
(c) 3R / 16
(d) R / 4

14 The equivalent resistance between the points A and B is

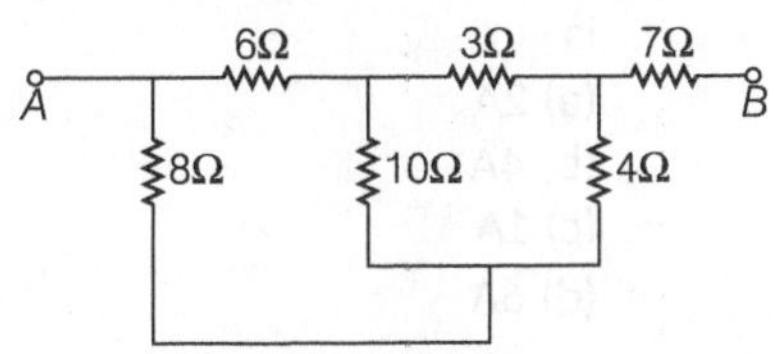

(a) $\frac{36}{7}$ Ω
(b) 10 Ω
(c) $\frac{85}{7}$ Ω
(d) None of the above

15 In the figure shown, find the equivalent resistance between A and B.

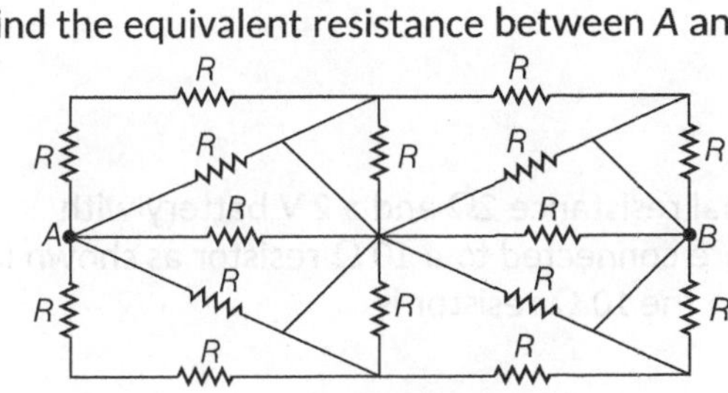

(a) 9R / 10 (b) 9R / 20
(c) 3R / 10 (d) 3R / 20

16 As the switch S is closed in the circuit shown in figure, current passing through it is

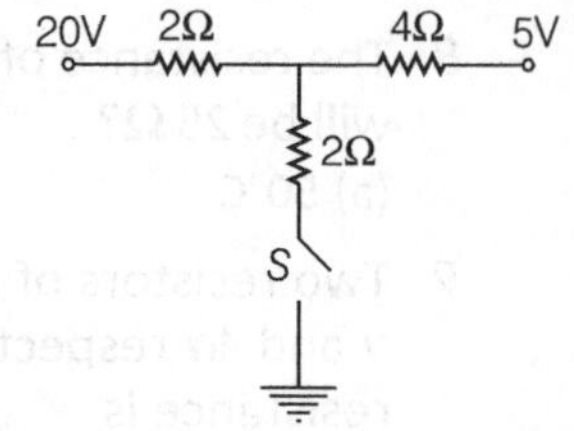

(a) 4.5 A
(b) 6.0 A
(c) 3.0 A
(d) zero

17 The potential difference between points A and B in a section of a circuit as shown in figure is

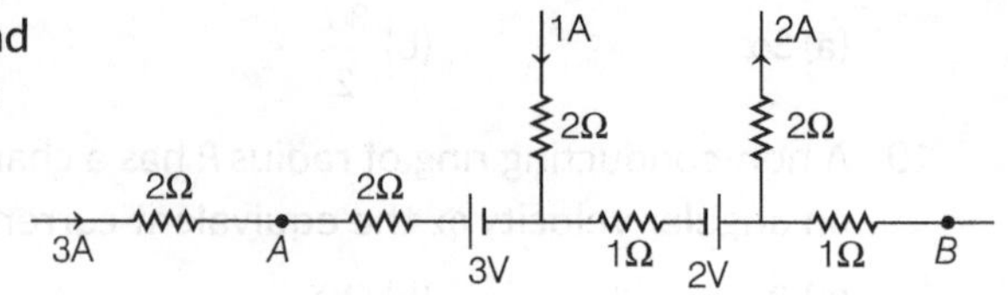

(a) 5 V
(b) −13 V
(c) 0 V
(d) +13 V

18 In the circuit shown, each resistance is 2Ω. The potential V_1 as indicated in the circuit, is equal to

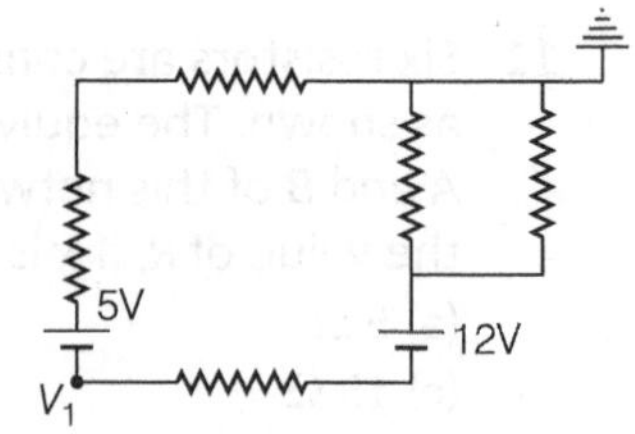

(a) 11 V
(b) −11 V
(c) 9 V
(d) −9 V

19 Figure shows a portion of circuit. The potential of point D is

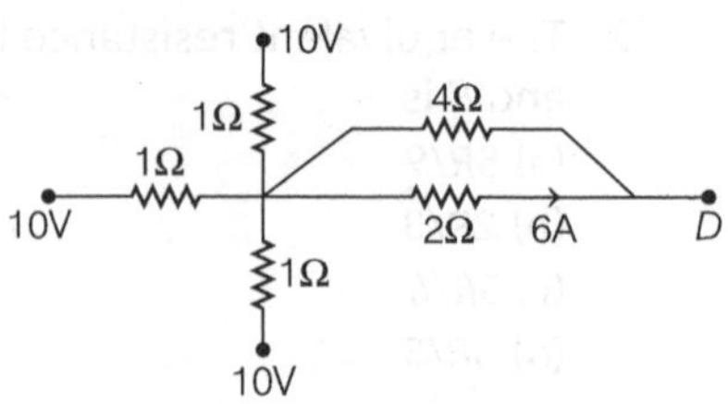

(a) −5 V
(b) −10 V
(c) 5 V
(d) 10 V

20 In the figure shown the current flowing through $2R$ is

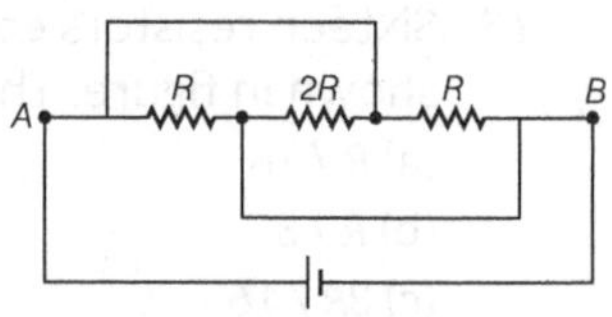

(a) from left to right
(b) from right to left
(c) no current
(d) None of the above

21 In the circuit shown in figure, the total current supplied by the battery is

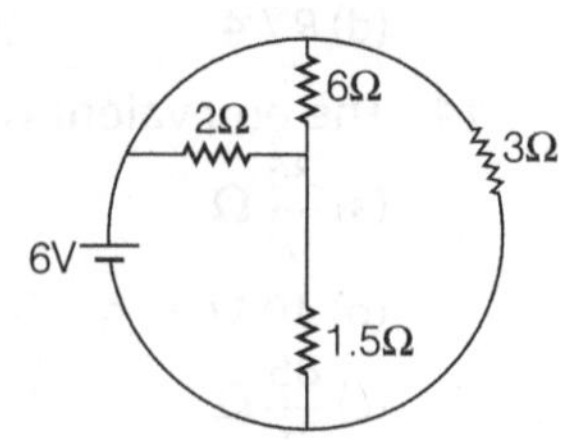

(a) 2A
(b) 4A
(c) 1A
(d) 6A

22 In the circuit shown, what is the potential difference V_{PQ}?

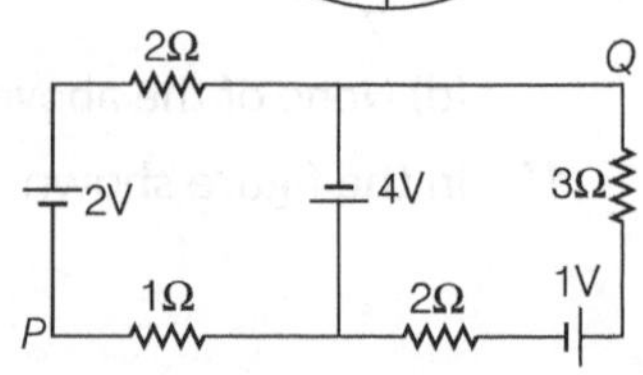

(a) +3V
(b) +2V
(c) −2V
(d) None of the above

23 A 5V battery with internal resistance 2Ω and a 2 V battery with internal resistance 1Ω are connected to a 10 Ω resistor as shown in the figure. The current in the 10 Ω resistor is

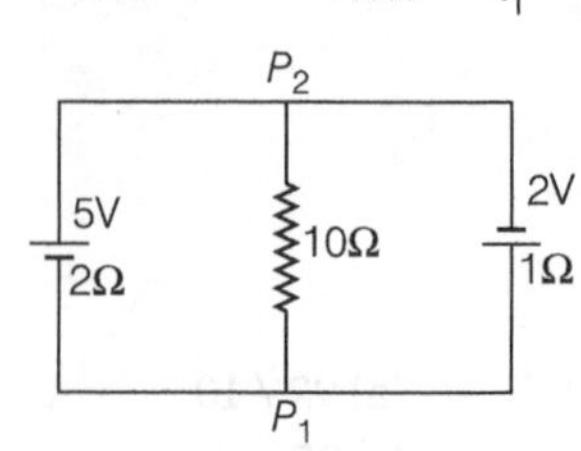

(a) 0.27 A, P_1 to P_2
(b) 0.03 A, P_2 to P_1
(c) 0.27 A, P_2 to P_1
(d) 0.03 A, P_1 to P_2

24 In the circuit as shown, a wire is connected between points A and B. How much current will flow through the wire AB?

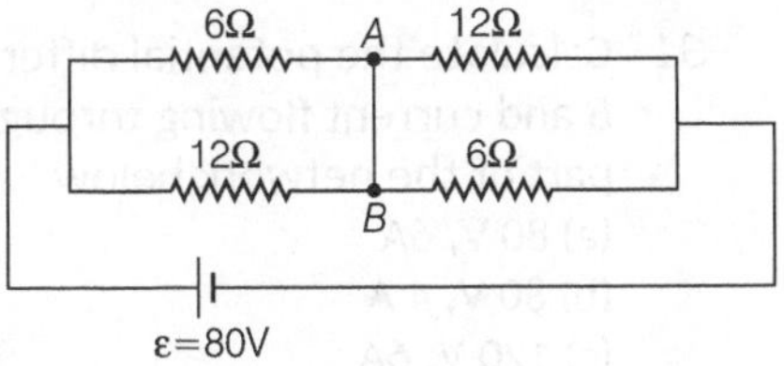

(a) Zero
(b) (10/3) A
(c) (20/3) A
(d) (5/3) A

25 In the given circuit, $R_B = 6\ \Omega, R_C = 3\ \Omega, R_D = 2\ \Omega$. The ratio of current through resistors A, B, C and D will be in the ratio

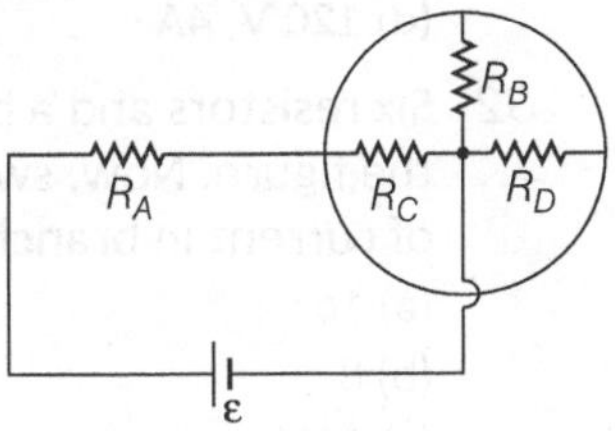

(a) 3 : 2 : 1 : 6
(b) 6 : 3 : 2 : 1
(c) 1 : 2 : 3 : 6
(d) 6 : 1 : 2 : 3

26 In the network shown in figure, the potential difference V_{ab} when the switch S is open, is

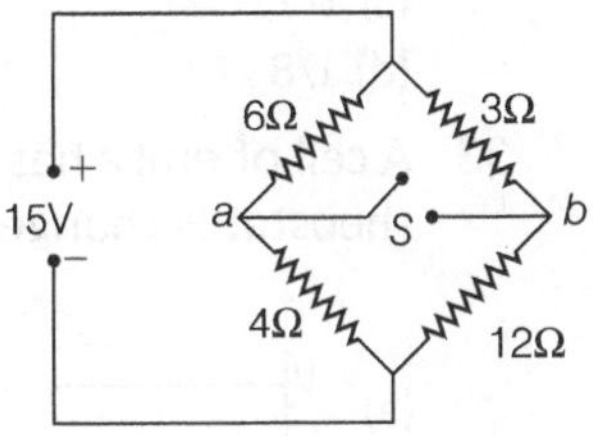

(a) 3 V
(b) – 6 V
(c) 9 V
(d) 12 V

27 In the circuit shown, the value of R in ohm, that will result in no current through the 30 V battery, is

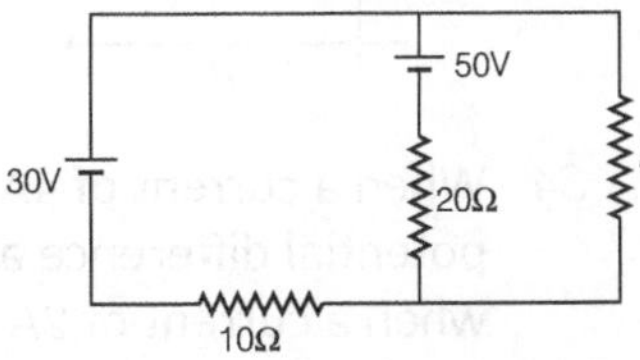

(a) 10 Ω
(b) 25 Ω
(c) 30 Ω
(d) 40 Ω

28 Find the current through 10 V battery.

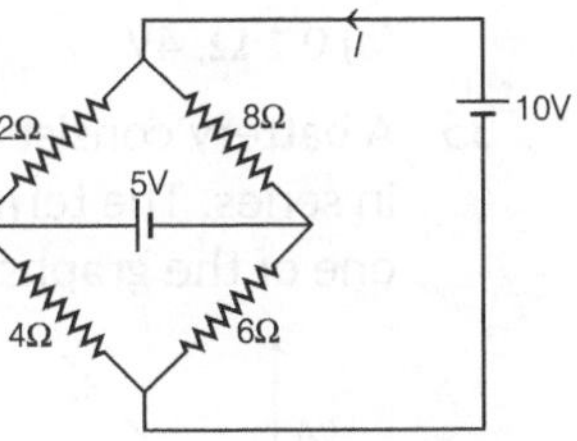

(a) 2.25 A
(b) 2.5 A
(c) 3 A
(d) 5.5 A

29 Three carbon resistors R_1, R_2 and R_3 are joined as shown. Their colour code reads as

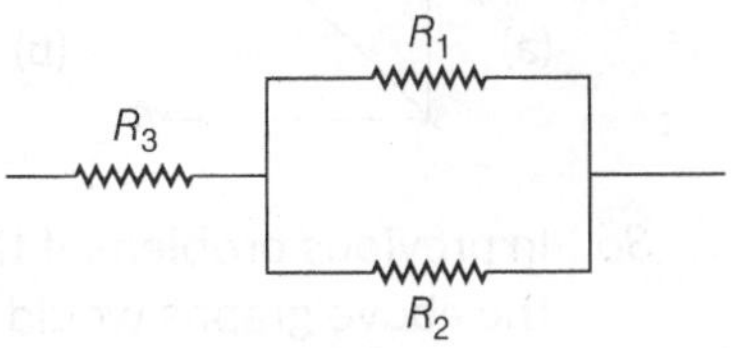

R_1 : Brown, Red, Brown
R_2 : Blue, Black, Black
R_3 : Orange, Black, Black
What is their equivalent resistance?

(a) 210 Ω
(b) 180/7 Ω
(c) 40 Ω
(d) 70 Ω

30 The circuit is as given in figure. The current through the branch BE of the circuit is

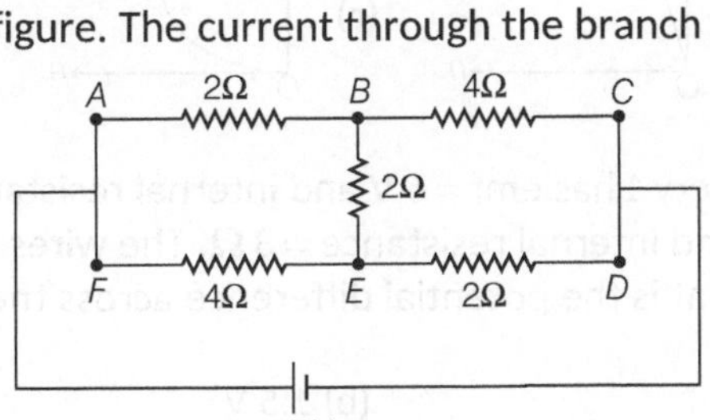

(a) 2A
(b) 5 A
(c) 7A
(d) 10 A

31 Calculate the potential difference between points A and B and current flowing through the 6 Ω resistor in the part of the network below.

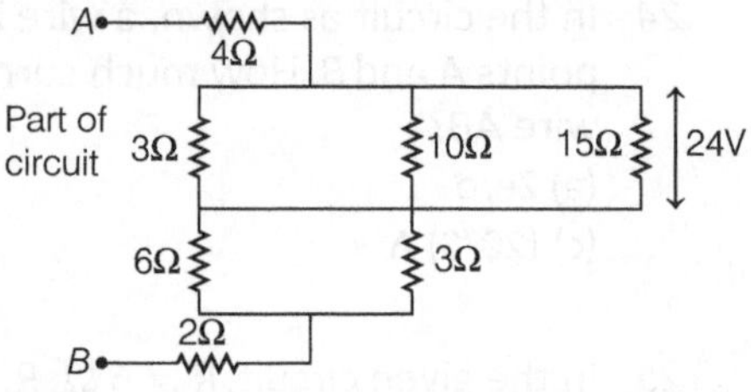

(a) 80 V, 6A
(b) 80 V, 4 A
(c) 120 V, 6A
(d) 120 V, 4A

32 Six resistors and a battery and connected as shown in the figure. Now, switch S is closed, then find the ratio of current in branch CD to current in branch AF.

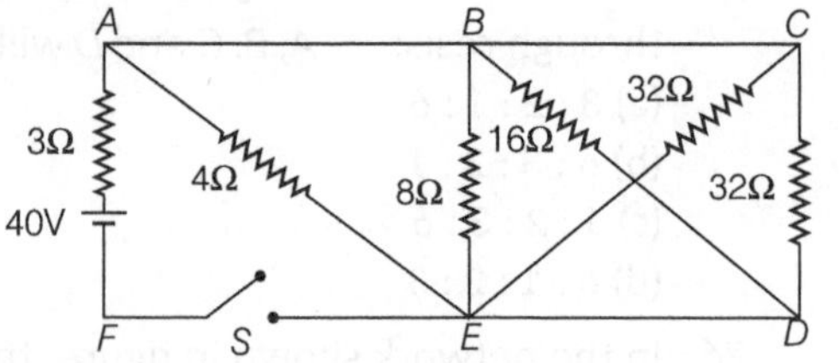

(a) 16
(b) 8
(c) 1/16
(d) 1/8

33 A cell of emf ε has an internal resistance r and is connected to rheostat. When resistance R of rheostat is changed, correct graph of potential difference across it is

(a) 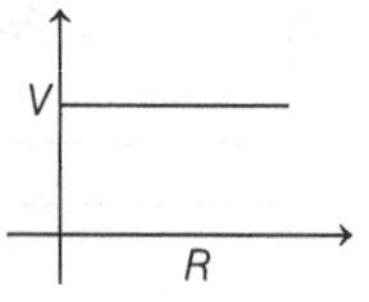 (b) 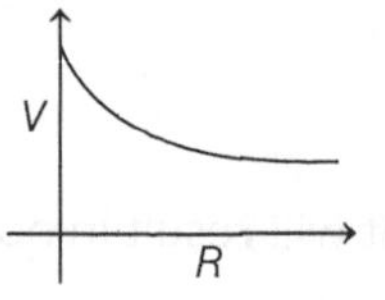 (c) 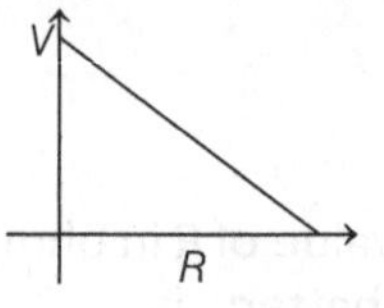 (d) 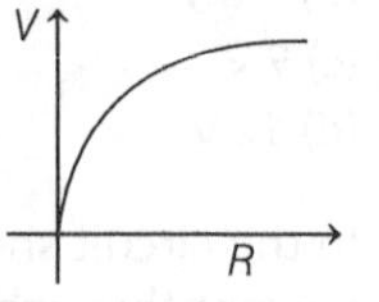

34 When a current of 4A flows within a battery from its positive to negative terminal, the potential difference across the battery is 12 V. The potential difference across the battery is 9 V when a current of 2A flows within it from its negative to its positive terminal. The internal resistance and the emf of the battery are

(a) 0.1 Ω, 4V (b) 0.2 Ω 5V (c) 0.5 Ω, 10V (d) 0.7 Ω, 10V

35 A battery consists of a variable number n of identical cells having internal resistance connected in series. The terminals of the battery are short-circuited and the current I measured. Which one of the graphs below shows the relationship between I and n?

(a) 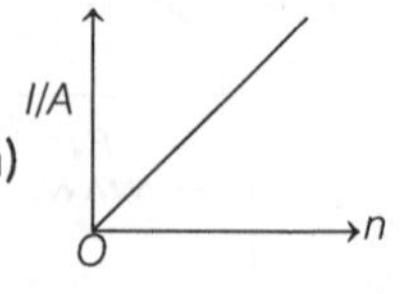 (b) 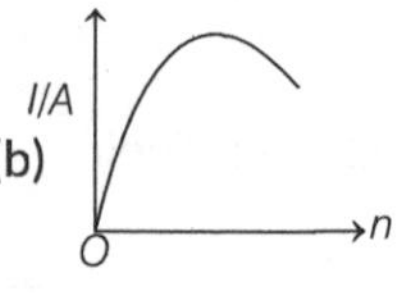 (c) 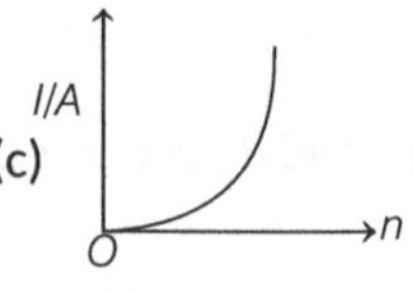 (d) 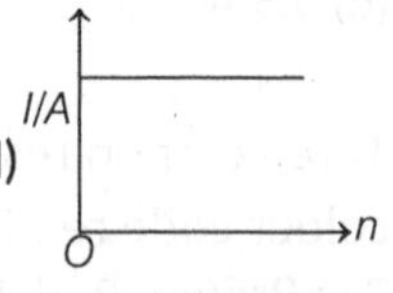

36 In previous problem, if the cells had been connected in parallel (instead of in series), which of the above graphs would have shown the relationship between total current I and n?

(a) 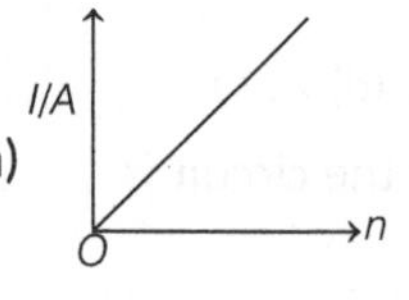 (b) 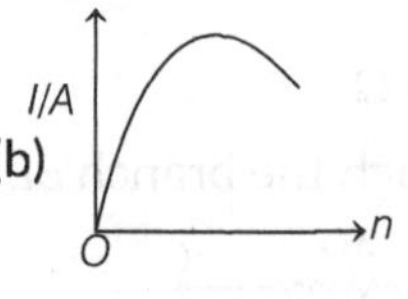 (c) 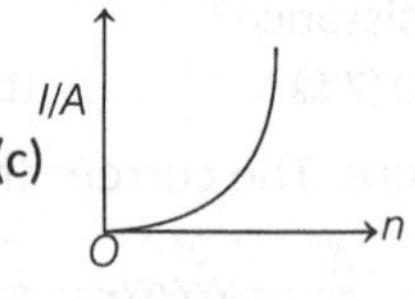 (d) 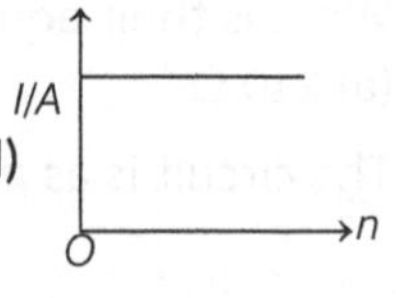

37 In the figure shown, battery 1 has emf = 6 V and internal resistance = 1 Ω. Battery 2 has emf = 2V and internal resistance = 3 Ω. The wires have negligible resistance. What is the potential difference across the terminals of battery 2?

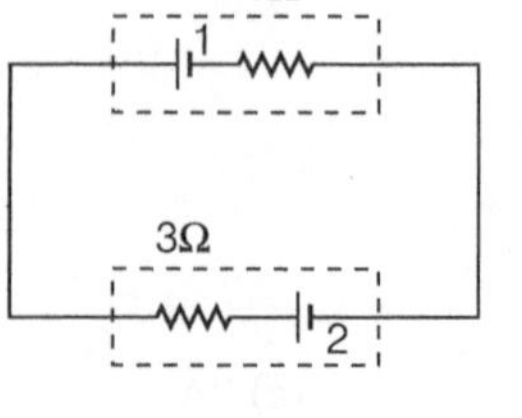

(a) 4 V (b) 1.5 V
(c) 5 V (d) 0.5 V

38 A circuit is comprised of eight identical batteries and a resistor $R = 0.8\ \Omega$. Each battery has an emf of 1.0 V and internal resistance of $0.2\ \Omega$. The voltage difference across any of the battery is

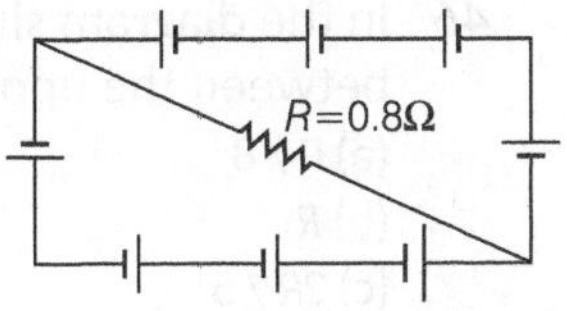

(a) 0.5 V (b) 1.0 V
(c) 0 (d) 2 V

39 Two batteries one of the emf 3 V, internal resistance $1\ \Omega$ and the other of emf 15 V, internal resistance $2\ \Omega$ are connected in series with a resistance R as shown. If the potential difference between a and b is zero, then resistance R in ohm is

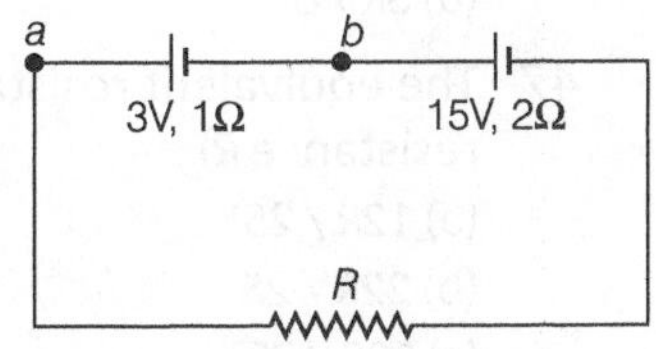

(a) 5 (b) 7
(c) 3 (d) 1

40 You are given 48 cells each of 2 V and internal resistance $1\ \Omega$. How will you connect them so that the current through an external resistance of 3Ω is the maximum?
(a) 8 cells in series, 6 such groups in parallel (b) 12 cells in series, 4 such groups in parallel
(c) 16 cells in series, 3 such groups in parallel (d) 24 cells in series, 2 such groups in parallel

41 The value of current through $20\ \Omega$ resistor is

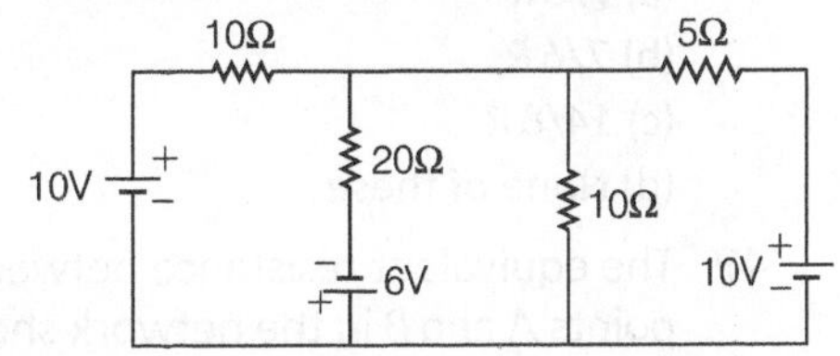

(a) 0.6 A
(b) 0.8 A
(c) 1.2 A
(d) 1.8 A

42 In order to determine the emf of a storage battery, it was connected in series with a standard cell in a certain circuit and a current I_1 was obtained. When the battery is connected to the same circuit, opposite to the standard cell, a current I_2 flow in the external circuit from the positive pole of the storage battery was obtained. What is the emf ε_1 of the storage battery? The emf of the standard cell is ε_2.

(a) $\varepsilon_1 = \frac{I_1 + I_2}{I_1 - I_2}\varepsilon_2$ (b) $\varepsilon_1 = \frac{I_1 + I_2}{I_2 - I_1}\varepsilon_2$ (c) $\varepsilon_1 = \frac{I_1 - I_2}{I_1 + I_2}\varepsilon_2$ (d) $\varepsilon_1 = \frac{I_2 - I_1}{I_1 + I_2}\varepsilon_2$

43 The battery in the diagram is to be charged by the generator G. The generator has a terminal voltage of 120 V when the charging current is 10 A. The battery has an emf of 100 V and an internal resistance of $1\ \Omega$. In order to charge the battery at 10 A charging current, the resistance R should be set at

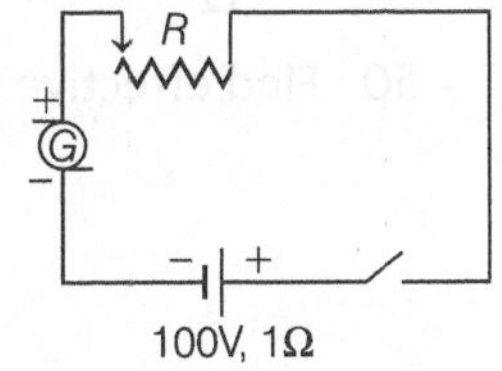

(a) $0.1\ \Omega$ (b) $0.5\ \Omega$
(c) $1.0\ \Omega$ (d) $5.0\ \Omega$

44 In the circuit shown, the equivalent resistance between the points A and B is

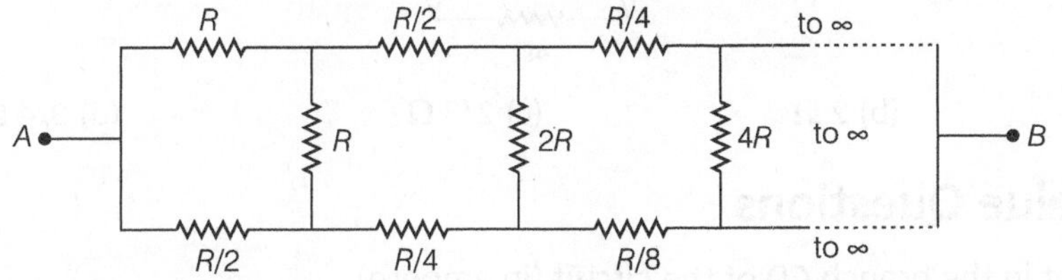

(a) R (b) $R/2$ (c) $4R/3$ (d) $2R/3$

45 Eight identical resistance r each are connected as shown. The equivalent resistance between AD is

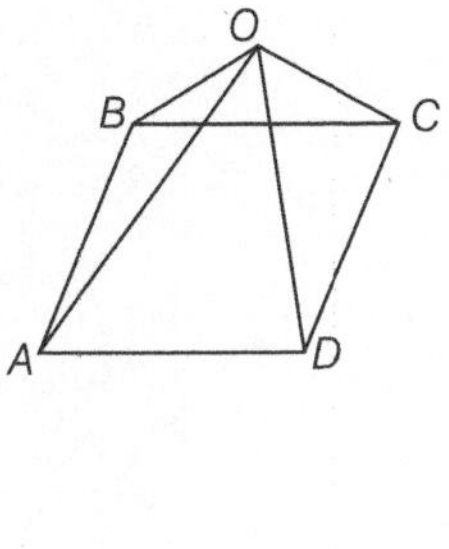

(a) $8R/15$
(b) $4R/15$
(c) $4R/5$
(d) $7R/10$

46 In the diagram shown, all the wires have resistance R. The equivalent resistance between the upper and lower dots shown in the diagram is

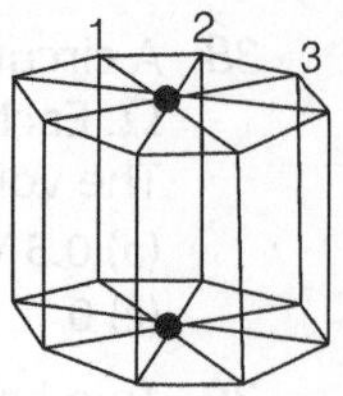

(a) $R/8$
(b) R
(c) $2R/5$
(d) $3R/8$

47 The equivalent resistance between the points A and B is (Each branch has resistance R)

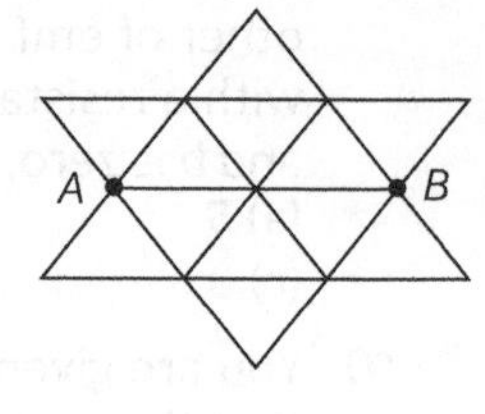

(a) $12R/25$
(b) $22R/25$
(c) $12R/35$
(d) $22R/35$

48 The resistance of all the wires between any two adjacent dots is R. Then, equivalent resistance between A and B as shown in figure is

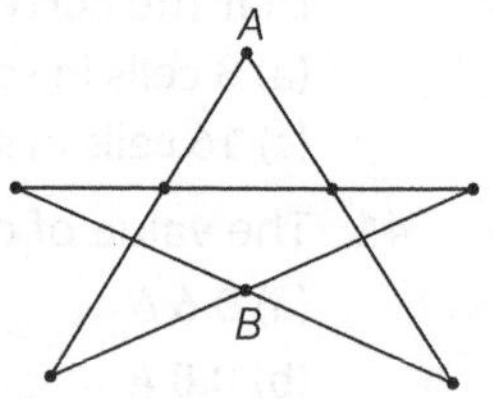

(a) $7/3\,R$
(b) $7/6\,R$
(c) $14/8\,R$
(d) None of these

49 The equivalent resistance between the terminal points A and B in the network shown in figure is

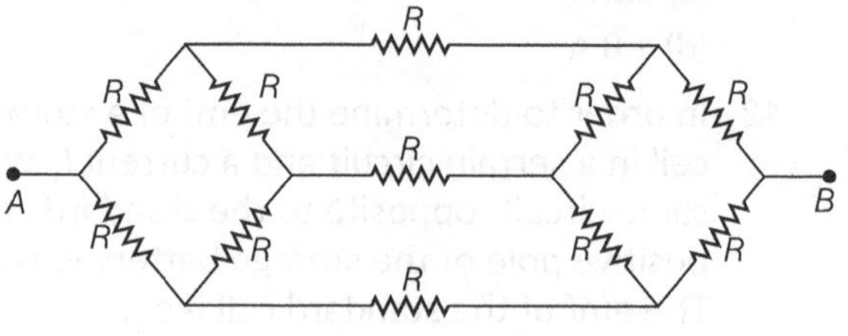

(a) $\frac{7R}{5}$
(b) $\frac{5R}{6}$
(c) $\frac{7R}{12}$
(d) $\frac{5R}{12}$

50 Find effective resistance between A and B.

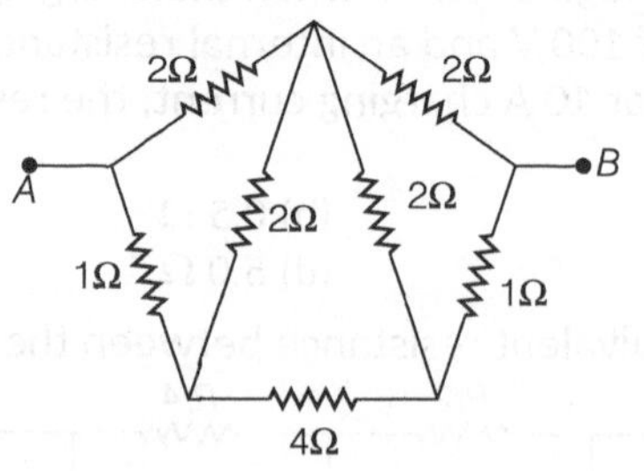

(a) 1 Ω (b) 2 Ω (c) 2/3 Ω (d) 3/4 Ω

Numerical Value Questions

51 Find current in the branch CD of the circuit (in ampere).

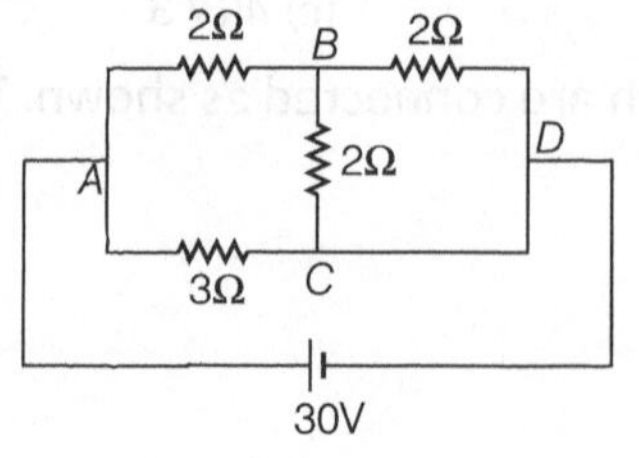

52 In the circuit shown in figure, $(V_A - V_B) = 16$ V. Find the current (in ampere) passing through 2 Ω resistor.

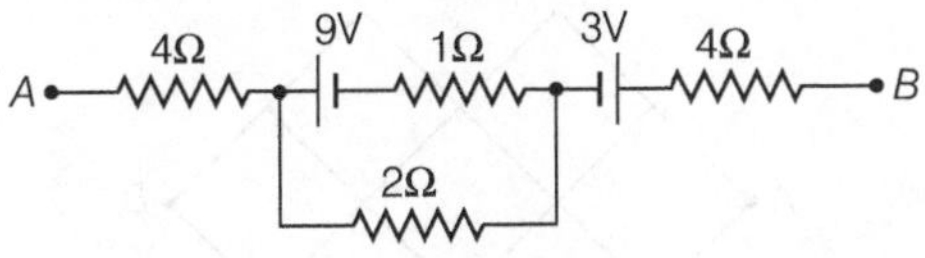

53 There is a cylindrical wire, whose temperature coefficient of resistivity is 6×10^{-3} /°C and coefficient of linear expansion is 1×10^{-3} /°C. The value of temperature coefficient of resistance is found be $N \times 10^{-3}$ /°C. Find the value of N.

54 An ionisation chamber with parallel conducting plates as anode and cathode, has 5×10^7 electrons and the same number of singly charged positive ions per cm^3. The electrons are moving towards the anode with velocity 0.4 m/s. The current density from anode to cathode is 4μ A/m^2. Find the velocity (in cm/s) of positive ions moving towards cathode.

55 Five 10Ω resistances are connected as shown in the figure. The resistance in the conducting wires (fully drawn lines) is negligible. Determine the resulting resistance R (in ohm) between A and B.

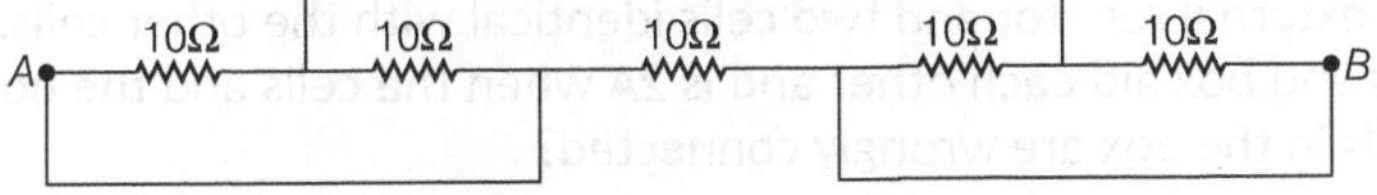

56 Find the effective resistance across terminals PQ in ohm.

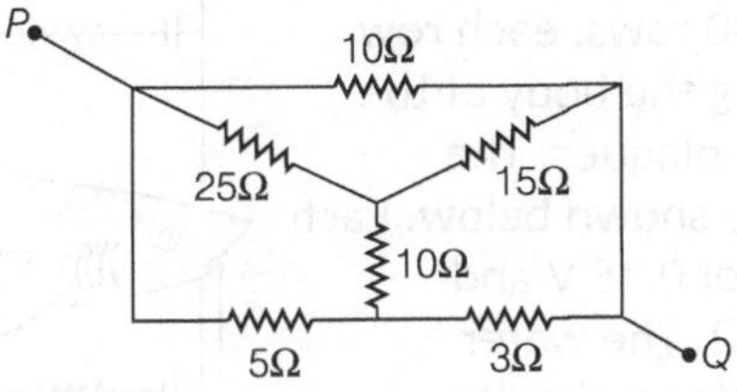

57 In the circuit shown, current through the resistance 2Ω is i_1 and current through the resistance 30Ω is i_2. Find the ratio $\frac{i_1}{i_2}$.

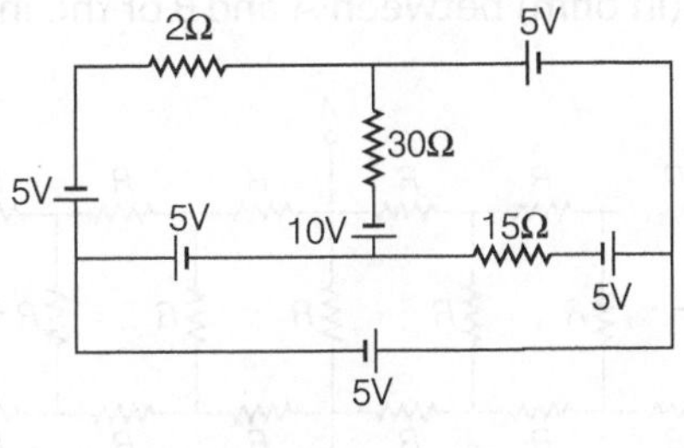

58 In the circuit, the electric current through the battery is .... ampere.

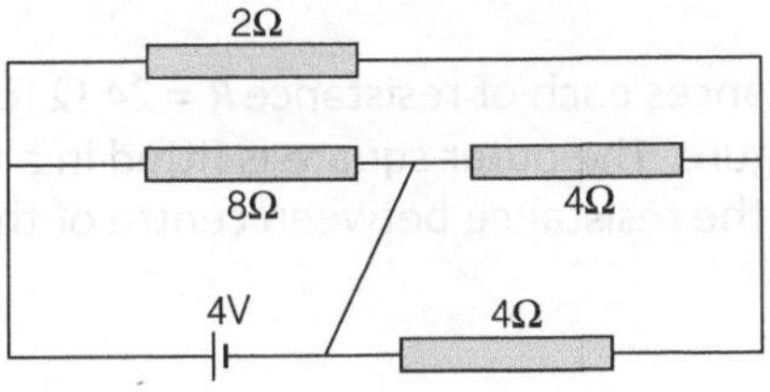

59 Each branch in the following circuit has a resistance R. The equivalent resistance of the circuit between the points A and B is x. Find the value of x/R.

60 The equivalent resistance between the points A and B in the following circuit is $x\ \Omega$, then x is

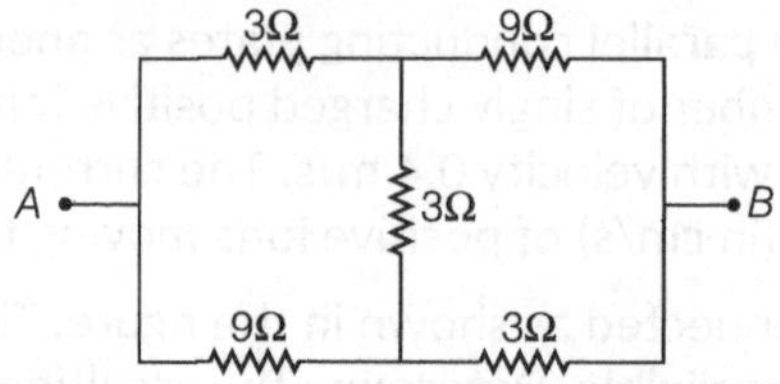

61 12 cells each of negligible internal resistance and having the same emf are connected in series and are kept in a closed box. Some of the cells are wrongly connected. This box is connected in series with an external resistor and two cells identical with the other cells. The current is 3A when the cells and box aid each other and is 2A when the cells and the box oppose each other. How many cells in the box are wrongly connected?

62 Eels are able to generate current with biological cells called electroplaques. The electroplaques in an Eel are arranged in 100 rows, each row stretching horizontally along the body of the fish containing 5000 electroplaques. The arrangement is suggestively shown below. Each electroplaques has an emf of 0.15 V and internal resistance of 0.25 Ω. The water surrounding the Eel completes a circuit between the head and its tail. If the water surrounding it has a resistance of 500 Ω, find the current (in ampere upto 2 decimal places), which an Eel can produce in water.

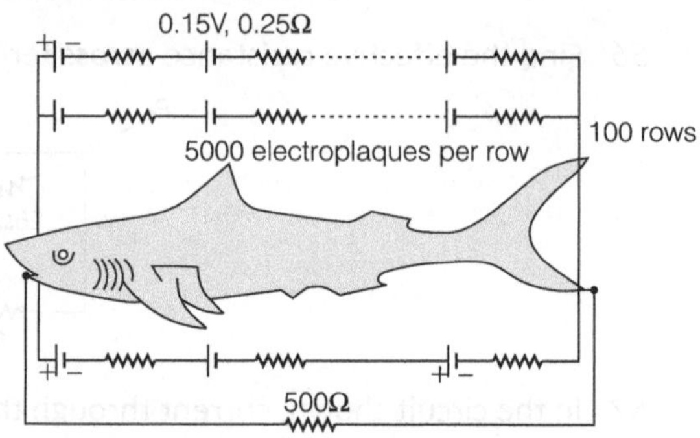

63 Find equivalent resistance (in ohm) between A and B of the infinite network shown. Resistance of each $R = 100\sqrt{3}\ \Omega$.

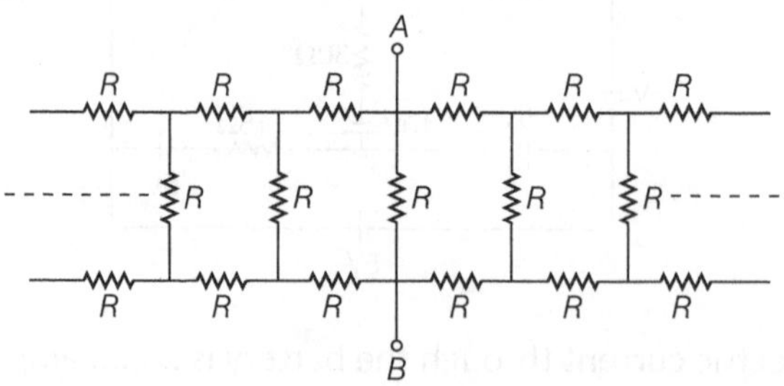

64 A network of twelve resistances each of resistance $R = 24\ \Omega$ form a square of squares as shown in the figure. The outer square is fitted in a metal ring of negligible resistance. Find the resistance between centre of the square and the ring.

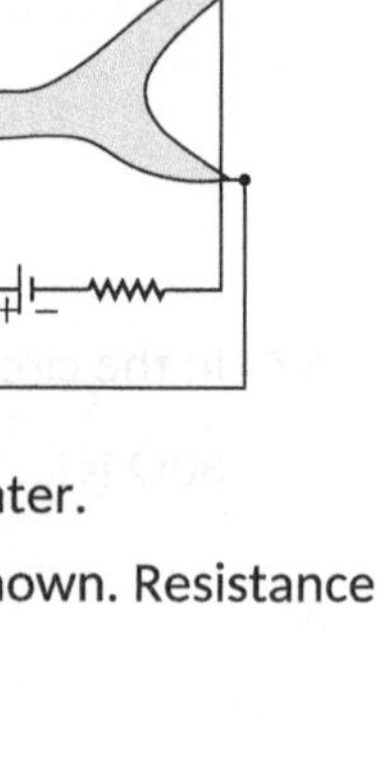

65 In the given network of resistors, each resistor has resistance $R = 12\ \Omega$. Find equivalent resistance (in ohm) across A and B.

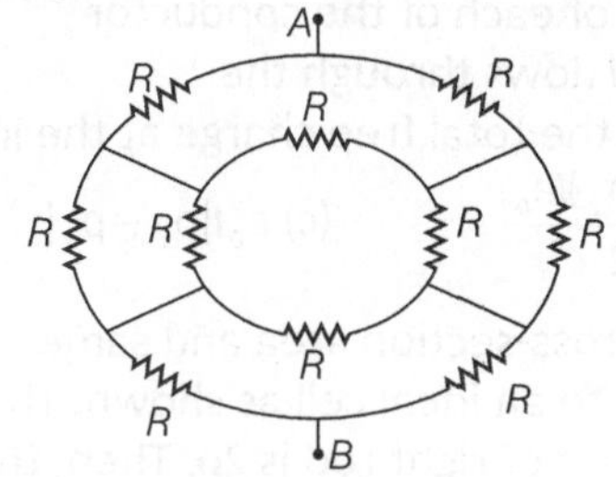

Answers

1. (c)	2. (c)	3. (a)	4. (b)	5. (a)	6. (b)	7. (c)	8. (d)	9. (c)	10. (c)
11. (d)	12. (a)	13. (c)	14. (c)	15. (b)	16. (a)	17. (d)	18. (d)	19. (a)	20. (b)
21. (b)	22. (b)	23. (b)	24. (b)	25. (d)	26. (b)	27. (c)	28. (a)	29. (d)	30. (b)
31. (d)	32. (c)	33. (d)	34. (c)	35. (d)	36. (a)	37. (c)	38. (c)	39. (c)	40. (b)
41. (a)	42. (a)	43. (c)	44. (d)	45. (a)	46. (d)	47. (d)	48. (b)	49. (a)	50. (b)
51. (15)	52. (3.5)	53. (5)	54. (10)	55. (5)					
56. (4)	57. (9)	58. (1.5)	59. (2.00)	60. (5)					
61. (1)	62. (1.46)	63. (100)	64. (9)	65. (15)					

For Detailed Solutions
Scan the code

LEVEL 03 \ JEE Advanced Level

Single Choice Correct Questions

1 Suppose a current carrying cone shaped wire made up of a uniform material has a cross-sectional area that gradually becomes smaller along the wire as shown in figure.

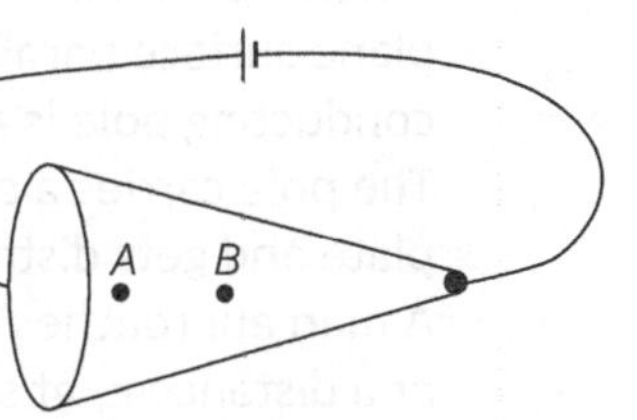

(a) Electric current is different in different portions of wire
(b) Current density at A is more than that of at point B
(c) Electric field at point A is lesser than that of point B
(d) Drift speed of electrons at point A is same as that of at point B

2 A cylindrical solid rod of length L and radius a is connected across a source of emf V and negligible internal resistance as shown in the figure. The resistivity of the rod at point P at a distance x from left end is given by $\rho = bx$ (where, b is a positive constant). Find the electric field at point P.

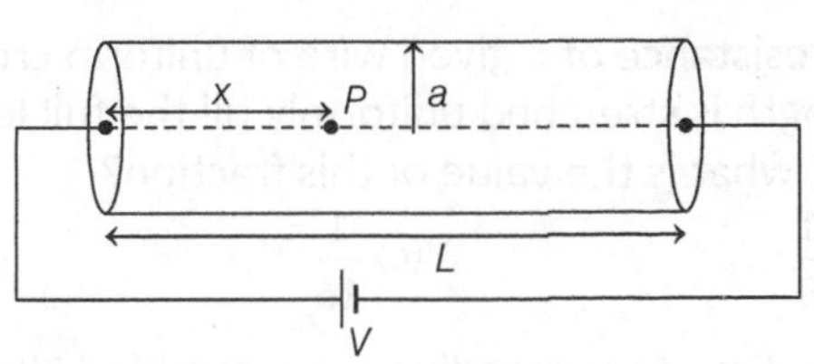

(a) $\frac{2V}{L^2}x$ (b) $\frac{V}{2L^2}x$ (c) $\frac{4V}{L^2}x$ (d) $\frac{8V}{L^2}x$

3 There are two concentric metallic spherical shells of radii 5 cm and 15 cm. The shells are being maintained at a constant potential difference of 10 V. The medium between the shells is filled with a material of resistivity $10^{-2}\ \Omega$-m. At a distance 10 cm from the centre, the electric field is

(a) 7.5 N/C (b) 75 N/C (c) 0.75 N/C (d) 750 N/C

4 Two long straight cylindrical conductors with resistivities ρ_1 and ρ_2 respectively are joined together as shown in figure. The radius of each of the conductor is a. If a uniform total current I flows through the conductors, the magnitude of the total free charge at the interface of the two conductor is

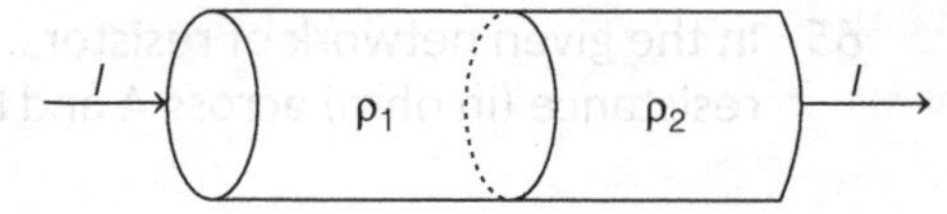

(a) zero (b) $\frac{(\rho_1 - \rho_2)I\varepsilon_0}{2}$ (c) $\varepsilon_0 I|\rho_1 - \rho_2|$ (d) $\varepsilon_0 I(\rho_1 + \rho_2)$

5 Two cylindrical rods of same cross-section area and same length are connected in series to an ideal cell as shown. The resistivity of left rod is ρ and that of right rod is 2ρ. Then, the variation of potential at any point P distant x from left end of combined rod system is given by

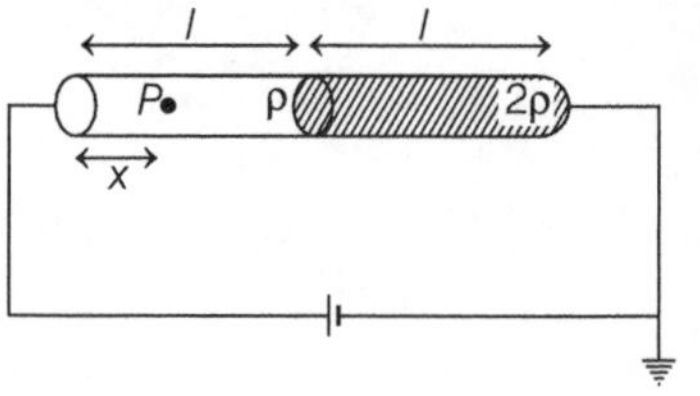

(a) 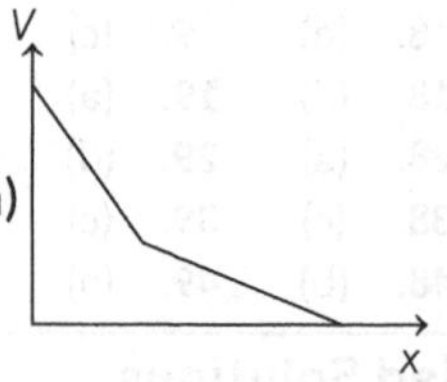

(b) 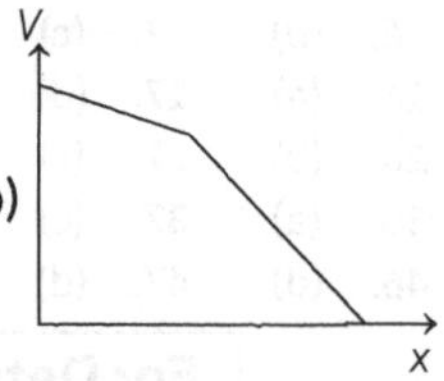

(c) 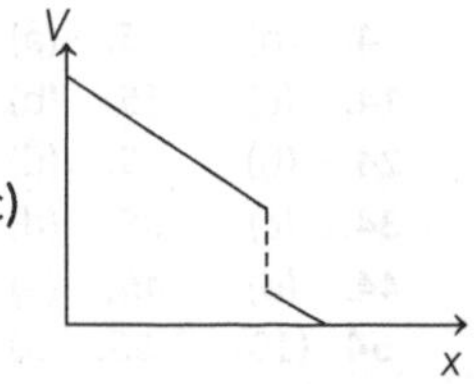

(d) 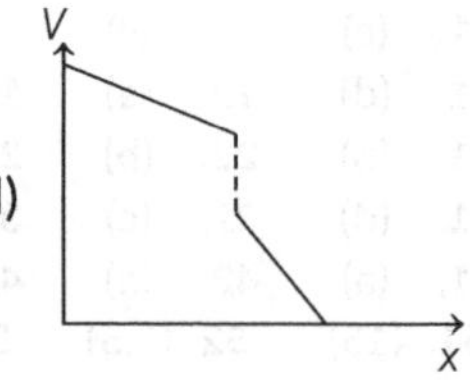

6 Current density in a cylindrical wire of radius R is given as $J = \begin{cases} J_0\left(\frac{x}{R} - 1\right) \text{for } 0 \le x < \frac{R}{2}. \\ J_0 \frac{x}{R} \text{for} \frac{R}{2} \le x \le R \end{cases}$

The current flowing in the wire is

(a) $\frac{7}{24}\pi J_0 R^2$ (b) $\frac{1}{6}\pi J_0 R^2$ (c) $\frac{7}{12}\pi J_0 R^2$ (d) $\frac{5}{12}\pi J_0 R^2$

7 A hemispherical metallic plate is buried in the earth with its plane surface parallel to the ground and touching the ground. A conducting pole is attached to the plate as shown in the figure. The pole carries a current I as shown. This current enters the plate and gets distributed uniformly on the surface of the plate. A man approaches the pole. If his steps are of length L and he is at a distance r_0 at some instant. Find the voltage, which will appear across his feet due to current I. Assume that the conductivity of the earth is σ.

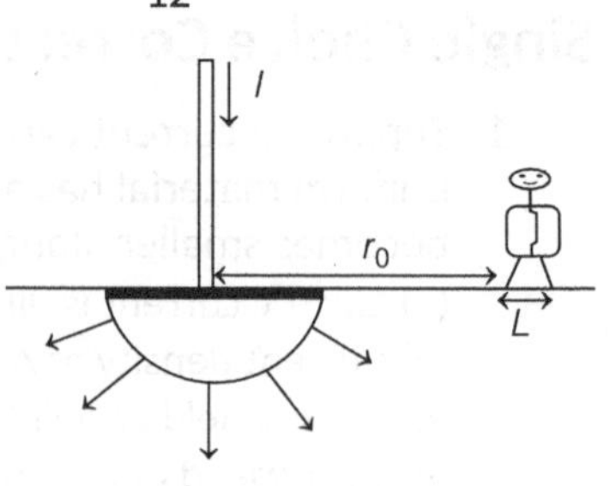

(a) $\frac{IL}{2\pi\sigma r_0(r_0 + L)}$ (b) $\frac{IL}{4\pi\sigma r_0(r_0 + L)}$ (c) $\frac{I}{2\pi\sigma L}\ln\left(\frac{r_0 + L}{r_0}\right)$ (d) $\frac{I}{4\pi\sigma L}\ln\left(\frac{r_0 + L}{r_0}\right)$

8 In order to increase the resistance of a given wire of uniform cross-section to four times its value, a fraction of its length is stretched uniformly till the full length of the wire becomes 1.5 times the original length, what is the value of this fraction?

(a) $\frac{1}{4}$ (b) $\frac{1}{8}$ (c) $\frac{1}{16}$ (d) $\frac{1}{6}$

9 A brass disc and a carbon disc of same radius are assembled alternatively to make a cylindrical conductor. The resistance of the cylinder is independent of the temperature. The ratio of thickness of the brass disc to that of the carbon disc is [α is temperature coefficient of resistance and neglect linear expansion]

(a) $\left|\frac{\alpha_C \rho_C}{\alpha_B \rho_B}\right|$ (b) $\left|\frac{\alpha_C \rho_B}{\alpha_B \rho_C}\right|$ (c) $\left|\frac{\alpha_B \rho_C}{\alpha_C \rho_B}\right|$ (d) $\left|\frac{\alpha_B \rho_B}{\alpha_C \rho_C}\right|$

10 A conducting rod of square cross-section is bent in a semi-circular shape of inner radius r as shown in figure. Length of side of cross-section is equal to a. Resistivity of material is ρ. The resistance of rod across its end will be (a is not small in comparison of r)

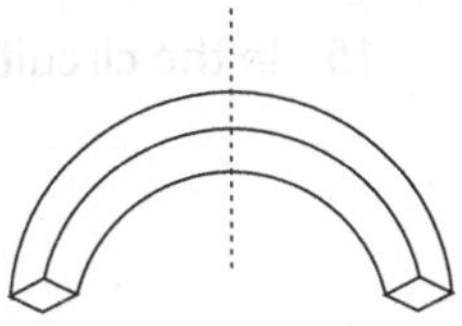

(a) $\dfrac{\pi\rho}{a\ln(1+a/r)}$ (b) $\dfrac{2\pi\rho}{a\ln(1+a/r)}$

(c) $\dfrac{\pi\rho(r+a)}{a^2}$ (d) $\dfrac{2\pi r\rho}{a^2}$

11 Maximum and minimum radii of given cylindrical conductor PQ are r_1 and r_2, then current through the battery will be (ρ = resistivity)

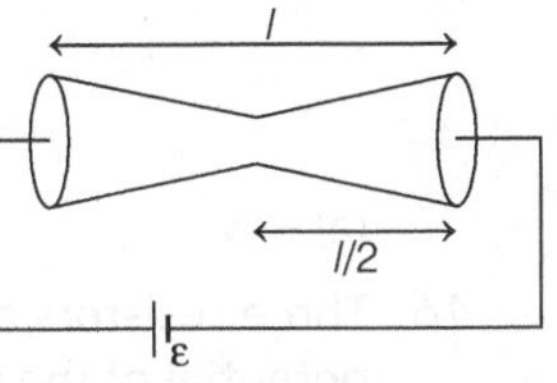

(a) $\dfrac{\pi\varepsilon r_1 r_2}{\rho l}$ (b) $\dfrac{2\pi\varepsilon r_1 r_2}{\rho l}$

(c) $\dfrac{\pi\varepsilon r_1 r_2}{2\rho l}$ (d) None of these

12 A metal rod of radius a is concentric with a metal cylindrical shell of radius b and length L. The space between rod and cylinder is tightly packed with a high resistance material of a resistivity ρ. A battery having a terminal voltage V is connected across the combination as shown. Neglect resistance of rod and cylinder. If I is the total current in the circuit, then I is

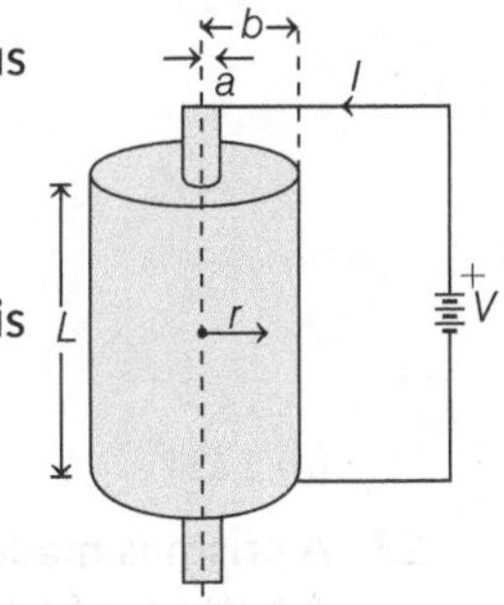

(a) $\dfrac{LV}{\rho}$ (b) $\dfrac{2\pi LV}{\rho(\ln b-\ln a)}$

(c) $\dfrac{4\pi LV}{\rho(\ln b-\ln a)}$ (d) $\dfrac{LV}{4\pi\rho(\ln b-\ln a)}$

13 Two current elements P and Q have current-voltage characteristics as shown below

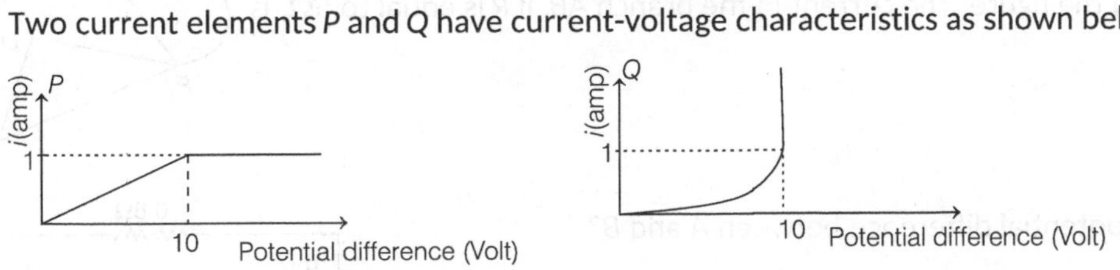

Which of the graphs given below represents current-voltage characteristics when P and Q are in series.

(a)

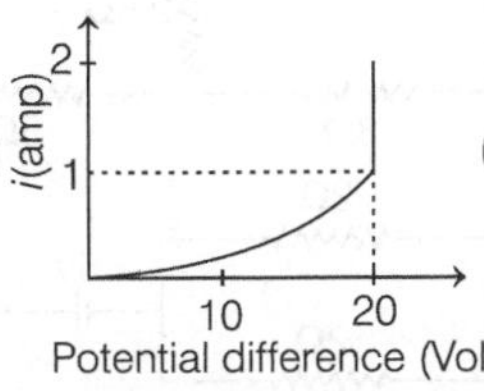

(b)

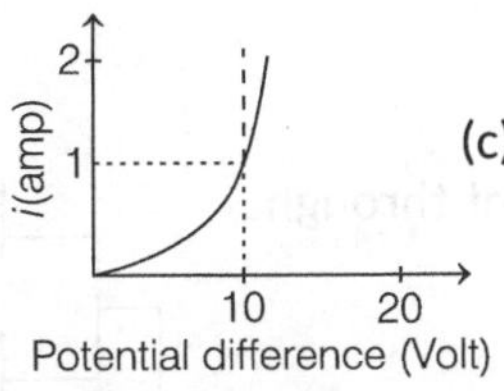

(c)

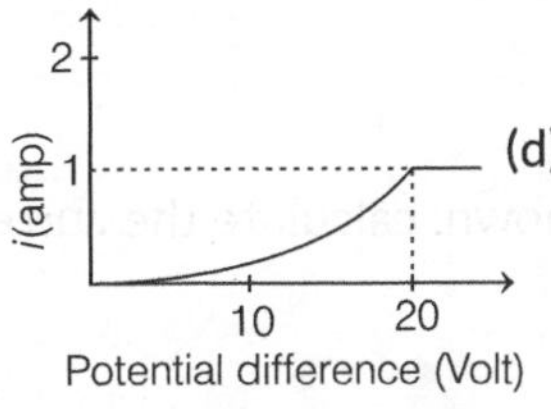

(d)

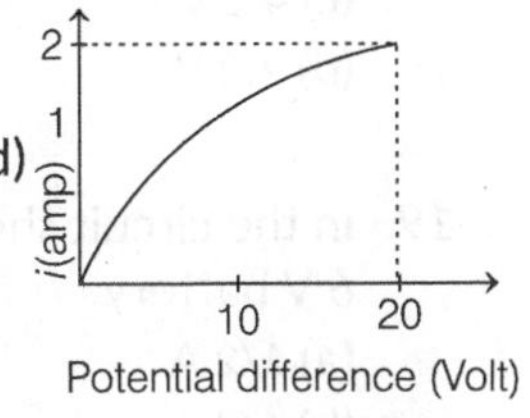

14 The two ends of a uniform conductor are joined to a cell of emf ε and some internal resistance. Starting from the mid-point P of the conductor, we move in the direction of the current and return to P. The potential V at every point on the path is plotted against the distance covered (x). Which of the following best represents the resulting curve?

(a)

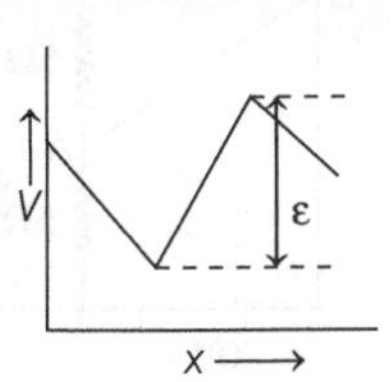

(b)

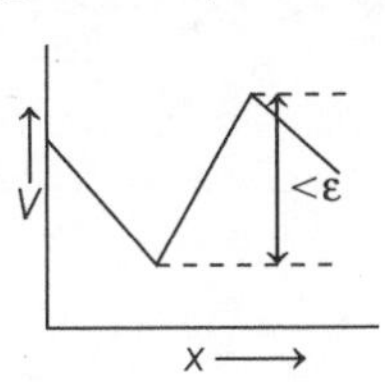

(c)

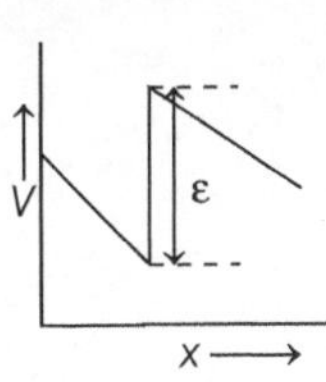

(d)

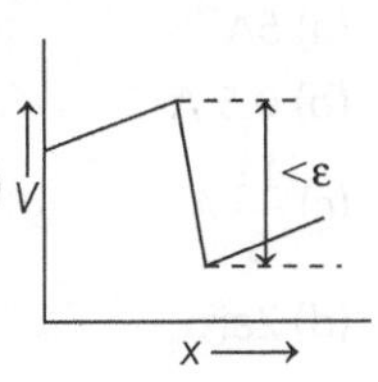

15 In the circuit given, find the potential of the point P.

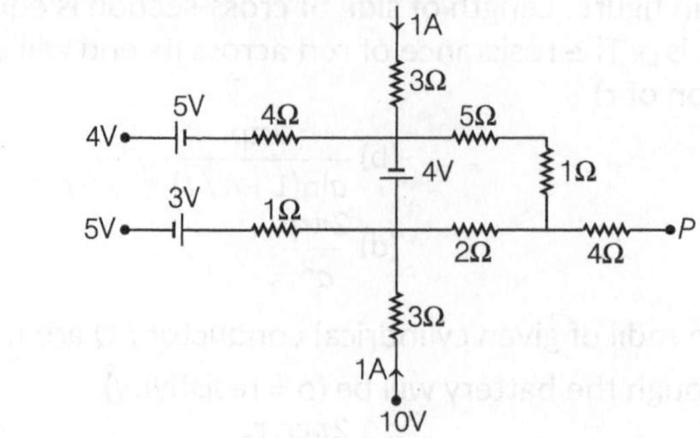

(a) –4V (b) –5V (c) –6V (d) –8V

16 Three resistors and two batteries are connected as shown in figure. Find the value of R, if the potential of the point a is greater than that of b by 7V.

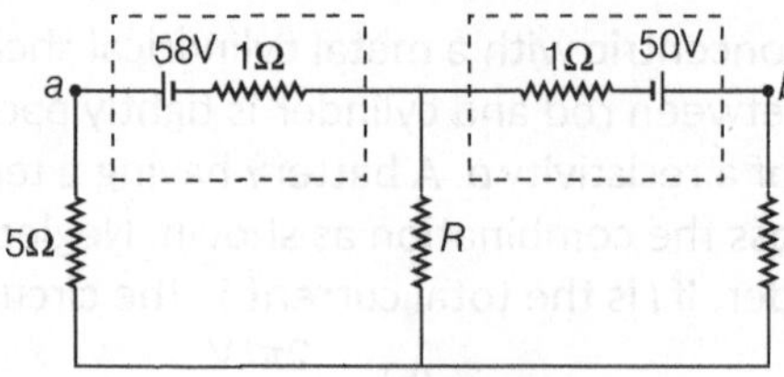

(a) 1 Ω (b) 5 Ω (c) 8 Ω (d) 9 Ω

17 A prism is made of wire mesh with each side having equal resistance R. A battery of 6 V and zero resistance is connected across E and F as shown in the figure. The current in the branch AB, if R is equal to 5Ω, is

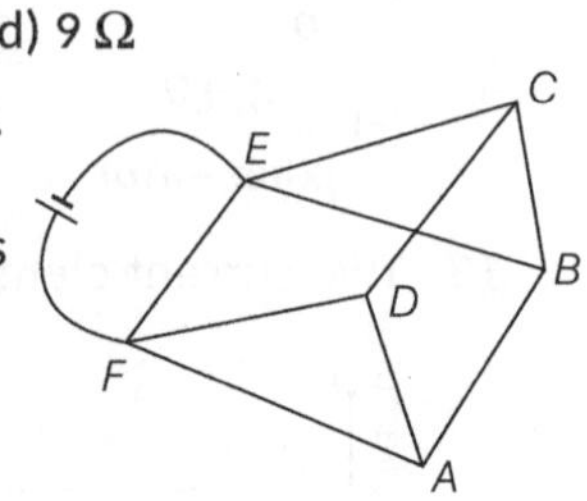

(a) 0.6 A
(b) 0.8 A
(c) 0.4 A
(d) 0

18 What is potential difference between A and B?

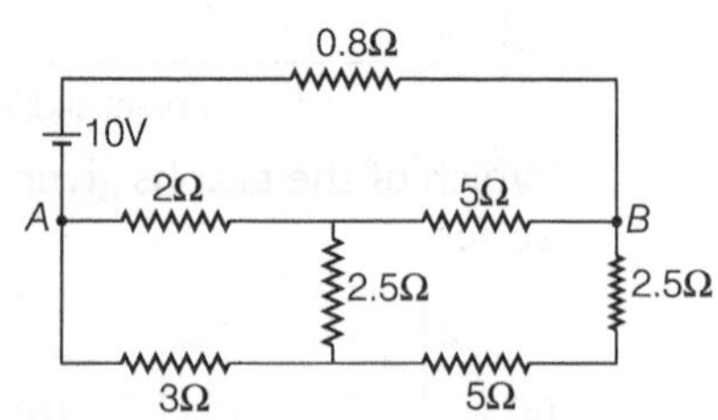

(a) 8.4 V
(b) 2.4 V
(c) 4.2 V
(d) 7.2 V

19 In the circuit shown, calculate the current through 6 V battery.

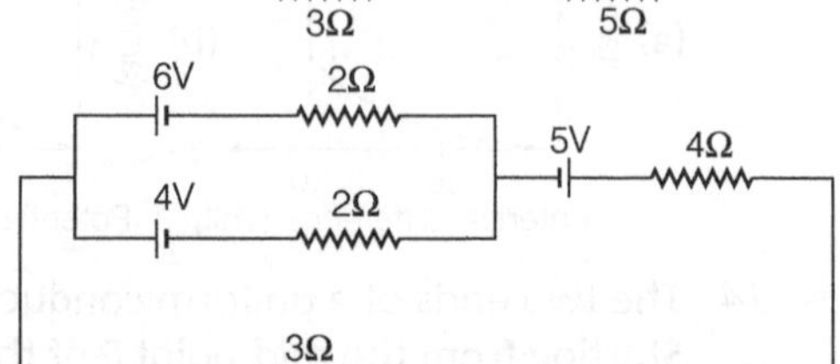

(a) 1/2 A
(b) 1/4 A
(c) 1/8 A
(d) None of the above

20 In the circuit shown, find the current in branch AB of the circuit.

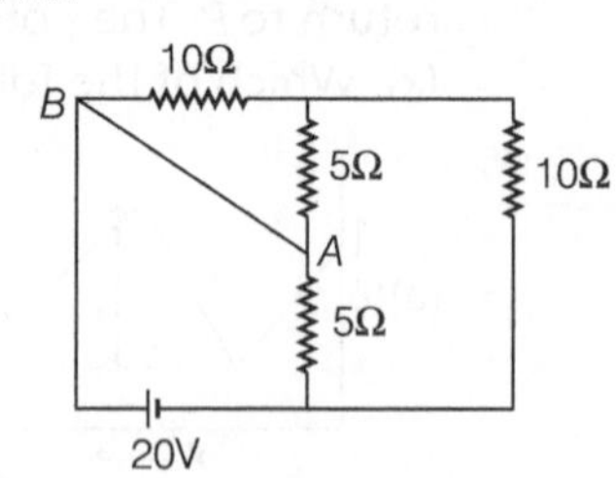

(a) 5A
(b) 0.5 A
(c) $\frac{11}{3}$ A
(d) Zero

21 In the circuit shown in the figure, the potential difference between points a and b is $V_a - V_b = 4\text{V}$. Find the emf E of the battery .

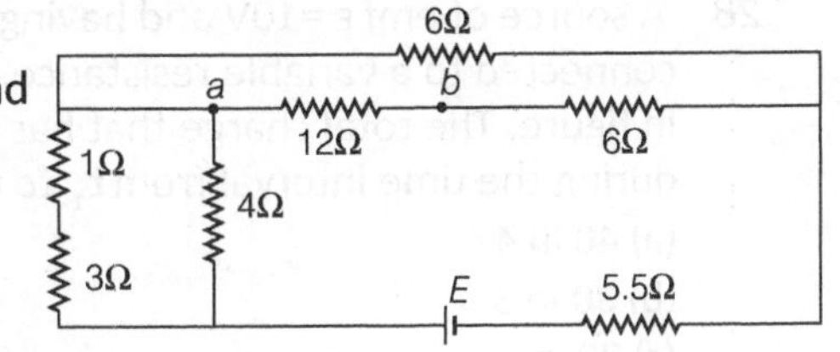

(a) 12 V
(b) 16 V
(c) 20 V
(d) 24 V

22 In the given circuit, three resistors of 3 Ω each, 3 cells ε_1 and ε_2 and ε_3 with internal resistance 3Ω each are connected. Currents in some branches are shown in the circuit. Find the emf of cell ε_3.

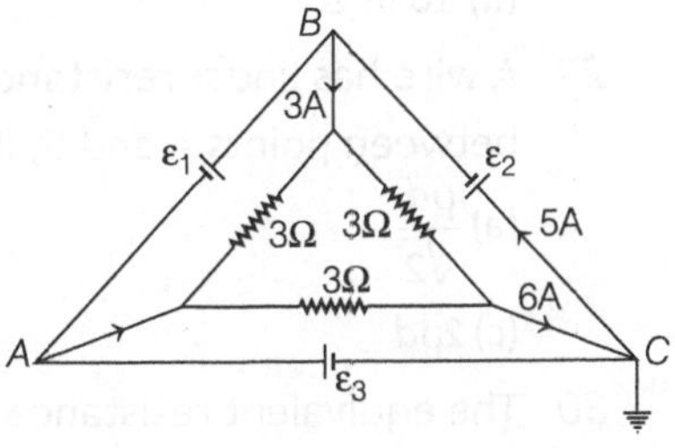

(a) 6 V
(b) 9 V
(c) 12 V
(d) 15 V

23 Under what condition, current passing through the resistance R can be increased by short-circuiting the battery of emf ε_2. The internal resistances of the two batteries are r_1 and r_2, respectively.

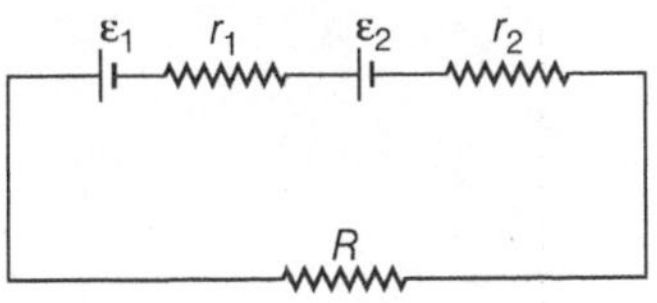

(a) $\varepsilon_2 r_1 > \varepsilon_1 (R + r_2)$
(b) $\varepsilon_1 r_2 > \varepsilon_2 (R + R_1)$
(c) $\varepsilon_2 r_2 > \varepsilon_1 (R + r_2)$
(d) $\varepsilon_1 r_1 > \varepsilon_2 (R + r_1)$

24 A long conductor of circular cross-section has radius r and length l as shown in the figure. The conductivity of the material near the axis is σ_1 and increases linearly with the distance from axis and becomes σ_2 near the surface. Find the current through the conductor, if a potential difference V is applied across the flat faces.

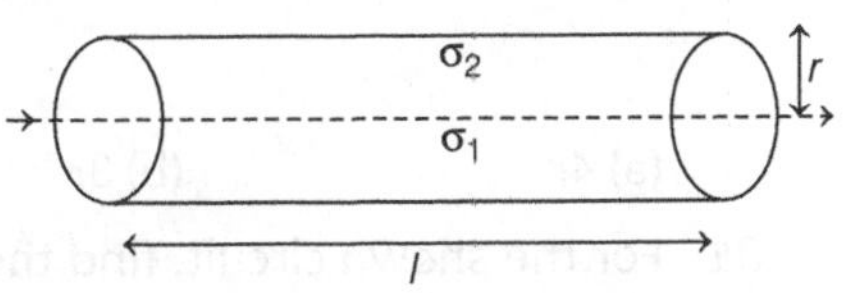

(a) $\dfrac{\pi r^2 V(\sigma_1 + \sigma_2)}{6l}$
(b) $\dfrac{2\pi r^2 V(\sigma_1 + \sigma_2)}{3l}$
(c) $\dfrac{\pi r^2 V(2\sigma_1 + \sigma_2)}{3l}$
(d) $\dfrac{\pi r^2 V(\sigma_1 + 2\sigma_2)}{3l}$

25 $ABCD$ is a square, where each side is a uniform wire of resistance 1Ω. A point E lies on CD such that, if a uniform wire of resistance 1Ω is connected across AE and constant potential difference is applied across A and C, then B and E are equipotential.

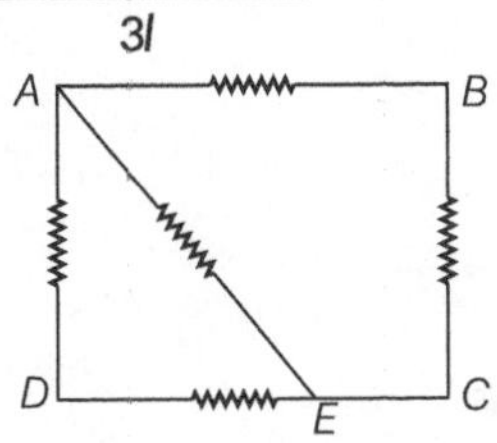

Then,

(a) $\dfrac{CE}{ED} = 1$
(b) $\dfrac{CE}{ED} = 2$
(c) $\dfrac{CE}{ED} = \dfrac{1}{\sqrt{2}}$
(d) $\dfrac{CE}{ED} = \sqrt{2}$

26 If the potential difference between A and D is V, what will be potential difference between F and C. Each of the resistance is R.

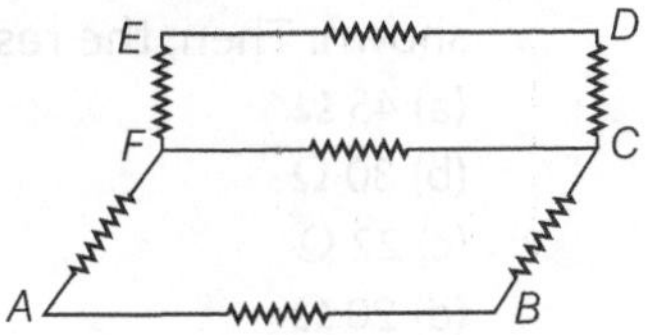

(a) $V/3$
(b) $V/5$
(c) $V/7$
(d) $V/8$

27 The current enters at A and comes out at D. Some of the resistance are shown. What should be resistance of wire CB, so that it draws double of the current that enters the wire BF?

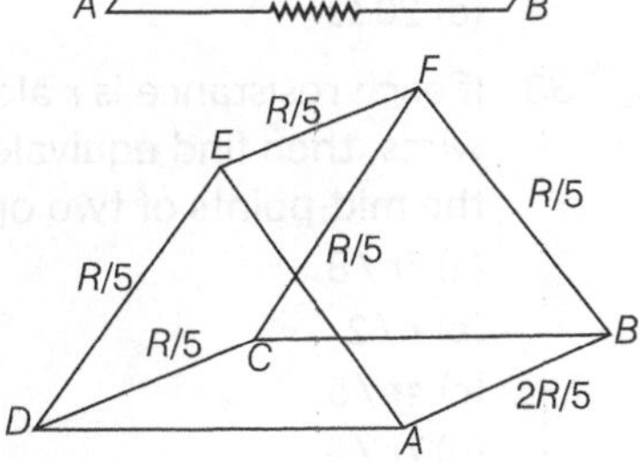

(a) $\dfrac{3R}{20}$
(b) $\dfrac{7R}{9}$
(c) $\dfrac{8R}{9}$
(d) $\dfrac{9R}{10}$

28 A source of emf $\varepsilon = 10\text{V}$ and having negligible internal resistance is connected to a variable resistance. The resistance varies as shown in figure. The total charge that has passed through the resistor R during the time interval from t_1 to t_2 is

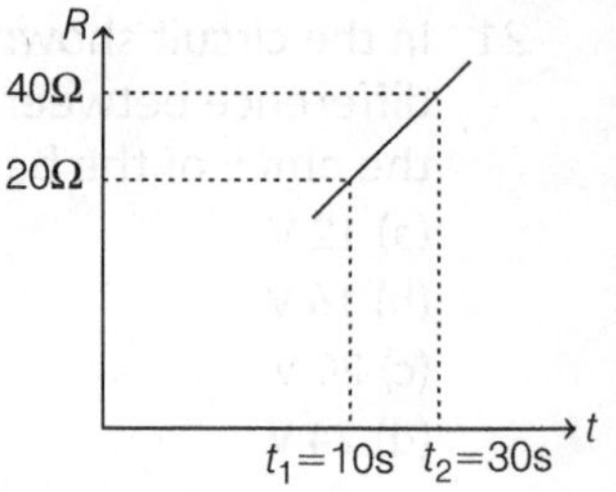

(a) 40 ln 4
(b) 30 ln 3
(c) 20 ln 2
(d) 10 ln 2

29 A wire has linear resistance ρ (in ohm/m). Find the equivalent resistance between points A and B, if the side of the big square is d.

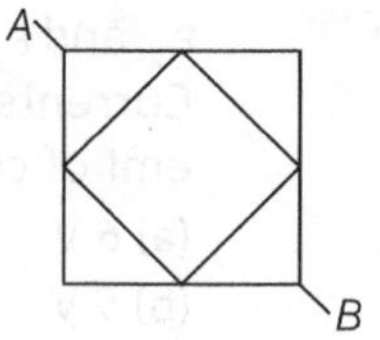

(a) $\frac{\rho d}{\sqrt{2}}$ (b) $\sqrt{2}\rho d$
(c) $2\rho d$ (d) None of these

30 The equivalent resistance between A and B in the arrangement of resistances as shown, is

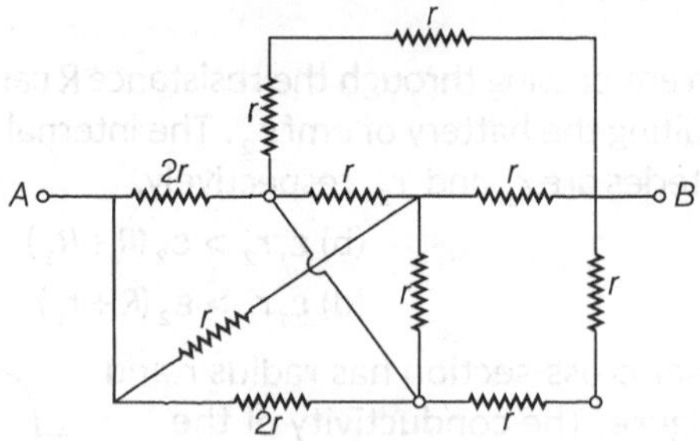

(a) $4r$ (b) $3r$ (c) $2.5\,r$ (d) r

31 For the shown circuit, find the effective resistance between the points A and B.

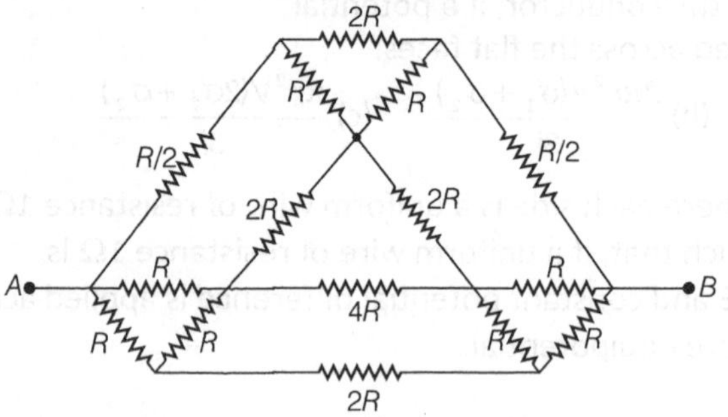

(a) R (b) $2R$ (c) $3R$ (d) $4R$

32 12 resistors (40 Ω each) are used to make a prism with hexagonal base as shown. Then, the resistance between two adjacent corners of the base is

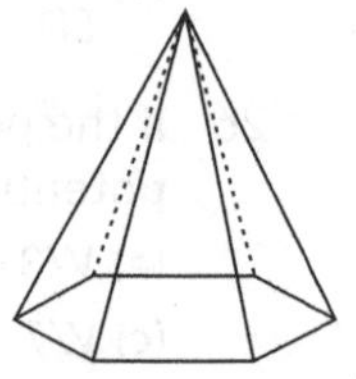

(a) 45 Ω
(b) 30 Ω
(c) 22 Ω
(d) 20 Ω

33 If each resistance is r along the 12 edges of a cubical skeleton of uniform wires, then find equivalent resistance between X and Y, where X and Y are the mid-points of two opposite edges of a face of the cube.

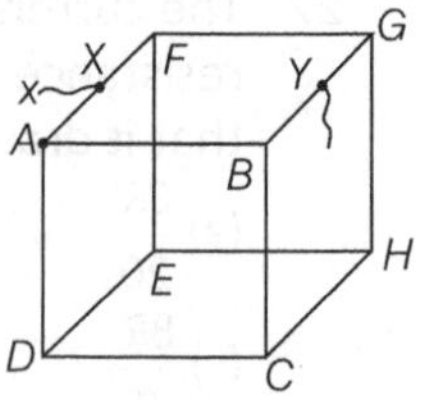

(a) $7r/8$
(b) $r/2$
(c) $4r/5$
(d) $7r/5$

34 Find the equivalent resistance of the network shown in figure between the points a and b.

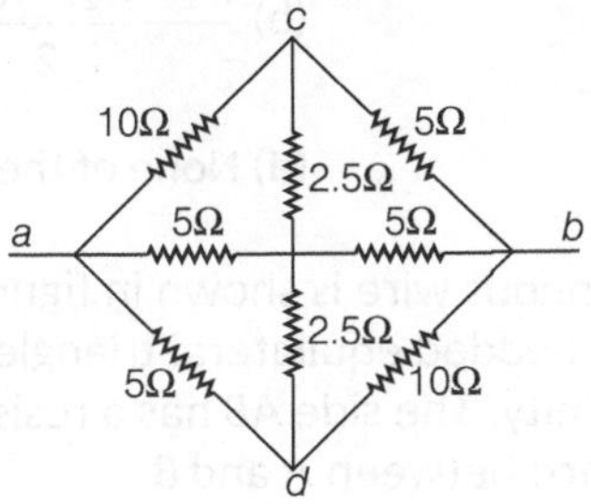

(a) $\frac{50}{13}\,\Omega$ (b) $\frac{50}{23}\,\Omega$ (c) $\frac{70}{29}\,\Omega$ (d) $\frac{70}{17}\,\Omega$

35 A hemispherical network of radius a is made by using a conducting wire of resistance per unit length r. The equivalent resistance across OP is

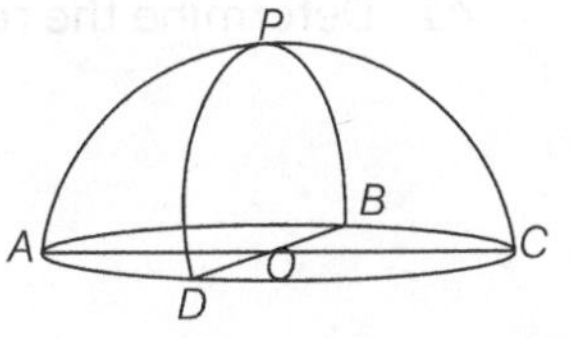

(a) $\frac{8ar}{(2+\pi)}$ (b) $\frac{8ar}{(2-\pi)}$

(c) $\frac{(2+\pi)ar}{8}$ (d) $\frac{(2-\pi)ar}{8}$

36 Find the equivalent resistance of the following group of resistance between A and B. Each resistance of the circuit is R.

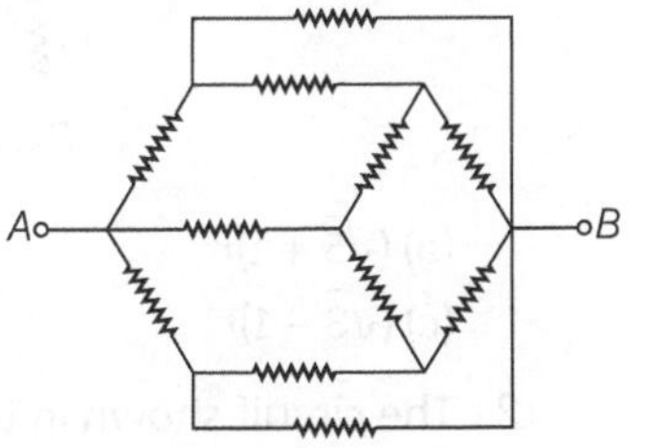

(a) $5R/14$

(b) $9R/14$

(c) $11R/14$

(d) $17R/14$

37 Equivalent resistance of the network given below between A and B will be (all resistances are $R\,\Omega$)

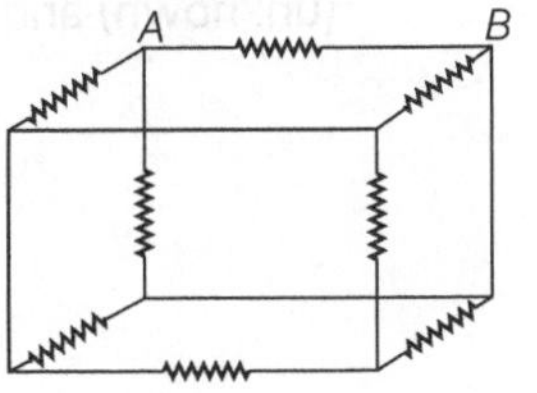

(a) $7R/30$

(b) $11R/30$

(c) $13R/30$

(d) $17R/30$

38 At what value of the resistance R_x in the circuit shown in figure will the total resistance between points A and B be independent of the number of loops? $[R = 2(\sqrt{3}+1)\,\Omega]$

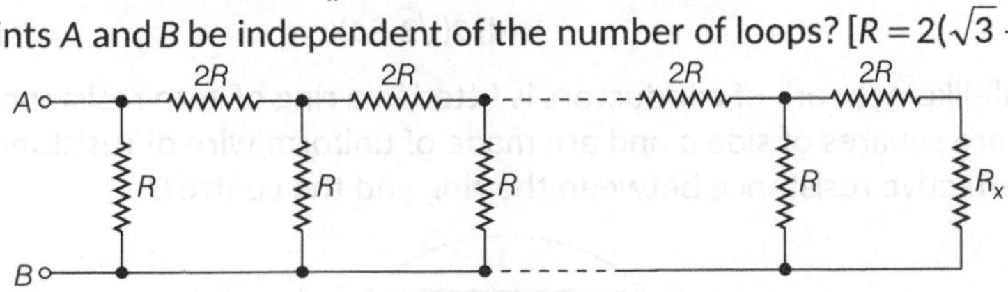

(a) $2\,\Omega$ (b) $4\,\Omega$ (c) $8\,\Omega$ (d) $16\,\Omega$

39 The circuit diagram shown consists of a large number of elements (each element has two resistors R_1 and R_2). The resistance of the resistors in each subsequent element differs by a factor of $K = 1/2$ from the resistance of the resistors in the previous elements. The equivalent resistance between A and B shown in figure is

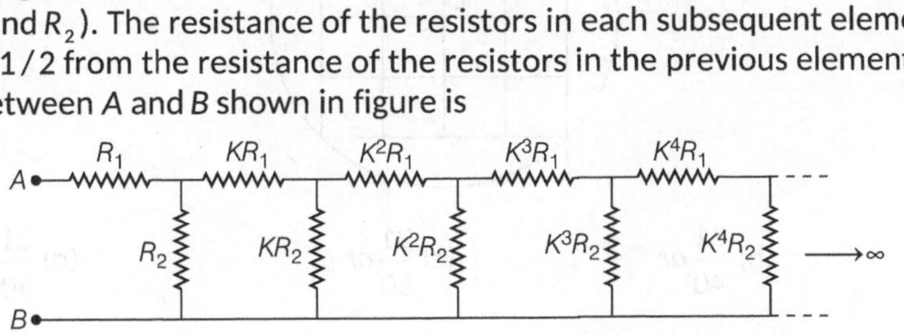

(a) $\frac{R_1 - R_2}{2}$ (b) $\frac{(R_1 - R_2) + \sqrt{6R_1R_2}}{2}$

(c) $\frac{(R_1 - R_2) + \sqrt{R_1^2 + R_2^2 + 6R_1R_2}}{2}$ (d) None of these

40 A frame made of thin homogeneous wire is shown in figure. Assume that the number of successively embedded equilateral triangle with sides decreasing by half tends to infinity. The side AB has a resistance $R = (\sqrt{7} + 1)$ Ω. Find the equivalent resistance between A and B.

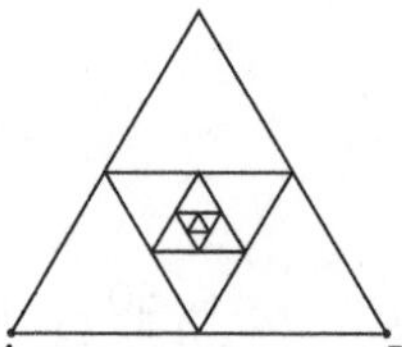

(a) $2\,\Omega$ (b) $3\,\Omega$

(c) $4\,\Omega$ (d) $6\,\Omega$

41 Determine the resultant resistance of the infinite circuit made of r resistance between A and B.

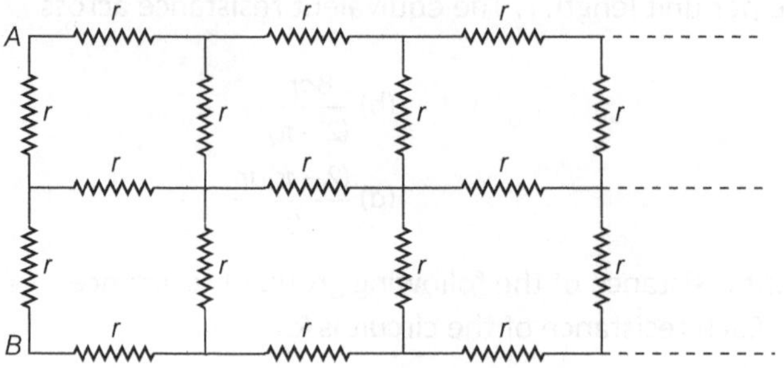

(a) $(\sqrt{5} + 1)r$ (b) $(\sqrt{5} - 1)r$

(c) $(\sqrt{3} - 1)r$ (d) $(\sqrt{3} + 1)r$

42 The circuit shown in the diagram extends to the right into infinity. Each battery has emf ε (unknown) and internal resistance r. Each resistor has the resistance $4r$. Value of emf of battery is

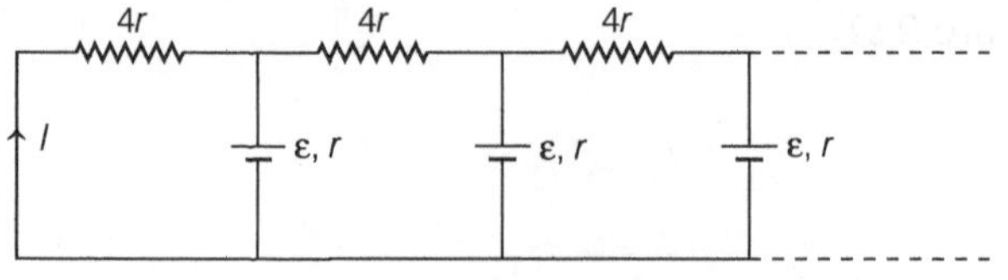

(a) $(\sqrt{2} - 1)Ir$ (b) $(\sqrt{2} + 1)Ir$

(c) $2(\sqrt{2} - 1)Ir$ (d) $2(\sqrt{2} + 1)Ir$

43 A window grill-like network of conductors is fitted in a ring of zero resistance. The interstices of the network are squares of side a and are made of uniform wire of resistance R per unit length. What is the effective resistance between the ring and the centre O.

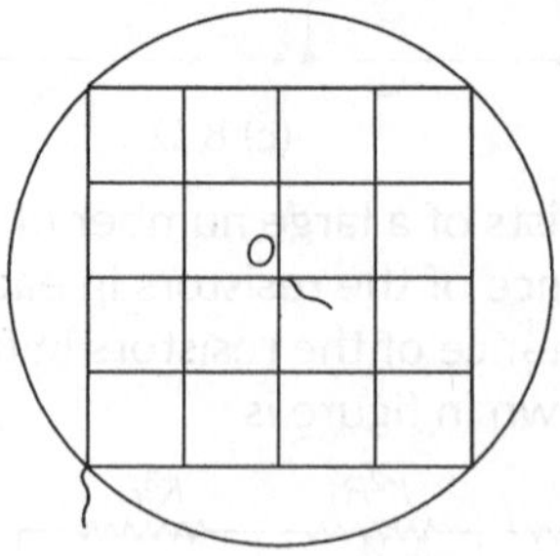

(a) $\frac{21}{40}ar$ (b) $\frac{31}{40}ar$ (c) $\frac{21}{50}ar$ (d) $\frac{31}{50}ar$

Multiple Choice(s) Correct Questions

44 Figure shows a solid resistance wire of uniform material but non-uniform cross-section area. A, B and C are the three sections. Current flowing is i, electric field at a section is E, potential is V and flux of electric field passing through a cross-section is ϕ, then

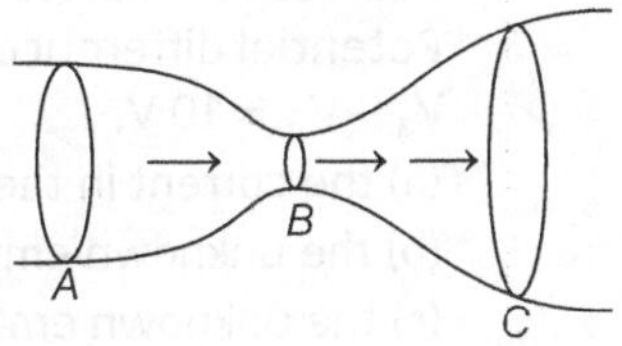

(a) $i_B > i_A > i_C$
(b) $E_B > E_A > E_C$
(c) $\phi_A = \phi_B = \phi_C$
(d) $V_B > V_A > V_C$

45 The resistivity of a cylindrical conductor carrying steady current along its length varies linearly with the distance from the current carrying end as given by $\rho = \rho_0\left(1+\frac{x}{l}\right)$, where l is the length of the conductor and x is the distance from the current entry end. ρ_0 is a positive constant.

(a) Electric field varies linearly with x
(b) Electric potential difference across the length x varies linearly with x
(c) Volume charge density in the conductor is zero
(d) Volume charge density in the conductor is non-zero

46 In the circuit shown in diagram,

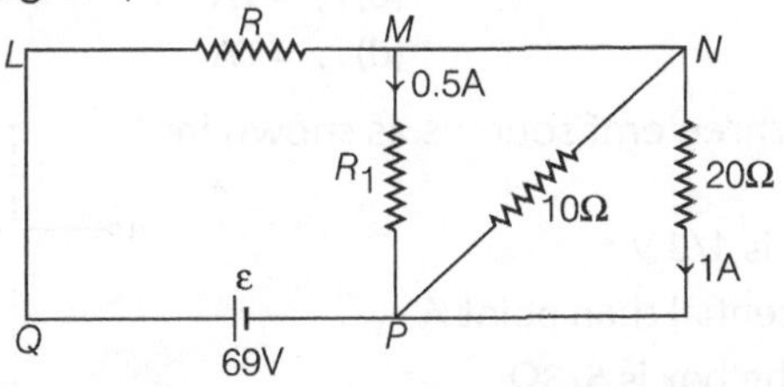

(a) the current through 10 Ω is 0.5 A
(b) the value of $R_1 = 40\ \Omega$
(c) the value of $R = 14\ \Omega$
(d) the potential difference across $R = 49$ V

47 A part of circuit is shown in figure. Choose the correct option(s).

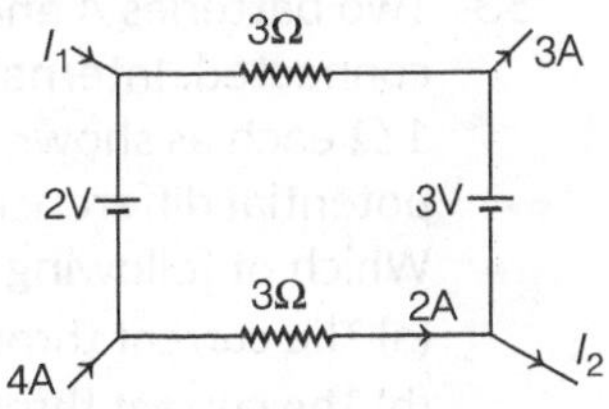

(a) $I_1 = 1/3$ A
(b) $I_2 = 2/3$ A
(c) $I_1 = -1/3$ A
(d) $I_2 = -2/3$ A

48 In the given circuit, the point A is 9V higher than point B.

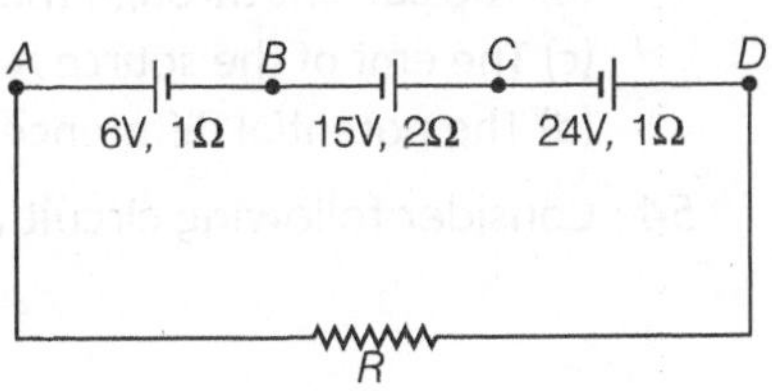

(a) $R = 1\ \Omega$
(b) $R = 7\ \Omega$
(c) Potential difference between B and D is 30 V
(d) Potential difference between B and C is 15 V

49 In the given diagram, choose correct options.

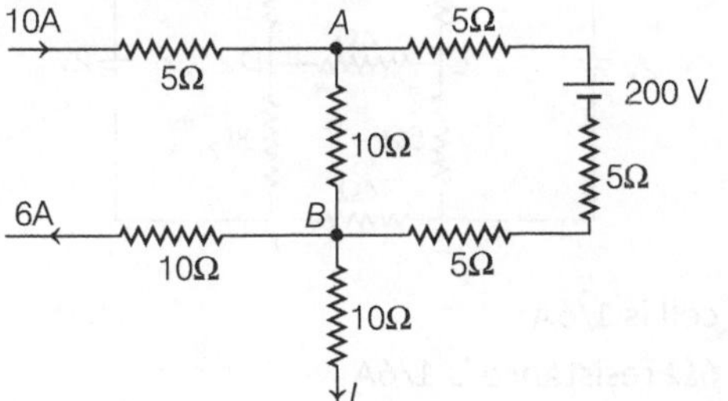

(a) The value of I is 4A
(b) The current from the cell is 4A
(c) Voltage drop across AB is 140 V
(d) The current from the cell is 10 A

50 In the circuit shown in figure, ε_1 and ε_2 are two ideal sources of unknown emfs. Some currents are shown. Potential difference appearing across $6\ \Omega$ resistance is $V_A - V_B = 10$ V,

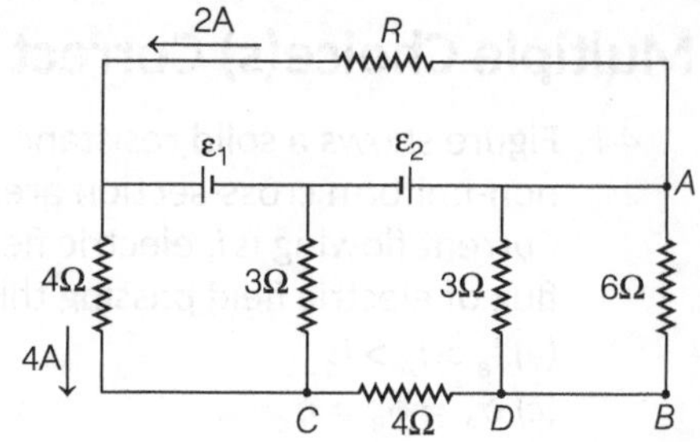

(a) the current in the $4\ \Omega$ resistor in between C and D is 5 A
(b) the unknown emf ε_1 is 43 V
(c) the unknown emf ε_2 is 57 V
(d) the resistance R is equal to $7\ \Omega$

51 For the circuit shown in figure,

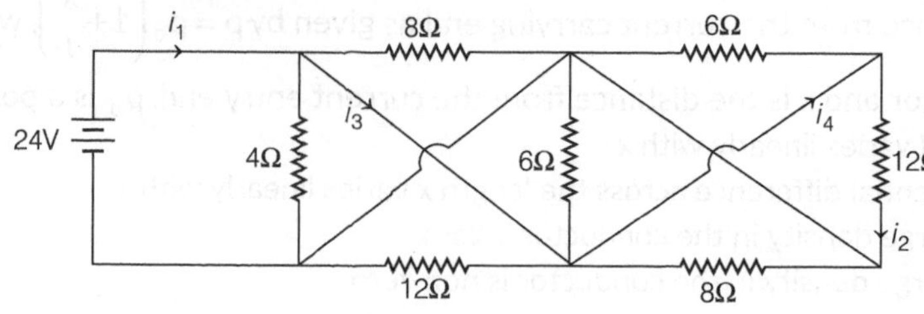

(a) $i_1 = 24$A
(b) $i_2 = 2$A
(c) $i_3 = 15$A
(d) $i_4 = 6$A

52 An electric box contains three emf sources as shown in the figure. Then,

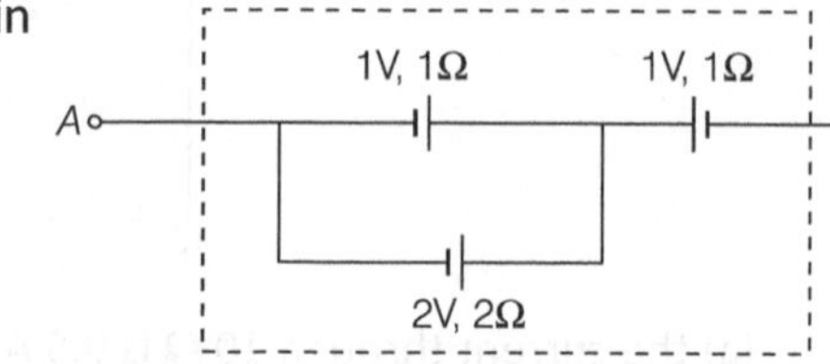

(a) emf of the electric box is 1/3 V
(b) point B is at higher potential than point A
(c) internal resistance of the box is $5/3\Omega$
(d) terminal voltage of 2V source is 4/3V

53 Two batteries A and B and three resistors are connected. Internal resistance of both batteries is $1\ \Omega$ each as shown. EMF of battery B is 5 V. The potential difference between P and Q is zero. Which of following is/are true?

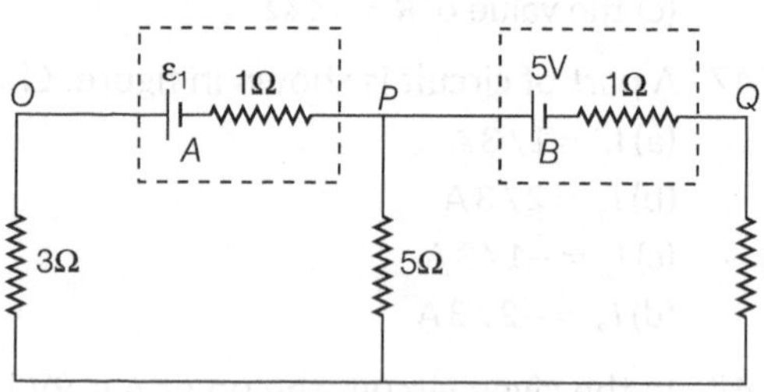

(a) The current through $5\ \Omega$ is 3 A
(b) The current through the battery A is 8 A
(c) The emf of the source A is 47 V
(d) The potential difference between O and P is 8V

54 Consider following circuit and select correct alternative(s).

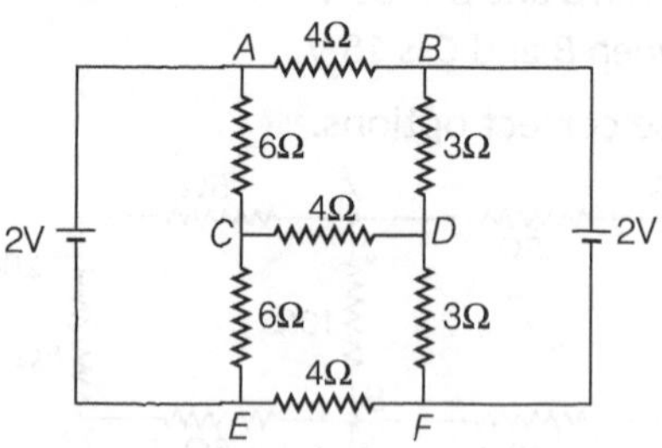

(a) Current through each cell is 1/6A
(b) Current through each 6Ω resistance is 1/6A
(c) Current through each 3Ω resistance is 1/3A
(d) Current through each 4Ω resistance is 1/2A

55 A battery of emf ε and internal resistance r is connected across a resistance R. Resistance R can be adjusted to any value greater than or equal to zero. A graph is plotted between the current (i) passing through the resistance and potential difference (V) across it. Select the correct alternative(s).

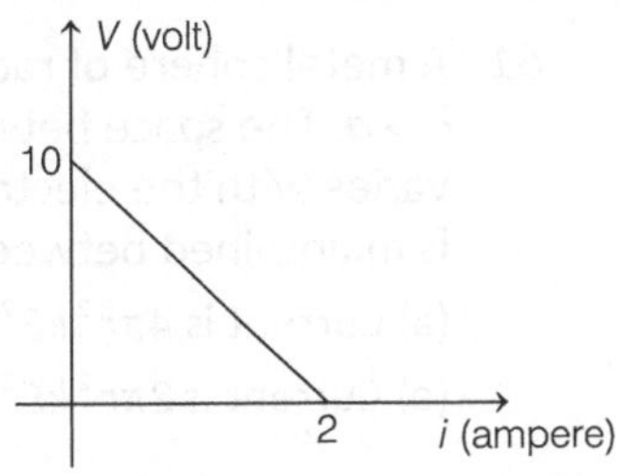

(a) internal resistance of the battery is 5Ω
(b) emf of the battery is 10 V
(c) Maximum current which can be taken from the battery is 2A
(d) V-i graph can never be a straight line as shown in figure

56 In the circuit shown in figure,

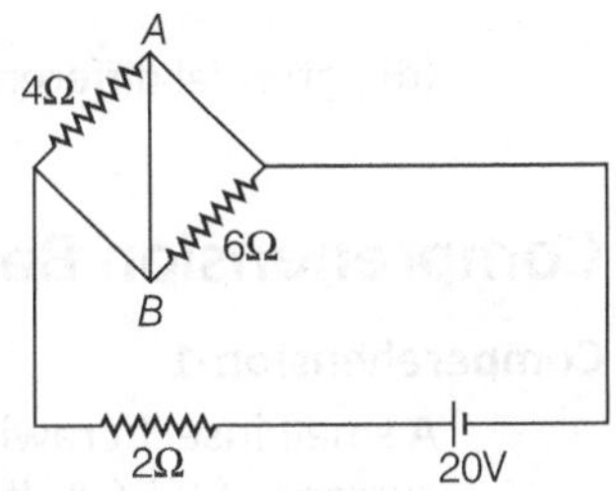

(a) power supplied by the battery is 200 W
(b) current flowing in the circuit is 5A
(c) potential difference across 4 Ω resistance is equal to the potential difference across 6Ω resistance
(d) current in wire AB is zero

57 In the circuit shown E,F,G and H are cells of emf 2V, 1V, 3V and 1V respectively and their internal resistances are 2Ω, 1Ω, 3Ω and 1Ω, respectively.

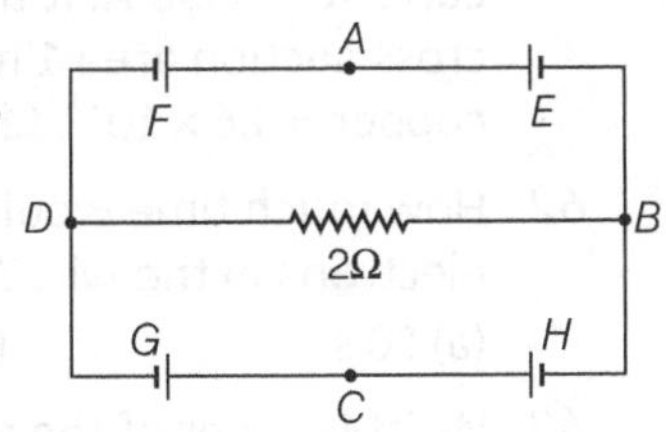

(a) $V_D - V_B = -2/13\text{V}$
(b) $V_D - V_B = 2/13\text{V}$
(c) $V_G = 21/13\text{V} =$ potential difference across cell G
(d) $V_H = 19/13\text{V} =$ potential difference across cell H

58 Which of the following options is/are correct regarding the below diagram?

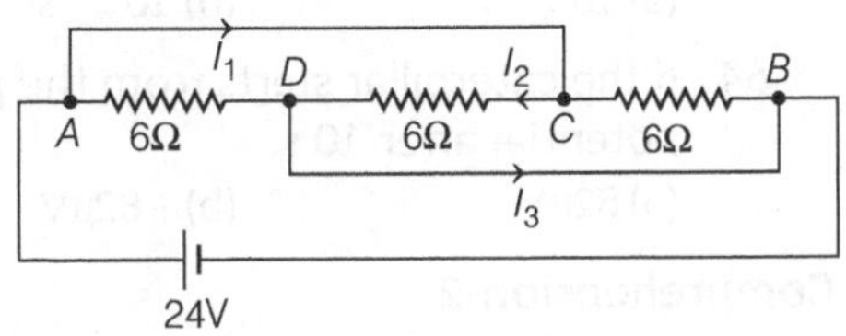

(a) $I_1 = 8\text{ A}$
(b) $I_2 = 4\text{ A}$
(c) $I_1 = 0$
(d) $I_3 = 8\text{ A}$

59 Consider 12 resistors arranged symmetrically in shape of bipyramid $ABCDEF$. The resistance of each resistor is R. The effective resistance between

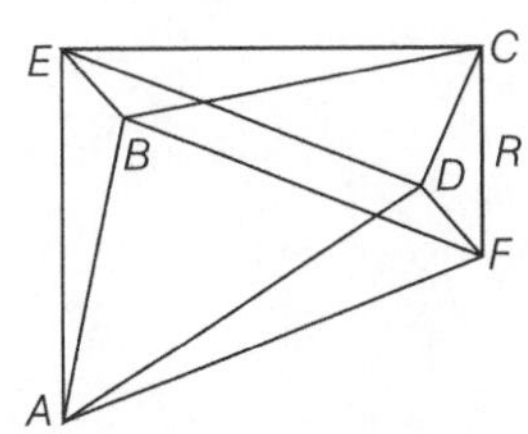

(a) E and F is $R/2$
(b) D and E is $5R/12$
(c) E and F is R
(d) A and C is $R/2$

60 Two circuits (shown below) are called 'circuit A' and circuit B, The equivalent resistance of circuit A is x and that of circuit B is y between 1 and 2.

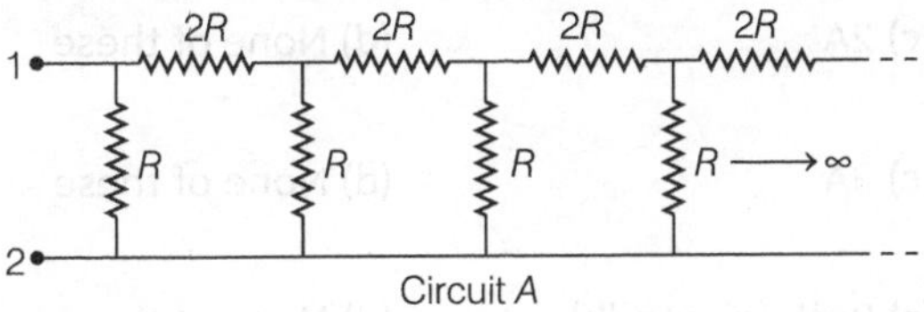

Circuit A

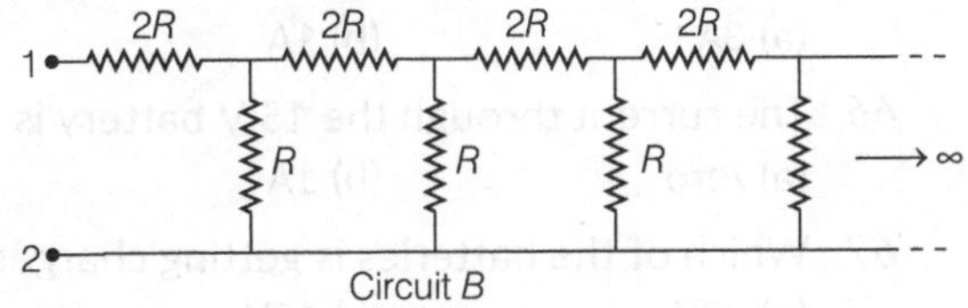

Circuit B

(a) $y > x$
(b) $y = (\sqrt{3}+1)R$
(c) $xy = 2R^2$
(d) $x - y = 2R$

61 A metal sphere of radius is surrounded by a concentric metal sphere of inner radius of b, where $b > a$. The space between the spheres is filled with a material, whose electrical conductivity σ varies with the electric field strength E as $\sigma = kE$, where k is a constant. A potential difference V is maintained between spheres.

(a) Current is $4\pi r^2 kE^2$, where $(a < r < b)$

(b) Current is $2\pi r^2 kE^2$, where $(a < r < b)$

(c) Potential difference between spheres is $V = \sqrt{\frac{1}{4\pi k}} \ln\left(\frac{b}{a}\right)$, where I is total current

(d) Potential difference between spheres is $V = \sqrt{\frac{1}{2\pi k}} \ln\left(\frac{b}{a}\right)$, where I is total current

Comprehension Based Questions

Comperehension-1

A small insect crawls in the direction of electron drift along bare copper wire that carries a current of 2.56 A. It travels with the drift speed of the electron in the wire of uniform cross-section area 1 mm^2. Number of free electrons for copper $= 8 \times 10^{22}$/cc and resistivity of copper $= 1.6 \times 10^{-8}$ Ω-m.

62 How much time would the caterpillar take to crawl 1.0 cm, if it crawls at the drift speed of the electrons in the wire?

(a) 50 s (b) 5 s (c) 5000 s (d) None of these

63 What is order of the average time of collision for free electrons of copper?

(a) 10^{-14} s (b) 10^{-16} s (c) 10^{-11} s (d) 10^{-8} s

64 If the caterpillar starts from the point of zero potential at $t = 0$, it reaches a point of .... potential after 10 s.

(a) 82μV (b) –82μV (c) 41μV (d) –41μV

Comprehension-2

The figure shows a network of five resistances and two batteries

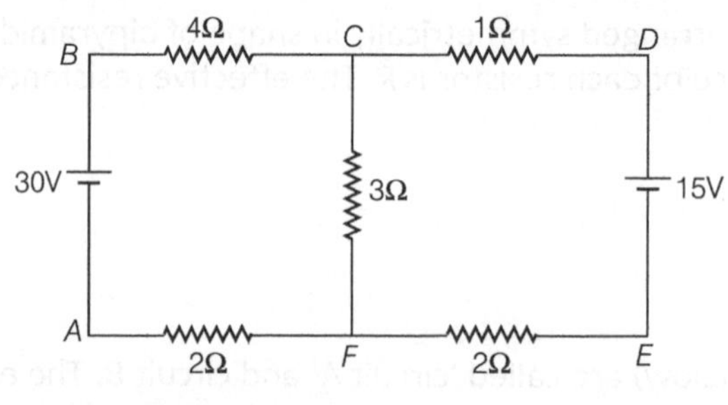

65 The current through the 30 V battery is

(a) 3A (b) 1A (c) 2A (d) None of these

66 The current through the 15 V battery is

(a) zero (b) 1A (c) 3A (d) None of these

67 Which of the batteries is getting charged?

(a) 30V (b) 15V (c) Both (a) and (b) (d) None of these

68 The total electrical power consumed by the circuit is

(a) 15W (b) 75 W (c) 105 W (d) 90 W

Comprehension-3

In the shown circuit, all cells are ideal.

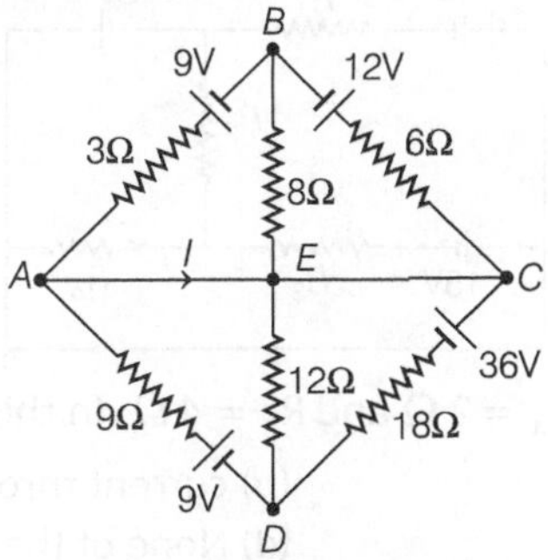

69 The magnitude of current through 8 Ω resistor connected between points *B* and *E* is equal to

(a) 0.25 A (b) 0.5 A

(c) 1 A (d) 2 A

70 The magnitude of current *I* through conducting wire connected between *A* and *E* is equal to

(a) $\frac{1}{3}$ A (b) $\frac{4}{3}$ A

(c) 1 A (d) zero

71 Let potential at points *B* and *D* be V_B and V_D, respectively. Then, $V_B - V_D$ is equal to

(a) 4V (b) −4 V

(c) 20 V (d) −20 V

Comprehension-4

The given figure shows a network of resistances and a battery.

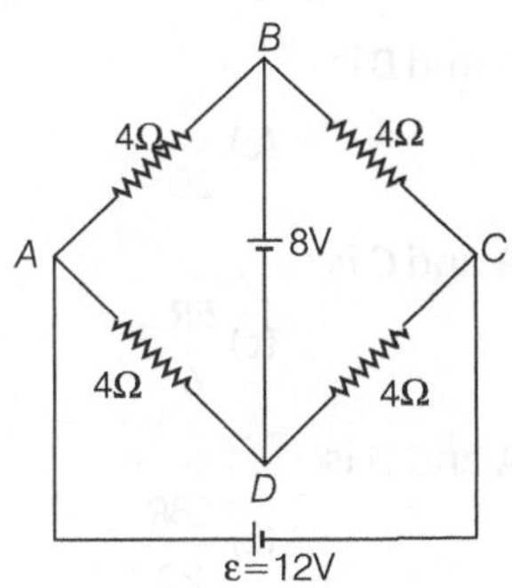

72 Identify the correct statement.

(a) The circuit satisfies the condition of a balanced Wheatstone bridge

(b) $V_B - V_D = 0$

(c) $V_B - V_D = 8$

(d) No current flows in the branch *BD*

73 Which of the two batteries is getting charged?

(a) 8 V battery (b) 12V battery

(c) None of these (d) Cannot be said

74 Choose the correct statement(s).

(a) The current coming out of the 8 V battery is 2A.

(b) The current coming out of the 12 V battery is 3A.

(c) The current flowing in all the 4Ω branches is same.

(d) The current flowing in the diagonally opposite branches is same.

Comprehension-5

In the circuit shown,

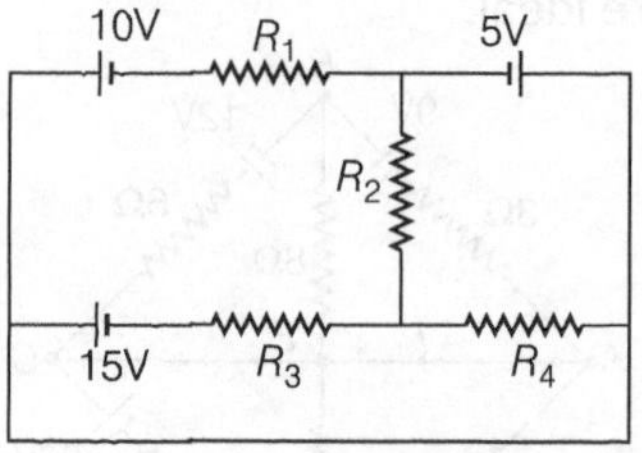

75 Current through R_2 is zero, if $R_4 = 2\,\Omega$ and $R_3 = 4\,\Omega$. In this case,

(a) current through R_3 is 2.5 A
(b) current through R_4 is 3A
(c) Both (a) and (b)
(d) None of the above

76 For what ratio $\frac{R_2}{R_4}$, current through R_3 will be zero?

(a) 1 : 1 (b) 1 : 2 (c) 1 : 3 (d) Not possible

Comprehension-6

Twelve wires of equal resistance R are placed in a given arrangement as shown in figure. Then,

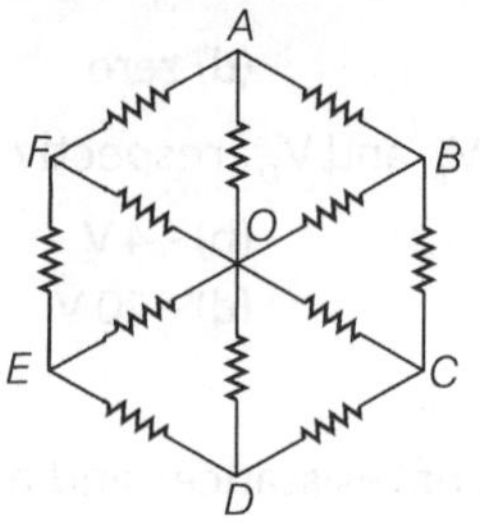

77 Equivalent resistance between A and B is

(a) $\frac{11R}{20}$ (b) $\frac{10R}{20}$ (c) $\frac{R}{20}$ (d) $\frac{3R}{2}$

78 Equivalent resistance between A and C is

(a) $\frac{3R}{4}$ (b) $\frac{R}{4}$ (c) $\frac{5R}{6}$ (d) $\frac{3R}{5}$

79 Equivalent resistance between A and O is

(a) $\frac{19R}{41}$ (b) $\frac{20R}{41}$ (c) $\frac{18R}{23}$ (d) $\frac{9R}{20}$

Comprehension-7

A network of resistances is constructed with R_1 and R_2 as shown in figure. The potential at the points 1, 2, 3 ...... N are $V_1, V_2, V_3 .... V_N$ respectively, each having a potential k times smaller than the previous one.

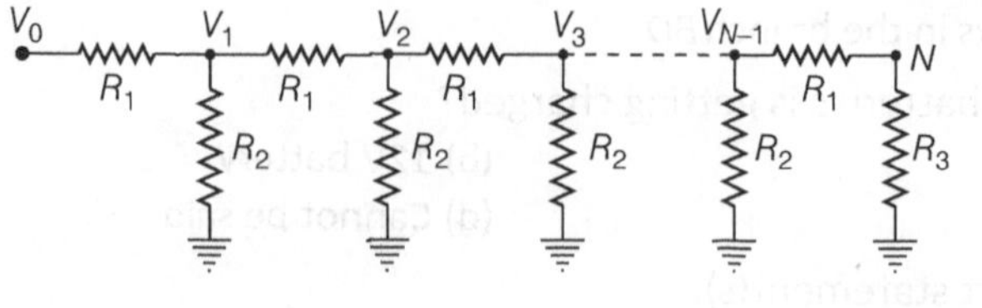

80 The ratio $\frac{R_1}{R_2}$ is

(a) $k^2 - \frac{1}{k}$ (b) $\frac{k}{k-1}$ (c) $k - \frac{1}{k^2}$ (d) $\frac{(k-1)^2}{k}$

81 The ratio $\frac{R_2}{R_3}$ is

(a) $\frac{(k-1)^2}{k}$ (b) $k^2-\frac{1}{k}$ (c) $\frac{k}{k-1}$ (d) $k-\frac{1}{k^2}$

82 The current that passes through the resistance R_2 nearest to the V_0 is

(a) $\frac{(k-1)^2}{k}\frac{V_0}{R_3}$ (b) $\frac{(k+1)^2}{k}\frac{V_0}{R_3}$ (c) $\left(k+\frac{1}{k^2}\right)\frac{V_0}{R_3}$ (d) $\left(\frac{k-1}{k^2}\right)\frac{V_0}{R_3}$

Comprehension-8

In the circuit given below, both batteries are ideal. emf ε_1 of battery 1 has a fixed value, but emf ε_2 of battery 2 can be varied between 1.0 V and 10.0 V. The graph gives the currents through the two batteries as a function of ε_2, but are not marked as which plot corresponds to which battery. But for both plots, current is assumed to be negative when the direction of the current through the battery is opposite the direction of that battery's emf. (Direction of emf is from negative to positive)

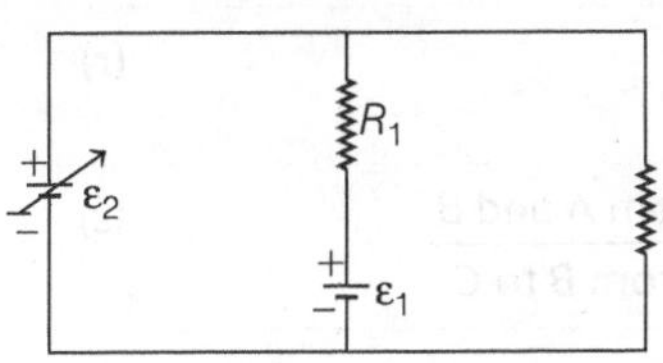

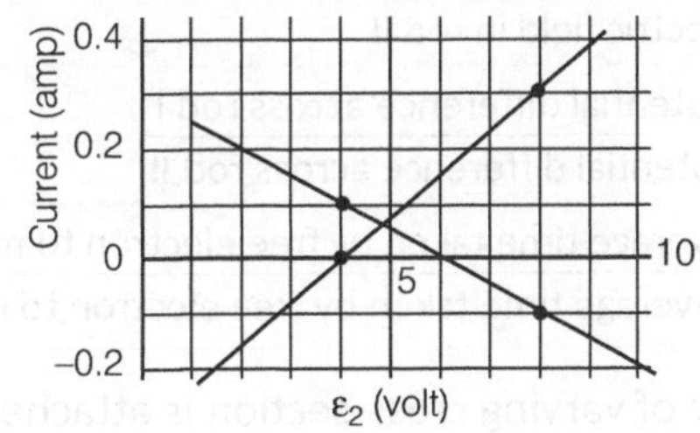

83 The value of emf ε_1 is

(a) 8 V (b) 6 V (c) 4 V (d) 2 V

84 The resistance R_1 has value

(a) 10Ω (b) 20 Ω (c) 30 Ω (d) 40 Ω

85 The resistance R_2 is equal to

(a) 10Ω (b) 20 Ω (c) 30 Ω (d) 40 Ω

Match the Columns Questions

86 Column I gives physical quantities based on a situation in which an ideal cell of emf V is connected across a cylindrical rod of uniform cross-section area and conductivity (σ) as shown in figure. E, J, ϕ and i are electric field at, current density through, electric flux through and current through shaded cross-section respectively as shown in figure. Physical quantities in Column II are related to those in Column I. Match the expression in Column I with the statements in Column II.

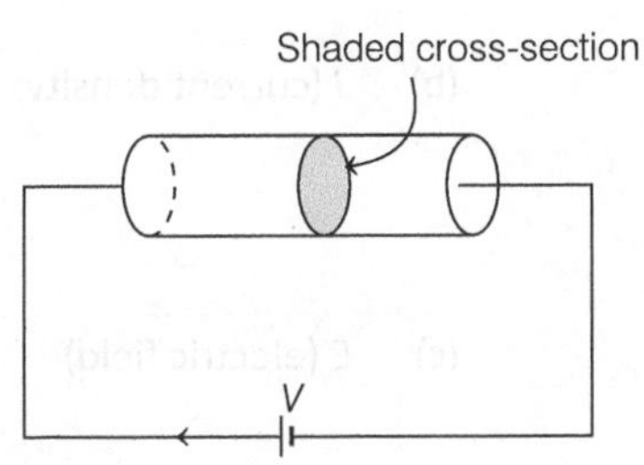

	Column I		Column II
(a)	$\frac{\phi}{i}$	(p)	Conductivity of rod
(b)	$\frac{E}{J}$	(q)	Resistance of rod
(c)	$\sigma\phi V$	(r)	Resistivity of rod
(d)	$\frac{V}{\sigma\phi}$	(s)	Power delivered to rod

87 Column I gives physical quantities of a situation in which a current i passes through two rods I and II of equal length that are joined in series. The ratio of free electron density (n), resistivity (ρ) and cross-section area (A) of both are in ratio $n_1 : n_2 = 2:1$, $\rho_1 : \rho_2 = 2:1$ and $A_1 : A_2 = 1:2$, respectively. Column II gives corresponding results. Match the ratios in Column I with the values in Column II.

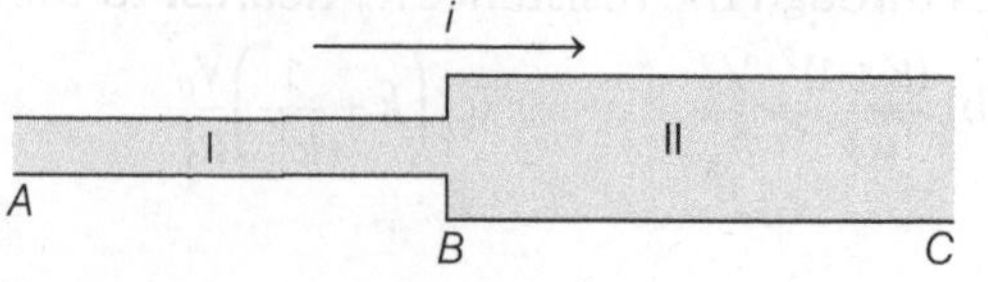

	Column I		Column II
(a)	$\dfrac{\text{Drift velocity of free electron in rod I}}{\text{Drift velocity of free electron in rod II}}$	(p)	0.5
(b)	$\dfrac{\text{Electric field in rod I}}{\text{Electric field in rod II}}$	(q)	1
(c)	$\dfrac{\text{Potential difference across rod I}}{\text{Potential difference across rod II}}$	(r)	2
(d)	$\dfrac{\text{Average time taken by free electron to move from } A \text{ and } B}{\text{Average time taken by free electron to move from } B \text{ to } C}$	(s)	4

88 A resistor of varying cross-section is attached with a battery as shown. Variation of area from both ends are similar and areas are same at ends and maximum at centre. Match the approximate graphical representation of various physical properties in columns, where X-axis represents position from end A of resistor and Y-axis various physical properties.

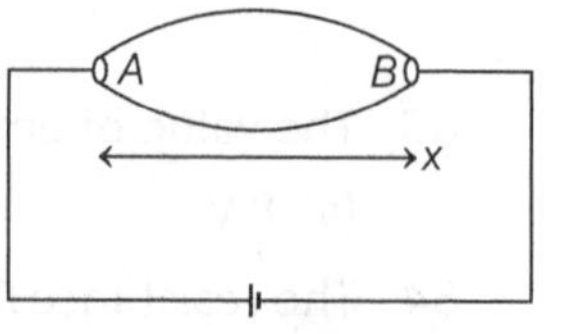

	Column I		Column II
(a)	i (current)	(p)	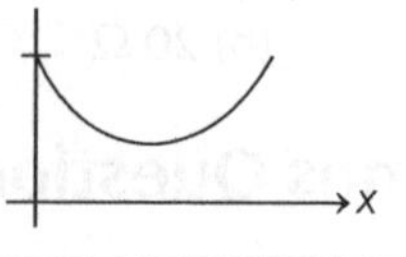
(b)	J (current density)	(q)	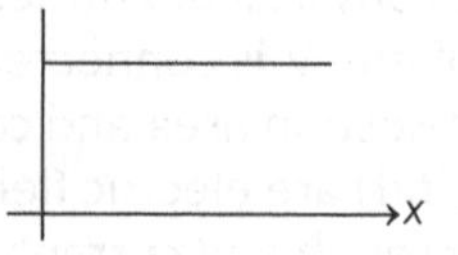
(c)	E (electric field)	(r)	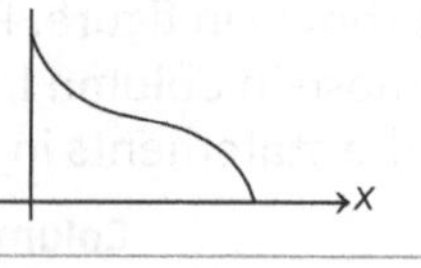
(d)	V_d (potential)	(s)	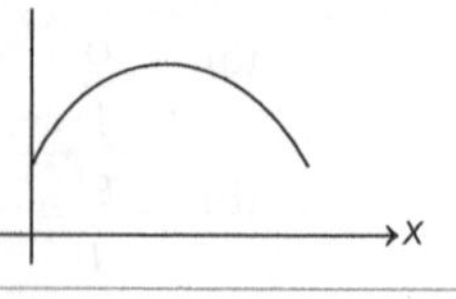
		(t)	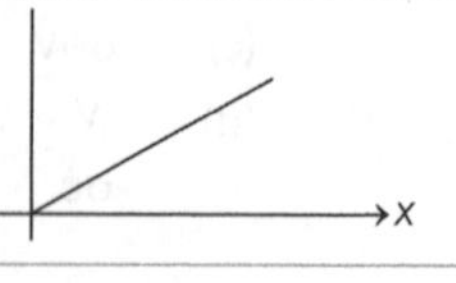

89 Six batteries of increasing emf and increasing internal resistance are as shown in figure. Match the following

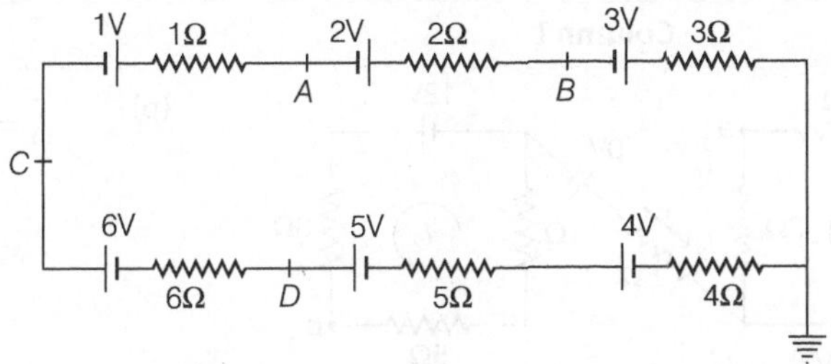

	Column I		Column II
(a)	Potential of point *A*	(p)	zero
(b)	Potential of point *B*	(q)	2 V
(c)	Potential of point *C*	(r)	4 V
(d)	Potential of point *D*	(s)	6 V
		(t)	None

90 In the given circuit *A*, *B*, *C* represent circuit elements. It may be an ideal battery or resistor. And +, – sign indicates high potential and low potential, respectively.

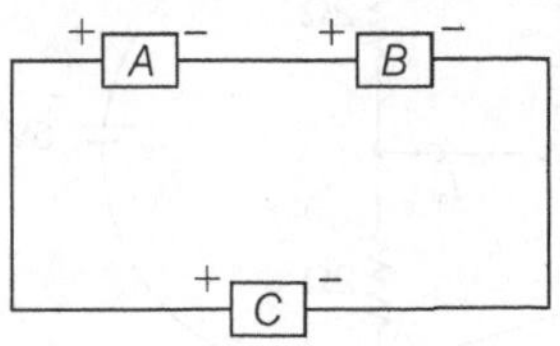

	Column I		Column II
(a)	If 'A' is resistor	(p)	Current clockwise
(b)	If 'A' is battery	(q)	'B' may be a battery
(c)	If 'B' is resistor	(r)	'C' must be a battery
(d)	If 'C' is resistor	(s)	Current anti-clockwise
		(t)	'B' must be battery

91 A circuit is shown in figure *R* is a non-zero variable with finite resistance. ε is some unknown emf with polarities are shown, then

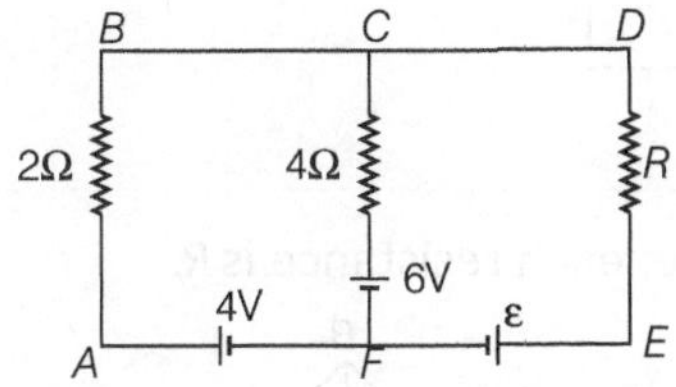

	Column I		Column II
(a)	Current passing through 4Ω resistance can be zero	(p)	possible, if $\varepsilon = 6$ V
(b)	Current passing through 4 Ω resistance can be from *F* to *C* direction	(q)	possible, if $\varepsilon > 6$ V
(c)	Current passing through 4Ω resistance can be from *C* to *F* direction	(r)	possible, if $\varepsilon < 6$ V
(d)	Current passing through 2Ω resistance will be from *B* to *A* direction	(s)	possible for any values of ε from zero to infinity

92 Consider the electric circuits given in Column I and Column II gives values of some quantities from electric circuits. Match the corresponding quantities with correct circuit.

	Column I		Column II
(a)	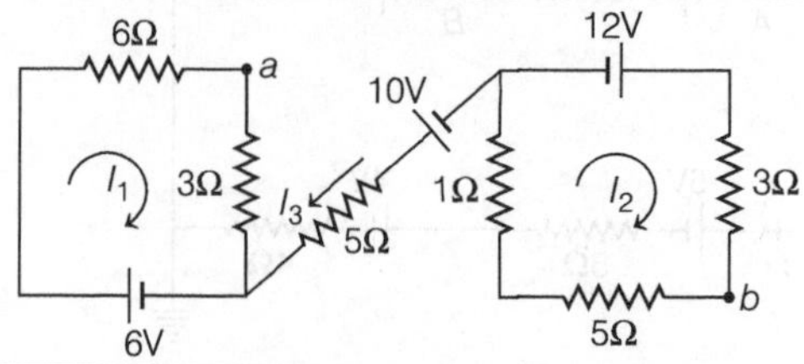	(p)	$I_1 = \frac{1}{55}$ A
(b)	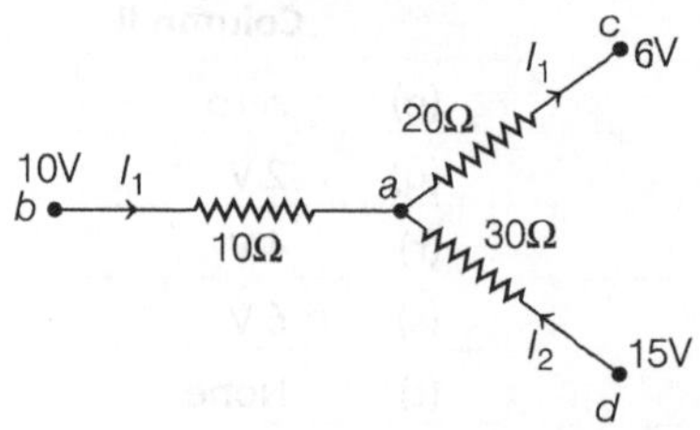	(q)	$I_2 = \frac{-1}{121}$ A
(c)	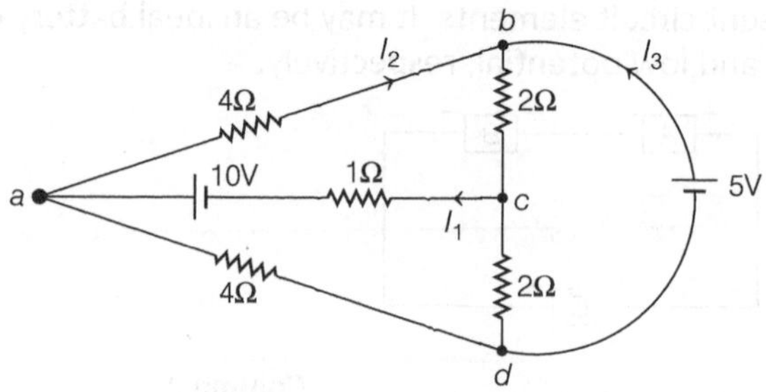	(r)	$I_3 = \frac{15}{8}$ A
(d)	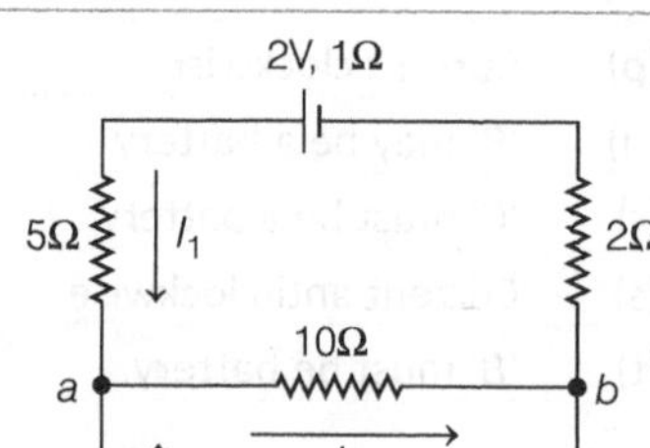	(s)	$V_a - V_b = 4$V

93 In the network shown in figure, each resistance is R.

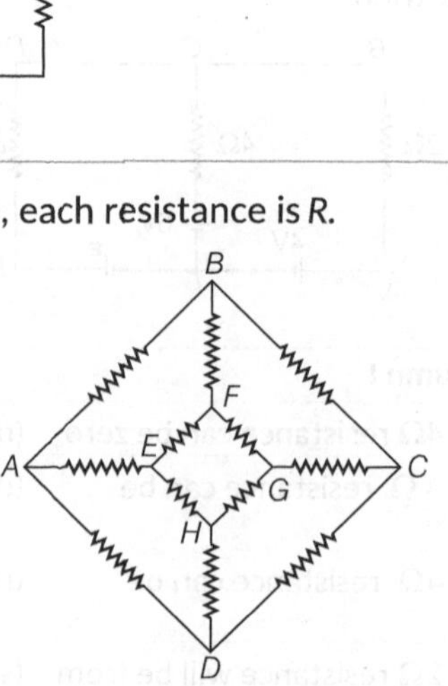

	Column I		Column II
(a)	Equivalent resistance between points *A* and *E*	(p)	7*R* / 12
(b)	Equivalent resistance between points *A* and *G*	(q)	5*R* / 6
(c)	Equivalent resistance between points *A* and *C*	(r)	3*R* / 4
		(s)	7*R* / 8

Numerical Value Questions

94 Each element in the finite chain of resistors shown in the figure is 1Ω. A current of 1 A flows through the final element. The potential difference *V* across the input terminals of the chain is *N* volt. Find *N*.

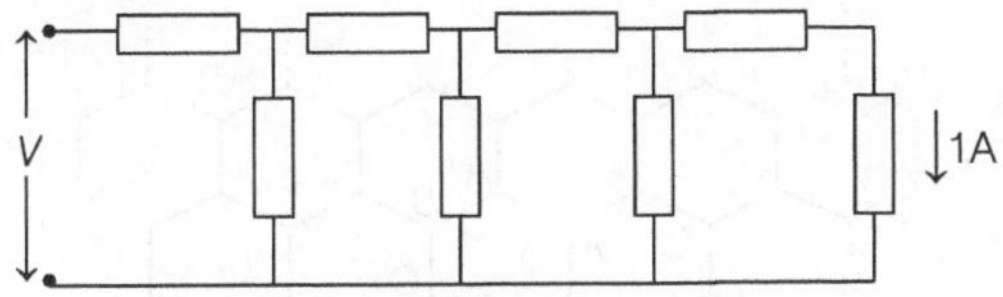

95 In the circuit diagram shown, find the current (in A) through the $1\ \Omega$ resistor.

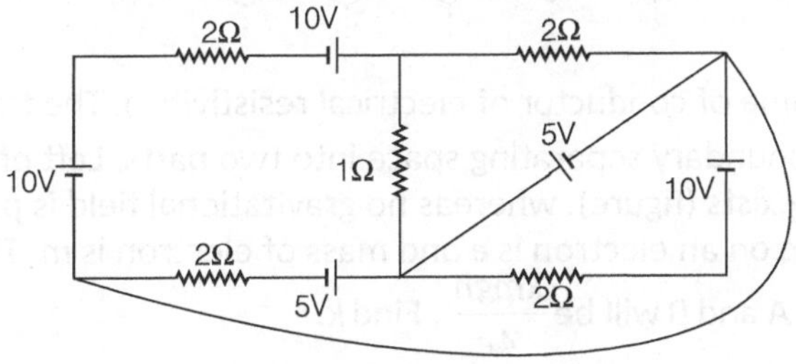

96 If the potential difference $V_A - V_B$ for the given circuit is $x/9$ V, then find *x*. All the cells are ideal and emf of each cell is 1 V.

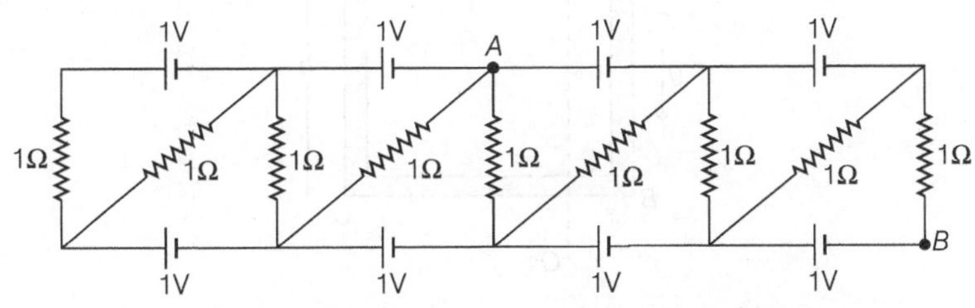

97 Figure shows a cell in which unit positive charge experience a constant non-electric force of 10N and a constant electric force of 8N in directions shown in the figure. Find the internal resistance (in ohm) of the cell.

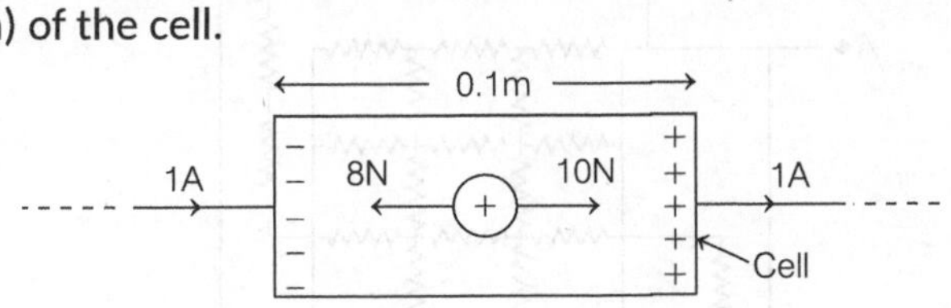

98 Find the current (in A) in the resistance R. Each resistance is of 2Ω.

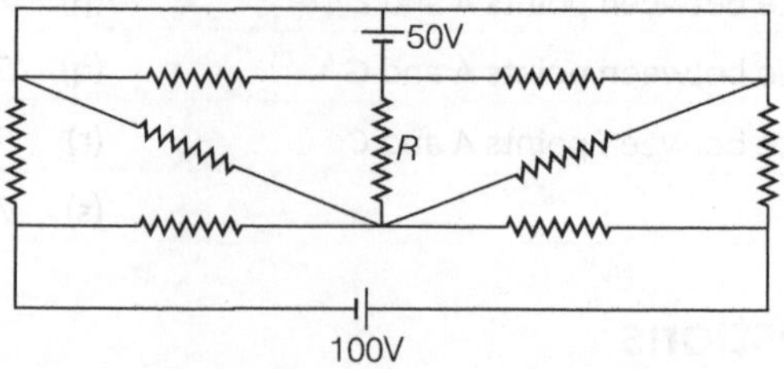

99 Consider an infinite mesh as shown in figure, where each side of the mesh has resistance $R = 24\Omega$. Consider hexagonal part $ABCDEF$ of infinite mesh and find equivalent resistance between A and B points of hexagon.

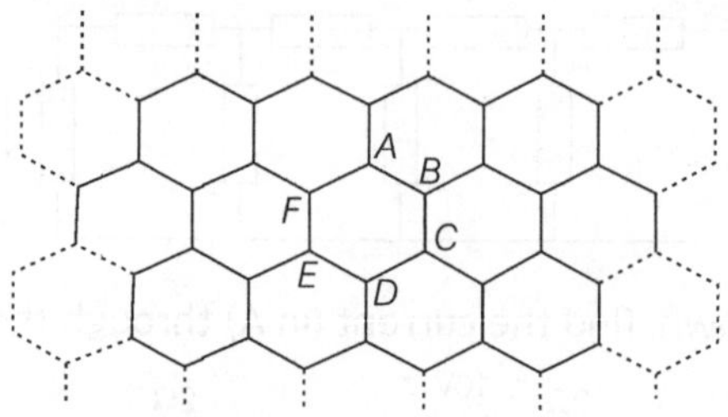

100 $ABCD$ is a square frame of conductor of electrical resistivity ρ. The frame lies in a vertical plane. PQ is an imaginary boundary separating space into two parts. Left of PQ, a uniform gravitational field **g** exists (figure), whereas no gravitational field is present right of PQ. Magnitude of charge on an electron is e and mass of electron is m. The electrical potential difference between A and B will be $\frac{kmgh}{4e}$. Find k.

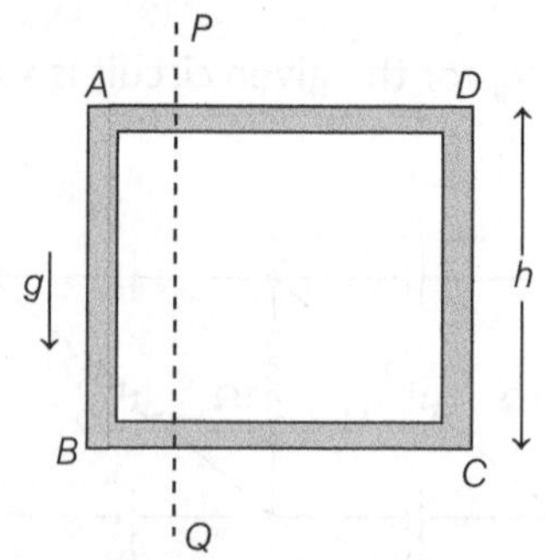

101 32 resistors are connected in a circuit as shown in the figure. The resistance of each resistor is 8Ω. Find equivalent resistance (in Ω) across AB.

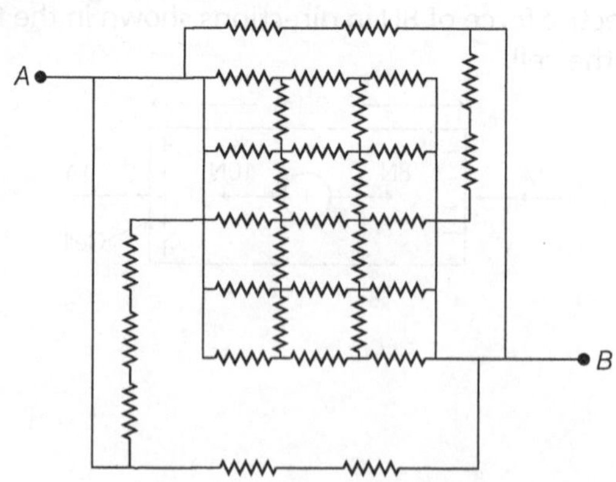

102 A beam of fast moving electrons having cross-sectional area $A = 1\,cm^2$ falls normally on a flat surface. The electrons are absorbed by the surface and the average pressure exerted by the electrons on this surface is found to be $p = 9.1\,Pa$. If the electrons are moving with a speed $v = 8 \times 10^7$ m/s, then find the effective current (in A) through any cross-section of the electron beam. (mass of electron $= 9.1 \times 10^{-31}$ kg)

Answers

1. (c)	**2.** (a)	**3.** (b)	**4.** (c)	**5.** (b)	**6.** (d)	**7.** (a)	**8.** (b)	**9.** (a)	**10.** (a)
11. (a)	**12.** (b)	**13.** (c)	**14.** (b)	**15.** (b)	**16.** (c)	**17.** (c)	**18.** (a)	**19.** (a)	**20.** (a)
21. (b)	**22.** (c)	**23.** (b)	**24.** (d)	**25.** (d)	**26.** (c)	**27.** (a)	**28.** (d)	**29.** (a)	**30.** (d)
31. (a)	**32.** (c)	**33.** (a)	**34.** (d)	**35.** (c)	**36.** (b)	**37.** (b)	**38.** (b)	**39.** (c)	**40.** (a)
41. (b)	**42.** (d)	**43.** (a)	**44.** (bc)	**45.** (ad)	**46.** (bcd)	**47.** (bc)	**48.** (bc)	**49.** (abc)	**50.** (abcd)
51. (abcd)	**52.** (abcd)	**53.** (abc)	**54.** (bc)	**55.** (abc)	**56.** (ac)	**57.** (acd)	**58.** (abd)	**59.** (abd)	**60.** (abc)
61. (ac)	**62.** (a)	**63.** (a)	**64.** (a)	**65.** (a)	**66.** (b)	**67.** (d)	**68.** (c)	**69.** (c)	**70.** (d)
71. (a)	**72.** (c)	**73.** (c)	**74.** (abd)	**75.** (a)	**76.** (d)	**77.** (a)	**78.** (a)	**79.** (d)	**80.** (d)
81. (c)	**82.** (d)	**83.** (b)	**84.** (b)	**85.** (d)	**86.** (a) → r; (b) → r; (c) → s; (d) → q				

87. (a) → q; (b) → s; (c) → s; (d) → q
88. (a) → q; (b) → p; (c) → p; (d) → r
89. (a) → p; (b) → p; (c) → p; (d) → p
90. (a) → pqr; (b) → q ; (c) → pr; (d) → st
91. (a) → q; (b) → pqr; (c) → q; (d) → pqrs
92. (a) → s; (b) → p; (c) → r ; (d) → q
93. (a) → p; (b) → q; (c) → r
94. (34)
95. (2.5) **96.** (22) **97.** (0.2) **98.** (18.75)
99. (16) **100.** (3) **101.** (3) **102.** (2)

For Detailed Solutions
Scan the code

Previous Years' Questions

JEE Main

1 A square shaped wire with resistance of each side 3Ω is bent to form a complete circle. The resistance between two diametrically opposite points of the circle in unit of Ω will be **[JEE Main 2021]**

2 The equivalent resistance of the given circuit between the terminals A and B is **[JEE Main 2021]**

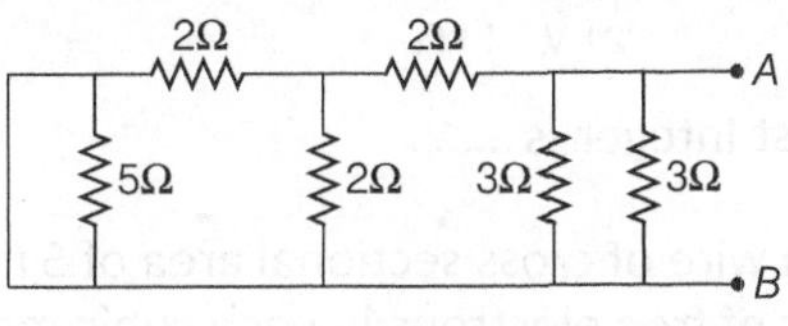

(a) 0
(b) $3\,\Omega$
(c) $9/2\,\Omega$
(d) $1\,\Omega$

3 The colour coding on a carbon resistor is shown in the given figure. The resistance value of the given resistor is **[JEE Main 2021]**

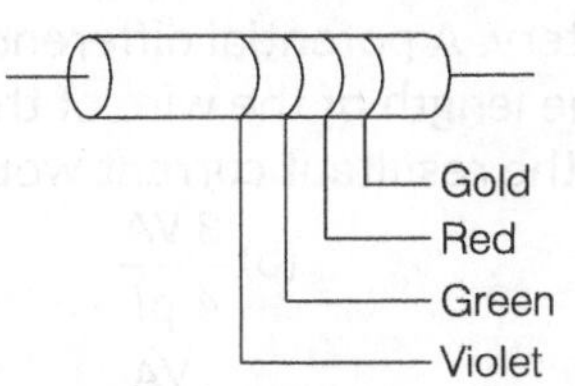

(a) $(5700 \pm 285)\,\Omega$
(b) $(7500 \pm 750)\,\Omega$
(c) $(5700 \pm 375)\,\Omega$
(d) $(7500 \pm 375)\,\Omega$

4 The resistance of a conductor at 15°C is 16 Ω and at 100°C is 20 Ω. What will be the temperature coefficient of resistance of the conductor? **[JEE Main 2021]**

(a) $0.010°C^{-1}$ (b) $0.033°C^{-1}$ (c) $0.003°C^{-1}$ (d) $0.042°C^{-1}$

5 A 16 Ω wire is bend to form a square loop. A 9 V supply having internal resistance of 1 Ω is connected across one of its sides. The potential drop across the diagonals of the square loop is ...... $\times 10^{-1}$ V. **[JEE Main 2021]**

6 A copper (Cu) rod of length 25 cm and cross-sectional area 3 mm^2 is joined with a similar aluminium (Al) rod as shown in figure. Find the resistance of the combination between the ends *A* and *B*.

(Take, resistivity of copper $= 1.7 \times 10^{-8}$ Ω-m, resistivity of aluminium $= 2.6 \times 10^{-8}$ Ω-m) **[JEE Main 2021]**

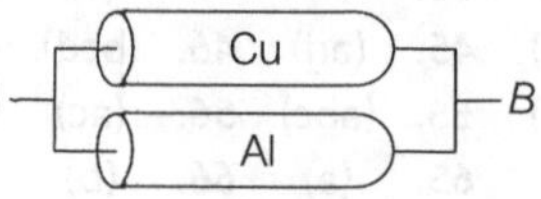

(a) 2.170 m Ω (b) 1.420 m Ω

(c) 0.0858 m Ω (d) 0.858 m Ω

7 In the given figure, switches S_1 and S_2 are in open condition. The resistance across *ab* when the switches S_1 and S_2 are closed is ...... Ω. **[JEE Main 2021]**

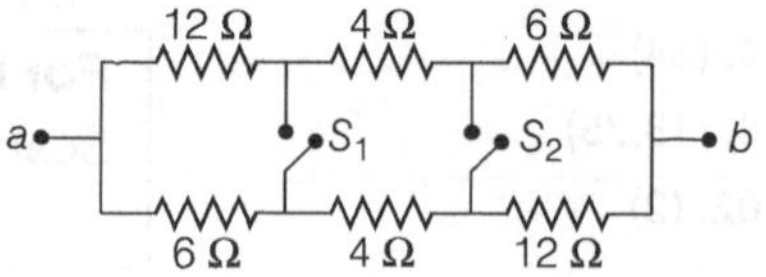

8 In the figure given, the electric current flowing through the 5 kΩ resistor is *x* mA. **[JEE Main 2021]**

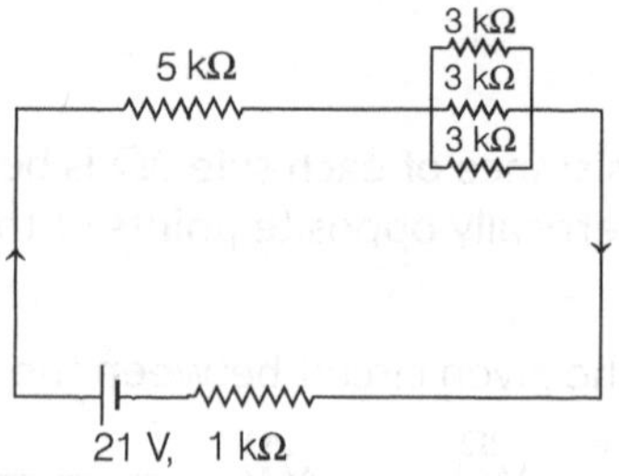

The value of *x* to the nearest Integer is ....

9 A current of 10 A exists in a wire of cross sectional area of 5 mm^2 with a drift velocity of 2×10^{-3} ms^{-1}. The number of free electrons in each cubic metre of the wire is **[JEE Main 2021]**

(a) 2×10^6 (b) 625×10^{25}

(c) 2×10^{25} (d) 1×10^{23}

10 A conducting wire of length *l*, area of cross-section *A* and electric resistivity ρ is connected between the terminals of a battery. A potential difference *V* is developed between its ends, causing an electric current. If the length of the wire of the same material is doubled and the area of cross-section is halved, the resultant current would be **[JEE Main 2021]**

(a) $\frac{1}{4}\frac{VA}{\rho l}$ (b) $\frac{3}{4}\frac{VA}{\rho l}$

(c) $\frac{1}{4}\frac{\rho l}{VA}$ (d) $4\frac{VA}{\rho l}$

11 A current of 6 A enters one corner P of an equilateral triangle PQR having three wires of resistance 2 Ω each and leaves by the corner R. The currents i_1 in ampere is .... **[JEE Main 2021]**

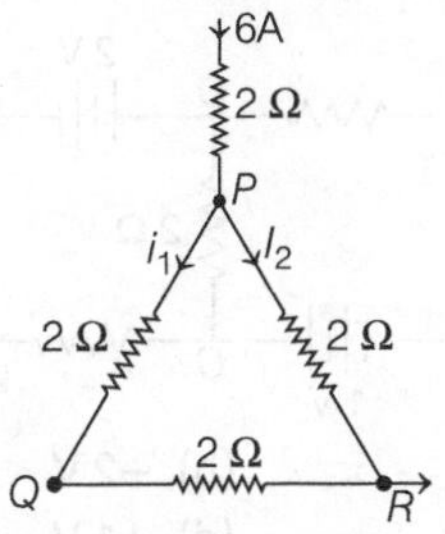

12 Five identical cells each of internal resistance 1 Ω and emf 5V are connected in series and in parallel with an external resistance R. For what value of R, current in series and parallel combination will remain the same ? **[JEE Main 2021]**

(a) 1 Ω (b) 25 Ω
(c) 5 Ω (d) 10 Ω

13 In an electrical circuit, a battery is connected to pass 20 C of charge through it in a certain given time. The potential difference between two plates of the battery is maintained at 15 V. The work done by the battery is .... J. **[JEE Main 2021]**

14 Five equal resistances are connected in a network as shown in figure. The net resistance between the points A and B is **[JEE Main 2021]**

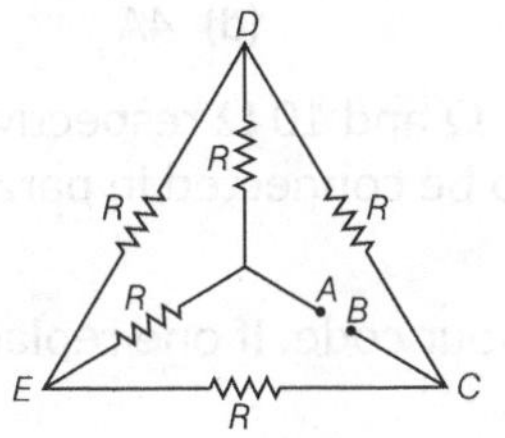

(a) $2R$ (b) $\frac{R}{2}$
(c) $\frac{3R}{2}$ (d) R

15 Consider four conducting materials copper, tungsten, mercury and aluminium with resistivity ρ_C, ρ_T, ρ_M and ρ_A, respectively. Then, **[JEE Main 2020]**

(a) $\rho_A > \rho_T > \rho_C$ (b) $\rho_A > \rho_M > \rho_C$
(c) $\rho_C > \rho_A > \rho_T$ (d) $\rho_M > \rho_A > \rho_C$

16 The value of current i_1 flowing from A to C in the circuit diagram is **[JEE Main 2020]**

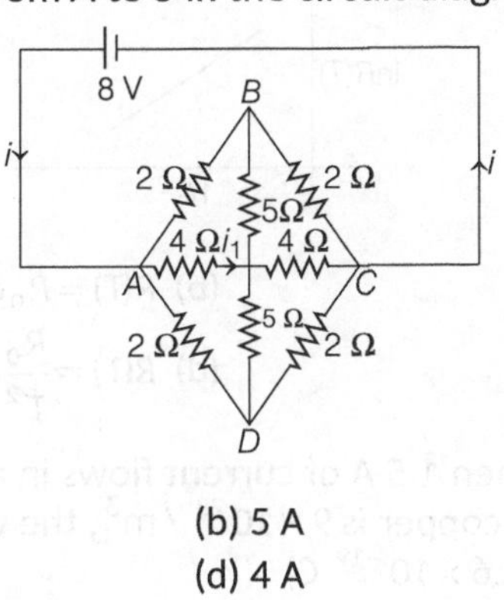

(a) 1 A (b) 5 A
(c) 2 A (d) 4 A

17 In the given circuit, currents in different branches and value of one resistor are shown. Then, potential at point B with respect to the point A is **[JEE Main 2020]**

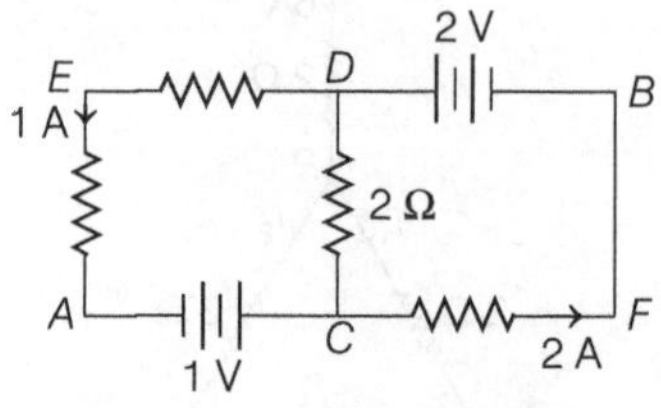

(a) +2 V
(b) −2 V
(c) −1 V
(d) +1 V

18 In the given circuit diagram, a wire is joining points B and D. The current in this wire is **[JEE Main 2020]**

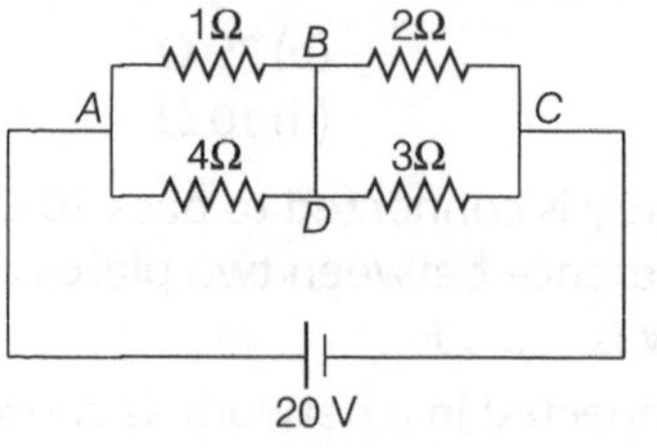

(a) 0.4 A
(b) zero
(c) 2A
(d) 4A

19 Four resistances of 15 Ω, 12 Ω, 4 Ω and 10 Ω respectively in cyclic order to form Wheatstone's network. The resistance that is to be connected in parallel with the resistance of 10 Ω to balance the network is .... Ω . **[JEE Main 2020]**

20 A 200 Ω resistor has a certain colour code. If one replaces the red colour by green in the code, the new resistance will be **[JEE Main 2019]**
(a) 100 Ω
(b) 400 Ω
(c) 300 Ω
(d) 500 Ω

21 In a conductor, if the number of conduction electrons per unit volume is 8.5×10^{28} m^{-3} and mean free time is 25 fs (femto second), it's approximate resistivity is (Take, $m_e = 9.1\times10^{-31}$ kg) **[JEE Main 2019]**
(a) 10^{-7} Ω-m
(b) 10^{-5} Ω-m
(c) 10^{-6} Ω-m
(d) 10^{-8} Ω-m

22 In an experiment, the resistance of a material is plotted as a function of temperature (in some range). As shown in the figure, it is a straight line. **[JEE Main 2019]**
One may conclude that

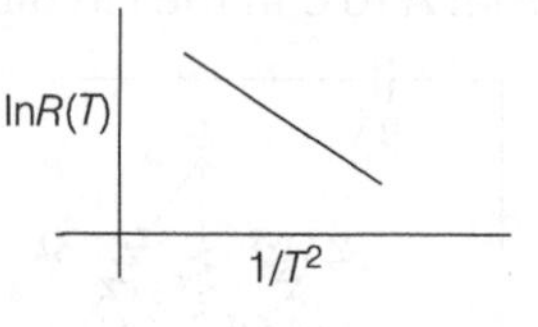

(a) $R(T) = R_0 e^{-T^2/T_0^2}$
(b) $R(T) = R_0 e^{T^2/T_0^2}$
(c) $R(T) = R_0 e^{-T_0^2/T^2}$
(d) $R(T) = \frac{R_0}{T^2}$

23 Drift speed of electrons, when 1.5 A of current flows in a copper wire of cross-section 5 mm^2 is v. If the electron density in copper is 9×10^{28} $/m^3$, the value of v (in mm/s) is close to (Take, charge of electron to be $= 1.6\times10^{-19}$ C) **[JEE Main 2019]**
(a) 0.02
(b) 0.2
(c) 2
(d) 3

24 A copper wire is stretched to make it 0.5% longer. The percentage change in its electrical resistance, if its volume remains unchanged is **[JEE Main 2019]**

(a) 2.0% (b) 1.0% (c) 0.5% (d) 2.5%

25 A carbon resistance has a following color code. What is the value of the resistance? **[JEE Main 2019]**

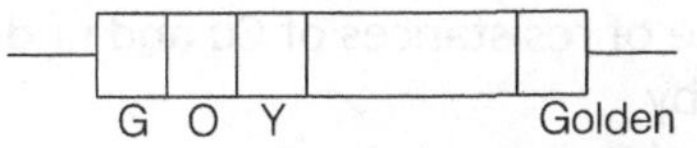

(a) 5.3 M$\Omega \pm 5\%$ (b) 64 k$\Omega \pm 10\%$

(c) 6.4 M$\Omega \pm 5\%$ (d) 530 k$\Omega \pm 5\%$

26 A uniform metallic wire has a resistance of 18 Ω and is bent into an equilateral triangle. Then, the resistance between any two vertices of the triangle is **[JEE Main 2019]**

(a) 12 Ω (b) 8 Ω

(c) 2 Ω (d) 4 Ω

27 In the figure shown, what is the current (in ampere) drawn from the battery? You are given $R_1 = 15\ \Omega, R_2 = 10\ \Omega, R_3 = 20\ \Omega, R_4 = 5\ \Omega, R_5 = 25\ \Omega, R_6 = 30\ \Omega, E = 15$ V **[JEE Main 2019]**

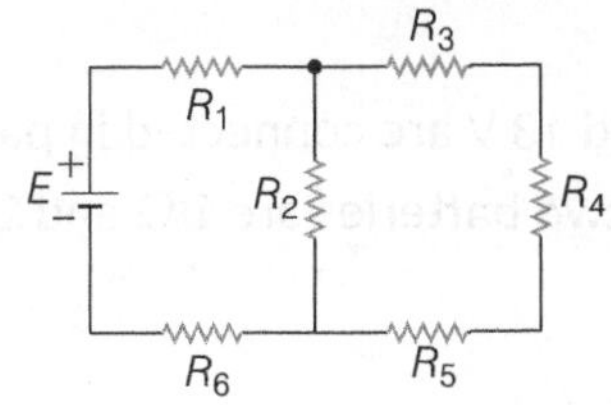

(a) 13/24 (b) 7/18

(c) 20/3 (d) 9/32

28 When the switch *S* in the circuit shown is closed, then the value of current *i* will be **[JEE Main 2019]**

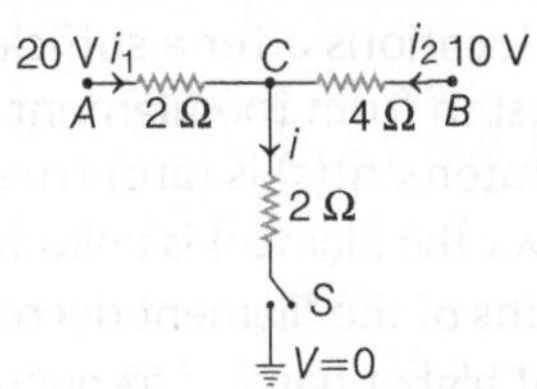

(a) 4A (b) 3A

(c) 2A (d) 5A

29 The Wheatstone bridge shown in figure here, gets balanced when the carbon resistor is used as R_1 has the color code (orange, red, brown). The resistors R_2 and R_4 are 80 Ω and 40 Ω, respectively.

Assuming that the color code for the carbon resistors gives their accurate values, the color code for the carbon resistor is used as R_3 would be **[JEE Main 2019]**

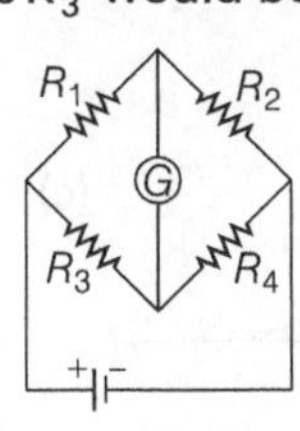

(a) brown, blue, black (b) brown, blue, brown

(c) grey, black, brown (d) red, green, brown

30 Two batteries with emf 12 V and 13 V are connected in parallel across a load resistor of 10 Ω. The internal resistances of the two batteries are 1 Ω and 2 Ω, respectively. The voltage across the load lies between **[JEE Main 2018]**

(a) 11.6 V and 11.7 V (b) 11.5 V and 11.6 V
(c) 11.4 V and 11.5 V (d) 11.7 V and 11.8 V

31 The temperature dependence of resistances of Cu and undoped Si in the temperature range 300-400 K, is best described by **[JEE Main 2016]**

(a) linear increase for Cu, linear increase for Si
(b) linear increase for Cu, exponential increase for Si
(c) linear increase for Cu, exponential decrease for Si
(d) linear decrease for Cu, linear decrease for Si

32 When 5V potential difference is applied across a wire of length 0.1m, the drift speed of electrons is $2.5 \times 10^{-4}\ ms^{-1}$. If the electron density in the wire is $8 \times 10^{28}\ m^{-3}$ the resistivity of the material is close to **[JEE Main 2015]**

(a) $1.6 \times 10^{-8}\ \Omega m$ (b) $1.6 \times 10^{-7}\ \Omega m$
(c) $1.6 \times 10^{-6}\ \Omega m$ (d) $1.6 \times 10^{-5}\ \Omega m$

JEE Advanced

33 Two batteries with emf 12 V and 13 V are connected in parallel across a load resistor of 10 Ω. The internal resistances of the two batteries are 1 Ω and 2 Ω, respectively. The voltage across the load lies between **[JEE Advanced 2016]**

(a) 11.7 V and 11.8 V
(b) 11.6 V and 11.7 V
(c) 11.5 V and 11.6 V
(d) 11.4 V and 11.5 V

34 An incandescent bulb has a thin filament of tungsten that is heated to high temperature by passing an electric current. The hot filament emits black-body radiation. The filament is observed to break up at random locations after a sufficiently long time of operation due to non-uniform evaporation of tungsten from the filament. If the bulb is powered at constant voltage, which of the following statement(s) is (are) true? **[JEE Advanced 2016]**

(a) The temperature distribution over the filament is uniform.
(b) The resistance over small sections of the filament decreases with time.
(c) The filament emits more light at higher band of frequencies before it breaks up.
(d) The filament consumes less electrical power towards the end of the life of the bulb.

35 An infinite line charge of uniform electric charge density λ lies along the axis of an electrically conducting infinite cylindrical shell of radius R. At time $t=0$, the space inside the cylinder is filled with a material of permittivity ε and electrical conductivity σ. The electrical conduction in the material follows Ohm's law. Which one of the following graphs best describes the subsequent variation of the magnitude of current density j (t) at any point in the material? **[JEE Advanced 2016]**

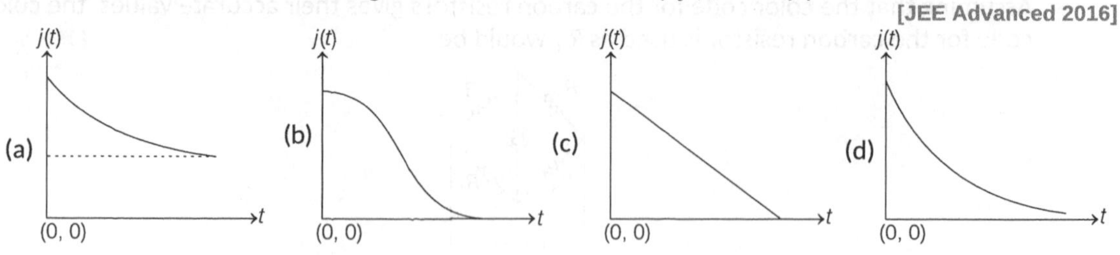

36 In the following circuit, the current through the resistor $R(=2\,\Omega)$ is I amperes. The value of I is

[JEE Advanced 2015]

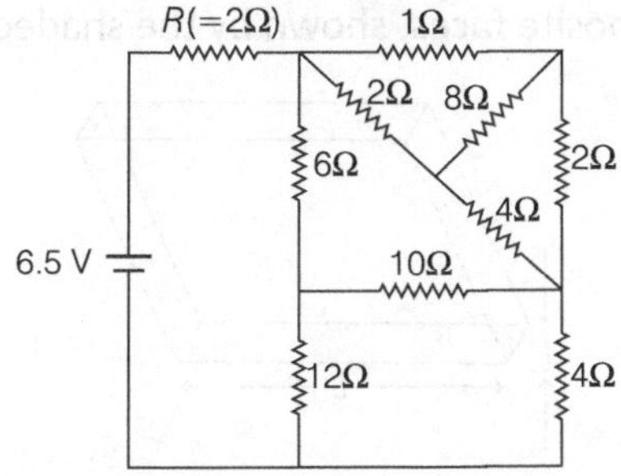

37 In an aluminium (Al) bar of square cross section, a square hole is drilled and is filled with iron (Fe) as shown in the figure. The electrical resistivities of Al and Fe are 2.7×10^{-8} Ωm and 1.0×10^{-7} Ωm, respectively. The electrical resistance between the two faces P and Q of the composite bar is

[JEE Advanced 2015]

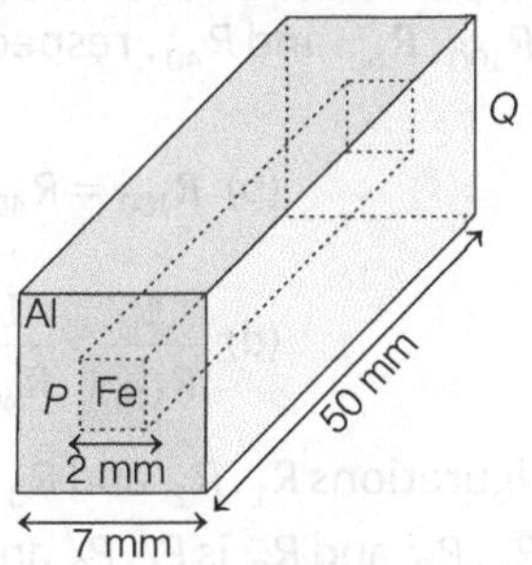

(a) $\dfrac{2475}{64}\,\mu\Omega$ (b) $\dfrac{1875}{64}\,\mu\Omega$

(c) $\dfrac{1875}{49}\,\mu\Omega$ (d) $\dfrac{2475}{132}\,\mu\Omega$

38 For the resistance network shown in the figure, choose the correct option(s).

[JEE Advanced 2012]

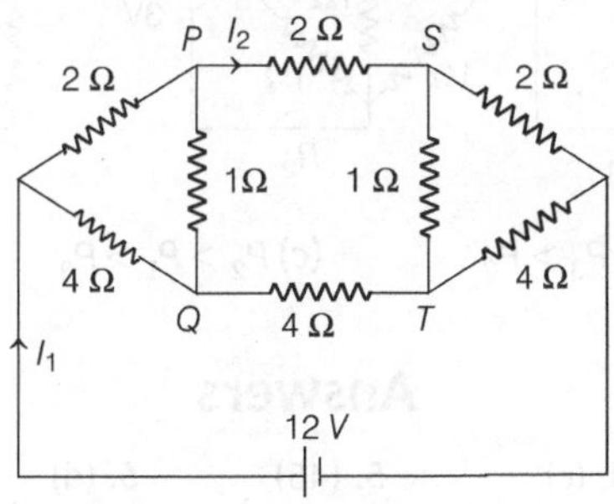

(a) The current through PQ is zero

(b) $I_1 = 3\,\text{A}$

(c) The potential at S is less than that at Q

(d) $I_2 = 2\,\text{A}$

39 Two batteries of different emfs and different internal resistances are connected as shown. The voltage across AB in volt is

[JEE Advanced 2011]

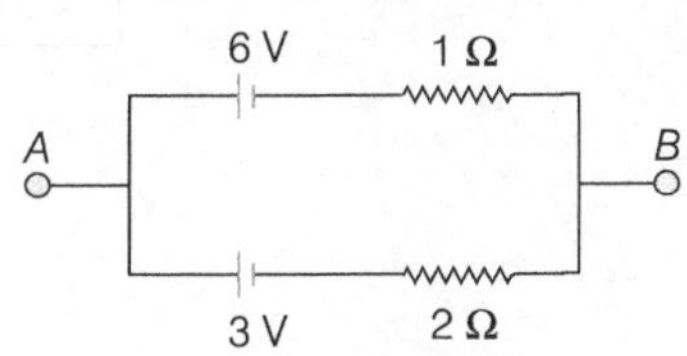

40 Consider a thin square sheet of side L and thickness t, made of a material of resistivity ρ. The resistance between two opposite faces, shown by the shaded areas in the figure is **[JEE Advanced 2010]**

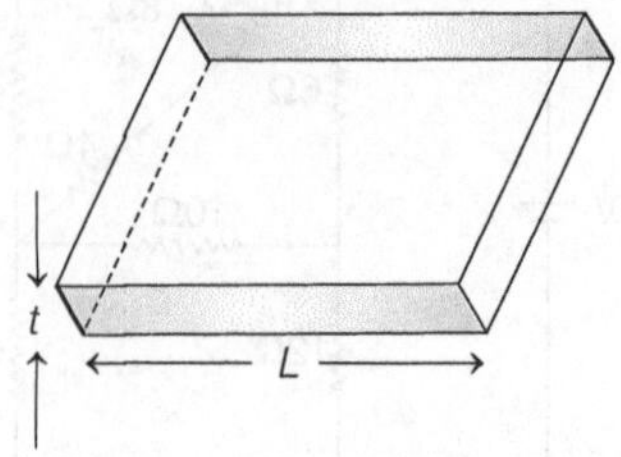

(a) directly proportional to L (b) directly proportional to t
(c) independent of L (d) independent of t

41 Incandescent bulbs are designed by keeping in mind that the resistance of their filament increases with the increase in temperature. If at room temperature, 100 W, 60 W and 40 W bulbs have filament resistances R_{100}, R_{60} and R_{40}, respectively, the relation between these resistances is **[JEE Advanced 2010]**

(a) $\frac{1}{R_{100}} = \frac{1}{R_{40}} + \frac{1}{R_{60}}$ (b) $R_{100} = R_{40} + R_{60}$

(c) $R_{100} > R_{60} > R_{40}$ (d) $\frac{1}{R_{100}} > \frac{1}{R_{60}} > \frac{1}{R_{40}}$

42 Figure shows three resistor configurations R_1, R_2 and R_3 connected to 3 V battery. If the power dissipated by the configuration R_1, R_2 and R_3 is P_1, P_2 and P_3, respectively, then **[JEE Advanced 2008]**

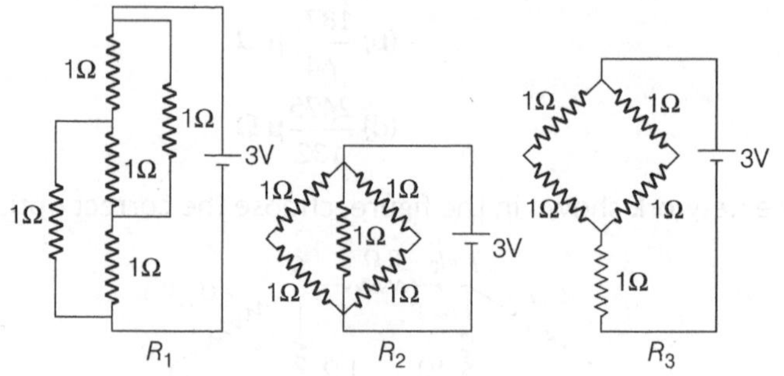

(a) $P_1 > P_2 > P_3$ (b) $P_1 > P_3 > P_2$ (c) $P_2 > P_1 > P_3$ (d) $P_3 > P_2 > P_1$

Answers

1. (3)	**2.** (d)	**3.** (d)	**4.** (c)	**5.** (45)	**6.** (d)	**7.** (10)	**8.** (3)	**9.** (b)
10. (a)	**11.** (2)	**12.** (a)	**13.** (300)	**14.** (d)	**15.** (d)	**16.** (a)	**17.** (d)	**18.** (c)
19. (10)	**20.** (d)	**21.** (d)	**22.** (c)	**23.** (a)	**24.** (b)	**25.** (d)	**26.** (d)	**27.** (d)
28. (d)	**29.** (b)	**30.** (b)	**31.** (c)	**32.** (d)				
33. (c)	**34.** (c,d)	**35.** (d)	**36.** (d)	**37.** (b)				
38. (a,b,c,d)		**39.** (5)	**40.** (c)	**41.** (a or c)				
42. (c)								

For Detailed Solutions
Scan the code

CHAPTER

07

ELECTRICAL INSTRUMENTS

LEVEL 01 \ Starter Level

1.1 Galvanometer, Ammeter and Voltmeter

1 What is the value of shunt which passes 10% of the main current through a galvanometer of 99 Ω?

2 The deflection of a moving coil galvanometer falls from 60 divisions to 12 divisions for the same value of current in the circuit, when a shunt of 12 Ω is connected. Find the resistance of the galvanometer.

3 A galvanometer has a resistance of 30 Ω and a full scale deflection current of 20 mA. What is the resistance needed and how is it to be connected to convert the galvanometer

(a) into an ammeter of 10 A range and

(b) into a voltmeter of 100 V range?

4 A moving coil galvanometer of resistance 20Ω gives a full scale deflection when a current of 1 mA is passed through it. It is to be converted into an ammeter reading 20A on full scale, but the shunt of 0.005 Ω only is available. What resistance should be connected in series with the galvanometer coil?

5 A galvanometer with a scale divided into 100 equal divisions, has a current sensitivity of 10 division per mA and voltage sensitivity of 2 division per mV. What adoptions are required to use it (a) to read 5A full scale and (b) 1 division per volt ?

6 A galvanometer having 50 divisions provided with a variable shunt S is used to measure the current when connected in series with a resistance of 90 Ω and a battery of internal resistance 10 Ω. It is observed that when the shunt resistances are 10Ω, 50Ω respectively, the deflections are 9 and 30 divisions respectively. What is the resistance of the galvanometer?

Further if the full scale deflection of the galvanometer movement is 300 mA, find the emf of the cell.

1.2 Measurement using Ammeter and Voltmeter

7 (a) What is the reading of the ammeter in the circuits (a) and (b)?

(b) What is the percentage error in measuring the current through the ammeter in circuit (b)?

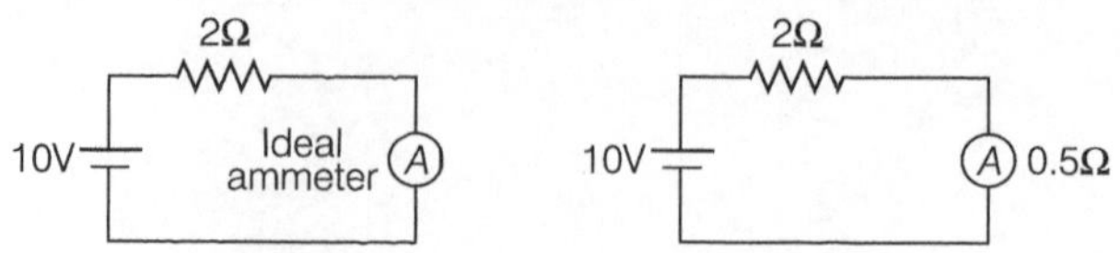

(a) (b)

8 Find the reading of ammeter. Is this the current through 6Ω?

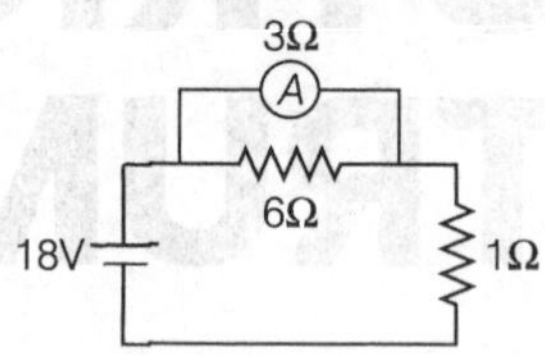

9 A voltmeter of resistance 400Ω is used to measure the potential difference across the 100Ω resistor in the circuit shown in the figure. (a) What will be the reading of the voltmeter ? (b) What was the potential difference across 100 Ω before the voltmeter was connected ?

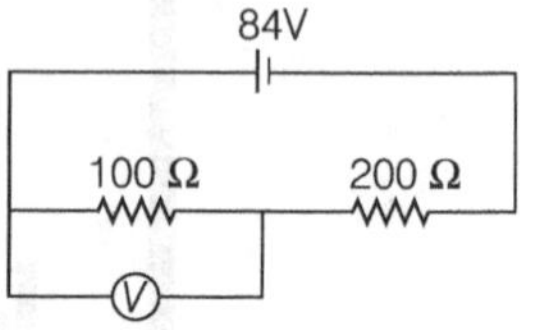

10 A cell of emf 3.4 V and internal resistance 3Ω connected to an ammeter having resistance 2Ω and to an external resistance of 100Ω. When a voltmeter is connected across the 100Ω resistance, the ammeter reading is 0.04 A. Find the voltage read by the voltmeter and its resistance. Had the voltmeter been an ideal one, what would have been its reading?

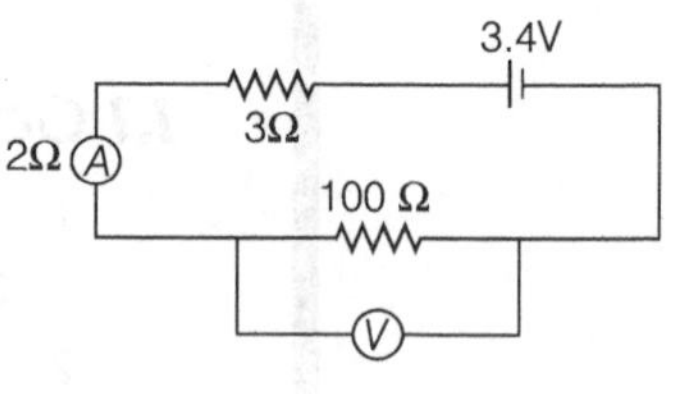

11 Find the reading of the ideal voltmeter as shown in figure.

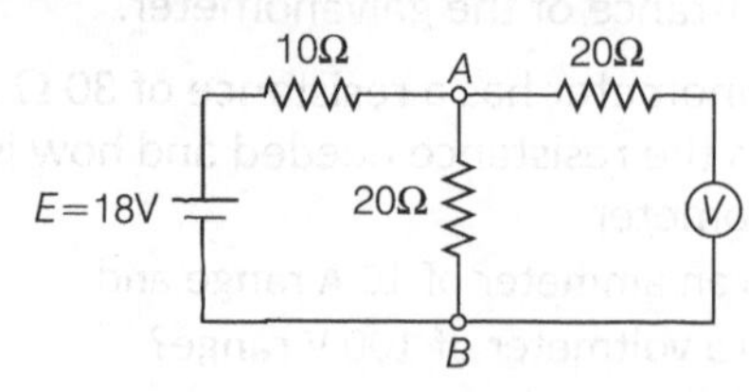

12 The emf E and the internal resistance r of the battery shown in figure are 50 V and 5.0 Ω, respectively. The external resistance R is 50 Ω. The resistances of the ammeter and voltmeter are 5.0Ω and 200Ω, respectively.

(a) Find the readings of the two meters.

(b) The switch is thrown to the other side. What will be the readings of the two meters now?

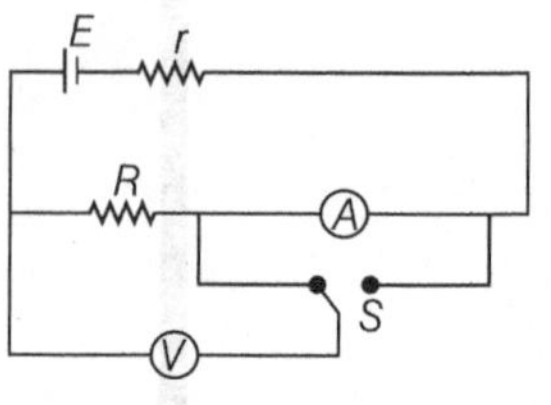

1.3 Meter Bridge

13 A meter bridge set-up with null deflection for the galvanometer is as shown in the figure. Find the value of the unknown resistor R.

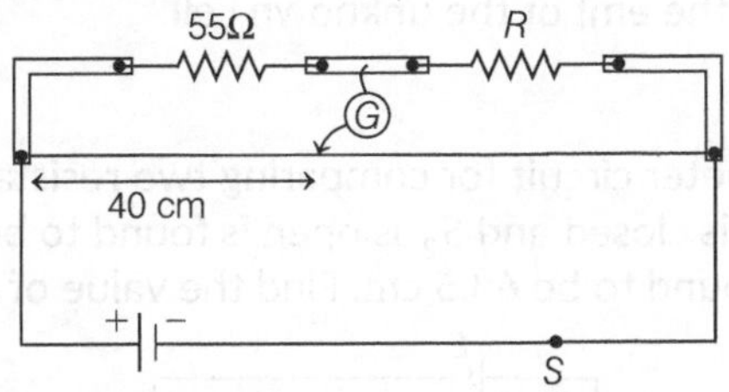

14 In the figure shown for which values of R_1 and R_2, the balance point for jockey is at 40 cm from A. When R_2 is shunted by a resistance of 10 Ω, balance shifts to 50 cm. Find R_1 and R_2. ($AB = 1$ m).

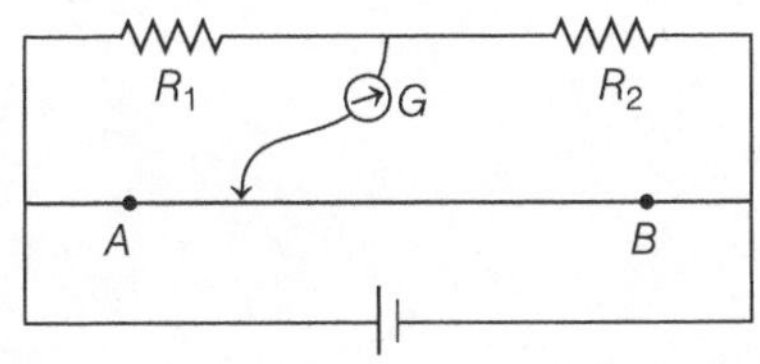

15 A meter bridge is set-up as shown, to determine an unknown resistance X using a standard 10Ω resistor. The galvanometer shows null point, when tapping key is at 52 cm mark. The end-corrections are 1 cm and 2 cm respectively for the ends A and B. Find value of X.

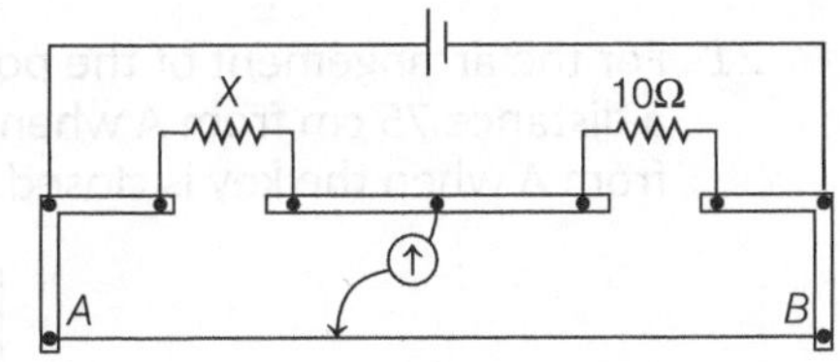

16 An unknown resistance is to be determined using resistances R_1, R_2 and R_3. If their corresponding null points are A, B and C, which of them will give most accurate reading?

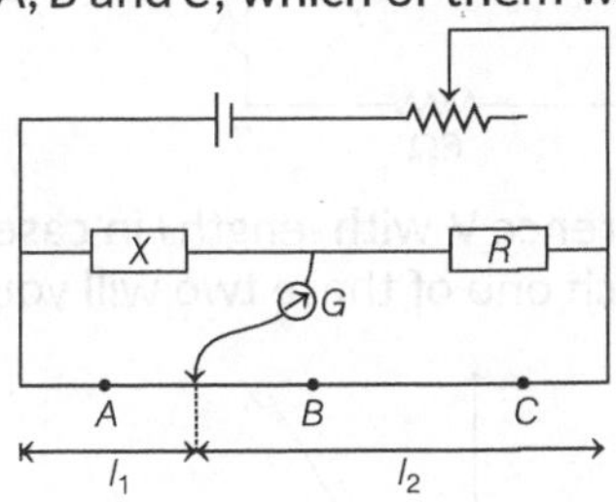

1.4 Potentiometer

17 In the primary circuit of a potentiometer as shown in figure, a potentiometer wire of length $L = 10$ m and resistance 10 Ω is joined in series with a resistance 20 Ω across a cell of emf $\varepsilon = 15$V. Find

(a) current in primary circuit.

(b) potential drop across potentiometer wire AB.

(c) potential gradient.

(d) maximum potential difference which can be measured with the potentiometer.

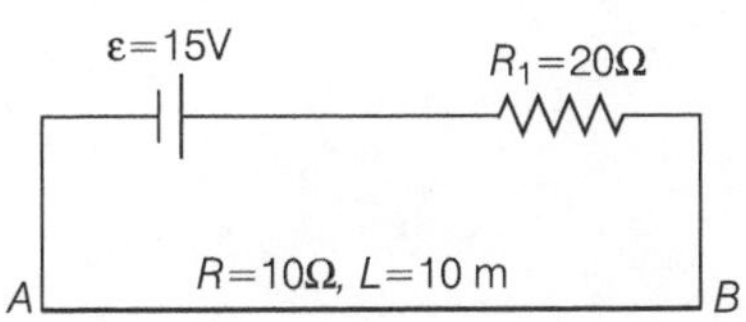

18 In the figure, the potentiometer wire of length $l = 450$ cm and resistance 9 Ω is joined to a cell of emf $E_1 = 10$ V and internal resistance $r_1 = 1$ Ω. Another cell of emf $E_2 = 5$ V and internal resistance $r_2 = 2$ Ω is connected as shown. The galvanometer G shows no deflection when the jockey is at C. Find the length AC.

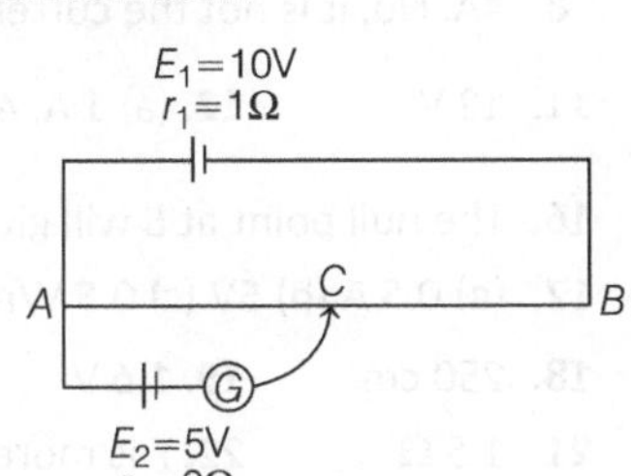

19 In an experiment to determine the emf of an unknown cell, its emf is compared with a standard cell of known emf $\varepsilon_1 = 1.12$ V. The balance point is obtained at 56 cm with standard cell and 80 cm with the unknown cell. Determine the emf of the unknown cell.

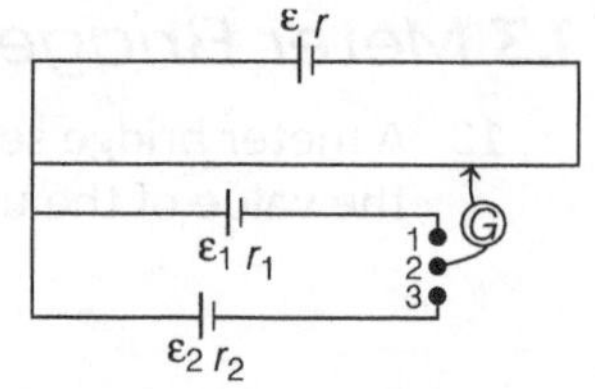

20 Figure shows a potentiometer circuit for comparing two resistances. The balance point with $R = 10\ \Omega$, when switch S_1 is closed and S_2 is open is found to be 58.3 cm, while that when S_2 is closed and S_1 is open is found to be 68.5 cm. Find the value of X.

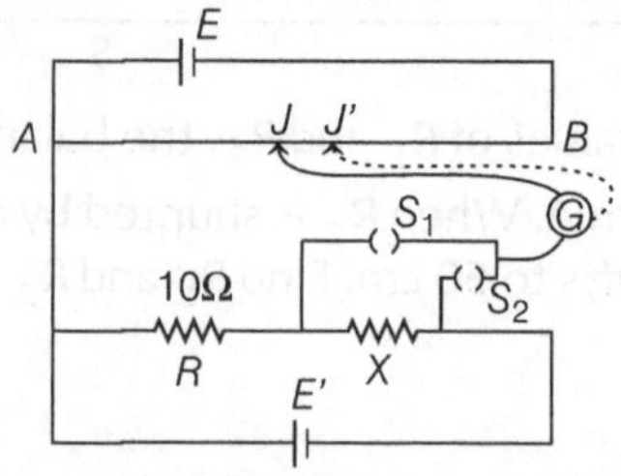

21 For the arrangement of the potentiometer shown in the figure, the balance point is obtained at a distance 75 cm from A when the key K is open. The second balance point is obtained at 60 cm from A when the key is closed. Find the internal resistance of the battery E_1.

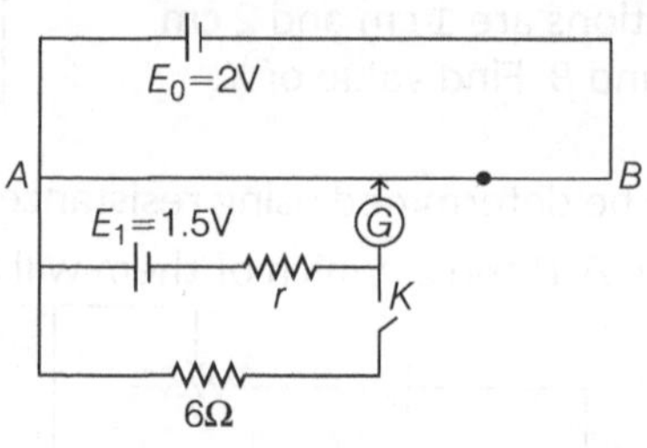

22 The variation of potential difference V with length l in case of two potentiometers X and Y is as shown in the given figure. Which one of these two will you prefer for comparing emf's of two cells and why?

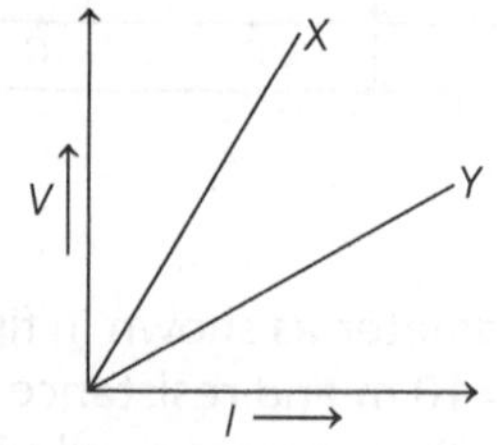

Answers

1. 11 Ω **2.** 48 Ω **3.** (a) 0.06 Ω in parallel (b) 4970 Ω in series **4.** 79.995 Ω
5. (a) 0.01 Ω in parallel (b) 9.995kΩ in series **6.** 233.3 Ω; 144 V **7.** (a) 5A, 4A (b) 20%
8. 4A, No, it is not the current through the 6 Ω resistor. **9.** (a) 24 V (b) 28 V **10.** 400 Ω, 3.2 V, 3.238 V
11. 12 V **12.** (a) 1 A, 40 V (b) 0.814 A, 44.8 V **13.** 82.5 Ω **14.** $R_1 = \frac{10}{3}\ \Omega$, $R_2 = 5\ \Omega$ **15.** 10.6 Ω
16. The null point at B will give most accurate reading
17. (a) 0.5 A (b) 5V (c) 0.5 V/m (d) 5 V
18. 250 cm **19.** 1.6 V **20.** 1.7 Ω
21. 1.5 Ω **22.** Y is more sensitive than X

For Detailed Solutions
Scan the code

LEVEL 02 \ JEE Main Level

Single Choice Correct Questions

1 The net resistance of a voltmeter should be large to ensure that
(a) it does not get overheated
(b) it does not draw excessive current
(c) it can measure large potential differences
(d) it does not appreciably change the potential difference to be measured

2 A milliammeter of range 10 mA and resistance 9 Ω is joined in a circuit as shown. The metre gives full-scale deflection for current I when A and B are used as its terminals, i.e. current enters at A and leaves at B (C is left isolated). The value of I is

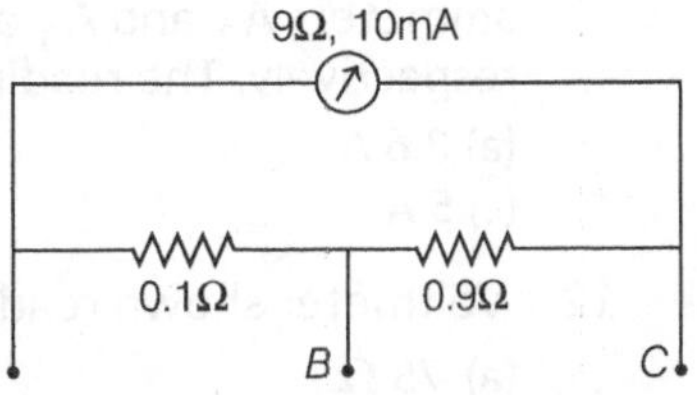

(a) 100 mA (b) 900 mA
(c) 1 A (d) 1.1 A

3 When a galvanometer is shunted with a 4Ω resistance, the deflection is reduced to one-fifth. If the galvanometer is further shunted with a 2Ω wire, determine current in galvanometer now if initially current in galvanometer is I_0 (given main current remain same).
(a) $\frac{I_0}{13}$ (b) $\frac{I_0}{5}$ (c) $\frac{I_0}{8}$ (d) $\frac{5I_0}{13}$

4 A galvanometer has a resistance of 20 Ω and reads full-scale when 0.2 V is applied across it. To convert it into a 10 A ammeter, the galvanometer coil should have a
(a) 0.01 Ω resistor connected across it (b) 0.02 Ω resistor connected across it
(c) 200 Ω resistor connected in series with it (d) 2000 Ω resistor connected in series with it

5 A galvanometer coil has a resistance 90 Ω and full scale deflection current 10 mA. A 910 Ω resistance is connected in series with the galvanometer to make a voltmeter. If the least count of the voltmeter is 0.1 V, the number of divisions on its scale is
(a) 90 (b) 91 (c) 100 (d) None of these

6 An ammeter has resistance R_0 and range I. What resistance should be connected in parallel with it to increase its range to nI?
(a) $\frac{R_0}{n}$ (b) $R_0(n+1)$ (c) $\frac{R_0}{n+1}$ (d) $\frac{R_0}{n-1}$

7 The galvanometer deflection, when key K_1 is closed but K_2 is open, equals θ_0 (see figure). On closing K_2 also and adjusting R_2 to 5 Ω, the deflection in galvanometer becomes $\frac{\theta_0}{5}$. The resistance of the galvanometer is given by (Neglect the internal resistance of battery)

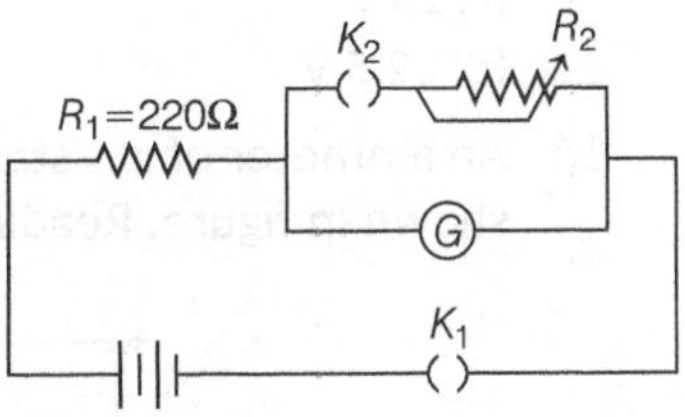

(a) 12 Ω (b) 25 Ω (c) 5 Ω (d) 22Ω

8 Two scales on a voltmeter measure voltage up to 20.0 V and 30.0 V. The resistance connected in series with the galvanometer is 1680 Ω for the 20.0 V scale and 2930 Ω for the 30.0V scale. The resistance of the galvanometer and the full scale current respectively are
(a) 320 Ω, 8 mA (b) 70 Ω, 10 mA (c) 820 Ω, 10 mA (d) 820 Ω, 8 mA

9 A moving coil galvanometer has 150 equal divisions. Its current sensitivity is 10 divisions per milliampere and voltage sensitivity is 2 divisions per milli volt. In order that each division reads 1 V, the resistance in ohm needed to be connected in series, with the coil will be
(a) 99995 (b) 9995 (c) 1000 (d) 10000

10 The ammeter has range 1 A without shunt. The range can be varied by using different shunt resistances. The graph between shunt resistance and range will have the nature

(a) *P*
(b) *Q*
(c) *R*
(d) *S*

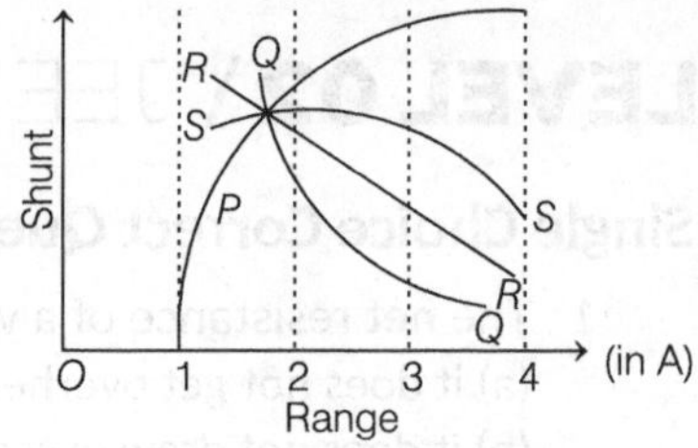

11 The reading of ammeter A_1 in figure is 2.4 A. Ammeter A_1 is ideal while the resistances of the ammeters A_2 and A_3 are 1 Ω and 3 Ω, respectively. The reading of ammeter A_3 will be

(a) 3.6 A (b) 4.8 A
(c) 5 A (d) 5.4 A

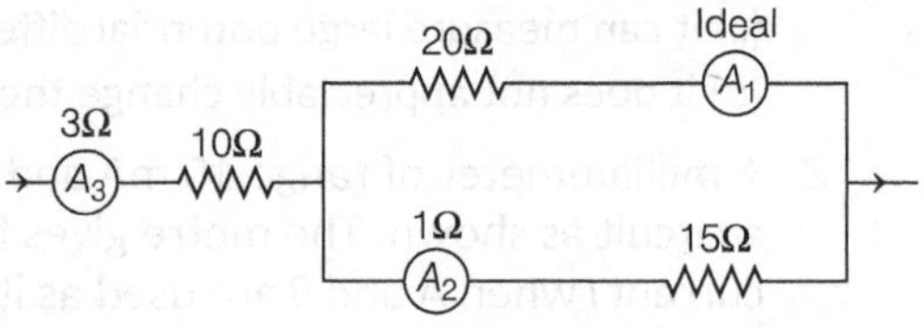

12 Voltmeter shown reads 18 V. The resistance of voltmeter, is

(a) 75 Ω (b) 120 Ω
(c) $\frac{800}{7}$ Ω (d) $\frac{900}{7}$ Ω

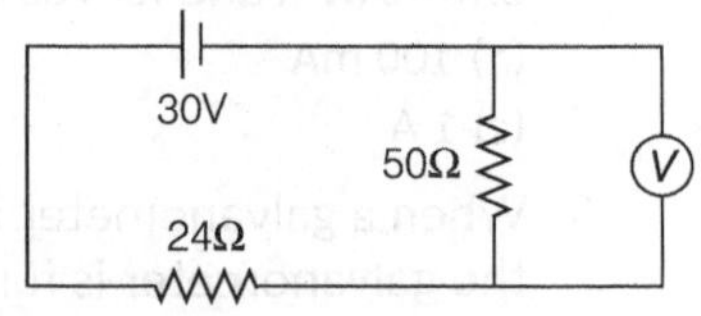

13 By error, a student places moving coil voltmeter V (nearly ideal) in series with the resistance in a circuit in order to read the current, as shown in figure. The voltmeter reading will

(a) 0 (b) 4V
(c) 6V (d) 12V

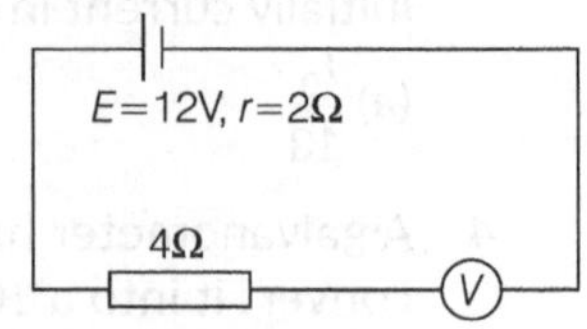

14 Two resistances of 400 Ω and 800 Ω are connected in series with 6 V battery of negligible internal resistance. A voltmeter of resistance 10000 Ω is used to measure the potential difference across 400 Ω. The error in the measurement of potential difference in volts approximately is

(a) 0.01 (b) 0.02 (c) 0.03 (d) 0.05

15 In the circuit, a voltmeter reads 30 V when it is connected across 400 Ω resistance. What the same voltmeter will read when it is connected across the 300 Ω resistance?

(a) 20 V
(b) 22.5 V
(c) 25 V
(d) 32.5 V

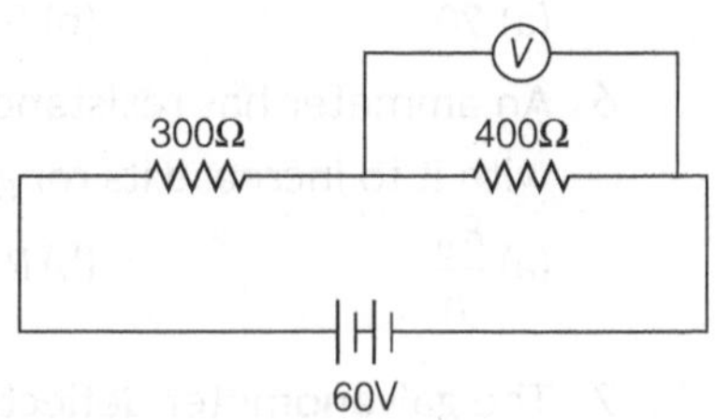

16 An ammeter of resistance 10 Ω and a voltmeter of resistance 20 Ω are connected in a circuit as shown in figure. Reading of ammeter and voltmeter are respectively

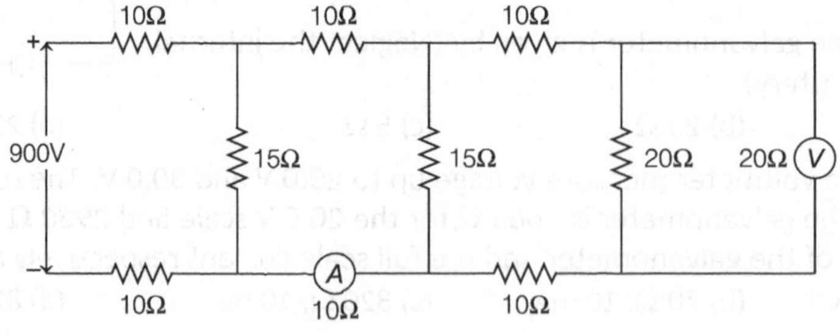

(a) $\frac{10}{3}$ A and $\frac{100}{3}$ V (b) 10 A and 100 V
(c) $\frac{10}{3}$ A and 100 V (d) 10 A and $\frac{100}{3}$ V

17 The reading of the ideal voltmeter in the adjoining diagram will be

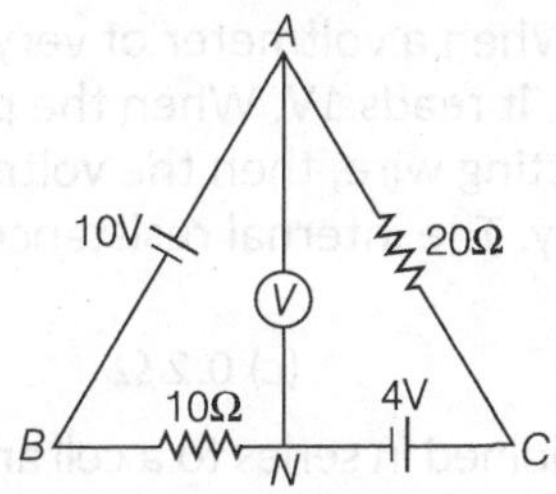

(a) 4 V (b) 8 V (c) 12 V (d) 14 V

18 In the circuit shown, the galvanometer shows zero current. The value of resistance R is

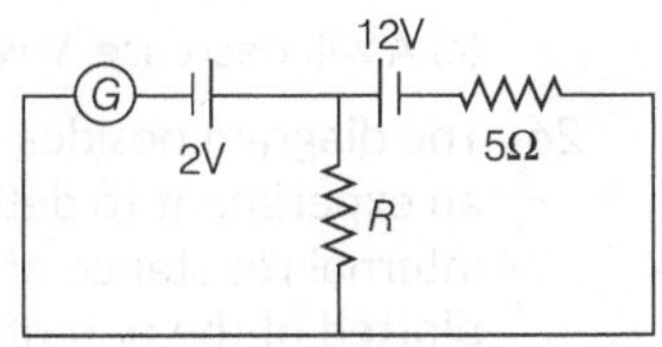

(a) 1 Ω
(b) 2 Ω
(c) 4 Ω
(d) 9 Ω

19 In the circuit shown, cells are of equal emf E but of different internal resistances $r_1 = 6\ \Omega$ and $r_2 = 4\ \Omega$. Reading of the ideal voltmeter connected across cell 1 is zero. Value of the external resistance R in ohm is equal to

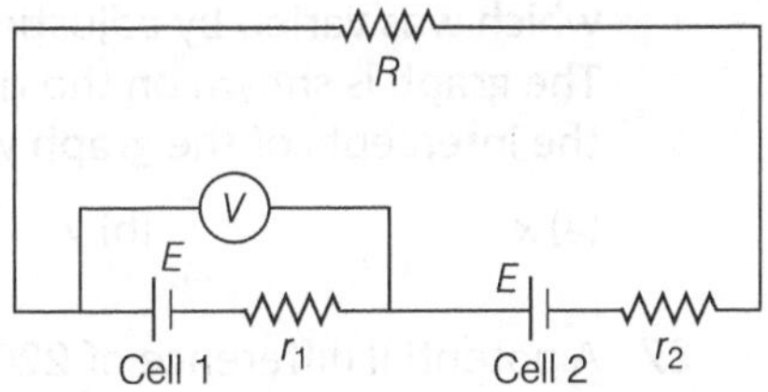

(a) 2
(b) 2.4
(c) 10
(d) 24

20 In the circuit shown the resistance of voltmeter is 10000 Ω and that of ammeter is 20 Ω. The ammeter reading is 0.10 A and voltmeter reading is 12 V, then R is equal to

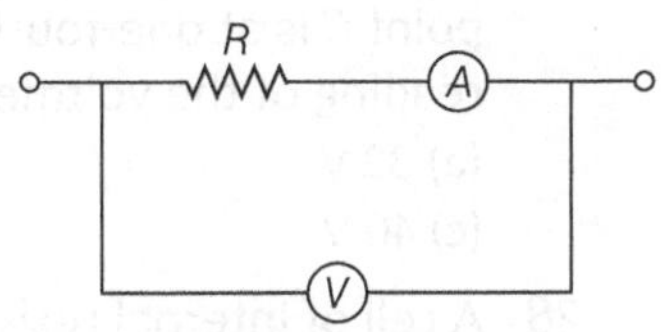

(a) 122 Ω (b) 140 Ω
(c) 116 Ω (d) 100 Ω

21 Which of the following wiring diagrams could be used to experimentally determine R using Ohm's law? Assume an ideal voltmeter and an ideal ammeter.

(a) 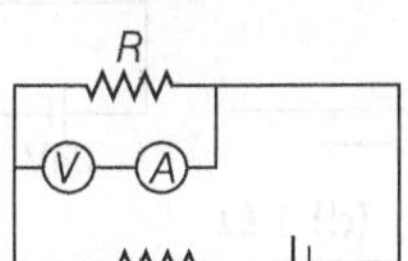

(b) 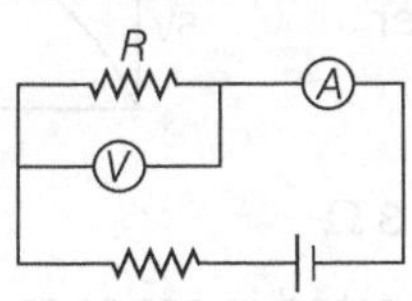

(c) 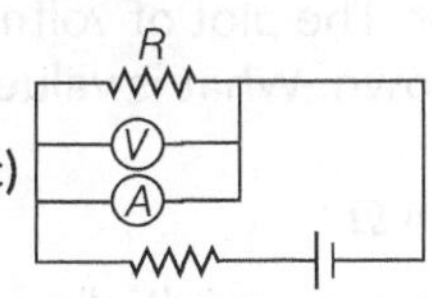

(d) 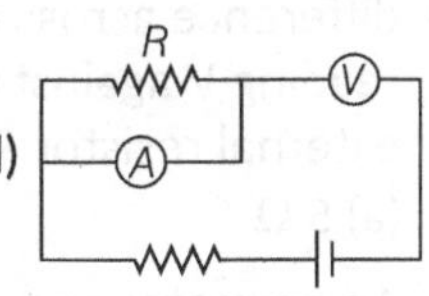

22 In the circuit shown, the reading of the ammeter is doubled after the switch is closed. Each resistor has a resistance = 1 Ω and the ideal cell has an emf = 10 V, then the ammeter has a coil resistance equal to

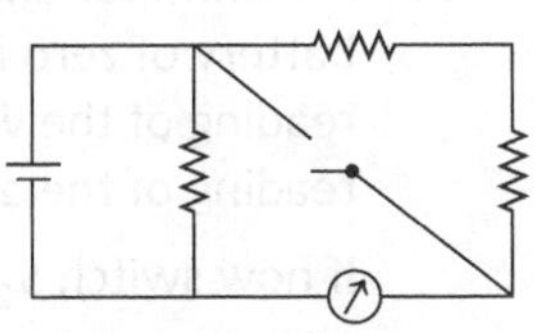

(a) 2 Ω (b) 1 Ω
(c) 2.5 Ω (d) None of these

23 In the circuit shown in the figure, reading of voltmeter is V_1 when only S_1 is closed, reading of voltmeter is V_2 when only S_2 is closed & reading of voltmeter is V_3 when both S_1 and S_2 are closed. Then,

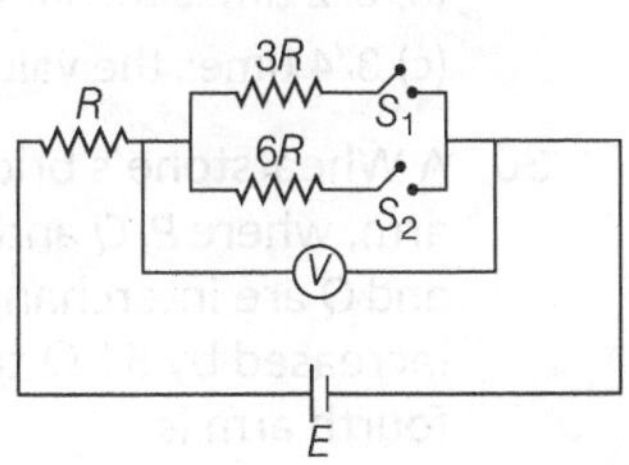

(a) $V_3 > V_2 > V_1$
(b) $V_2 > V_1 > V_3$
(c) $V_3 > V_1 > V_2$
(d) $V_1 > V_2 > V_3$

24 When an ammeter of negligible internal resistance is inserted in series with circuit, it reads 1 A. When a voltmeter of very large resistance is connected across X, it reads 1V. When the points A and B are short-circuited by a conducting wire, then the voltmeter measures 10 V across the battery. The internal resistance of the battery is

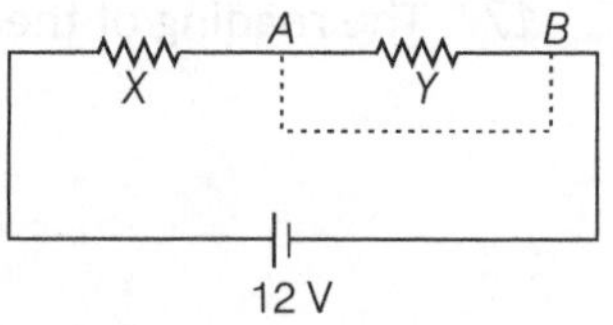

(a) 0 (b) 0.1 Ω (c) 0.2 Ω (d) 0.52 Ω

25 An ammeter and a voltmeter are joined in series to a cell and their readings are A and V ,respectively. If a resistance is now joined in parallel with the voltmeter

(a) both A and V will increase (b) both A and V will decrease

(c) A will decrease, V will increase (d) A will increase, V will decrease

26 The diagram besides shows a circuit used in an experiment to determine the emf and internal resistance of the cell C. A graph was plotted of the potential difference V between the terminals of the cell against the current I, which was varied by adjusting the rheostat. The graph is shown on the right; x and y are the intercepts of the graph with the axes as shown. What is the internal resistance of the cell?

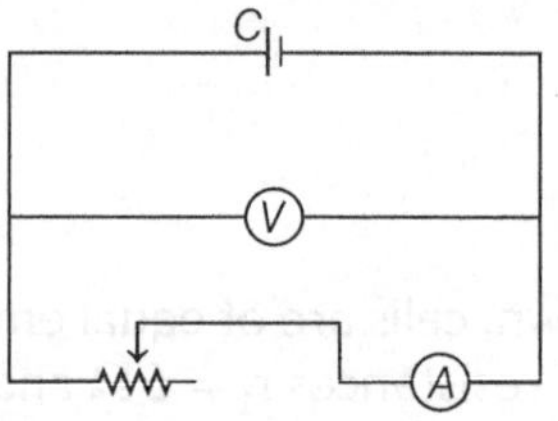

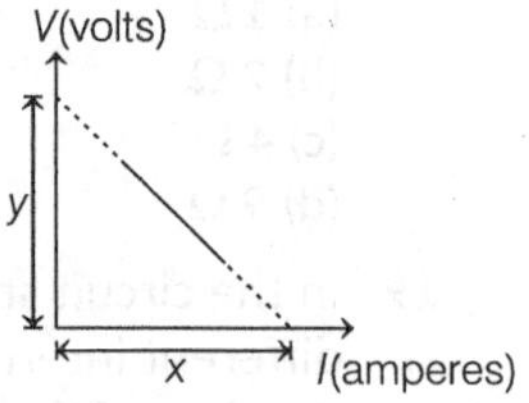

(a) x (b) y (c) $\frac{x}{y}$ (d) $\frac{y}{x}$

27 A potential difference of 220 V is maintained across a 12000 Ω rheostat, as shown in the figure. The voltmeter has a resistance of 6000 Ω and point C is at one-fourth of the distance from A to B. Therefore, the reading of the voltmeter will be

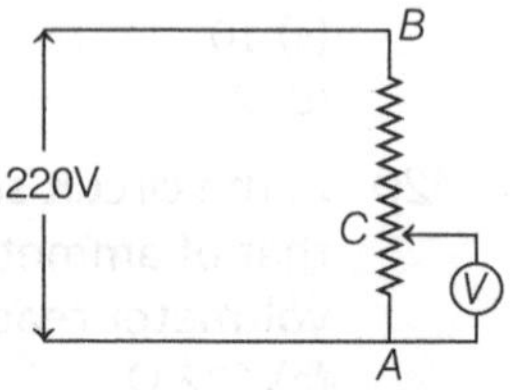

(a) 32 V (b) 36 V

(c) 40 V (d) 42 V

28 A cell of internal resistance 1 Ω is connected across a resistor. A voltmeter having variable resistance G is used to measure potential difference across resistor. The plot of voltmeter reading V against G is shown. What is value of external resistor R?

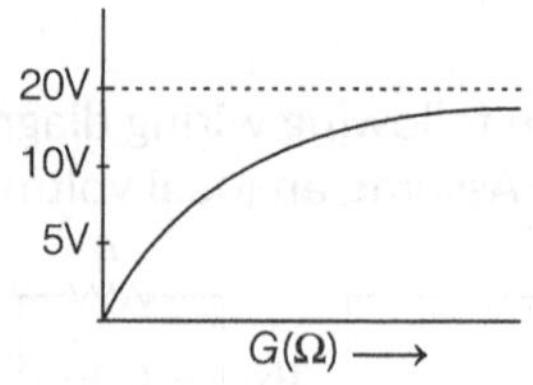

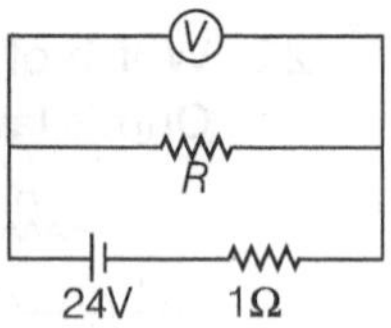

(a) 5 Ω (b) 4 Ω (c) 3 Ω (d) 1 Ω

29 An ammeter and a voltmeter are initially connected in series to a battery of zero internal resistance. When switch S_1 is closed, the reading of the voltmeter becomes half of the initial, whereas the reading of the ammeter becomes double.

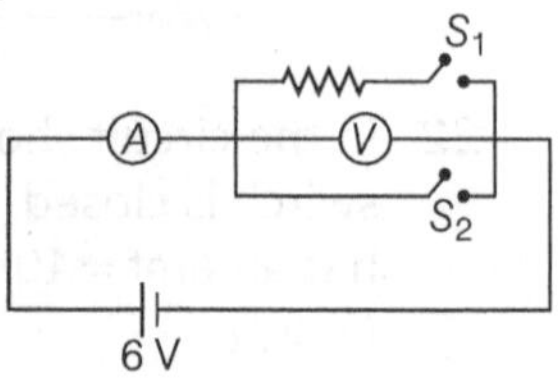

If now switch S_2 is also closed, then reading of ammeter becomes

(a) 3/2 times the initial value (b) 3/2 times the value after closing S_1

(c) 3/4 times the value after closing S_1 (d) 3/4 times the initial value

30 A Wheatstone's bridge is balanced with a resistance of 625 Ω in the third arm, where P, Q and S are in the 1st, 2nd and 4th arm, respectively. If P and Q are interchanged, the resistance in the third arm has to be increased by 51 Ω to secure balance. The unknown resistance in the fourth arm is

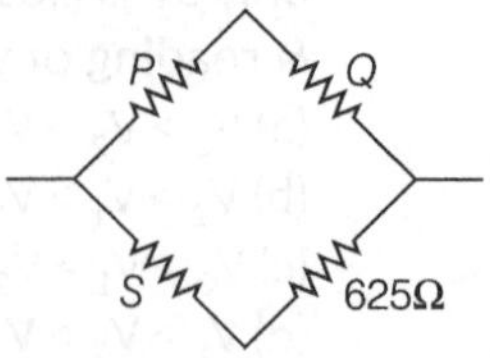

(a) 625 Ω (b) 650 Ω

(c) 676 Ω (d) 600 Ω

31 In a balanced Wheatstone bridge, current in the galvanometer is zero. It remains zero, when

1. battery emf is increased
2. all resistances are increased by 10 Ω
3. all resistances are made five times
4. the battery and the galvanometer are interchanged

(a) Only 1 is correct (b) 1, 2 and 3 are correct
(c) 1, 3 and 4 are correct (d) 1 and 3 are correct

32 Which one of the following is an incorrect statement?

(a) The Wheatstone bridge is most sensitive, when all the four resistances are of the same order.
(b) Sensitivity of a potentiometer is directly proportional to its potential gradient.
(c) Potentiometer is a better device than voltmeter for measuring potential difference.
(d) Rheostat can be used as a potential divider.

33 Two resistances are connected in the two gaps of a meter bridge. The balance point is 20 cm from the zero end. When a resistance 15 Ω is connected in series with the smaller of two resistances, the null point shifts to 40 cm. The smaller of the two resistances has the value

(a) 3 Ω (b) 6 Ω (c) 9 Ω (d) 12 Ω

34 In the shown arrangement of the experiment of the meter bridge, if *AC* corresponding to null deflection of galvanometer is *x*. What would be its value, if the radius of the wire *AB* is doubled?

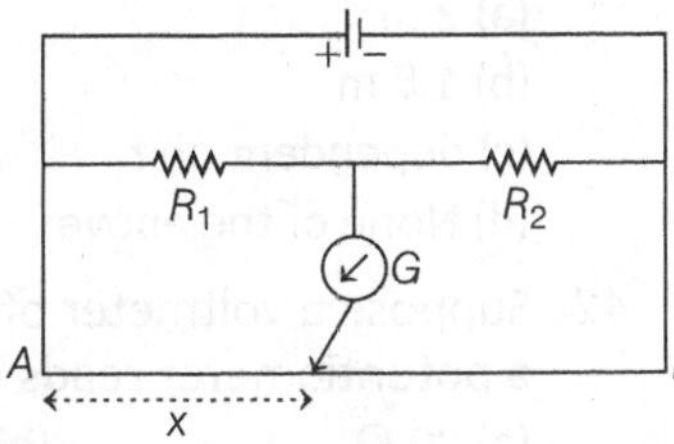

(a) x (b) $\frac{x}{4}$
(c) $4x$ (d) $2x$

35 In a meter bridge experiment, resistance X is connected in the right gap. When R_1 and R_2 are connected in the left gap separately, the balance points are 40 cm and 50 cm respectively, from the left end. Now, if both R_1 and R_2 are connected in series in the left gap with X in the right gap, then the new balance point from the left end will be

(a) 50 cm (b) 62.5 cm
(c) 75 cm (d) Data insufficient for calculation

36 A meter bridge is set up as shown to determine end corrections at *A* and *B*. When a resistance of 15 Ω is used in left gap and of 20 Ω in right gap, then null point comes at a distance 42 cm from *A*. When these resistances are interchanged, null point comes at a distance 57 cm from *A*. Values of end corrections are

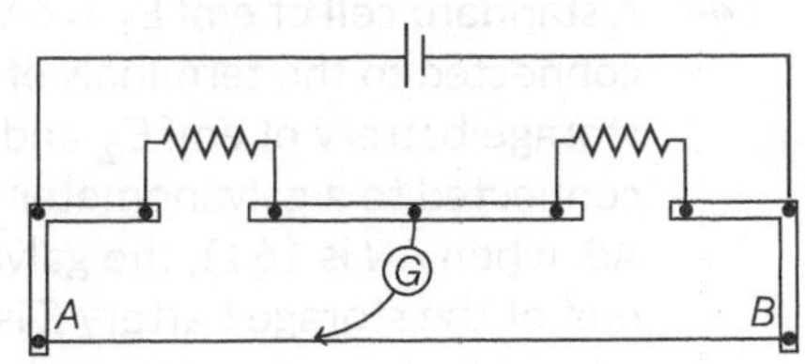

(a) 1 cm, 2 cm (b) 2 cm, 3 cm
(c) 3 cm, 4 cm (d) 3 cm, 2 cm

37 The diagram shows a modified meter bridge, which is used for measuring two unknown resistances at the same time. When the first galvanometer is used for obtaining the balance point, it is found at point *C*. Now the first galvanometer is removed and the second galvanometer is used, which gives balance point at *D*. Using the details given in the diagram, find out the values of R_1 and R_2. $AB = L$; $AC = \frac{L}{4}$ and $AD = \frac{2L}{3}$.

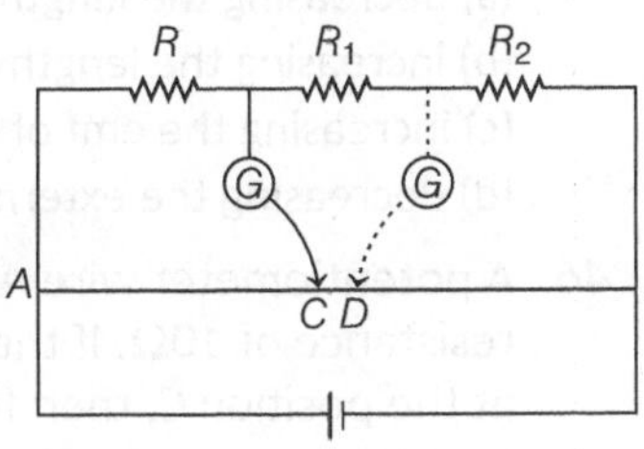

(a) $\frac{5R}{3}, \frac{4R}{3}$ (b) $\frac{3R}{5}, \frac{3R}{4}$ (c) $\frac{4R}{3}, \frac{5R}{3}$ (d) $\frac{3R}{4}, \frac{3R}{5}$

38 The resistance of a potentiometer wire 8 m long is 8 Ω. A high resistance box and a 2 V accumulator are connected in series with it. What should be the value of the resistance in the box, if it is desired to have a potential gradient of 1 μV/mm?

(a) 992 Ω (b) 1992 Ω (c) 19992 Ω (d) None of these

39 Figure shows a simple potentiometer circuit for measuring a small emf produced by a thermocouple. The meter wire PQ has a resistance of 5 Ω and length 1 m. If a balance point is obtained 0.6 m along PQ when measuring an emf of 6 mV, what is the value of resistance R?

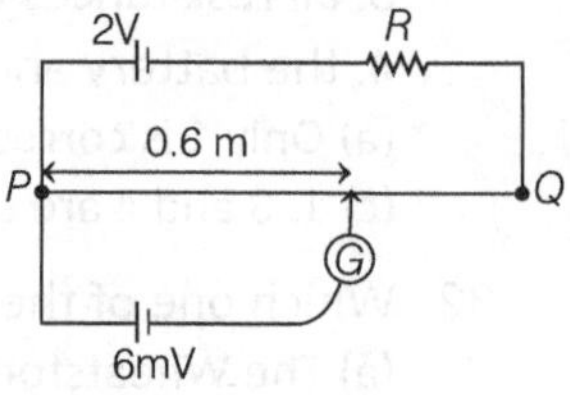

(a) 95 Ω (b) 995Ω

(c) 195 Ω (d) 1995 Ω

40 In a potentiometer experiment, n identical cells each of emf E are used in series and the balancing length is found to be 58 cm of the wire. If the polarities of 2 cells out of n cells is reversed, then the balancing length becomes 29 cm. Value of n is

(a) 8 (b) 12 (c) 16 (d) 24

41 In the given potentiometer circuit, length of the wire AB is 3 m and resistance is $R = 4.5\ \Omega$. The length AC for no deflection in galvanometer is

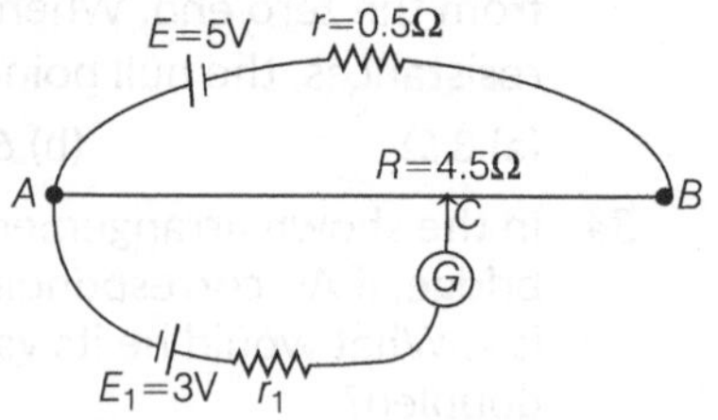

(a) 2 m

(b) 1.8 m

(c) dependent on r_1

(d) None of the above

42 Suppose a voltmeter of resistance 660 Ω reads the voltage of a very old cell to be 1.32 V while a potentiometer reads its voltage to be 1.44 V. The internal resistance of the cell is

(a) 30 Ω (b) 60 Ω (c) 6 Ω (d) 0.6 Ω

43 The length of a potentiometer wire is l. A cell of emf E is balanced at a length $l/3$ from the positive end of the wire. If the length of the wire is increased by $l/2$, at what distance will the same cell give a balance point?

(a) $\frac{2l}{3}$ (b) $\frac{l}{2}$ (c) $\frac{l}{6}$ (d) $\frac{4l}{3}$

44 A standard cell of emf $E_1 = 6$ V and negligible internal resistance is connected to the terminals of a potentiometer of resistance 20 Ω. A storage battery of emf E_2 and negligible internal resistance is connected to a galvanometer as shown. On moving the jockey along AB, when AN is 16 Ω, the galvanometer shows null deflection. The emf of the storage battery E is

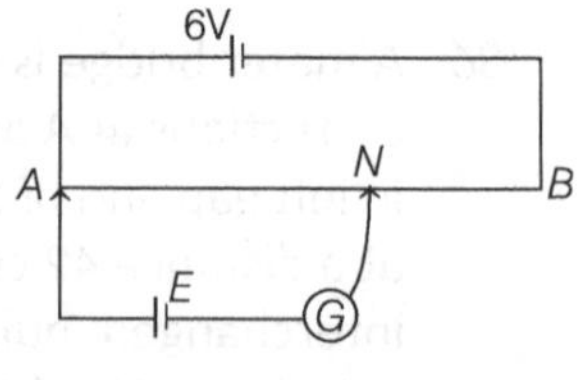

(a) 4.2 V (b) 4.5 V (c) 4.8 V (d) 5.1 V

45 The sensitivity of a potentiometer can be increased by

(a) decreasing the length of potentiometer wire

(b) increasing the length of potentiometer wire

(c) increasing the emf of the cell in primary circuit

(d) decreasing the external resistance connected in primary circuit

46 A potentiometer wire AB is 100 cm long and has a total resistance of 10Ω. If the galvanometer shows zero deflection at the position C, then find the value of unknown resistance R.

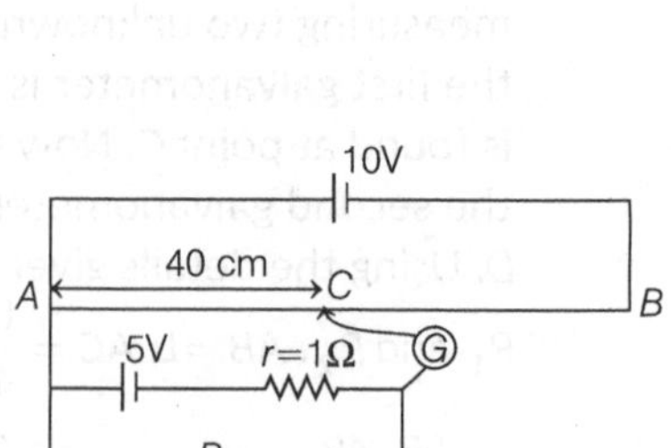

(a) 1 Ω

(b) 2 Ω

(c) 4 Ω

(d) 8 Ω

47 Circuit for the measurement of resistance by potentiometer is shown. The galvanometer is first connected at point A and zero deflection is observed at length $PJ = 10$ cm. In second case, it is connected at point C and zero deflection is observed at a length 30 cm from P. Then, the unknown resistance X is

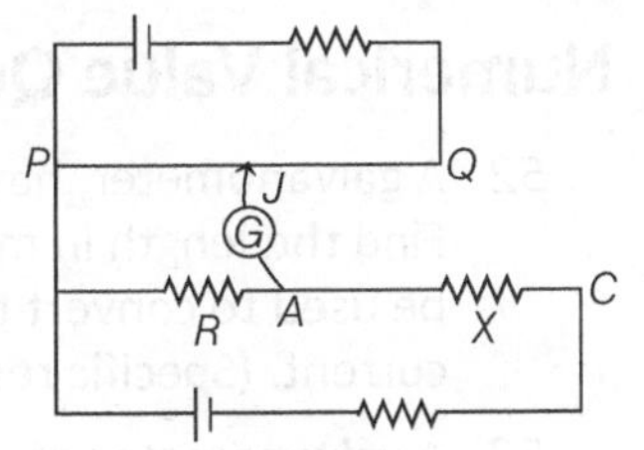

(a) $2R$
(b) $\frac{R}{2}$
(c) $\frac{R}{3}$
(d) $3R$

48 In the arrangement shown in figure when the switch S_2 is open, the galvanometer shows no deflection for $l = \frac{L}{2}$. When the switch S_2, is closed, the galvanometer shows no deflection for $l = \frac{5L}{12}$. The internal resistance r of 6 V cell and the emf E of the other battery are respectively

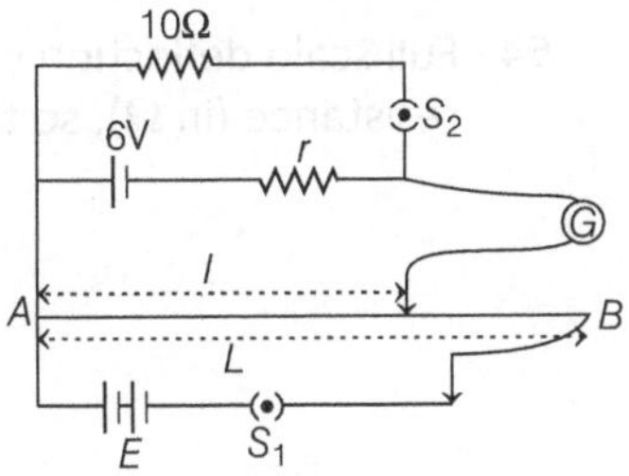

(a) 3 Ω, 8 V
(b) 2 Ω, 12 V
(c) 2 Ω, 24 V
(d) 3 Ω, 12 V

49 Consider the potentiometer circuit arranged as in figure. The potentiometer wire (resistance $15r$) is 600 cm long. If the jockey touches the wire at a distance of 560 cm from A, what will be the current in the galvanometer?

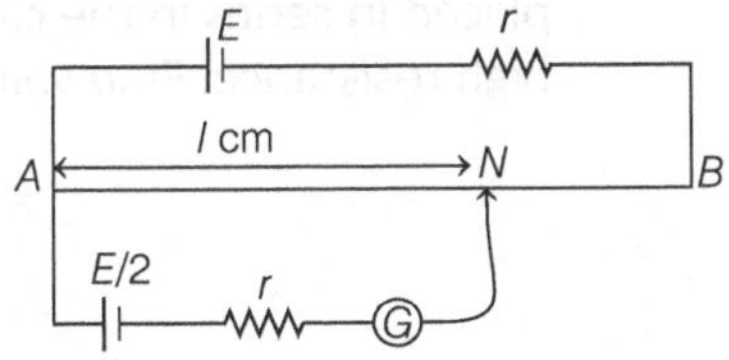

(a) $\frac{3E}{11r}$
(b) $\frac{5E}{22r}$
(c) $\frac{3E}{22r}$
(d) $\frac{5E}{11r}$

50 In the circuit shown in figure, XY is a potentiometer wire 100cm long. The circuit is connected as shown. With switches S_2 and S_3 open, a balance point is found at Z. After switch S_1 has remained closed for some time, it is found that contact at Z must be moved towards Y to maintain a balance. Which of the following is the most likely reason for this?

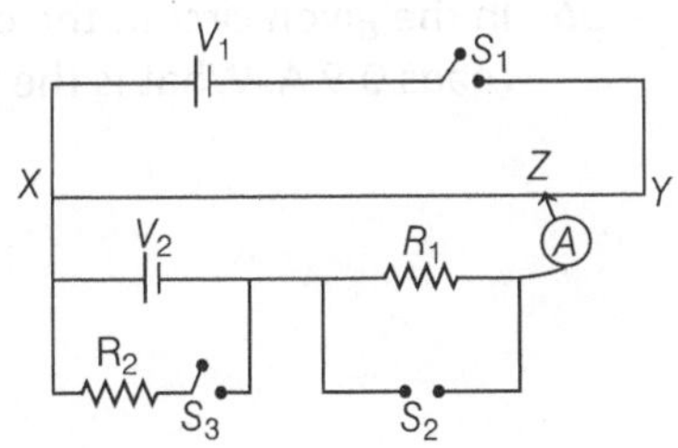

(a) The cell V_1 is running down
(b) The cell V_2 is running down
(c) The wire XZ is getting warm and its resistance is increasing
(d) The resistor R_1 is getting warm and increasing in value

51 A battery of emf $E_0 = 12$ V is connected across a 4m long uniform wire having resistance 4Ω/m. The cells of small emf $\varepsilon_1 = 2$ V and $\varepsilon_2 = 4$ V having internal resistances 2Ω and 6Ω respectively, are connected as shown in the figure. If galvanometer shows no deflection at the point N, the distance of point N from the point A is equal to

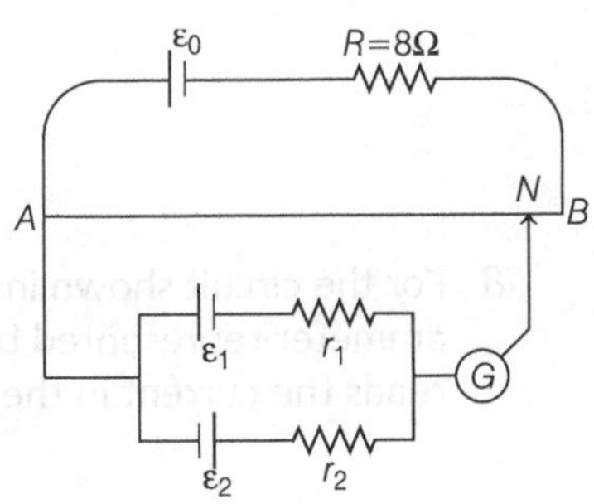

(a) $\frac{1}{6}$ m
(b) $\frac{1}{3}$ m
(c) 25 cm
(d) 50 cm

Numerical Value Questions

52 A galvanometer, having a resistance of 50 Ω, gives a full scale deflection for a current of 0.05 A. Find the length in meter of a resistance wire of area of cross-section 2.97×10^{-2} cm that can be used to convert the galvanometer into an ammeter which can read a maximum of 5 A current. (Specific resistance of the wire $= 5 \times 10^{-7}$ Ω-m)

53 A galvanometer of resistance 40 Ω, shunted by a resistance of 50 Ω gives a deflection of 50 divisions when joined in series with a resistance of (1000/9) Ω and 2 V battery. What is the current senstivity of galvanometer (in div/mA)?

54 Full scale deflection current for galvanometer is 1 A. What should be the value of shunt resistance (in Ω), so that galvanometer shows half scale deflection.

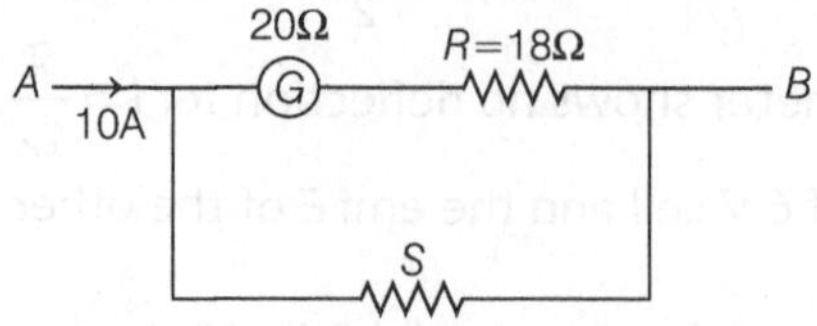

55 DC supply of 120V is connected to a large resistance X. A 100 V voltmeter of resistance 10 kΩ placed in series in the circuit reads 4V. This is an unusual use of voltmeter for measuring very high resistance. Find value of X (in kΩ).

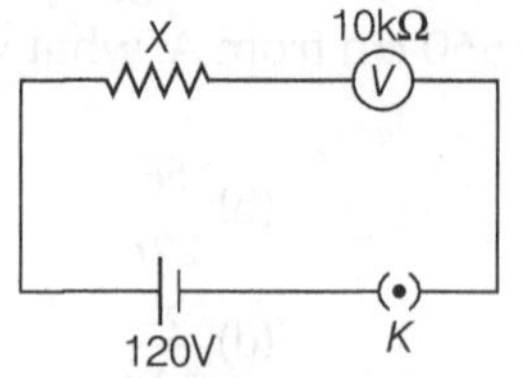

56 In the given circuit, the current flowing through the resistance 20Ω is 0.3 A while the ammeter reads 0.9 A. What is the value of R_1 (in Ω)?

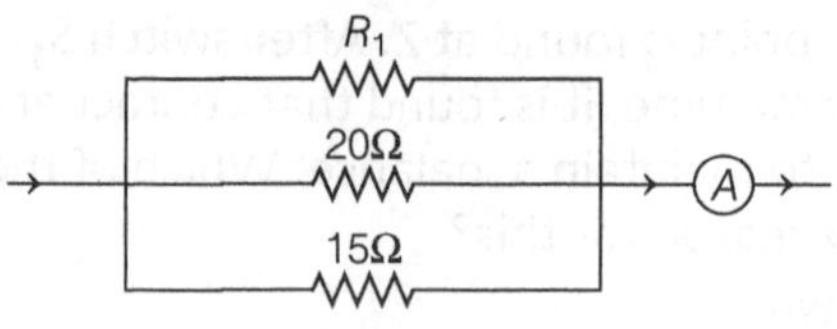

57 In the circuit shown, the coil resistance of the ammeter is 1Ω and that of the voltmeter is 20Ω. The main branch current is (19/20) A. What is the reading of the voltmeter(in volt)?

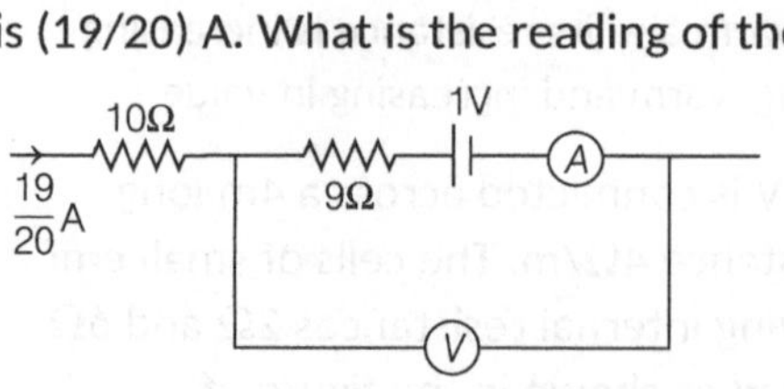

58 For the circuit shown in the figure. Find the emf (in mV) of the unknown battery *E*. The ammeter represented by the symbol A can be considered perfect (i.e. no resistance) and it reads the current in the wire to be 0.5 A in the direction shown.

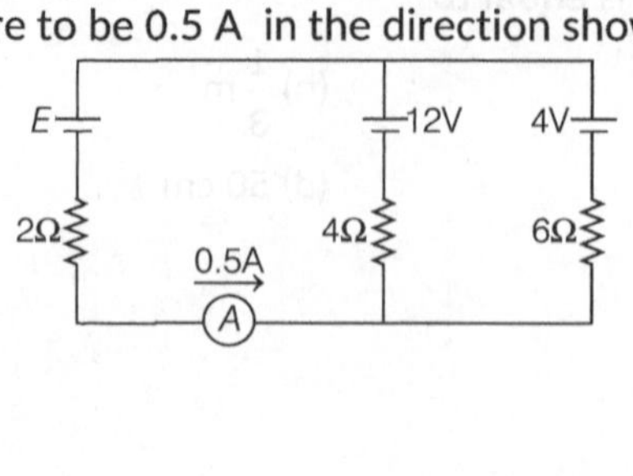

59 Two cells A and B of emf 1.3V and 1.5V respectively are arranged as shown in figure. The voltmeter reads 1.45 V. The voltmeter is assumed to be ideal. Find the ratio of internal resistances $\left(\frac{r_1}{r_2}\right)$ of the cells.

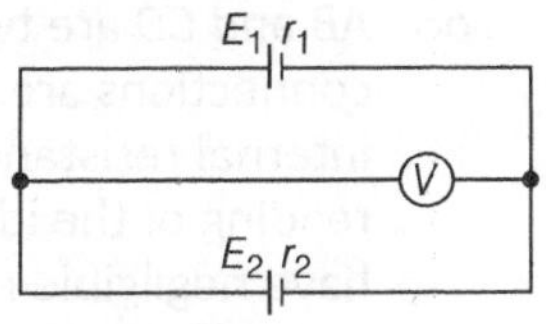

60 In the circuit shown, all voltmeters and ammeter are ideal. The internal resistance of battery is 20 Ω. The readings of voltmeter and ammeter are marked in the figure. What is resistance R_2 (in Ω)?

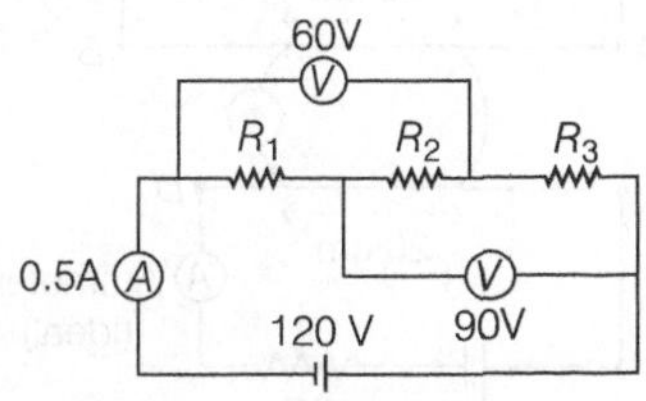

61 A resistance of 2Ω is connected across one gap of a meter bridge (the length of the wire is 100 cm) and an unknown resistance, greater than 2Ω, is connected across the other gap. When these resistances are interchanged, the balanced point shifts by 20 cm. Neglecting any end corrections, the unknown resistance is .... Ω.

62 The resistance of uniform wire AB is 9Ω. Find the resistance of part BC (in Ω) for which galvanometer G will show zero deflection.

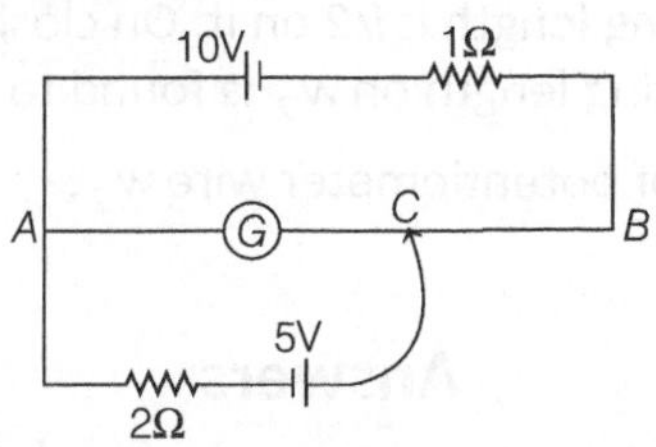

63 A 10 m long potentiometer wire has a resistance of 20Ω. It is connected in series with a battery of emf 3V and a resistance of 10 Ω. The internal resistance of cell is negligible. If the length can be read accurately up to 1 mm, the potentiometer can read voltage with an accuracy β mV. Then, what is the value of β?

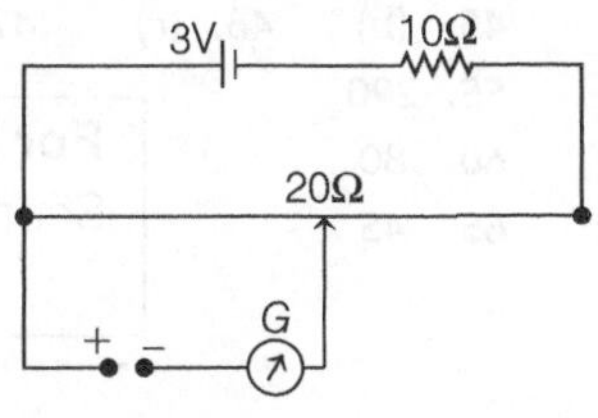

64 An ideal cell having a steady emf of 2 V is connected across the potentiometer wire of length 10 m. The potentiometer wire is of magnesium and having resistance of 11.5 Ω/m. An another cell gives a null point at 6.9 m. If a resistance of 5Ω is put in series with potentiometer wire, find the new position of the null point (in m).

65 A potentiometer circuit is arranged as shown in figure. AB is potentiometer wire of length 1m. Internal resistances of cells are shown in figure. Now, when only key K_1 is closed, balanced length AJ is 40 cm and when only K_2 closed, balanced length AJ becomes 55 cm. Find the balanced length AJ (in cm) when both the keys are closed.

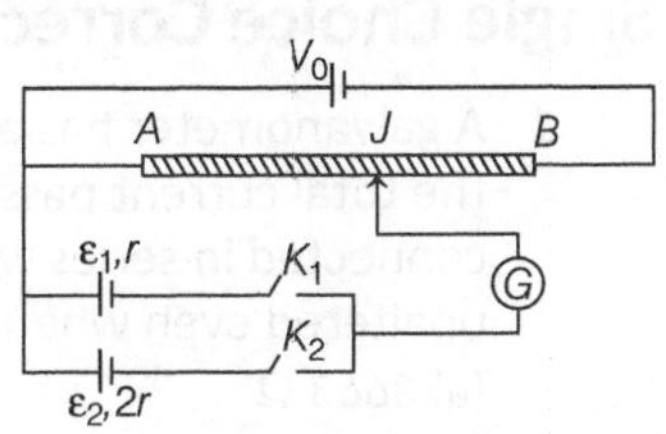

66 AB and CD are two uniform resistance wires of lengths 100 cm and 80 cm, respectively. The connections are shown in the figure. The cell of 5V is ideal while the other cell of emf E has internal resistance 2 Ω. A length of 20 cm of wire CD is balanced by 40 cm of wire AB. If the reading of the ideal ammeter is 2A, then find the emf E in volt. The other connecting wires have negligible resistance.

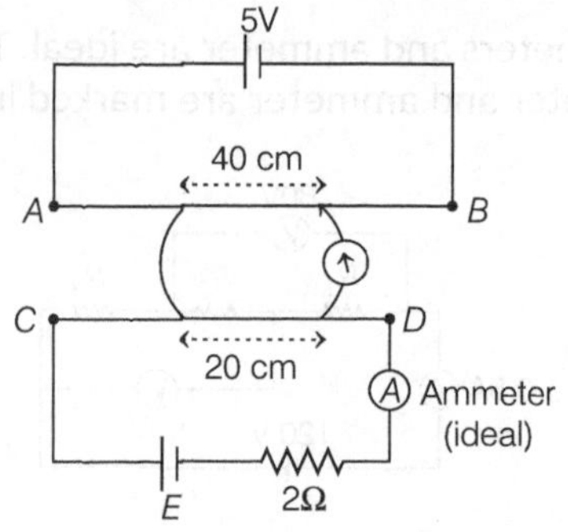

67 Two potentiometer wires w_1 and w_2 of equal length l, connected to a battery of emf ε_P and internal resistance 1Ω through two switches S_1 and S_2. A battery of emf ε is balanced on these potentiometer wires one by one. Potentiometer wire w_1 is of resistance 2 Ω. When only S_1 is closed and S_2 is open, balancing length is $l/2$ on it. On closing S_2 and opening S_1, the balancing length on w_2 is found to be $\frac{2l}{3}$. Find the resistance (in Ω) of potentiometer wire w_2.

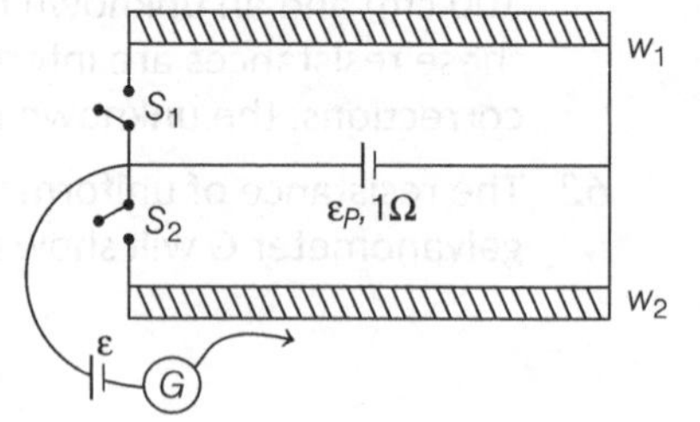

Answers

1. (d)	2. (c)	3. (a)	4. (b)	5. (c)	6. (d)	7. (d)	8. (d)	9. (b)	10. (b)
11. (d)	12. (d)	13. (d)	14. (d)	15. (b)	16. (d)	17. (b)	18. (a)	19. (a)	20. (d)
21. (b)	22. (a)	23. (b)	24. (c)	25. (d)	26. (d)	27. (c)	28. (a)	29. (b)	30. (b)
31. (c)	32. (b)	33. (c)	34. (a)	35. (b)	36. (d)	37. (a)	38. (b)	39. (c)	40. (a)
41. (d)	42. (b)	43. (b)	44. (c)	45. (b)	46. (c)	47. (a)	48. (b)	49. (c)	50. (a)
51. (c)	52. 3	53. 6	54. 2	55. 290					
56. 30	57. 7	58. 6600	59. 3	60. 80					
61. 3	62. 3	63. 0.2	64. 7.2	65. 45					
66. 12	67. 1								

For Detailed Solutions
Scan the code

LEVEL 03 \ JEE Advanced Level

Single Choice Correct Questions

1 A galvanometer has a resistance of 3663 Ω. A shunt S is connected across it such that (1/34) of the total current passes through the galvanometer. The external resistance which must be connected in series with the main circuit, so that the total current in the main circuit remain unaltered even when the galvanometer is shunted is

(a) 3663 Ω
(b) 111 Ω
(c) 107.7 Ω
(d) 3555.3 Ω

2 To find the resistance of a galvanometer by the half deflection method, the following circuit is used with resistance $R_1 = 9970\ \Omega$, $R_2 = 30\ \Omega$ and $R_3 = 0$. The deflection in the galvanometer is d. With $R_3 = 107\ \Omega$, the deflection changed to $d/2$. The galvanometer resistance is approximately

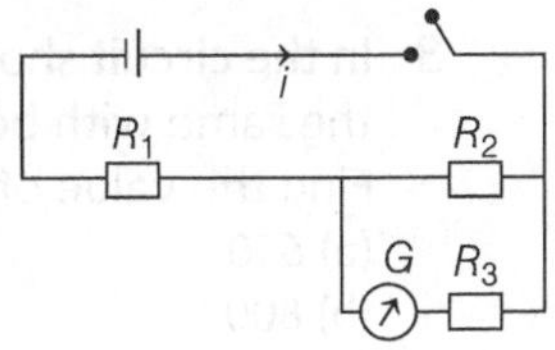

(a) $107\ \Omega$
(b) $137\ \Omega$
(c) $\frac{107}{2}\ \Omega$
(d) $77\ \Omega$

3 We are aware that a galvanometer can be converted into an ammeter as well as voltmeter by using appropriate resistances. We have a galvanometer whose maximum allowable potential drop increases n times by connecting a $27\ \Omega$ resistor in series. When we shunt the same galvanometer with a 3Ω resistor, its range of current increases to n times the original value. When two such galvanometers are used to construct an ammeter and a voltmeter and are connected as shown, the current in galvanometer making up ammeter and voltmeter respectively are

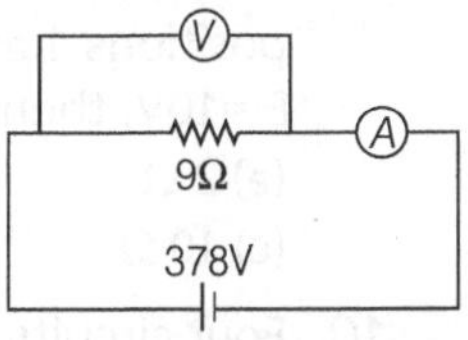

(a) 10 A and 8 A
(b) 30 A and 10 A
(c) 32 A and 30 A
(d) 8 A and 30 A

4 Two ammeters 1 and 2 have different internal resistances r_1 (known) and r_2 (unknown). Each ammeter has scale such that the angular deviation of the needle from zero is proportional to the current. Initially, the ammeters are connected in series and then to a source of constant voltage. The deviations of the needles of the ammeters are θ_1 and θ_2, respectively. The ammeters are then connected in parallel and then to the same voltage source. This time, the deviations of the needles are θ'_1 and θ'_2, respectively. r_2 in terms of $r_1, \theta_1, \theta_2, \theta'_1$ and θ'_2 is

(a) $r_1 \frac{\theta_1\theta'_1}{\theta_2\theta'_2}$
(b) $r_1 \frac{\theta_2\theta'_1}{\theta_1\theta'_2}$
(c) $r_1 \frac{\theta_2\theta'_2}{\theta_1\theta'_1}$
(d) $r_1 \frac{\theta_1\theta_2}{\theta'_1\theta'_2}$

5 In a circuit, voltmeter reads 3V and the ammeter reads 2A. Then, which of the following is incorrect?

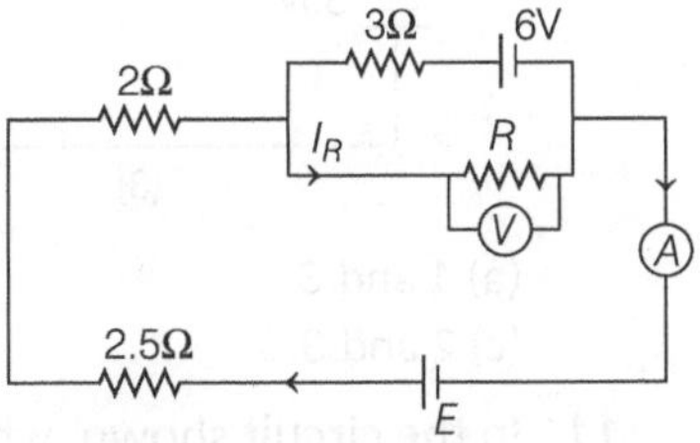

(a) The resistance R is $1\ \Omega$
(b) The emf E is 9V
(c) The current in $3\ \Omega$ resistor is 1A
(d) 6 V battery is discharging

6 In the circuit shown, when switch S_1 is closed and S_2 is open, the ideal voltmeter shows a reading of 18V. When switch S_2 is closed and S_1 is open, the reading of voltmeter is 24V. When S_1 and S_2 both are closed, the voltmeter reading will be

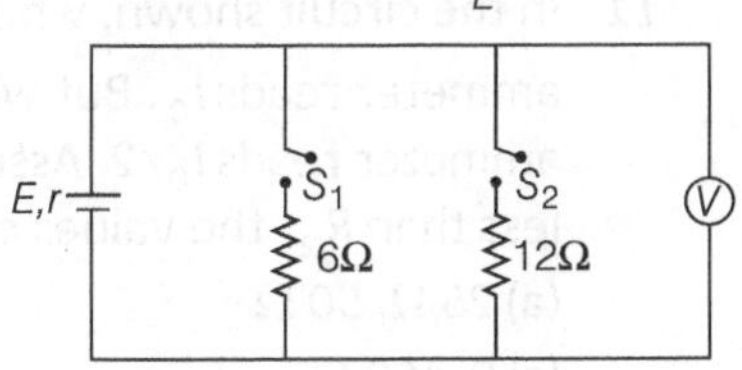

(a) 14.4 V
(b) 20.6 V
(c) 24.2 V
(d) 10.8 V

7 In the circuit shown in the figure, no current flows through the ideal ammeter. If the internal resistance of the cell is negligible, then find value of unknown resistance R.

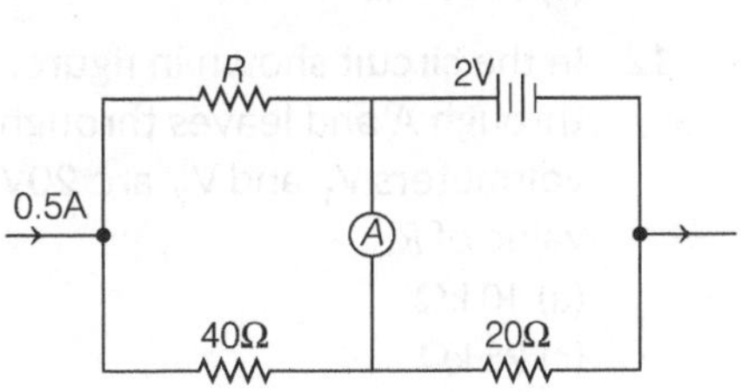

(a) $5\ \Omega$
(b) $10\ \Omega$
(c) $15\ \Omega$
(d) $20\ \Omega$

8 In the circuit shown, the reading of the ammeter (ideal) is the same with both switches open as with both closed. Find the value of resistance R in Ω.

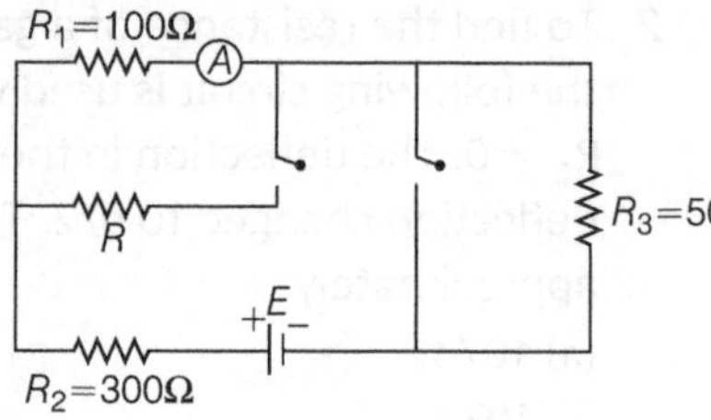

(a) 600
(b) 800
(c) 400
(d) 300

9 In the circuit shown, the variable resistance is adjusted to a particular value, so that the ammeter reading is same in both the positions 1 and 2 of the key. The reading of ammeter is 2A. If E =10V, then value of x is

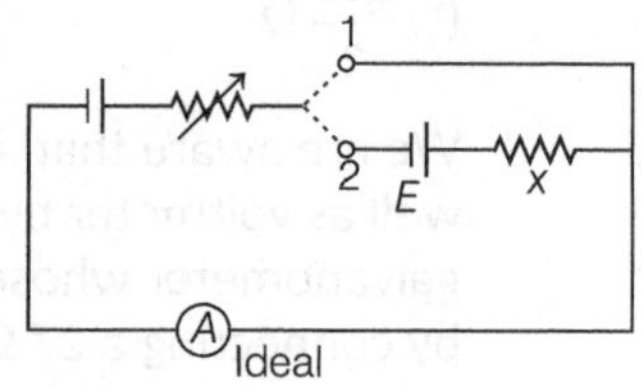

(a) 2 Ω
(b) 5 Ω
(c) 10 Ω
(d) 20 Ω

10 Four circuits are given in the options below. In all the circuits voltmeters, ammeter and cells are ideal. The readings of each ammeter and voltmeter are shown in the figure. In which pair of circuits shown below, readings of voltmeters V_1 and V_2 and ammeter A may be correct?

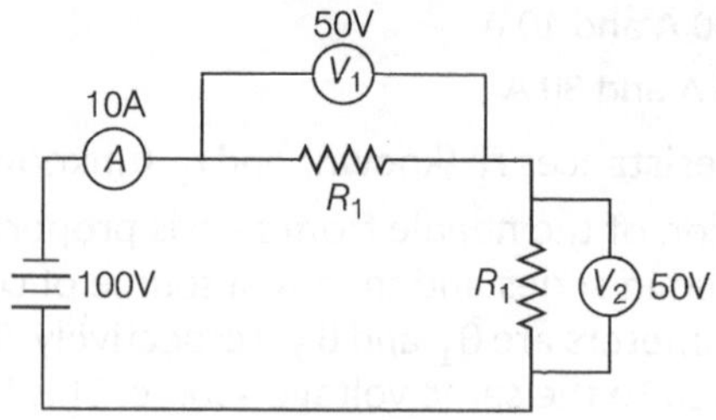

(1)

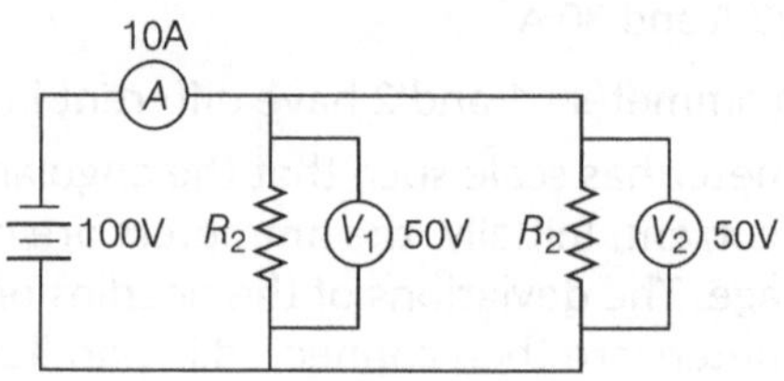

(2)

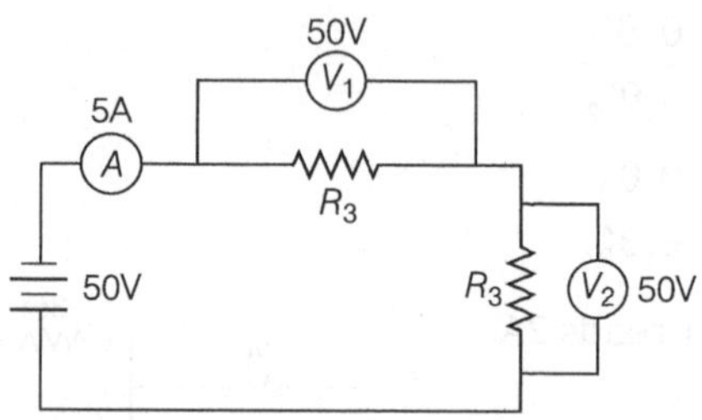

(3)

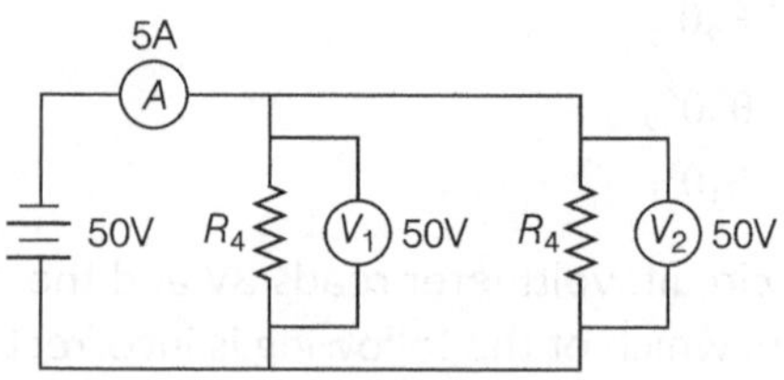

(4)

(a) 1 and 3
(b) 2 and 4
(c) 2 and 3
(d) 1 and 4

11 In the circuit shown, when keys K_1 and K_2 both are closed, the ammeter reads I_0. But when K_1 is open and K_2 is closed, the ammeter reads $I_0/2$. Assuming that, ammeter resistance is much less than R_2, the values of r and R_1 are

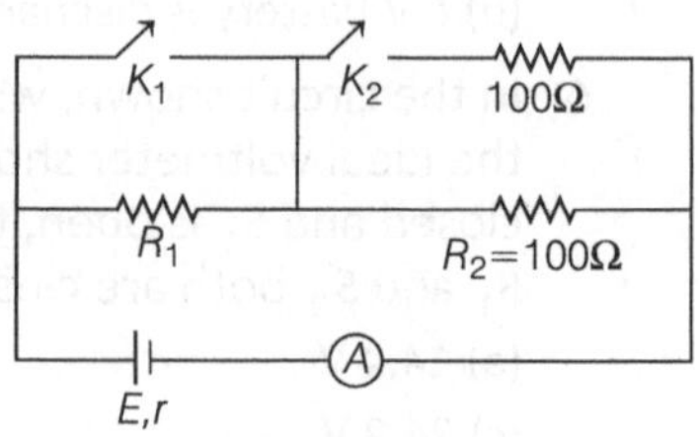

(a) 25 Ω, 50 Ω
(b) 25 Ω,100Ω
(c) 0, 100 Ω
(d) 0,50 Ω

12 In the circuit shown in figure, a current $I = 600\ \mu A$ enters through A and leaves through B. Reading of the identical voltmeters V_1 and V_2 are 20V and 30V respectively. Find the value of R.

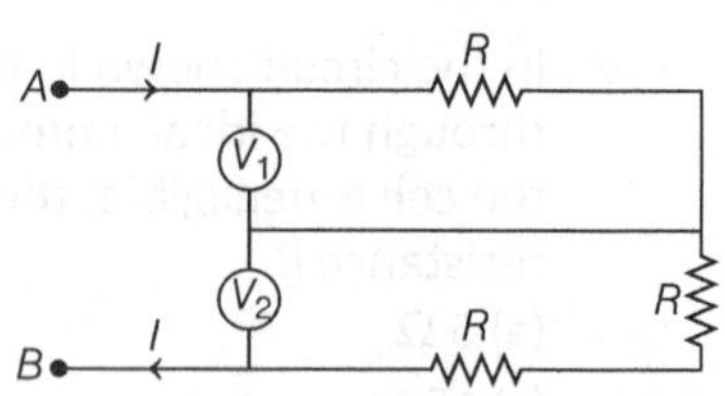

(a) 10 kΩ
(b) 20 kΩ
(c) 25 kΩ
(d) 50 kΩ

13 In the circuit shown in the figure, two resistors R_1 and R_2 have been connected in series to an ideal cell. When a voltmeter is connected across R_1, its reading is $V_1 = 4.0$ V and when the same voltmeter is connected across R_2, its reading is $V_2 = 6.0$ V. The reading of the voltmeter, when it is connected across the cell is $V_3 = 12.0$ V. Find the actual potential difference across R_1 in the circuit.

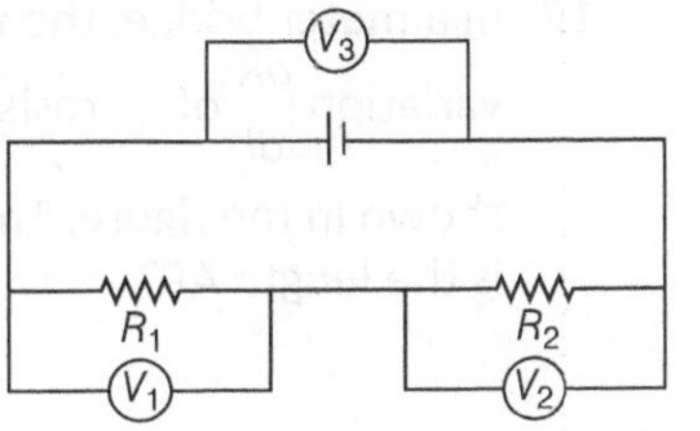

(a) 3V (b) 4V
(c) 4.8 V (d) 5.5 V

14 A galvanometer of coil resistance 1 Ω is converted into voltmeter by using a resistance of 5Ω in series and another identical galvanometer is converted into ammeter by using a shunt of 1 Ω. Now, ammeter and voltmeter are connected in circuit as shown, the readings of voltmeter and ammeter respectively, are

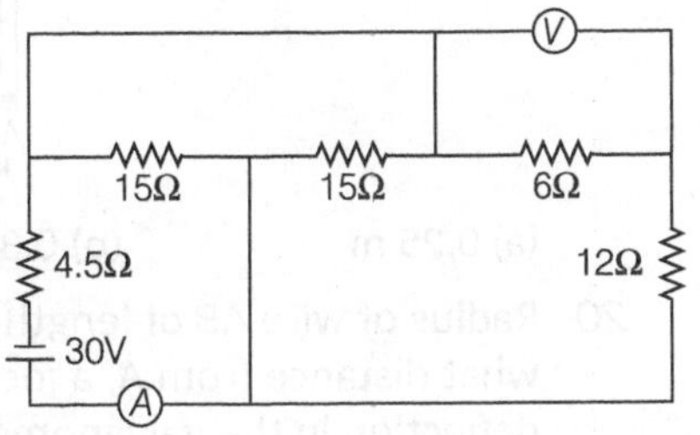

(a) 3V, 3 A (b) 6V, 3A
(c) 2 V, 2A (d) 4V, 2A

15 Four ammeters with identical internal resistance r and a resistor of resistance R are connected to a current source as shown in figure. It is known that, the reading of the ammeter A_1 is $I_1 = 3$A and the reading of the ammeter A_2 is $I_2 = 5$A. Determine the ratio of the resistances R/r.

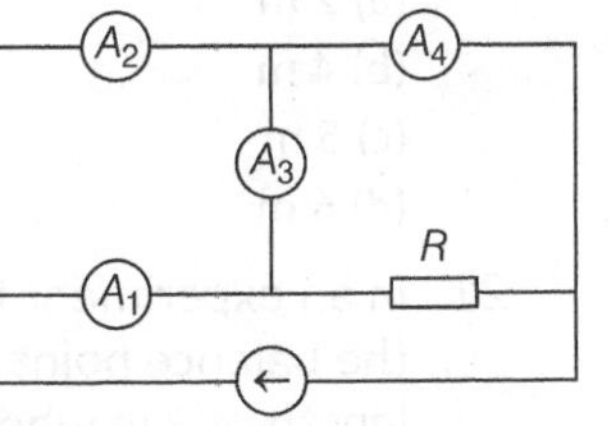

(a) 9 (b) 3
(c) $\frac{1}{9}$ (d) $\frac{1}{3}$

16 The circuit shown consists of three identical resistances and three identical voltmeters A, B and C. If the first voltmeter A reads $V_A = 22$ V and the third voltmeter C reads $V_C = 8$V, calculate reading of the second voltmeter B.

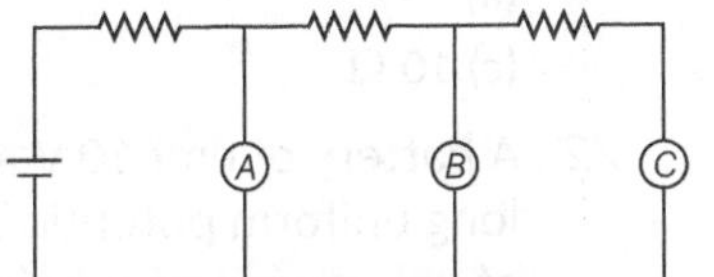

(a) 10 V (b) 12 V
(c) 14 V (d) 15 V

17 An ideal voltmeter is connected between C and D gives reading 140 V. When we replace it with an ideal ammeter, its reading is 21 A. Now , we connect a resistor r between C and D. The current flowing through r is

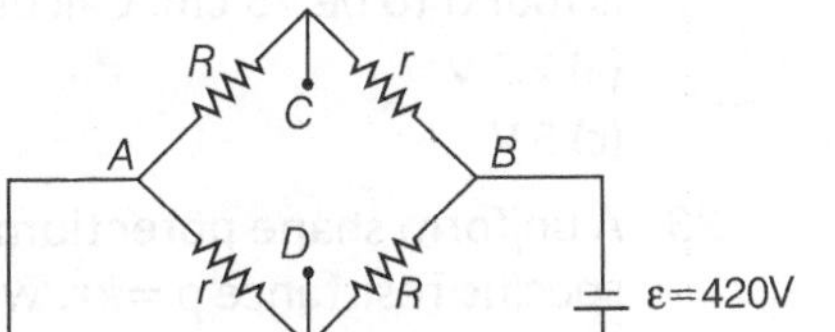

(a) 8A
(b) 10 A
(c) 12 A
(d) 14 A

18 Figure shows a meter bridge, wire AC has uniform cross-section. The length of wire AC is 100 cm. X is a standard resistor of 4 Ω and Y is a coil. When Y is immersed in melting ice, the null point is at 40 cm from point A. When the coil Y is heated to 100°C, a 78 Ω resistor has to be connected in parallel with Y in order to keep the bridge balanced at the same point. Temperature coefficient of resistance of the coil is

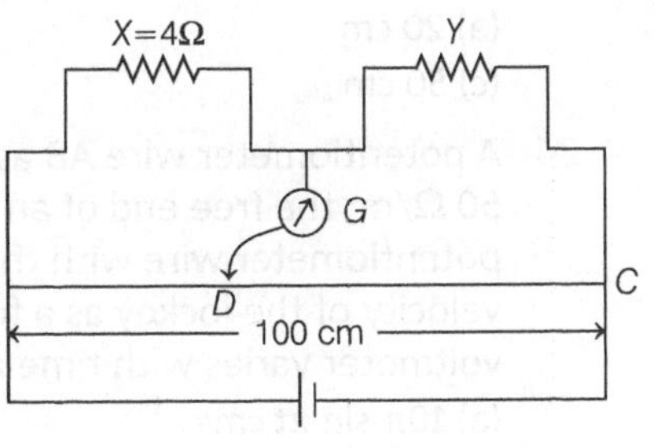

(a) 6.3×10^{-4} K^{-1} (b) 4.3×10^{-4}K^{-1}
(c) 8.3×10^{-4} K^{-1} (d) 2.3×10^{-4} K^{-1}

19 In a meter bridge, the wire of length 1 m has a non-uniform cross-section such that, the variation $\frac{dR}{dl}$ of its resistance R with length l is $\frac{dR}{dl} \propto \frac{1}{\sqrt{l}}$. Two equal resistances are connected as shown in the figure. The galvanometer has zero deflection when the jockey is at point P. What is the length AP?

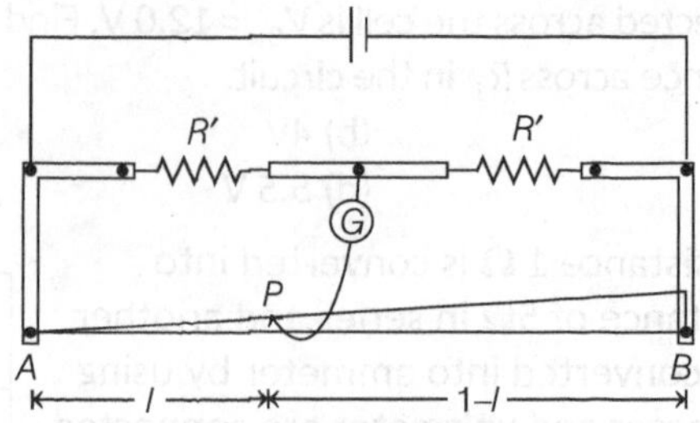

(a) 0.25 m (b) 0.3 m (c) 0.35 m (d) 0.2 m

20 Radius of wire AB of length 10 m changes from r to $2r$. At what distance from A, a jockey be contacted to give no deflection in the galvanometer?

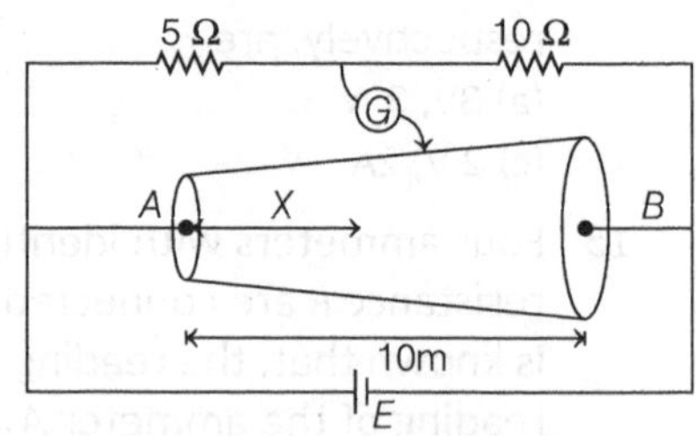

(a) 2 m
(b) 4 m
(c) 5 m
(d) 6 m

21 In an experiment to measure the internal resistance of a cell by potentiometer, it is found that the balance point is at a length of 2m when the cell is shunted by a 5 Ω resistance, and is at a length of 3 m when the cell is shunted by a 10 Ω resistance. The internal resistance of the cell is

(a) 1 Ω (b) 1.5 Ω
(c) 10 Ω (d) 15 Ω

22 A battery of emf 10 V and internal resistance 2Ω is connected in primary circuit with a 100 cm long uniform potentiometer wire and a rheostat whose resistance is fixed at 998 Ω. A battery of unknown emf is being balanced on this potentiometer wire and balancing length is found to be 50 cm. When area of cross-section of potentiometer wire is doubled, then balancing length is found to be 75 cm. Calculate emf of the battery.

(a) 2.5 V (b) 4 V
(c) 5 V (d) 6 V

23 A uniform shape potentiometer wire AB (100 cm long) has specific resistance $\rho = kx$, where x is the position of point from starting end A. Then, at which position (in cm) null point of wire will come from the left end A for the given circuit ? Resistance of wire AB is 40 Ω.

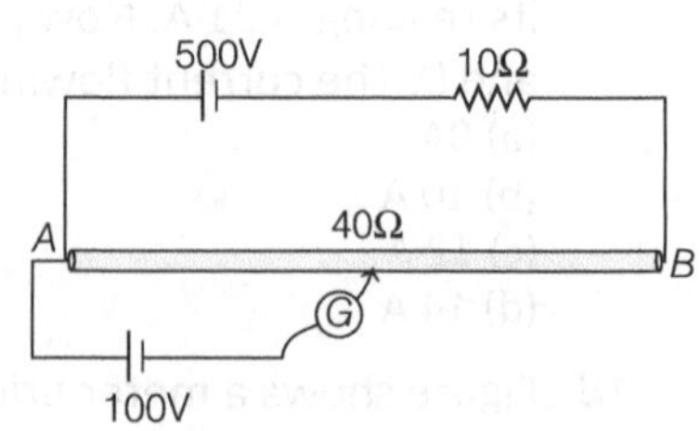

(a) 20 cm (b) 25 cm
(c) 50 cm (d) 80 cm

24 A potentiometer wire AB as shown is 40 cm long of resistance 50 Ω/m; the free end of an ideal voltmeter is touching the potentiometer wire with the help of a jockey. What should be the velocity of the jockey as a function of time, so that reading of voltmeter varies with time as $(2 \sin \pi t)$ V.

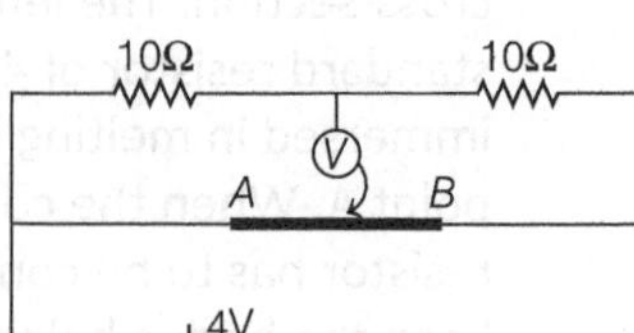

(a) $10\pi \sin \pi t \text{ cms}^{-1}$ (b) $10\pi \cos \pi t \text{ cms}^{-1}$
(C) $20\pi \sin \pi t \text{ cms}^{-1}$ (d) $20\pi \cos \pi t \text{ cms}^{-1}$

25 In the laboratory, the high emf of a battery is measured by using potentiometer and two resistances R_1 and R_2, where $R_1 << R_2$ as shown in the figure.

$R_1 = (100 \pm 0.10)\ \Omega$ and $R_2 = (9900 \pm 9.90)\ \Omega$, $AB = 1$ m.

The voltage across R_1, when switches S_1 and S_2 are connected to point 1 and point 2 is balanced against $l_1 = (60 \pm 0.06)$ cm. When the switches S_1 and S_2 are shifted to point 3 and point 4 as shown in the given figure, the potential difference of standard $E_0 = 2$ V is balanced against length $l_2 = (75 \pm 0.075)$ cm.

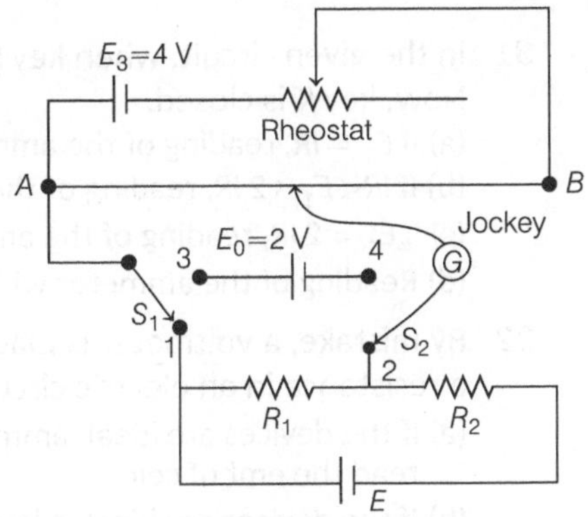

The maximum error in emf of battery E is

(a) 0.32 V (b) 0.50 V (c) 0.64 V (d) 0.80 V

Multiple Choice(s) Correct Questions

26 A galvanometer can be used as a voltmeter or an ammeter. While it was used as a voltmeter, a resistance of 72 Ω was used to increase its range to n times. While using it as an ammeter, a shunt resistance of 2 Ω was used which increased its range to n times.

(a) Resistance of galvanometer is 12 Ω (b) Resistance of ammeter is 1.7 Ω

(c) Resistance of voltmeter is 84 Ω (d) Resistance of ammeter is 3.4 Ω

27 The galvanometer shown in the figure has resistance 10 Ω. It is shunted by a series combination of a resistance $S = 1\ \Omega$ and an ideal cell of emf 2V. A current 2A passes as shown

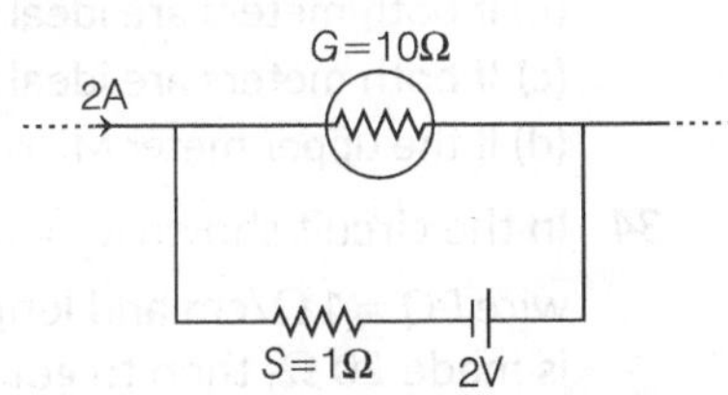

(a) The reading of the galvanometer is 1 A

(b) The reading of the galvanometer is zero

(c) The potential difference across the resistance S is 1.5 V

(d) The potential difference across the resistance S is 2 V

28 Voltmeter and ammeter shown in circuit diagram are ideal, then the correct statements.

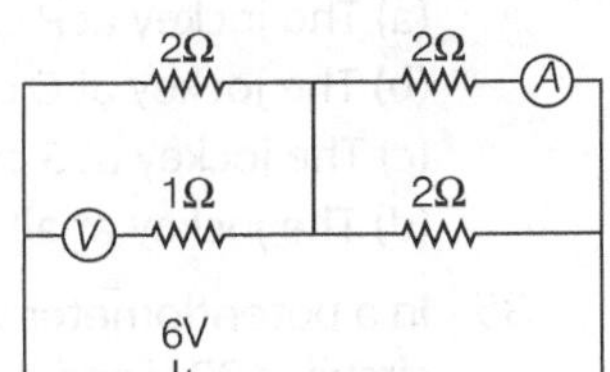

(a) Reading of voltmeter is 4 V.

(b) Reading of ammeter is 1.0 A.

(c) Current through 1 Ω resistance is 1.0 A.

(d) Current through 1 Ω resistance is zero.

29 In the figure shown, two ideal voltmeters and an ammeter which are connected across the various circuit elements. If the voltmeter connected across 9 Ω resistance reads 4.5 V, then

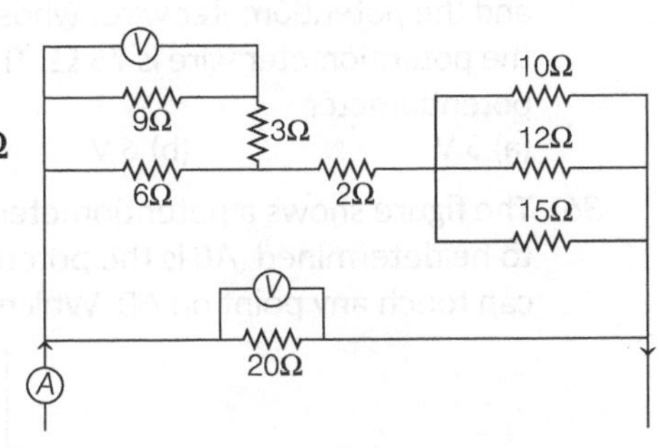

(a) the current through 12 Ω resistance is 0.5A

(b) the reading of the voltmeter connected across 20 Ω resistance, is 20 V

(c) the reading of the voltmeter connected across 20 Ω resistance, is 15 V

(d) the reading of the ammeter is 2.25 A

30 In the circuit shown, the battery is ideal. It has a voltage of 21 V. The galvanometer G_1 shows 0.42 A. When another identical galvanometer G_2 is connected across R, the reading of G_1 increases to 0.50 A.

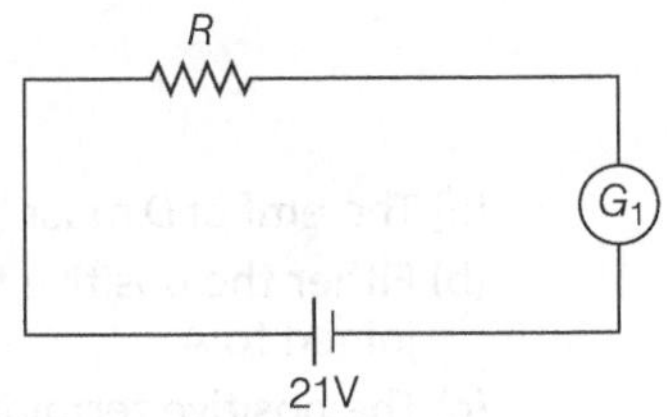

(a) The resistance of galvanometer is 30 Ω.

(b) The resistance R is 40 Ω.

(c) The reading of G_2 is 0.20A.

(d) If instead of this, G_2 was connected parallel to G_1, their reading would be same 0.21 A.

31 In the given circuit, when key K is open, reading of ammeter is I. Now, key K is closed.

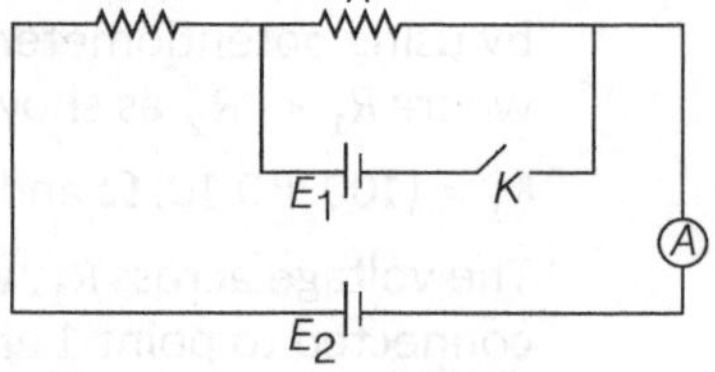

(a) If $E_1 = IR$, reading of the ammeter is I.
(b) If $IR < E_1 < 2IR$, reading of the ammeter is greater than I.
(c) If $E_1 = 2IR$, reading of the ammeter will be zero.
(d) Reading of the ammeter will not change.

32 By mistake, a voltmeter is placed in series and an ammeter is parallel with a resistance in an electric circuit, with a cell in series.

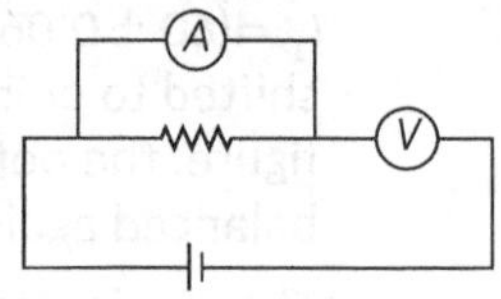

(a) If the devices are ideal, ammeter will read zero current and voltmeter will read the emf of cell.
(b) If the devices are ideal, a large current will flow through the ammeter and it will be damaged.
(c) The main current in the circuit will be very low and almost all current will flow through the ammeter, if resistance of ammeter is much smaller than the resistance in parallel.
(d) The devices may get damaged, if emf of the cell is very high and the meters are non-ideal.

33 In the given circuit, the potential difference between a and b is 300 V. The resistors are of 100 Ω each.

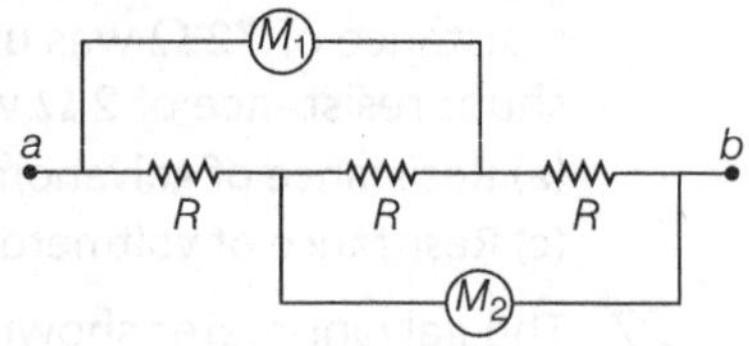

(a) If the upper meter M_1 is ideal voltmeter and lower is ideal ammeter M_2, reading of voltmeter is 300 V.
(b) If both meters are ideal voltmeters, they show 200 V each.
(c) If both meters are ideal ammeters, they read 6A each.
(d) If the upper meter M_1 is ideal voltmeter and lower M_2 is ideal ammeter, reading of ammeter is 3A.

34 In the circuit shown $R_1 = R_2 = 10\ \Omega$ and resistance per unit length of wire $PQ = 1\ \Omega/\text{cm}$ and length $PQ = 10$ cm. S is mid-point of wire PQ. If R_2 is made 20 Ω, then to get zero deflection in galvanometer.

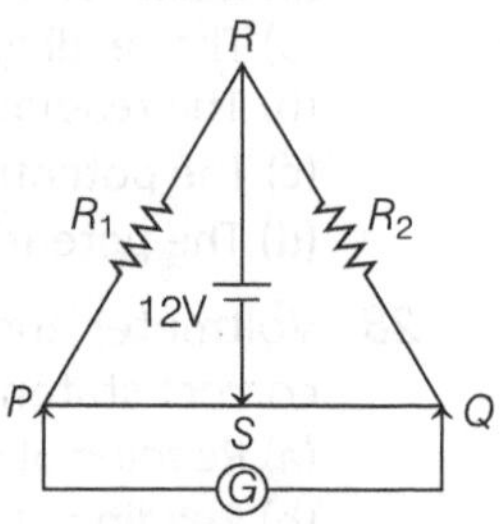

(a) The jockey at P can be moved towards S by 2 cm.
(b) The jockey at Q can be moved towards S by 2 cm.
(c) The jockey at S can be moved towards P by (5/3) cm.
(d) The jockey at all positions fixed and R_1 should be made 20 Ω.

35 In a potentiometer wire experiment, the emf of a battery in the primary circuit is 20 V and its internal resistance is 5 Ω. There is a resistance box (in series with the battery and the potentiometer wire) whose resistance can be varied from 120 Ω to 170 Ω. Resistance of the potentiometer wire is 75 Ω. The following potential differences can be measured using this potentiometer

(a) 5 V (b) 6 V (c) 7 V (d) 8 V

36 The figure shows a potentiometer arrangement. D is the driving cell and C is the cell whose emf is to be determined. AB is the potentiometer wire and G is galvanometer. J is sliding contact which can touch any point on AB. Which of the following are essential conditions for obtaining balance?

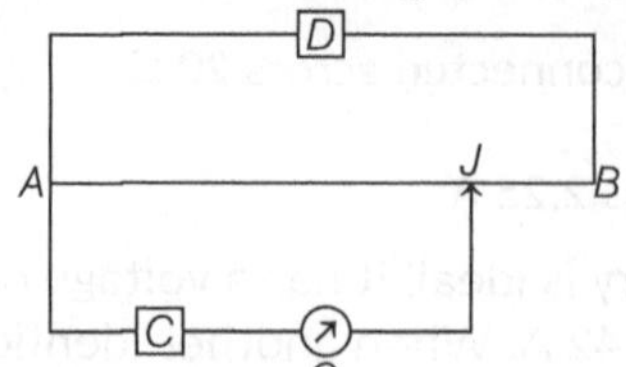

(a) The emf of D must be greater than the emf of C.
(b) Either the positive terminals of both D and C or the negative terminals of both D and C must be joined to A.
(c) The positive terminals of D and C must be joined.
(d) The resistance of G must be less than resistance of AB.

37 A student was given the experiment to measure the emf of a unknown cell using potentiometer. He cheated result and wrote the observation arbitrarily and was caught. Which reading helped the teacher to arrive at the conclusion that he cheated?

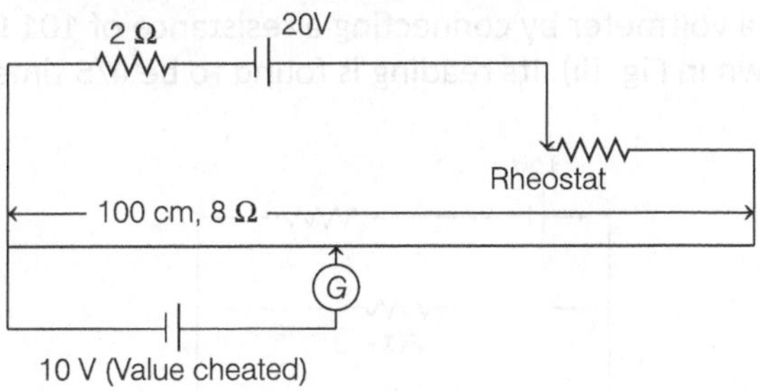

S.No.	Value of rheostat	Null point
1.	R_1	50 cm
2.	R_2	60 cm
3.	R_3	70 cm
4.	R_4	80 cm

(a) 1
(b) 2
(c) 3
(d) 4

38 In the potentiometer circuit of given figure, the galvanometer reveals a current in the direction shown wherever the sliding contact touches the wire. This could be caused by

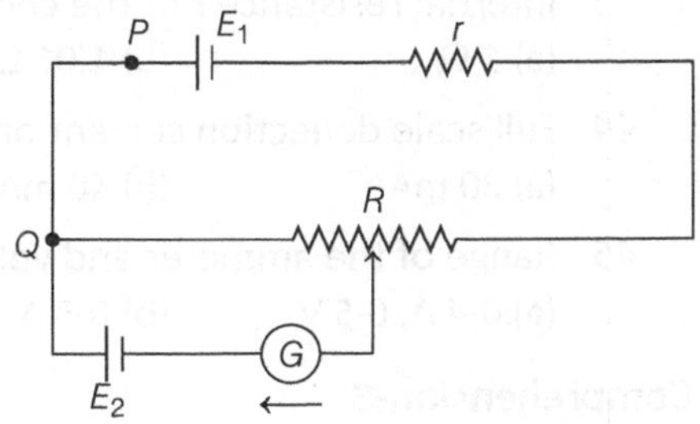

(a) E_1 being too low
(b) r being too high
(c) a break in PQ
(d) E_2 being too low

39 Figure shows a potentiometer circuit. The length of potentiometer wire AB is 120 cm and its resistance is 2Ω. The internal resistance of 2V cell is negligible and the internal resistance of 6V cell is 1 Ω. Initially the switches S_1 and S_2 are open and S_3 is closed.

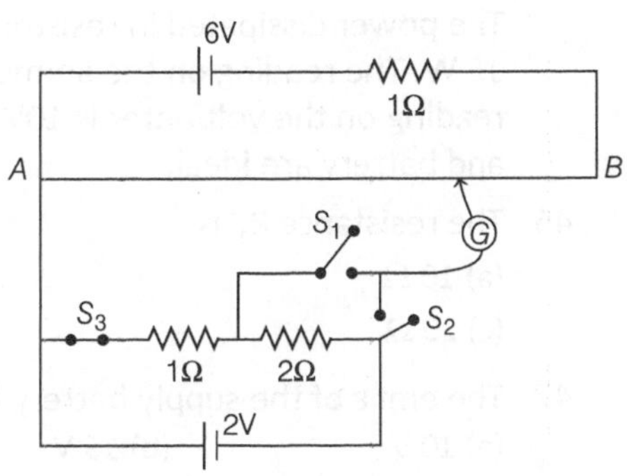

(a) When only S_2 is open, balanced length is 20 cm.
(b) When only S_1 is open, balanced length is 60 cm.
(c) When all three switches are closed, balanced length is 60 cm.
(d) When S_1 and S_2 are open, but S_3 is closed, balance length is 60 cm.

Comprehension Based Questions

Comprehension-1

An ammeter and a voltmeter are connected in series to a battery with an emf of 10V. When a certain resistance is connected in parallel with the voltmeter, the reading of the voltmeter decreases three times, whereas the reading of the ammeter increases the two times.

40 Find the voltmeter reading after the connection of the resistance.
(a) 1 V (b) 2 V (c) 3 V (d) 4 V

41 If resistance of the ammeter is 2 Ω, then resistance of the voltmeter is
(a) 1 Ω (b) 2 Ω (c) 3 Ω (d) 4 Ω

42 If resistance of the ammeter is 2 Ω, then resistance of the resistor which is added in parallel to the voltmeter is
(a) $\frac{3}{5}$ Ω (b) $\frac{2}{7}$ Ω (c) $\frac{3}{7}$ Ω (d) None of these

Comprehension-2

A galvanometer (coil resistance 99 Ω) is converted into a ammeter using a shunt of 1 Ω and connected as shown in the Fig. (i). The ammeter reads 3A.

The same galvanometer is converted into a voltmeter by connecting a resistance of 101 Ω in series. This voltmeter is connected as shown in Fig. (ii). Its reading is found to be 4/5 times of the full scale reading.

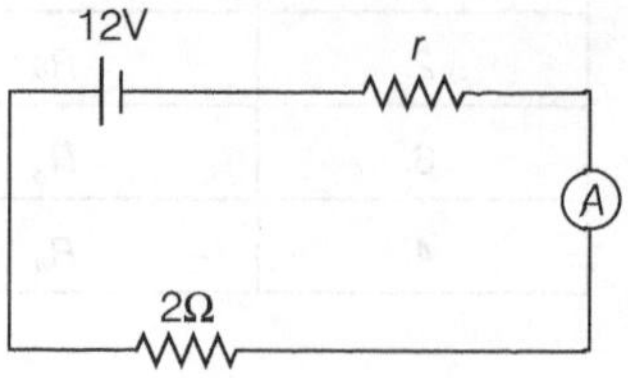

(i)

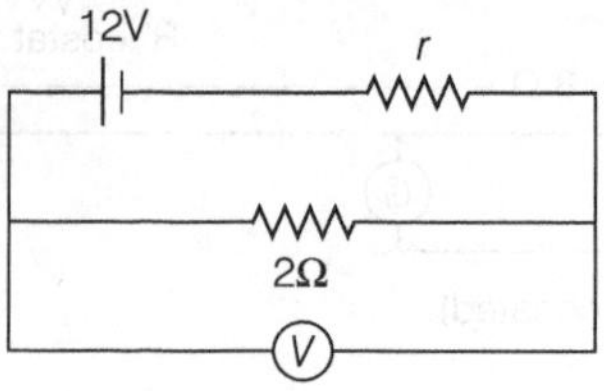

(ii)

43 Internal resistance r of the cell is
(a) 1 Ω (b) 1.01 Ω (c) 1.05 Ω (d) 1.1 Ω

44 Full scale deflection current of the galvanometer is
(a) 30 mA (b) 40 mA (c) 50 mA (d) 100 mA

45 Range of the ammeter and voltmeter respectively are
(a) 0-4 A, 0-5 V (b) 0-5 A, 0-10 V (c) 0-8 A, 0-15 V (d) 0-10 A, 0-20 V

Comprehension-3

The power dissipated in resistor R_3 shown in the figure is 15 W. The reading on the ammeter is 500 mA and the reading on the voltmeter is 10V. Ammeter, voltmeter and battery are ideal.

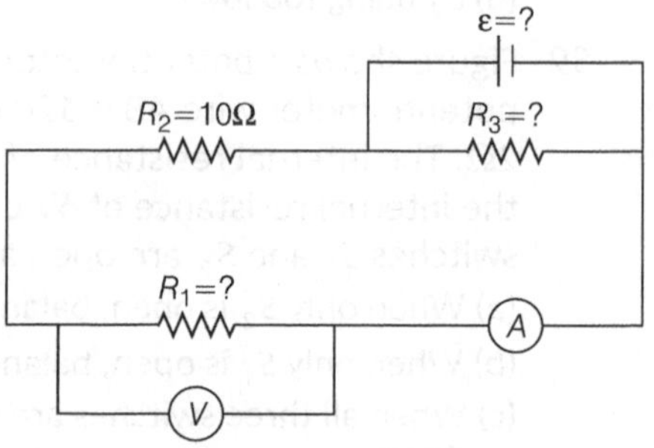

46 The resistance R_1 is
(a) 10 Ω (b) 15 Ω
(c) 20 Ω (d) 25 Ω

47 The emf ε of the supply battery is
(a) 10 V (b) 15 V (c) 20 V (d) 25 V

48 The resistance R_3 is
(a) 10 Ω (b) 15 Ω (c) 20 Ω (d) 25 Ω

Comprehension-4

In a potentiometer circuit, two wires of same material of resistivity ρ, one of radius of cross-section a and other of radius of cross-section $2a$ are joined in series. They are of length L and $2L$, respectively. This combination acts as the potentiometer wire of length $3L$. The emf of the cell in the primary circuit is ε and internal resistance is $\rho L/2\pi a^2$. This cell is connected to the potentiometer wire by a conducting wire of negligible resistance with positive terminal of the cell connected to one end (call it A) of longer wire.

49 The maximum voltage which can be balanced on the potentiometer wire is
(a) ε (b) $\frac{\varepsilon}{2}$ (c) $\frac{3\varepsilon}{4}$ (d) $\frac{2\varepsilon}{5}$

50 The balancing length, measured from point A, obtained in measurement of emf of cell of emf $\frac{\varepsilon}{2}$ is
(a) L (b) $\frac{L}{2}$ (c) $\frac{2L}{5}$ (d) $\frac{5L}{2}$

51 If positive terminal of another cell of emf $\frac{\varepsilon}{2}$ and internal resistance $\rho L / 2\pi a^2$ is connected to point A and other terminal of this cell is joined to other end of two wires, then the current through this cell is

(Take, $R = \rho L / 2\pi a^2$)

(a) $\frac{\varepsilon}{2R}$ (b) $\frac{\varepsilon}{3R}$ (c) $\frac{\varepsilon}{5R}$ (d) $\frac{\varepsilon}{7R}$

Comprehension-5

In the circuit shown, the internal resistance of the cell is negligible. The distance of the jockey (slider) from left hand end of the wire is l. The adjoining graphs shows the variation of current I (marked in figure) with length l of the slide wire.

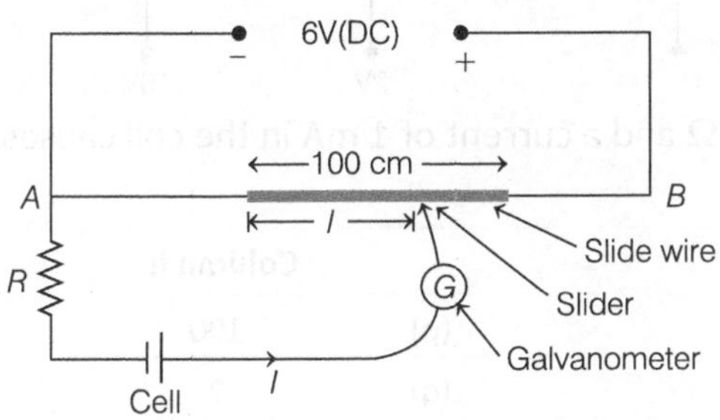

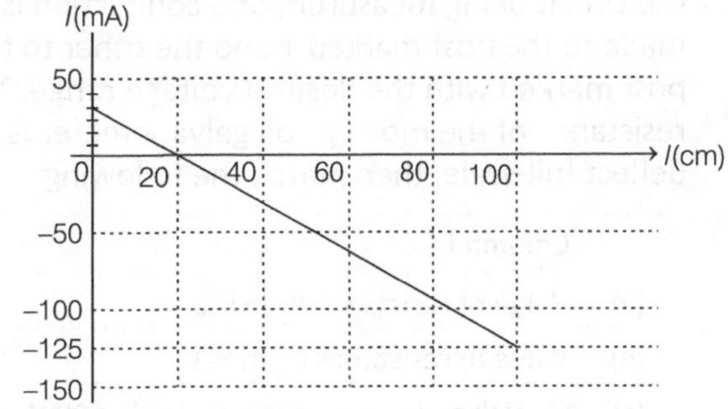

52 For balancing condition of the instrument, value of l is equal to

(a) 40 cm (b) 20 cm (c) 100 cm (d) None of these

53 Value of the emf of cell is

(a) 0.98 V (b) 1.20 V (c) 1.86 V (d) 3 V

54 Value of the resistance R is

(a) 30 Ω (b) 40 Ω (c) 38 Ω (d) 45 Ω

Comprehension-6

A wire AB (of length 1 m, area of cross-section πm^2) is used in potentiometer experiment to calculate emf and internal resistance (r) of battery.
The emf and internal resistance of driving battery are 15 V and 3Ω, respectively. The resistivity of wire AB varies as $\rho = \rho_0 x$ (where x is distance from A in metres and $\rho_0 = 24\pi\ \Omega$)
The distance of null point from A is obtained at $\sqrt{\frac{2}{3}}$ m when switch S is open and at $\frac{1}{\sqrt{2}}$ m when switch is closed.

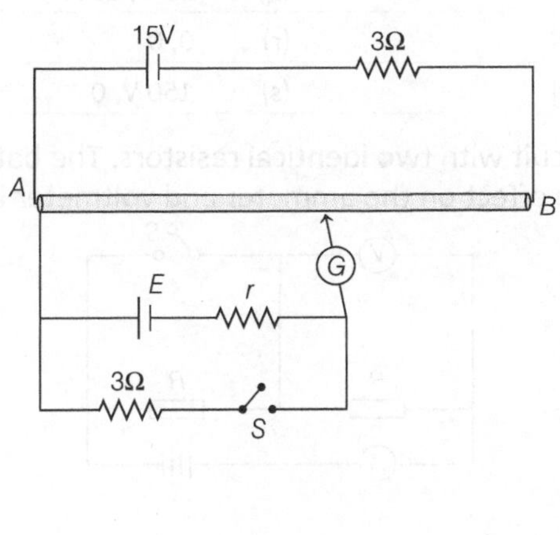

55 The resistance of whole wire AB is

(a) $6\,\Omega$ (b) $12\,\Omega$ (c) $18\,\Omega$ (d) $24\,\Omega$

56 The emf E of battery is

(a) 4 V (b) 6 V (c) 8 V (d) 10 V

57 Internal resistance of battery E is

(a) $4\,\Omega$ (b) $3\,\Omega$ (c) $2\,\Omega$ (d) $1\,\Omega$

Match the Columns Questions

58 Figure shows the internal wiring of a three range voltmeter whose binding posts are marked +, 2 V, 10 V and 100 V. When the meter is connected to the circuit being measured, one connection is made to the post marked + and the other to the post marked with the desired voltage range. The resistance of the moving coil galvanometer is $40\,\Omega$ and a current of 1 mA in the coil causes it to deflect full-scale, then match the following.

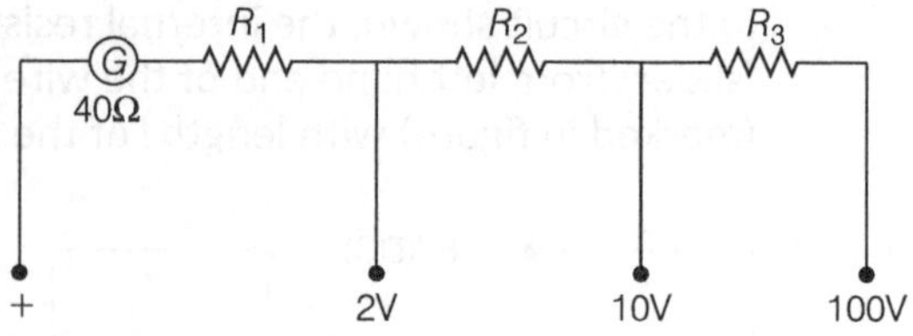

	Column I		Column II
(a)	Value of resistance R_1 in kΩ	(p)	100
(b)	Value of resistance R_3 in kΩ	(q)	2
(c)	Overall resistance of the meter in 100 V range in kΩ	(r)	1.96
(d)	Overall resistance of the meter in 2 V range in kΩ	(s)	90

59 Voltmeter of resistance $10\,\Omega$ and ammeter of resistance $10\,\Omega$ are attached in the following circuit with switches S_1 and S_2.

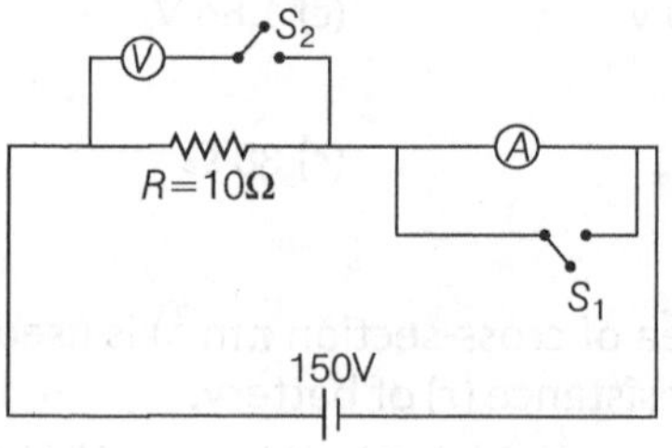

	Column I (switch position)		Column II (reading of voltmeter and ammeter)
(a)	S_1 open, S_2 closed	(p)	0, 7.5 A
(b)	S_1 closed, S_2 open	(q)	50 V, 10 A
(c)	S_1 open, S_2 open	(r)	0, 0
(d)	S_1 closed, S_2 closed	(s)	150 V, 0

60 The diagram shows a circuit with two identical resistors. The battery has a negligible internal resistance. What will the effect on the ammeter and voltmeter be if the switch S is closed?

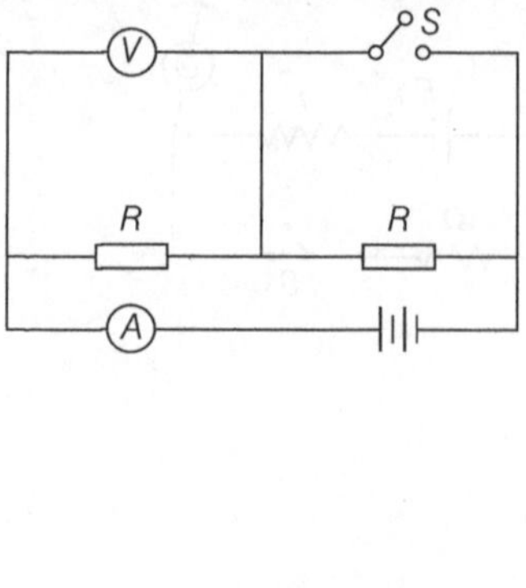

	Column I		Column II
(a)	Ammeter reading	(p)	Increases
(b)	Voltmeter reading	(q)	Decreases
(c)	Equivalent resistance of circuit	(r)	does not change
(d)	Power dissipated across R in right branch	(s)	becomes zero

61 In the circuit shown in the figure, the electromotive force of the battery is 9V and its internal resistance is 15 Ω. The two identical voltmeters and the ammeter can all be considered ideal (i.e. voltmeter is of very large resistance and ammeter is of negligible resistance).

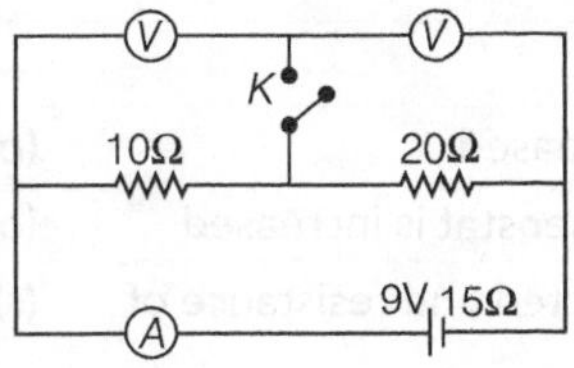

	Column-I		Column-II
(a)	Voltmeter (left) reading (in volts), when K is open.	(p)	4.5
(b)	Voltmeter (right) reading (in volts), when K is closed.	(q)	3
(c)	Ammeter reading (in amperes), when K is open.	(r)	4
(d)	Ammeter reading (in amperes), when K is closed.	(s)	0.2
		(t)	0

62 Column I has four circuits each having an ammeter. Column II has four values of current in the ammeter. The ammeter has zero resistance. The voltmeter, in (b) has infinite resistance and a reading 8V. The resistance R has not been specified. Match the circuit with its correct ammeter reading.

Column-I — **Column-II**

(a)

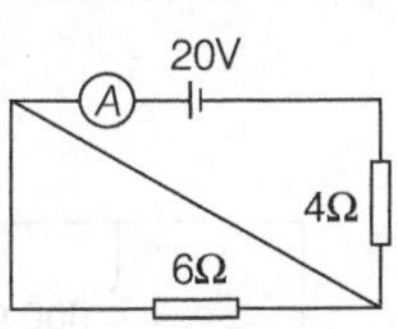

(p) 0

(b)

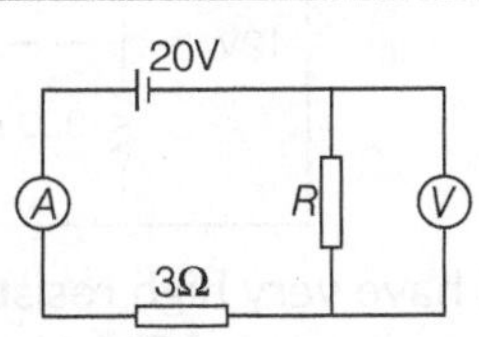

(q) 2 A

(c)

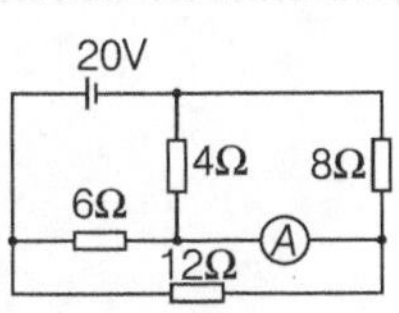

(r) 4 A

(d)

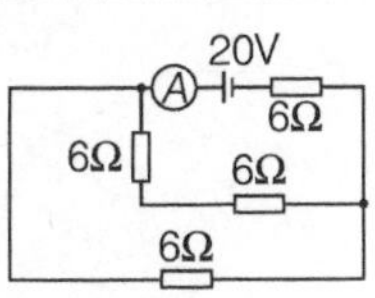

(s) 5 A

63 A potentiometer experiment set-up is shown in the given diagram. R is the point when there is no deflection in the galvanometer. E_1 is the source. Match the following.

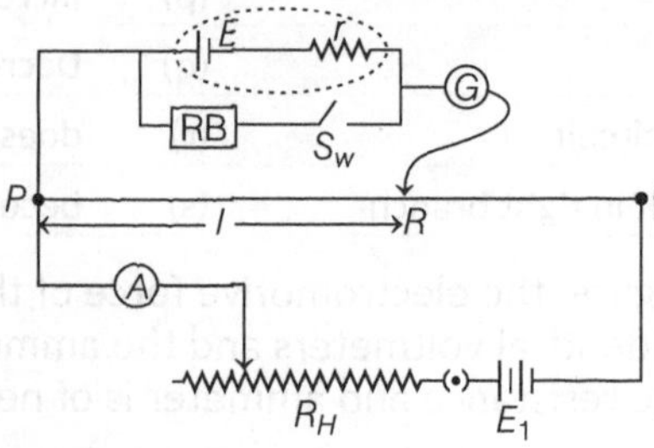

	Column I		Column II
(a)	If only emf of battery E_1 is increased	(p)	Point R will shift to left
(b)	If only resistance offered by rheostat is increased	(q)	Point R will shift to right
(c)	If resistance box (RB) is connected and resistance of RB is increased	(r)	Point R will remain unchanged
(d)	If an identical battery is connected in parallel to E	(s)	Point R may shift to left or right
		(t)	Reading of ammeter A changes

Numerical Value Questions

64 In the circuit shown, voltmeter is ideal but ammeter has a resistance 1 Ω. If the reading of ammeter and voltmeter are x A and yV respectively, then find value of $\frac{y}{x}$.

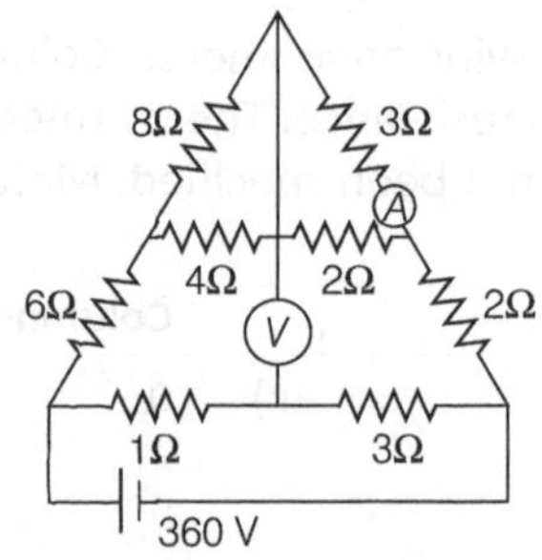

65 In the figure current in the galvanometer I is 10 mA, while the voltmeter reads a potential difference of 3 V. What is the current in galvanometer II (in mA)? The galvanometers are identical, the internal resistance of the battery is negligible.

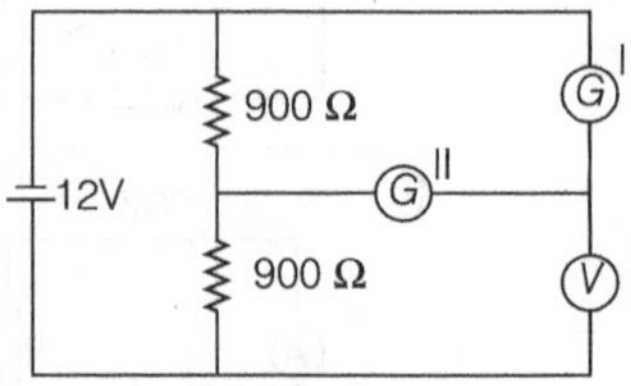

66 In the circuit shown below, all the voltmeters are identical and have very high resistance. Each resistor has the same resistance. The voltage of the ideal battery shown is 27 V. Find the reading of voltmeter V_3 (in volts).

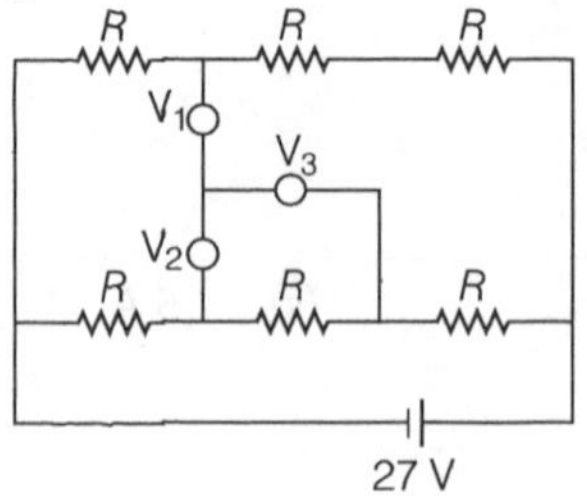

67 In the given circuit, the ammeters A_1 and A_2 are ideal and the ammeter A_3 has a resistance of $1.9 \times 10^{-3}\ \Omega$. The reading in ampere of the ammeter A_2 is $\left(\frac{34}{K}\right)$. What is the value of K?

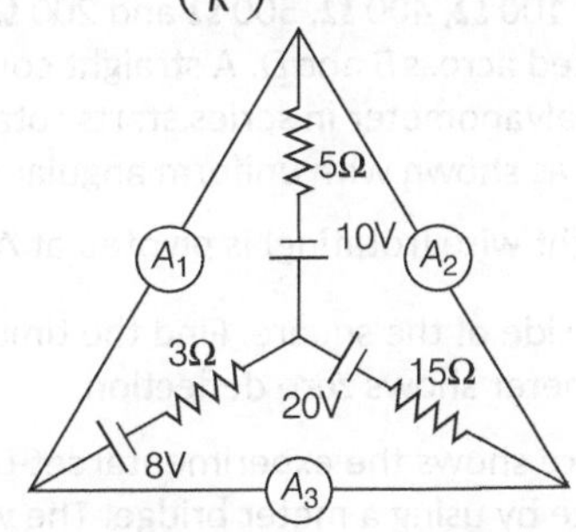

68 The figure shows a portion of an electric circuit. Resistors are known and are indicated on the diagram and the voltmeters are identical. If the voltmeters V_1 and V_2 read 7.5 V and 5.0 V, respectively. Find the reading of voltmeter V_3 (in V).

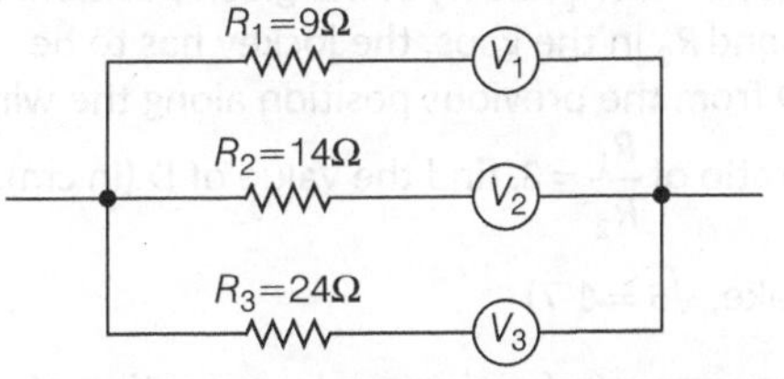

69 A constant voltage $V_0(= 12\ \text{V})$ is applied to a potential divider of resistance $R(= 4\Omega)$, connected to an ideal ammeter. A constant resistor $r\ (= 2\ \Omega)$ is connected to the sliding contact of the potential divider (as shown). Find the minimum current (in A) measured by ammeter.

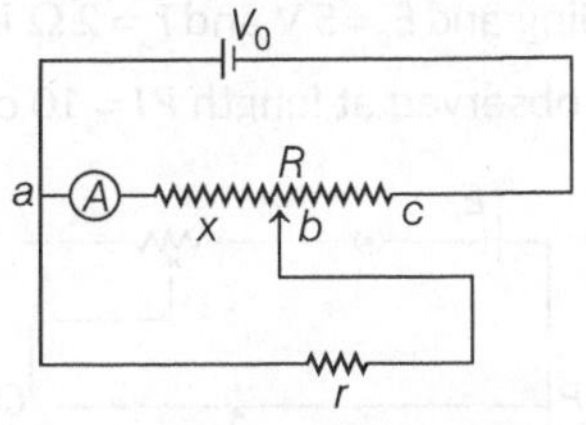

70 The circuit of a carbon arc lamp is shown in the figure (to ignite the lamp the carbon electrodes are touched together, then they are slowly drawn apart). The voltage across the gap between the carbon rods as a function of the current can be given by the following formula $V = A + B/I$, where A and B are constants.

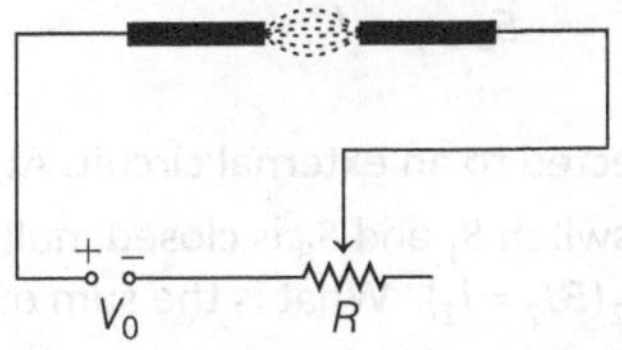

If the voltage of the battery V_0 can be considered constant, what is the maximum resistance(in Ω) of the rheostat R_{max}, that should be connected in series to ignite the lamp. Data $A = 40$ V, $B = 20$ W, $V_0 = 100$ V

71 Four wires of equal length 2 m are arranged as shown in diagram. Wires 2, 3 and 4 are of equal cross-sectional area and wire 1 is of half the cross-section of these wires. Jockey is moved through distance x cm on wire 1 to get null point, then find x.

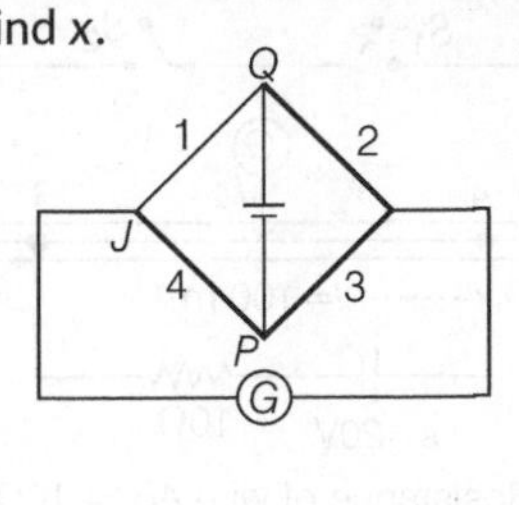

Problems in Physics

72 $ABCD$ is a square frame made from different wires of same length and each having different uniform resistance per unit length. Resistances of wires forming sides AB, BC, CD and DA are 100 Ω, 400 Ω, 500 Ω and 200 Ω, respectively. An ideal cell is connected across B and D. A straight conducting wire containing a resistance and a galvanometer in series starts rotating about pivoted point A from initial position as shown with uniform angular velocity $\omega = \frac{\pi}{360}$ rad/s. One end of the straight wire (rotating) is pivoted at A and other end always in sliding contact with a side of the square. Find the time (in seconds) after start, when galvanometer shows zero deflection.

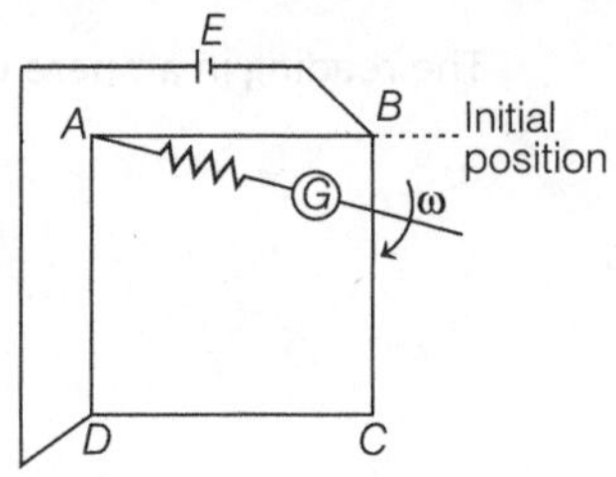

73 The circuit diagram given in the figure shows the experimental set-up for the measurement of unknown resistance by using a meter bridge. The wire connected between the points P and Q has non-uniform resistance such that resistance per unit length varies directly as the distance from the point P. Null point is obtained with the jockey J with R_1 and R_2 in the given position. On interchanging the positions R_1 and R_2 in the gaps, the jockey has to be displaced through a distance D from the previous position along the wire to establish the null point. If the ratio of $\frac{R_1}{R_2} = 3$, find the value of D (in cm).

Ignore any end corrections. (Take, $\sqrt{3} = 1.7$)

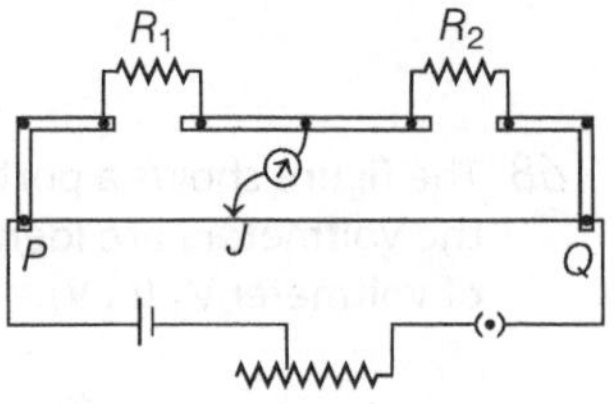

74 One of the circuits for the measurement of resistance by potentiometer is shown. The galvanometer is connected at point A and zero deflection is observed at length $PJ = 30$ cm. In second case, the secondary cell is changed.

Take, $E_s = 10$ V and $r = 1\,\Omega$ in 1st reading and $E_s = 5$ V and $r = 2\,\Omega$ in 2nd reading.

In second case, the zero deflection is observed at length $PJ = 10$ cm. What is the resistance R (in Ω)?

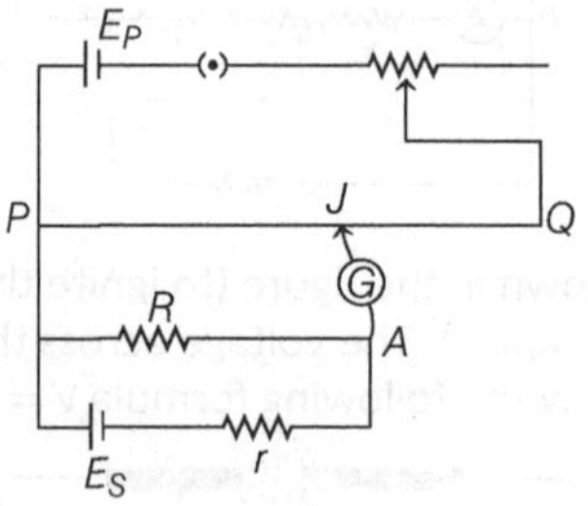

75 Figure shows a potentiometer connected to an external circuit. At an instant, either switch S_1 and S_3 is closed or S_2 and S_4 is closed. When switch S_1 and S_3 is closed, null point is attained at $J_1 (AJ_1 = l_1)$ and when S_2 and S_4 is closed, it is attained at $J_2 (BJ_2 = l_2)$. What is the sum of l_1 and l_2 in metre.

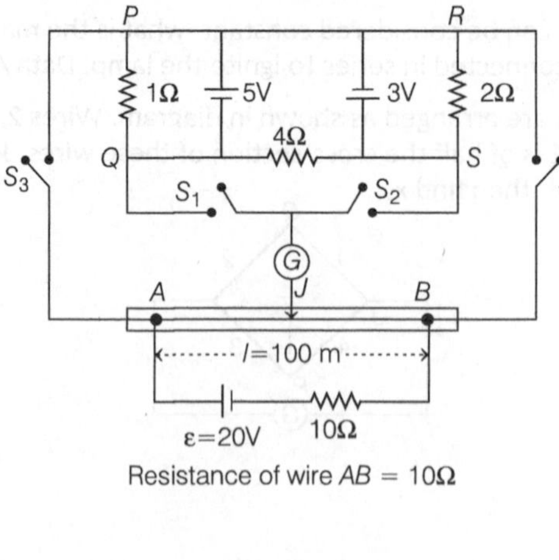

Resistance of wire AB = 10Ω

76 In the potentiometer circuit shown here, a potentiometer wire AB of length 100 cm and resistance 8 Ω is used. While measuring potential difference between points X and Y, balancing length AC is obtained to be equal to 25 cm. What is internal resistance r (in Ω) of the driving circuit?

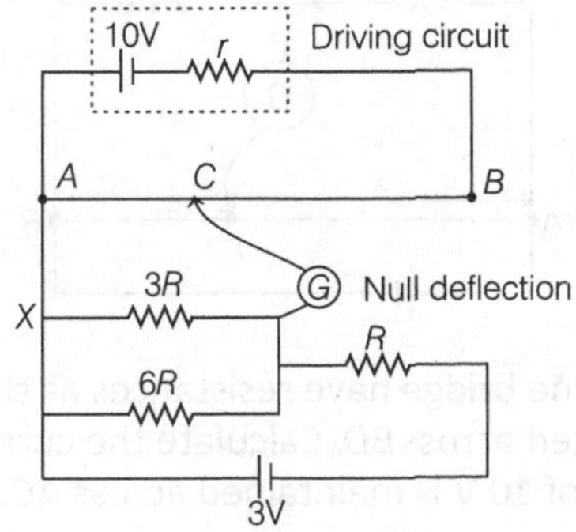

Answers

1. (d)	2. (d)	3. (a)	4. (b)	5. (b)	6. (a)	7. (b)	8. (a)	9. (b)	10. (d)
11. (d)	12. (d)	13. (c)	14. (a)	15. (a)	16. (b)	17. (c)	18. (c)	19. (a)	20. (a)
21. (c)	22. (a)	23. (c)	24. (d)	25. (c)	26. (abc)	27. (bd)	28. (abd)	29. (acd)	30. (ac)
31. (ac)	32. (acd)	33. (abcd)	34. (acd)	35. (abc)	36. (ab)	37. (ab)	38. (abc)	39. (abcd)	40. (b)
41. (c)	42. (a)	43. (b)	44. (c)	45. (b)	46. (c)	47. (b)	48. (b)	49. (c)	50. (d)
51. (d)	52. (b)	53. (b)	54. (a)	55. (b)	56. (c)	57. (d)			

58. a → r; b → s; c → p; d → q 59. a → q; b → r; c → p; d → s 60. a → p; b → p; c → q; d → qs
61. a → q; b → r; c → s; d → s 62. a → s; b → r; c → p; d → q 63. a → pt; b → qt; c → q; d → r

64. 17	65. 2.22	66. 6	67. 27
68. 3	69. 2	70. 45	
71. 50	72. 74	73. 35	
74. 1	75. 80	76. 2	

For Detailed Solutions
Scan the code

Previous Years' Questions

JEE Main

1 In the given potentiometer circuit arrangement, the balancing length AC is measured to be 250 cm. When the galvanometer connection is shifted from point (1) to point (2) in the given diagram, the balancing length becomes 400 cm. The ratio of the emf of two cells, $\frac{\varepsilon_1}{\varepsilon_2}$ is [JEE Main 2021]

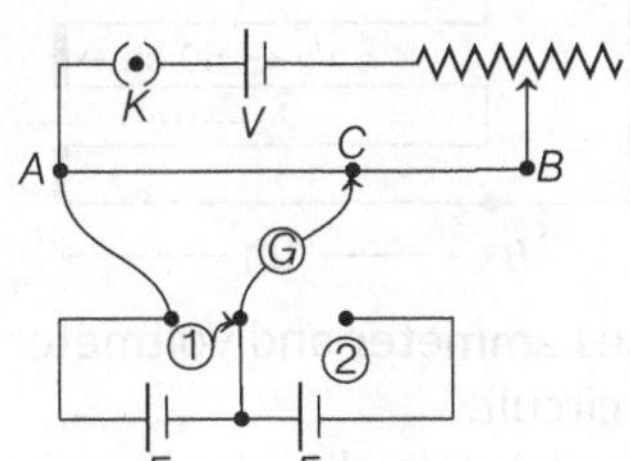

(a) 5/3 (b) 8/5
(c) 4/3 (d) 3/2

Problems in Physics

2 Consider a 72 cm long wire AB as shown in the figure. The galvanometer jockey is placed at P on AB at a distance x cm from A. The galvanometer shows zero deflection.

The value of x, to the nearest integer, is .... **[JEE Main 2021]**

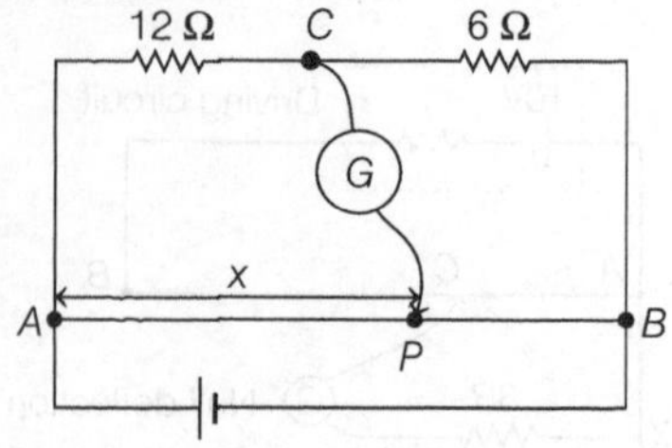

3 The four arms of a Wheatstone bridge have resistances as shown in the figure. A galvanometer of 15 Ω resistance is connected across BD. Calculate the current through the galvanometer when a potential difference of 10 V is maintained across AC. **[JEE Main 2021]**

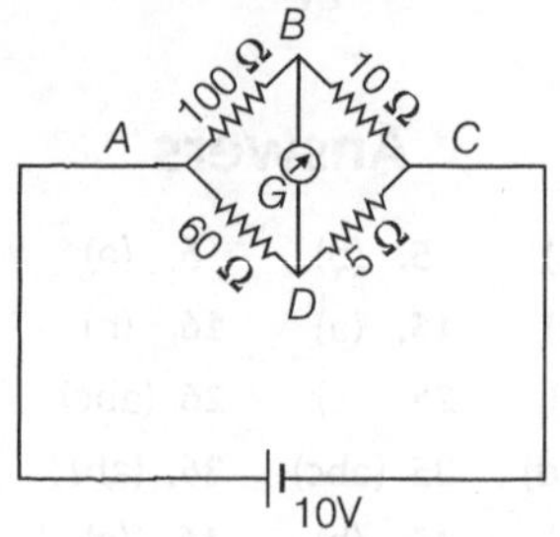

(a) 2.44 μA (b) 2.44 mA (c) 4.87 mA (d) 4.87 μA

4 In the given circuit of potentiometer, the potential difference E across AB (10 m length) is larger than E_1 and E_2 as well. For key K_1 (closed), the jockey is adjusted to touch the wire at point J_1, so that there is no deflection in the galvanometer. Now, the first battery (E_1) is replaced by second battery (E_2) for working by making K_1 open and K_2 closed. The galvanometer gives then null deflection at J_2. The value of $\frac{E_1}{E_2}$ is $\frac{a}{b}$, where a = ... **[JEE Main 2021]**

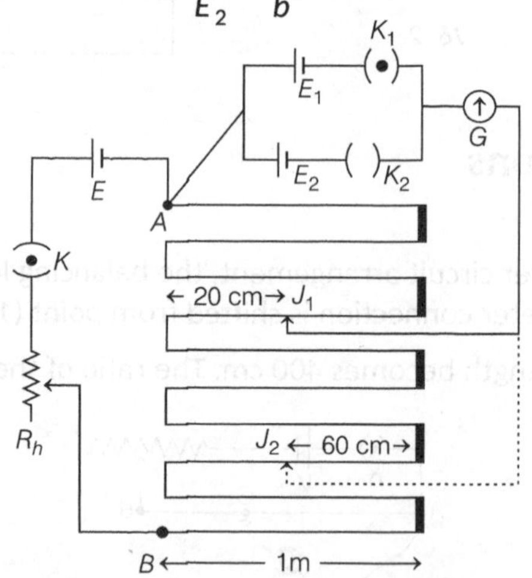

5 A circuit to verify Ohm's law uses ammeter and voltmeter in series or parallel connected correctly to the resistor. In the circuit, **[JEE Main 2020]**

(a) ammeter is always used in parallel and voltmeter in series
(b) Both ammeter and voltmeter must be connected in parallel
(c) ammeter is always connected in series and voltmeter in parallel
(d) Both ammeter and voltmeter must be connected in series

6 A potentiometer wire PQ of 1 m length is connected to a standard cell E_1. Another cell E_2 of emf 1.02 V is connected with a resistance r and switch S (as shown in figure). With switch S open, the null position is obtained at a distance of 49 cm from Q. The potential gradient in the potentiometer wire is **[JEE Main 2020]**

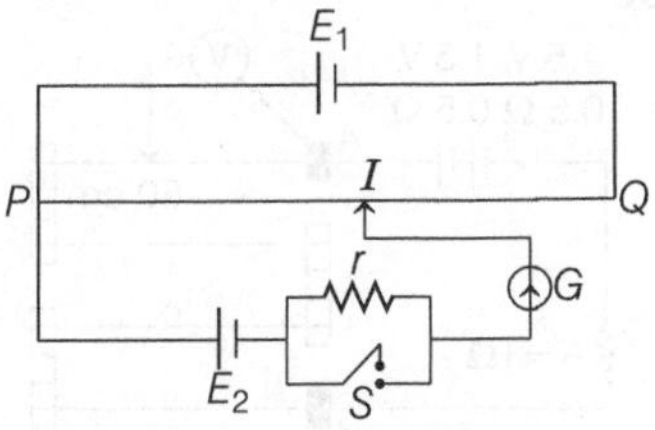

(a) 0.02 V/cm (b) 0.01 V/cm (c) 0.03 V/cm (d) 0.04 V/cm

7 In the given circuit, an ideal voltmeter connected across the 10 Ω resistance reads 2 V. The internal resistance r, of each cell is **[JEE Main 2019]**

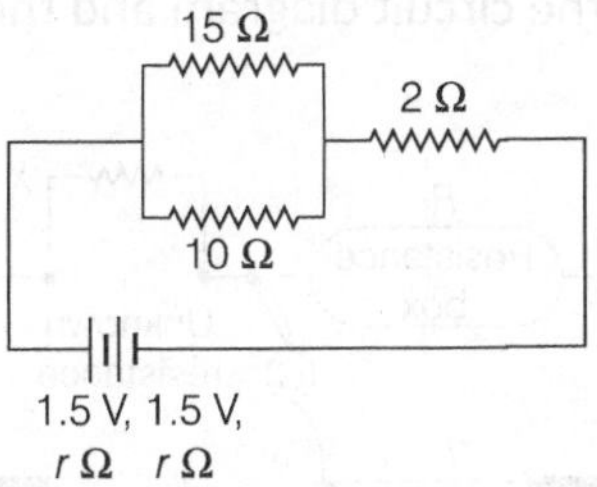

(a) 1.5 Ω (b) 0.5 Ω (c) 1 Ω (d) 0 Ω

8 To verify Ohm's law, a student connects the voltmeter across the battery as shown in the figure. The measured voltage is plotted as a function of the current and the following graph is obtained **[JEE Main 2019]**

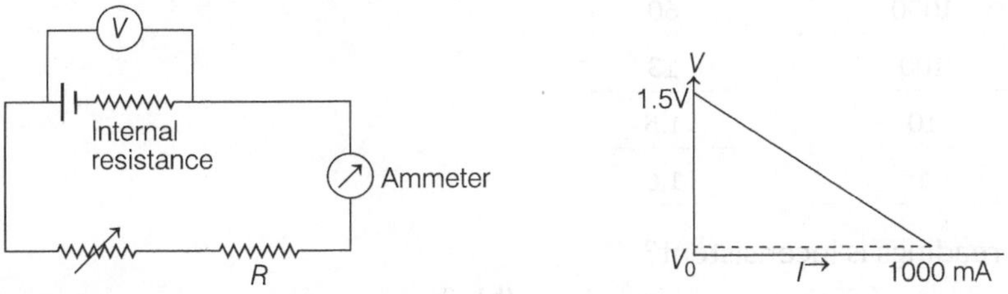

If V_0 is almost zero, then identify the correct statement.

(a) The emf of the battery is 1.5 V and its internal resistance is 1.5 Ω

(b) The value of the resistance R is 1.5 Ω

(c) The potential difference across the battery is 1.5 V when it sends a current of 1000 mA

(d) The emf of the battery is 1.5 V and the value of R is 1.5 Ω

9 In the given circuit, the internal resistance of the 18 V cell is negligible. If $R_1 = 400\ \Omega$, $R_3 = 100\ \Omega$ and $R_4 = 500\ \Omega$ and the reading of an ideal voltmeter across R_4 is 5 V, then the value of R_2 will be **[JEE Main 2019]**

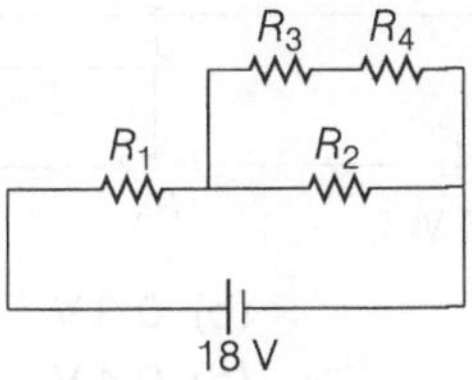

(a) 550 Ω (b) 230 Ω (c) 300 Ω (d) 450 Ω

10 In the circuit shown, a four-wire potentiometer is made of a 400 cm long wire, which extends between A and B. The resistance per unit length of the potentiometer wire is $r = 0.01\ \Omega$/cm. If an ideal voltmeter is connected as shown with jockey J at 50 cm from end A, the expected reading of the voltmeter will be **[JEE Main 2019]**

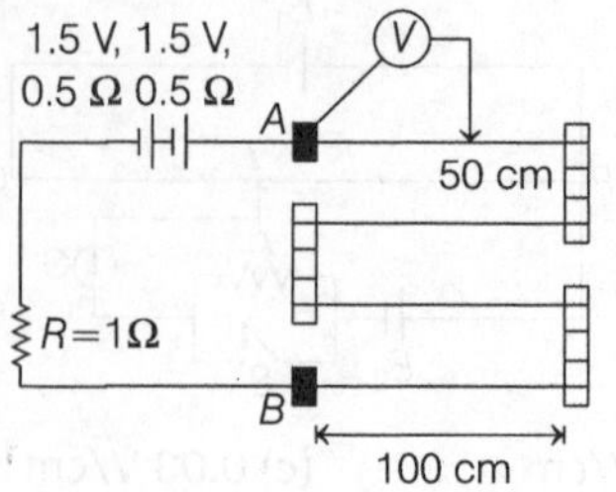

(a) 0.20 V (b) 0.75 V
(c) 0.25 V (d) 0.50 V

11 In a meter bridge experiment, the circuit diagram and the corresponding observation table are shown in figure **[JEE Main 2019]**

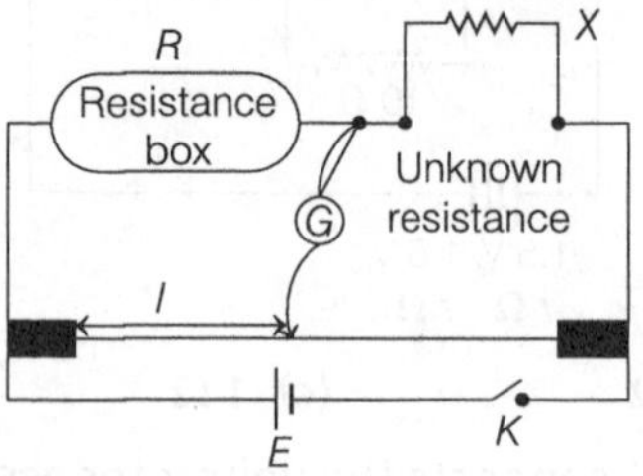

S. No.	R (Ω)	l (cm)
1.	1000	60
2.	100	13
3.	10	1.5
4.	1	1.0

Which of the readings is inconsistent?

(a) 3 (b) 2
(c) 1 (d) 4

12 The resistance of the meter bridge AB in given figure is $4\ \Omega$. With a cell of emf $\varepsilon = 0.5$ V and rheostat resistance $R_h = 2\ \Omega$. The null point is obtained at some point J. When the cell is replaced by another one of emf $\varepsilon = \varepsilon_2$, the same null point J is found for $R_h = 6\ \Omega$. The emf ε_2 is **[JEE Main 2019]**

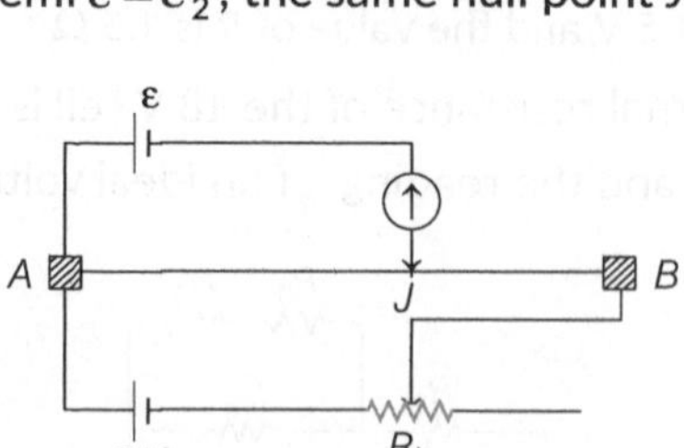

(a) 0.6 V (b) 0.3 V
(c) 0.5 V (d) 0.4 V

13 In the experimental set up of meter bridge shown in the figure, the null point is obtained at a distance of 40 cm from A. If a 10 Ω resistor is connected in series with R_1, the null point shifts by 10 cm.
The resistance that should be connected in parallel with $(R_1 + 10)$ Ω such that the null point shifts back to its initial position is **[JEE Main 2019]**

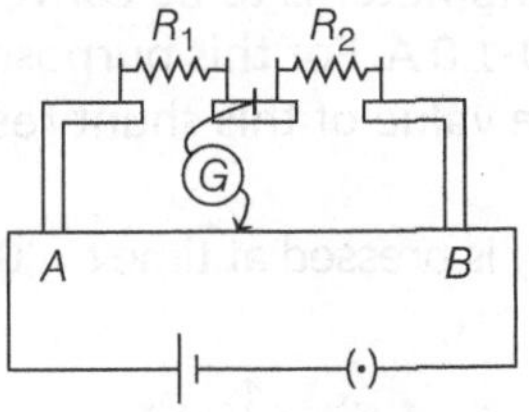

(a) 60 Ω (b) 20 Ω
(c) 30 Ω (d) 40 Ω

14 An ideal battery of 4 V and resistance *R* are connected in series in the primary circuit of a potentiometer of length 1 m and resistance 5 Ω. The value of *R* to give a potential difference of 5 mV across 10 cm of potentiometer wire is **[JEE Main 2019]**
(a) 395 Ω (b) 495 Ω (c) 490 Ω (d) 480 Ω

15 A potentiometer wire *AB* having length *L* and resistance $12r$ is joined to a cell *D* of emf ε and internal resistance *r*. A cell *C* having emf $\frac{\varepsilon}{2}$ and internal resistance $3r$ is connected. The length *AJ* at which the galvanometer as shown in figure shows no deflection is **[JEE Main 2019]**

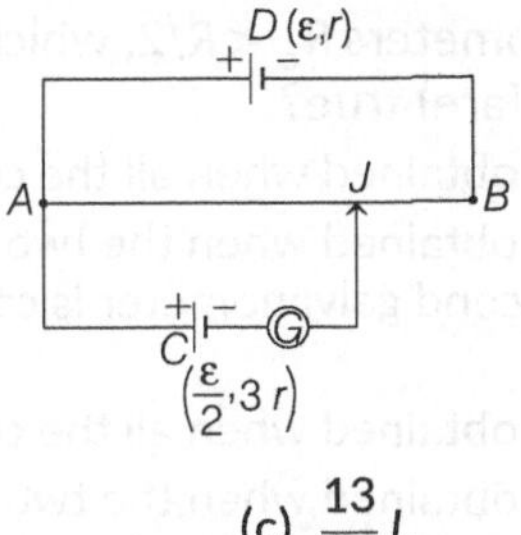

(a) $\frac{5}{12}L$ (b) $\frac{11}{12}L$ (c) $\frac{13}{24}L$ (d) $\frac{11}{24}L$

16 In a potentiometer experiment, it is found that no current passes through the galvanometer when the terminals of the cell are connected across 52 cm of the potentiometer wire. If the cell is shunted by a resistance of 5 Ω, a balance is found when the cell is connected across 40 cm of the wire. Find the internal resistance of the cell. **[JEE Main 2018]**
(a) 1 Ω (b) 1.5 Ω (c) 2 Ω (d) 2.5 Ω

17 On interchanging the resistances, the balance point of a meter bridge shifts to the left by 10 cm. The resistance of their series combination is 1 kΩ. How much was the resistance on the left slot before interchanging the resistances? **[JEE Main 2018]**
(a) 990 Ω (b) 505 Ω (c) 550 Ω (d) 910 Ω

18 Which of the following statement is incorrect? **[JEE Main 2017]**
(a) In a balanced Wheatstone bridge, if the cell and the galvanometer are exchanged, the null point is disturbed.
(b) A rheostat can be used as a potential divider.
(c) Kirchhoff's second law represents energy conservation.
(d) Wheatstone bridge is the most sensitive when all the four resistances are of the same order of magnitude.

JEE Advanced

19 A moving coil galvanometer has 50 turns and each turn has an area 2×10^{-4} m^2. The magnetic field produced by the magnet inside the galvanometer is 0.02 T. The torsional constant of the suspension wire is 10^{-4} N-m rad^{-1}. When a current flows through the galvanometer, a full scale deflection occurs, if the coil rotates by 0.2 rad. The resistance of the coil of the galvanometer is 50 Ω. This galvanometer is to be converted into an ammeter capable of measuring current in the range 0-1.0 A. For this purpose, a shunt resistance is to be added in parallel to the galvanometer. The value of this shunt resistance in ohms, is .... **[JEE Advanced 2019]**

20 In the circuit shown below, the key is pressed at time $t = 0$. Which of the following statement(s) is (are) true? **[JEE Advanced 2016]**

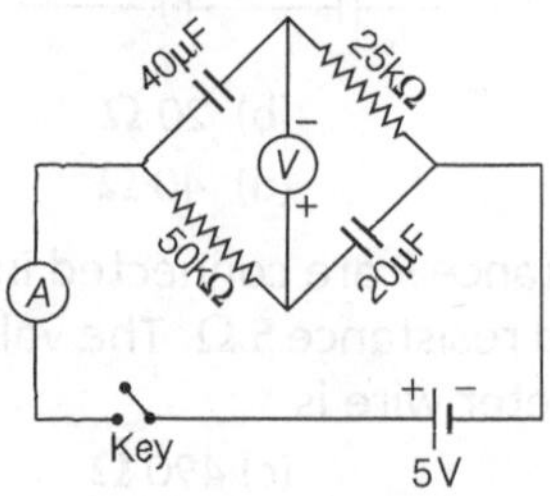

(a) The voltmeter display -5 V as soon as the key is pressed and displays $+5$ V after a long time.
(b) The voltmeter will display 0 V at time $t = \ln 2$ s.
(c) The current in the ammeter becomes $1/e$ of the initial value after 1 s.
(d) The current in the ammeter becomes zero after a long time.

21 Consider two identical galvanometers and two identical resistors with resistance R. If the internal resistance of the galvanometers $R_c < R/2$, which of the following statement(s) about any one of the galvanometers is (are) true? **[JEE Advanced 2016]**
(a) The maximum voltage range is obtained when all the components are connected in series
(b) The maximum voltage range is obtained when the two resistors and one galvanometer are connected in series, and the second galvanometer is connected in parallel to the first galvanometer
(c) The maximum current range is obtained when all the components are connected in parallel
(d) The maximum current range is obtained when the two galvanometers are connected in series, and the combination is connected in parallel with both the resistors

22 During an experiment with a meter bridge, the galvanometer shows a null point when the jockey is pressed at 40.0 cm using a standard resistance of 90 Ω, as shown in the scale used in the meter bridge is 1 mm. The unknown resistance is **[JEE Advanced 2014]**

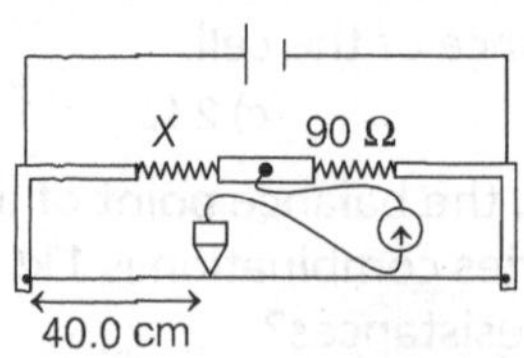

(a) (60 ± 0.15) Ω (b) (135 ± 0.56) Ω
(c) (60 ± 0.25) Ω (d) (135 ± 0.23) Ω

23 A galvanometer gives full scale deflection with 0.006 A current. By connecting it to a 4990 Ω resistance, it can be converted into a voltmeter of range 0-30 V. If connected to a $\frac{2n}{249}$ Ω resistance, it becomes an ammeter of range 0-1.5 A. The value of n is .... **[JEE Advanced 2014]**

24 A meter bridge is set-up as shown in figure, to determine an unknown resistance X using a standard 10 Ω resistor. The galvanometer shows null point when tapping-key is at 52 cm mark. The end-corrections are 1 cm and 2 cm respectively for the ends A and B. The determined value of X is **[JEE Advanced 2011]**

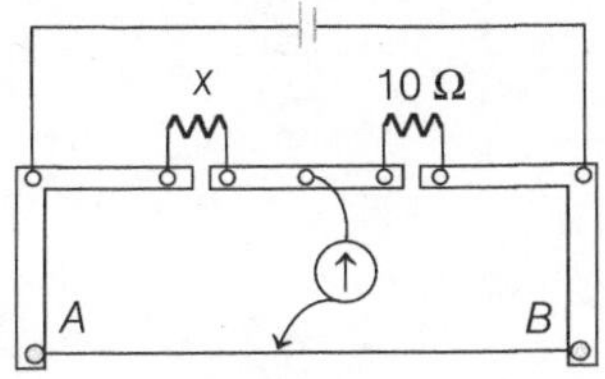

(a) 10.2 Ω (b) 10.6 Ω (c) 10.8 Ω (d) 11.1 Ω

25 To verify Ohm's law, a student is provided with a test resistor R_T, a high resistance R_1, a small resistance R_2, two identical galvanometers G_1 and G_2, and a variable voltage source V. The correct circuit to carry out the experiment is **[JEE Advanced 2010]**

(a)
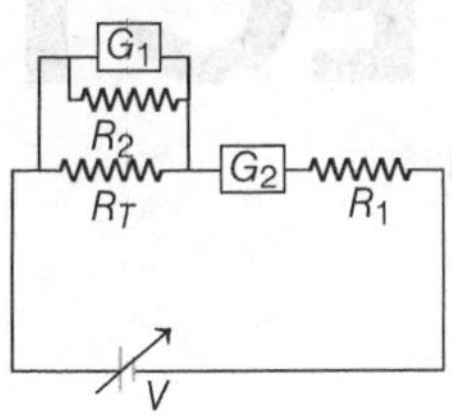

(b)
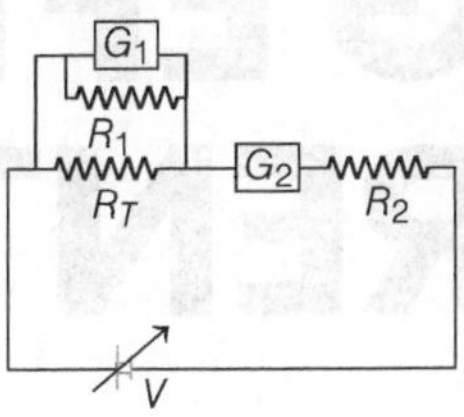

(c)
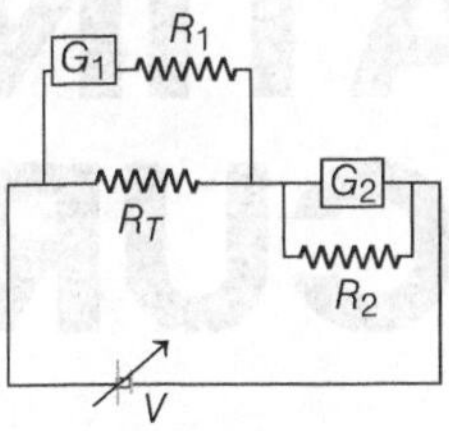

(d)
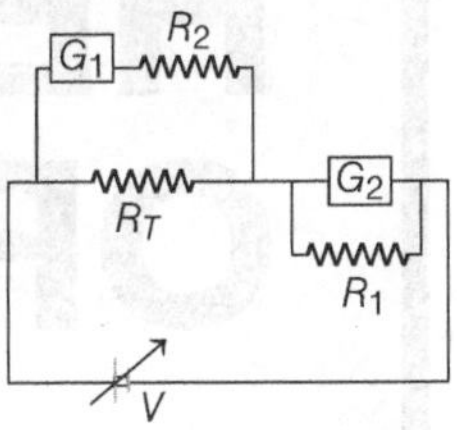

Answers

1. (a)	**2.** 48	**3.** (c)	**4.** 1	**5.** (c)	**6.** (a)	**7.** (b)	**8.** (a)	**9.** (c)
10. (c)	**11.** (d)	**12.** (b)	**13.** (a)	**14.** (a)	**15.** (c)	**16.** (b)	**17.** (c)	**18.** (a)
19.5.55	**20.** (abcd)	**21.** (a,c)	**22.** (c)					
23. 5	**24.** (b)	**25.** (c)						

For Detailed Solutions
Scan the code

CHAPTER

08

HEATING EFFECT OF CURRENT

LEVEL 01 \ Starter Level

1.1 Heating Effect of Current

1 Two wires of same mass, having ratio of the lengths 1 : 2, densities 1 : 3 and resistivities 2: 1. They are connected one by one to the same voltage supply. The rate of heat dissipation in the first wire is found to be 10 W. Find the rate of heat dissipation in the second wire.

2 In the following circuit, 5Ω resistor develops 45 J/s due to current flowing through it. Find the power developed across 12 Ω resistor.

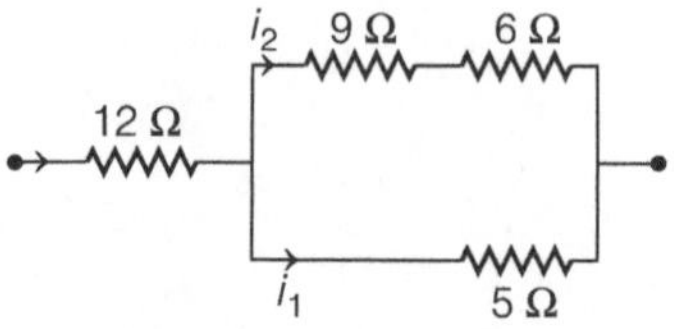

3 If bulb rating is 100 W and 220 V, then determine

(i) resistance of filament,

(ii) current through filament, if it is operated at a potential difference for which it is designed.

(iii) If bulb operates at 110 V power supply, then find power consumed by bulb.

4 A 500 W heater is designed to operate from a 200 V line. By what percentage will its heat output drop, if the line voltage drops to 160 V? Also, find the heat produced by it in 10 min.

5 How much time heater will take to increase the temperature of 100 g water by 50°C, if resistance of heating coil is 400Ω and supply voltage is 200 V?

6 An electric kettle taking 3 A at 210 V brings 1 L of water from 20°C to 80°C in 10 min. Find its efficiency.

7 A fuse of lead wire has an area of cross-section 0.2 mm^2. On short-circuiting, the current in the fuse wire reaches 30 A. How long after the short-circuiting, will the fuse begin to melt? For lead, take its specific heat 0.032 cal/g-°C, melting point 327°C, density 11.34g/cm^3 and the resistivity $= 22 \times 10^{-6}$ Ω-cm. The initial temperature of wire is 20°C. Neglect heat losses.

8 A storage battery with emf $E = 3.6$ V loaded with an external resistance produces a current $I = 2.0$ A. In this case, the potential difference between the terminals of the storage battery equals $V = 2.0$ V. Find the thermal power generated in the battery and the power supplied by the battery.

9 A series battery by 6 lead accumulators each of emf 2.0 V and internal resistance 0.5Ω is charged by a 100 V DC supply. What series resistance should be used in the charging circuit in order to limit the current to 8 A? Using the required resistor, calculate

(i) the power supplied by the DC source

(ii) the power dissipated as heat and

(iii) chemical energy stored in the battery in 15 min.

1.2 Series and Parallel Combinations of the Heating Elements

10 An electric heater and an electric bulb are rated 500 W, 220 V and 100 W, 220 V, respectively. Both are connected in series to a 220 V supply. Calculate power consumed by (i) heater (ii) bulb.

11 In the figure shown B_1, B_2 and B_3 are three bulbs rated as (200 V, 100 W), (200 V, 50W) and (200 V, 25 W), respectively.

(i) Which bulb is the brightest and also find power delivered by it?

(ii) What is the total power delivered by all the three bulbs?

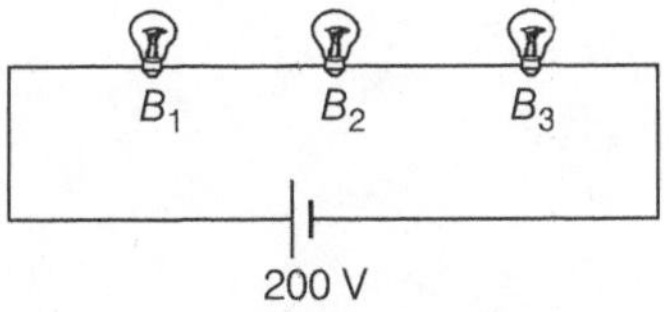

12 Two bulbs are rated 30 W-200 V and 60 W-200V. They are connected with a 400 V power supply. Find which bulb is likely to get fused, if they are connected in (i) series and (ii) parallel.

13 A 100 W bulb B_1 and two 60 W bulbs B_2 and B_3 are connected to a 250 V source as shown in the figure. Now, W_1, W_2 and W_3 are the output powers of the bulbs B_1, B_2 and B_3, respectively. Arrange W_1, W_2 and W_3 in the increasing or decreasing order.

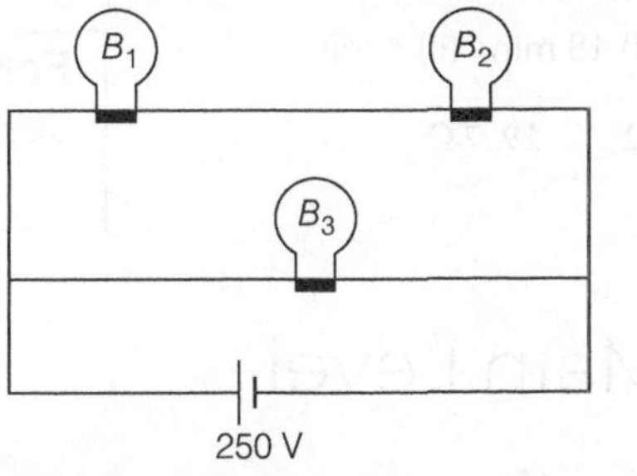

14 Four identical bulbs each rated 100 W, 220 V are connected across a battery as shown. What is total electric power consumed by the bulbs?

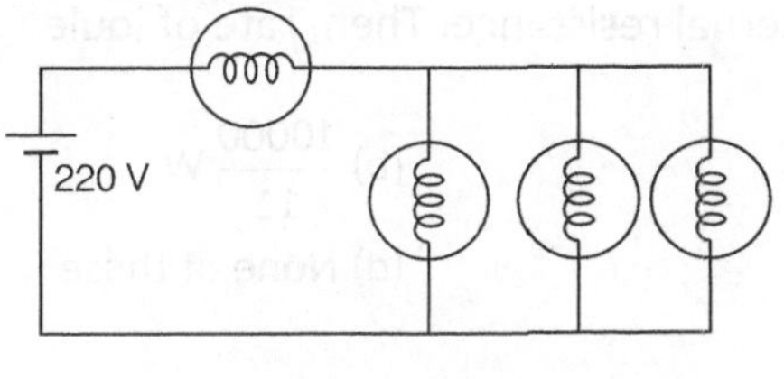

15 An electric tea kettle has two heating coils. When one of the coils is switched on, boiling begins in 12 min. When the other coil is switched on, the boiling begins in 6 min. In what time, will the boiling begin, if both coils are switched on simultaneously

(i) in series (ii) in parallel

16 A 100 W bulb designed to operate on 200 V is to be connected across a 500 V source. What resistance should be put in series, so that the bulb consumes 100 W?

1.3 Maximum Power Transfer Theorem

17 In the circuit as shown in figure, find value of external resistor R, so that the power dissipated in it is maximum. Also, find the maximum power dissipated in it.

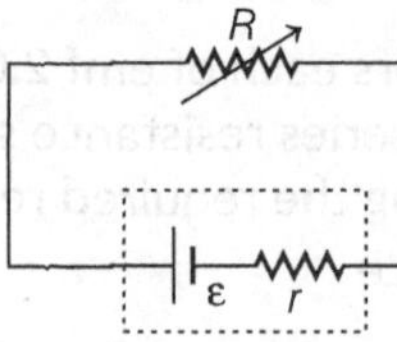

18 A battery has an open circuit potential difference of 36 V between its terminals. When a load resistance of 10Ω is connected across the battery, the power dissipated within the battery is 18 W. What should be the load resistance R, so that maximum power will be dissipated in R.

19 In the figure shown find R, so that the power generated in R is maximum.

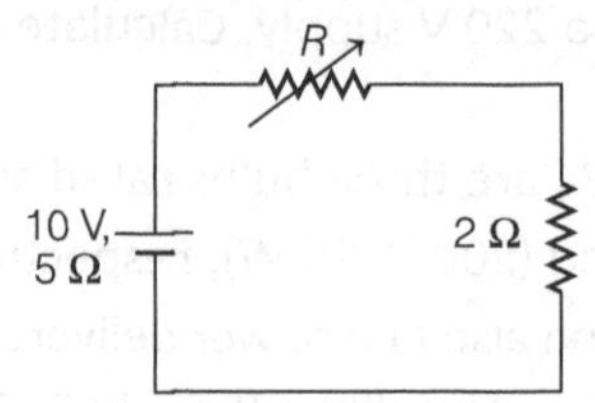

Answers

1. $\frac{5}{3}$ W 2. 192 W 3. (i) $R = 484\,\Omega$ (ii) $\frac{5}{11}$ A (iii) 25 W 4. 36%, 192 kJ 5. 210 s

6. 66.67% 7. 0.095 s 8. 3.2 W, 4W 9. 8Ω (i) 800W (ii) 704W (iii) 86400 J

10. (i) 13.88 W (ii) 69.44 W 11. (i) 8.16 W (ii) 14.28 W 12. (i) 30 W bulb (ii) both the bulbs

13. $W_3 > W_2 > W_1$ 14. 75 W 15. (i) 18 min (ii) 4 min

16. 600Ω 17. $r, \frac{\varepsilon^2}{4r}$ 18. 2Ω 19. 7Ω

For Detailed Solutions
Scan the code

LEVEL 02 \ JEE Main Level

Single Choice Correct Questions

1 In the figure shown each resistor is of 20 Ω and the cell has emf 10 volt with negligible internal resistance. Then, rate of joule heating in the circuit is

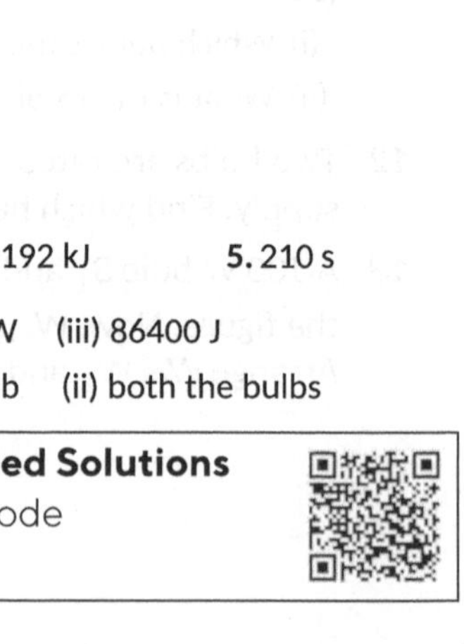

(a) $\frac{100}{11}$ W (b) $\frac{10000}{11}$ W

(c) 11 W (d) None of these

2 Find the resistor in which the dissipated power is the greatest.

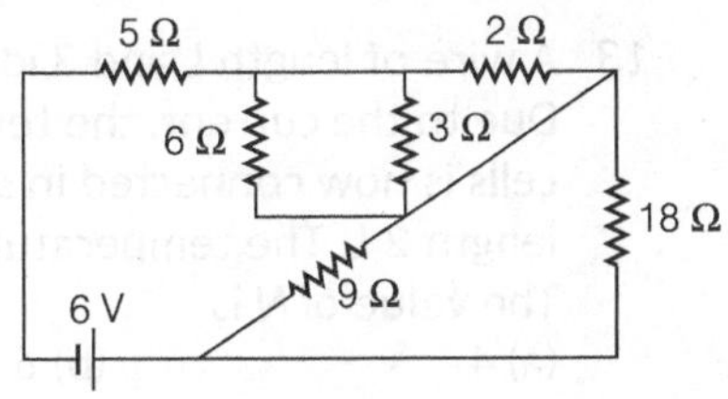

(a) 6Ω
(b) 9Ω
(c) 18Ω
(d) 5Ω

3 The ratio of powers dissipated respectively in R and $3R$, as shown is

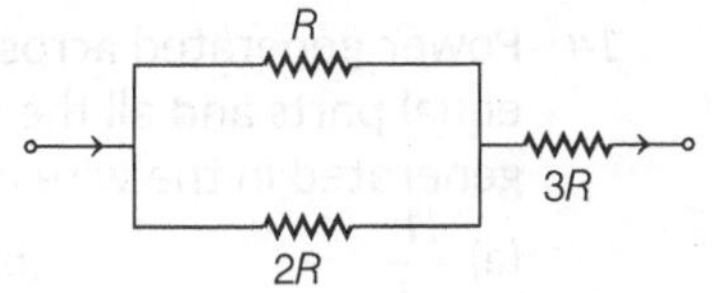

(a) 9
(b) $\frac{27}{4}$
(c) $\frac{4}{9}$
(d) $\frac{4}{27}$

4 In the circuit shown in figure, the heat produced in the 5Ω resistor due to the current flowing through it is 10 calories per second. The heat generated in the 4Ω resistor is

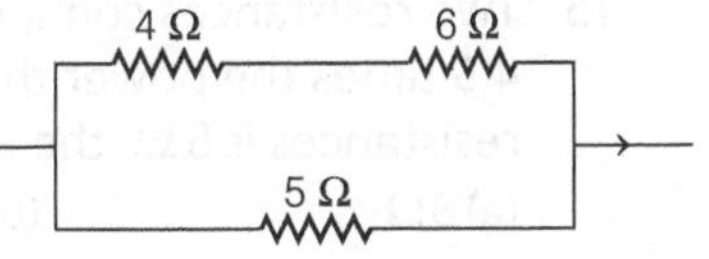

(a) 1 cal/s
(b) 2 cal/s
(c) 3 cal/s
(d) 4 cal/s

5 The charge flowing through a resistance R varies with time as $Q = 2t - 8t^2$. The total heat produced in the resistance is $\left(\text{for } 0 \le t \le \frac{1}{8}\right)$

(a) $\frac{R}{6}$ J
(b) $\frac{R}{3}$ J
(c) $\frac{R}{2}$ J
(d) R J

6 What amount of heat will be generated in a coil of resistance R due to a charge q passing through it, if the current in the coil decreases to zero uniformly during a time interval Δt?

(a) $\frac{4}{3}\frac{q^2R}{\Delta t}$
(b) $\ln\frac{q^2R}{2\Delta t}$
(c) $\frac{2q^2R}{3\Delta t}$
(d) $\ln\frac{(2\Delta t)}{q^2R}$

7 Masses of three wires of same material are in the ratio 1 : 3 : 5. Their lengths are in the ratio 5 : 3 : 1. When they are connected in series to an external source, the amount of heat produced in them are in the ratio

(a) 125 : 15 :1
(b) 1 : 15 : 125
(c) 5 : 3 : 1
(d) 1 : 3 : 5

8 Rate of dissipation of Joule's heat in resistance per unit volume is (symbols have their usual meanings)

(a) σE
(b) σJ
(c) JE
(d) None of these

9 The resistance of each arm in the circuit shown in figure has same value (i.e. $R_{12} = R_{10} = R_{13} = R_{20} = R_{24} = \ldots\ldots$ etc.). The ratio $\frac{Q_{12}}{Q_{34}}$ of the amounts of heat liberated per unit time in conductors 1-2 and 3-4 is

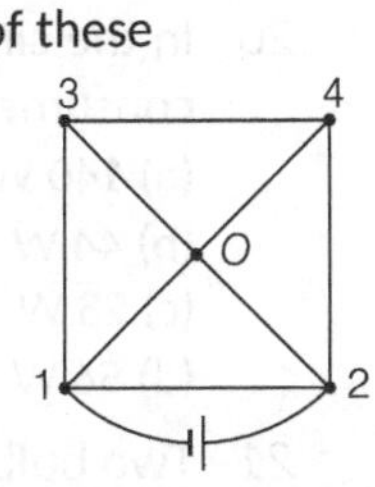

(a) 4 : 1
(b) 8 : 1
(c) 16 : 1
(d) None of these

10 A resistance coil of 60Ω is immersed in 42 kg of water. A current of 7A is passed through it. The rise in temperature of water per minute is

(a) 4°C
(b) 8°C
(c) 1.0°C
(d) 12°C

11 A coil of wire of resistance 50Ω is embedded in a block of ice and a potential difference of 210 V is applied across it. The amount of ice which melts in 1 s is

(a) 0.262 g
(b) 2.62 g
(c) 26.2 g
(d) 0.0262 g

12 If the length of the filament of a heater is reduced by 10%, the power of the heater will

(a) increase by about 9%
(b) increase by about 11%
(c) increase by about 19%
(d) decrease by about 10%

13 A wire of length L and 3 identical cells of negligible internal resistances are connected in series. Due to the current, the temperature of the wire is raised by ΔT in time t. N number of similar cells is now connected in series with a wire of the same material and cross-section but of length $2L$. The temperature of the wire is raised by the same amount ΔT in the same time t. The value of N is

(a) 4 (b) 6 (c) 8 (d) 9

14 Power generated across a uniform wire connected across a supply is H. If the wire is cut into n equal parts and all the parts are connected in parallel across the same supply, the total power generated in the wire is

(a) $\frac{H}{n^2}$ (b) n^2H (c) nH (d) $\frac{H}{n}$

15 Two resistances connected in parallel across a cell of negligible internal resistance, use 4.5 times the power that they would use when in series across the same cell. If one of the resistances is $8\,\Omega$, the other is

(a) 8Ω (b) 12Ω (c) 16Ω (d) 36Ω

16 A battery is supplying power to a tape-recorder by cable of resistance of $0.02\ \Omega$. If the battery is generating 50 W power at 5 V, then power received by tape-recorder is (neglect internal resistance of battery)

(a) 50 W (b) 45 W (c) 30 W (d) 48 W

17 When electric bulb of same power, but different marked voltage are connected in series across the power line, their brightness will be

(a) proportional to their marked voltage
(b) inversely proportional to their marked voltage
(c) proportional to the square of their marked voltage
(d) inversely proportional to the square of their marked voltage

18 Two bulbs rated (25 W - 220 V) and (100 W - 220 V) are connected in series to a 440 V line. Which one is likely to fuse?

(a) 25 W bulb (b) 100 W bulb (c) Both bulbs (d) None of these

19 The wattage rating of a light bulb indicates the power dissipated by the bulb, if it is connected across 110 V DC potential difference. If 50 W and 100 W bulbs are connected in series to a 110 V DC source, how much power will be dissipated in the 50 W bulb?

(a) 50 W (b) 100 W (c) 22 W (d) 11 W

20 In the circuit shown, the battery is ideal. What is the total power consumed by the circuit approximately?

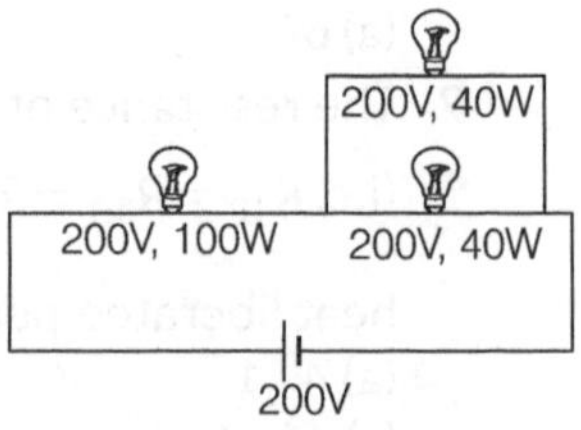

(a) 140 W
(b) 44 W
(c) 28 W
(d) 56 W

21 Two bulbs of 500 W and 200 W are manufactured to operate on 220 V line. The ratio of heats produced in 500 W and 200 W in two cases when first they are joined in series and secondly in parallel, will be

(a) $\frac{5}{2}, \frac{2}{5}$ (b) $\frac{5}{2}, \frac{5}{2}$ (c) $\frac{2}{5}, \frac{5}{2}$ (d) $\frac{2}{5}, \frac{2}{5}$

22 The circuit given is made up using identical light bulbs. The light bulbs of maximum brightness of the following will be

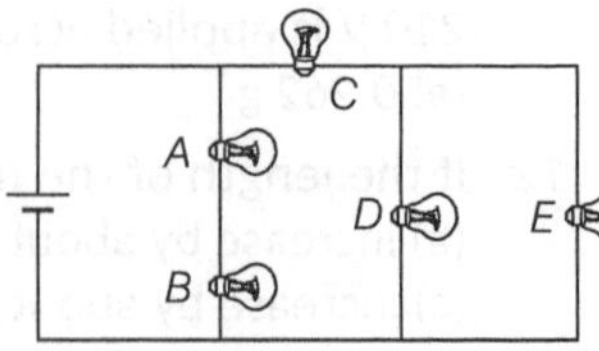

(a) bulb A (b) bulb C
(c) bulb D (d) bulb E

23 In the circuit given, ammeter A reads 0.5 A. Bulbs L_1 and L_2 are brightly lit, but L_3 is not lit. What is the reason for L_3 not being lit?

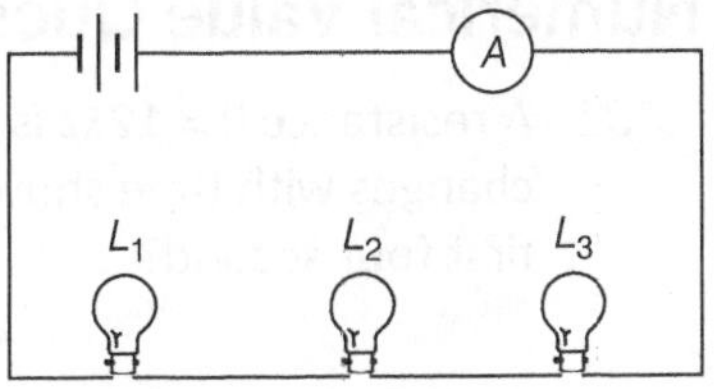

(a) The ammeter is faulty
(b) The filament of L_3 is broken
(c) The resistance of L_3 is much lower than that of L_1 and L_2
(d) There is a break in the connecting wire between L_2 and L_3

24 If X, Y and Z in the figure are identical lamps, which of the following changes to the brightness of the lamps occur when switch S is closed?

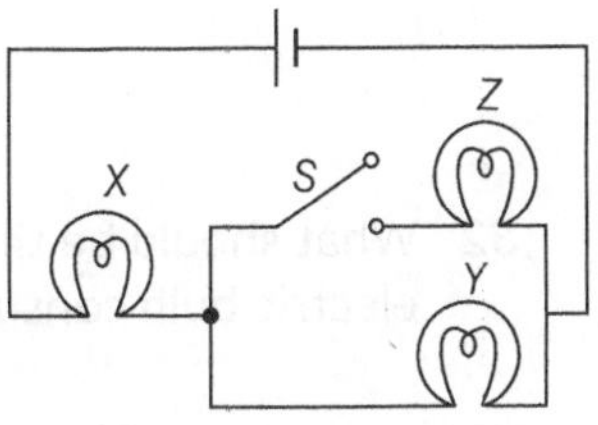

(a) X stays the same, Y decreases
(b) X increases, Y decreases
(c) X increases, Y stays the same
(d) X decreases, Y increases

25 The electrical installation in a building is protected by a 15 A fuse. If the supply voltage is 220 V, the maximum number of 60W bulbs that can be used in the building is

(a) 44 (b) 55
(c) 88 (d) 66

26 An electric kettle has two coils. When one of these is switched on, the water in the kettle boils in 6 min. When the other coil is switched on, the water boils in 3 min. If the two coils are connected in parallel, the time taken to boil the water in the kettle is

(a) 2 min (b) 3 min
(c) 6 min (d) 9 min

27 40 electric bulbs are connected in series across a 220 V supply. After one bulb is fused, the remaining 39 are connected again in series across the same supply. Illumination of the bulbs will

(a) decrease by 6.25% (b) increase by 6.25%
(c) decrease by 2.5% (d) increase by 2.5%

28 In the figure shown the thermal power generated in Y is maximum when $Y = 4\Omega$. Then, X is

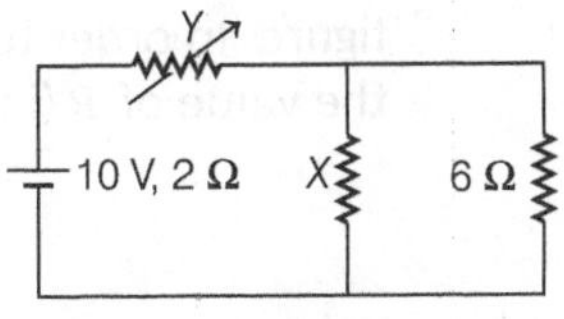

(a) 2Ω (b) 3Ω
(c) 1Ω (d) 6Ω

29 In a circuit shown in figure, resistances R_1 and R_2 are known, as well as emf's E_1 and E_2. The internal resistances of the sources are negligible. At what value of the resistance R will the thermal power generated in it be the highest?

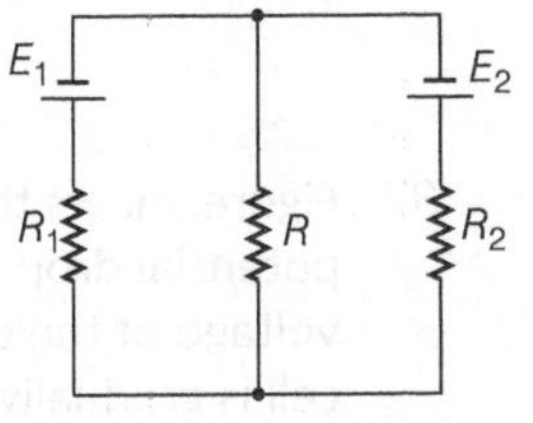

(a) $R_1 + R_2$ (b) $R_1 - R_2$
(c) $\sqrt{R_1 R_2}$ (d) $\dfrac{R_1 R_2}{R_1 + R_2}$

30 For maximum power from battery, the internal resistance of battery r is

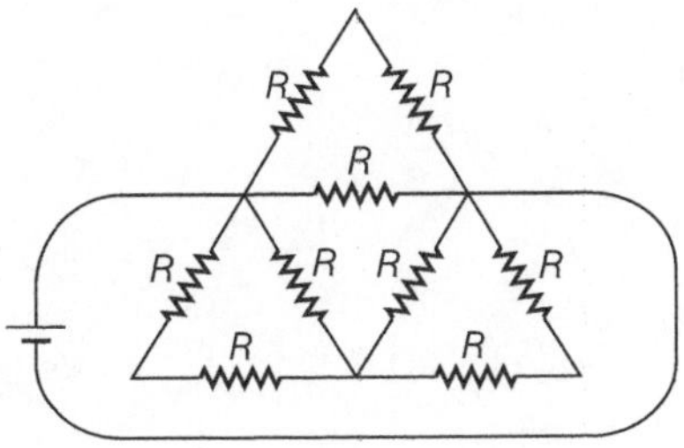

(a) $10R$ (b) $\dfrac{4R}{9}$

(c) $\dfrac{R}{8}$ (d) $\dfrac{10R}{9}$

Numerical Value Questions

31 A resistance $R = 12\,\Omega$ is connected across a source of varying emf as shown in figure. Its emf changes with time shown in the graph. What is the heat dissipated (in J) in the resistance in the first four second?

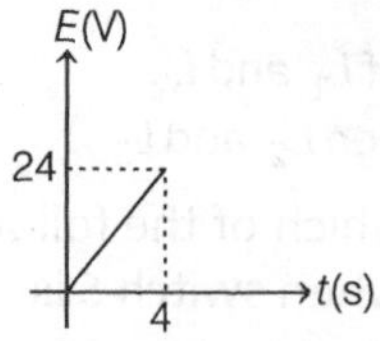

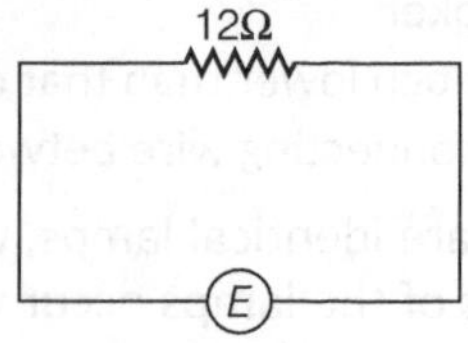

32 What should be the value of resistance R (in Ω) in the circuit shown in figure, so that the electric bulb consumes the rated power?

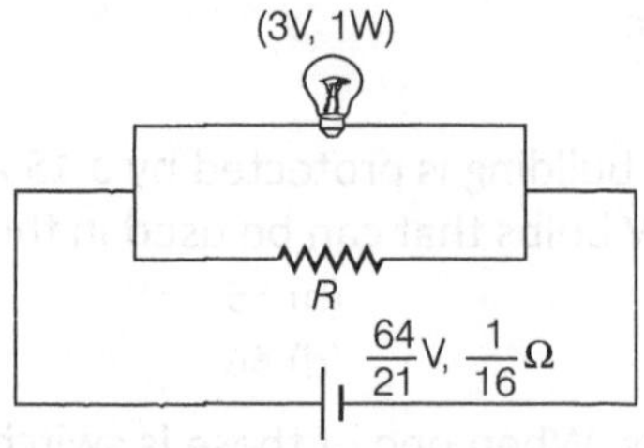

33 Two electric bulbs 40 W, 200 V and 100 W, 200 V are connected in series. Find the maximum voltage (in volts) that can be applied across the combination, without fusing either bulb.

34 Electric bulb 50W-100 V glowing at full power are to be used in parallel with battery 120 V, $10\,\Omega$. What is the maximum number of bulbs that can be connected, so that they glow in full power?

35 A fuse with a circular cross-sectional radius of 0.15 mm blows at 15 A. If the radius of a fuse, made of the same material is 0.60 mm, then find the current (in A) at which it blows.

36 A battery of internal resistance $6\,\Omega$ is connected to the network of resistance as shown in figure. In order to that the maximum power can be delivered to the network, what should be the value of R (in Ω)?

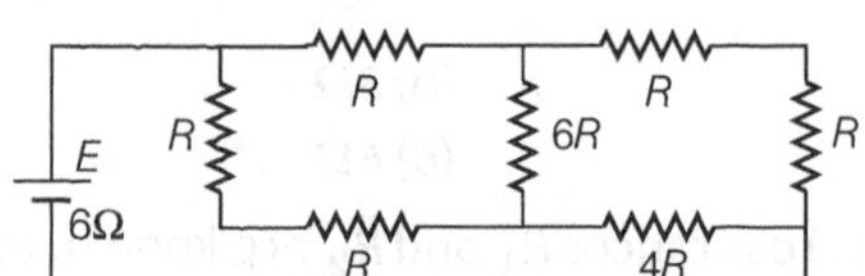

37 Figure shows three bulbs connected in series. Their voltage rating shows that they will fuse, if potential drop across them is greater than 100 V. Voltage of the cell can be varied. In each step voltage of the cell increases by 20 V. Initially, the voltage of the cell is 100 V. If voltage of the cell is gradually increased and it is found that after Nth step current in the circuit becomes zero. Find the value of N.

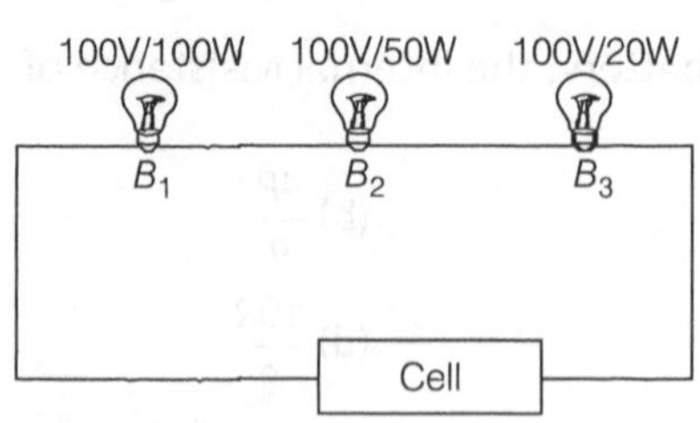

38 Circuit shows seven identical bulbs (A to G) connected through a battery of emf 200 V. The bulbs are rated at 200 V, 100 W. Find the power (in W) dissipated in the bulb which glow brightest.

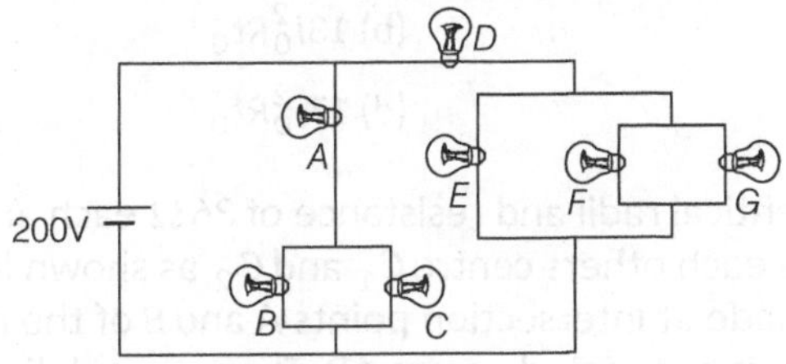

39 In a house, there are 3 lamps of 40 W each, 8 lamps of 60 W each, a radio of 40 W and a TV of 160 W. The lamps are in operation on an average for 2 h a day, the radio for 4 h a day and the TV for 1 hour a day. On Sundays, an electric iron of 750 W is used for 1 hour and the TV for an extra 3 h. Calculate the electricity bill (in ₹) for the month of february of a leap year at the rate of ₹ 5 per unit. The first Sunday falls on 3rd February.

40 The coil of a calorimeter C has a resistance of $R_1 = 60\,\Omega$. The coil R_1 is connected to the circuit as shown in figure. How many degrees (°C) will 240 g of water poured into the calorimeter be heated in 7 min during which a current flows through the coil, if the ammeter shows 3 A? The resistance of R_2 is $30\,\Omega$. Disregard the resistances of the battery and the ammeter; and the heat losses and heat capacity of the calorimeter and the resistor.

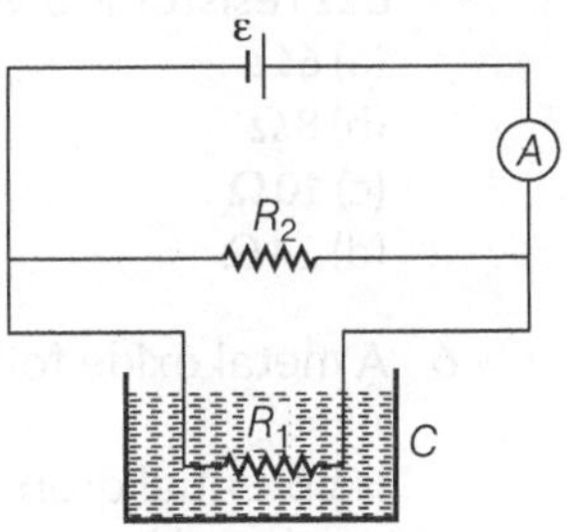

Answers

1. (c)	**2.** (d)	**3.** (d)	**4.** (b)	**5.** (a)	**6.** (a)	**7.** (a)	**8.** (c)	**9.** (c)	**10.** (c)
11. (b)	**12.** (b)	**13.** (b)	**14.** (b)	**15.** (c)	**16.** (d)	**17.** (c)	**18.** (a)	**19.** (c)	**20.** (b)
21. (c)	**22.** (b)	**23.** (c)	**24.** (b)	**25.** (b)					
26. (a)	**27.** (d)	**28.** (b)	**29.** (d)	**30.** (b)					
31. 64	**32.** 7	**33.** 280	**34.** 4	**35.** 120					
36. 3	**37.** 3	**38.** 56.25	**39.** 245	**40.** 25					

For Detailed Solutions
Scan the code

LEVEL 03 \ JEE Advanced Level

Single Choice Correct Questions

1 The charge supplied by source varies with time t as $Q = at - bt^2$. The total heat produced in the resistor $2R$ is (Assume that direction of current is not changing)

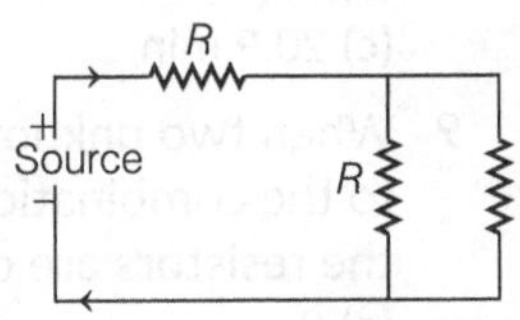

(a) $\dfrac{a^3R}{6b}$ (b) $\dfrac{a^3R}{27b}$

(c) $\dfrac{a^3R}{3b}$ (d) None of these

2 A total charge Q flows across a resistor R during a time interval, T in such a way that the current *versus* time graph for 0 to T is like the loop of a sine curve in the range 0 to π. The total heat generated in the resistor is

(a) $\dfrac{Q^2\pi^2R}{8T}$ (b) $\dfrac{2Q^2\pi^2R}{T}$ (c) $\dfrac{2Q^2\pi R}{T}$ (d) $\dfrac{Q^2v^2R}{2T}$

3 A time varying current i is passed through a resistance R as shown in figure. Total heat produced in the resistor is

(a) $11 i_0^2 R t_0$
(b) $13 i_0^2 R t_0$
(c) $15 i_0^2 R t_0$
(d) $17 i_0^2 R t_0$

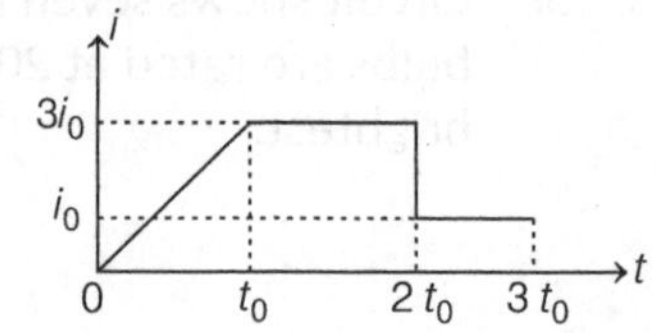

4 Two circular rings of identical radii and resistance of 36 Ω each are placed in such a way that cross each others centre C_1 and C_2 as shown in figure. Conducting joints are made at intersection points A and B of the rings. An ideal cell of emf 20 volts is connected across AB. The power delivered by cell is

(a) 80 W
(b) 100 W
(c) 120 W
(d) 200 W

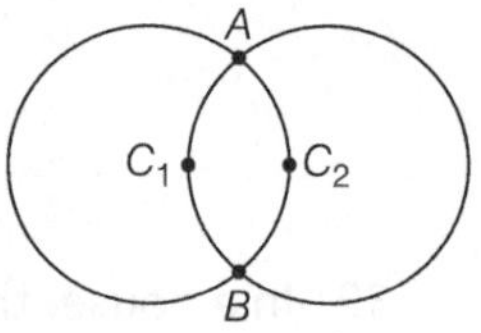

5 In the circuit shown in figure, the power which is dissipated as heat in the 6 Ω resistor is 6 W. What is the value of resistance R in the circuit?

(a) 6 Ω
(b) 8 Ω
(c) 10 Ω
(d) 24 Ω

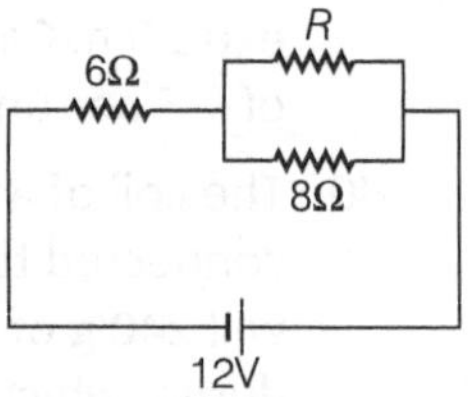

6 A metal oxide follows the relation $V = a\sqrt{I}$, where $a = 1 \dfrac{\text{volt}}{(\text{ampere})^{1/2}}$, V is potential drop across the oxide and I is current through it. If heat produced in resistor and metal oxide is same, then value of R is

(a) 3 Ω
(b) $\frac{1}{3}$ Ω
(c) 6 Ω
(d) 9 Ω

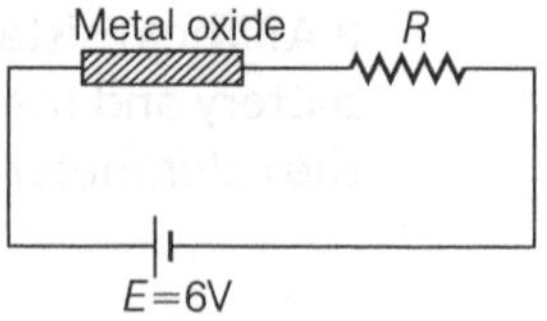

7 A heater is designed to operate with a power of 1000 W in a 100 V line. It is connected in combination with a resistance of 10 Ω and a resistance R, to a 100 V mains as shown in figure. What will be the value of R, so that the heater operates with a power of 62.5 W?

(a) 1 Ω
(b) 3 Ω
(c) 5 Ω
(d) 15 Ω

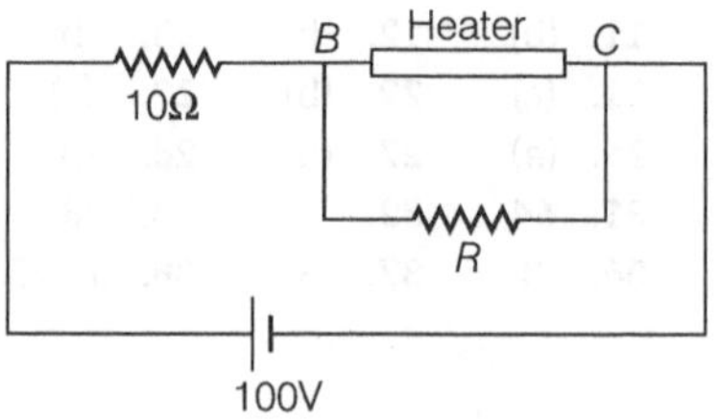

8 It takes 16 min to boil some water in an electric kettle. Due to some defect, it becomes necessary to remove 10% turns of the heating coil of the kettle. After repairs, how much time will it take to boil the same mass of water?

(a) 17.7 min
(b) 14.4 min
(c) 20.9 min
(d) 13.7 min

9 When two unknown resistors are connected in series with a battery, 225 W power is delivered to the combination with a total current 5A. For the same total current 50W is delivered when the resistors are connected in parallel. Find the ratio of two resistances.

(a) 9
(b) 12
(c) 2
(d) 4

10 Three 200 Ω resistors are connected as shown in figure. The maximum power that can be dissipated in anyone of the resistor is 50W. Find the maximum voltage that can be applied to the terminals A and B.

(a) 100 V
(b) 150 V
(c) 200 V
(d) 250 V

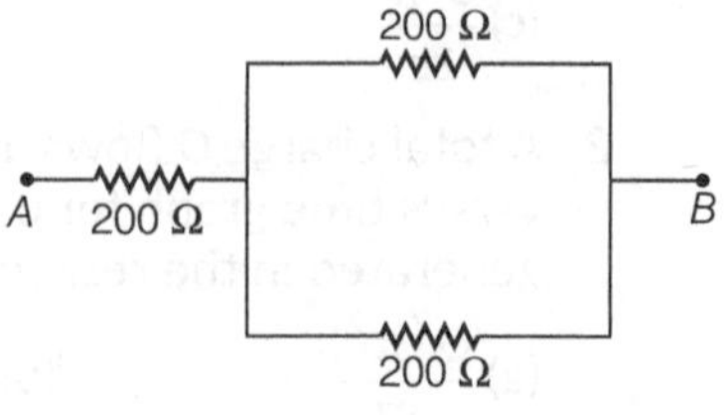

11 Four resistors carrying a current shown in figure are immersed in a box containing ice at 0°C. How much ice must be put in the box every 10 min to keep the average quantity of ice in the box constant? Latent heat of ice is 80 cal/g.

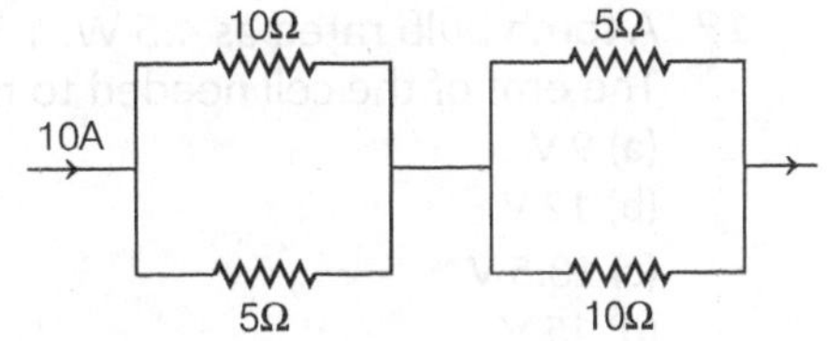

(a) 1.190 kg (b) 3.20 kg
(c) 4.2 kg (d) 0.25 kg

12 Six resistors, each of resistance R, are connected at their ends so as to form tetrahedral frame. Now, two cells of negligible internal resistance, and emf E is connected to the combination. One cell between any two corners and other cell between the remaining two corners. Thermal power developed will be

(a) $\frac{E^2}{R}$ (b) $\frac{2E^2}{R}$
(c) $\frac{4E^2}{R}$ (d) $\frac{8E^2}{R}$

13 A wire of length 1 m and radius 10^{-3} m is carrying a heavy current and is assumed to radiate as a black body. At equilibrium, its temperature is 900 K while that of the surroundings is 300 K. The resistivity of the material of the wire at 300 K is $\pi^2 \times 10^{-8}$ Ω-m and its temperature coefficient of resistance is 7.8×10^{-3} per °C. (Stefan's constant $= 5.68 \times 10^{-8}$ W-m^{2}-K^{4}). So, the current in the wire is

(a) 25A (b) 15A (c) 21 A (d) 36 A

14 A resistance coil wired to an external battery is placed inside a thermally insulated cylinder fitted with a frictionless piston and containing an ideal gas. A current $i = 240$ mA flows through the coil which has resistance $R = 490\,\Omega$. At what constant speed v must the piston of mass $m = 12$ kg move upward in order that the temperature of the gas remains unchanged? [Assume that 60% energy is utilised to do work by the gas, and neglect the work done by atmospheric pressure]

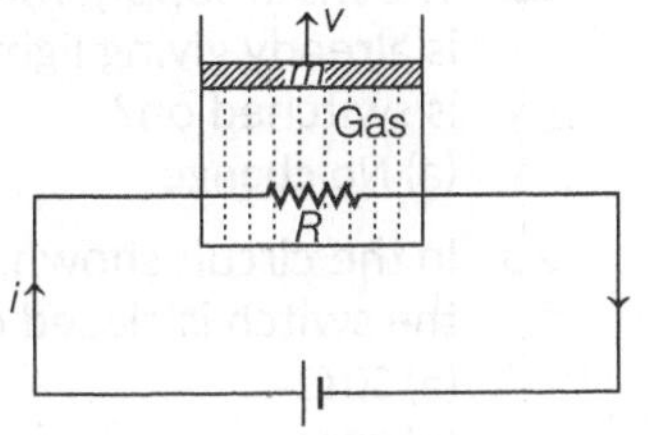

(a) 0.144 m/s (b) 1.41 m/s (c) 0.24 m/s (d) 2.4 m/s

15 n identical light bulbs, each designed to draw P power from a certain voltage supply are joined in series and that combination is connected across that supply. The power consumed by one bulb will be

(a) nP (b) P (c) $\frac{P}{n}$ (d) $\frac{P}{n^2}$

16 The operating temperature of the filament of lamp is 2000°C. The temperature coefficient of the material of the filament is 0.005°C^{-1}. If the atmospheric temperature is 0°C, then the current in the 100W-200V lamp when it is switched on is nearest to

(a) 2.5A (b) 3.5A (c) 4.5A (d) 5.5A

17 Three light bulbs of 40W, 60W and 100 W are connected in series with a 240 V source. Then,
(a) the potential difference will be maximum across the 40 W bulb
(b) the current will be maximum in 100 W bulb
(c) the resistance of the 40 W bulb is minimum
(d) the current through the 60 W bulb will be 0.1 A

18 A house is fitted with certain number of 100 W, 230 V incandescent lamps. The power to the house is fed by a generator producing the power at 240 V. The resistance of the wires from the generator to the house is 2 Ω. The maximum number of lamps that can be illuminated, so that the voltage across none of the lamps drops below 230 V is

(a) 10 (b) 11 (c) 12 (d) 15

19 A torch bulb rated as 4.5 W, 1.5 V is connected as shown in the figure. The emf of the cell needed to make the bulb glow at full intensity is

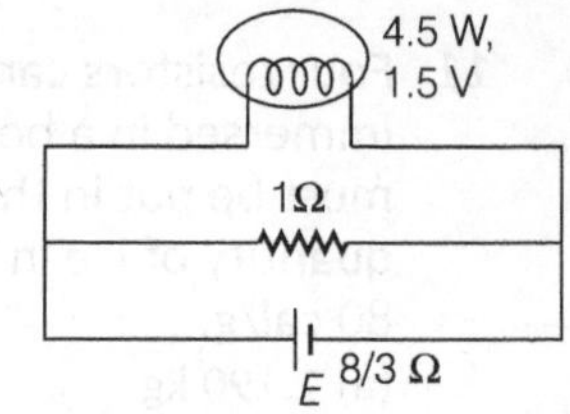

(a) 9 V
(b) 12 V
(c) 13.5 V
(d) 15 V

20 The circuit shown in figure, contains a battery, a rheostat and two identical lamps. What will happen to the brightness of the lamps, if the resistance of the rheostat is increased?

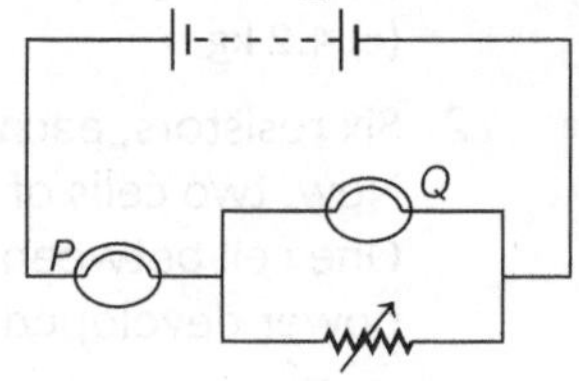

	Lamp P	**Lamp Q**
(a)	Less brighter	Brighter
(b)	Less brighter	Less brighter
(c)	Brighter	Less brighter
(d)	No change	Brighter

21 A circuit is connected as shown. All light bulbs are identical. When the switch in the circuit is closed illuminating bulb 4, which other bulb(s) also become brighter?

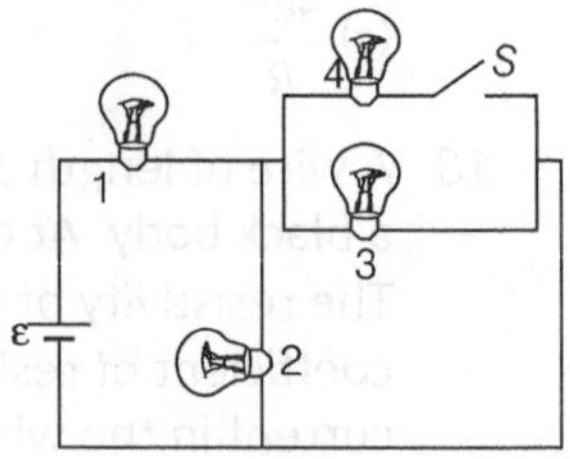

(a) Bulb 1 only
(b) Bulb 2 only
(c) Bulbs 2 and 3 only
(d) Bulbs 1, 2 and 3

22 The main supply voltage to a room is 120 V. The resistance of the lead wires is 6 Ω. A 60W bulb is already giving light. What is the decrease in the voltage across the bulb when a 240 W heater is switched on?
(a) No change (b) 10 V (c) 20 V (d) More than 10 V

23 In the circuit shown, the bulb B glows equally bright whether the switch is closed or open. The resistance of bulb is

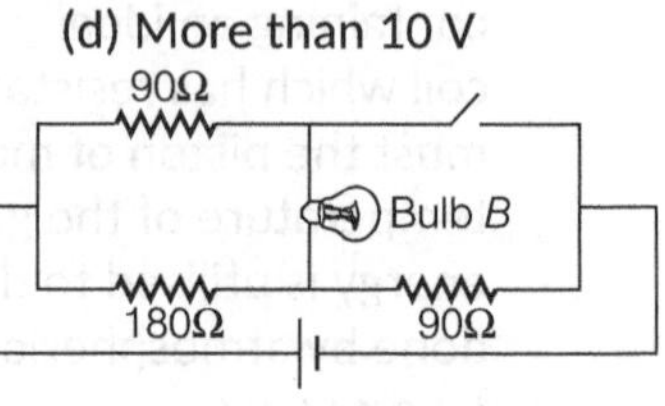

(a) 30 Ω
(b) 20 Ω
(c) 60 Ω
(d) 120 Ω

24 Assuming all bulbs are identical, rank the brightnesses of the bulbs from brightest to dimmest.

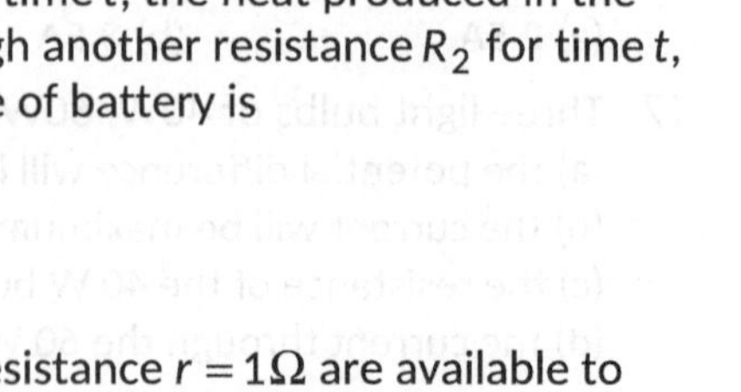

(a) $A > D = H > B = C > E = F = G$
(b) $A > B = C > D = H > E = F = G$
(c) $A > D = H > E = F = G > B = C$
(d) All have equal brightness

25 When a battery sends current through a resistance R_1 for time t, the heat produced in the resistor is Q. When the same battery sends current through another resistance R_2 for time t, the heat produced in R_2 is again Q. The internal resistance of battery is
(a) $\sqrt{R_1 R_2}$ (b) $R_1 + R_2$
(c) $R_1 - R_2$ (d) $\frac{R_1 + R_2}{2}$

26 Two identical batteries each of emf $E = 2$ V and internal resistance $r = 1\Omega$ are available to produce heat in an external circuit. What is the maximum rate of production of heat that can be obtained in the external circuit?
(a) 1 W (b) 2 W
(c) 4 W (d) 8 W

27 In the circuit shown the cell will deliver maximum power to the network, if the value of R is

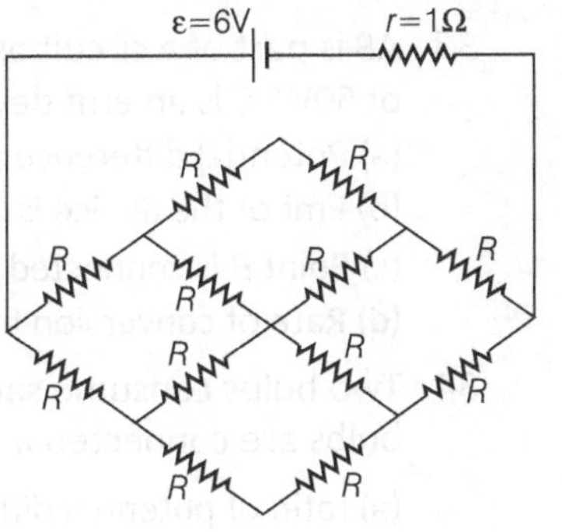

(a) $\frac{2}{3}\Omega$

(b) $\frac{4}{3}\Omega$

(c) $\frac{1}{3}\Omega$

(d) $\frac{5}{3}\Omega$

Multiple Choice(s) Correct Questions

28 A variable current flows through a 1Ω resistor for 2 s. Time dependence of the current is shown in the graph.

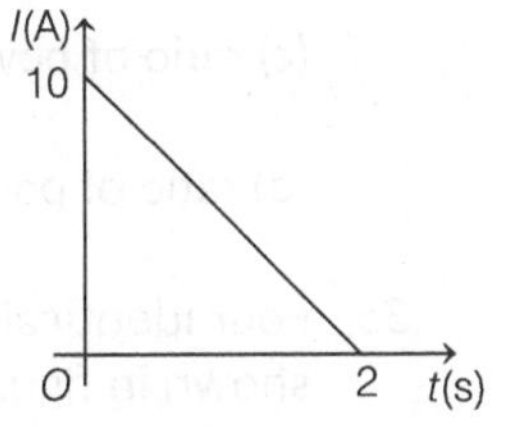

(a) Total charge flown through the resistor is 10 C
(b) Average current through the resistor is 5A
(c) Total heat produced in the resistor is 50 J
(d) Maximum power during the flow of current is 100 W

29 A conductor is made of an isotropic material and has shape of a truncated cone. A battery of constant emf is connected across it and its left end is earthed as shown in figure. If at a section distant x from left end, electric field intensity, potential and the rate of generation of heat per unit length are E, V and H respectively, which of the following graph(s) is/are correct?

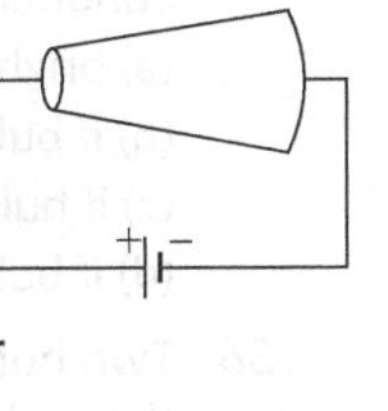

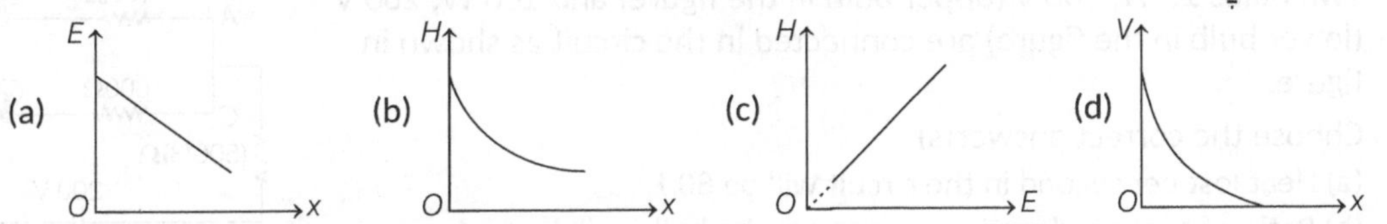

30 Square copper plates of equal thickness and having, area 1 cm^2 and 1 m^2, are connected in series as shown in the figure.

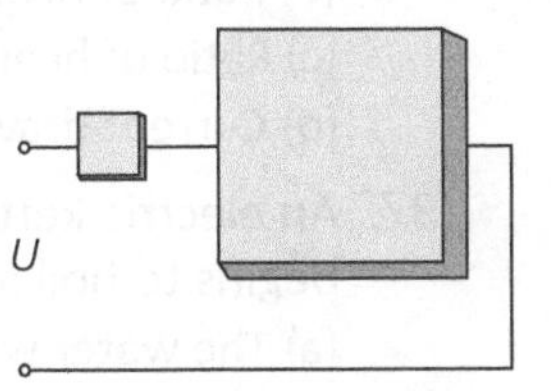

(a) The resistance of the both the plates are equal
(b) The potential drop across both the plates are equal
(c) The current in both the plates are equal
(d) The total heat dissipated in both the plates per unit time are equal

31 Two bars of radii r and $2r$ but of same material are kept in contact as shown. An electric current I is the passed through the bars. Then,

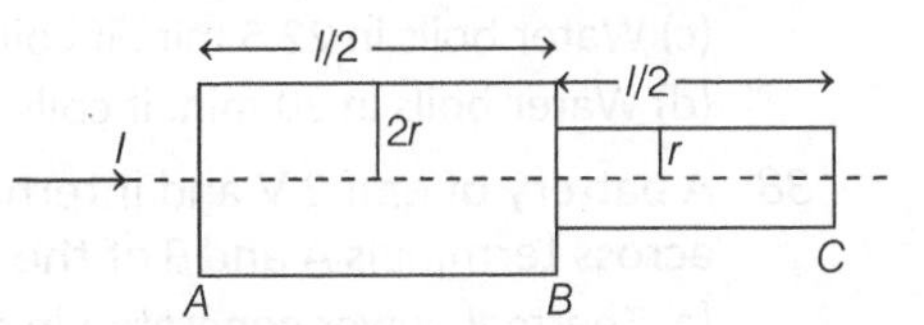

(a) heat produced in bar BC is 4 times the heat produced in bar AB
(b) electric field in both halves is same
(c) current density across AB is double that of across BC
(d) potential drop across BC is 4 times that of across AB

32 In the circuit as shown in figure,

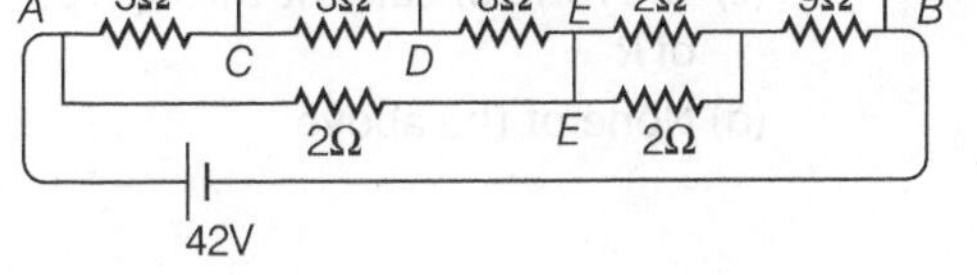

(a) power disputed in 5Ω resistor is 10 W
(b) power disputed in 5Ω resistor is 5 W
(c) power disputed in 8Ω resistor is zero
(d) power disputed in 8Ω resistor is 2 W

33 AB is part of a circuit as shown, that absorbs energy at a rate of 50W. E is an emf device that has no internal resistance.

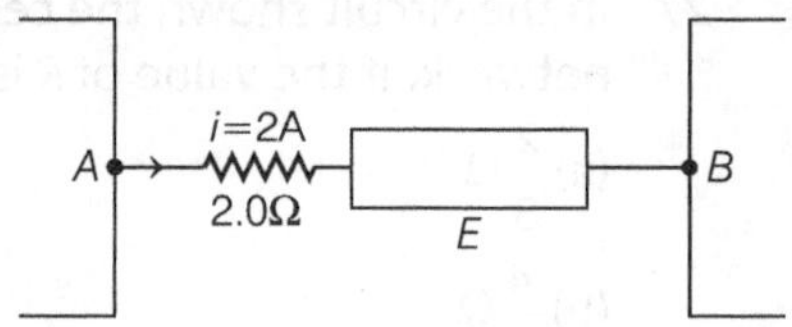

(a) Potential difference across AB is 17V
(b) Emf of the device is 21 V
(c) Point B is connected to the positive terminal of E
(d) Rate of conversion from electrical to chemical energy is 42 W in device E

34 Two bulbs consume same power when operated at 200 V and 300 V, respectively. When the bulbs are connected in series across a DC source of 500 V, then
(a) ratio of potential difference across them is $\frac{3}{2}$
(b) ratio of potential difference across them is $\frac{4}{9}$
(c) ratio of power consumed across them is $\frac{4}{9}$
(d) ratio of power consumed across them is $\frac{2}{3}$

35 Four identical bulbs A, B, C and D are connected in a circuit as shown in figure. Now, whenever any bulb fails, then it cannot conduct current through it. Then,

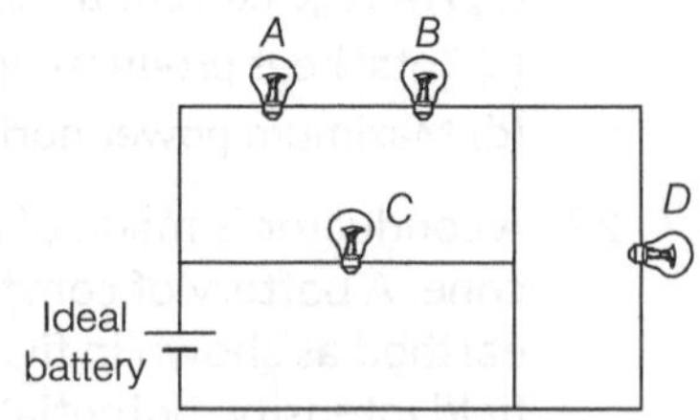

(a) brightness of bulb C is highest
(b) if bulb C fails, brightness of bulb D increases
(c) if bulb C fails, brightness of all bulbs remain same
(d) if bulb A fails, B will not glow

36 Two bulbs 25 W, 100 V (upper bulb in the figure) and 100 W, 200 V (lower bulb in the figure) are connected in the circuit as shown in figure.

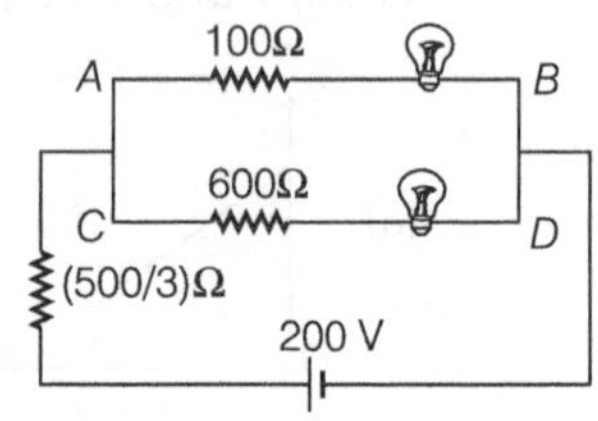

Choose the correct answer(s).
(a) Heat lost per second in the circuit will be 80 J
(b) Ratio of heat produced per second in the bulbs will be 1 : 1
(c) Ratio of heat produced in branch AB to that produced in branch CD will be 1 : 2
(d) Current drawn from the cell is 0.4A

37 An electric kettle heater has two coils. When one coil is switches on, the water in the kettle begins to boil after 15 min. When the other coil is switched on, the water boils in 30 min.
(a) The water will boil in 45 min, if the coils are connected in series.
(b) The water will boil in 10 min, if the coils are connected in parallel.
(c) Water boils in 22.5 min, if coils are connected in series.
(d) Water boils in 20 min, if coils are connected in parallel.

38 A battery of emf 2 V and internal resistance 1Ω is connected across terminals A and B of the circuit as shown in the given figure.

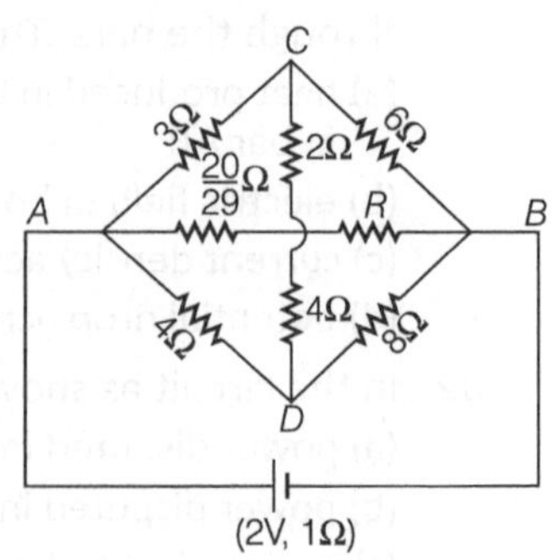

(a) Thermal power generated in external circuit will be maximum possible when $R = \frac{16}{29}\Omega$
(b) Maximum possible thermal power generated in external circuit is equal to 4 W
(c) The ratio of current through 6Ω to that through 4Ω is independent of R
(d) None of the above

39 N cells each of emf E and identical resistance r are grouped into sets of K cells connected in series. The (N/K) sets are connected in parallel to a load of resistance R, then

(a) maximum power is delivered to the load, if $K = \sqrt{\dfrac{NR}{r}}$

(b) maximum power is delivered to the load, if $K = \sqrt{\dfrac{r}{NR}}$

(c) maximum power delivered to the load is $\dfrac{NE^2}{4r}$

(d) maximum power delivered to the load is $\dfrac{E^2}{4Nr}$

Comprehension Based Questions

Comprehension-1

A car battery with a 12 V emf and an internal resistance of 0.04 Ω is being charged with a current of 50 A.

40 The potential difference V across the terminals of the battery are
(a) 10 V (b) 12 V (c) 14 V (d) 16 V

41 The rate at which energy is being dissipated as heat inside the battery is
(a) 100 W (b) 500 W (c) 600 W (d) 700 W

42 The rate of energy conversion from electrical to chemical is
(a) 100 W (b) 500 W (c) 600 W (d) 700 W

Comprehension-2

All bulbs consume same power. The resistance of bulb 1 is 36 Ω.

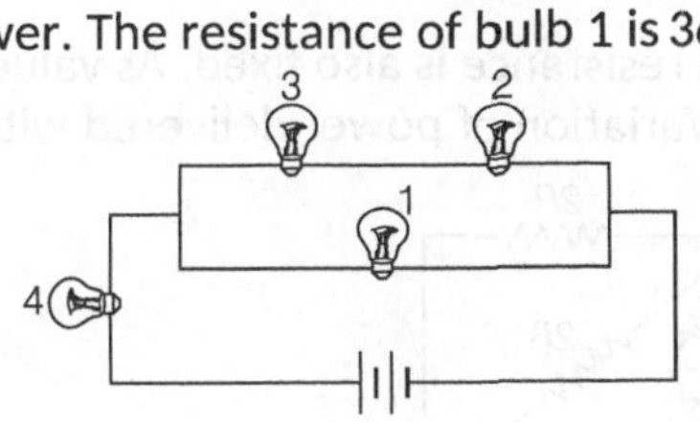

43 What is the resistance of bulb 3?
(a) 4Ω (b) 9Ω (c) 12Ω (d) 18Ω

44 What is the resistance of bulb 4?
(a) 4Ω (b) 9Ω (c) 12Ω (d) 18Ω

45 What is the voltage output of the battery, if the power of each bulb is 4W?
(a) 12 V (b) 16 V (c) 24 V (d) None of these

Comprehension-3

Important aspect of fuse wire and battery

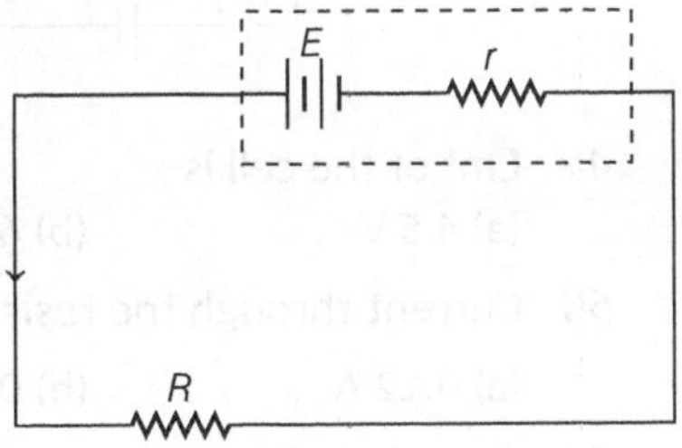

Electric fuse is a protective device used in series with an electric circuit or an electric appliance to save it from damage due to overheating produced by strong current in the circuit or appliance. Fuse wire is generally made from an alloy of lead and tin which has high resistance and low melting point. It is connected in series in an electric installation. If a circuit gets accidentally short-circuited, a large current flows, then fuse wire melts away which causes a break in the circuit.

The power through fuse (P) is equal to heat energy lost per unit area per unit time (h) (neglecting heat loses from ends of the wire).

$P = I^2R = h \times 2\pi r l \ldots \left[R = \dfrac{\rho l}{\pi r^2}\right]$, where r and l are the length and radius of fuse wire, respectively. A battery is described by its emf (E) and internal resistance (r), efficiency of battery (η) is defined as the ratio of the output power and the input power.

$$\eta = \frac{\text{Output power}}{\text{Input power}} \times 100\%$$

But $I = \dfrac{E}{R+r}$, input power $= EI$, output power $= EI - I^2 r$, then

$$\eta = \left(\frac{EI - I^2 r}{EI}\right) \times 100 = \left(1 - \frac{Ir}{E}\right) \times 100 = 1 - \left(\frac{E}{R+r}\right)\left(\frac{r}{E}\right) \times 100 = \left(\frac{R}{R+r}\right) \times 100$$

We know that, output power of a source is maximum when the external resistance is equal to internal resistance, i.e. $R = r$.

46 Two fuse wires of same material are having length ratio 1 : 2 and radius ratio 4 : 1. Then, respective ratio of their current rating will be

(a) 8 : 1 (b) 2 : 1 (c) 1 : 8 (d) 4 : 1

47 The maximum power of a 20.0 Ω fuse wire is 2.0 kW, then this fuse wire can be connected safely to DC source (negligible internal resistance) of

(a) 300 V (b) 190 V (c) 250 V (d) 220 V

48 Efficiency of a battery (non-ideal) when delivering maximum power is

(a) 100% (b) 50% (c) 90% (d) 40%

Comprehension-4

Cell emf is fixed and internal resistance is also fixed. As value of R changes, power delivered to circuit by cell also changes. Variation of power delivered with R is shown in the graph

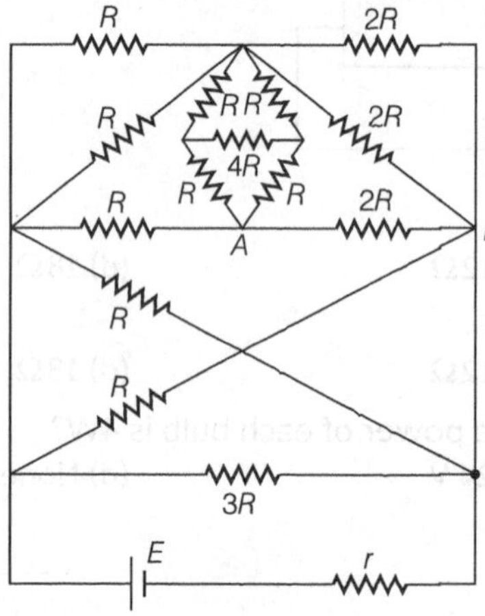

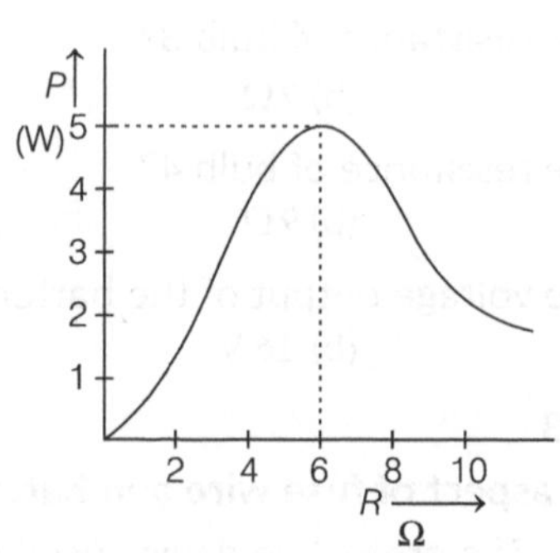

49 Emf of the cell is

(a) 4.5 V (b) 6 V (c) 9 V (d) 11.5 V

50 Current through the resistance $3R$ is nearly (where, $R = 6\,\Omega$)

(a) 0.12 A (b) 0.17 A (c) 0.37 A (d) 0.49 A

51 Potential difference across each resistance $2R$, i.e. between A and B is nearly (where, $R = 6\,\Omega$)

(a) 1.2 V (b) 1.6 V (c) 2 V (d) 3.5 V

Comprehension-5

A high tension wire is at a high potential with respect to a wire which is well grounded called earth wire. You must have seen such wires stretched parallel to roads. There is a high tension wire between two points A and B, 1 km apart. The distance between high tension wire and earth wire is 1 m. The resistance of the high tension (and also the earth wire) is 1Ω/m. This wire is at a potential of 11 kV at point A w.r.t. earth wire and is carrying 1A current which returns back to the generator through the earth wire. This wire is quite a thick wire. There is a sign board at a pole over which this wire is stretched reading 'DANGER, 11 kV'. You might think what would happen, if one touched this wire. Will one feel a shock or not. Well! it depends on whether the current through our body exceeds a particular value, which we may call critical current.

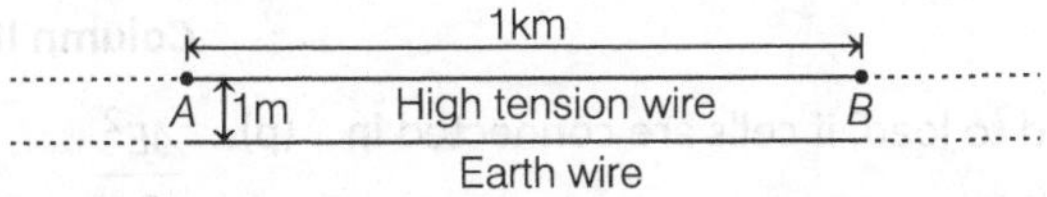

Bird

52 Consider a bird having effective resistance $10\,\Omega$ between its feet sitting on this high tension wire. The distance between its feet is 10 cm. The current through the bird is approximately

(a) 10A (b) 1A (c) 0.01A (d) 0.005A

53 If the 'critical current' for the bird is 0.1 A, find the maximum power at 11 kV can be transmitted at point A, so that the bird may not get shock. Assume that the distance between the feet is 10 cm.

(a) 111 kW (b) 11 kW (c) 101 kW (d) 110011 kW

Match the Columns Questions

54 For the circuit shown in figure, 4 cells are arranged. In Column I, the cell number is given while in Column II, some statements related to cells are given. Match the entries of Column I with the entries of Column II.

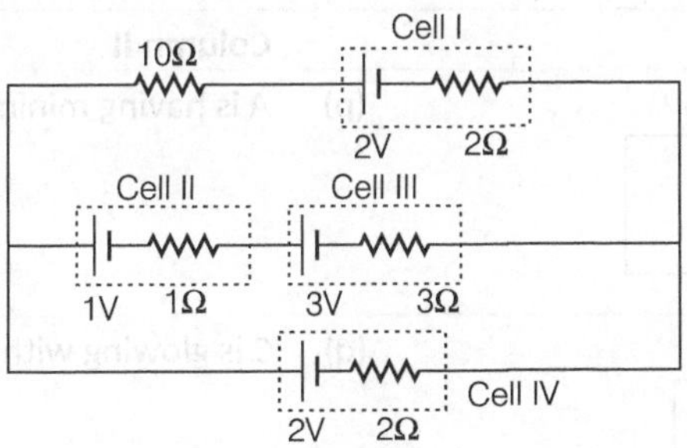

	Column I		Column II
(a)	Cell I	(p)	Chemical energy of cell is decreasing
(b)	Cell II	(q)	Chemical energy of cell is increasing
(c)	Cell III	(r)	Work done by cell is positive
(d)	Cell IV	(s)	Thermal energy developed in cell is positive

55 Referring to the circuit shown in figure match Column I with Column II

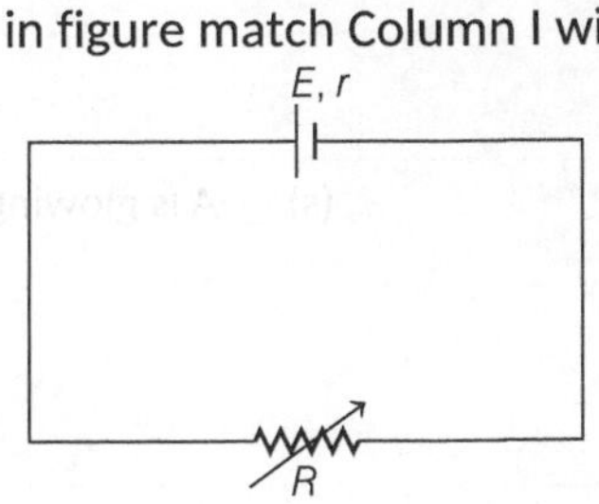

	Column I		Column II
(a)	Terminal potential difference across the cell will be maximum	(p)	$r > R$
(b)	Power transferred to resistance R is less than the maximum possible	(q)	$r < R$
(c)	Power dissipated in the cell is maximum	(r)	$R = \infty$
(d)	Faster drift of ions in the electrolyte in the cell will be for	(s)	$R = 0$

56 Consider two identical cells each of emf E and internal resistance r connected to a load resistance R.

	Column I		Column II
(a)	Maximum power transferred to load, if cells are connected in series	(p)	$\frac{4E^2}{9r}$
(b)	Maximum power transferred to load, if cells are connected in parallel	(q)	$\frac{E^2}{2r}$
(c)	Power transferred to load, if cells are connected in series and $R = r$	(r)	$E_{eq} = E, r_{eq} = \frac{r}{2}$
(d)	Power transferred to load, if cells are connected in parallel and $R = r$	(s)	$E_{eq} = 2E, r_{eq} = 2r$

57 Three bulbs A, B and C are having rated power P_A, P_B and P_C respectively, each bulb is designed to operate at rated voltage V. It is given that $P_A > P_B > P_C$. In Column I, the three bulbs are arranged in different configuration, while in Column II, the information about intensities of bulbs are mentioned. Match the entries of Column I with the entries of Column II. Neglect the variation, if resistance due to change in temperature.

	Column I		Column II
(a)	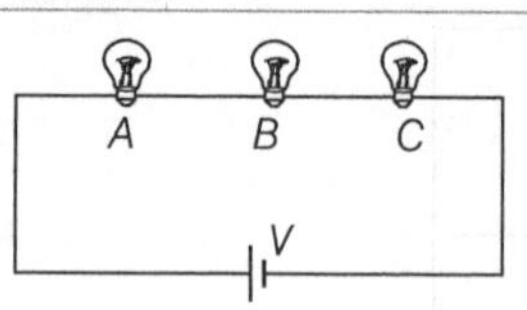	(p)	A is having minimum intensity
(b)	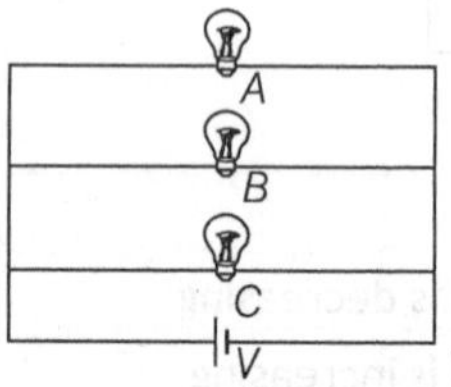	(q)	C is glowing with maximum brightness
(c)	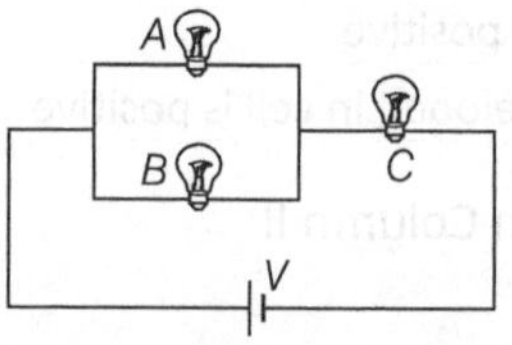	(r)	B is glowing with minimum brightness
(d)	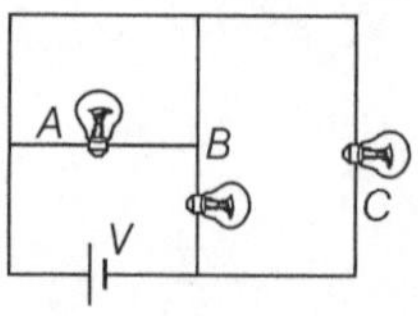	(s)	A is glowing with maximum brightness

58 Match the Column I with Column II.

Column I	Column II
(a) 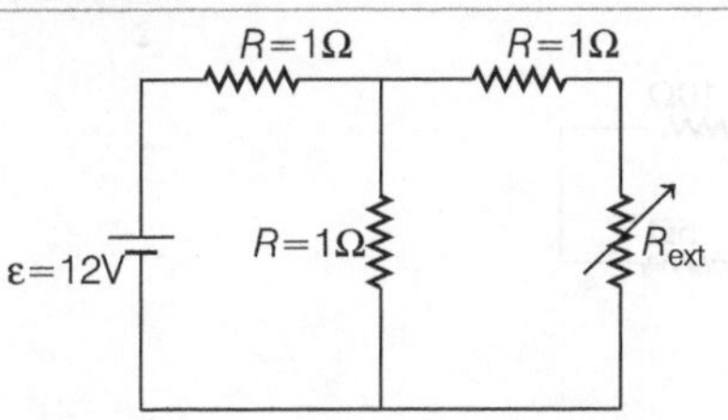 Numerical value of maximum power (in watt) delivered to R_{ext} is	(p) $\frac{1}{2}$
(b) 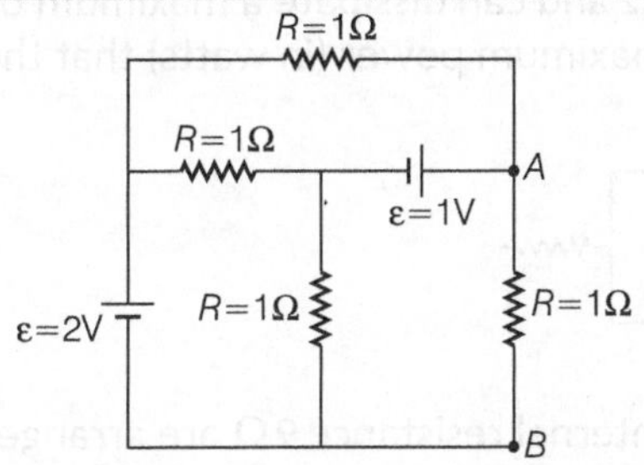 Numerical value of potential difference (in volt) between points A and B is	(q) $\frac{3}{2}$
(c) 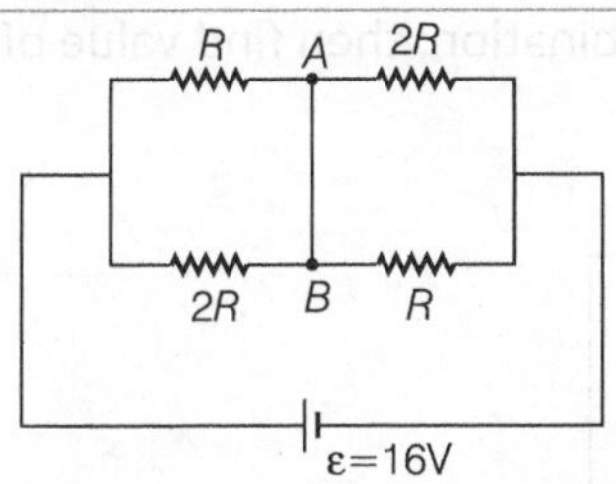 $R = 1\Omega$ Wire AB has negligible resistance. Numerical value of current (in ampere) through AB is	(r) 6
(d) 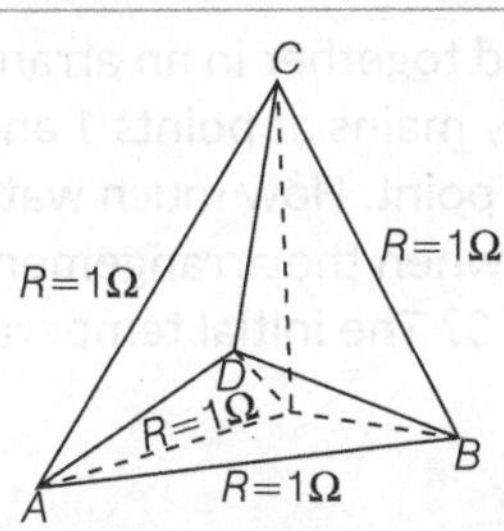 Resistances are arranged in tetrahedron configuration. All resistances are equal. Numerical value of equivalent resistance (in Ω) between A and B is	(s) 4

Numerical Value Questions

59 In the given figure i - t graph, it is revealed that the current decreases by 50% in each time interval t. The heat dissipated in the resistor R is $\frac{Q^2R}{nT}$. The value of n is (Assuming total charge flows is Q)

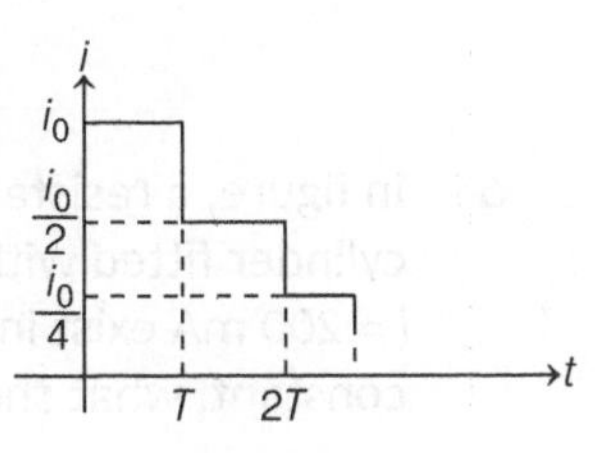

60 Figure shows a battery circuit with an unknown resistance x. If the value of maximum power dissipation in unknown resistance x is $\frac{5}{n}$ W, then find n.

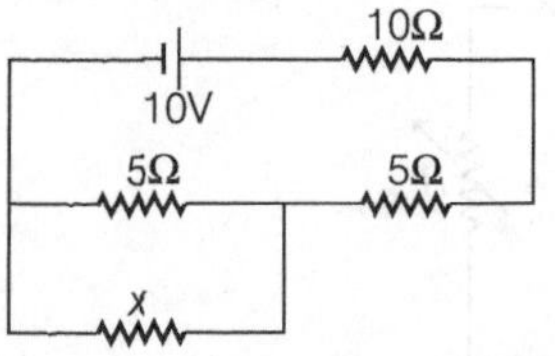

61 Each of 3 resistors in figure has a resistance of 2 Ω and can dissipate a maximum of 18 W without becoming excessively heated. Find the maximum power (in watts) that the circuit can dissipate.

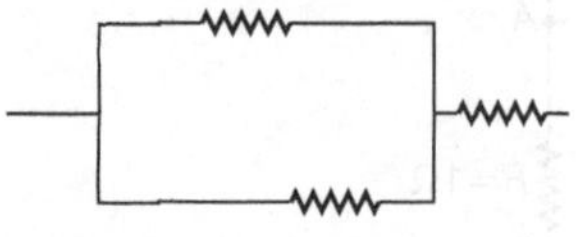

62 324 numbers of identical galvanic cells, each of internal resistance 9 Ω are arranged as several in series group of cells connected in parallel. The arrangement has been laid out, so that power output in an externally connected resistance of value 4 Ω is maximum. If n number of cells are connected in every series group that form parallel combination, then find value of n.

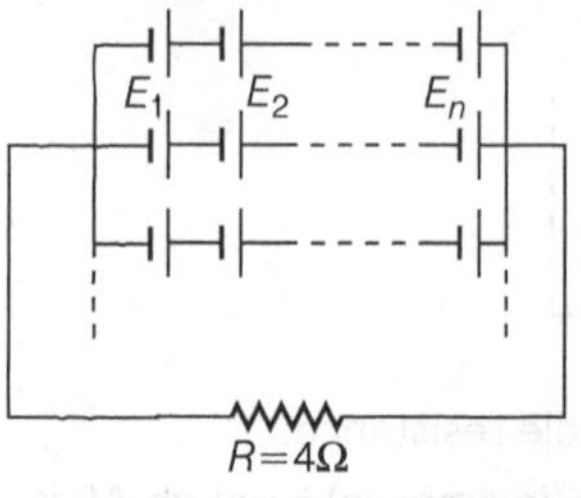

63 The resistances in a burner of an electric stove are connected together in an arrangement shown in figure. This arrangement is connected to the supply mains at points 1 and 2, and after a certain time, 500 grams of water are heated to the boiling point. How much water (in gram) can be heated to the boiling point in the same time interval when the arrangement of resistances is connected to the supply mains at points 1 and 3? The initial temperature of the water is the same in both cases. Neglect all heat loses.

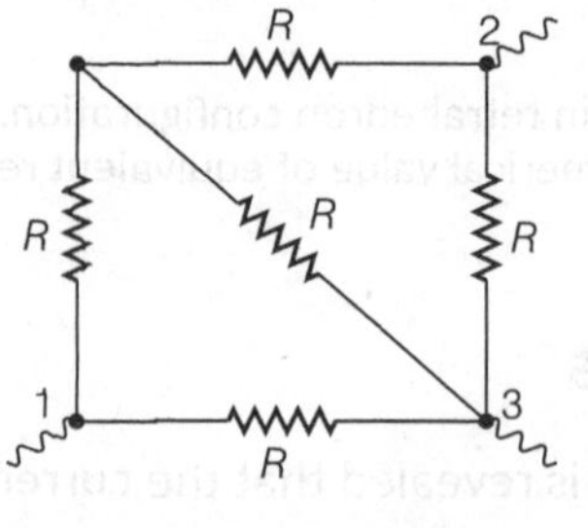

64 In figure, a resistance coil wired to an external battery, is placed inside a thermally insulated cylinder fitted with a frictionless piston and containing an ideal diatomic gas. A current $i = 200$ mA exist in the coil, which has a resistance $R = 350\,\Omega$. If the pressure of the gas remains constant, what should be the speed v

(in cm/s) of the piston, of mass $m = 10$ kg?
(Take $p_{atm} = 10^5$ Pa, $A = 10$ cm^2)

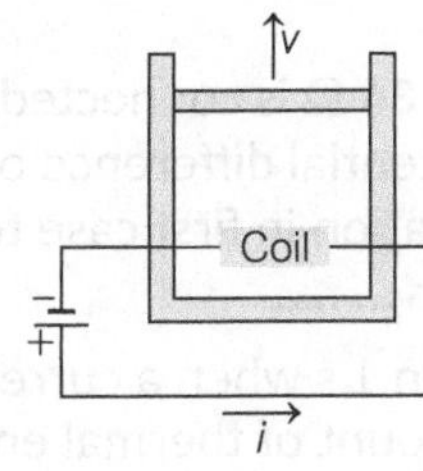

65 Figure shows an automobile circuit. How much power (in watt) is dissipated by the automobile circuit when switches *A*, *B*, *C* and *D* are closed.

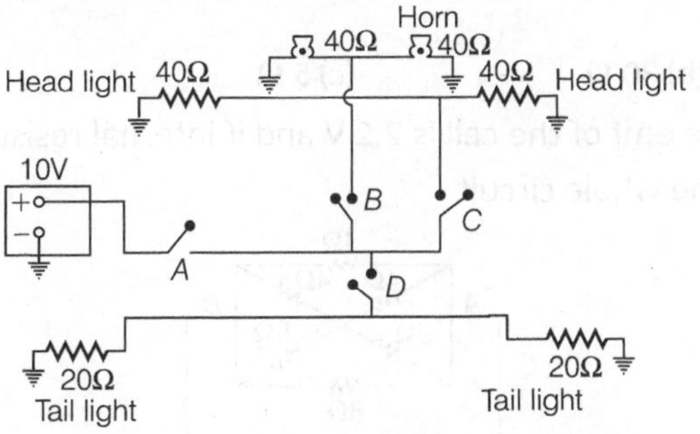

66 A battery of emf ε fitted with a resistor *R* in series supplies current to the resistors between *A* and *B* in the upper part of the circuit. If the power dissipated in the network between *A* and *B* as shown by rounded dotted line, does not get affected by closing the key *K*, find the value of *R* (in Ω), if $R' = 20\sqrt{2}\ \Omega$.

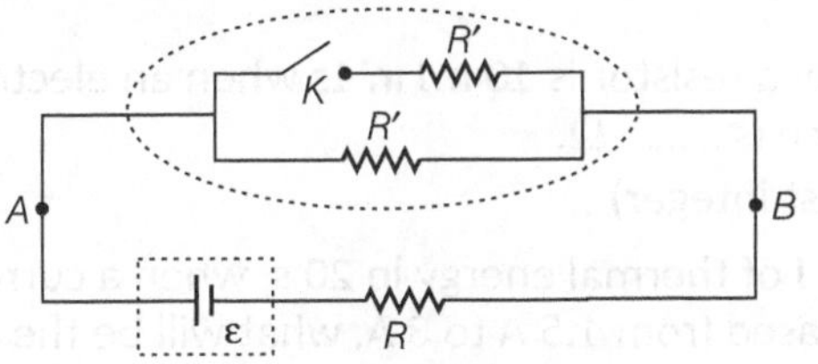

Answers

1. (b)	**2.** (a)	**3.** (b)	**4.** (b)	**5.** (d)	**6.** (b)	**7.** (c)	**8.** (b)	**9.** (c)	**10.** (b)
11. (a)	**12.** (c)	**13.** (d)	**14.** (a)	**15.** (d)	**16.** (d)	**17.** (a)	**18.** (b)	**19.** (c)	**20.** (a)
21. (a)	**22.** (d)	**23.** (a)	**24.** (a)	**25.** (a)	**26.** (b)	**27.** (a)	**28.** (abd)	**29.** (bc)	**30.** (abcd)
31. (ad)	**32.** (bc)	**33.** (bd)	**34.** (bc)	**35.** (acd)	**36.** (ad)	**37.** (ab)	**38.** (ac)	**39.** (ac)	**40.** (c)
41. (a)	**42.** (c)	**43.** (b)	**44.** (a)	**45.** (b)	**46.** (a)	**47.** (b)	**48.** (b)	**49.** (b)	**50.** (b)

51. (c) **52.** (c) **53.** (a) **54.** (a) → qs; (b) → prs; (c) → prs; (d) → qs

55. (a) → r; (b) → pqrs; (c) → s; (d) → s **56.** (a) → qs; (b) → qr; (c) → ps; (d) → pr

57. (a) → pq; (b) → s; (c) → qr; (d) → p

58. (a) → r; (b) → q; (c) → s; (d) → p

59. 3	**60.** 12	**61.** 27	**62.** 12
63. 800	**64.** 2	**65.** 20	**66.** 20

For Detailed Solutions
Scan the code

Previous Years' Questions

JEE Main

1 A uniform heating wire of resistance 36 Ω is connected across a potential difference of 240 V. The wire is then cut into half and potential difference of 240 V is applied across each half separately. The ratio of power dissipation in first case to the total power dissipation in the second case would be 1 : *x*, where *x* is .... **[JEE Main 2021]**

2 A resistor dissipates 192 J of energy in 1 s when a current of 4 A is passed through it. Now, when the current is doubled, the amount of thermal energy dissipated in 5 s is .... J. **[JEE Main 2021]**

3 An electric bulb of 500 W at 100 V is used in a circuit having a 200 V supply. Calculate the resistance *R* to be connected in series with the bulb, so that the power delivered by the bulb is 500 W. **[JEE Main 2021]**

(a) 20 Ω (b) 30 Ω (c) 5 Ω (d) 10 Ω

4 In the given figure, the emf of the cell is 2.2 V and if internal resistance is 0.6 Ω. Calculate the power dissipated in the whole circuit **[JEE Main 2021]**

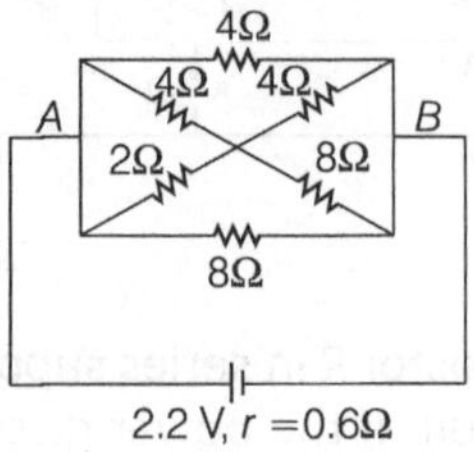

(a) 1.32 W (b) 0.65 W
(c) 2.2 W (d) 4.4 W

5 The energy dissipated by a resistor is 10 mJ in 1s when an electric current of 2 mA flows through it. The resistance is .... Ω.
(Round off to the nearest integer) **[JEE Main 2021]**

6 A resistor develops 500 J of thermal energy in 20 s, when a current of 1.5 A is passed through it. If the current is increased from 1.5 A to 3 A, what will be the energy developed in 20 s? **[JEE Main 2021]**

(a) 1500 J (b) 1000 J (c) 500 J (d) 2000 J

7 A torch battery of length *l* is to be made up of a thin cylindrical bar of radius *a* and a concentric thin cylindrical shell of radius *b* is filled in between with an electrolyte of resistivity ρ (see figure). If the battery is connected to a resistance *R*, the maximum joule's heating in *R* will takes place for **[JEE Main 2020]**

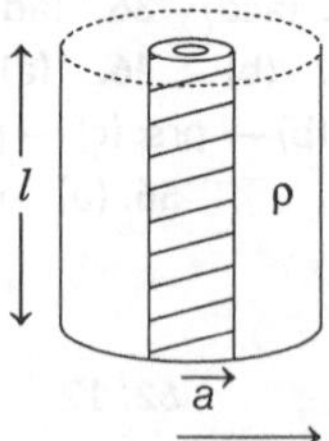

(a) $R = \frac{\rho}{2\pi l}\left(\frac{b}{a}\right)$ (b) $R = \frac{\rho}{2\pi l}\ln\left(\frac{b}{a}\right)$

(c) $R = \frac{\rho}{\pi l}\ln\left(\frac{b}{a}\right)$ (d) $R = \frac{2\rho}{\pi l}\ln\left(\frac{b}{a}\right)$

8 A battery of 3.0 V is connected to a resistor dissipating 0.5 W of power. If the terminal voltage of the battery is 2.5 V, the power dissipated within the internal resistance is **[JEE Main 2020]**

(a) 0.072 W (b) 0.125 W
(c) 0.50 W (d) 0.10 W

9 In a building, there are 15 bulbs of 45 W, 15 bulbs of 100 W, 15 small fans of 10 W and 2 heaters of 1 kW. The voltage of electric main is 220 V. The minimum fuse capacity (rated value) of the building will be **[JEE Main 2020]**

(a) 25 A (b) 10 A
(c) 20 A (d) 15 A

10 The resistive network shown below is connected to a DC source of 16 V. The power consumed by the network is 4 W. The value of R is **[JEE Main 2019]**

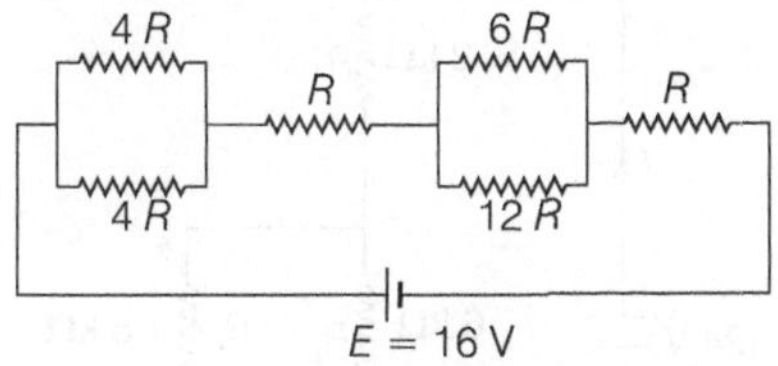

(a) 6 Ω (b) 8 Ω
(c) 1 Ω (d) 16 Ω

11 A current of 2 mA was passed through an unknown resistor which dissipated a power of 4.4 W. Dissipated power when an ideal power supply of 11 V is connected across it is **[JEE Main 2019]**

(a) 11×10^{-4} W (b) 11×10^{-5} W
(c) 11×10^{5} W (d) 11×10^{-3} W

12 Two equal resistances when connected in series to a battery consume electric power of 60 W. If these resistances are now connected in parallel combination to the same battery, the electric power consumed will be **[JEE Main 2019]**

(a) 60 W (b) 30 W
(c) 240 W (d) 120 W

13 Two electric bulbs rated at 25 W, 220 V and 100 W, 220 V are connected in series across a 220 V voltage source. If the 25 W and 100 W bulbs draw powers P_1 and P_2 respectively, then **[JEE Main 2019]**

(a) $P_1 = 16$ W, $P_2 = 4$W (b) $P_1 = 4$ W, $P_2 = 16$W
(c) $P_1 = 9$ W, $P_2 = 16$W (d) $P_1 = 16$ W, $P_2 = 9$W

14 A cell of internal resistance r drives current through an external resistance R. The power delivered by the cell to the external resistance will be maximum when **[JEE Main 2019]**

(a) $R = 2r$ (b) $R = r$
(c) $R = 0.001\, r$ (d) $R = 1000\, r$

15 In a large building, there are 15 bulbs of 40 W, 5 bulbs of 100 W, 5 fans of 80 W and 1 heater of 1 kW. The voltage of the electric mains is 220 V. Then, the minimum current should be **[JEE Main 2014]**

(a) 8 A (b) 10 A
(c) 12 A (d) 14 A

16 The supply voltage to room is 120 V. The resistance of the lead wires is 6 Ω. A 60 W bulb is already switched on. What is the decrease of voltage across the bulb, when a 240 W heater is switched on in parallel to the bulb? **[JEE Main 2013]**

(a) Zero (b) 2.9 V
(c) 13.3 V (d) 10.04 V

JEE Advanced

17 When two identical batteries of internal resistance 1 Ω each are connected in series across a resistor R, the rate of heat produced in R is J_1. When the same batteries are connected in parallel across R, the rate is J_2. If $J_1 = 2.25 J_2$ then the value of R in Ω is **[JEE Advanced 2010]**

18 For the circuit shown in the figure **[JEE Advanced 2009]**

(a) the current I through the battery is 7.5 mA

(b) the potential difference across R_L is 18 V

(c) ratio of powers dissipated in R_1 and R_2 is 3

(d) if R_1 and R_2 are interchanged, magnitude of the power dissipated in R_L will decrease by a factor of 9

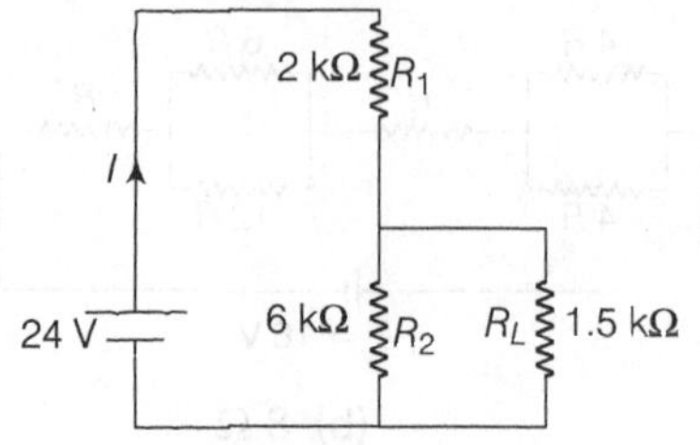

Answers

1. (4)	2. (3840)	3. (a)	4. (c)	5. (2500)	6. (d)	7. (b)	8. (d)
9. (c)	10. (b)	11. (b)	12. (c)	13. (a)	14. (b)	15. (c)	16. (d)
17. (4)	18. (ad)						

For Detailed Solutions
Scan the code

CHAPTER

09

CAPACITANCE

LEVEL 01 \ Starter Level

1.1 Capacitance of an Isolated Conductor

1 If 30μC charge is given to an isolated conductor of capacitance 5μF, then find

(i) potential of the conductor and

(ii) energy stored in the electric field of conductor.

2 What is the capacitance of the earth? (Take, radius of the earth = 6400 km)

3 *n* identical spherical mercury drops each of capacitance *C* coalesce to form a bigger drop. Find the capacitance of the bigger drop.

4 Two conductors *A* and *B* of capacitances 3μF and 6μF have charges 6μC and 3μC, respectively. They are placed at a large distance from each other. When they are joined by a conducting wire, find

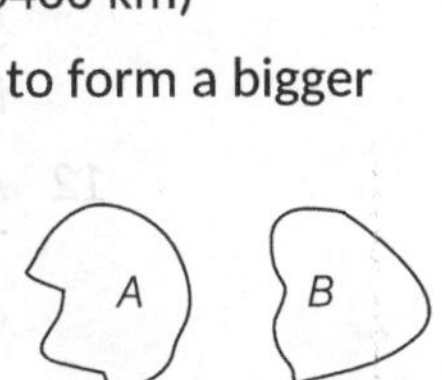

(i) the final charges on *A* and *B*,

(ii) the common potential and

(iii) the heat produced during the process of flow of charges.

5 Two metallic spheres of radii 40cm and 10cm carry charges of 100 μC and – 20μC, respectively. The two spheres are placed at a large distance from each other. If they are now connected by a metallic wire, then find

(i) the final charges on them

(ii) the charge flown through the metallic wire.

1.2 Capacitor

6 When 2.0×10^{12} electrons are transferred from one conductor to another, a potential difference of 10V appears between the conductors. Calculate the capacitance of the system of the conductors.

7 Charges $+q$ and $-q$ are given to plates of a parallel plate air capacitor having area of each plate A and separation between them is d, then find

(i) capacitance of the system,

(ii) electric field between the plates,

(iii) potential difference between the plates and

(iv) energy stored between the plates.

8 Two parallel conducting plates of a capacitor of capacitance C containing charges Q and $-2Q$ are at a distance d apart. Find

(i) potential difference between the plates and

(ii) energy stored in the capacitor

9 A parallel plate air capacitor is made using two square plates, each of edge 0.1m, spaced 1cm apart. It is connected to a 100 V battery.

(i) What is the capacitance?

(ii) What is the charge on each plate?

(iii) What is the energy stored in the capacitor?

(iv) What is the electric field between the plates?

(v) If the battery is disconnected and then the plates are pulled apart to a separation of 2 cm, what are the answers to the above parts?

10 The plates S and T of an uncharged parallel plate capacitor are connected across a battery. The battery is then disconnected and the charged plates are now connected in a system as shown in the figure. The system shown is in equilibrium. All the strings are insulating and massless. Find the magnitude of charge on one of the capacitor plates. (Area of plates = A)

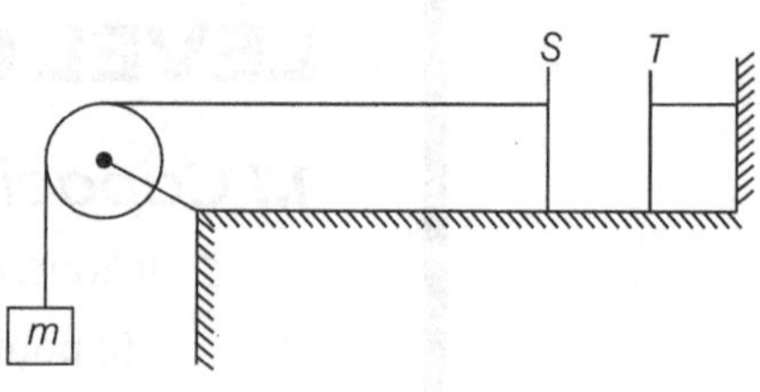

11 The stratosphere acts as a conducting layer for the earth. If the stratosphere extends beyond 50 km from the surface of earth, then calculate the capacitance of the spherical capacitor formed between stratosphere and earth's surface. (Take, radius of earth as 6400 km)

12 A cylindrical capacitor has two co-axial cylinders of length 15 cm and radii 1.5 cm and 1.4 cm. The outer cylinder is earthed and the inner cylinder is given a charge of 3.5 μC. Determine the capacitance of the system and the potential of the inner cylinder.

1.3 Applications of KCL and KVL

13 Initially, all capacitors are uncharged. Find the charges on all the capacitors in the following circuits using Kirchhoff's junction law.

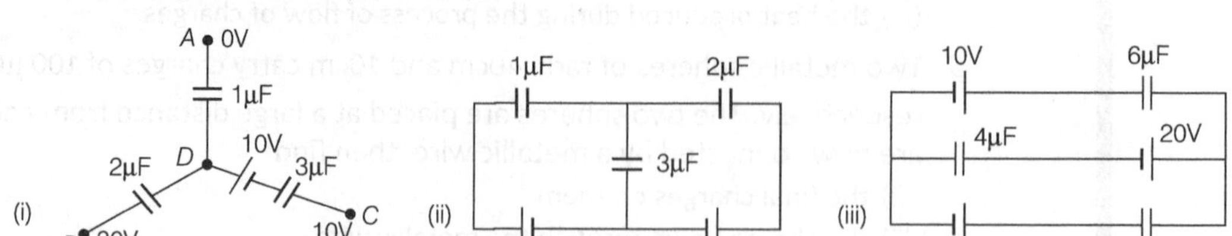

14 After closing the switch, find the charges on all the capacitors in the steady state in the following circuits using Kirchhoff s voltage law.

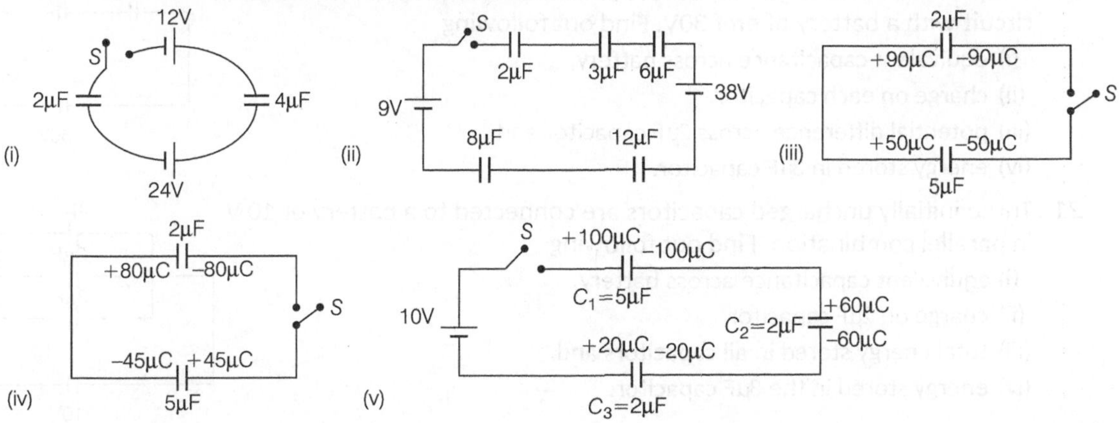

1.4 Redistribution of Charges

15 A capacitor of capacitance C which is initially uncharged is connected across a battery of emf E. Find work done by battery, energy stored in the capacitor and heat dissipated in the circuit during the process of charging.

16 In the following circuits, capacitor is initially charged as shown in figures. The switch is closed at $t = 0$, find heat produced in each circuit after a long time.

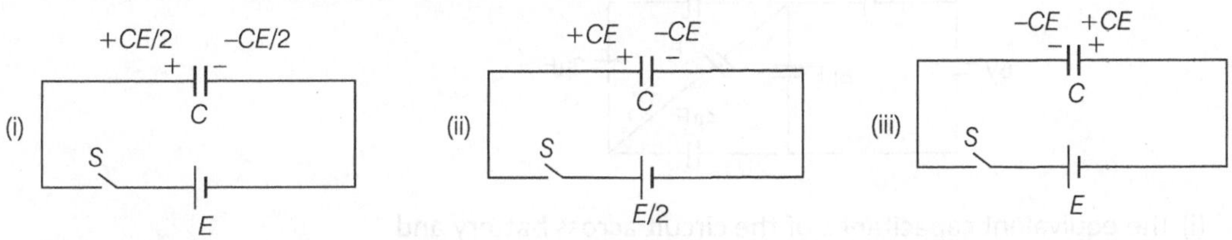

17 A 3 μF capacitor is charged to a potential of 100 V and is connected in parallel with an uncharged 2μF capacitor.

(i) What is the final potential difference across each capacitor?

(ii) What is the heat produced in the circuit?

18 Two capacitors of capacitances 2μF and 3μF are charged to potential difference 20 V and 10 V respectively as shown in figure.

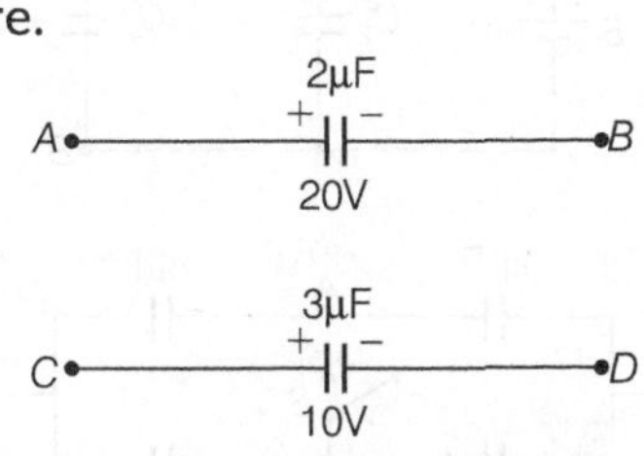

(i) If A is connected with C and B is connected with D, how much heat is produced in the circuit?

(ii) If A is connected with D and B is connected with C, how much heat is produced in the circuit?

19 Find the energy dissipated in the circuit after the switch is flipped from position 1 to 2.

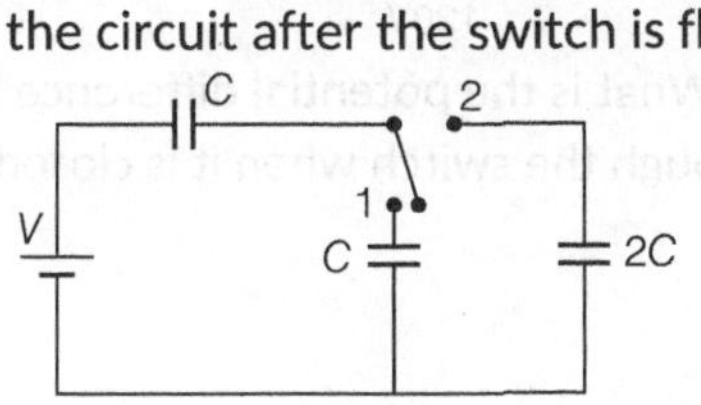

1.5 Series and Parallel Combinations

20 Three initially uncharged capacitors are connected in series as shown in circuit with a battery of emf 30V. Find out following

(i) equivalent capacitance across battery,
(ii) charge on each capacitor,
(iii) potential difference across 2μF capacitor and
(iv) energy stored in 3μF capacitor.

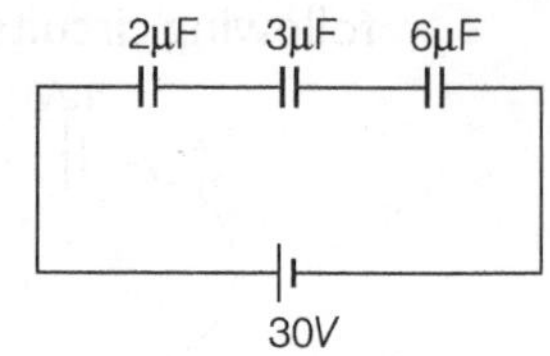

21 Three initially uncharged capacitors are connected to a battery of 10 V in parallel combination. Find out following

(i) equivalent capacitance across battery,
(ii) charge on 2μF capacitor,
(iii) total energy stored in all capacitors and
(iv) energy stored in the 3μF capacitor.

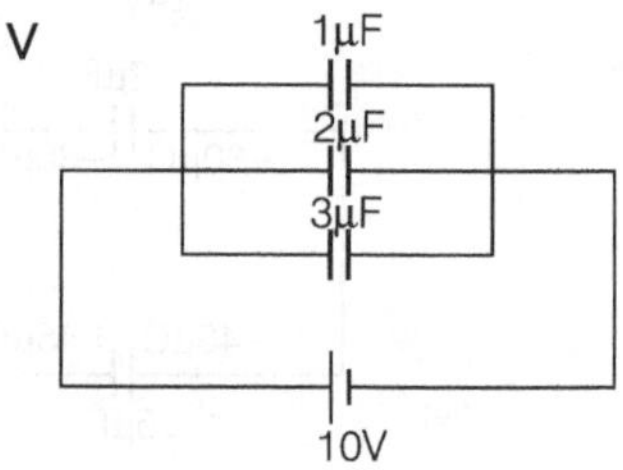

22 In the given circuit, find out

(i) charges on each capacitor,
(ii) potential difference across 1μF capacitor and
(iii) energy stored in 2μF capacitor.

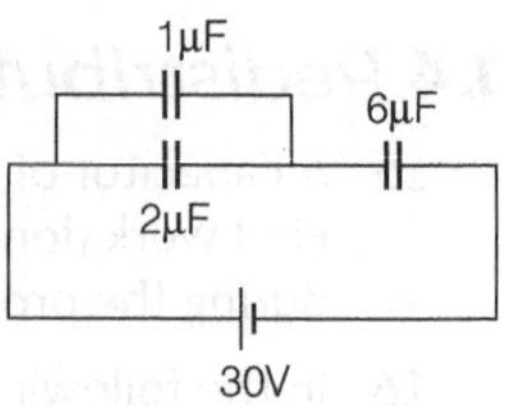

23 In the circuit shown, find

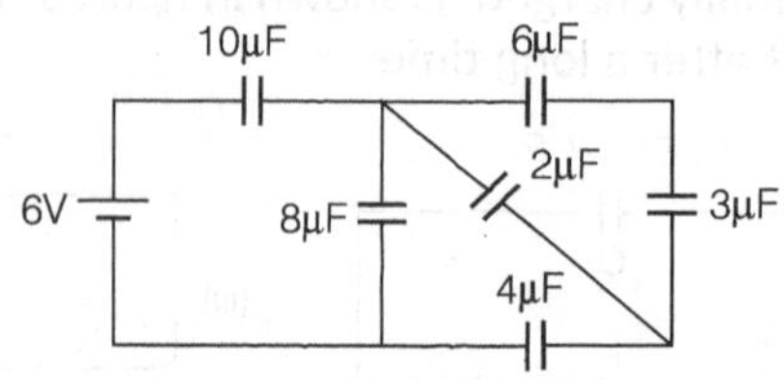

(i) the equivalent capacitance of the circuit across battery and
(ii) charge on each capacitor.

24 Find the potential difference between points A and B of the system shown in figure, if the emf is equal to $E = 110$ V and the capacitance ratio $C_2/C_1 = 2.0$.

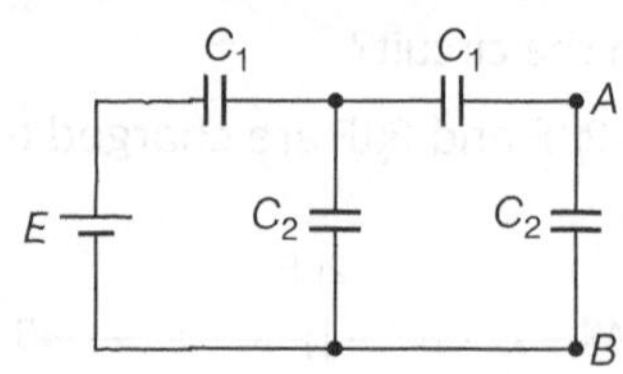

25

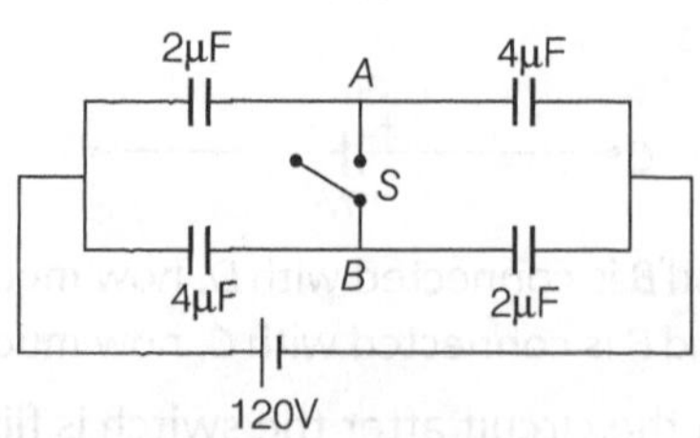

(i) Initially, switch S is open. What is the potential difference between A and B?
(ii) What charge will flow through the switch when it is closed?

26 Three capacitors each having capacitance $C = 2\mu F$ are connected with a batten of emf 30 V as shown in figure. When the switch S is closed, find

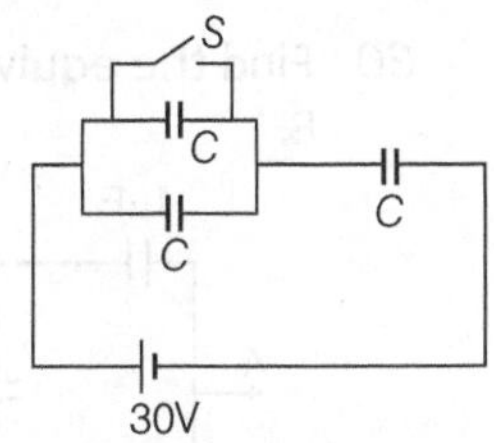

(i) the amount of charge flown through the battery,
(ii) the amount of charge flown through the switch S,
(iii) the energy supplied by the battery and
(iv) the heat generated in the circuit.

27 A capacitor of capacity $C_1 = 1\,\mu F$ withstands the maximum voltage $V_1 = 6$ kV while another capacitor of capacitance $C_2 = 2\,\mu F$ withstands the maximum voltage $V_2 = 4$ kV. What maximum voltage will the system of these two capacitors withstand, if they are connected in series?

28 From a supply of identical capacitors rated 8 μF, 200V, what is the minimum number of capacitors required to form a composite 16μF, 1000 V?

1.6 Equivalent Capacitance

29 Find the equivalent capacitance between the points A and B of the given assemblies.

(i)

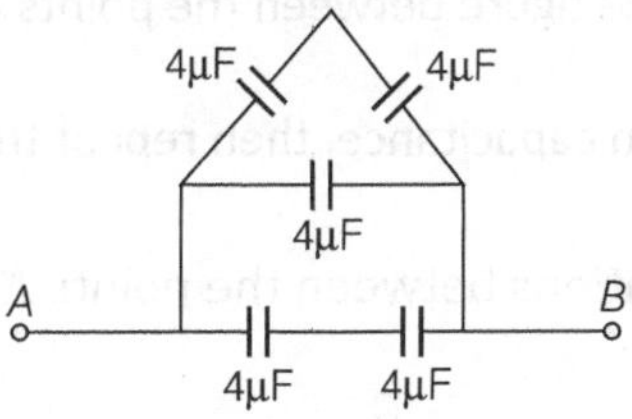

(ii)

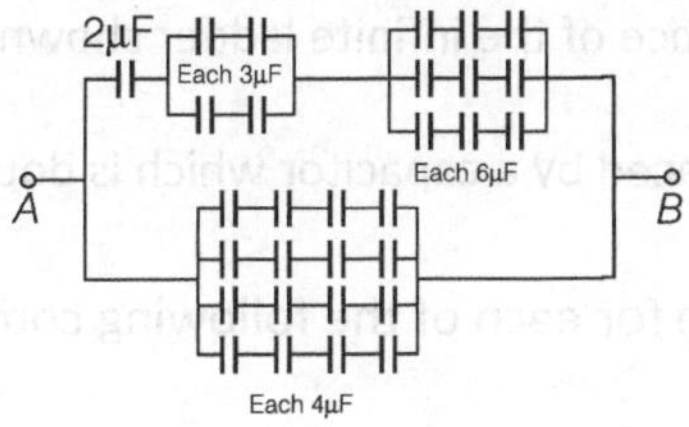

(iii)

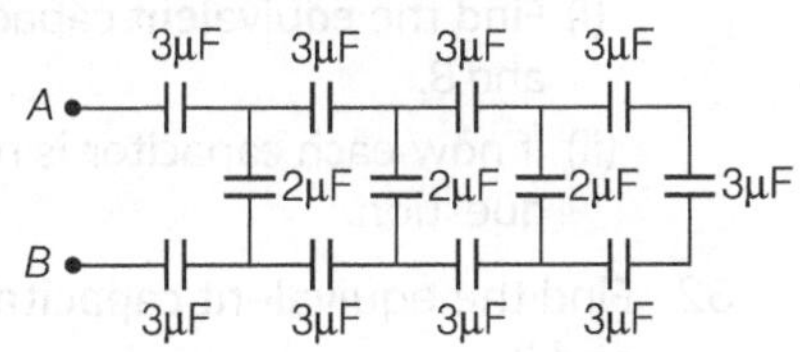

(iv)

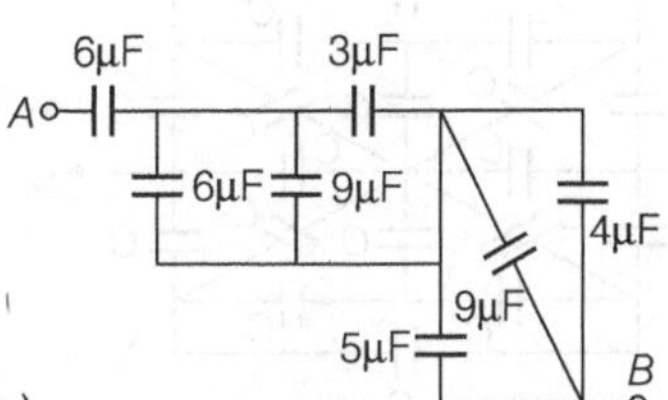

(v)

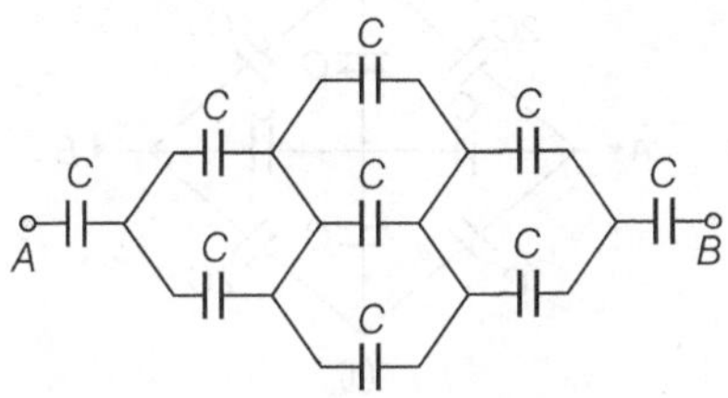

(vi)

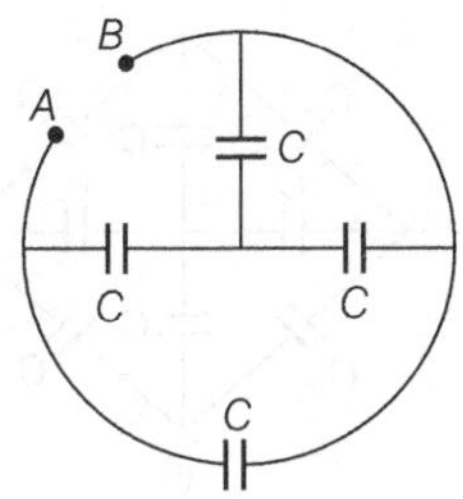

(vii)

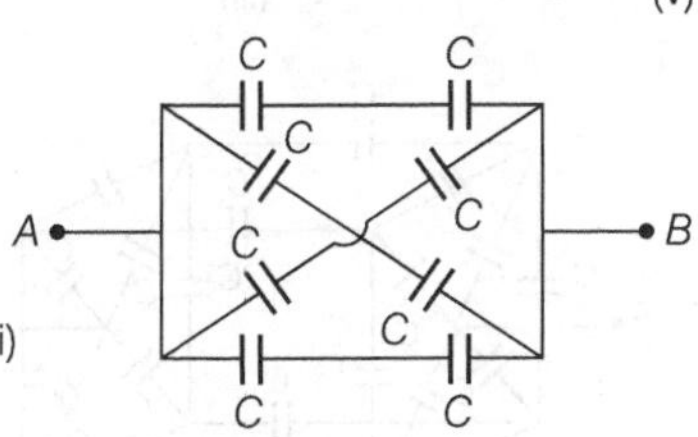

(viii)

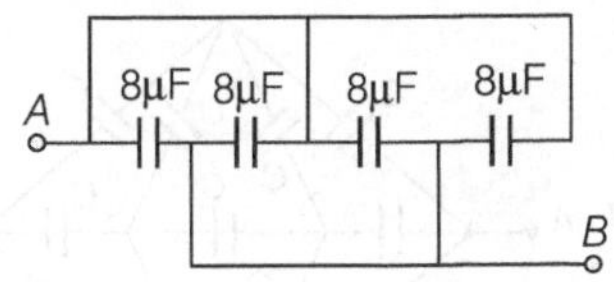

(ix)

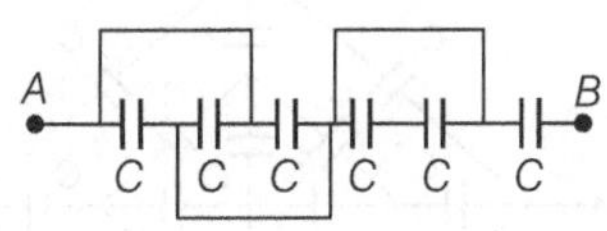

(x)

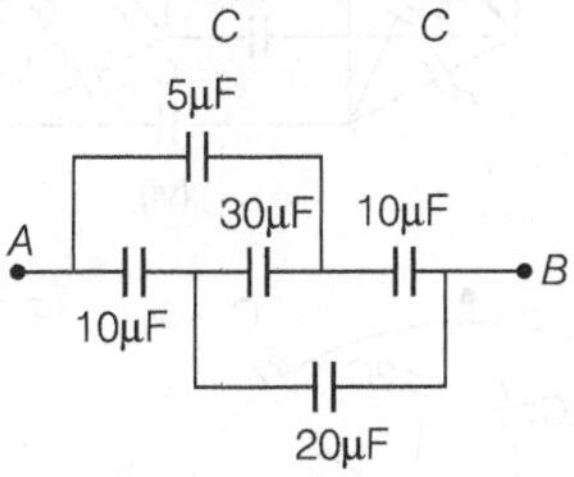

(xi)

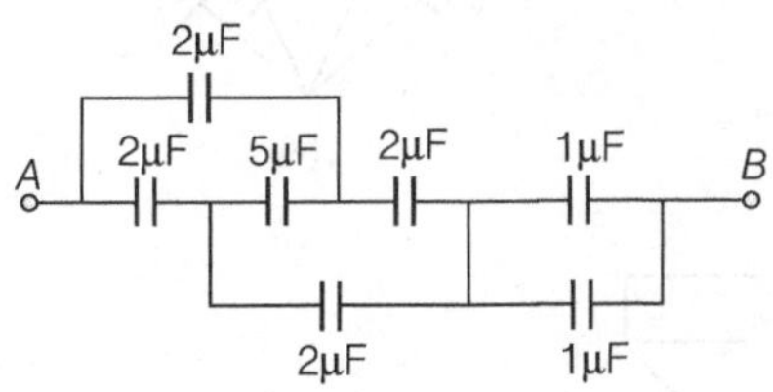

(xii)

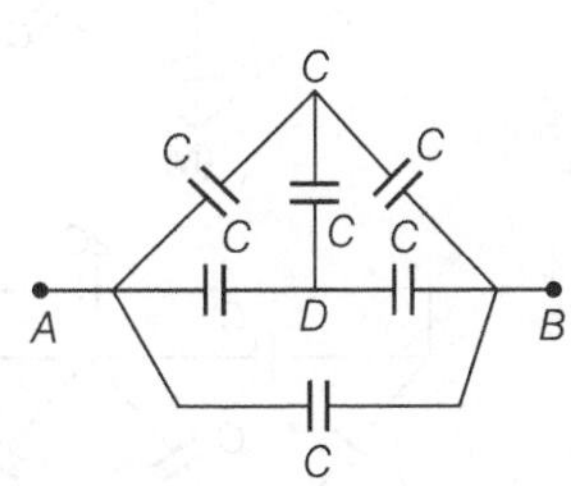

(xiii)

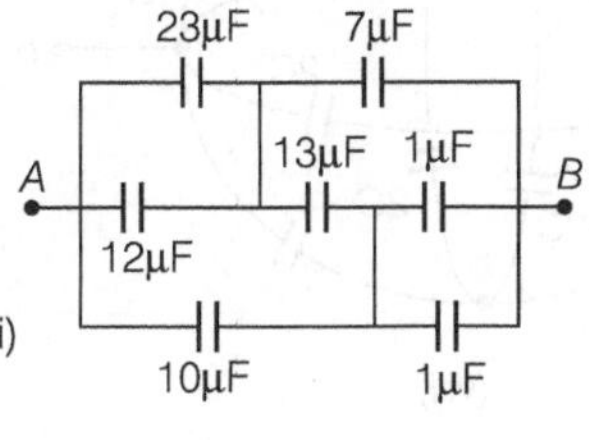

(xiv)

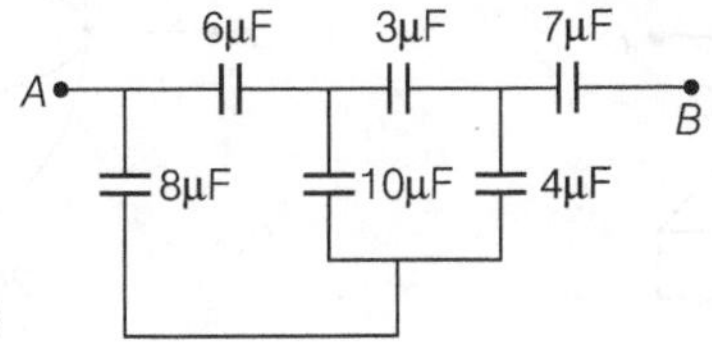

(xv)

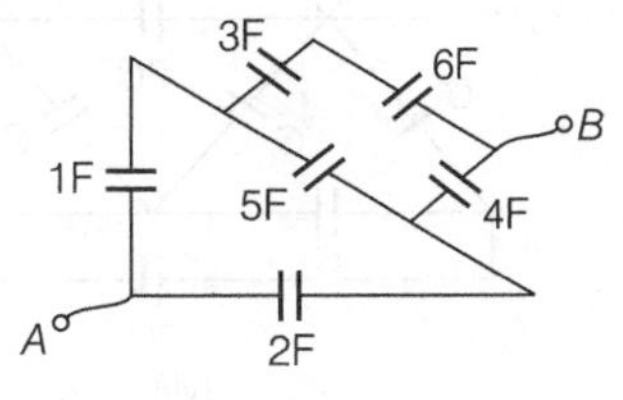

30 Find the equivalent capacitance of each of the following combination between the points A and B.

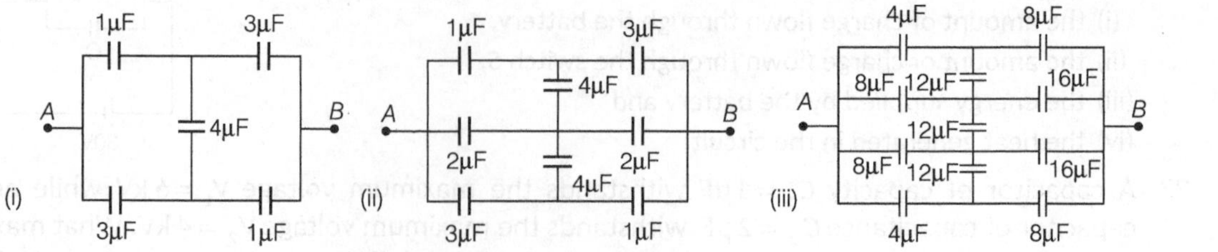

31 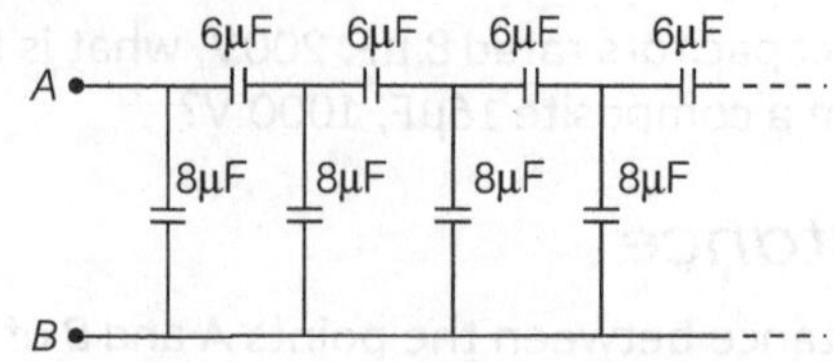

(i) Find the equivalent capacitance of the infinite ladder shown in the figure between the points A and B.

(ii) If now each capacitor is replaced by a capacitor which is double in capacitance, then repeat the question.

32 Find the equivalent capacitance for each of the following combinations between the points A and B.

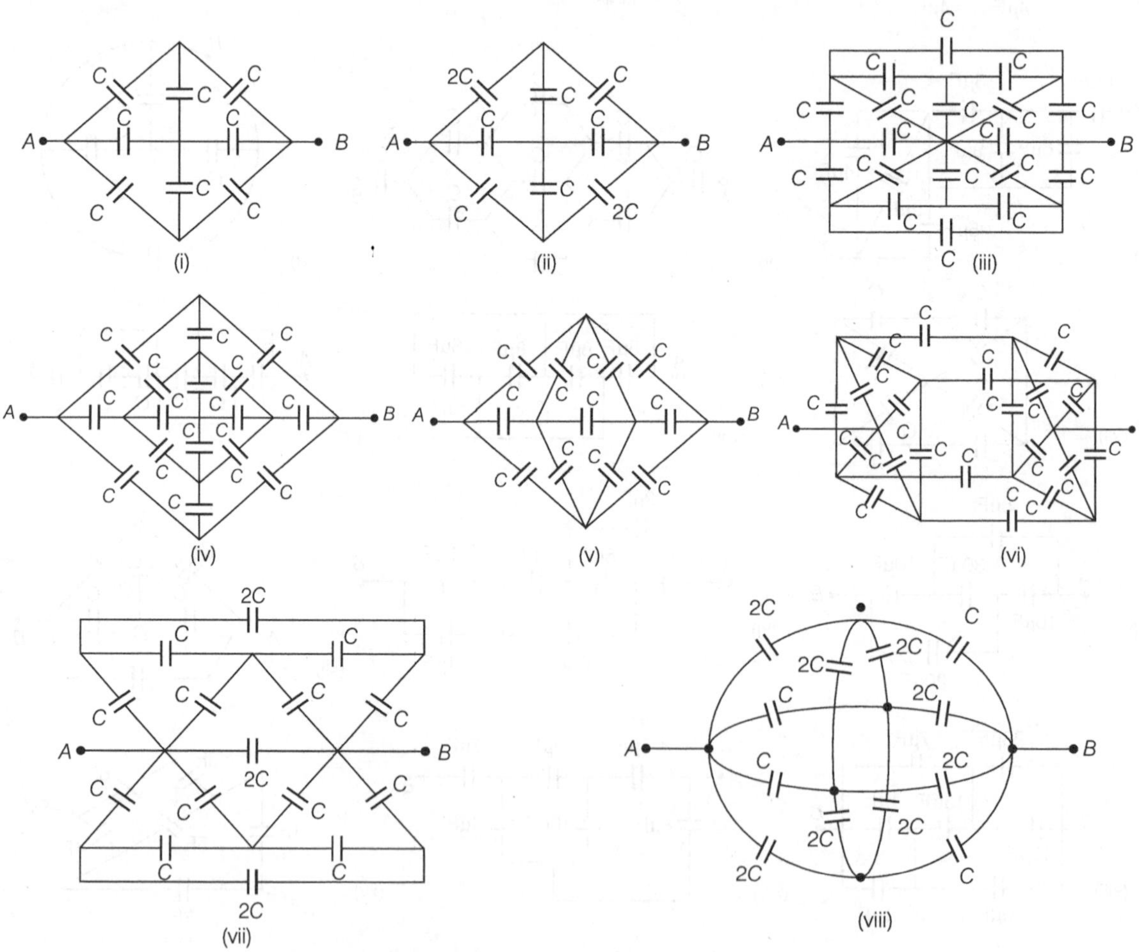

1.7 Combination of Parallel Plates

33 Find out equivalent capacitance between A and B. (Take, each plate area = A and distance between two successive plates is d)

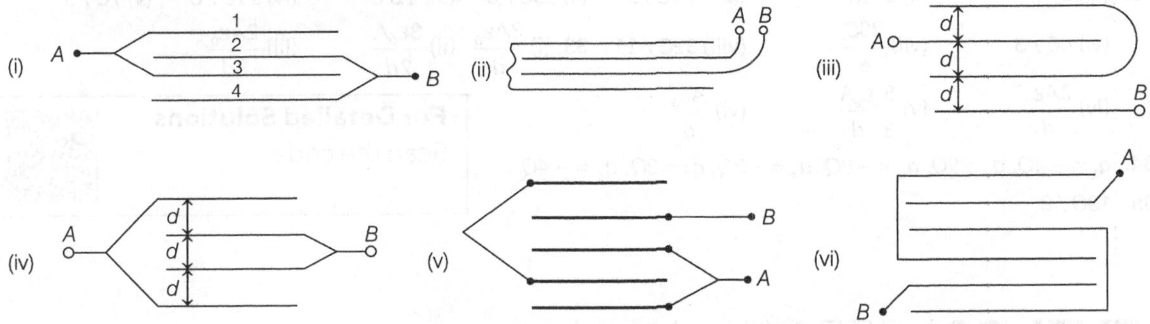

34 Three identical very large conducting plates A, B and C are placed parallel to each other. Plates A and C are connected by thin conducting wire. A charge + 4Q is given to plate A and − 12Q is given to plate B. Find the final charges on each surfaces of plates. Here, separation between A and B is L and between B and C is 3L.

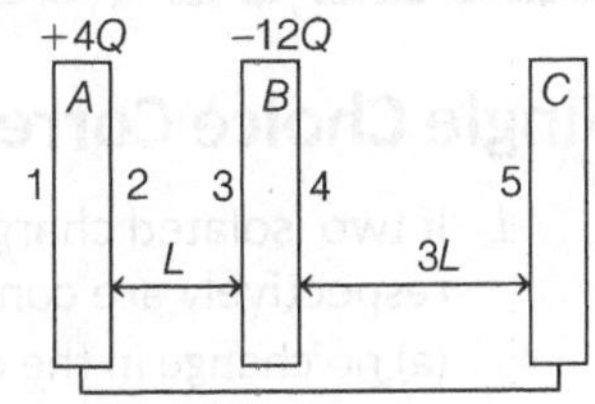

35 The metal plate on the left carries charge + Q and on the right has charge 4Q respectively as shown in figure. The central plate has initially charge equal to 5Q.
When both S_1 and S_2 switches are closed simultaneously, then find the charge flown through S_1 upto steady state.

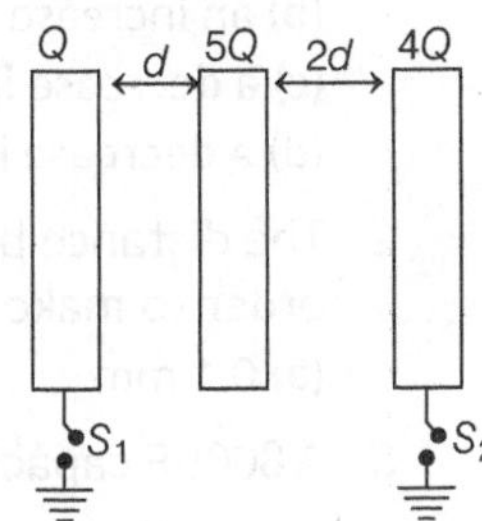

Answers

1. (i) 6V (ii) 90 μJ **2.** 711 μF **3.** $n^{1/3}C$ **4.** (i) $q_A = 3\mu C$, $q_B = 6\mu C$ (ii) 1V (iii) 2.25 μJ

5. (i) + 64 μC and + 16 μC (ii) 36 μC **6.** 32 nF **7.** (i) $\frac{A\varepsilon_0}{d}$ (ii) $\frac{q}{A\varepsilon_0}$ (iii) $\frac{qd}{A\varepsilon_0}$ (iv) $\frac{q^2 d}{2A\varepsilon_0}$

8. (i) $V = \frac{3Q}{2C}$ (ii) $\frac{9Q^2}{8C}$ **9.** (i) 8.85×10^{-12}F (ii) 8.85×10^{-10}C (iii) 4.425×10^{-8} J (iv) 10^4 V/m

(v) 4.425×10^{-12}F, 8.85×10^{-12}C, 8.85×10^{-10}J, 10^4 V/m **10.** $\sqrt{2mgA\varepsilon_0}$ **11.** 0.092 F

12. 1.21×10^{-10}F, 2.89×10^4V **13.** (i) $q_{1\mu F} = 20\mu C$, $q_{2\mu F} = 20\mu C$, $q_{3\mu F} = 0$ (ii) $q_{1\mu F} = 3\mu C$, $q_{2\mu F} = 6\mu C$, $q_{3\mu F} = 9\mu C$

(iii) $q_{4\mu F} = 8\mu C$, $q_{6\mu F} = 72\mu C$, $q_{10\mu F} = 80\mu C$ **14.** (i) 16μC on each (ii) 24μC on each

(iii) $q_{2\mu F} = 40\mu C$, $q_{5\mu F} = 100\mu C$ (iv) $q_{2\mu F} = 10\mu C$, $q_{5\mu F} = 25\mu C$ (v) $q_{C_1} = 75\mu C$, $q_{C_2} = 35\mu C$, $q_{C_3} = 45\mu C$

15. $CE^2, \frac{CE^2}{2}, \frac{CE^2}{2}$ **16.** (i) $\frac{CE^2}{8}$ (ii) $\frac{CE^2}{8}$ (iii) $2CE^2$ **17.** (i) 60 V (ii) 6 mJ **18.** (i) 60 μJ (ii) 540 μJ

19. $\frac{CV^2}{12}$ **20.** (i) 1μF (ii) 30μC (iii) 15 V (iv) 150 μJ **21.** (i) 6 μF (ii) 20 μC (iii) 300 μJ (iv) 150 μJ

22. (i) $q_{1\mu F} = 20\mu C$, $q_{2\mu F} = 40\mu C$, $q_{6\mu F} = 60\mu C$ (ii) 20 V (iii) 400 μJ

23. $C_{eq} = 5\mu F$, $q_{2\mu F} = q_{3\mu F} = q_{6\mu F} = 3\mu C$, $q_{4\mu F} = 6\mu C$, $q_{8\mu F} = 24\mu C$, $q_{10\mu F} = 30\mu C$ **24.** 10 V

25. (i) $V_A - V_B = -40$ V (ii) 120 μC **26.** (i) 20 μC (ii) 60 μC (iii) 0.6 mJ (iv) 0.3 mJ

27. 9 kV **28.** 50

29. (i) $8\,\mu F$ (ii) $5\,\mu F$ (iii) $1\,\mu F$ (iv) $3.6\,\mu F$ (v) $3C/10$ (vi) $5C/3$ (vii) $2C$ (viii) $32\,\mu F$ (ix) $3C/4$ (x) $10\,\mu F$ (xi) $1\,\mu F$ (xii) $2C$ (xiii) $7.5\,\mu F$ (xiv) $2.8\,\mu F$ (xv) $2F$

30. (i) $11/6\,\mu F$ (ii) $11/4\,\mu F$ (iii) $16\,\mu F$

31. (i) $12\,\mu F$ (ii) $24\,\mu F$

32. (i) $3C/2$ (ii) $15C/8$ (iii) $1.3\,C$ (iv) $11C/8$ (v) $7C/5$ (vi) $4C/3$ (vii) $\frac{23C}{6}$ (viii) $32C/11$

33. (i) $\frac{3A\varepsilon_0}{d}$ (ii) $\frac{3\varepsilon_0 A}{2d}$ (iii) $\frac{2A\varepsilon_0}{3d}$ (iv) $\frac{2A\varepsilon_0}{d}$ (v) $\frac{5}{3}\frac{\varepsilon_0 A}{d}$ (vi) $\frac{A\varepsilon_0}{d}$

34. $q_1 = -4Q, q_2 = 9Q, q_3 = -9Q, q_4 = -3Q, q_5 = 3Q, q_6 = -4Q$

35. $13Q/3$

For Detailed Solutions
Scan the code

LEVEL 02 \ JEE Main Level

Single Choice Correct Questions

1 If two isolated charged metallic spheres of radii R_1 and R_2 having charges Q_1 and Q_2 respectively are connected to each other, then there is
(a) no change in the electrical energy of the system
(b) an increase in the electrical energy of the system
(c) a decrease in the electrical energy of the system in any case
(d) a decrease in electrical energy of the system, if $Q_1R_2 \neq Q_2R_1$

2 The distance between the plates of a circular parallel plate condenser of diameter 40 mm, in order to make its capacity equal to that of a metallic sphere of radius 1m, will be
(a) 0.1 mm (b) 10 mm (c) 0.01 mm (d) 1 mm

3 A $800\mu F$ capacitor is charged at a constant rate of 50×10^{-6} A. The time after which the voltage across the capacitor becomes 10 V is
(a) 160 s (b) 50 s (c) 10 s (d) 500 s

4 If the charge on a capacitor is increased by 2C, the energy stored in it increases by 21%. The original charge on the capacitor (in C) is
(a) 10 (b) 20 (c) 30 (d) 40

5 The plates of a parallel plate condenser are being moved away with a constant speed v. If the plate separation at any instant of time is d, then the rate of change of capacitance with time is proportional to
(a) $\frac{1}{d}$ (b) $\frac{1}{d^2}$ (c) d^2 (d) d

6 If the plates of a parallel plate capacitor are moved further apart to a distance $2d$ from initial distance d with a battery connected across, which of the following statements is correct?
(a) The charge on each plate doubles.
(b) The electric field in the region between the plates reduces to half of its initial value.
(c) The voltage across the plates doubles.
(d) The charge on each plate remains unaffected.

7 A parallel plate capacitor of $C = 12\mu F$ is charged to a potential difference of 10V and the battery disconnected. Now, the plates are pulled further apart to double the initial distance. The work done by the external agent in slowly pulling the plates further apart is equal to
(a) $-600\,\mu J$ (b) $300\,\mu J$ (c) $600\,\mu J$ (d) $-300\,\mu J$

8 A photographic flash unit consists of a Xenon-filled flash tube energised by the discharge of a capacitor, previously charged by a 1000 V source. The average power delivered to the tube is 2000 W in a time of 0.04 s. The capacitance of the capacitor can be estimated as

(a) 40 μF (b) 80 μF (c) 160 μF (d) None of these

9 Capacity of an isolated sphere is increased n times when it is enclosed by an earthed concentric sphere. The ratio of their radii would be

(a) $\frac{n^2}{n-1}$ (b) $\frac{n}{n-1}$ (c) $\frac{2n}{n+1}$ (d) $\frac{2n+1}{n+1}$

10 One plate of the capacitor is held fixed and the other plate is connected to a spring as shown in the figure. The capacitor has $C = 10\mu F$ and distance between the plates $d = 1$ mm when the switch was open. Now, with the switch closed, the distance is reduced to $0.8d$. Assume the spring was initially in a relaxed state. If $E = 4V$, then the spring constant k is equal to

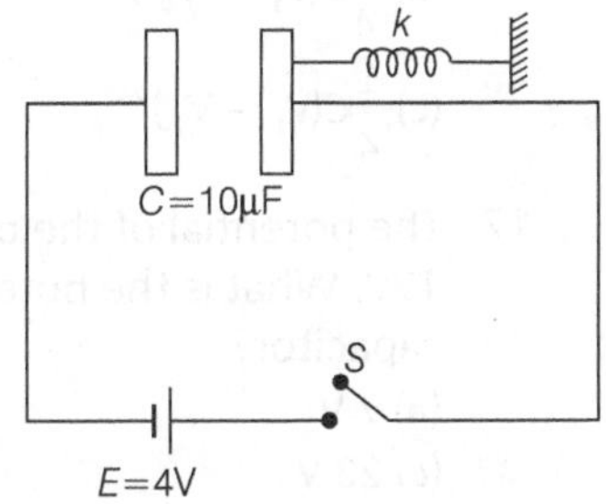

(a) 1500 N/m (b) 35 N/m

(c) 625 N/m (d) 125 N/m

11 A capacitor of capacitance C is charged to a potential difference V from a cell and then disconnected from it. A charge $+Q$ is now given to its positive plate. The potential difference across the capacitor is

(a) V (b) $V+\frac{Q}{C}$ (c) $V+\frac{Q}{2C}$ (d) $V-\frac{Q}{C}$, if $Q < CV$

12 In the figure shown, the plates of a parallel plate capacitor have unequal charges. Its capacitance is C and P is a point outside the capacitor and close to the plate of charge $-Q$. The distance between the plates is d, then which of the following statement is incorrect?

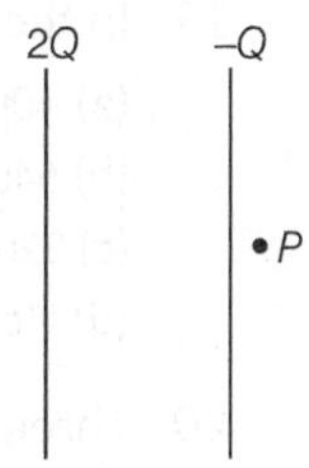

(a) A point charge at point P will experience electric force due to the capacitor.

(b) The potential difference between the plates will be $\frac{3Q}{2C}$.

(c) The energy stored in the electric field in the region between the plates is $\frac{9Q^2}{8C}$.

(d) The force on one plate due to the other plate is $\frac{9Q^2}{8\pi\varepsilon_0 A}$.

13 A capacitor having capacitance C is charged by a battery having emf $2V$ (battery is disconnected after charging). Now, this charged capacitor is connected to another battery having emf V. The positive plate of capacitor is connected with negative terminal of battery and *vice-versa*. The amount of heat produced after connecting it with battery

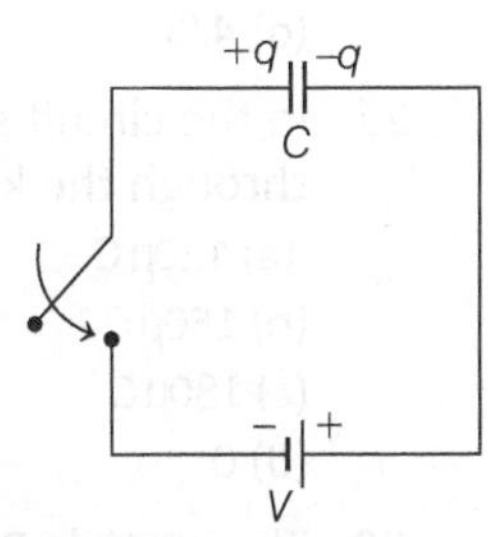

(a) $4.5\, CV^2$ (b) $1.5CV^2$

(c) $4\, CV^2$ (d) None of these

14 A $4\mu F$ capacitor is charged by a 200 V supply. It is then disconnected from the supply and is connected to another uncharged $2\mu F$ capacitor. What percentage of electrostatics energy of the first capacitor is dissipated in the form of heat and electromagnetic radiation?

(a) 20% (b) 25 % (c) 33.33 % (d) 40 %

15 A parallel plate capacitor of capacitance C is connected to a battery and is charged to potential difference V. Another capacitor of capacitance $2C$ is similarly charged to a potential difference $2V$. The charging battery is then disconnected and the capacitors are connected in parallel to

each other in such a way that the positive terminal of one is connected to the negative terminal of the other. The final energy of the configuration is

(a) zero (b) $\frac{1}{2}CV^2$ (c) $\frac{25}{6}CV^2$ (d) $\frac{9}{2}CV^2$

16 Two identical capacitors, have the same capacitance C. One of them is charged to potential V_1 and the other to V_2. The negative ends of the capacitors are connected together. When the positive ends are also connected, the decrease in energy of the combined system is

(a) $\frac{1}{4}C(V_1^2 - V_2^2)$ (b) $\frac{1}{4}C(V_1^2 + V_2^2)$

(c) $\frac{1}{4}C(V_1 - V_2)^2$ (d) $\frac{1}{4}C(V_1 + V_2)^2$

17 The potential of the point A is greater than that of B by 19V. What is the potential difference across the 3μF capacitor?

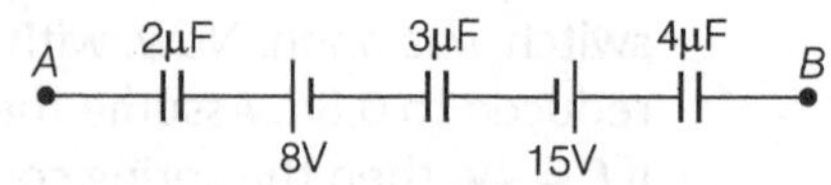

(a) 7 V (b) 8 V

(c) 23 V (d) 4 V

18 For the given circuit, the charge that appears across the 2μF capacitor is

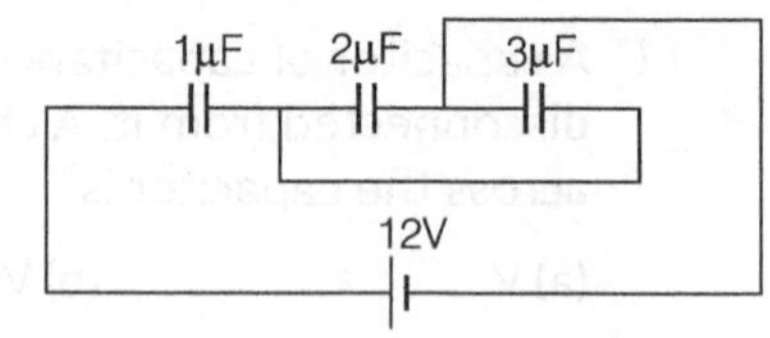

(a) 1μC (b) 4μC

(c) 3μC (d) 8μC

19 In the circuit shown, the energy stored in 1μF capacitor is

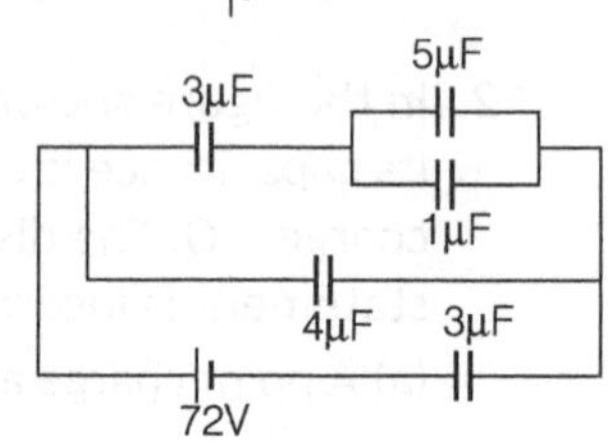

(a) 40μJ

(b) 64μJ

(c) 32μJ

(d) None of the above

20 Three capacitors A, B and C are connected in a circuit as shown in figure. What is the charge (in μC) on the capacitor B?

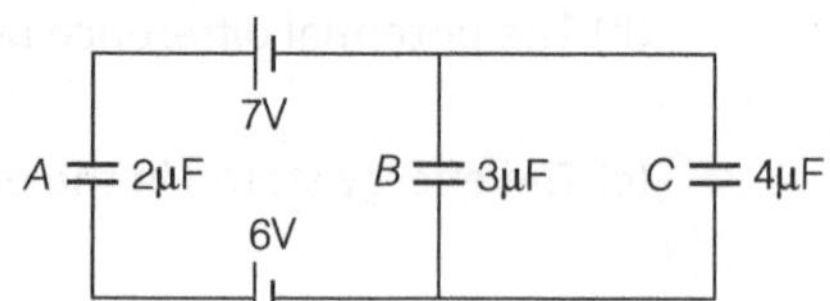

(a) 1/3

(b) 2/3

(c) 1

(d) 4/3

21 In the circuit shown in figure, the amount of charge that will flow through the key when it is closed is

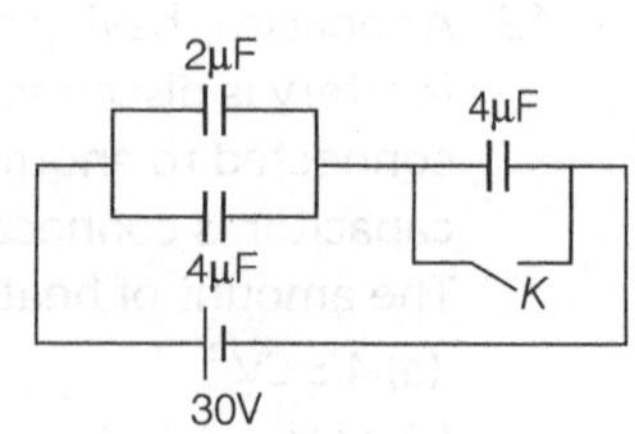

(a) 120μC

(b) 150μC

(c) 180μC

(d) 0

22 The particle P shown in figure has a mass of 10 mg and a charge of – 0.01 μC. The plates are horizontal and each plate has a surface area 100 cm^2. What potential difference V should be applied to the combination to hold the particle P in equilibrium?

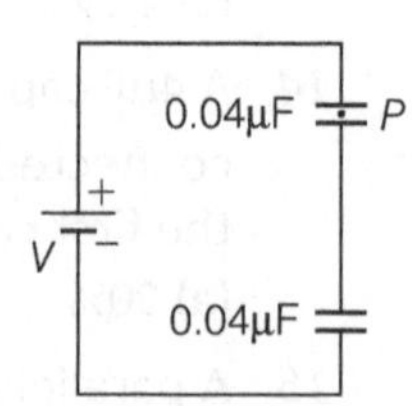

(a) 13 mV (b) 23 mV

(c) 43 mV (d) 93 mV

23 Figure shows two capacitors C_1 and C_2 connected with 10 V battery and terminals *A* and *B* are earthed. The graph shows the variation of potential as one moves from left to right, then the ratio of $\frac{C_1}{C_2}$ is

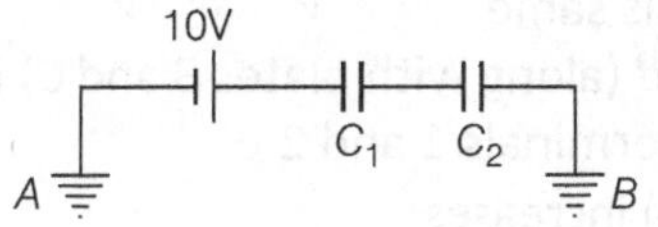

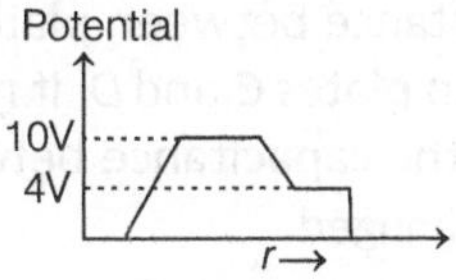

(a) $\frac{5}{2}$ (b) $\frac{2}{3}$ (c) $\frac{2}{5}$ (d) $\frac{3}{2}$

24 In the given circuit, if point *C* is connected to the earth and a potential of + 2000 V is given to the point *A*, then the potential at *B* is (assume circuit is complete)

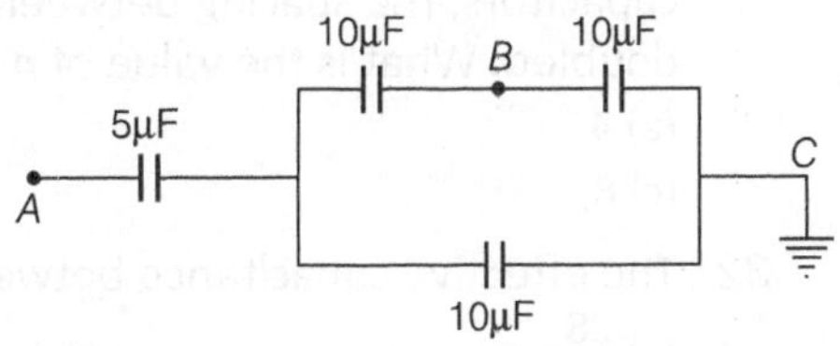

(a) 250 V (b) 500 V
(c) 750 V (d) 200 V

25 In the following figure, the charge and potential difference across 8μF capacitance will be respectively (approximately)

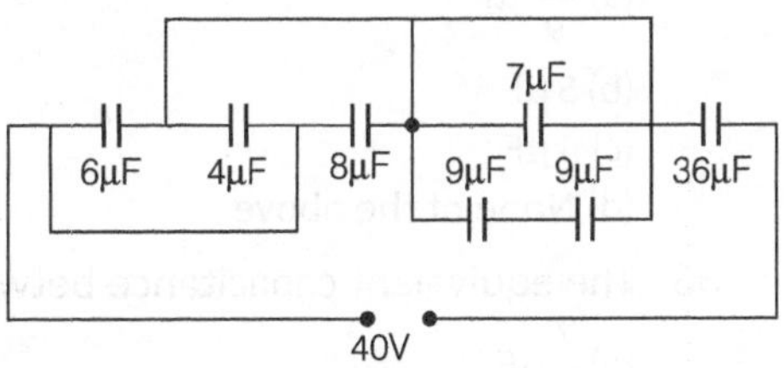

(a) 320 μC, 40 V
(b) 420 μC, 50 V
(c) 214 μC, 27 V
(d) 360 μC, 45 V

26 Four capacitors and two sources of emf are connected as shown in the figure. The potential difference between the points *a* and *b* is

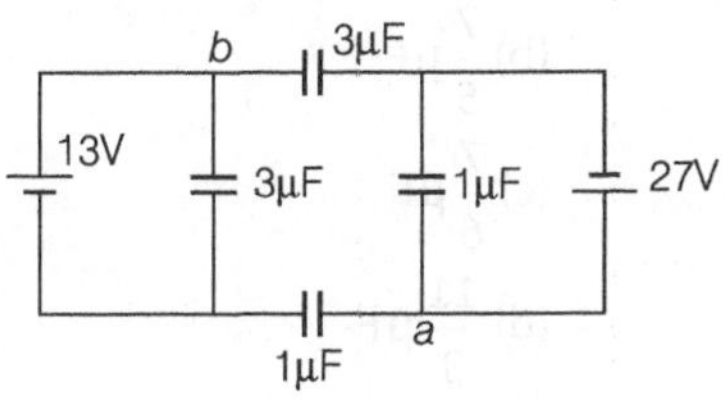

(a) zero
(b) 13 V
(c) 17 V
(d) 27 V

27 In the circuit when switch *S* is closed, the amount of charge passing through the switch is xμC, then the value of *x* is

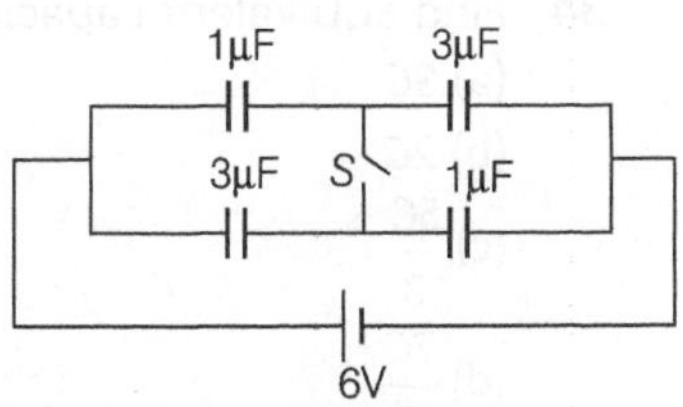

(a) 12
(b) 6
(c) 3
(d) 1

28 Find the charge flown through the switch when it is closed.

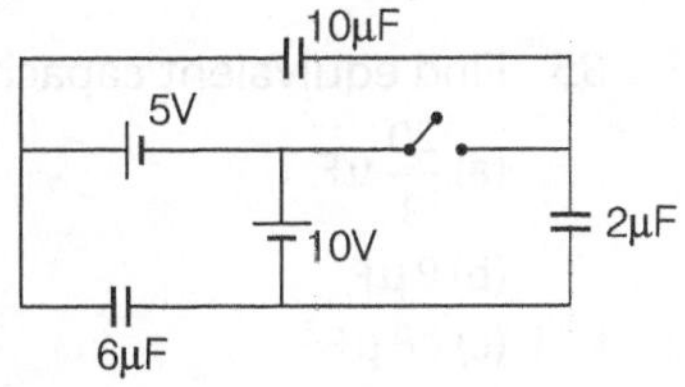

(a) 20 μC
(b) 30 μC
(c) 42 μC
(d) 60 μC

29 The circuit involves two ideal cells connected to a 1 μF capacitor *via* a key *K*. Initially, the key *K* is in position 1 and the capacitor is charged fully by 2V cell. The key is then pushed to position 2. The heat produced in circuit (in μJ) after pushing the key to position 2 is

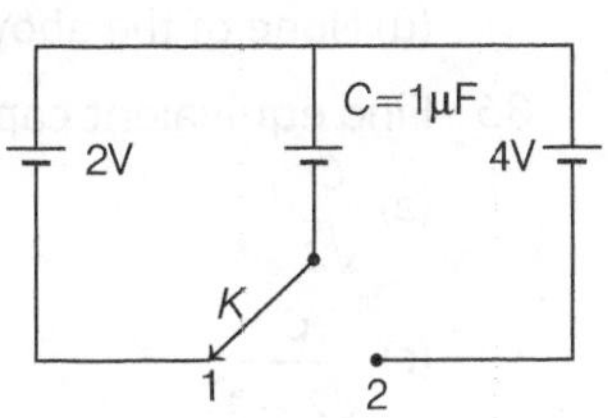

(a) 2 (b) 6
(c) 8 (d) 16

30 A, B, C and D are large conducting plates kept parallel to each other. A and D are fixed. Plates B are C connected to each other by a rigid conducting rod can slide over frictionless rails as shown in figure. Initially, the distance between plates A and B is same as that between plates C and D. If now the rod (along with plates B and C) is slightly moved towards right, the capacitance between the terminals 1 and 2

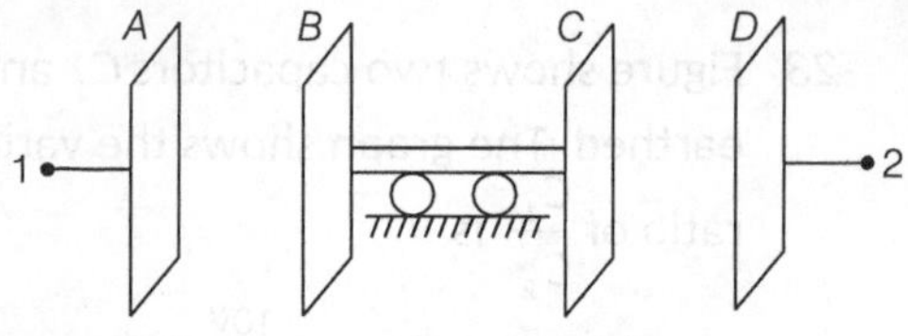

(a) remains unchanged
(b) increases
(c) decreases
(d) Nothing can be said

31 Twelve identical capacitors are connected in series between two points. Out of these for n capacitors, the spacing between the plates is reduced to half and for the remaining it is doubled. What is the value of n for which the effective capacitance remains unchanged?
(a) 4
(b) 6
(c) 8
(d) 2

32 The effective capacitance between points A and B will be

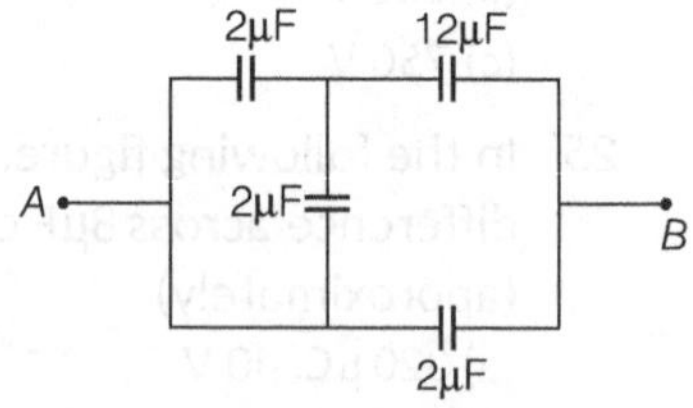

(a) $\frac{28}{9}$ μF
(b) 5 μF
(c) 4 μF
(d) None of the above

33 The equivalent capacitance between points A and B is

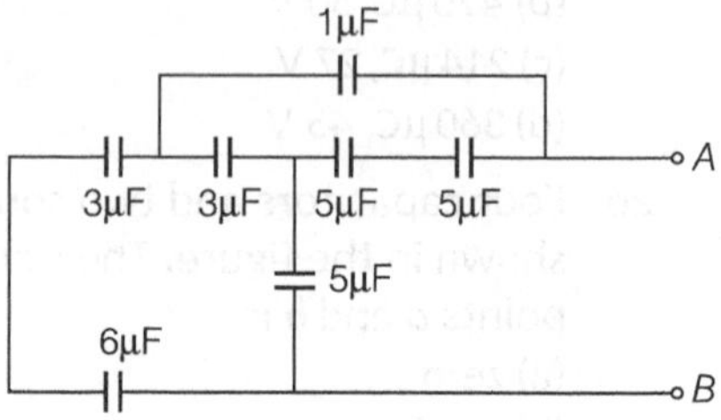

(a) $\frac{7}{3}$ μF
(b) $\frac{7}{5}$ μF
(c) $\frac{7}{6}$ μF
(d) $\frac{11}{3}$ μF

34 Find equivalent capacitance between A and B.

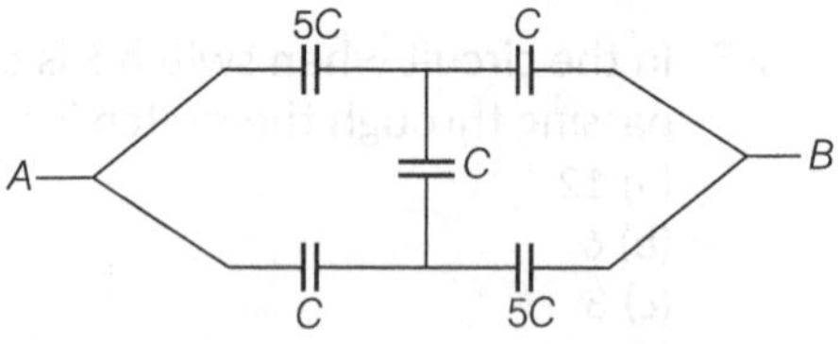

(a) 3C
(b) 2C
(c) $\frac{5C}{3}$
(d) $\frac{7C}{5}$

35 Find equivalent capacitance across AB (all capacitances in μF).

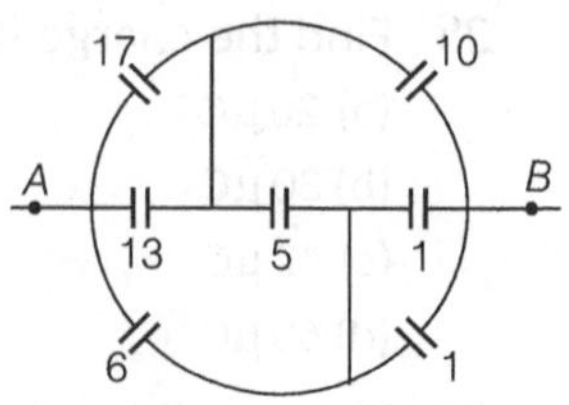

(a) $\frac{20}{3}$ μF
(b) 9 μF
(c) 48 μF
(d) None of the above

36 Find equivalent capacitance between points A and B.

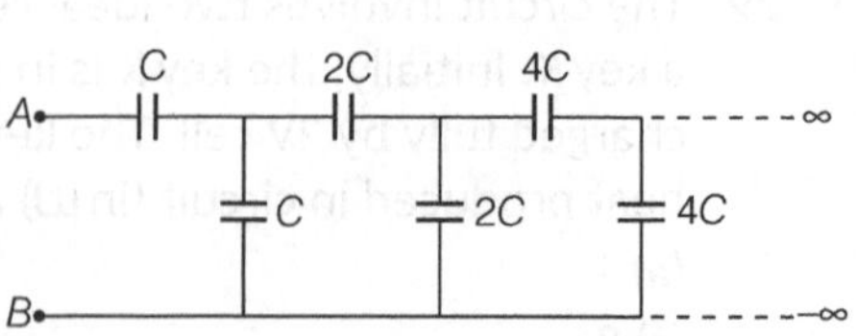

(a) $\frac{C}{\sqrt{2}}$
(b) $\frac{C}{\sqrt{2}+1}$
(c) $\frac{C}{\sqrt{2}-1}$
(d) $\frac{C}{2\sqrt{2}}$

37 Each edge of the cube contains a capacitance C. The equivalent capacitance between the points A and B will be

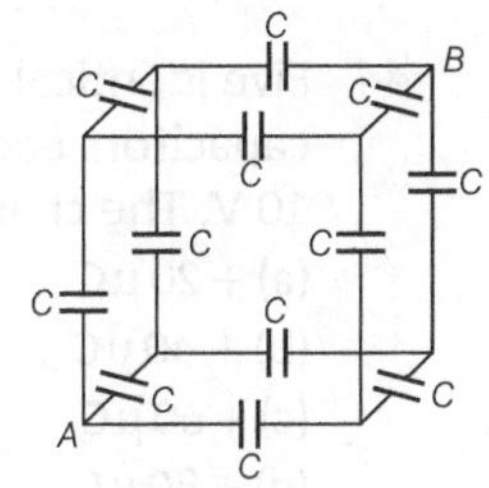

(a) $\frac{6C}{5}$ (b) $\frac{5C}{6}$

(c) $\frac{12C}{7}$ (d) $\frac{7C}{12}$

38 Three capacitors 2 μF, 3 μF and 5 μF can withstand voltages to 3V, 2V and 1V, respectively. Their series combination can withstand a maximum voltage equal to

(a) 5 V (b) 6 V (c) (26/5)V (d) (31/6) V

39 A circuit is made using five identical capacitors, each of which can withstand a maximum voltage of 2 V. The maximum voltage which can be safely applied across points A and B is

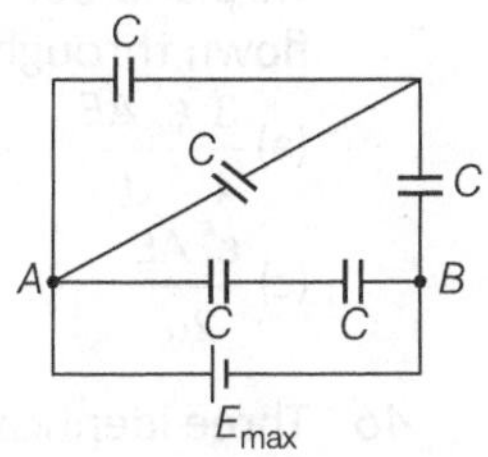

(a) 1 V
(b) 2 V
(c) 3 V
(d) 4V

40 Two spherical conductors A_1 and A_2 of radii r_1 and r_2 are placed concentrically in air. The two are connected by a copper wire as shown in figure, then the equivalent capacitance of the system is

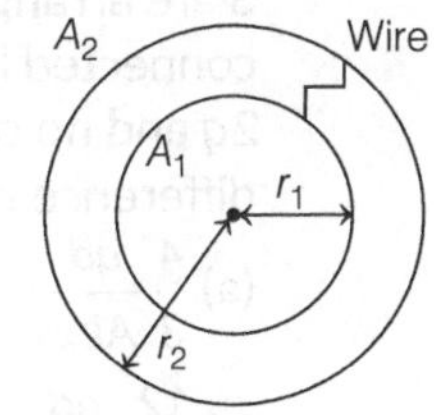

(a) $\frac{4\pi\varepsilon_0 r_1 r_2}{(r_2 - r_1)}$ (b) $4\pi\varepsilon_0 (r_1 + r_2)$

(c) $4\pi\varepsilon_0 r_2$ (d) $4\pi\varepsilon_0 r_1$

41 A capacitor is made of a flat plate of area A and a second plate having a stair like structure as shown in figure. The capacitance of the arrangement is

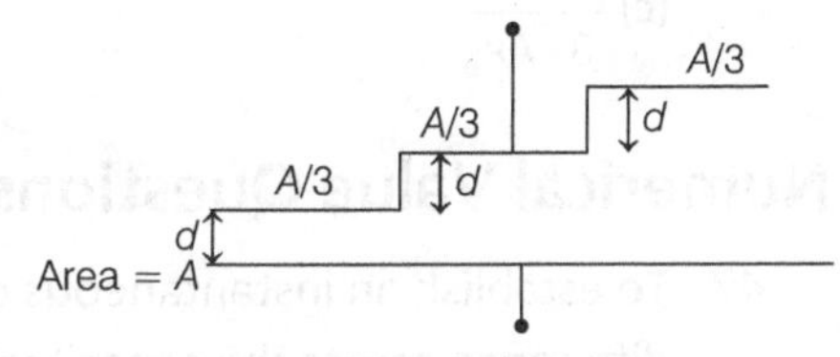

(a) $\frac{\varepsilon_0 A}{d}$ (b) $\frac{18}{11}\frac{\varepsilon_0 A}{d}$

(c) $\frac{11}{18}\frac{\varepsilon_0 A}{d}$ (d) $\frac{\varepsilon_0 A}{3d}$

42 Two conducting large plates P_1 and P_2 are placed parallel to each other at very small, separation d. The plate area of either face of plate is A. A charge $+2Q$ is given to plate P_1 and $-Q$ to the plate P_2 (neglect ends effects). If plates P_1 and P_2 are now connected by a conducting wire, then total amount of heat produced is

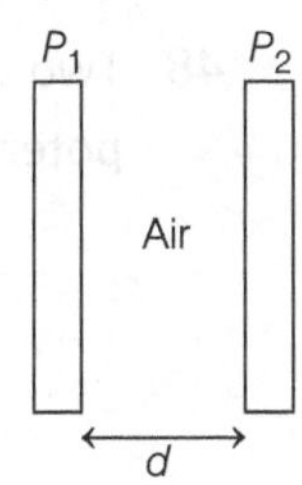

(a) $\frac{4Q^2 d}{3\varepsilon_0 A}$ (b) $\frac{9Q^2 d}{8\varepsilon_0 A}$

(c) $\frac{3Q^2 d}{4\varepsilon_0 A}$ (d) None of these

43 Find equivalent capacitance between A and B. [Assume, each conducting plate is having same dimensions and neglect the thickness of the plate, $\frac{\varepsilon_0 A}{d} = 7\mu F$, where A is area of plates and $A >> d$].

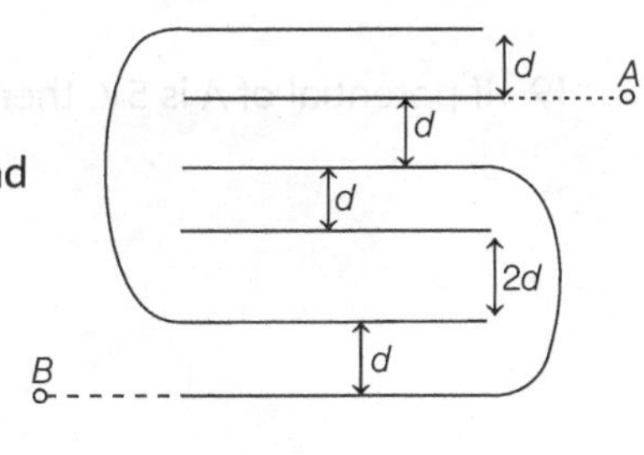

(a) 7 μF (b) 11 μF

(c) 12 μF (d) 13 μF

44 Five identical capacitor plates are arranged such that they make capacitors each of 2 μF. The plates are connected to a source of emf 10 V. The charge on plate *C* is

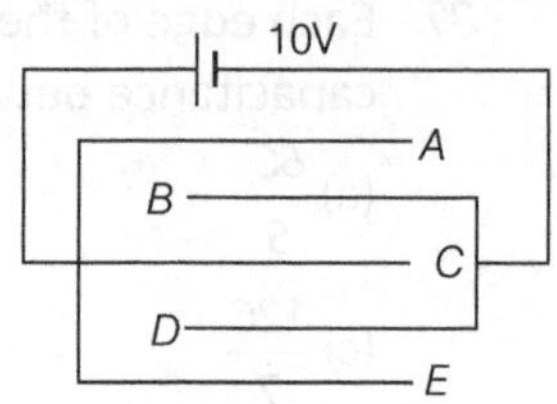

(a) + 20 μC
(b) + 40 μC
(c) + 60 μC
(d) + 80 μC

45 Five conducting parallel plates having area *A* and separation between them *d*, are placed as shown in the figure. Plate numbers 2 and 4 are connected wire and between point *A* and *B*, a cell of emf *E* is connected. The charge flown through the cell is

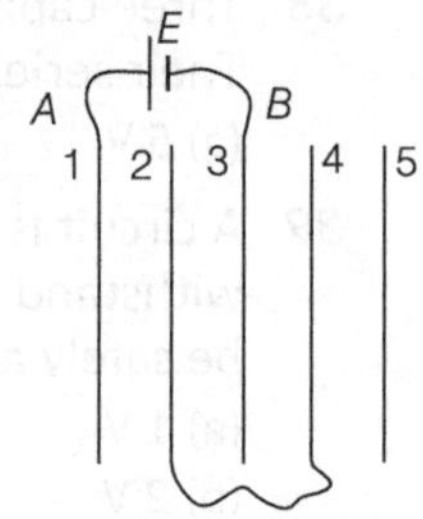

(a) $\frac{3}{4}\frac{\varepsilon_0 AE}{d}$
(b) $\frac{4\varepsilon_0 AE}{d}$
(c) $\frac{\varepsilon_0 AE}{2d}$
(d) $\frac{2}{3}\frac{\varepsilon_0 AE}{d}$

46 Three identical metal plates of large area *A* numbered 1,2 and 3 are arranged as shown in the figure. The plates 1 and 3 are connected by a conducting wire. The charge given to plate 2 is 2*q* and no other plate is given any charge. Find the potential difference between the plates 2 and 3.

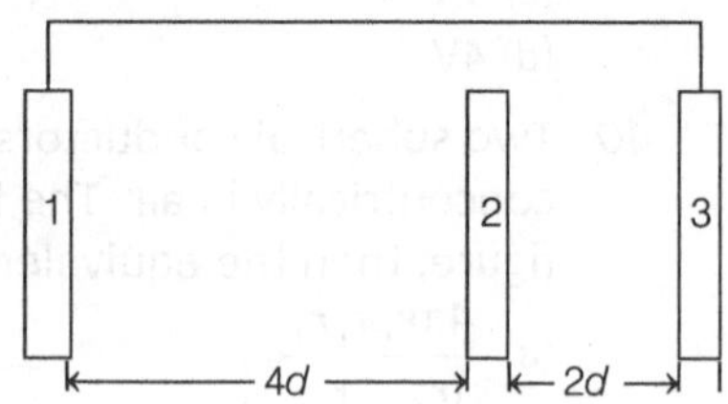

(a) $\frac{4}{3}\frac{qd}{A\varepsilon_0}$
(b) $\frac{8}{3}\frac{qd}{A\varepsilon_0}$
(c) $\frac{12}{3}\frac{qd}{A\varepsilon_0}$
(d) $\frac{2}{3}\frac{qd}{A\varepsilon_0}$

Numerical Value Questions

47 To establish an instantaneous current of 2mA through a 50 μF capacitor, the potential difference across the capacitor plates should be changed at the rate of β volt/second. Find the value of β.

48 Two uncharged capacitors are connected across a battery as shown in figure. V_A, V_B and V_C are potentials at the points *A*, *B* and *C*, respectively. What is the value of $V_A + V_B + V_C$ (in volts)?

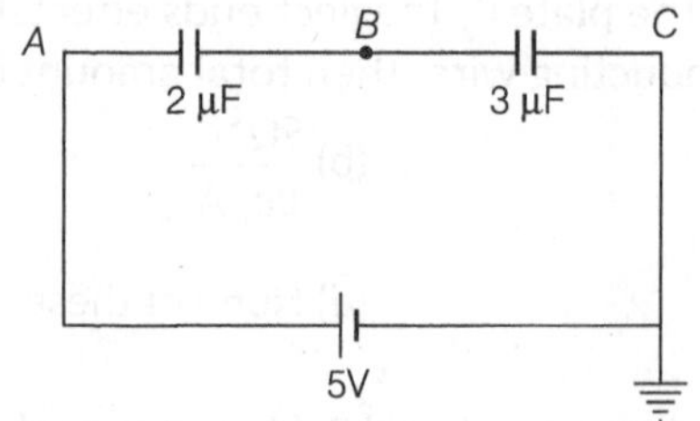

49 If potential of *A* is 5V, then find the potential of *B* (in volt).

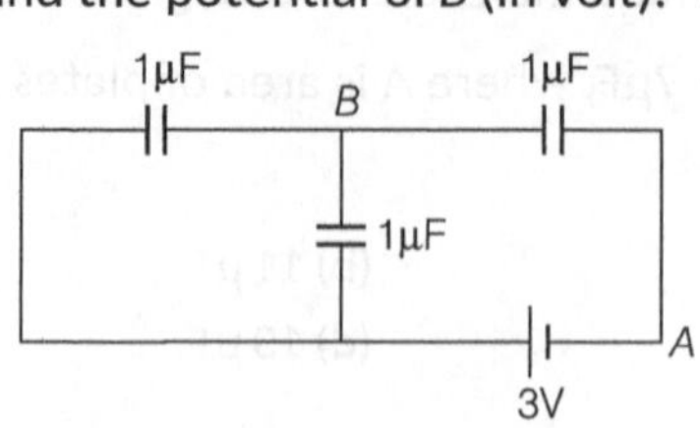

50 In the circuit as shown in figure, the potential difference between the points *a* and *b* is 4 V. Find the emf *E* (in V) of the battery.

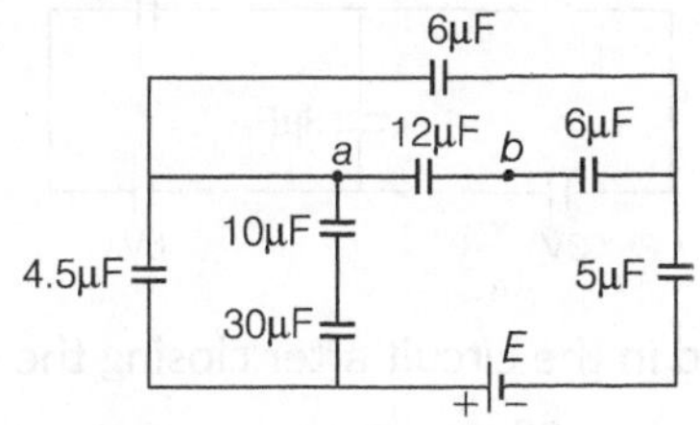

51 Figure shows a network of capacitors. Find the value of *C* (in μF), if the effective capacitance between the points *A* and *B* is 12μF.

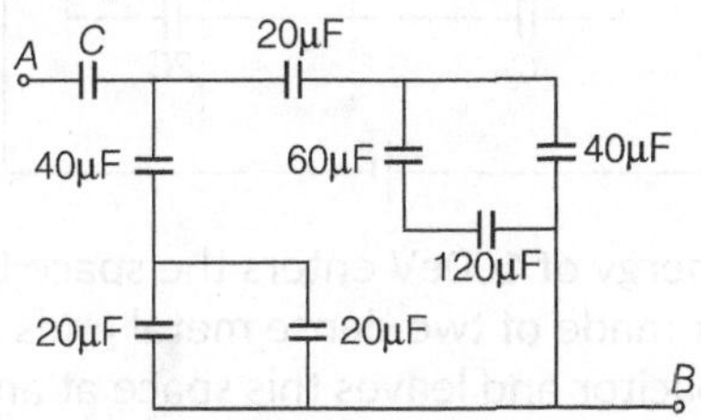

52 Find the equivalent capacitance (in μF) between points *A* and *B* of the arrangements shown in figure. Each capacitor has capacitance $C = 2\ \mu F$.

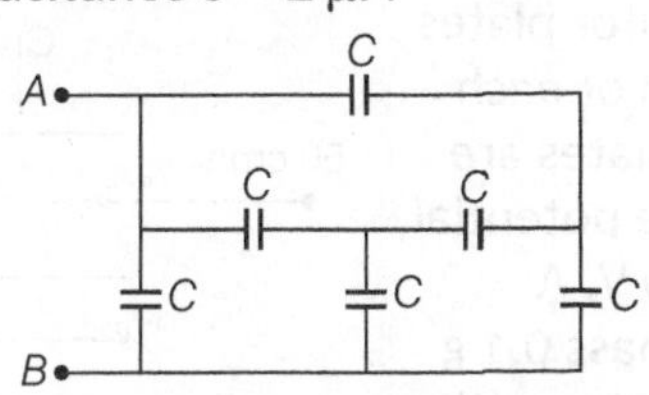

53 In figure $C_1 = 2\ \mu F$, $C_2 = 6\ \mu F$ and $C_3 = 3\ \mu F$. The breakdown voltages of the individual capacitors are $V_1 = 90$ V, $V_2 = 60$V and $V_3 = 150$V, respectively. Find the maximum voltage(in V) that can be applied across points *a* and *b*, so that no capacitor breaks down.

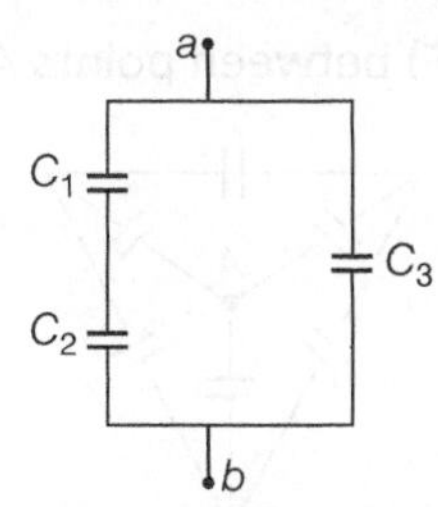

54 Plate 1 of a capacitor having capacitance 1.5 μF is given a charge of 30 μC while the other plate is uncharged. The capacitor is now connected with a battery of potential 40 V with plate 1 connected to the positive terminal. Work done by battery is ________ μJ.

55 A 8μF capacitor is charged by a 125V battery and then disconnected from the battery. When this capacitor is connected to a second (initially uncharged) capacitor C_0, the final voltage on each capacitor is 20V. What is the value of C_0 (in μF)?

56 Five plates of dimension 5 cm × 8 m are placed at separation of 8.85 mm from each other. If they are connected through a battery which provides constant potential difference of 20V as shown in figure, find the charge (in nC) flown through the battery.

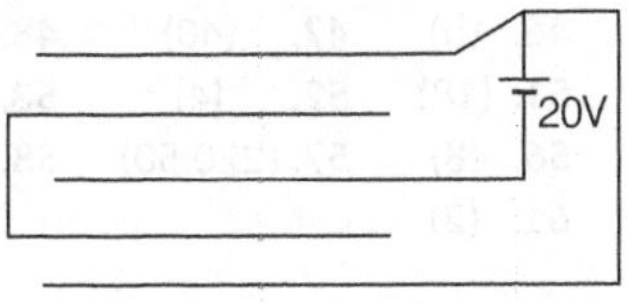

57 Calculate the heat produced (in μJ) in the circuit when key is closed.

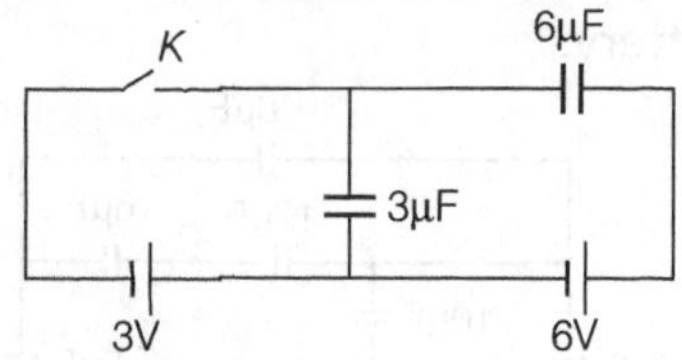

58 The amount of heat produced in the circuit after closing the switch S is $\frac{C\varepsilon^2}{n}$. Find the value of n.

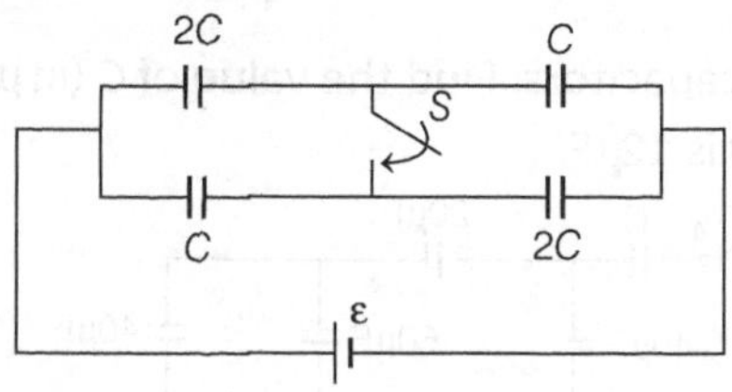

59 An electron with a kinetic energy of 100eV enters the space between the plates of plane capacitor made of two dense metal grids at an angle of 30° with the plates of capacitor and leaves this space at an angle of 45° with the plates. What is the potential difference (in V) of the capacitor?

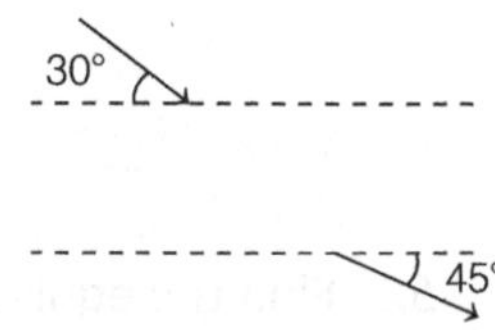

60 In the figure shown, the capacitor plates have a gap = 4 mm. The length of each plate is 10 cm. The capacitor plates are connected to a cell, so that the potential difference between them is 10 V. A particle of charge – 1 nC and mass 0.1 g enters the gap between the plates with a velocity is 50 cm/s as shown, exactly in the middle of the gap. What is the distance (in mm) of the point, where the particle hits the screen, from point O?

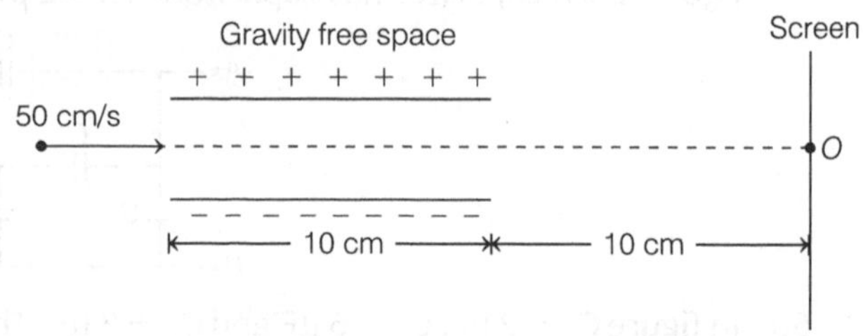

61 Find the effective capacitance(in μF) between points A and B in the networks shown below. All capacitors are of capacitance = 1μF.

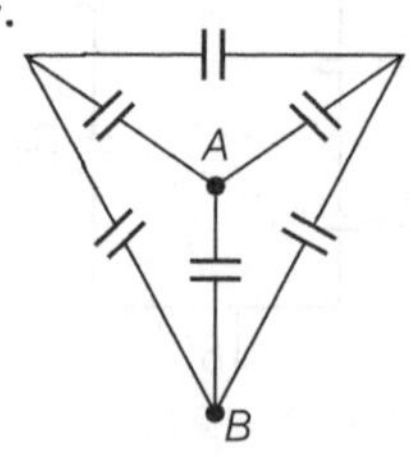

Answers

1. (d)	**2.** (a)	**3.** (a)	**4.** (b)	**5.** (b)	**6.** (b)	**7.** (c)	**8.** (c)	**9.** (b)	**10.** (c)
11. (c)	**12.** (d)	**13.** (a)	**14.** (c)	**15.** (b)	**16.** (c)	**17.** (b)	**18.** (b)	**19.** (c)	**20.** (b)
21. (c)	**22.** (c)	**23.** (b)	**24.** (a)	**25.** (c)	**26.** (c)	**27.** (b)	**28.** (b)	**29.** (a)	**30.** (a)
31. (c)	**32.** (b)	**33.** (a)	**34.** (b)	**35.** (b)	**36.** (a)	**37.** (a)	**38.** (d)	**39.** (c)	**40.** (c)
41. (c)	**42.** (b)	**43.** (b)	**44.** (b)	**45.** (d)					
46. (b)	**47.** (40)	**48.** (7)	**49.** (7)	**50.** (46)					
51. (18)	**52.** (4)	**53.** (120)	**54.** (1800)	**55.** (42)					
56. (8)	**57.** (220.50)	**58.** (12)	**59.** (50)	**60.** (1.50)					
61. (2)									

For Detailed Solutions
Scan the code

LEVEL 03 \ JEE Advanced Level

Single Choice Correct Questions

1 A capacitor has square plates, each of side l making an angle of θ with each other as shown in figure. The capacitance of the arrangement for small θ is

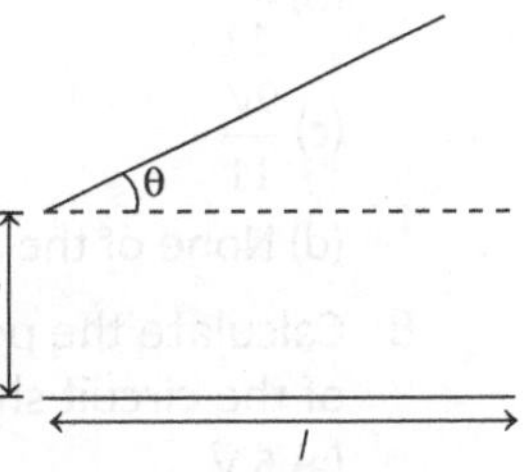

(a) $\frac{\varepsilon_0 l^2}{2d}\left(1+\frac{\theta l}{d}\right)$ (b) $\frac{\varepsilon_0 l^2}{d}\left(1+\frac{\theta l}{2d}\right)$

(c) $\frac{\varepsilon_0 l^2}{2d}\left(1-\frac{\theta l}{d}\right)$ (d) $\frac{\varepsilon_0 l^2}{d}\left(1-\frac{\theta l}{2d}\right)$

2 An air filled parallel plate capacitor with the plate area A is connected to a battery with an emf E and small internal resistance. One of the plates vibrates, so that the distance between plates varies as $d = d_0 + a\cos\omega t (a << d_0)$. The capacitor breaks down when the instantaneous current in the circuit reaches the value I. Maximum possible amplitude of vibration a is

(a) $\frac{Id_0^2}{EA\omega\varepsilon_0}$ (b) $\frac{Id_0^2}{2EA\omega\varepsilon_0}$ (c) $\frac{IA}{2Ed_0^2\omega\varepsilon_0}$ (d) $\frac{IA}{Ed_0^2\omega\varepsilon_0}$

3 The gap between the plates of a parallel plate capacitor is filled with glass of resistivity ρ. The capacitance of the capacitor without the glass equals C. The leakage current of the capacitor when a voltage V is applied to it is

(a) $\frac{V\rho}{C\varepsilon_0}$ (b) $\frac{CV}{\rho\varepsilon_0}$ (c) $\frac{V\varepsilon_0}{C\rho}$ (d) $\frac{CV\rho}{\varepsilon_0}$

4 The left plate of the capacitor shown in the figure above carries a charge $+Q$ while the right plate is uncharged at $t = 0$. The total charge on the right plate after closing the switch will be

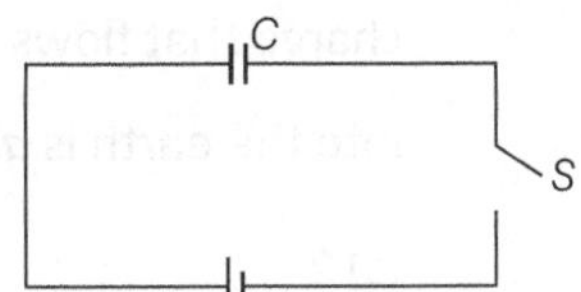

(a) $\frac{Q}{2} + C\varepsilon$ (b) $\frac{Q}{2} - C\varepsilon$

(c) $-\frac{Q}{2}$ (d) $-C\varepsilon$

5 Find the potential difference across point a and b as shown in figure.

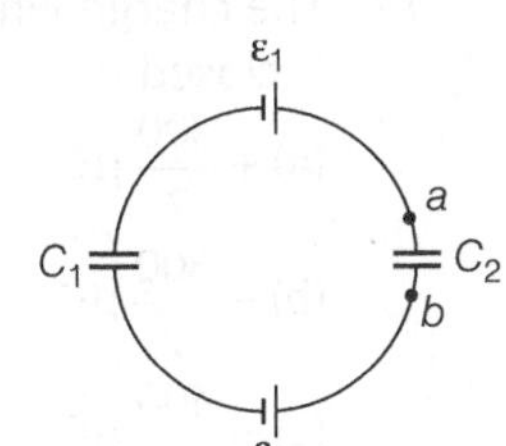

(a) $\left(\frac{\varepsilon_1+\varepsilon_2}{C_1+C_2}\right)C_1$ (b) $\left(\frac{\varepsilon_1+\varepsilon_2}{C_1+C_2}\right)C_2$

(c) $\left(\frac{\varepsilon_1-\varepsilon_2}{C_1+C_2}\right)C_1$ (d) $\left(\frac{\varepsilon_1-\varepsilon_2}{C_1+C_2}\right)C_2$

6 Three capacitors of 2μF, 3μF and 5μF are independently charged with batteries of emf's 5V, 20V and 10V, respectively. After disconnecting from the voltage sources, these capacitors are connected as shown in figure with their positive polarity plates connected to A and negative polarity is earthed. Now, a battery of 20V and an uncharged capacitor of 4μF capacitance are connected to the junction A as shown with a switch S. When switch is closed, the potential of the junction A is

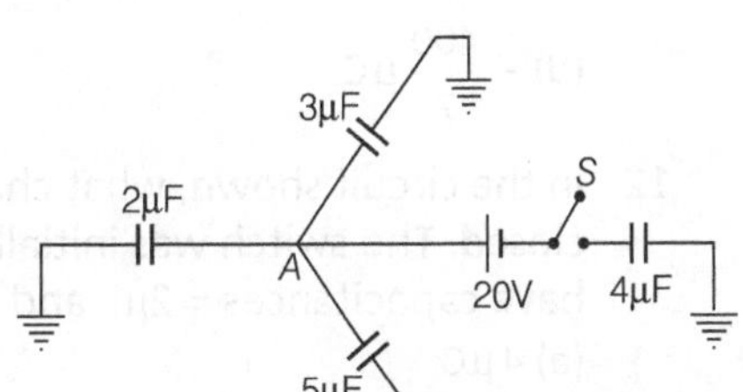

(a) $\frac{50}{7}$ V (b) $\frac{100}{7}$ V

(c) $\frac{150}{7}$ V (d) $\frac{200}{7}$ V

7 In the given circuit, find the potential at point *O*.

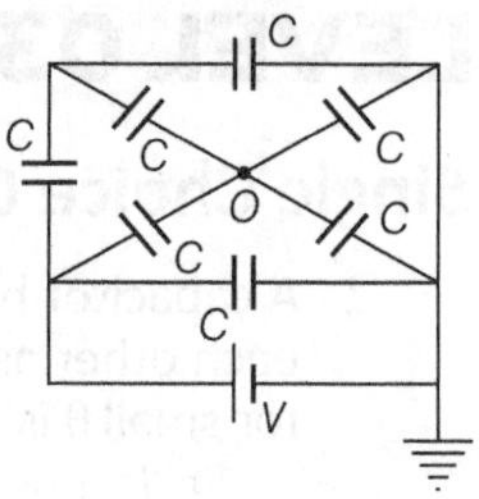

(a) $\frac{4V}{11}$

(b) $\frac{3V}{11}$

(c) $\frac{9V}{11}$

(d) None of the above

8 Calculate the potential difference between the points *a* and *b* of the circuit shown.

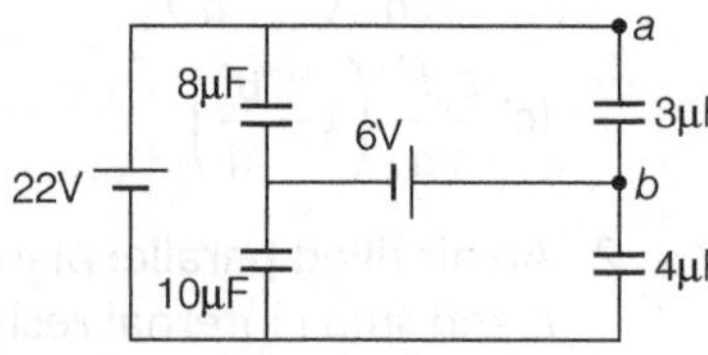

(a) 5 V
(b) 7 V
(c) 8 V
(d) 14 V

9 Find the amount by which the total energy stored in the capacitors will increase in the circuit shown in the figure after switch *K* is closed. (Take, $C = 3\mu F$ and $V = 10$ V)

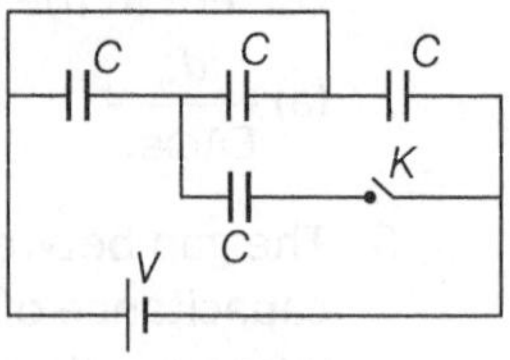

(a) 50 μJ
(b) 100 μJ
(c) 150 μJ
(d) 200 μJ

10 In the circuit shown, the potential of the topmost wire is maintained at 100 V and the lowermost wire is grounded. After closing the switch, the charge that flows through the switch is q_1 and the charge that flows into the earth is q_2. The ratio of $\left|\frac{q_1}{q_2}\right|$ is

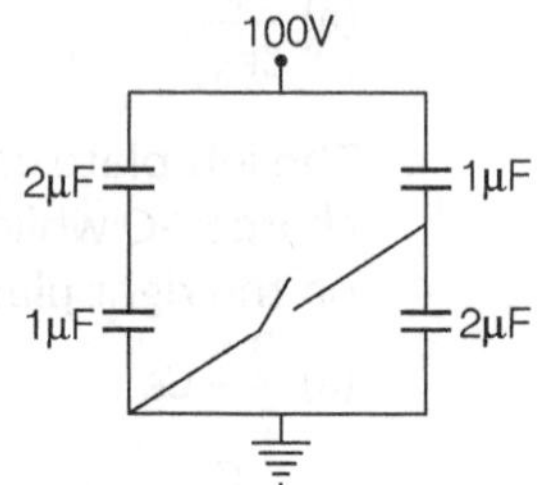

(a) 2 (b) 3
(c) 4 (d) 6

11 The charge which flows from point *A* to *B*, when switch is closed is

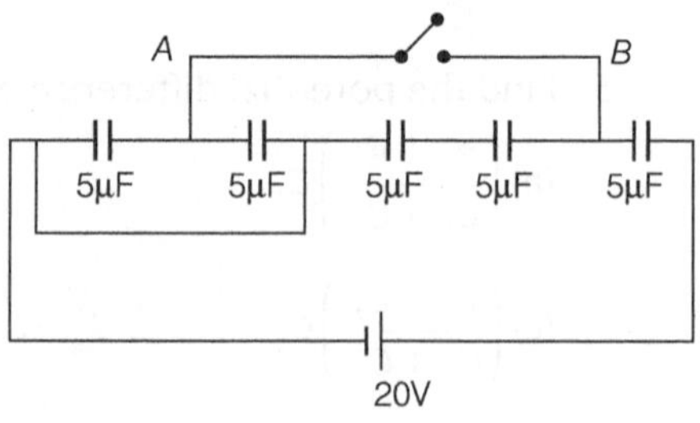

(a) $+\frac{300}{7}$ μC

(b) $-\frac{300}{7}$ μC

(c) $+\frac{400}{7}$ μC

(d) $-\frac{400}{7}$ μC

12 In the circuit shown, what charge will flow through the switch after it is closed. The switch was initially open since a long time. All capacitors have capacitances = 2μF and all cells have emf = 10 V.

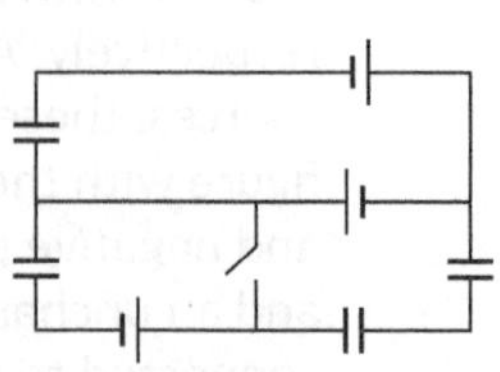

(a) 4 μC (b) 5 μC
(c) 8 μC (d) 10 μC

13 In the following circuit composed of identical capacitors, across which terminals would you connect a battery in order for all capacitors to charge up?

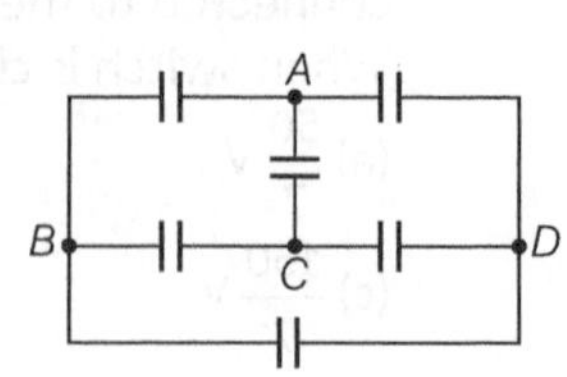

(a) *AB* (b) *AC*
(c) *BD* (d) None of these

14 Two parallel plate capacitors are placed horizontally and connected to a battery E as shown in figure. After charging the capacitors, the switch K is opened. A charged particle is placed in the left capacitor and is at rest. If one reduces the distance between the plates of the right capacitor, the particle will move

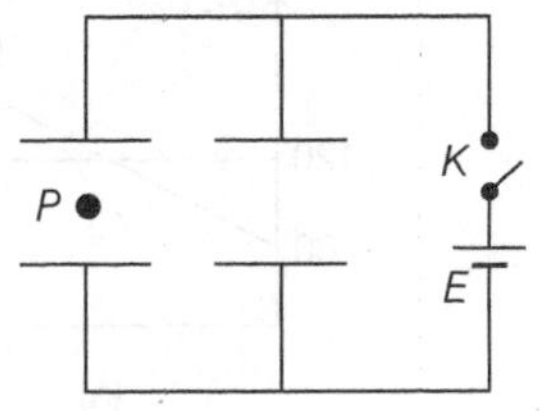

(a) horizontally (b) upwards
(c) downwards (d) in circle

15 The two plates of a capacitor of capacitance 20 μF are given charges 100 μC and 300 μC respectively as shown in figure. The amount of heat produced in the circuit, when switch S is closed is

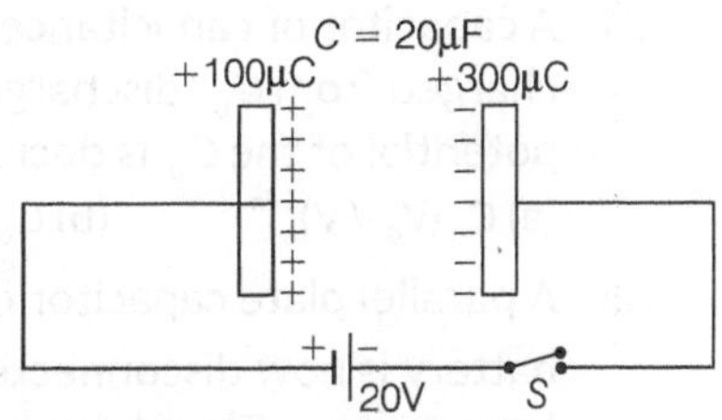

(a) 4.25 mJ (b) 6.75 mJ
(c) 6.25 mJ (d) 8.75 mJ

16 An insulated conductor initially free from charge is charged by repeated contacts with a plate which after each contact has a charge Q due to some mechanism. If q is charge on the conductor after the first operation, what is the maximum charge which can be given to the conductor?

(a) Q (b) $Q+q$ (c) $\frac{Qq}{Q+q}$ (d) $\frac{Qq}{Q-q}$

17 In the given circuit, all the capacitors are initially uncharged. After closing the switch S_1 for a long time, suddenly S_2 is also closed and kept closed for a long time. Total heat produced after closing S_2 will be

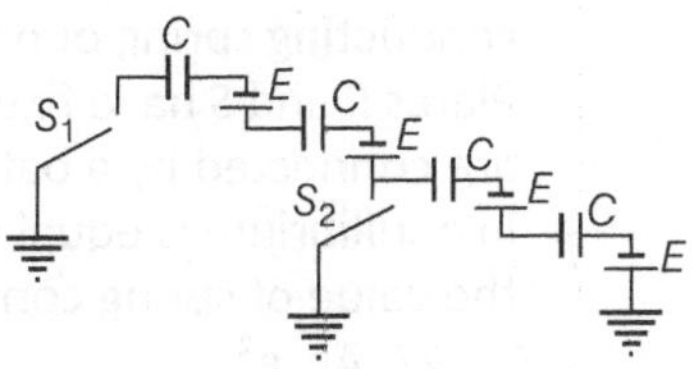

(a) $4\,CE^2$ (b) $\frac{1}{2}CE^2$
(c) $2\,CE^2$ (d) 0

18 The amount of heat liberated when the switch S is shifted from position 1 to 2 is

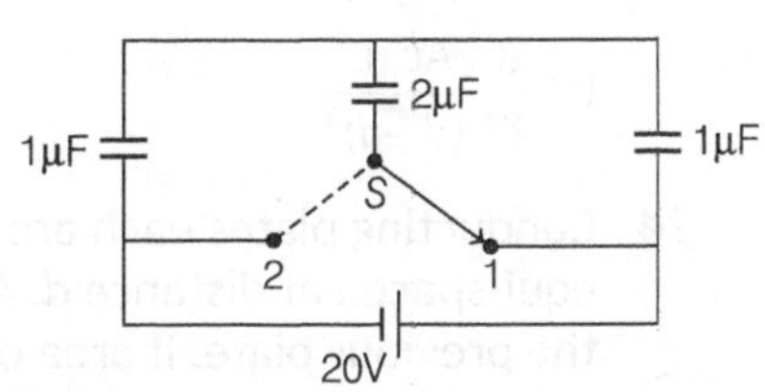

(a) 180 μJ
(b) 200 μJ
(c) 220 μJ
(d) 210 μJ

19 Four capacitors are connected as shown, connected with a battery. Breakdown voltages of each capacitor C_1, C_2, C_3 and C_4 are 60V, 50V, 90V and 150V, respectively. If potential difference of battery is increased from zero, then the capacitor which will breakdown first will be (Take, $C_1 = 1\mu F$, $C_2 = 2\,\mu F$, $C_3 = 3\,\mu F$, $C_4 = 4\,\mu F$)

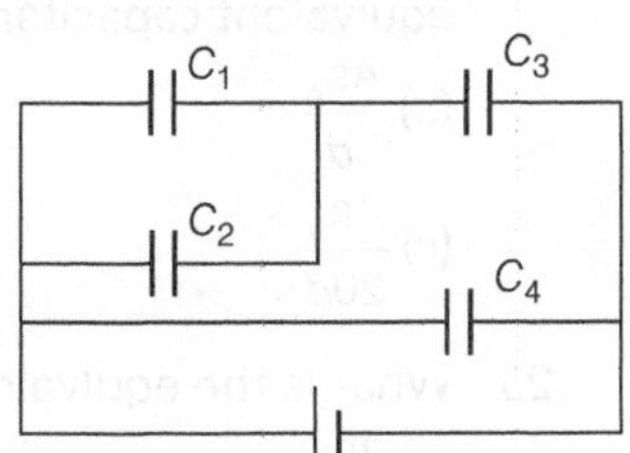

(a) C_1 (b) C_2
(c) C_3 (d) C_4

20 In the figure, switch S is closed and steady state is reached. Now at $t = 0$, voltage V_2 is increased according to equation $V = t + 10$; upto the breakdown stage. The correct plot of charges q_1 and q_2 with respect to t will be

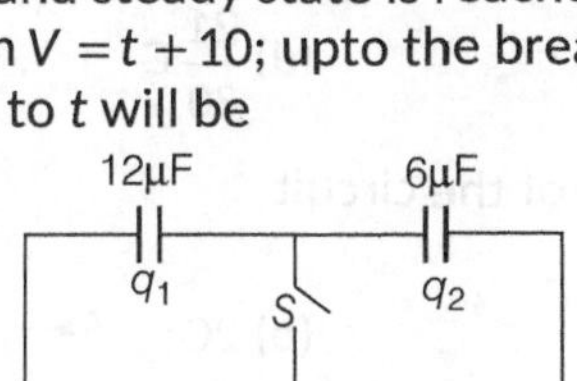

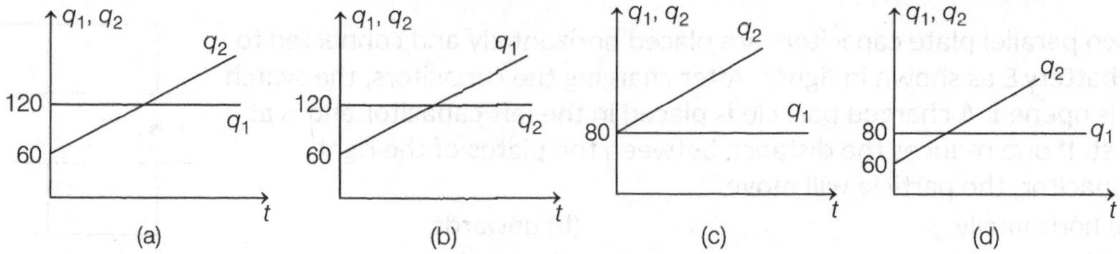

21 A capacitor of capacitance C_0 is charged to a voltage V_0 and then isolated. A capacitor C is then charged from C_0, discharged and charged again; the process is repeated n times. Due to this, potential of the C_0 is decreased to V, then value of C is

(a) $C_0(V_0/V)^{1/n}$ (b) $C_0[(V_0/V)^{1/n}-1]$ (c) $C_0[(V_0/V)-1]$ (d) $C_0[(V/V_0)^n+1]$

22 A parallel plate capacitor of capacitance 200 μF is charged by a battery of emf 100V. The battery is now disconnected and temperature of the plates is equal to atmospheric temperature. The plates are now connected by a thin wire of negligible heat capacity. Assume, 50% of their stored energy increases their temperature till the capacitor gets completely discharged and energy equally distributes over the plates. If thermal capacity of each plate is 0.5 JK^{-1} and coefficient of linear expansion is $2\times10^{-5}\,^\circ C^{-1}$, percentage increase in the volume of the plates is

(a) 0.001% (b) 0.002% (c) 0.003% (d) 0.004%

23 Two parallel plate capacitors with area A are connected through a conducting spring of natural length l in series as shown in figure. Plates P and S have fixed positions at separation d. Now, the plates are connected by a battery of emf ε. If the extension in the spring in equilibrium is equal to the separation between the plates, find the value of spring constant k.

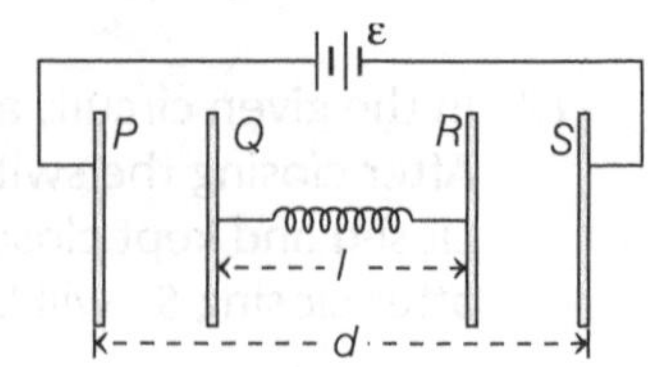

(a) $\frac{27}{8}\frac{A\varepsilon_0\varepsilon^2}{(d-l)^3}$ (b) $\frac{27}{16}\frac{A\varepsilon_0\varepsilon^2}{(d-l)^3}$

(c) $\frac{8}{27}\frac{A\varepsilon_0\varepsilon^2}{(d-l)^3}$ (d) $\frac{16}{27}\frac{A\varepsilon_0\varepsilon^2}{(d-l)^3}$

24 Conducting plates each are placed face-to-face and equi-spaced at distance d. Area of each plate is half of the previous plate. If area of first plate is A, then the equivalent capacitance of the system shown is

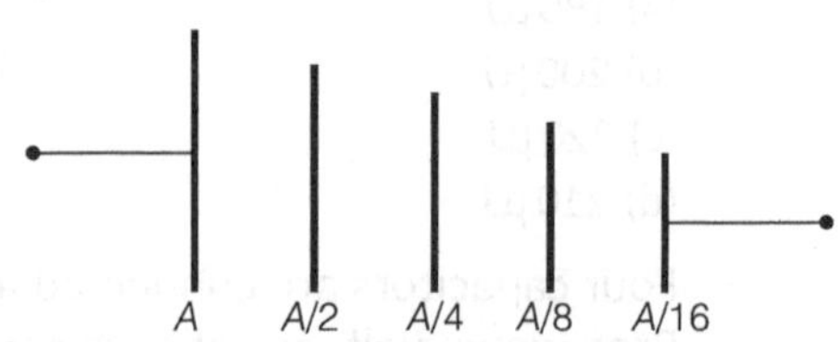

(a) $\frac{A\varepsilon_0}{d}$ (b) $\frac{A\varepsilon_0}{10d}$

(c) $\frac{A\varepsilon_0}{20d}$ (d) $\frac{A\varepsilon_0}{30d}$

25 What is the equivalent capacitance across the battery?

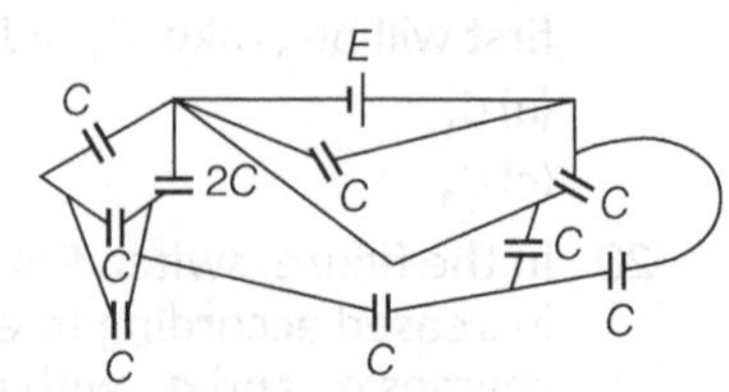

(a) $\frac{79}{30}C$ (b) $\frac{59}{30}C$

(c) $\frac{41}{30}C$ (d) $\frac{21}{30}C$

26 Find the equivalent capacitance of the circuit between point A and B.

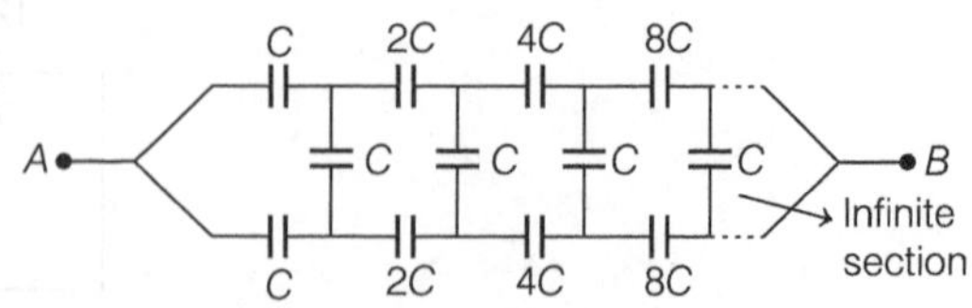

(a) C (b) $2C$

(c) $\frac{C}{2}$ (d) $\frac{C}{4}$

27 Find the equivalent capacitance of the combination shown in the figure. Each capacitor is of capacitance $C = 15\,\mu F$.

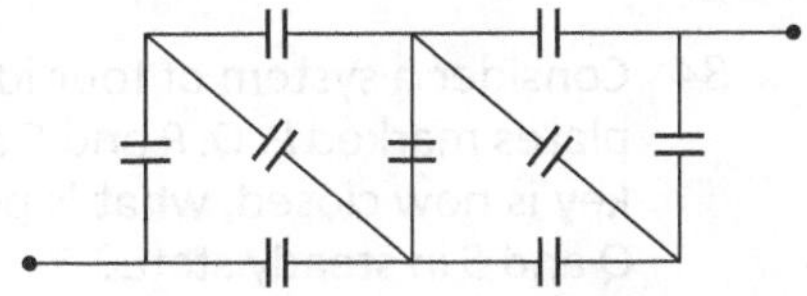

(a) $7\,\mu F$
(b) $11\,\mu F$
(c) $13\,\mu F$
(d) $19\,\mu F$

28 A finite ladder is constructed by connecting several sections of $2\mu F$ and $4\mu F$ capacitor combinations as shown in the figure. It is terminated by a capacitor of capacitance C. What value should be chosen for C, such that equivalent capacitance of the ladder between the point A and B becomes independent of the number of sections in between?

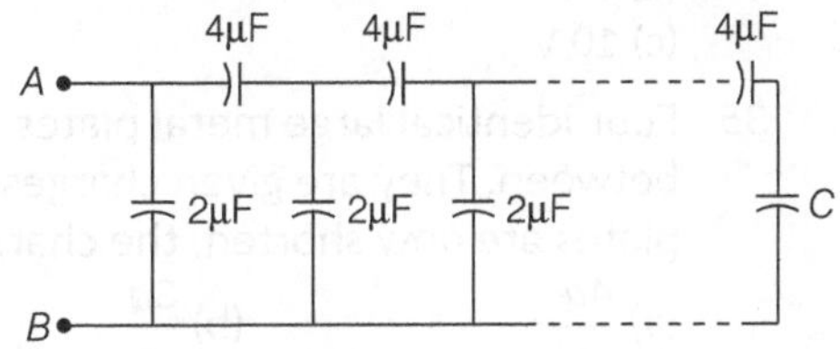

(a) $16\,\mu F$
(b) $8\,\mu F$
(c) $4\,\mu F$
(d) $2\,\mu F$

29 The potential difference between the points A and B and that between E and F of the circuit shown in figure, respectively are

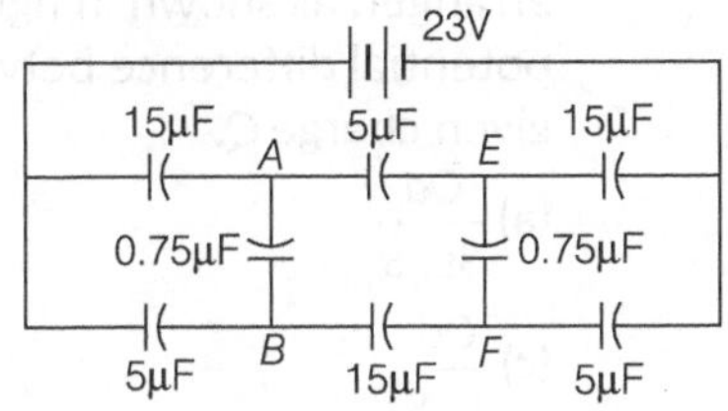

(a) 5V, 5V
(b) 10 V, 5V
(c) 15 V, 5V
(d) 0V, 0V

30 Find equivalent capacitance between A and B.

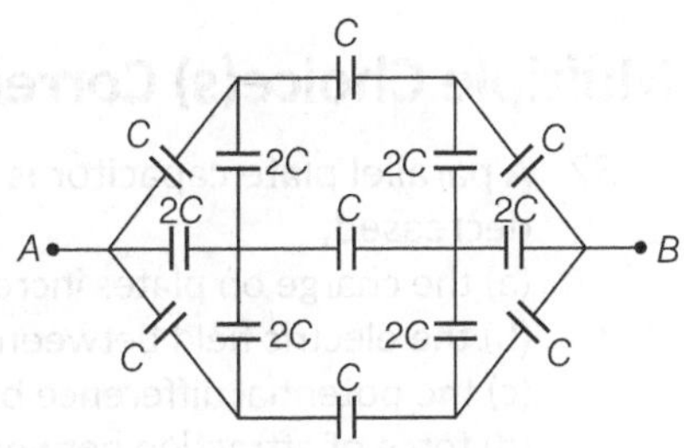

(a) $\frac{9C}{8}$
(b) $\frac{11C}{8}$
(c) $\frac{13C}{16}$
(d) $\frac{19C}{16}$

31 Three long concentric conducting cylindrical shells have radii R, $2R$ and $2\sqrt{2}R$. Inner and outer shells are connected to each other. The capacitance across middle and inner shells per unit length is

(a) $\frac{\varepsilon_0}{3\ln 2}$
(b) $\frac{6\pi\varepsilon_0}{\ln 2}$
(c) $\frac{\pi\varepsilon_0}{2\ln 2}$
(d) None of these

32 In the arrangement shown, all plates have equal area equal to A. The amount of spacing between plates is mentioned. Find the equivalent capacitance of the system between A and B, if $C = \frac{A\varepsilon_0}{L}$.

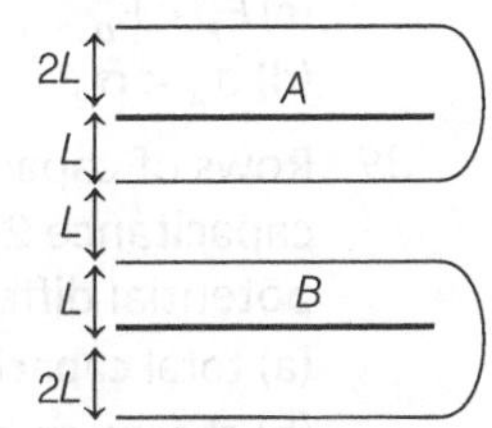

(a) (5/7) C
(b) (3/7) C
(c) (1/7) C
(d) (2/7) C

33 Three identical large metal plates of area A are at distances d and $2d$ from each other as shown in figure. Metal plate A is uncharged, while metal plates B and C have respective charges $+q$ and $-q$. Metal plates A and C are connected by switch K through a wire. How much energy is lost when switch is closed?

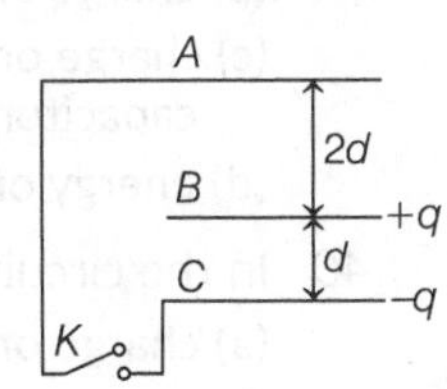

(a) $\frac{2q^2 d}{3\varepsilon_0 A}$
(b) $\frac{q^2 d}{6\varepsilon_0 A}$
(c) $\frac{q^2 d}{3\varepsilon_0 A}$
(d) None of these

34 Consider a system of four identical uncharged infinite metal plates marked P, Q, R and S arranged as shown in figure. The key is now closed, what is potential difference between plates Q and S in steady state?

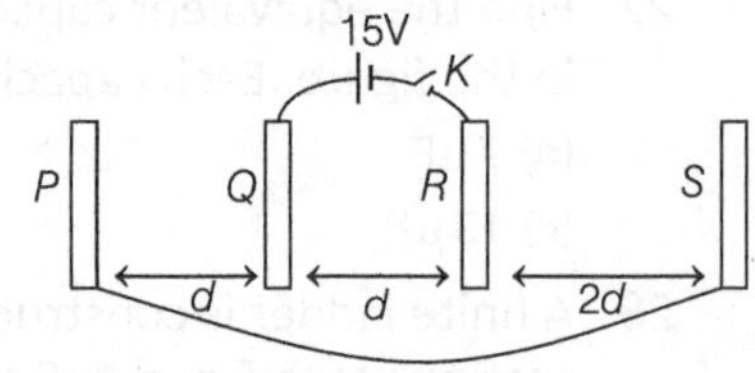

(a) 5 V (b) 8 V
(c) 10 V (d) 12 V

35 Four identical large metal plates are arranged parallel to each other, with equal gaps in between. They are given charges q, $2q$, $3q$ and $4q$ (counting from any one side). The outermost plates are now shorted, the charge which flows through the shorting wire is

(a) $\frac{4q}{3}$ (b) $\frac{3q}{4}$ (c) $\frac{5q}{3}$ (d) None of these

36 Four parallel large plates separated by equal distance d are arranged as shown in figure. The area of the plate is S. Find the potential difference between the plates B and C, if the plate B is given charge Q.

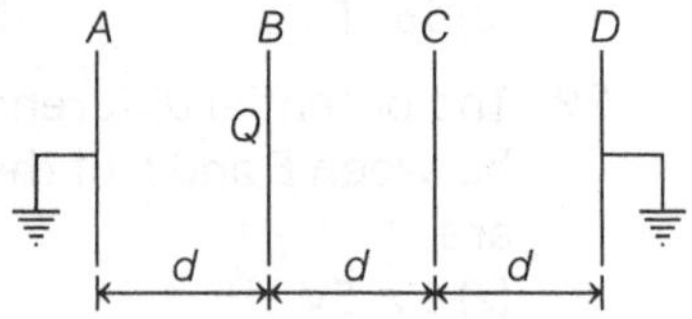

(a) $\frac{Qd}{3\varepsilon_0 S}$ (b) $\frac{2Qd}{3\varepsilon_0 S}$
(c) $\frac{Qd}{\varepsilon_0 S}$ (d) $\frac{2Qd}{\varepsilon_0 S}$

Multiple Choice(s) Correct Questions

37 A parallel plate capacitor is connected to a battery. If the distance between the plates is decreased,
(a) the charge on plates increases
(b) the electric field between the plates increases
(c) the potential difference between the plates increases
(d) force of attraction between the plates increases

38 In the given figure, a capacitor of non-parallel plates is as shown in figure. The plates of capacitor are connected by a cell of emf V_0. If σ denotes surface charge density and E denotes electric field, then

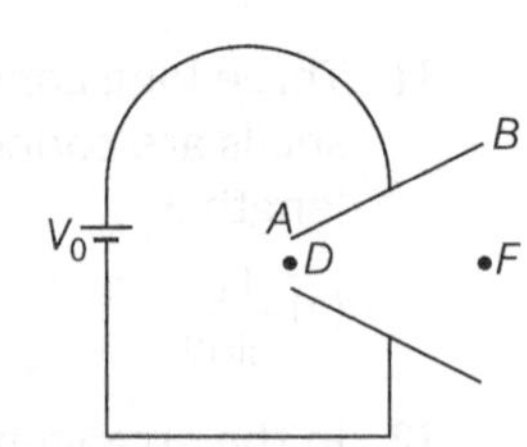

(a) $\sigma_A > \sigma_B$
(b) $E_F > E_D$
(c) $E_F < E_D$
(d) $\sigma_A < \sigma_B$

39 Rows of capacitors containing 1,2,4,8, ... ∞ capacitors, each of capacitance 2 μF, are connected in parallel as shown in figure. The potential difference across $AB = 10$ V, then

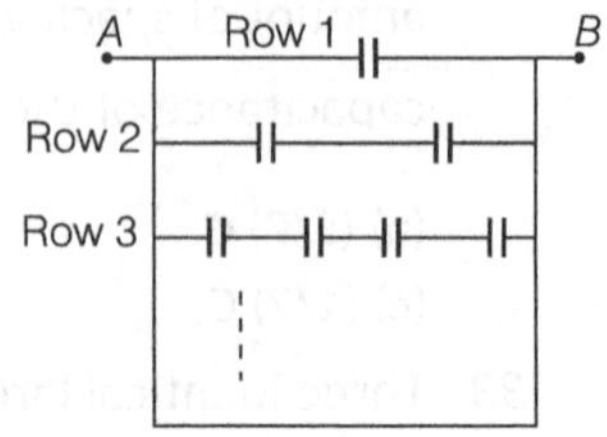

(a) total capacitance across AB is 4 μF
(b) charge on each capacitor will be same
(c) charge on the capacitor in the first row is more than on any other capacitor
(d) energy of all the capacitors is 100 μJ

40 In the circuit shown, in steady state

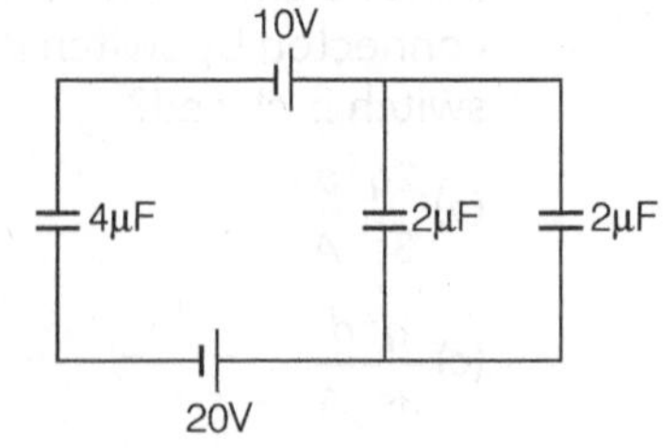

(a) charge on 4μF capacitor is 20μC
(b) charge on 2μF capacitor is 20μC
(c) the potential difference across 4μF capacitor is 5 V
(d) the potential difference across 4μF capacitor is 10 V

41 In the circuit diagram shown below,

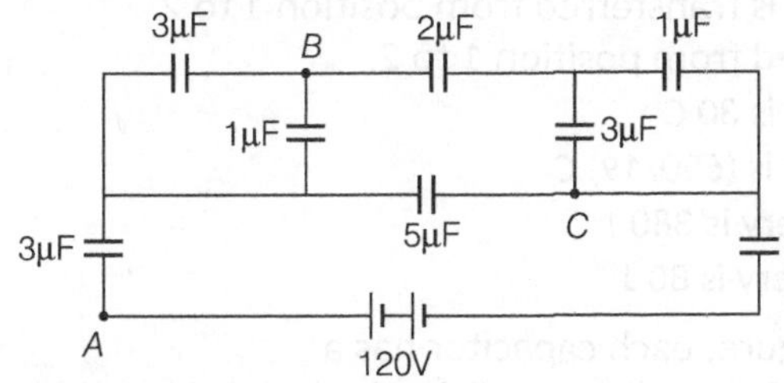

(a) the energy stored in all the capacitors is 7.2 mJ
(b) energy stored in 5μF capacitor is 1 mJ
(c) the potential difference between A and B is 45 V
(d) the potential difference between B and C is 15 V

42 Two identical plates P_1 and P_2 of cross-sectional area A are separated at a very small distance d, where $\frac{A\varepsilon_0}{d} = 2$ SI unit. Initially, charge on the plate P_1 is -10 C and charge on the plate P_2 is $+30$ C. Now, plates P_1 and P_2 are connected to positive and negative terminals of battery of potential difference 10 V with the help of switches S_1 and S_2, respectively. Choose the correct option(s).

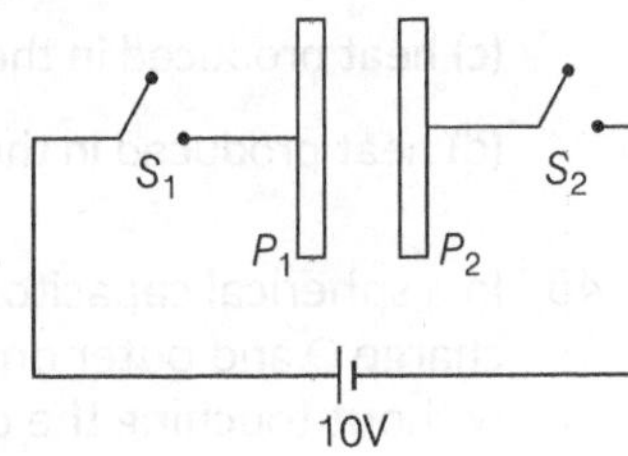

(a) The work done by the battery is 400 J.
(b) The net charge on the plate P_2 is -10C.
(c) The charge on the left side of the plate P_1 is 10C.
(d) Heat produced in the circuit is 200 J.

43 Two capacitors of capacitances 1μ F and 3μF are charged to the same voltages 5V. They are connected in parallel with oppositely charged plates connected together, then
(a) final common voltage will be 5V
(b) final common voltage will be 2.5 V
(c) heat produced in the circuit will be zero
(d) heat produced in the circuit will be 37.5 μJ

44 Negative plate of a 2 μF capacitor charged to 150 V is connected to the oppositely charged plate of 3 μF capacitor charged to 120 V. The other plates of each capacitor are connected to free wires. An uncharged capacitor of 1.5 μF is dropped on to free ends as shown in figure. After steady state, select the correct statements.

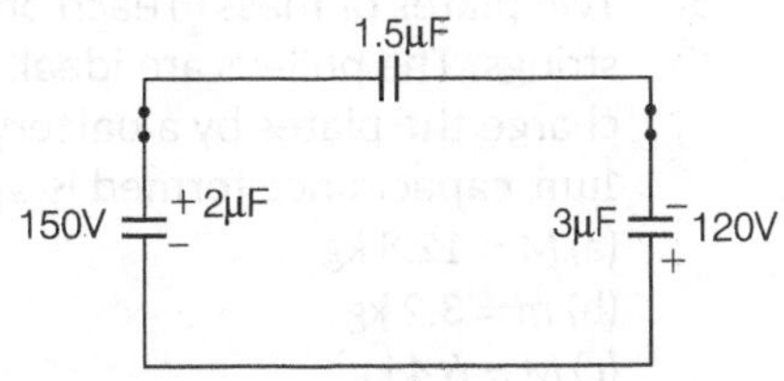

(a) Potential drop across 2 μF capacitor is 60V.
(b) Potential drop across 1.5 μF capacitor is 120 V.
(c) Potential drop across each capacitor is 90V.
(d) Charge on 3 μF capacitor is 180 μC .

45 Three capacitors $C_1 = 2\mu F$, $C_2 = 2\mu F$ and $C_3 = 3\mu F$ having initial charges 4 μC, zero and 1μC connected through a battery of emf 2V as shown in figure. On closing the switches S_1 and S_2,

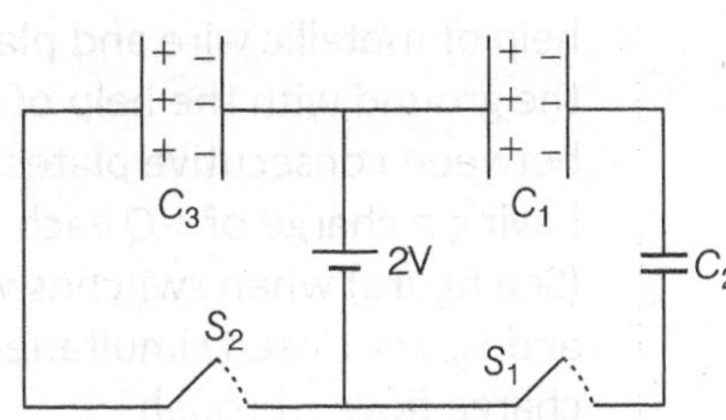

(a) charge on capacitor C_1 is 4 μC
(b) charge on capacitor C_2 is 0
(c) charge on capacitor C_3 is 6 μC
(d) charge flown through S_2 is 5 μC

46 When the steady state is reached by keeping switch in position-1, then switch S is transferred from position 1 to 2. Then, after S is transferred from position 1 to 2,

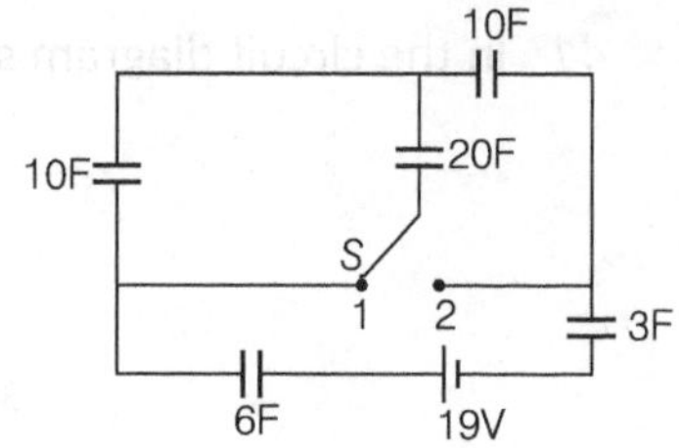

(a) charge on 3F capacitor is 30 C
(b) charge on 3F capacitor is (650/19) C
(c) work done by the battery is 380 J
(d) work done by the battery is 80 J

47 In the circuit shown in figure, each capacitor has a capacitance C. The emf of the cell is E and circuit already in steady state. If the switch S is closed,

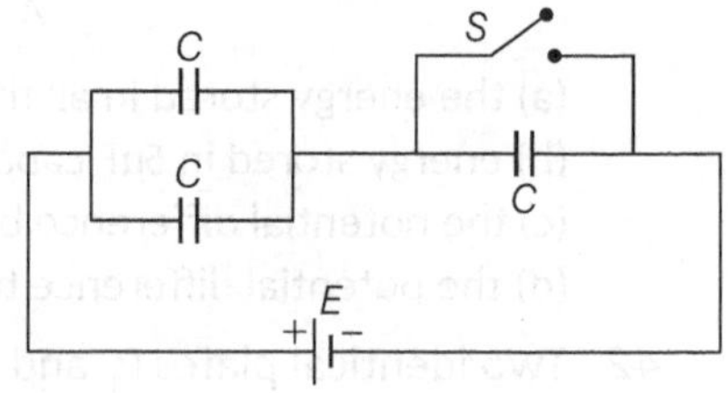

(a) the amount of charge flowing through the switch is $2CE$
(b) the amount of charge flowing through the cell is $\left(\frac{4}{3}\right)CE$
(c) heat produced in the circuit is $2CE^2$
(d) heat produced in the circuit is $\left(\frac{2}{3}\right)CE^2$

48 In a spherical capacitor, we have two concentric spherical shells, the inner one carrying a charge Q and outer one carrying a charge of $-Q$. If the inner shell is displaced from the centre without touching the outer shell,
(a) the capacitance of the capacitor will increase
(b) the capacitance of the capacitor will remain same
(c) the energy of capacitor with decrease
(d) the potential difference between inner and outershell will increase

49 Consider capacitance circuit consisting 12 identical capacitors each of capacitance $1\mu F$ as shown in the figure. The equivalent capacitance between the terminals

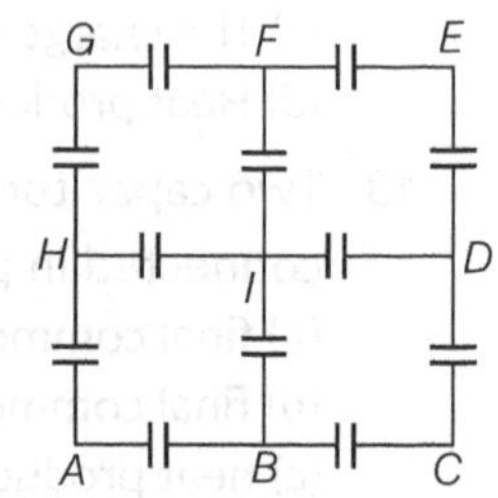

(a) A and E is $\frac{2}{3}\mu F$
(b) A and I is $\frac{8}{7}\mu F$
(c) H and I is $\frac{12}{7}\mu F$
(d) H and B is $\frac{5}{7}\mu F$

50 Two plates of mass m each are connected to mass M by two light inextensible strings. The pulleys are ideal. To hold system in equilibrium, we need to charge the plates by a battery of at least 8 volts. If distance between plates is $1\mu m$, capacitance formed is $1\mu F$, then $(g = 10 \text{ m/s}^2)$

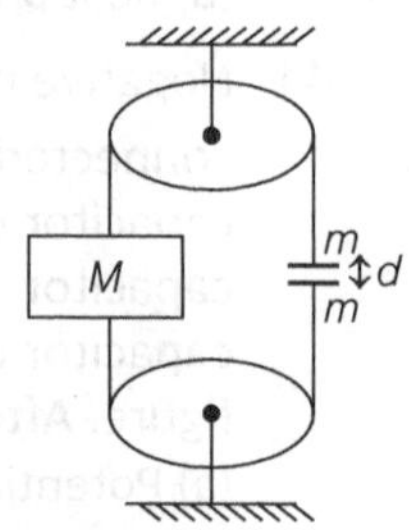

(a) $M = 12.8$ kg
(b) $m = 3.2$ kg
(c) $M = 6.4$ kg
(d) $m = 1.6$ kg

51 There is a system of four conducting parallel plates A, B, C and D. Plates A and C are connected to each other with the help of metallic wire and plates B and D can be connected to the ground with the help of switches S_1 and S_2. Distance between consecutive plates is d. Initially plates A and B are having a charge of $+Q$ each and plate D is having charge $+2Q$ (See figure) when switches were open. At $t = 0$, switches S_1 and S_2 are closed simultaneously. Then, the magnitude of charge flown through

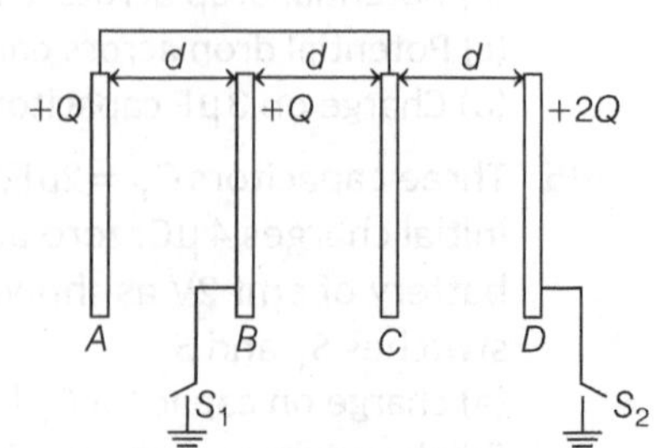

(a) S_1 is $5Q/3$
(b) S_1 is Q
(c) S_2 is $7Q/3$
(d) S_2 is $2Q$

52 Figure shows an arrangement of four identical rectangular plates A, B, C and D each of area S. Plates A and C are given charges Q_1 and Q_2, respectively. Ignore the separation between the plates in comparison to the plate dimensions.

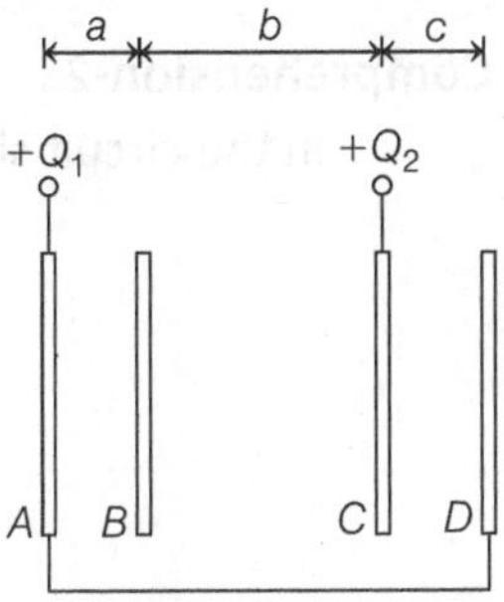

(a) Potential difference between plates A and B is independent of Q_1.
(b) Potential difference between plates C and D is independent of Q_1.
(c) Potential difference between plates A and B is independent of Q_2.
(d) Potential difference between plates C and D is independent of Q_2.

53 Two fixed very large conducting plates A and B of area S carry charges $-Q$ and q respectively, where $Q > q > 0$. A third identical plate C carrying a charge Q is released at distance d from B. Third plate collides with B. Assume collision is elastic and time of collision is sufficient to re-distribute charge among B and C (Neglect gravity).

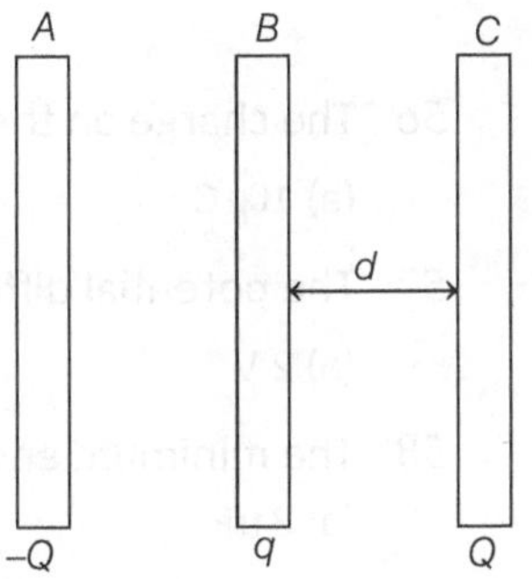

(a) Force acting on plate C before collision is $\dfrac{(Q-q)Q}{2\varepsilon_0 S}$
(b) Force acting on plate C after collision is $\dfrac{(q/2)^2}{2\varepsilon_0 S}$
(c) Subsequent to collision, kinetic energy of plate C increases
(d) Subsequent to collision, electric field between the plates B and C is zero

Comprehension Based Questions

Comprehension-1

Two conductors carry charges $+Q$ and $-Q$. They create a non-uniform electric field as shown in figure. A coordinate system is placed on the positive conductor, and the electric field strength along the X-axis is found to vary according to the function $E = \dfrac{\beta Q}{\varepsilon_0}\left(1+\dfrac{x^2}{d^2}\right)$, where β is a constant and d is the least distance the conductors, answer the following equations.

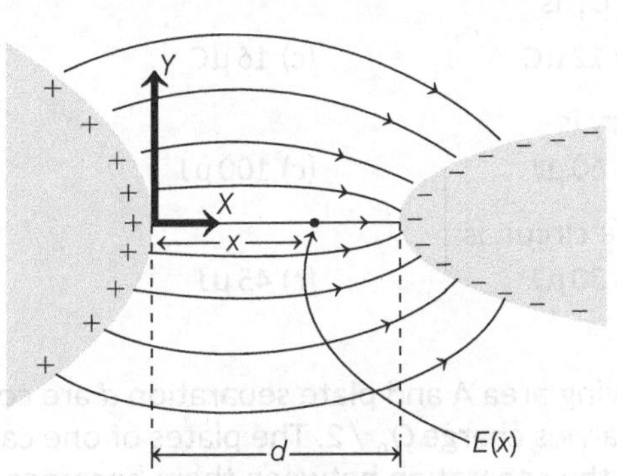

54 The potential difference between the conductors is
(a) $\dfrac{\beta Qd}{3\varepsilon_0}$ (b) $\dfrac{3\beta Qd}{2\varepsilon_0}$ (c) $\dfrac{4\beta Qd}{3\varepsilon_0}$ (d) None of these

55 The capacitance of the system is
(a) $\dfrac{3\varepsilon_0}{\beta d}$ (b) $\dfrac{2\varepsilon_0}{3\beta d}$ (c) $\dfrac{3\varepsilon_0}{4\beta d}$ (d) None of these

Comprehension-2

In the circuit shown in figure, five capacitors are connected to a battery.

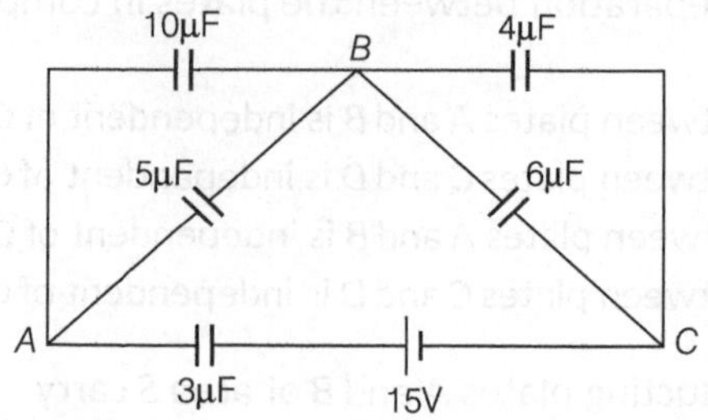

56 The charge on the $6\,\mu F$ capacitor is

(a) $10\mu C$ (b) $12\,\mu C$ (c) $18\,\mu C$ (d) $24\,\mu C$

57 The potential difference across the $5\,\mu F$ capacitor is

(a) 2 V (b) 3 V (c) 5 V (d) 10 V

58 The minimum energy is stored in the capacitor of

(a) $3\,\mu F$ (b) $6\,\mu F$ (c) $5\,\mu F$ (d) $4\,\mu F$

Comprehension-3

Following figure shows the initial charges on the capacitor. After the switch S is closed,

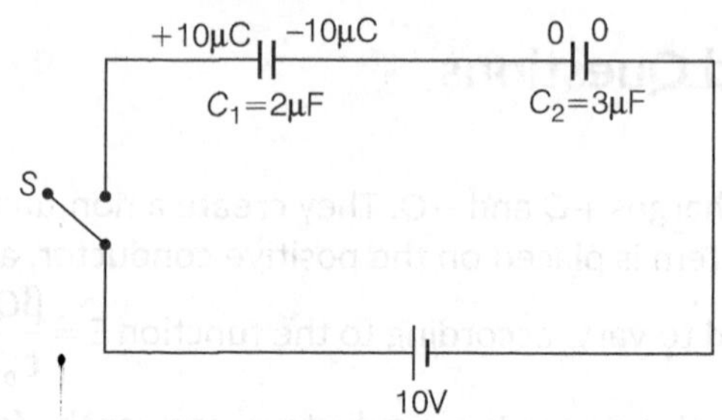

59 the charge on capacitor C_1 is

(a) $8\,\mu C$ (b) $12\,\mu C$ (c) $16\,\mu C$ (d) None of these

60 the work done by battery is

(a) $20\,\mu J$ (b) $60\,\mu J$ (c) $100\,\mu J$ (d) None of these

61 the heat produced in the circuit is

(a) $15\,\mu J$ (b) $30\,\mu J$ (c) $45\,\mu J$ (d) None of these

Comprehension-4

Two capacitors each having area *A* and plate separation *d* are connected as shown in the circuit. Each capacitor carries charge $Q_0/2$. The plates of one capacitor are slowly pulled apart by an external agent till the separation between them becomes $2d$. The other capacitor is not disturbed.

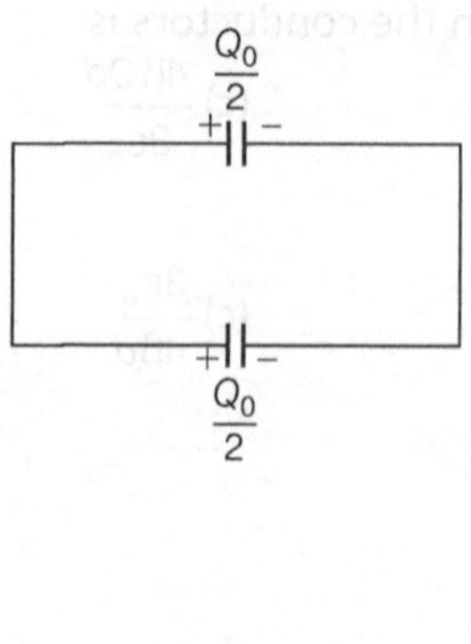

62 The work done by the external agent in this process will be

(a) $\frac{Q_0^2 d}{12A\varepsilon_0}$ (b) $\frac{Q_0^2 d}{6A\varepsilon_0}$ (c) $\frac{Q_0^2 d}{18A\varepsilon_0}$ (d) $\frac{Q_0^2 d}{9A\varepsilon_0}$

63 The potential difference across the capacitors when the separation between the plates is x, is

(a) $\frac{Q_0 xd}{2A\varepsilon_0 (x+d)}$ (b) $\frac{Q_0 xd}{A\varepsilon_0 (x+d)}$ (c) $\frac{Q_0 xd}{4A\varepsilon_0 (x+d)}$ (d) $\frac{Q_0 d}{2A\varepsilon_0 x}$

64 Heat produced in this process will be equal to

(a) $\frac{5Q_0^2 d}{72A\varepsilon_0}$ (b) $\frac{5Q_0^2 d}{36A\varepsilon_0}$ (c) $\frac{1Q_0^2 d}{2A\varepsilon_0}$ (d) 0

Comprehension-5

The figure shows a diagonal symmetric arrangement of capacitors and a battery.

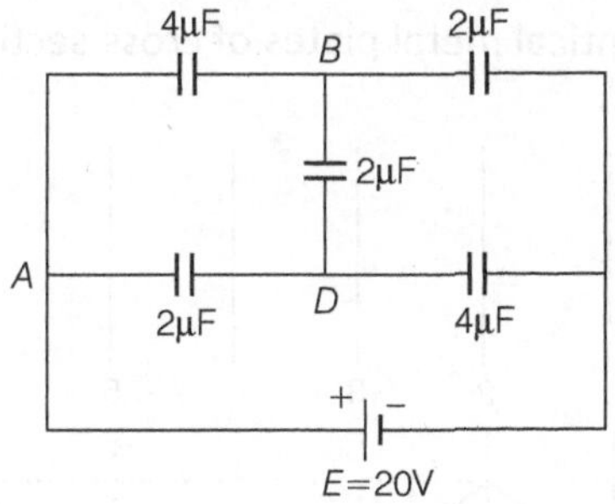

65 Identify the correct statements.
(a) Both the 4 μF capacitors carry equal charges in opposite sense
(b) Both the 4 μF capacitors carry equal charges in same sense
(c) $V_B - V_D > 0$
(d) $V_D - V_B > 0$

66 If the potential of C is zero, then identify the correct statements.

(a) $V_A = +20$ V (b) $4(V_A - V_B) + 2(V_D - V_B) = 2V_B$
(c) $2(V_A - V_D) + 2(V_B - V_D) = 4V_D$ (d) $V_A = V_B + V_D$

67 The potential of the points B and D are

(a) $V_B = 8$V (b) $V_B = 12$V
(c) $V_D = 8$V (d) $V_D = 12$V

68 The value of charges q_1, q_2 and q_3 as shown in the figure are

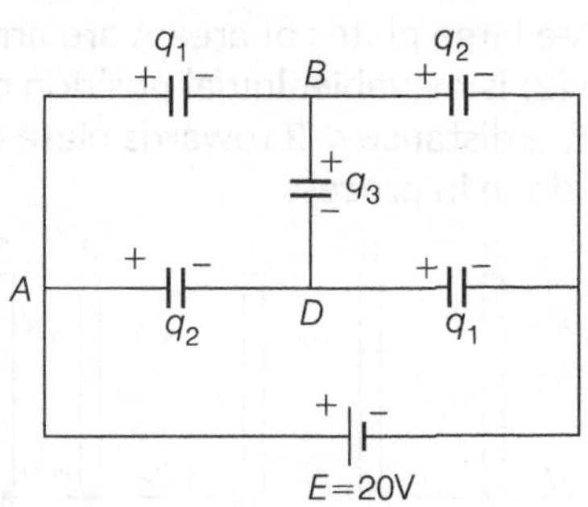

(a) $q_1 = 32\,\mu C$; $q_2 = 24\,\mu C$; $q_3 = -8\,\mu C$ (b) $q_1 = 48\,\mu C$; $q_2 = 16\,\mu C$; $q_3 = +8\,\mu C$
(c) $q_1 = 32\,\mu C$; $q_2 = 24\,\mu C$; $q_3 = +8\,\mu C$ (d) $q_1 = 3\,\mu C$; $q_2 = 4\,\mu C$; $q_3 = +2\,\mu C$

Comprehension-6

A circuit consists of two capacitors of capacitance C each and ideal cell of emf E volts as shown in figure. Initially, both the capacitors are uncharged. The key is pushed to position 1 and then to position 2. This operation is repeated infinite times.

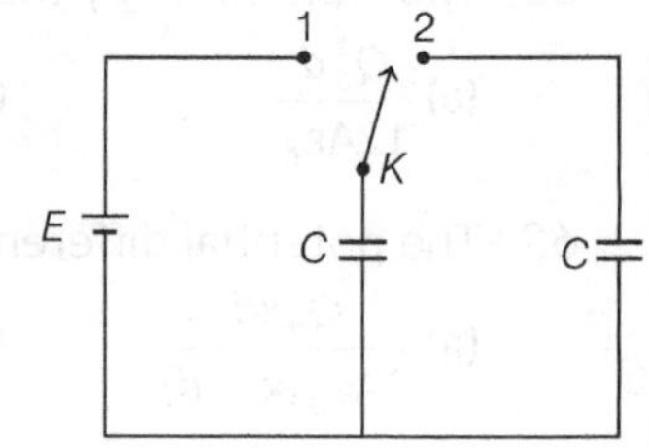

69 The magnitude of net charge that crosses the cell is

(a) CE (b) $1.5\,CE$
(c) $2\,CE$ (d) 0

70 The magnitude of net work done by the cell is

(a) CE^2 (b) $1.5\,CE^2$ (c) $2\,CE^2$ (d) 0

71 The net heat produced in the circuit is

(a) CE^2 (b) $1.5\,CE^2$ (c) $2\,CE^2$ (d) 0

Comprehension-7

The V *versus* x plot for six identical metal plates of cross-sectional area A is shown. The plates are placed with separation 'd'.

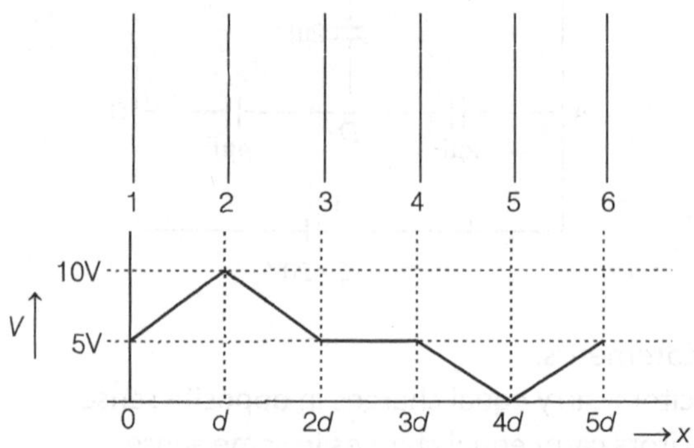

72 Equivalent capacitance between 2 and 5 is

(a) $\frac{2\varepsilon_0 A}{d}$ (b) $\frac{\varepsilon_0 A}{d}$ (c) $\frac{3\varepsilon_0 A}{d}$ (d) $\frac{4\varepsilon_0 A}{d}$

73 Total charge on plate 2 is

(a) $\frac{10\varepsilon_0 A}{d}$ (b) $\frac{5}{3}\frac{\varepsilon_0 A}{d}$ (c) $\frac{4}{3}\frac{\varepsilon_0 A}{d}$ (d) None of these

74 Ratio of charge on plate 2 to plate 5 i.e. $|Q_2 : Q_5|$, is

(a) 2 : 1 (b) 3 : 1 (c) 1 : 1 (d) None of these

Comprehension-8

A battery of voltage V and three large plates of area A are arranged as shown in diagram. Plates (1) and (3) are fixed and plate (2) is movable. Initial position of middle plates shown in Fig (1). Now, middle plate is shifted by a distance $d/2$ towards plate (3) and final position is shown in Fig.(2). Neglect external work done in process.

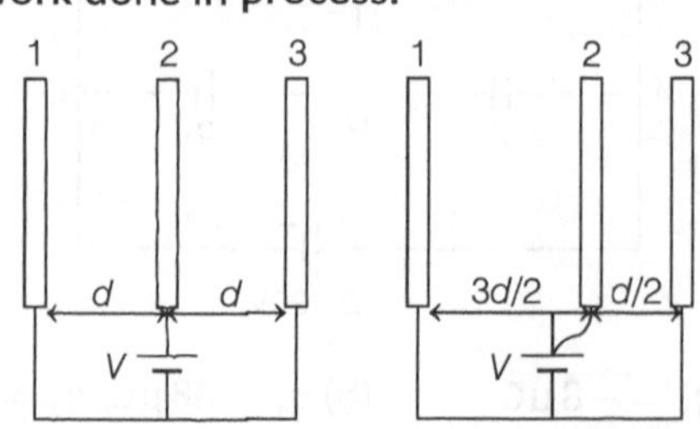

Fig. (1) Fig. (2)

75 If total charge on plate (2) initially is x and total charge on plate (2) finally is y then

(a) $x = \frac{\varepsilon_0 A}{2d} V$ (b) $x = \frac{2\varepsilon_0 A}{d} V$ (c) $y = \frac{8}{3}\frac{\varepsilon_0 A}{d} V$ (d) $y = \frac{3}{8}\frac{\varepsilon_0 A}{d} V$

76 If work done by the battery in shifting the plate (2) from initial position to final position is x and the amount of heat produced in doing so is y, then

(a) $x = \frac{2}{3}\frac{\varepsilon_0 A}{d} V^2$ (b) $x = \frac{3}{8}\frac{\varepsilon_0 A}{d} V^2$ (c) $y = \frac{\varepsilon_0 AV^2}{3d}$ (d) $y = \frac{\varepsilon_0 AV^2}{d}$

Comprehension-9

A parallel plate capacitor with air as a dielectric is arranged horizontally. The lower plate is fixed and the other connected with a vertical spring. The area of each plate is A. In the steady position, the distance between the plates is d_0. When the capacitor is connected with an electric source with the voltage V, a new equilibrium appears, with the distance between the plates as d_1. Mass of the upper plate is m.

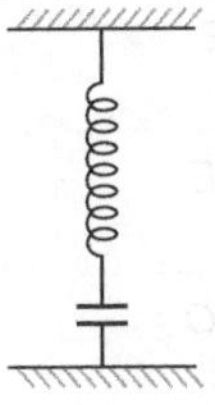

77 Spring constant k is

(a) $\frac{\varepsilon_0 AV^2}{(d_0 - d_1)^2 d_0}$ (b) $\frac{\varepsilon_0 AV^2}{2d_1^2 (d_0 - d_1)}$ (c) $\frac{\varepsilon_0 AV^2}{2(d_0 - d_1)^2 d_0}$ (d) $\frac{\varepsilon_0 AV^2}{3(d_0 - d_1)^2 d_0}$

78 Maximum voltage for a given k in which an equilibrium is possible, is

(a) $\sqrt{\frac{k}{A\varepsilon_0}\left(\frac{2}{3}d_0\right)^{3/2}}$ (b) $\sqrt{\frac{k}{2A\varepsilon_0}\left(\frac{2d_0}{3}\right)^{3/2}}$ (c) $\sqrt{\frac{2k}{A\varepsilon_0}\left(\frac{2}{3}d_0\right)^{3/2}}$ (d) $\sqrt{\frac{k}{A\varepsilon_0}\left(\frac{2d_0}{3}\right)^{1/2}}$

79 What is the angular frequency of the oscillating system around the equilibrium value d_1? (Amplitude of the oscillation << d_1)

(a) $\sqrt{\frac{k}{m}\left(\frac{2d_1 - 3d_0}{d_1}\right)}$ (b) $\sqrt{\frac{k}{m}\left(\frac{3d_1 - 2d_0}{d_1}\right)}$ (c) $\sqrt{\frac{k}{m}\left(\frac{d_1 - d_0}{d_1}\right)}$ (d) $\sqrt{\frac{k}{m}\left(\frac{2d_1 - d_0}{d_1}\right)}$

Match the Columns Questions

80 In the circuit, both capacitors are identical. Column I indicates action done on capacitor 1 and Column II indicates effect on capacitor 2.

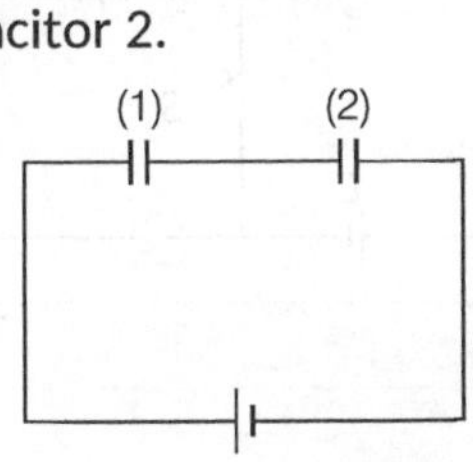

	Column I		Column II
(a)	Plates are moved further apart	(p)	Numerical value of charge on left plate increases
(b)	Area increased	(q)	Potential difference increases
(c)	Left plate is earthed	(r)	Numerical value of charge on right plate decreases
(d)	Its plates are short-circuited	(s)	None of the above effects

81 Match Column I against Column II referring to the adjacent figure.

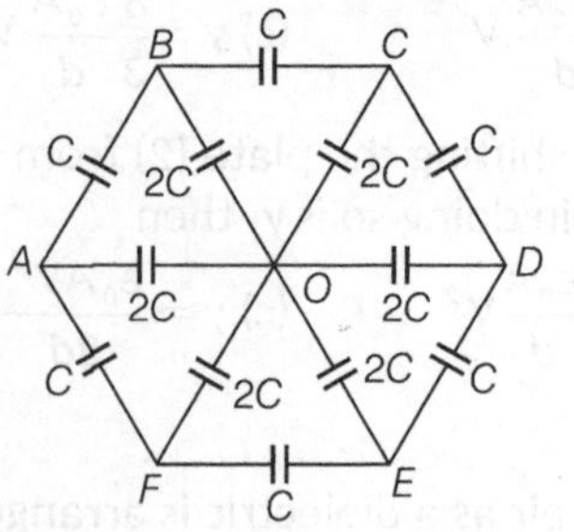

Column I	Column II
(a) Effective capacitance between *A* and *D*	(p) $\frac{45}{13}C$
(b) Effective capacitance between *F* and *E*	(q) $\frac{15}{8}C$
(c) Effective capacitance between *A* and *E*	(r) $\frac{45}{19}C$
(d) Effective capacitance between *A* and *O*	(s) $\frac{9}{5}C$

82 The circuit involves two ideal cells connected to a 1 μF capacitor *via* a key *K*. Initially the key *K* is in position 1 and the capacitor is charged fully by 2V cell. The key is then pushed to position 2. Column I gives physical quantities involving the circuit after the key is pushed from position 1 to position 2. Column II gives corresponding results. Match the statements in Column I with the corresponding values in Column II.

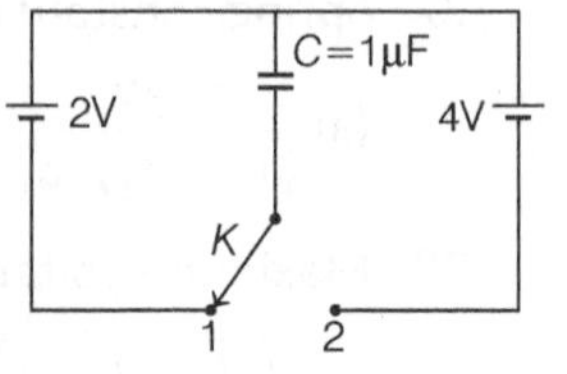

Column I	Column II
(a) The net charge crossing the 4 V cell in μC is	(p) 2
(b) The magnitude of work done by 4 V cell in μJ is	(q) 6
(c) The gain in potential energy of capacitor in μJ is	(r) 8
(d) The net heat produced in circuit in μJ is	(s) 16

83 Consider the situation shown. The switch *S* is open for a long time and then closed. Then, after closing *S*,

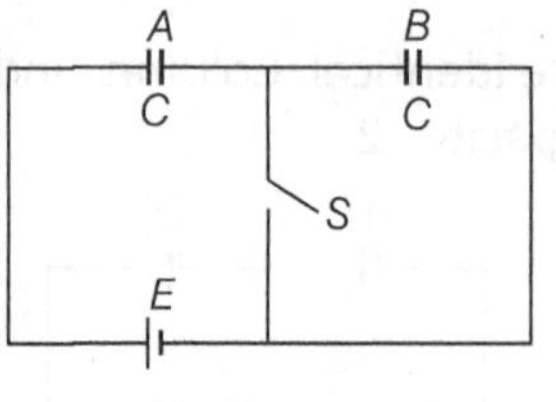

Column I	Column II
(a) Charge flown through battery	(p) $\frac{CE^2}{2}$
(b) Work done by battery	(q) $\frac{CE}{2}$
(c) Charge on capacitor *A*	(r) $\frac{CE^2}{4}$
(d) Heat developed in the system	(s) CE

84 Three conducting bodies A (solid sphere), B (hollow sphere) and C(hollow sphere with thickness R) are arranged as shown in the figure. A charge q is given to the inner sphere. Match the statements from Column I with C the values in Column II.

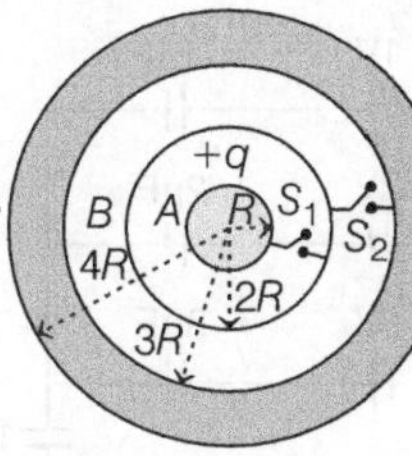

	Column I		Column II
(a)	When S_1, S_2 both are open, equivalent capacity of the system is	(p)	$16\pi\varepsilon_0 R$
(b)	When S_1 is closed but S_2 is open, equivalent capacity of the system is	(q)	$\frac{16\pi\varepsilon_0 R}{3}$
(c)	When S_2 is closed but S_1 is opened, equivalent capacity of the system is	(r)	$\frac{48\pi\varepsilon_0 R}{5}$
(d)	When both S_1 and S_2 are closed, equivalent capacity of the system is	(s)	$\frac{48\pi\varepsilon_0 R}{11}$

85 The equivalent capacitance across A and B for the plates connected as shown in Column I, are given in Column II. Match the items in Column II with the items in Column I.

	Column I		Column II
(a)	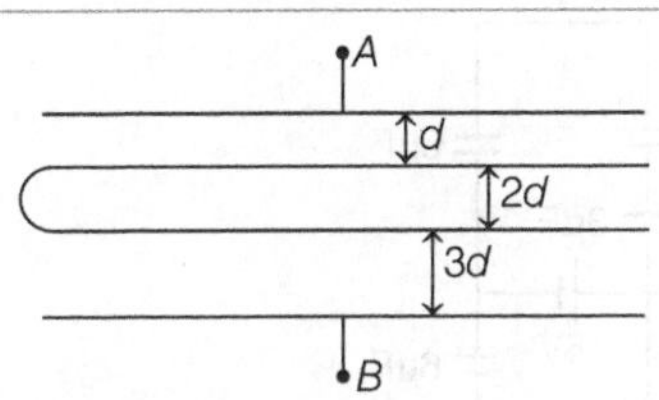 Area of each plate is S	(p)	$\frac{3\varepsilon_0 S}{4d}$
(b)	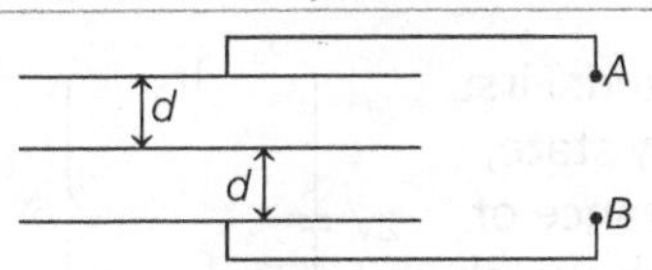 Area of each plate is S	(q)	$\frac{2\varepsilon_0 S}{3d}$
(c)	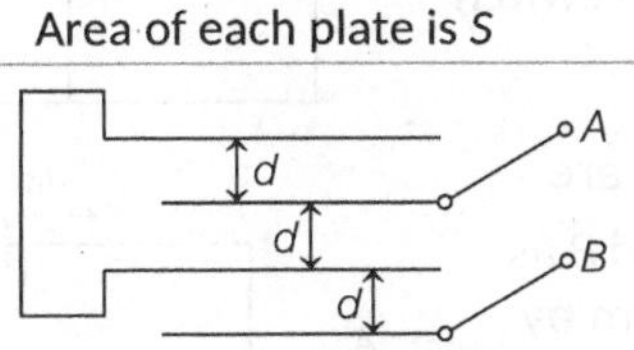 Area of each plate is S	(r)	$\frac{\varepsilon_0 S}{4d}$
(d)	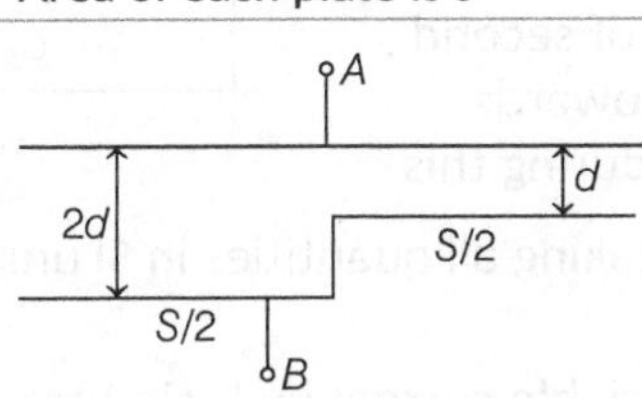 Area of upper plate is S	(s)	$\frac{\varepsilon_0 S}{2d}$
		(t)	$\frac{\varepsilon_0 S}{d}$

Numerical Value Questions

86 In the figure shown, the emf ε for which charge on $2\,\mu F$ capacitor is $4\,\mu C$ are ε_1 and ε_2. Find $\varepsilon_1 + \varepsilon_2$ (in volt).

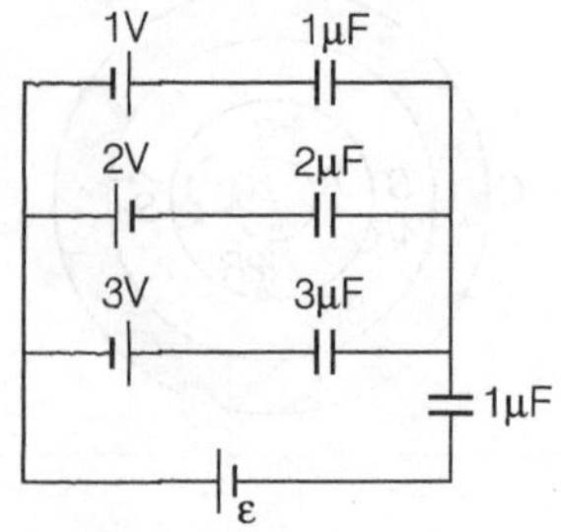

87 Find the potential difference (in V) across points A and B in the circuit shown in figure.

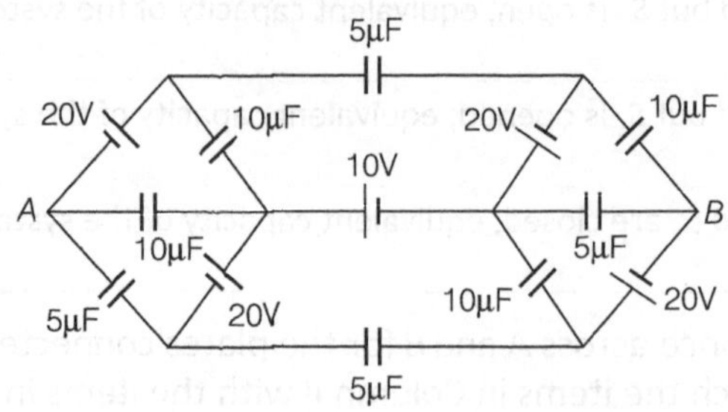

88 Find the charge (in μC) flown through the switch from A to B when it is closed.

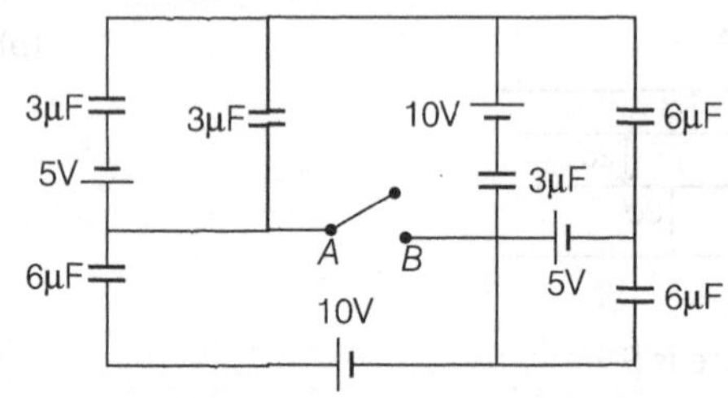

89 In given circuit, initially all capacitors were uncharged. First switch S_1 is closed and S_2 is kept open. After steady state, first S_1 is opened and then S_2 is closed. The capacitance of each capacitor is $6\mu F$. Find the charge flown through switch S_2 after it is closed in μC.

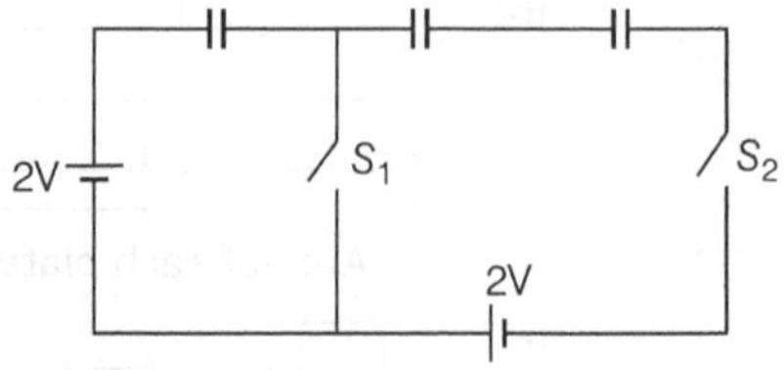

90 Two identical capacitors having plate separation d_0 are connected parallel to each other across points A and B as shown in figure. A charge Q is imparted to the system by connecting a battery across A and B and battery is removed. Now, first plate of first capacitor and second plate of second capacitor starts moving with constant velocity u_0 towards left. The magnitude of current flowing in the loop during this process at this instant is $\frac{Qu_0}{nd_0}$. Find the value of n, taking all quantities in SI units.

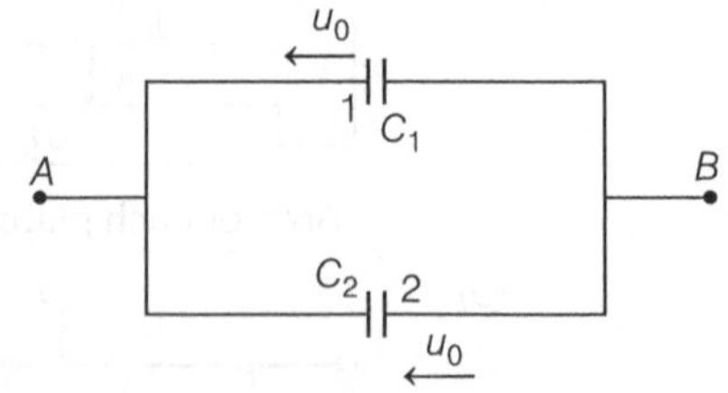

91 In a circuit, a branch carrying only capacitor has variable current $i = I_0 \sin 2\pi t$ and capacitor was uncharged at $t = 0$. The maximum magnitude of potential difference across capacitor is $\frac{2I_0}{k\pi C}$. Then, value of k will be

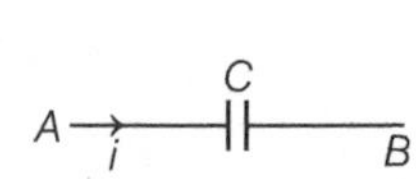

92 A charge of $8\,\mu C$ is given to point A and point B is earthed. If each capacitance is $1\,\mu F$, what is the total energy (in μJ) of the system ?

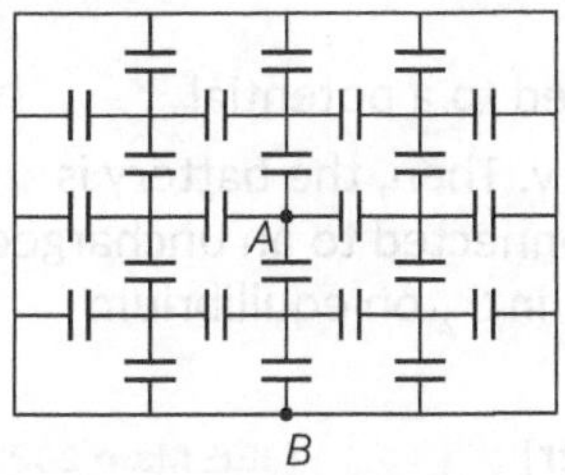

93 In the middle of two plates making a parallel plate capacitor, a plate was placed, having charge q. The plate is moved parallel to itself by a distance x. What is the charge (in μC) flowing through the external circuit of the capacitor, if the distance between its plates equal to d? (Take, $q = 8\mu C$, $x = 1$ mm, $d = 4$ mm)

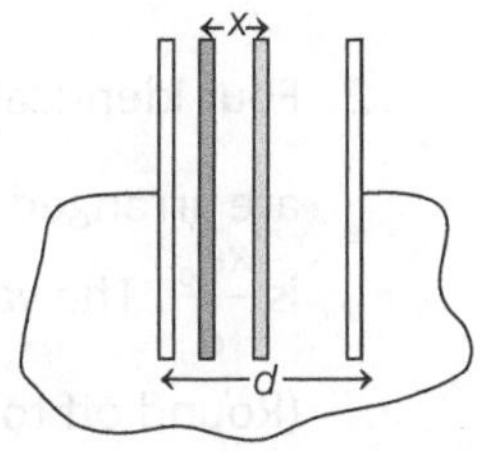

94 The figure shows four identical conducting plates each of area A. The separation between the consecutive plates is equal to L. When both the switches are closed, if charge present on the upper surface of the lowest plate from the top is written $\frac{xV_0\varepsilon_0 A}{L}$, then what is the value of x? Treat symbols as having usual meaning.

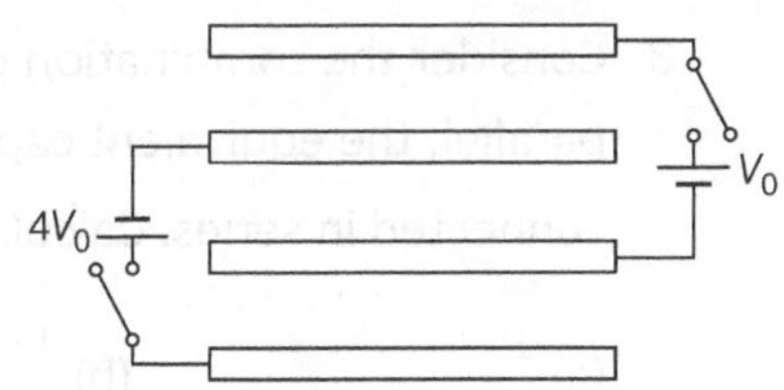

95 Four capacitors of capacitance $1\mu F$ each are connected with conducting wires between AB, BC, CD and DA, where $ABCD$ is a square. Considering the square to be horizontal, a point M is vertically above the centroid of the square. Four capacitors of capacitance $2\,\mu F$ each are connected between AM, BM, CM and DM, thus forming a pyramid kind of a frame as shown. If the effective capacitance between points A-C is C_1 and between A-B is C_2, then find $\frac{C_2}{C_1}$.

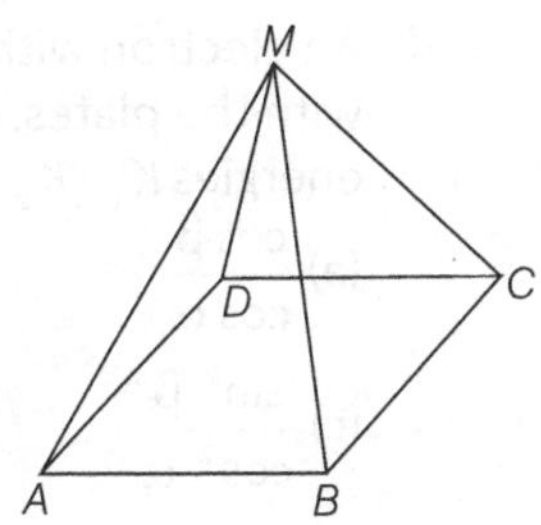

Answers

1. (d)	**2.** (a)	**3.** (b)	**4.** (b)	**5.** (c)	**6.** (b)	**7.** (a)	**8.** (c)	**9.** (b)	**10.** (b)
11. (d)	**12.** (d)	**13.** (d)	**14.** (c)	**15.** (c)	**16.** (d)	**17.** (d)	**18.** (b)	**19.** (b)	**20.** (a)
21. (b)	**22.** (c)	**23.** (a)	**24.** (d)	**25.** (a)	**26.** (a)	**27.** (b)	**28.** (c)	**29.** (a)	**30.** (d)
31. (b)	**32.** (b)	**33.** (b)	**34.** (a)	**35.** (c)	**36.** (a)	**37.** (abd)	**38.** (ac)	**39.** (ac)	**40.** (ac)
41. (abcd)	**42.** (abc)	**43.** (bd)	**44.** (abd)	**45.** (abcd)	**46.** (bd)	**47.** (abd)	**48.** (ac)	**49.** (abc)	**50.** (bc)
51. (ac)	**52.** (ab)	**53.** (abcd)	**54.** (c)	**55.** (c)	**56.** (c)	**57.** (a)	**58.** (c)	**59.** (c)	**60.** (b)
61. (a)	**62.** (a)	**63.** (b)	**64.** (d)	**65.** (bc)	**66.** (abcd)	**67.** (bc)	**68.** (c)	**69.** (c)	**70.** (c)
71. (a)	**72.** (b)	**73.** (a)	**74.** (c)	**75.** (bc)	**76.** (ac)	**77.** (b)	**78.** (a)	**79.** (b)	

80. (a) $\rightarrow$ r, (b) $\rightarrow$ pq, (c) $\rightarrow$ s, (d) $\rightarrow$ pq

81. (a) $\rightarrow$ s; (b) $\rightarrow$ r; (c) $\rightarrow$ q; (d) $\rightarrow$ p

82. (a) $\rightarrow$ p; (b) $\rightarrow$ r; (c) $\rightarrow$ q; (d) $\rightarrow$ p

83. (a) $\rightarrow$ q; (b) $\rightarrow$ p; (c) $\rightarrow$ s; (d) $\rightarrow$ r

84. (a) $\rightarrow$ s; (b) $\rightarrow$ r; (c) $\rightarrow$ q; (d) $\rightarrow$ p

85. (a) $\rightarrow$ r; (b) $\rightarrow$ s; (c) $\rightarrow$ q; (d) $\rightarrow$ p

86. (40) **87.** (10) **88.** (69) **89.** (4) **90.** (2)

91. (2) **92.** (12) **93.** (2) **94.** (3) **95.** (1.20)

For Detailed Solutions
Scan the code

Previous Years' Questions

JEE Main

1 A 2 μF capacitor C_1 is first charged to a potential difference of 10 V using a battery. Then, the battery is removed and the capacitor is connected to an uncharged capacitor C_2 of 8 μF. The charge in C_2 on equilibrium condition is .... μC.
(Round off to the nearest integer) **[JEE Main 2021]**

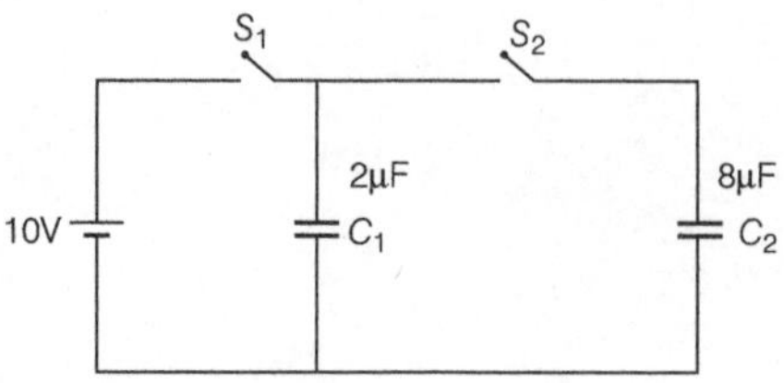

2 Four identical rectangular plates with length, $l = 2$ cm and breadth, $b = \frac{3}{2}$ cm are arranged as shown in figure. The equivalent capacitance between A and C is $\frac{x\varepsilon_0}{d}$. The value of x is ....
(Round off to the nearest integer) **[JEE Main 2021]**

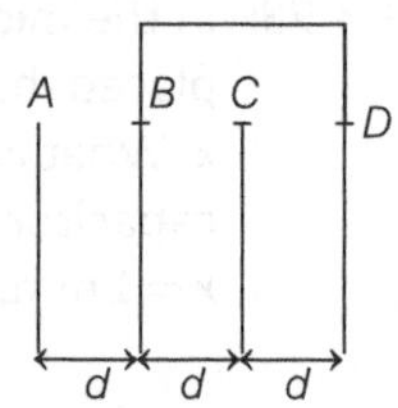

3 Consider the combination of two capacitors C_1 and C_2, with $C_2 > C_1$, when connected in parallel, the equivalent capacitance is 15/4 time the equivalent capacitance of the same connected in series. Calculate the ratio of capacitors $\frac{C_2}{C_1}$. **[JEE Main 2021]**

(a) $\frac{15}{11}$ (b) $\frac{111}{80}$ (b) $\frac{29}{15}$ (d) $\frac{15}{4}$

4 An electron with kinetic energy K_1 enters between parallel plates of a capacitor at an angle α with the plates. It leaves the plates at angle β with kinetic energy K_2. Then, the ratio of kinetic energies $K_1 : K_2$ will be **[JEE Main 2021]**

(a) $\frac{\cos\beta}{\cos\alpha}$ (b) $\frac{\cos\beta}{\sin\alpha}$
(c) $\frac{\sin^2\beta}{\cos^2\alpha}$ (d) $\frac{\cos^2\beta}{\cos^2\alpha}$

5 Two equal capacitors are first connected in series and then in parallel. The ratio of the equivalent capacities in the two cases will be **[JEE Main 2021]**

(a) 1 : 2 (b) 2 : 1
(c) 4 : 1 (d) 1 : 4

6 Two isolated conducting spheres S_1 and S_2 of radii $\frac{2}{3}R$ and $\frac{1}{3}R$, have charges 12 μC and – 3μC respectively, and are at a large distance from each other. They are now connected by a conducting wire. A long time after this is done, the charges on S_1 and S_2 respectively, are **[JEE Main 2020]**

(a) 4.5 μC on both (b) + 4.5 μC and – 4.5μC
(c) 3 μC and 6 μC (d) 6 μC and 3 μC

7 A capacitor is made of two square plates each of side a making a very small angle α between them, as shown in figure. The capacitance will be close to **[JEE Main 2020]**

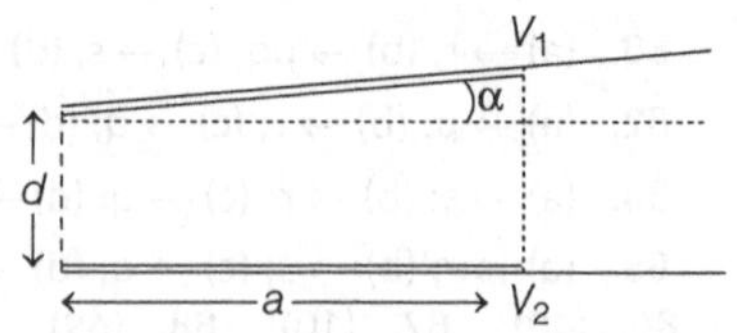

(a) $\frac{\varepsilon_0 a^2}{d}\left(1 - \frac{\alpha a}{4d}\right)$ (b) $\frac{\varepsilon_0 a^2}{d}\left(1 + \frac{\alpha a}{d}\right)$
(c) $\frac{\varepsilon_0 a^2}{d}\left(1 - \frac{\alpha a}{2d}\right)$ (d) $\frac{\varepsilon_0 a^2}{d}\left(1 - \frac{3\alpha a}{2d}\right)$

8 A 5 μF capacitor is charged fully by a 220V supply. It is then disconnected from the supply and is connected in series to another uncharged 2.5 μF capacitor. If the energy change during the charge redistribution is $\frac{X}{100}$ J, then value of X to the nearest integer is .... **[JEE Main 2020]**

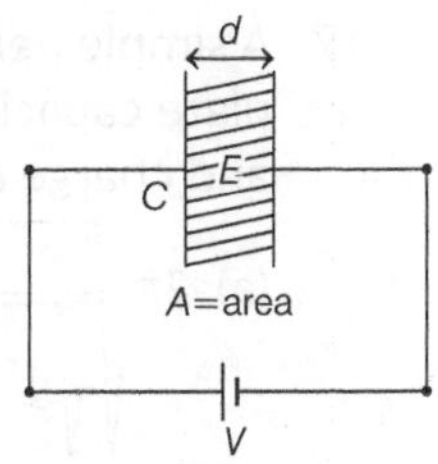

9 A 10 μF capacitor is fully charged to a potential difference of 50 V. After removing the source voltage it is connected to an uncharged capacitor in parallel. Now, the potential difference across them becomes 20 V. The capacitance of the second capacitor is **[JEE Main 2020]**

(a) 15 μF (b) 30 μF
(c) 20 μF (d) 10 μF

10 In the circuit shown in the figure, the total charge is 750 μC and the voltage across capacitor C_2 is 20 V. Then, the charge on capacitor C_2 is **[JEE Main 2020]**

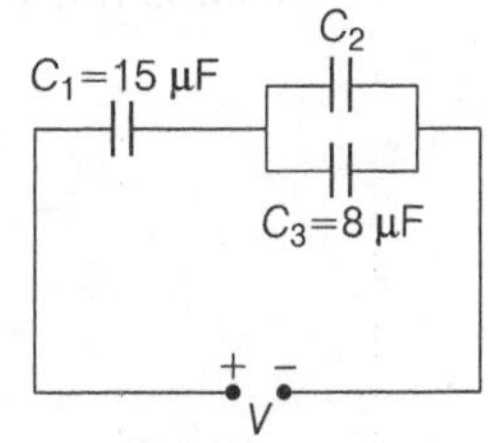

(a) 450 μC
(b) 590 μC
(c) 160 μC
(d) 650 μC

11 A capacitor C is fully charged with voltage V_0. After disconnecting the voltage source, it is connected in parallel with another uncharged capacitor of capacitance $\frac{C}{2}$. The energy loss in the process after the charge is distributed between the two capacitors is **[JEE Main 2020]**

(a) $\frac{1}{3}CV_0^2$ (b) $\frac{1}{6}CV_0^2$ (c) $\frac{1}{2}CV_0^2$ (d) $\frac{1}{4}CV_0^2$

12 Two capacitors of capacitances C and $2C$ are charged to potential differences V and $2V$ respectively. These are then connected in parallel in such a manner that the positive terminal of one is connected to the negative terminal of the other. The final energy of this configuration is **[JEE Main 2020]**

(a) $\frac{9}{2}CV^2$ (b) $\frac{3}{2}CV^2$ (c) $\frac{25}{6}CV^2$ (d) zero

13 In the circuit shown, charge on the 5μF capacitor is **[JEE Main 2020]**

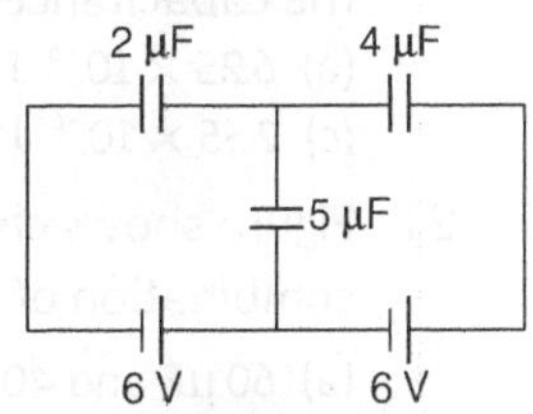

(a) 18.00 μC
(b) 10.90 μC
(c) 16.36 μC
(d) 5.45 μC

14 Effective capacitance of parallel combination of two capacitors C_1 and C_2 is 10 μF. When these capacitors are individually connected to a voltage source of 1 V, the energy stored in the capacitor C_2 is 4 times that of C_1. If these capacitors are connected in series, then their effective capacitance will be **[JEE Main 2020]**

(a) 4.2 μF (b) 3.2 μF (c) 1.6 μF (d) 8.4 μF

15 A 60 pF capacitor is fully charged by a 20 V supply. It is then disconnected from the supply and is connected to another uncharged 60 pF capacitor in parallel. The electrostatic energy that is lost in this process by the time, the charge is redistributed between them is (in nJ). **[JEE Main 2020]**

16 A parallel plate capacitor has 1μF capacitance. One of its two plates is given + 2μC charge and the other plate + 4μC charge. The potential difference developed across the capacitor is **[JEE Main 2019]**

(a) 1 V (b) 5 V (c) 2 V (d) 3 V

17 A simple pendulum of length L is placed between the plates of a parallel plate capacitor having electric field E, as shown in figure. Its bob has mass m and charge q. The time period of the pendulum is given by **[JEE Main 2019]**

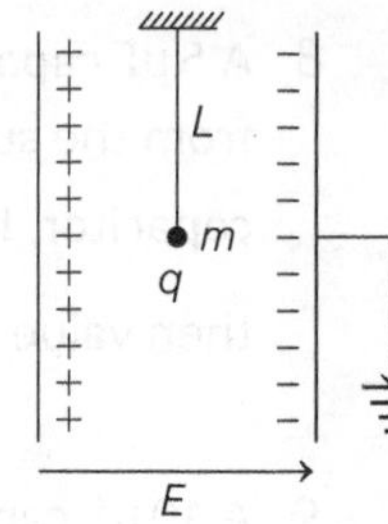

(a) $2\pi\sqrt{\dfrac{L}{\sqrt{g^2+\left(\dfrac{qE}{m}\right)^2}}}$ (b) $2\pi\sqrt{\dfrac{L}{\sqrt{g^2-\dfrac{q^2E^2}{m^2}}}}$

(c) $2\pi\sqrt{\dfrac{L}{\left(g+\dfrac{qE}{m}\right)}}$ (d) $2\pi\sqrt{\dfrac{L}{\left(g-\dfrac{qE}{m}\right)}}$

18 The charge on a capacitor plate in a circuit as a function of time is shown in the figure. What is the value of current at $t = 4$ s? **[JEE Main 2019]**

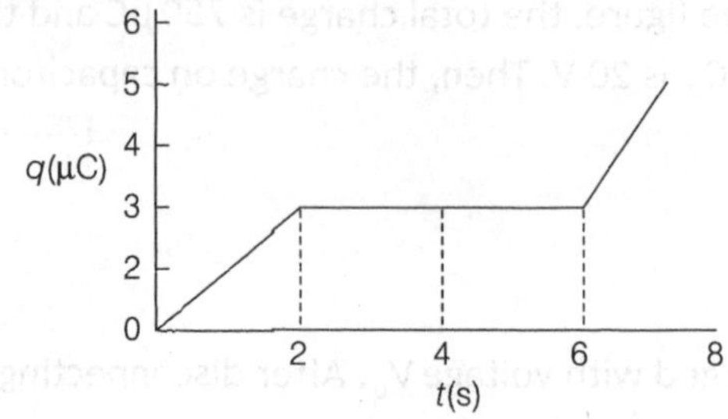

(a) 2 μA (b) 1.5 μA (c) Zero (d) 3 μA

19 A parallel plate capacitor with plates of area 1 m^2 each, are at a separation of 0.1 m. If the electric field between the plates is 100 N/C, the magnitude of charge on each plate is

$\left(\text{Take, } \varepsilon_0 = 8.85 \times 10^{-12} \dfrac{C^2}{N-m^2}\right)$ **[JEE Main 2019]**

(a) 9.85×10^{-10} C (b) 8.85×10^{-10} C

(c) 7.85×10^{-10} C (d) 6.85×10^{-10} C

20 A capacitor with capacitance 5 μF is charged to 5 μC. If the plates are pulled apart to reduce the capacitance to 2 μF, how much work is done? **[JEE Main 2019]**

(a) 6.25×10^{-6} J (b) 2.16×10^{-6} J

(c) 2.55×10^{-6} J (d) 3.75×10^{-6} J

21 Figure shows charge (q) *versus* voltage (V) graph for series and parallel combination of two given capacitors. The capacitances are **[JEE Main 2019]**

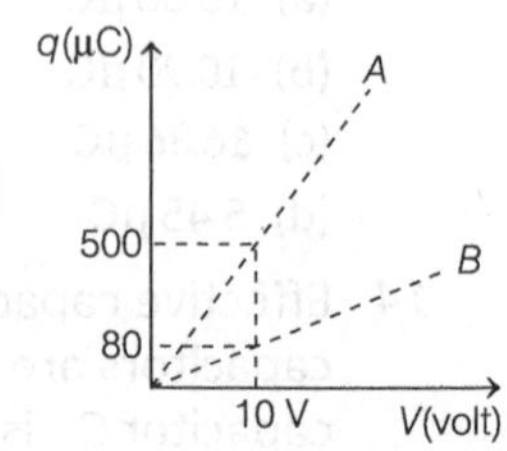

(a) 60 μF and 40 μF

(b) 50 μF and 30 μF

(c) 20 μF and 30 μF

(d) 40 μF and 10μF

22 In the given circuit, the charge on 4 μF capacitor will be **[JEE Main 2019]**

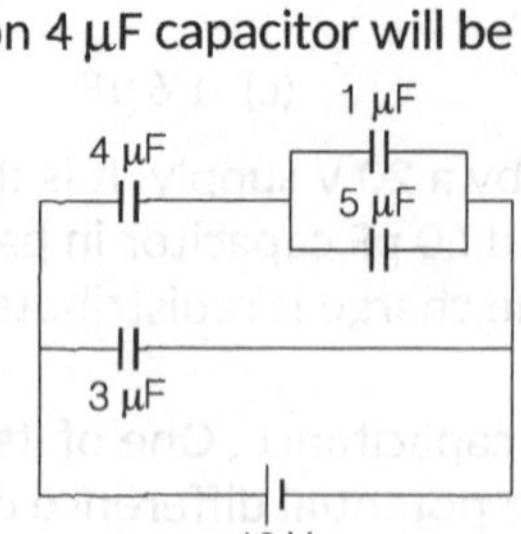

(a) 5.4 μC (b) 9.6 μC (c) 13.4 μC (d) 24 μC

23 In the figure shown below, the charge on the left plate of the 10 μF capacitor is – 30 μC. The charge on the right plate of the 6 μF capacitor is **[JEE Main 2019]**

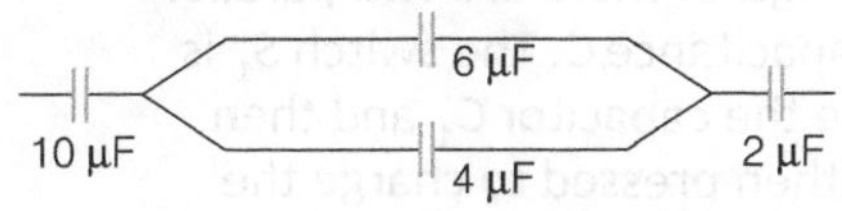

(a) + 12 μC
(b) + 18 μC
(c) – 12 μC
(d) – 18 μC

24 Seven capacitors, each of capacitance 2 μF are to be connected in a configuration to obtain an effective capacitance of $\left(\frac{6}{13}\right)$ μF. Which of the combinations shown in figures below will achieve the desired value? **[JEE Main 2019]**

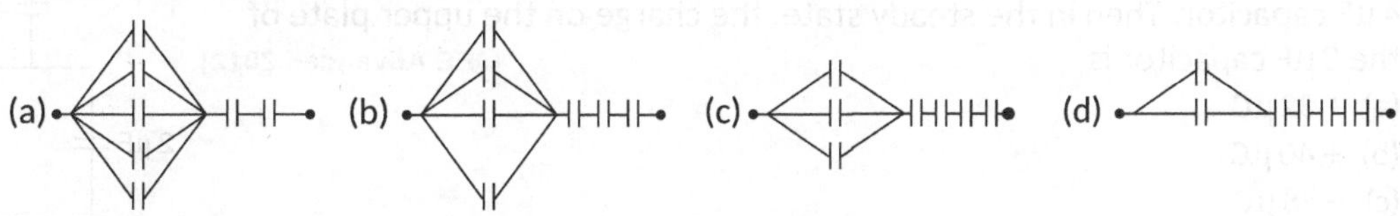

25 In the figure shown, after the switch '*S*' is turned from position '*A*' to position '*B*', the energy dissipated in the circuit in terms of capacitance '*C*' and total charge '*Q*' is **[JEE Main 2019]**

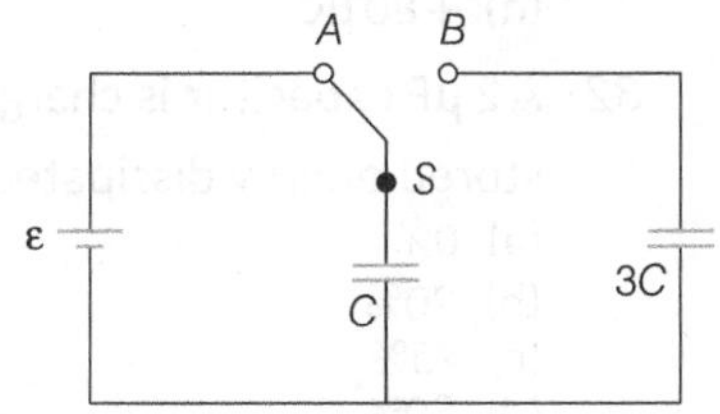

(a) $\frac{3}{4}\cdot\frac{Q^2}{C}$
(b) $\frac{5}{8}\cdot\frac{Q^2}{C}$
(c) $\frac{1}{8}\cdot\frac{Q^2}{C}$
(d) $\frac{3}{8}\cdot\frac{Q^2}{C}$

26 In the circuit shown, find *C* if the effective capacitance of the whole circuit is to be 0.5 μF. All values in the circuit are in μF. **[JEE Main 2019]**

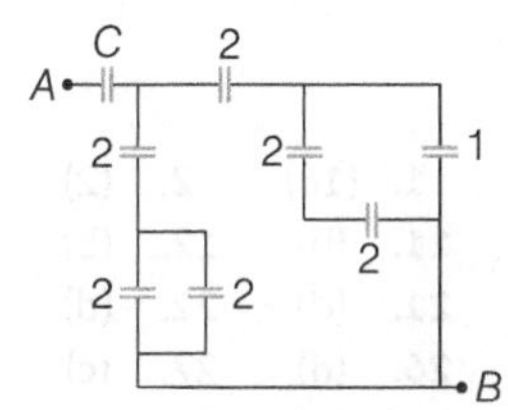

(a) $\frac{6}{5}$ μF
(b) 4 μF
(c) $\frac{7}{10}$ μF
(d) $\frac{7}{11}$ μF

27 A capacitance of 2 μF is required in an electrical circuit across a potential difference of 1kV. A large number of 1 μF capacitors are available which can withstand a potential difference of not more than 300 V. The minimum number of capacitors required to achieve this is **[JEE Main 2017]**

(a) 16
(b) 24
(c) 32
(d) 2

28 A combination of capacitors is set-up as shown in the figure. The magnitude of the electric field, due to a point charge *Q* (having a charge equal to the sum of the charges on the 4 μF and 9μF capacitors), at a point distance 30 m from it, would equal to **[JEE Main 2016]**

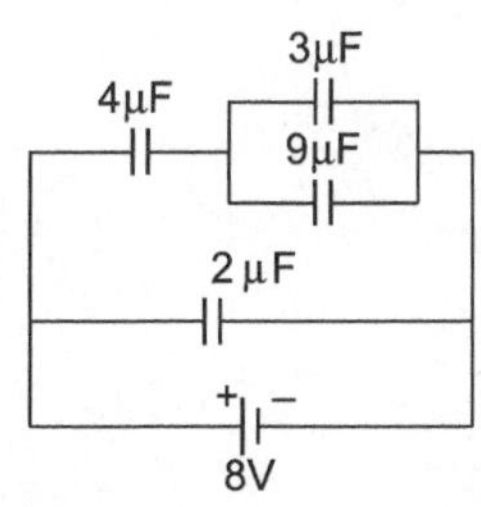

(a) 240 N/C
(b) 360 N/C
(c) 420 N/C
(d) 480 N/C

29 Two capacitors C_1 and C_2 are charged to 120 V and 200 V, respectively. It is found that by connecting them together, the potential on each one can be made zero. Then, **[JEE Main 2013]**

(a) $5C_1 = 3C_2$
(b) $3C_1 = 5C_2$
(c) $3C_1 + 5C_2 = 0$
(d) $9C_1 = 4C_2$

JEE Advanced

30 In the circuit shown in the figure, there are two parallel plate capacitors each of capacitance C. The switch S_1 is pressed first to fully charge the capacitor C_1 and then released. The switch S_2 is then pressed to charge the capacitor C_2. After some time, S_2 is released and then S_3 is pressed. After some time **[JEE Advanced 2013]**

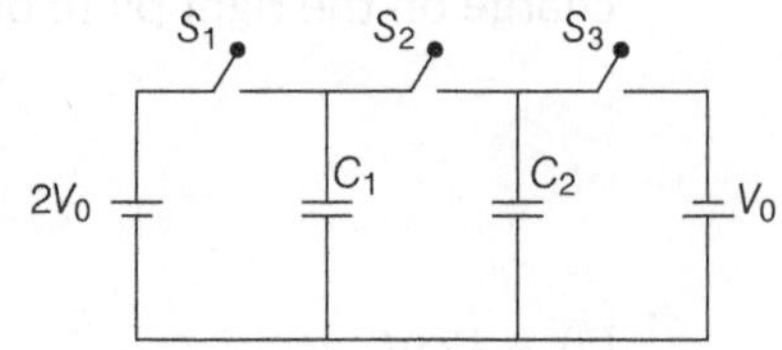

(a) the charge on the upper plate of C_1 is $2\,CV_0$
(b) the charge on the upper plate of C_1 is CV_0
(c) the charge on the upper plate of C_2 is 0
(d) the charge on the upper plate of C_2 is $-CV_0$

31 In the given circuit, a charge of +80 μC is given to the upper plate of the 4 μF capacitor. Then in the steady state, the charge on the upper plate of the 3 μF capacitor is **[JEE Advanced 2012]**

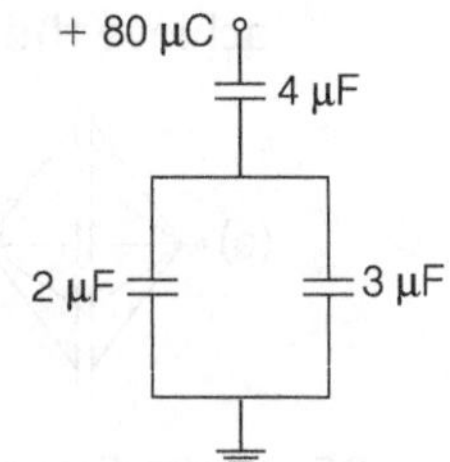

(a) + 32 μC
(b) + 40 μC
(c) + 48 μC
(d) + 80 μC

32 A 2 μF capacitor is charged as shown in the figure. The percentage of its stored energy dissipated after the switch S is turned to position 2 is **[JEE Advanced 2011]**

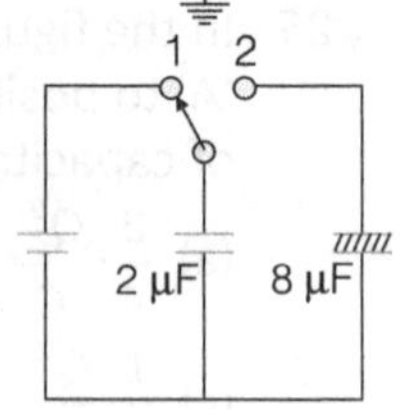

(a) 0%
(b) 20%
(c) 75%
(d) 80%

Answers

1. (16)	2. (2)	3. (*)	4. (d)	5. (d)	6. (d)	7. (c)	8. (4)	9. (a)	10. (b)
11. (b)	12. (b)	13. (c)	14. (c)	15. (6)	16. (a)	17. (a)	18. (c)	19. (b)	20. (d)
21. (d)	22. (d)	23. (b)	24. (c)	25. (d)					
26. (d)	27. (c)	28. (c)	29. (b)	30. (bd)					
31. (c)	32. (d)								

For Detailed Solutions
Scan the code

CHAPTER

10

DIELECTRICS

LEVEL 01 \ Starter Level

1.1 Capacitor Fully Filled with Dielectric

1 A parallel plate capacitor has plates of area A separated by distance d. The capacitor is connected across a cell of emf V.

(a) Find the capacitance (C_0), charge (q_0), electric field (E_0) and energy (U_0) of the capacitor.

(b) Find new capacitance (C), charge (q), electric field (E) and energy (U) of the capacitor in the following two cases.

Case-1 If battery is first disconnected and then a dielectric slab of dielectric constant K is inserted fully between the capacitor plates.

Case-2 If a dielectric slab of dielectric constant K is inserted fully between the capacitor plates while it is connected to battery.

2 A capacitor has two circular plates whose radii are 8 cm and distance between them is 1 mm. When mica (dielectric constant $= 6$) is placed between the plates, calculate the capacitance of this capacitor.

3 A parallel plate capacitor having plate area 100 cm^2 and separation 1.0 mm holds a charge of 0.12 μC when connected to a 120 V battery. Find the dielectric constant of the material filling the gap.

4 A parallel plate air capacitor has square plates of edge 10.0 cm separated by 1.0 mm. It is fully filled by a dielectric slab of dielectric constant 5.0 and connected across a 60.0 V battery. Then, find

(a) the charge supplied by the battery (b) the induced charge on the dielectric

5 A parallel plate capacitor is formed by two plates, each of area 100 cm^2, separated by a distance of 1 mm. A dielectric of dielectric constant 5.0 and dielectric strength 1.9×10^7 V/m is filled between the plates. Find the maximum charge that can be stored on the capacitor without causing any dielectric breakdown.

1.2 Capacitor Partially Filled by A Dielectric

6 The distance between the plates of a parallel plate condenser is 5 cm. A field of 3×10^4 V/m is applied between the plates. It is then disconnected from the battery and an uncharged metal plate of thickness 1 cm is inserted into the condenser parallel to its plates. Find the potential difference between the plates;

(i) before the introduction of the metal plate,

(ii) after its introduction.

(iii) What would be the potential difference, if the metal plate is replaced by a plate of dielectric constant 2?

7 A potential difference of 130 V is applied across a plane air capacitor spaced 1 cm apart. A plane parallel glass layer with a thickness of 0.5 cm and a plane parallel paraffin layer with a thickness of 0.3 cm are placed in the space between the capacitor plates. Find

(i) intensity of electric field in each layer,

(ii) the drop of potential in each layer,

(iii) the surface charge density of the induced charges on each layer.
Given that: $K_{\text{glass}} = 6, K_{\text{paraffin}} = 2$.

8 A parallel plate air capacitor of capacitance 5 μF is connected to a battery of emf 30 V. The separation between the plates is 2 mm. Now, a dielectric slab of thickness 1 mm and dielectric constant 5 is inserted into the gap to occupy the lower half of it.

(i) Find the capacitance of the new combination.

(ii) How much charge has flown through the battery after the slab is inserted?

9 A parallel plate air capacitor of capacitance 15 μF is connected to a battery of 100 V. Separation between the plates 1 mm. A dielectric slab of thickness 0.5 mm and dielectric constant 5 is inserted into the gap.

(i) Find the increase in electrostatic energy.

(ii) If the battery is now disconnected and the dielectric slab is taken out, find the further increase in energy.

(iii) Why does the energy increase in inserting the slab as well as in taking it out?

1.3 Equivalent Capacitance with Dielectric

10 Some dielectric slabs are inserted between the plates of a parallel plate air capacitor of plates area A and separation d as shown in the following figures. Calculate capacitance for each case.

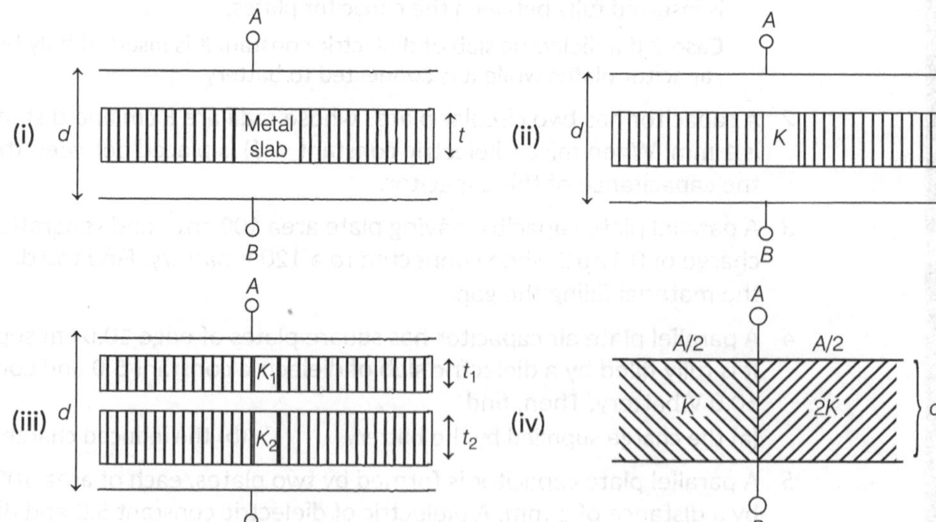

(v) 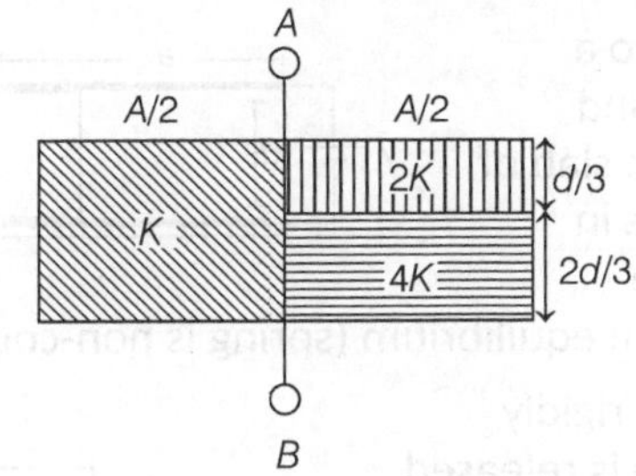

(vi) 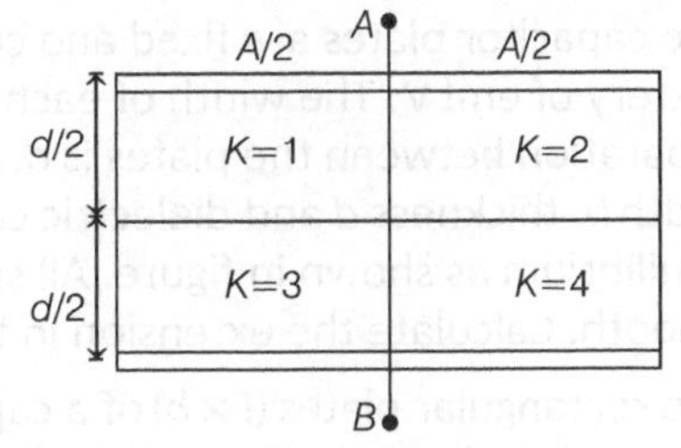

11 A parallel plate capacitor is having square plates of edge a and plate-separation d. The gap between the plates is filled with a dielectric. Calculate the capacitance for each case.

(i) 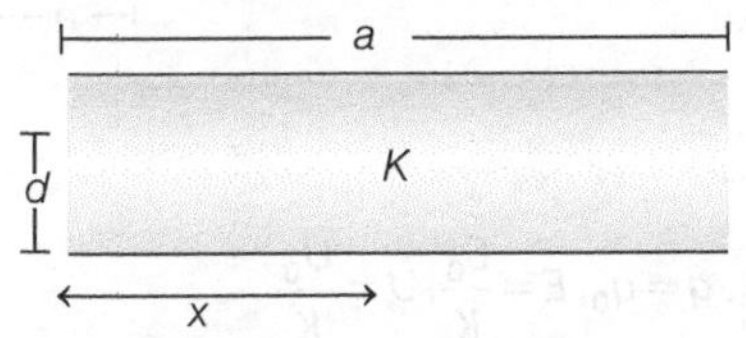

Dielectric constant K varies parallel to an edge as $K = K_0 + \lambda x$, where K_0 and λ are constants and x is the distance from the left end.

(ii) 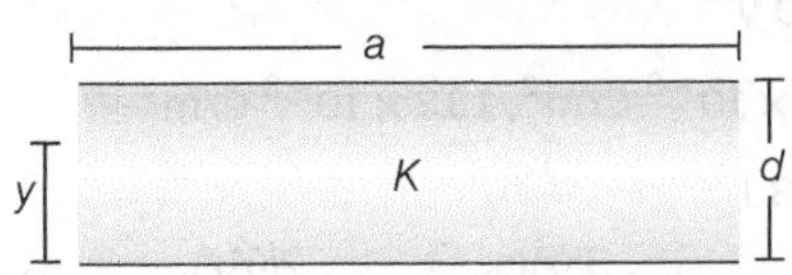

Dielectric constant K varies perpendicular to an edge as $K = K_0 + \lambda y$, where K_0 and λ are constants and y is the distance from an edge.

1.4 Capacitors with Dielectrics in Circuit

12 Two parallel plate air capacitors each of capacitance C were connected in series to a battery with emf. ε. Then, one of the capacitors was filled up with uniform dielectric with relative permittivity k.

(i) How many times did the electric field strength in that capacitor decrease?

(ii) What amount of charge flows through the battery?

13 Two parallel plate capacitors of capacitance 10 μF and 30 μF are connected in parallel and charged to a potential difference 300 V with the help of a battery. The battery is disconnected and the space between these two capacitors are filled with dielectric of constant 6 and 3 respectively. Find the voltage across the combination.

1.5 Force on Dielectric Slab

14 Figure shows a parallel plate capacitor with plates width b and length l. The separation between the plates is d. The plates are rigidly clamped and connected to a battery of e.m.f. V. Dielectric slab is of length l, width b, thickness d and dielectric constant K.

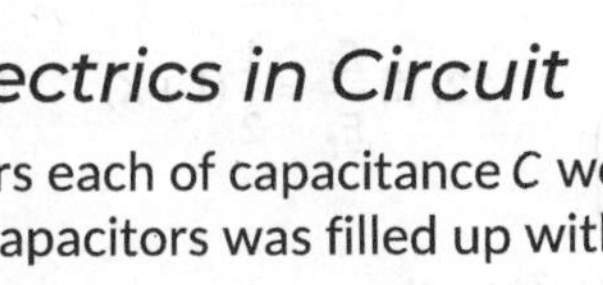

(i) Find the energy of the system when a length x of the slab is introduced into the capacitor.

(ii) What force is exerted on the slab by the capacitor?

15 The capacitor plates are fixed and connected to a battery of emf V. The width of each plate is b and separation between the plates is d. A dielectric slab of width b, thickness d and dielectric constant K is in equilibrium as shown in figure. All surfaces are smooth. Calculate the extension in the spring in equilibrium (spring is non-conducting).

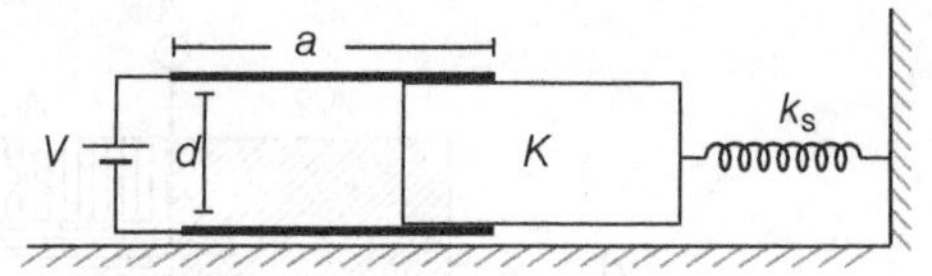

16 The rectangular plates ($l \times b$) of a capacitor are rigidly clamped as shown in figure. The dielectric slab is released from rest with a length a inside the capacitor. Neglecting friction and gravity, show that the slab will execute periodic motion and find its time period.

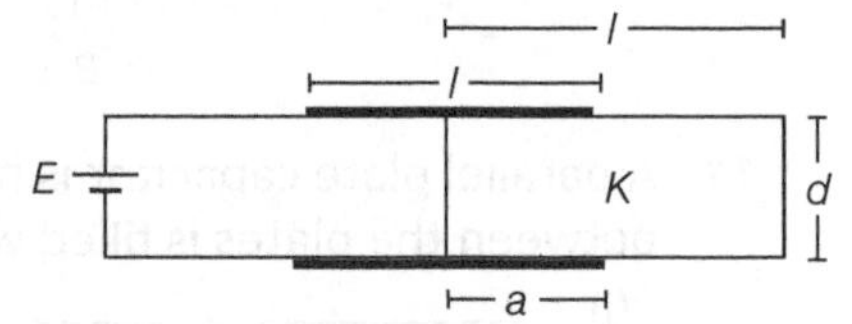

Answers

1. (i) $C_0 = \frac{A\varepsilon_0}{d}, q_0 = \frac{A\varepsilon_0 V}{d}, E_0 = \frac{V}{d}, U_0 = \frac{1}{2}\frac{A\varepsilon_0}{d}V^2$ (ii) $C = KC_0, q = q_0, E = \frac{E_0}{K}, U = \frac{U_0}{K}$

(iii) $C = KC_0, q = Kq_0, E = E_0, U = KU_0$ **2.** 1.067×10^{-9} F **3.** 11.3 **4.** (a) 26.6 nC, (b) 21.3 nC

5. 8.4 μC **6.** (i) 1500 V (ii) 1200 V (iii) 1350 V

7. (i) 5×10^3 V/m, 1.5×10^4 V/m (ii) 25 V, 45 V (iii) 22.2×10^{-8} C/m^2, 13.3×10^{-8} C/m^2

8. (a) 8.3 μF, (b) 99 μC **9.** (a) 0.05 J (b) 0.083 J

10. (i) $\frac{A\varepsilon_0}{d-t}$ (ii) $\frac{A\varepsilon_0}{d - t + \frac{t}{K}}$ (iii) $\frac{A\varepsilon_0}{(d - t_1 - t_2) + \frac{t_1}{K_1} + \frac{t_2}{K_2}}$ (iv) $\frac{3KA\varepsilon_0}{2d}$ (v) $\frac{2KA\varepsilon_0}{d}$ (vi) $\frac{25A\varepsilon_0}{12d}$

11. (i) $\frac{a^2\varepsilon_0}{2d}(2K_0 + \lambda a)$ (ii) $\frac{\lambda a^2 \varepsilon_0}{\ln\left(1 + \frac{\lambda d}{K_0}\right)}$ **12.** (i) $\frac{E_i}{E_f} = \frac{1}{2}(1 + K)$ (ii) $\Delta q = \frac{1}{2}C\varepsilon\frac{K-1}{K+1}$ **13.** 80 V

14. (i) $U = \frac{\varepsilon_0 bV^2}{2d}[(K-1)x + l]$ (ii) $F = \frac{\varepsilon_0 bV^2(K-1)}{2d}$

15. $\frac{\varepsilon_0 bV^2(K-1)}{2dK_s}$ **16.** $8\sqrt{\frac{md(l-a)}{b\varepsilon_0\varepsilon^2(K-1)}}$

For Detailed Solutions
Scan the code

LEVEL 02 \ JEE Main Level

Single Choice Correct Questions

1 A capacitor stores 50 μC charge when connected across a battery. When the gap between the plates is filled with a dielectric, a charge of 100 μC flows through the battery. The dielectric constant of the material is

(a) 2.5 (b) 2 (c) 4 (d) 3

2 The capacitance of an air filled parallel plate capacitor is 10 μF. The separation between the plates is doubled and the space between the plates is then filled with wax giving the capacitance a new value of 40 μF.The dielectric constant of wax is

(a) 4 (b) 8 (c) 10 (d) 12

3 A capacitor is charged by using a battery which is then disconnected. A dielectric slab is then slipped between the plates which results in

(a) reduction of charges on the plates and increase of potential difference across the plates

(b) increase in the potential difference across the plates, reduction in stored energy, but no change in the charge on the plates

(c) decrease in the potential difference across the plates, reduction in stored energy, but no change in the charge on the plates.

(d) None of the above

4 Two square plates of capacitor are of length 32 cm each. Between the plates, material having thickness 640 μm is sandwiched whose dielectric strength is 20 MV/m. The maximum charge that can be put on the capacitor is (Permittivity of free space $= 9 \times 10^{-12}$ SI unit and dielectric constant of material is 4)

(a) 4.6 μC (b) 5 μC (c) 73.7 μC (d) 18.4 μC

5 If dielectric constant and dielectric strength be denoted by K and x respectively, then a material suitable for use as a dielectric in a capacitor must have

(a) high K and high x (b) high K and low x (c) low K and low x (d) low K and high x

6 An isolated capacitor when filled with a dielectric $K = 3$ has charge Q_0, voltage V_0 and electric field E_0. If the dielectric is replaced with another one having $K = 9$, the new value of charge, voltage and field will be respectively

(a) $3Q_0, 3V_0, 3E_0$ (b) $Q_0, 3V_0, 3E_0$ (c) $Q_0, \frac{V_0}{3}, 3E_0$ (d) $Q_0, \frac{V_0}{3}, \frac{E_0}{3}$

7 In a capacitor of capacitance 20 μF, the distance between the plates is 2 mm. If a dielectric slab of width 1 mm and dielectric constant 2 is inserted between the plates, then the new capacitance will be

(a) 2 μF (b) 15.5 μF (c) 26.6 μF (d) 32 μF

8 Capacitance of a capacitor becomes 7/6 times its original value if a dielectric slab of thickness $t = \left(\frac{2}{3}\right)d$ is introduced between plates where d is the separation between plates. The dielectric constant of slab is

(a) $\frac{14}{11}$ (b) $\frac{14}{13}$ (c) $\frac{13}{7}$ (d) $\frac{11}{7}$

9 The capacitance of an air capacitor is 15 μF and the separation between the parallel plates is 6 mm. A copper plate of 2 mm thickness is introduced symmetrically between the plates. The capacitance now becomes

(a) 5 μF (b) 7.5 μF (c) 22.5 μF (d) 30 μF

10 Separation between the plates of a parallel-plate capacitor is 5 mm. This capacitor, having air as the dielectric medium between the plates, is charged to a potential difference 25 V using a battery. The battery is then disconnected and a dielectric slab of thickness 3 mm and dielectric constant $K = 10$ is placed between the plates, as shown. Potential difference between the plates after the dielectric slab has been introduced is

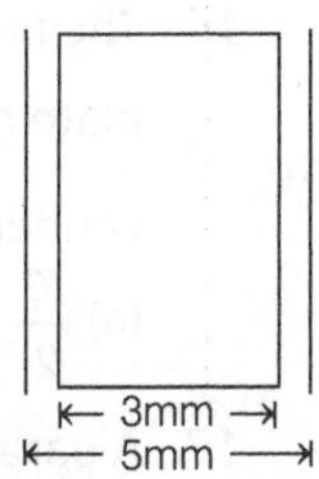

(a) 18.5 V (b) 13.5 V

(c) 11.5 V (d) 6.5 V

11 The plates of a parallel plate capacitor are charged upto 100 V. A 2 mm thick plate is inserted between the plates. To maintain the same potential difference, the distance between the capacitor plates is increased by 1.6 mm. The dielectric constant of the plate is

(a) 5 (b) 1.25

(c) 4 (d) 2.5

12 A parallel plate capacitor has two layers of dielectric as shown in figure. This capacitor is connected across a battery. The graph which shows the variation of electric field (E) and distance (x) from left plate.

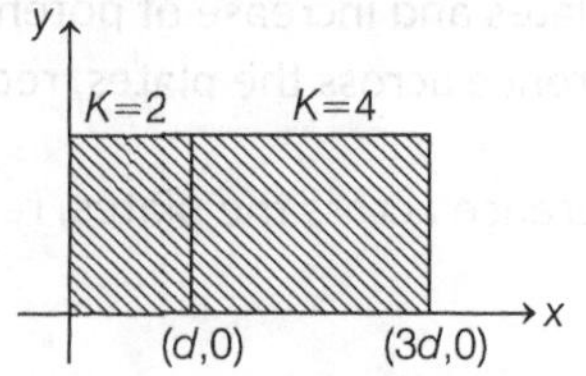

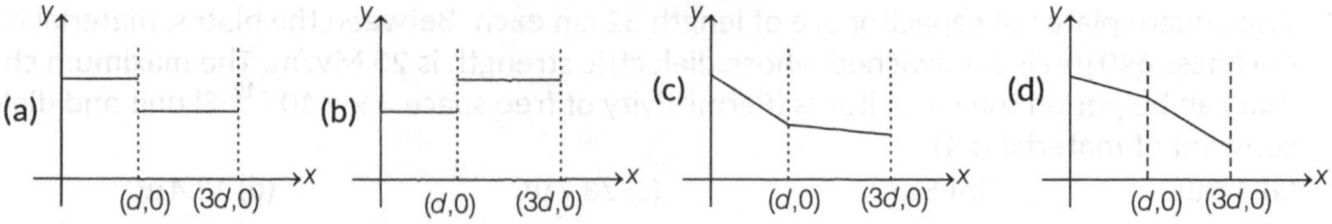

13 In the figure shown a parallel plate capacitor has a dielectric of width $d/2$ and dielectric constant $K = 2$. The other dimensions of the dielectric are same as that of the plates. The plates P_1 and P_2 of the capacitor have area A each. The energy of the capacitor is

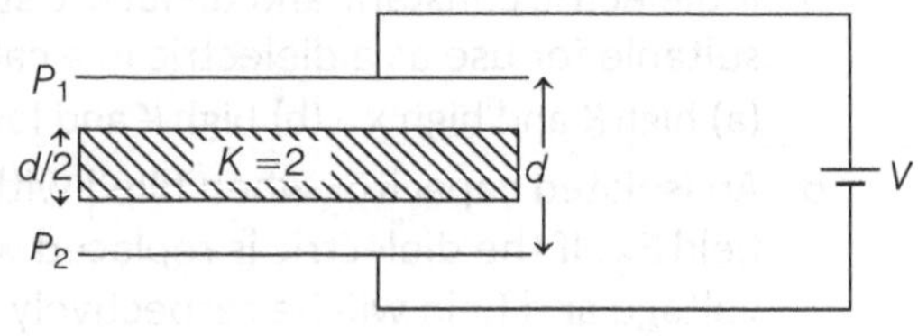

(a) $\dfrac{\varepsilon_0 AV^2}{3d}$ (b) $\dfrac{2\varepsilon_0 AV^2}{d}$ (c) $\dfrac{3}{2}\dfrac{\varepsilon_0 AV^2}{d}$ (d) $\dfrac{2\varepsilon_0 AV^2}{3d}$

14 A parallel plate capacitor is completely filled by an insulating material 3 mm thick and of relative permittivity 4. The distance between the plates is increased to allow the insertion of a second sheet 5 mm thick and of relative permittivity ε_r. If the capacitance of the capacitor so formed is one-half of the original capacitance, then the value of ε_r.

(a) $\dfrac{20}{3}$ (b) $\dfrac{16}{3}$ (c) $\dfrac{10}{3}$ (d) None of these

15 A parallel plate capacitor have distance between its plate is d and area of each plate is A. Now, half of capacitor is filled with a substance of dielectric constant K according to given figure. The capacity of the condenser, is

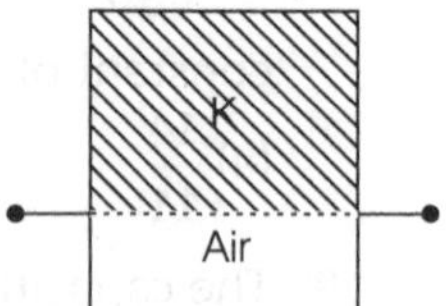

(a) $\dfrac{\varepsilon_0 A}{2d}(K+2)$ (b) $\dfrac{\varepsilon_0 A}{3d}(K+1)$

(c) $\dfrac{\varepsilon_0 A}{2d}(K+1)$ (d) $\dfrac{\varepsilon_0 A}{4d}(K+3)$

16 The two metallic plates of radius r are placed at a distance d apart and its capacity is C. If a plate of radius $\dfrac{r}{2}$ and thickness d of dielectric constant 6 is placed between the plates of the condenser, then its capacity will be

(a) $\dfrac{7C}{2}$ (b) $\dfrac{3C}{7}$ (c) $\dfrac{7C}{3}$ (d) $\dfrac{9C}{4}$

17 A parallel plate capacitor of capacitance C (without dielectrics) is filled by dielectric slabs as shown in figure. Then, the new capacitance of the capacitor is

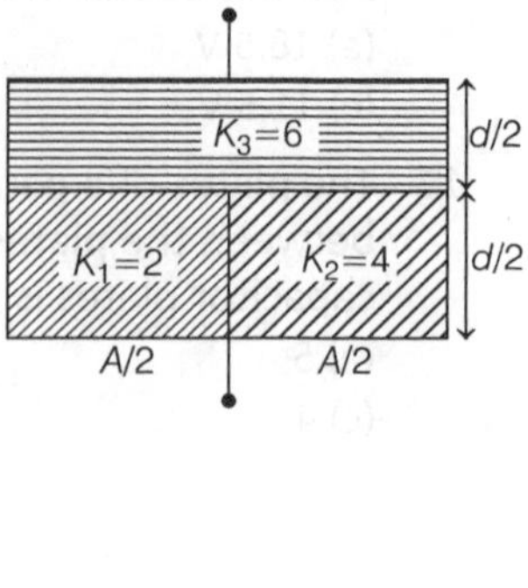

(a) 3.9 C
(b) 4 C
(c) 2.4 C
(d) 3 C

18 The area of the plates of a parallel plate capacitor is A and the gap between them is d. The gap is filled with a non-homogeneous dielectric whose dielectric constant varies with the distance 'y' from one plate as $K = \lambda \sec\left(\dfrac{\pi y}{2d}\right)$, where λ is a dimensionless constant. The capacitance of this capacitor is

(a) $\pi\varepsilon_0\lambda \dfrac{A}{2d}$ (b) $\pi\varepsilon_0\lambda \dfrac{A}{d}$

(c) $2\pi\varepsilon_0\lambda \dfrac{A}{d}$ (d) None of these

19 In the circuit shown below, capacitors A and B have identical geometry, but a material of dielectric constant 3 is present between the plates of B. The potential differences across A and B are respectively

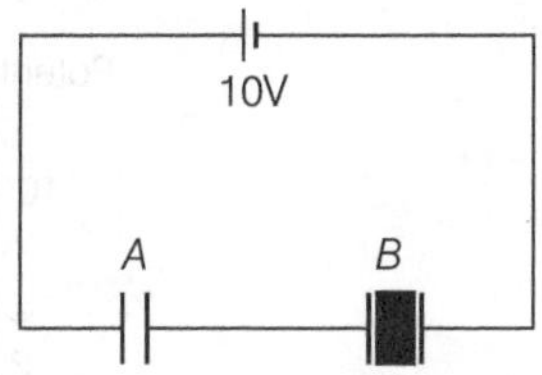

(a) 2.5 V, 7.5 V (b) 2 V, 8 V

(c) 7.5 V, 2.5 V (d) 8 V, 2 V

20 Two identical capacitors 1 and 2 are connected in series to a battery as shown in figure. Capacitor 2 contains a dielectric slab of dielectric constant K as shown. Q_1 and Q_2 are the charges stored in the capacitors. Now, the dielectric slab is removed and the corresponding charges are Q_1' and Q_2'. Then,

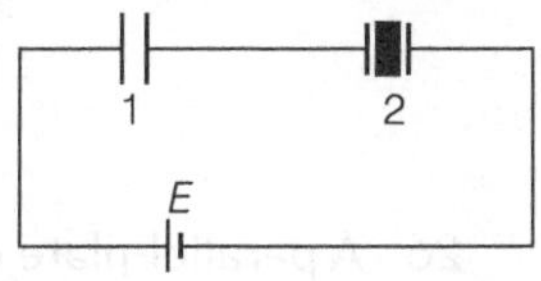

(a) $\dfrac{Q_1}{Q_1'} = \dfrac{K+1}{K}$ (b) $\dfrac{Q_2}{Q_2'} = \dfrac{K+1}{2}$

(c) $\dfrac{Q_2'}{Q_2} = \dfrac{K+1}{2K}$ (d) $\dfrac{Q_1'}{Q_1} = \dfrac{K}{2}$

21 Two parallel plate capacitors of capacitance C_0 and $2C_0$ are connected in parallel and charged to a potential difference V_0. The battery is disconnected and the region between the capacitor plates of capacitance C_0 is completely filled with a dielectric of dielectric constant K. The potential difference across the capacitors, now becomes

(a) $\dfrac{V_0}{(3K+1)}$ (b) $\dfrac{3V_0}{(K+2)}$ (c) $\dfrac{2V_0}{K}$ (d) $\dfrac{V_0}{(2K+1)}$

22 A parallel plate capacitor A of capacitance 1 μF is charged to a potential drop 100 volt and then disconnected from the battery. Now, the capacitor A is completely filled with a dielectric slab of dielectric constant $k = 4$. Another parallel plate capacitor B of capacitance 2 μF is charged to a potential drop 20 V and then disconnected from the battery. Now, the two capacitors A and B are connected with each other with opposite polarities. Then, the heat dissipated after connecting the capacitors is

(a) 1.15×10^{-3} J (b) 1.25×10^{-3} J (c) 1.35×10^{-3} J (d) 1.45×10^{-3} J

23 A smooth dielectric slab A of mass m and dielectric constant k is placed between the plates of a parallel plate capacitor and connected to another block B of equal mass m through an inextensible string and a light pulley arrangement as shown in the figure. The capacitor plates with separation d and width b are connected to a battery of emf E as shown in the figure. If $E = \sqrt{\dfrac{mgd}{b(K-1)\varepsilon_0}}$, then acceleration of the dielectric slab will be

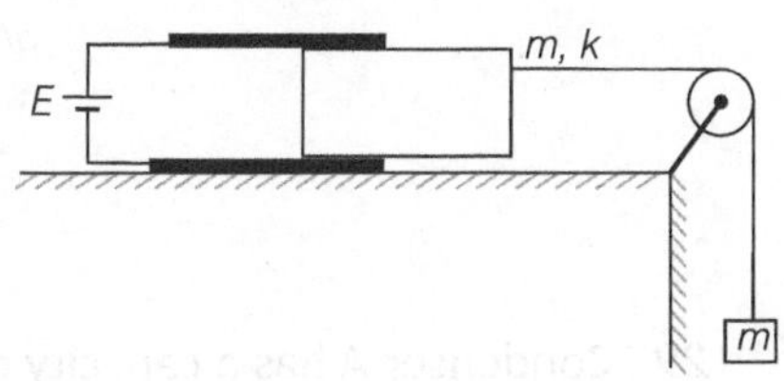

(a) g (b) $\dfrac{g}{2}$ (c) $\dfrac{g}{4}$ (d) $\dfrac{g}{8}$

Numerical Value Questions

24 A parallel plate capacitor has a capacitance of 120 pF (without dielectric) and plate area of 100 cm^2. The space between the plates is now filled with Mica ($k = 5.0$). The potential difference between the plates is 50 V. What is modulus of charge induced on either surface of mica (in nC)?

25 A capacitor filled partially with dielectric material of dielectric constant k. Its electric potential v/s position graph is as shown. Distance between the two plates is 4 mm. Find the dielectric constant of medium.

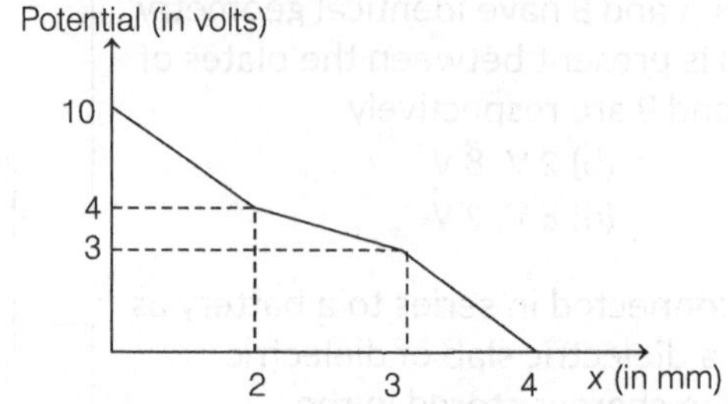

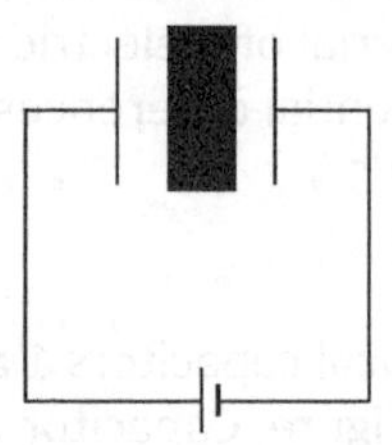

26 A parallel-plate capacitor is filled with a dielectric and a certain potential difference is applied to its plates. The energy of the capacitor is 4×10^{-5} J. After the capacitor is disconnected from the power source, the dielectric is extracted from the capacitor. The work performed against the forces of the electric field in extracting the dielectric is 12×10^{-5} J. Find the dielectric constant of the dielectric.

27 A parallel plate capacitor is to be designed which is to be connected across 1 kV potential difference. The dielectric material which is to be filled between the plates has dielectric constant $K = 6\pi$ and dielectric strength 10^7 V/m. For safety the electric field is never to exceed 10% of the dielectric strength. With such specifications, if we want a capacitor of capacitance 50pF, what minimum area (in mm^2) of plates is required for safe working? (use $\varepsilon_0 = \frac{1}{36\pi} \times 10^{-9}$ in SI unit)

28 If the equivalent capacitance between a and b is $\frac{nA\varepsilon_0}{15d}$, then find n.

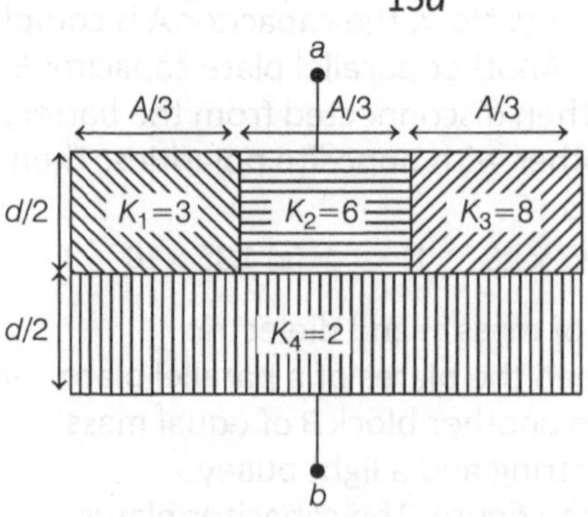

29 Condenser A has a capacity of 15 μF when it is filled with a medium of dielectric constant 15. Another condenser B has a capacity 1 μF with air between the plates. Both are charged separately by a battery of 100 V. After charging, both are connected in parallel without the battery and the dielectric material being removed. Find the common potential drop (in V) across each of them now.

30 Two conducting square plates with length l are arranged parallel to each other at a distance $d(<<l)$. The space between the plates is filled with a dielectric of relative permittivity $\varepsilon_r = 4$ up to a length $x = \frac{l}{3}$. The upper plate is given a charge Q and the lower plate is given a charge $-Q$ as shown in the figure. If $Q = 6\,\mu C$ then find how much charge (in μC) is present on the portion of upper plate which is in contact with the dielectric?

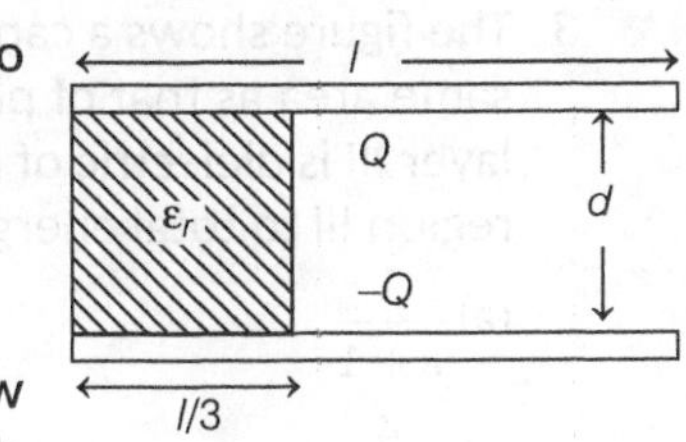

31 A parallel plate capacitor is made by fixing two large plates 2 mm apart inside of a container as shown in figure. The plates are connected to a battery of voltage 90 V. If at $t = 0$, the tap is opened from which a liquid of dielectric constant 5 starts filling in the container at a constant rate of $4\pi\ m^3/s$, find the current (in mA) in connecting wires as a function of time. Neglect any resistance in connecting wires.

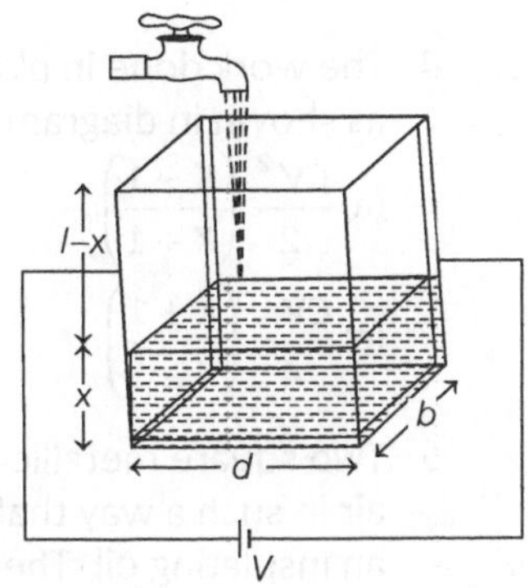

Answers

1. (d)	**2.** (b)	**3.** (c)	**4.** (c)	**5.** (a)	**6.** (d)	**7.** (c)	**8.** (a)	**9.** (c)	**10.** (c)
11. (a)	**12.** (a)	**13.** (d)	**14.** (a)	**15.** (c)	**16.** (d)	**17.** (a)	**18.** (a)	**19.** (c)	**20.** (c)
21. (b)	**22.** (c)	**23.** (c)	**24.** 24						
25. 3	**26.** 4	**27.** 300	**28.** 43						
29. 800	**30.** 4	**31.** 10							

For Detailed Solutions
Scan the code

LEVEL 03 \ JEE Advanced Level

Single Choice Correct Questions

1 When a dielectric slab is introduced inside a parallel plate capacitor then

(a) charge appearing on the surface of slab is due to free electrons inside the dielectric

(b) dielectric gets uniformly charged throughout the volume

(c) charge appearing at the surface is due to bound charge

(d) if σ is surface charge density at the surface of dielectric then electric field due to surface charge on dielectric will be $\frac{\sigma}{k\varepsilon_0}$

2 In a parallel plate capacitor, the distance between the plates is 10 cm. Two dielectric slabs of thickness 5 cm each and dielectric constants $K_1 = 2$ and $K_2 = 4$ respectively, are inserted between the plates. A potential of 100 V is applied across the capacitor as shown in the figure. The value of the net bound surface charge density at the interface of the two dielectrics is

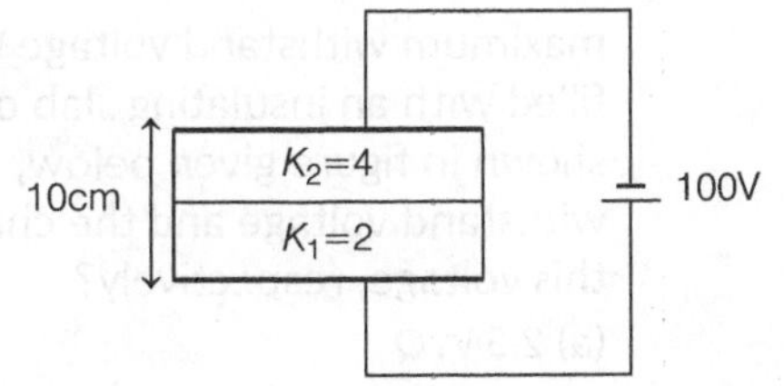

(a) $-\frac{2000}{3}\varepsilon_0$
(b) $-\frac{1000}{3}\varepsilon_0$
(c) $-250\varepsilon_0$
(d) $\frac{2000}{3}\varepsilon_0$

3 The figure shows a capacitor having three layers of equal thickness and same area as that of plate. Layer I is vacuum, layer II is conductor and layer III is dielectric of dielectric constant K. The ratio of energy stored in region III to total energy stored in capacitor is

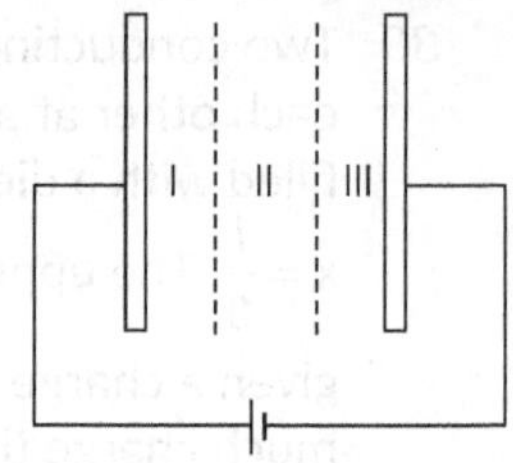

(a) $\frac{1}{K+1}$ (b) $\frac{3}{K+1}$

(c) $\frac{4}{K+3}$ (d) $\frac{4}{3K+1}$

4 The work done in placing the dielectric slab inside one of the capacitors as shown in diagram is

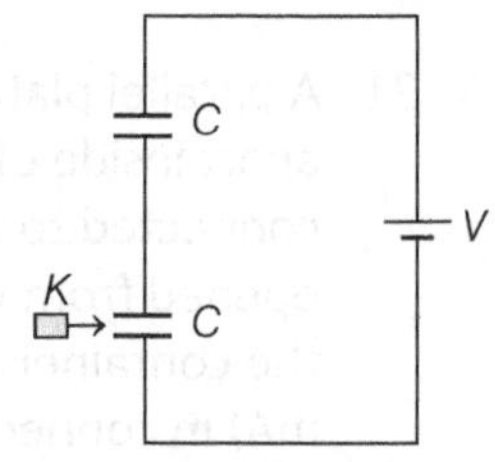

(a) $\frac{CV^2}{2}\left(\frac{K-1}{K+1}\right)$ (b) $\frac{CV^2}{4}\left(\frac{K-1}{K+1}\right)$

(c) $\frac{CV^2}{4}\left(\frac{K+1}{K-1}\right)$ (d) $\frac{CV^2}{2}\left(\frac{K+1}{K-1}\right)$

5 Two square metallic plates of 1 m side are kept 0.01 m apart, like a parallel plate capacitor, in air in such a way that one of their edges is perpendicular, to an oil surface in a tank filled with an insulating oil. The plates are connected to a battery of emf 500 volt. The plates are then lowered vertically into the oil at a speed of 0.001 m/s. The current drawn from the battery during the process is

[Dielectric constant of oil = 11, $\varepsilon_0 = 8.85 \times 10^{-12} \, C^2/N^2 \, m^2$]

(a) 8.85×10^{-9} A (b) 9.735×10^{-9} A

(c) 4.425×10^{-9} A (d) 1.77×10^{-9} A

6 In the given figure, two identical capacitors are shown, with one of them having charge Q and the other being uncharged.

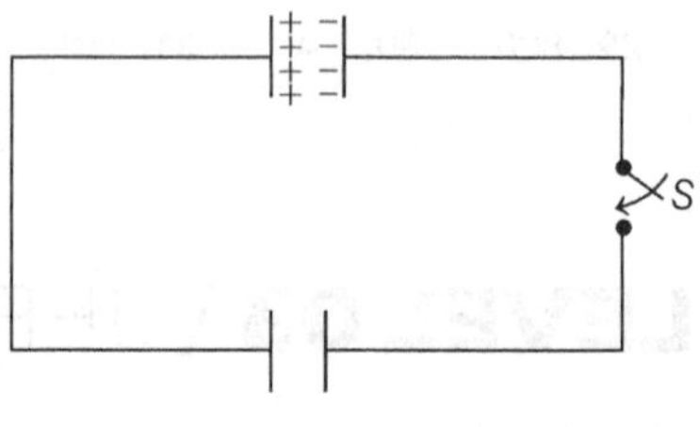

Process (A) A dielectric of dielectric constant k is introduced between the plates of the charged capacitor and then the switch S is closed.

Process (B) Switch S is closed without inserting the dielectric.

Find the Incorrect statement.

(a) Energy lost during process (A) is less as compared to process (B)

(b) Charge transferred to the uncharged capacitor in process (A) is more as compared to process (B)

(c) Common potential difference after redistribution of charge is more in process (B) as compared to process (A)

(d) In process (B), electric field between the plates, in both the capacitors, becomes same, after redistribution of charge.

7 A parallel plate air capacitor carries a charge Q at its maximum withstand voltage V. If the capacitor is half filled with an insulating slab of dielectric constant 4 as shown in figure given below, what are the maximum withstand voltage and the charge on the capacitor at this voltage, respectively?

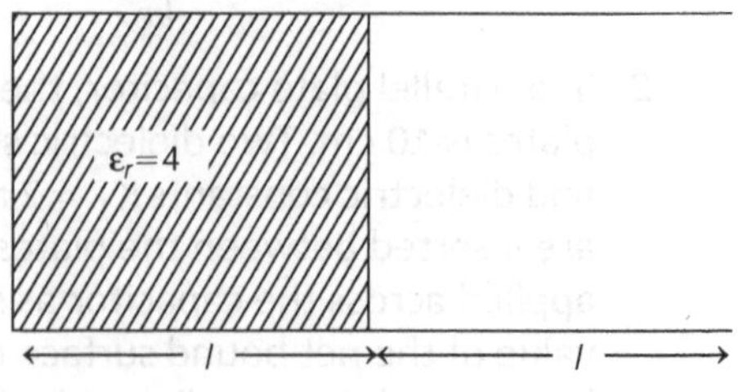

(a) 2.5 V, Q (b) 4 V, 2.5 Q

(c) V, 2.5 Q (d) $\frac{V}{4}$, Q

8 A thin metallic partition of negligible thickness is inserted between two shaded metallic plates as shown. The remaining ends are then packed with insulating plates to form a container like structure. 2 taps shown are opened at $t = 0$ and finally closed at $t = 5$ s. Find capacitance of system between A and B after closing taps. (Assume liquid to be non- conducting) Volumetric flow rates and dielectric constants of liquid are given.

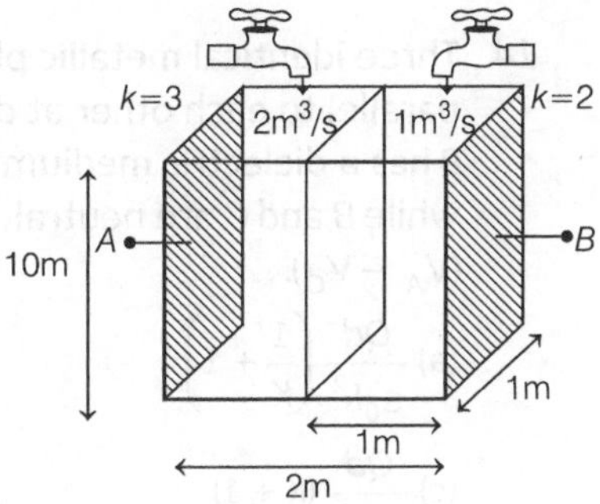

(a) 8.85×10^{-11} F
(b) 8.85×10^{-10} F
(c) 8.63×10^{-10} F
(d) 8.63×10^{-11} F

9 Find the capacitance between the inner and outer curved cylindrical conductor surface as shown in figure. Space between conductor surface is filled with dielectric of dielectric constant K. Consider $b << R$.

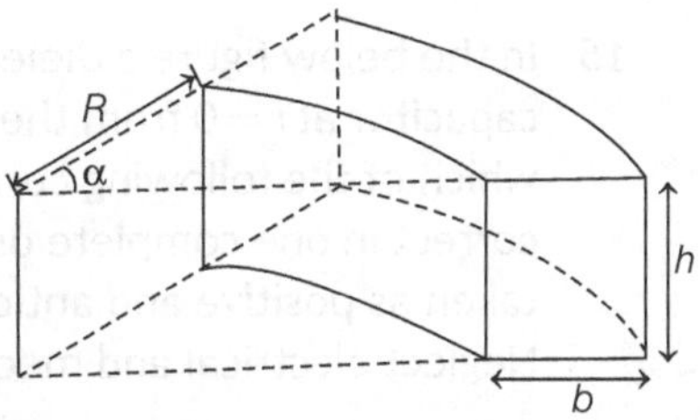

(a) $\dfrac{\varepsilon_0 KhR\alpha}{b}$
(b) $\dfrac{\varepsilon_0 KbR\alpha}{h}$
(c) $\dfrac{\varepsilon_0 KhR\alpha}{2b}$
(d) $\dfrac{\varepsilon_0 KbR\alpha}{2h}$

10 The gap between the plates of a parallel-plate capacitor is filled with a dielectric material whose relative permittivity ε varies linearly from ε_1 to ε_2 $(\varepsilon_2 > \varepsilon_1)$ in the direction perpendicular to the plates. The area of each plate equals to S and the separation between the plates is equal to d. The capacitance of the capacitor is

(a) $\dfrac{\varepsilon_0(\varepsilon_2 - \varepsilon_1)S}{d \ln\left(\dfrac{\varepsilon_2}{\varepsilon_1}\right)}$
(b) $\dfrac{\varepsilon_0(\varepsilon_2 + \varepsilon_1)S}{d \ln\left(\dfrac{\varepsilon_2}{\varepsilon_1}\right)}$
(c) $\dfrac{\varepsilon_0 S}{d \ln\left(\dfrac{\varepsilon_2}{\varepsilon_1}\right)}$
(d) None of these

11 A parallel plate capacitor is made of two square plates of side a, separated by a distance d ($d << a$). The lower triangular portion is filled with a dielectric of dielectric constant K, as shown in the figure. Capacitance of this capacitor is

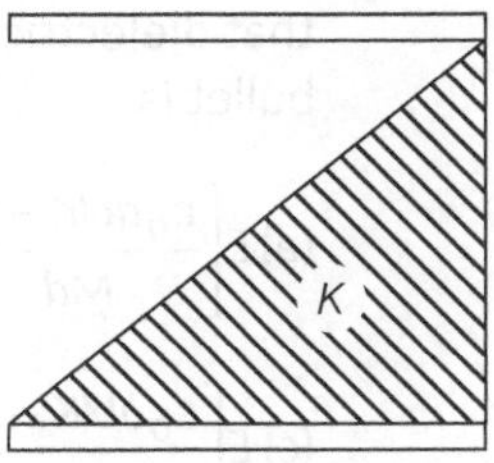

(a) $\dfrac{K\varepsilon_0 a^2}{2d}$
(b) $\dfrac{K\varepsilon_0 a^2}{d} \ln K$
(c) $\dfrac{K\varepsilon_0 a^2}{d(K-1)} \ln K$
(d) $\dfrac{K\varepsilon_0 a^2}{2d(K+1)}$

12 A spherical capacitor is made of two conducting spherical shells of radii a and $b = 3a$. The space between the shells is filled with a dielectric of dielectric constant $K = 3$ upto a radius $c = 2a$ as shown. If the capacitance of given arrangement is n times the capacitance of an isolated spherical conducting shell of radius a, then value of n is

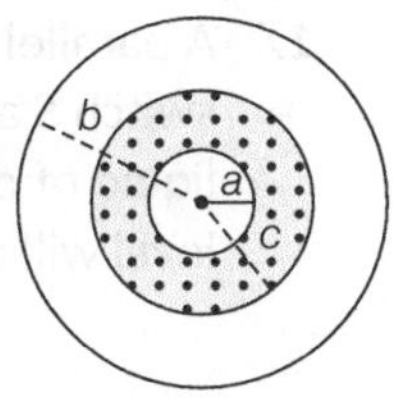

(a) 2
(b) 3
(c) 4
(d) 6

13 Four identical plates 1, 2, 3 and 4 are placed parallel to each other at equal distance as shown in the figure. Plates 1 and 4 are joined together and the space between 2 and 3 is filled with a dielectric of dielectric constant $k = 2$. The capacitance of the system between 1 and 3, 2 and 3 are C_1 and C_2 respectively. The ratio of $\dfrac{C_1}{C_2}$ is

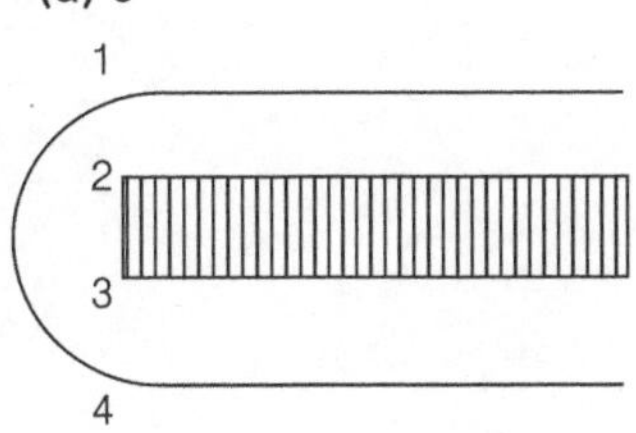

(a) 1
(b) $\dfrac{3}{2}$
(c) $\dfrac{2}{3}$
(d) $\dfrac{2}{5}$

14 Three identical metallic plates A, B and C of size $(l \times l)$ are placed parallel to each other at distances d. The region between A and B has a dielectric medium of constant k. If A is given charge $+Q$ while B and C are neutral, calculate the potential difference $(V_A - V_C)$.

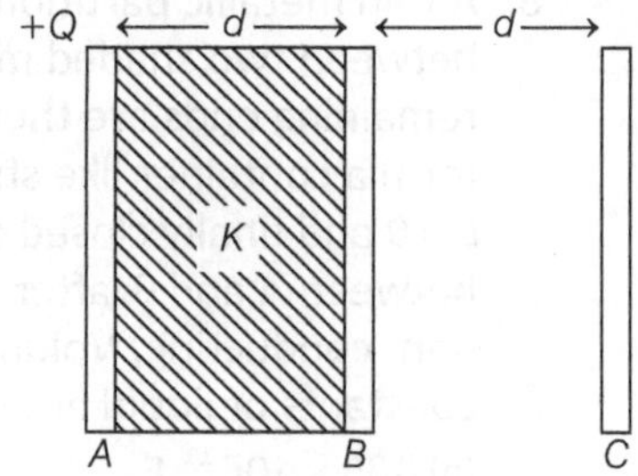

(a) $\frac{Qd}{\varepsilon_0 l^2}\left(\frac{1}{K}+1\right)$ (b) $\frac{Qd}{2\varepsilon_0 l^2}\left(\frac{1}{K}+1\right)$

(c) $\frac{Qd}{\varepsilon_0 l^2}(K+1)$ (d) $\frac{Qd}{2\varepsilon_0 l^2}(K+1)$

15 In the below figure a dielectric is released inside the capacitor at $t = 0$ from the end of the capacitor. Then which of the following current (i) *versus* time (t) graph is correct in one complete oscillation if clockwise current is taken as positive and anticlockwise as negative? ($a > c$). Neglect electrical and mechanical resistance.

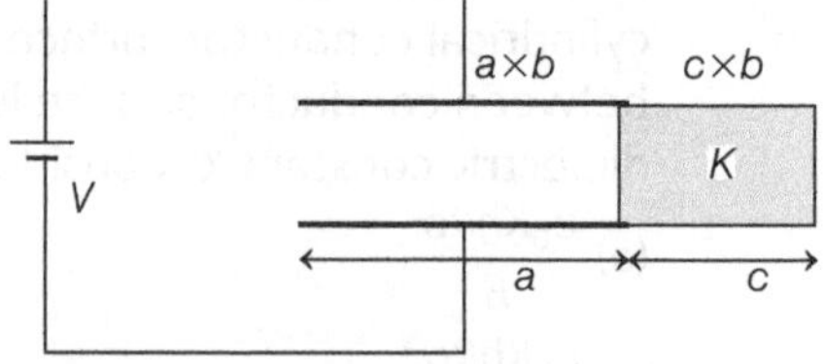

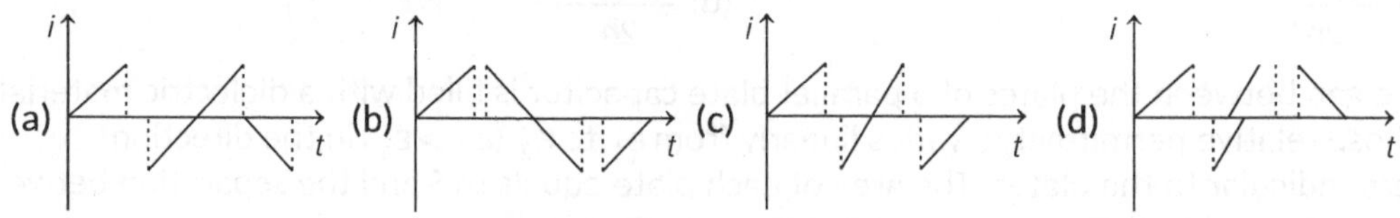

16 A parallel plate capacitor is half filled with a dielectric (K) of mass M. Capacitor is attached with a cell of emf E. Plates are held fixed on smooth insulating horizontal surface. A bullet of mass M hits the dielectric elastically and its found that dielectric just leaves out the capacitor. The speed of bullet is

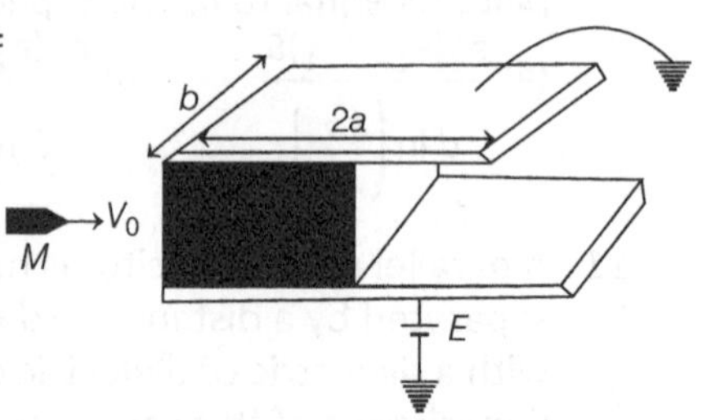

(a) $E\left[\frac{\varepsilon_0 ab(K-1)}{Md}\right]^{\frac{1}{2}}$ (b) $E\left[\frac{2\varepsilon_0 ab(K-1)}{Md}\right]^{\frac{1}{2}}$

(c) $E\left[\frac{\varepsilon_0 ab(K-1)}{2Md}\right]^{\frac{1}{2}}$ (d) None of these

17 A parallel plate capacitor of plate area $A = lb$ with separation d is connected to a battery via a switch S as shown in figure. Capacitor plates are kept vertical and touched on the surface of a liquid of density ρ as shown. If S is closed, find the height between plates to which the liquid level will rise.

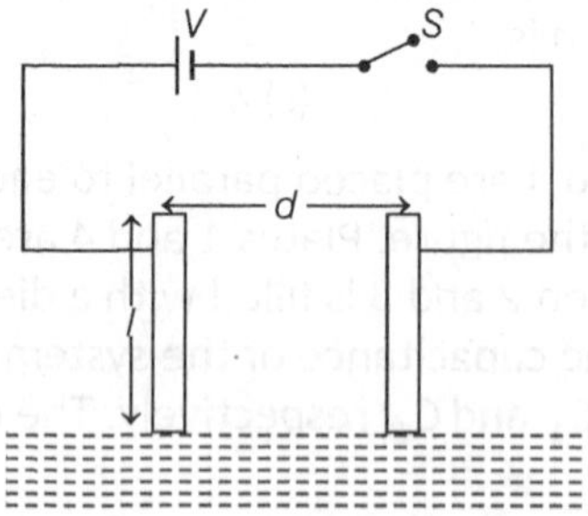

(a) $\frac{\varepsilon_0 V^2(K-1)}{d^2 \rho g}$ (b) $\frac{\varepsilon_0 V^2(K-1)}{2d^2 \rho g}$ (c) $\frac{2\varepsilon_0 V^2(K-1)}{d^2 \rho g}$ (d) $\frac{\varepsilon_0 V^2 K}{d^2 \rho g}$

18 A horizontal parallel-plate capacitor is lowered on a liquid surface in such a way that its lower plate is just submerged in the liquid of dielectric constant K as shown in figure. Find the height to which the liquid level will be raised between the plates, if the capacitor plates are given a surface charge density $+\sigma$ and $-\sigma$ on its plates.

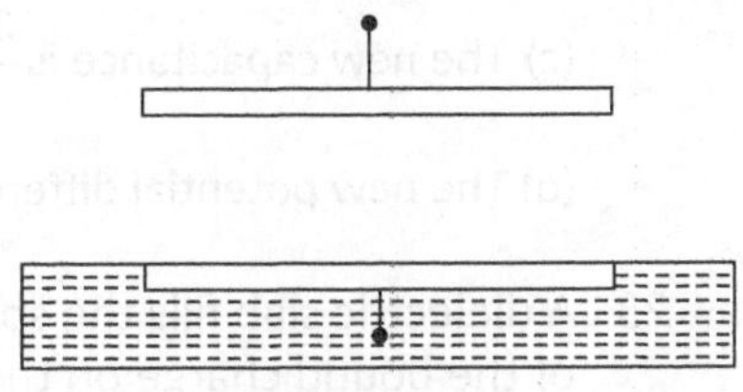

(a) $\dfrac{\sigma^2(K-1)}{\varepsilon_0 K\rho g}$ (b) $\dfrac{\sigma^2(K-1)}{2\varepsilon_0 K\rho g}$ (c) $\dfrac{\sigma^2(K^2-1)}{\varepsilon_0 K^2\rho g}$ (d) $\dfrac{\sigma^2(K^2-1)}{2\varepsilon_0 K^2\rho g}$

Multiple Choice(s) Correct Questions

19 A parallel plate capacitor of plate area A and plate separation d is charged to a potential difference V and then the battery disconnected. A slab of dielectric constant K is then inserted between the plates of the capacitor so as to fill the space between the plates. If Q, E and W denote respectively, the magnitude of the charge on each plate, the magnitude of the electric field between the plates (after the slab is inserted) and the magnitude of the work done on the system, in the process of inserting the slab, then

(a) $Q=\dfrac{\varepsilon_0 AV}{d}$ (b) $Q=\dfrac{\varepsilon_0 KAV}{d}$ (c) $E=\dfrac{V}{Kd}$ (d) $W=\dfrac{\varepsilon_0 AV^2}{2d}\left(1-\dfrac{1}{K}\right)$

20 A capacitor of capacity C_0 is connected to a battery of emf V_0. When steady state is attained, a dielectric slab of dielectric constant K is slowly introduced in the capacitor to fill the capacitor completely. Mark the correct statement(s), in final steady state.

(a) Magnitude of induced charge on the each surface of slab is $C_0V_0(K-1)$

(b) Electric force due to induced charges on any plate is zero.

(c) Force of attraction between plates of capacitor is $\dfrac{K(C_0V_0)^2}{2\varepsilon_0 A}$

(d) Field due to induced charges in dielectric slab is $\dfrac{(K-1)C_0V_0}{\varepsilon_0 A}$

21 On a parallel plate capacitor following operations can be performed.

P – connect the capacitor to a battery of emf V.

Q – disconnect the battery,

R – reconnect the battery with polarity reversed,

S – insert a dielectric slab in the capacitor.

(a) In PQR (perform P, then Q, then R), the stored electric energy remains unchanged and no thermal energy is developed.

(b) The charge appearing on the capacitor is greater after the action PSQ than after the action PQS.

(c) The electric energy stored in the capacitor is greater after the action SPQ than after the action PQS.

(d) The electric field in the capacitor after the action PS is the same as that after SP.

22 A parallel plate capacitor of area A and separation d is charged to potential difference V and removed from the charging source. A dielectric slab of constant $K=2$, thickness d and area $\dfrac{A}{2}$ is inserted, as shown in the figure. Let σ_1 be free charge density at the conductor-dielectric surface and σ_2 be the charge density at the conductor-vacuum surface.

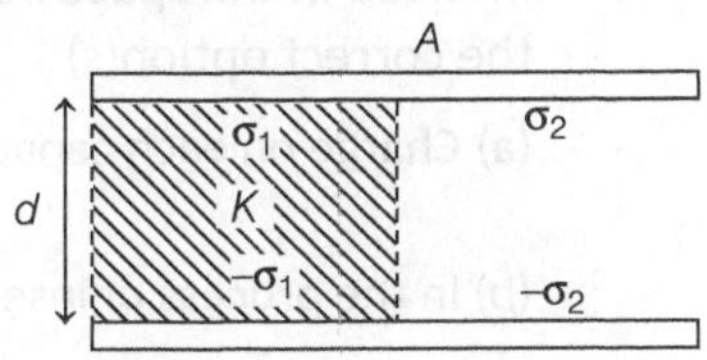

(a) The electric field has the same value inside the dielectric as that in the free space between the plates.

(b) The ratio $\dfrac{\sigma_1}{\sigma_2}$ is equal to $\dfrac{2}{1}$.

(c) The new capacitance is $\frac{3\varepsilon_0 A}{2d}$.

(d) The new potential difference between the plates is $\frac{2}{3}$ V.

23 A dielectric slab fills the space between the plates of a parallel-plate capacitor. The magnitude of the bound charge on the slab is 75% of the magnitude of the free charge on the plates. The capacitance is 480 μF and the maximum charge that can be stored on the capacitor is $240\,\varepsilon_0 L^2 E_{max}$, where E_{max} is the breakdown field.

(a) The dielectric constant for the dielectric slab is 4.

(b) Without the dielectric, the capacitance of the capacitor would be 360 μF.

(c) The plate area is $60\,L^2$.

(d) If the dielectric slab is having the same area as the capacitor plate but the width half that of the capacitor, the capacitance would be 192 μF.

24 In the figure shown

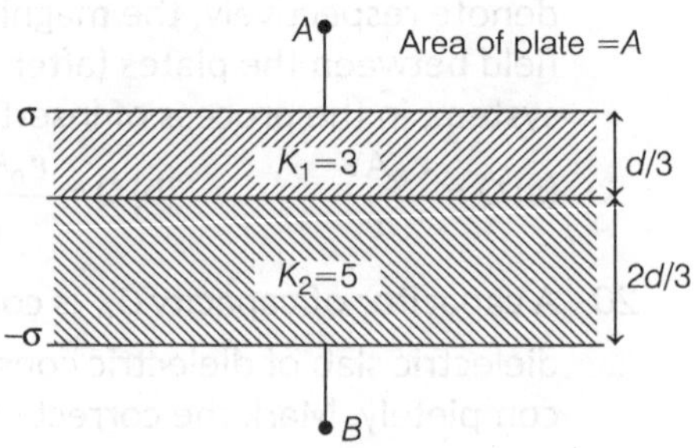

(a) The ratio of energy density in first dielectric to second dielectric is $\frac{5}{3}$.

(b) The ratio of energy density in first dielectric to second dielectric is $\frac{3}{5}$.

(c) Net induced surface charge density on the interface of the two dielectrics is $\frac{2\sigma}{15}$.

(d) Net induced surface charge density on the interface of the two dielectrics is $-\frac{2\sigma}{15}$.

25 Identical dielectric slabs are inserted into two identical capacitors A and B. These capacitors and a battery are connected as shown in figure. Now, the slab of capacitor B is pulled out with battery remaining connected.

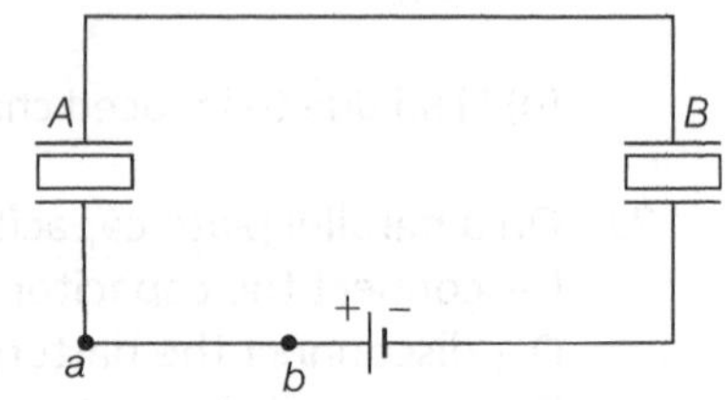

(a) During the process, positive charge flows from a to b.

(b) Finally charge on capacitor B will be less than that on capacitor A.

(c) During the process, a work is done by the external force F, which completely appears as heat in the circuit.

(d) During the process, internal energy of the battery increases.

26 In the figure shown capacitors A and B of capacitance C each are in steady state. A dielectric slab of dielectric constant $K = 2$ and dimensions equal to the inner dimensions of the capacitor is inserted in the space between the plates of the capacitor B. Choose the correct option(s)

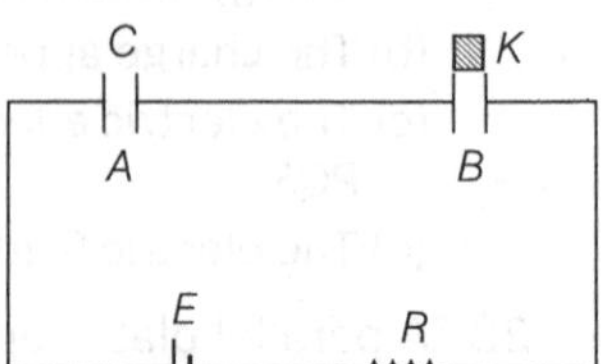

(a) Charge on each capacitor will increase by $\frac{CE}{6}$.

(b) In the process of inserting the dielectric, energy of the cell decreases by an amount of $\frac{CE^2}{6}$.

(c) In the process of inserting the dielectric, energy of the cell increases by an amount of $\frac{CE^2}{6}$.

(d) In the process of inserting the dielectric, energy in the capacitor A increases by an amount of $\frac{7CE^2}{72}$.

27 A parallel plate capacitor has a dielectric slab in it. The slab just fills the space inside the capacitor. The capacitor is charged by a battery and the battery is disconnected. Now, an external agent starts pulling the slab out uniformly at $t = 0$. If at time t, capacitance of the capacitor is C, potential difference across plate is V, and energy stored in it is U, then which of the following graph is/are correct?

(a) 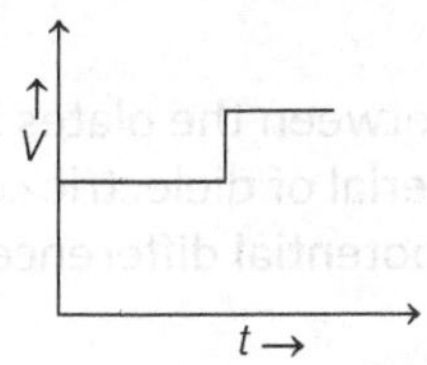

(b) 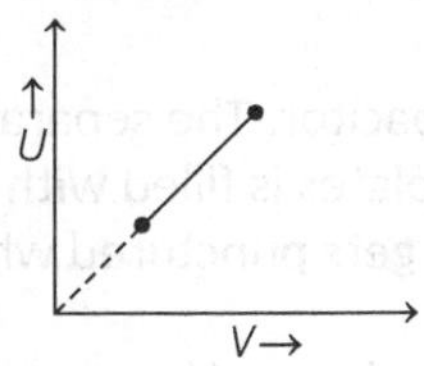

(c) 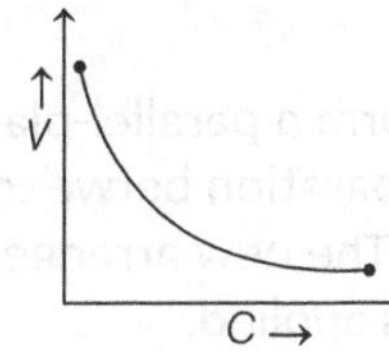

(d) 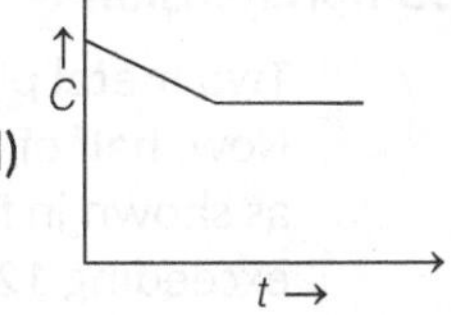

Comprehension Based Questions

Comprehension-1

Consider a series combination of capacitor C_1, C_2 and C_3. Capacitor C_1 and C_3 have air as the dielectric medium between the plates while C_2 has a medium of dielectric constant $K = 3$ filled uniformly between the plates. A potential difference V is applied across the combination. It is observed that potential difference across C_3 is 12 V.

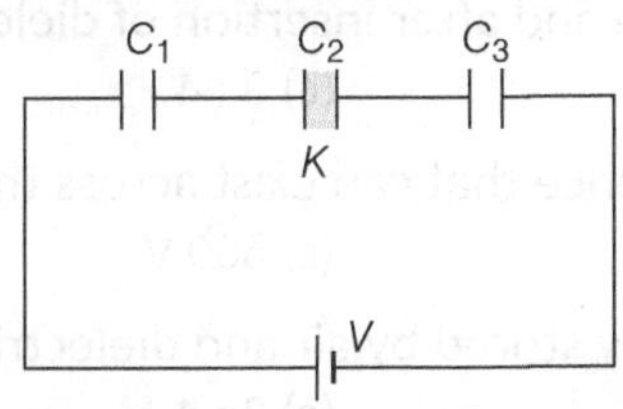

The dielectric material ($K = 3$) between the plates of C_2 is now removed. In this situation, potential difference across C_3 is found to change by 3 V. It is also found that potential differences across C_1 and C_2 are equal. (Capacitance of C_2 without dielectric is 2 μF).

28 Applied potential difference V is

(a) 10 V (b) 20 V (c) 30 V (d) 36 V

29 Capacitance C_3 is

(a) 8 μF (b) 6 μF (c) 3 μF (d) 2 μF

30 Capacitance C_1 is

(a) 8 μF (b) 6 μF (c) 3 μF (d) 2 μF

Comprehension-2

Figure shows a parallel plate capacitor having an insulator of dielectric constant $K = 2$. The plate area of the capacitor is A and separation between them is d, while the surface area of the dielectric is $\frac{A}{3}$ and its height is $\frac{d}{3}$. The capacitor is connected to a battery of potential difference V. Let q_1 be the free charge that appears at the upper surface of lower plate, just below the dielectric, while q_2 is the surface charge that appears on the other side as shown.

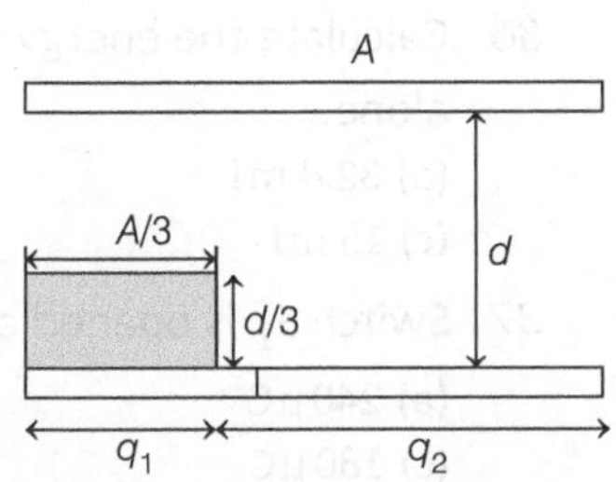

31 The equivalent capacitance of the system is

(a) $\frac{12}{11}\frac{\varepsilon_0 A}{d}$ (b) $\frac{11}{12}\frac{\varepsilon_0 A}{d}$ (c) $\frac{16}{15}\frac{\varepsilon_0 A}{d}$ (d) $\frac{15}{16}\frac{\varepsilon_0 A}{d}$

32 The ratio of charge $\frac{q_1}{q_2}$ on the lower plate of capacitor is

(a) 1 (b) $\frac{3}{5}$ (c) $\frac{3}{8}$ (d) $\frac{5}{3}$

Comprehension-3

Two metal plates form a parallel-plate capacitor. The separation between the plates is $2d$. Now, half of the separation between the plates is filled with a material of dielectric constant 2 as shown in figure. The new arrangement gets punctured when a potential difference just exceeding 1200 V is applied.

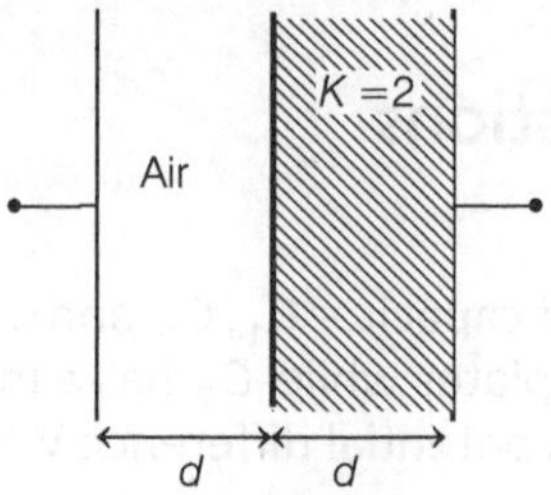

33 The ratio of capacitance before and after insertion of dielectric is

(a) 1 : 2 (b) 2 : 1 (c) 3 : 4 (d) 4 : 3

34 The maximum potential difference that can exist across the dielectric is

(a) 300 V (b) 400 V (c) 600 V (d) 800 V

35 The ratio of electrostatic energy stored by air and dielectric parts of capacitor is

(a) 1 : 2 (b) 2 : 1 (c) 3 : 4 (d) 4 : 3

Comprehension-4

In the circuit shown, capacitor A has capacitance $C_1 = 2\,\mu F$ when filled with dielectric slab $(k = 2)$. Capacitor B and C are air capacitors and have capacitances $C_2 = 3\,\mu F$ and $C_3 = 6\,\mu F$ respectively.

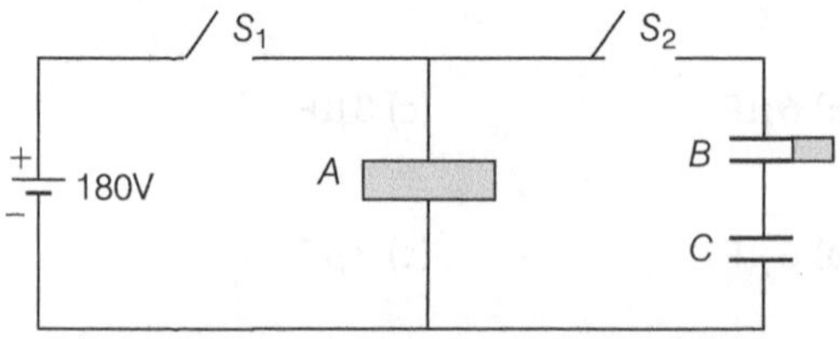

36 Calculate the energy supplied by battery during process of charging when switch S_1 is closed alone.

(a) 32.4 mJ (b) 64.8 mJ
(c) 15 mJ (d) 30 mJ

37 Switch S_1 is opened and S_2 is closed. The charge on capacitor B is

(a) 240 μC (b) 280 μC
(c) 180 μC (d) 200 μC

38 Now switch S_2 is opened, slab of A is removed. Another dielectric slab $k = 2$ which can just fill the space in B, is inserted into it and then switch S_2 is closed. The charge on capacitor B is

(a) 90 μC (b) 240 μC
(c) 180 μC (d) 270 μC

Comprehension-5

An isolated spherical capacitor has charge $+Q$ on its inner conductor of radius a and $-Q$ on its outer conductor of radius b. Half of volume between the two conductors is then filled with a liquid dielectric of constant K as shown in figure

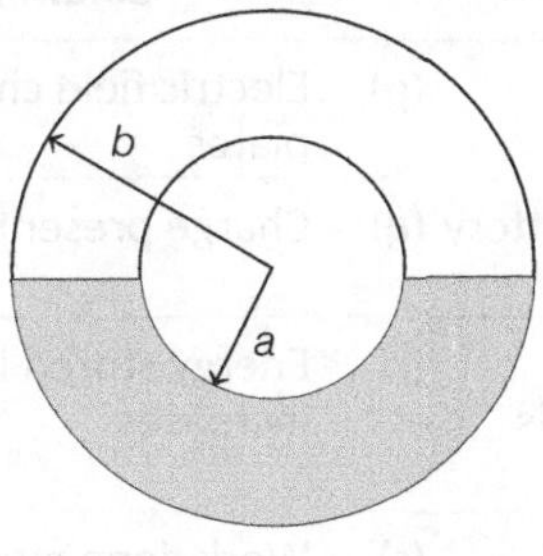

39 The capacity of half filled spherical capacitor is

(a) $\frac{4\pi\varepsilon_0 Kab}{b-a}$ (b) $\frac{2\pi\varepsilon_0 (K+1)ab}{b-a}$ (c) $\frac{2\pi\varepsilon_0 Kab}{b-a}$ (d) $\frac{4\pi\varepsilon_0 (K+1)ab}{b-a}$

40 The electric field between the two plates of capacitor as a function of distance r from centre is

(a) $\frac{Q}{2\pi\varepsilon_0 Kr^2}$ (b) $\frac{Q}{4\pi\varepsilon_0 Kr^2}$ (c) $\frac{Q}{2\pi\varepsilon_0 (K+1)r^2}$ (d) $\frac{Q}{4\pi\varepsilon_0 (K+1)r^2}$

41 Surface density of charge on lower half of inner conductor is

(a) $\frac{Q}{4\pi a^2(1+K)}$ (b) $\frac{Q}{4\pi b^2(1+K)}$ (c) $\frac{KQ}{2\pi a^2(1+K)}$ (d) $\frac{KQ}{4\pi b^2(1+K)}$

Comprehension-6

A block A of mass m kept on a rough horizontal surface is connected to a dielectric slab of mass $\frac{m}{6}$ and dielectric constant K by means of a light and inextensible string passing over a fixed pulley as shown in figure. The dielectric can completely fill the space between the parallel plate capacitor of plate area $l \times l$ and separation between the plates d kept in vertical position.

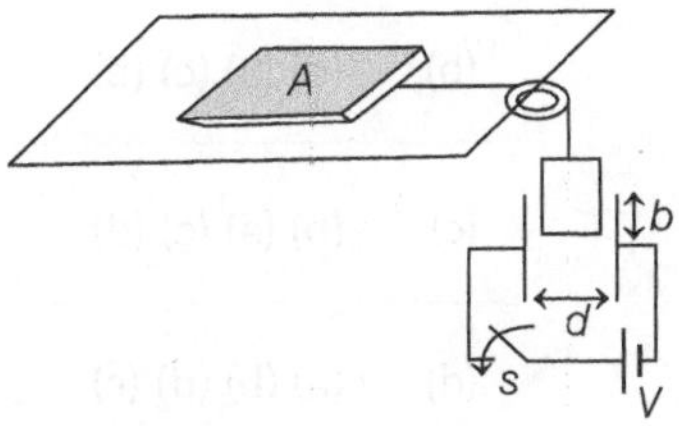

Initially switch S is open and length of the dielectric inside the capacitor is b. The coefficient of friction between the block A and the surface is $\mu = \frac{1}{4}$. Ignore any other friction.

42 The minimum value of the emf V_{min} of the battery so that after closing the switch the block A will move

(a) $\sqrt{\frac{mgd}{6\varepsilon_0 lK}}$ (b) $\sqrt{\frac{mgd}{12\varepsilon_0 lK}}$

(c) $\sqrt{\frac{mgd}{6\varepsilon_0 l(K-1)}}$ (d) $\sqrt{\frac{mgd}{12\varepsilon_0 l(K-1)}}$

43 If $V = 2V_{min}$ find the speed of the block A when the dielectric completely fills the space between the plates of the capacitor.

(a) $\sqrt{\frac{3}{7}g(l-b)}$ (b) $\sqrt{\frac{6}{7}g(l-b)}$

(c) $\sqrt{\frac{3}{11}g(l-b)}$ (d) $\sqrt{\frac{6}{11}g(l-b)}$

Match the Columns Questions

44 Some events related to a capacitor are listed in column-I. Match these events with their effect(s) in column-II.

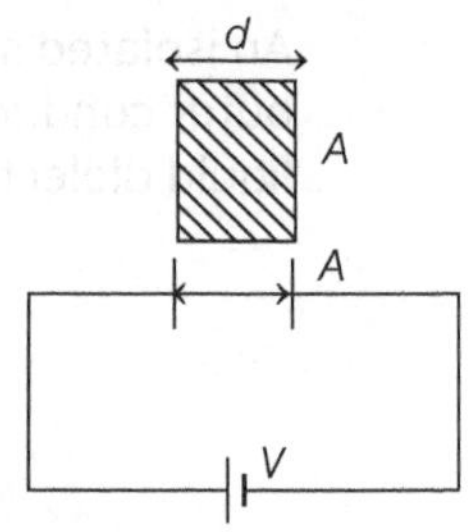

Column I (Events)		Column II (Effects)	
(a)	Insertion of dielectric while battery remains attached	(p)	Electric field changes between plates
(b)	Removal of dielectric while battery is not present	(q)	Charge present on plates changes
(c)	Slow decrease in separation between plates while battery is attached	(r)	Energy stored in capacitor increases
(d)	Slow increase of separation between plates while battery is not present	(s)	Work done by external agent is positive

45 On a capacitor of capacitance C_0 following steps are performed in the order as given in column-I.

(a) Capacitor is charged by connecting it across a battery of Emf V_0.

(b) Dielectric of dielectric constant k and thickness *d* is inserted.

(c) Capacitor is disconnected from battery.

(d) Separation between plates is doubled.

Column-I (Steps performed)		Column-II Final value of Quantity (Symbols have usual meaning)	
(a)	(a) (d) (c) (b)	(p)	$Q=\frac{C_0V_0}{2}$
(b)	(d) (a) (c) (b)	(q)	$Q=\frac{kC_0V_0}{k+1}$
(c)	(b) (a) (c) (d)	(r)	$C=\frac{kC_0}{k+1}$
(d)	(a) (b) (d) (c)	(s)	$V=\frac{V_0(k+1)}{2k}$

46 Two dielectric slabs of relative primitivities $\varepsilon_{r_1}=2$, $\varepsilon_{r_2}=3$ and thickness *d* and 2*d* respectively are filled in between the grounded parallel plates each of area *A*. If a battery of emf $\varepsilon=2$ volt is connected to the middle conducting plate find all the quantity in SI unit. (Assume initially uncharged).

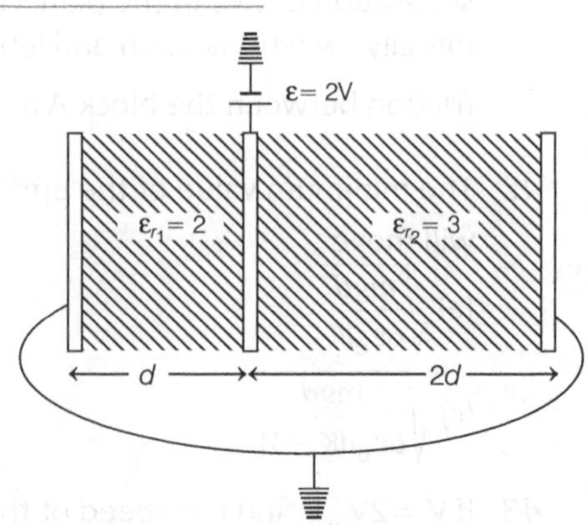

Column I		Column II	
(a)	Capacitance of the system	(p)	$\frac{5A\varepsilon_0}{d}$
(b)	Charge flown through the battery	(q)	$\frac{2A\varepsilon_0}{d}$
(c)	Induced charges on the left surface of slab of thickness *d*	(r)	$\frac{7A\varepsilon_0}{d}$
(d)	Energy stored in the system	(s)	$\frac{7A\varepsilon_0}{2d}$

Numerical Value Questions

47 A parallel plate capacitor is filled by a dielectric whose dielectric constant varies with the applied voltage according to the law $K = \alpha V$, where $\alpha = 1$ per volt. The same (but containing no dielectric) capacitor charged to a voltage $V = 156$ volt is connected in parallel to the first non-linear, uncharged capacitor. Find the final voltage V_f (in volt) across the capacitors.

48 Between the plates of a parallel-plate capacitor there is a metallic plate whose thickness takes up $\eta = 0.60$ of the capacitor gap. When that plate is absent the capacitor has a capacity $C = 20$ nF. The capacitor is connected to a dc voltage source $V = 100$ V. The metallic plate is slowly extracted from the gap. Find the mechanical work performed in the process of plate extraction. (in μJ)

49 A capacitor consists of two fixed semicircular plates of radius R and a movable slab of thickness d and dielectric constant K placed between them. The slab can freely rotate about the axis O and practically fills the entire gap between the fixed plate. A constant voltage V is maintained between the plates. If the torque acting on the dielectric slab when the slab is displaced to an angle θ as shown in the figure, due to electrostatic force is $\tau = \dfrac{(K-1)\varepsilon_0 R^2 V^2}{nd}$, find the value of n.

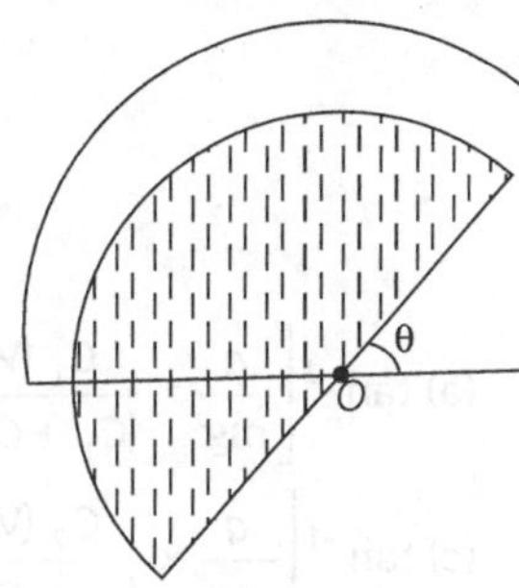

50 One end of an open coaxial cable is immersed vertically in a liquid dielectric of unknown permittivity. The cable is connected to a source of potential difference V. The electrostatic forces draw the liquid dielectric into the annular space of the cable for a height h above the level of the dielectric outside the cable. The radius of the inner conductor is a, that of the outer conductor is b ($b > a$), and the mass density of the liquid is ρ_m. Determine the relative permittivity of the liquid dielectric, if $V = 1000$ V, $h = 3.29$ cm, $a = 0.5$ mm, $b = 1.5$ mm, $\rho_m = 1\ \text{gm/cm}^3$ ($g = 10\ \text{ms}^{-2}$) (Assume that electrostatic force balances the weight of the liquid) (Take, $\ln 3 = 1.1$)

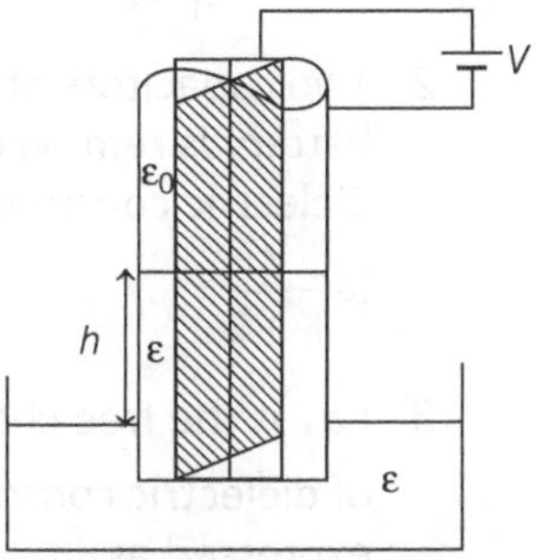

51 A parallel plate capacitor consists of two square plates of side length $a = 6$ cm. There is dielectric slab in the shape of a cylinder of parabolic cross-section between the plates, as shown in figure. The dielectric constant of the material is $K = 2$. The length of latus rectum of the parabola is equal to the side length of the plate. Then, the capacitance is given by $4\pi\varepsilon_0 N \times 10^{-2}$ farad, then find the value of N.

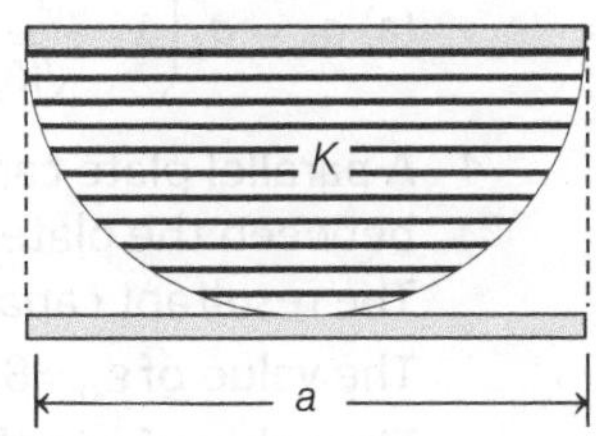

Answers

1. (c)	**2.** (a)	**3.** (a)	**4.** (b)	**5.** (c)	**6.** (b)	**7.** (c)	**8.** (d)	**9.** (a)	**10.** (a)
11. (c)	**12.** (b)	**13.** (c)	**14.** (b)	**15.** (c)	**16.** (a)	**17.** (b)	**18.** (d)	**19.** (acd)	**20.** (abd)
21. (bcd)	**22.** (abcd)	**23.** (acd)	**24.** (ad)	**25.** (ad)	**26.** (abd)	**27.** (bcd)	**28.** (d)	**29.** (c)	**30.** (d)
31. (c)	**32.** (b)	**33.** (c)	**34.** (b)	**35.** (b)	**36.** (b)	**37.** (c)	**38.** (d)	**39.** (b)	**40.** (c)
41. (c)	**42.** (c)	**43.** (a)							

44. (a) → qr, (b) → prs, (c) → pqr, (d) → rs

45. (a) → prs, (b) → prs (c) → r, (d) → qr

46. (a) → s, (b) → r, (c) → q, (d) → r

47. 12 **48.** 150 **49.** 4 **50.** 82.78 **51.** 3

For Detailed Solutions
Scan the code

Previous Years' Questions

JEE Main

1 A simple pendulum of mass m, length l and charge $+q$ suspended in the electric field produced by two conducting parallel plates as shown. The value of deflection of pendulum in equilibrium position will be **[JEE Main 2021]**

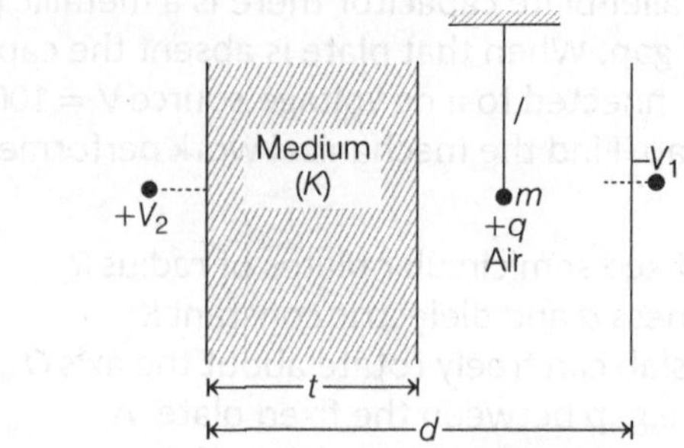

(a) $\tan^{-1}\left[\frac{q}{mg}\times\frac{C_1(V_2-V_1)}{(C_1+C_2)(d-t)}\right]$ (b) $\tan^{-1}\left[\frac{q}{mg}\times\frac{C_2(V_2-V_1)}{(C_1+C_2)(d-t)}\right]$

(c) $\tan^{-1}\left[\frac{q}{mg}\times\frac{C_2(V_1+V_2)}{(C_1+C_2)(d-t)}\right]$ (d) $\tan^{-1}\left[\frac{q}{mg}\times\frac{C_1(V_1+V_2)}{(C_1+C_2)(d-t)}\right]$

2 Two capacitors of capacities 2C and C are joined in parallel and charged upto potential V. The battery is removed and the capacitor of capacity C is filled completely with a medium of dielectric constant K. The potential difference across the capacitors will now be **[JEE Main 2021]**

(a $\frac{V}{K+2}$ (b) $\frac{V}{K}$ (c) $\frac{3V}{K+2}$ (d) $\frac{3V}{K}$

3 f q_f is the free charge on the capacitor plates and q_b is the bound charge on the dielectric slab of dielectric constant K placed between the capacitor plates, then bound charge q_b can be expressed as **[JEE Main 2021]**

(a) $q_b = q_f\left(1-\frac{1}{\sqrt{K}}\right)$ (b) $q_b = q_f\left(1-\frac{1}{K}\right)$ (c) $q_b = q_f\left(1+\frac{1}{\sqrt{K}}\right)$ (d) $q_b = q_f\left(1+\frac{1}{K}\right)$

4 A parallel plate capacitor has plate area $100\ \text{m}^2$ and plate separation of 10 m. The space between the plates is filled upto a thickness 5 m with a material of dielectric constant of 10. The resultant capacitance of the system is x pF.
The value of $\varepsilon_0 = 8.85\times10^{-12}\ \text{fm}^{-1}$.
The value of x to the nearest integer is .... **[JEE Main 2021]**

5 In the reported figure, a capacitor is formed by placing a compound dielectric between the plates of parallel plate capacitor. The expression for the capacity of the said capacitor will be (Take, area of plate $= A$) **[JEE Main 2021]**

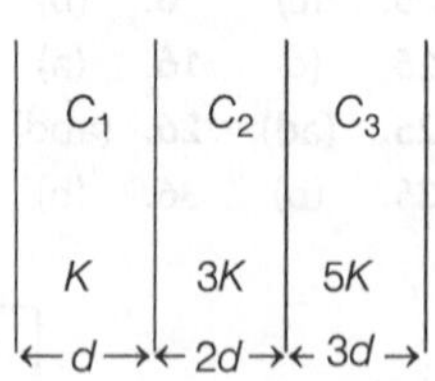

(a) $\frac{15}{34}\frac{K\varepsilon_0 A}{d}$ (b) $\frac{15}{6}\frac{K\varepsilon_0 A}{d}$

(c) $\frac{25}{6}\frac{K\varepsilon_0 A}{d}$ (d) $\frac{9}{6}\frac{K\varepsilon_0 A}{d}$

6 A parallel plate capacitor with plate area A and distance of separation d is filled with a dielectric. What is the capacity of the capacitor when permittivity of the dielectric varies as **[JEE Main 2021]**

$$\varepsilon(x) = \varepsilon_0 + kx, \text{ for } \left(0 < x \le \frac{d}{2}\right)$$

$$\varepsilon(x) = \varepsilon_0 + k(d - x), \text{ for } \left(\frac{d}{2} \le x \le d\right)$$

(a) $\left(\varepsilon_0 + \frac{kd}{2}\right)^{2/kA}$ (b) $\frac{kA}{2\ln\left(\frac{2\varepsilon_0 + kd}{2\varepsilon_0}\right)}$ (c) zero (d) $\frac{kA}{2}\ln\left(\frac{2\varepsilon_0}{2\varepsilon_0 - kd}\right)$

7 A parallel plate capacitor whose capacitance C is 14 **pF** is charged by a battery to a potential difference $V = 12$ V between its plates. The charging battery is now disconnected and a porcelain plate with $K = 7$ is inserted between the plates, then the plate would oscillate back and forth between the plates with a constant mechanical energy of .... pJ. (Assume no friction) **[JEE Main 2021]**

8 In a parallel plate capacitor set up, the plate area of capacitor is 2 m^2 and the plates are separated by 1m. If the space between the plates are filled with a dielectric material of thickness 0.5 m and area 2 m^2 (see figure) the capacitance of the set-up will be ...... ε_0.
(Dielectric constant of the material = 3.2)
(Round off to the nearest integer) **[JEE Main 2021]**

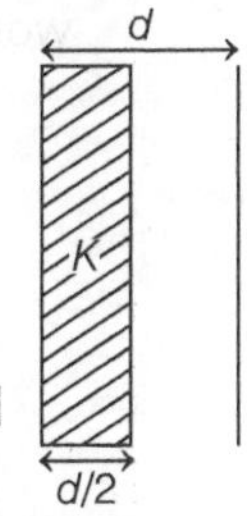

9 For changing the capacitance of a given parallel plate capacitor, a dielectric material of dielectric constant K is used, which has the same area as the plates of the capacitor. The thickness of the dielectric slab is $\frac{3}{4}d$, where d is the separation between the plates of parallel plate capacitor. The new capacitance (C') in terms of original capacitance (C_0) is given by the following relation **[JEE Main 2021]**

(a) $C' = \frac{3+K}{4K}C_0$ (b) $C' = \frac{4+K}{3}C_0$ (c) $C' = \frac{4K}{K+3}C_0$ (d) $C' = \frac{4}{3+K}C_0$

10 A parallel plate capacitor has plate of length l, width w and separation of plates is d. It is connected to a battery of emf V. A dielectric slab of the same thickness d and of dielectric constant $k = 4$ is being inserted between the plates of the capacitor. At what length of the slab inside plates, will the energy stored in the capacitor be two times the initial energy stored? **[JEE Main 2020]**

(a) $\frac{2l}{3}$ (b) $\frac{l}{3}$ (c) $\frac{l}{4}$ (d) $\frac{l}{2}$

11 A parallel plate capacitor has plates of area A separated by distance d between them. It is filled with a dielectric which has a dielectric constant that varies as $k(x) = K(1 + \alpha x)$, where x is the distance measured from one of the plates. If $(\alpha d) << 1$, the total capacitance of the system is best given by the expression **[JEE Main 2020]**

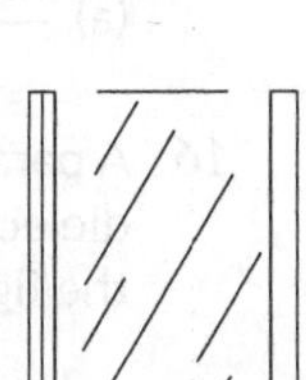

(a) $\frac{AK\varepsilon_0}{d}(1 + \alpha d)$ (b) $\frac{A\varepsilon_0 K}{d}\left(1 + \left(\frac{\alpha d}{2}\right)^2\right)$

(c) $\frac{AK\varepsilon_0}{d}\left(1 + \frac{\alpha d}{2}\right)$ (d) $\frac{A\varepsilon_0 K}{d}\left(1 + \frac{\alpha^2 d^2}{2}\right)$

12 Voltage rating of a parallel plate capacitor is 500 V. Its dielectric can withstand a maximum electric field of 10^6 V/m. The plate area is 10^{-4} m^2. What is the dielectric constant, if the capacitance is 15 pF?
(Take, $\varepsilon_0 = 8.86 \times 10^{-12}$ C^2 /N - m^2) **[JEE Main 2019]**
(a) 3.8 (b) 8.5 (c) 4.5 (d) 6.2

13 The parallel combination of two air filled parallel plate capacitors of capacitance C and nC is connected to a battery of voltage, V. When the capacitors are fully charged, the battery is removed and after that a dielectric material of dielectric constant K is placed between the two plates of the first capacitor. The new potential difference of the combined system is **[JEE Main 2019]**
(a) $\frac{(n+1)V}{(K+n)}$ (b) $\frac{nV}{K+n}$ (c) V (d) $\frac{V}{K+n}$

14 Two identical parallel plate capacitors of capacitance C each, have plates of area A, separated by a distance d. The space between the plates of the two capacitors, is filled with three dielectrics of equal thickness and dielectric constants K_1, K_2 and K_3. The first capacitor is filled as shown in Fig. I, and the second one is filled as shown in Fig. II. If these two modified capacitors are charged by the same potential V, the ratio of the energy stored in the two, would be (E_1 refers to capacitor (I) and E_2 to capacitor (II)) : **[JEE Main 2019]**

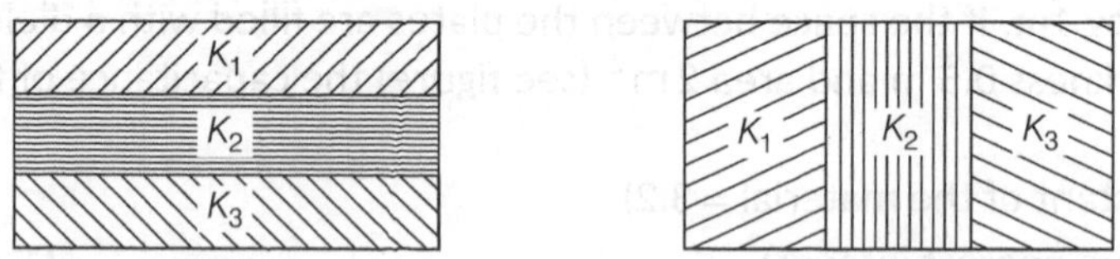

(I) (II)

(a) $\frac{E_1}{E_2} = \frac{K_1K_2K_3}{(K_1+K_2+K_3)(K_2K_3+K_3K_1+K_1K_2)}$ (b) $\frac{E_1}{E_2} = \frac{(K_1+K_2+K_3)(K_2K_3+K_3K_1+K_1K_2)}{K_1K_2K_3}$
(c) $\frac{E_1}{E_2} = \frac{9K_1K_2K_3}{(K_1+K_2+K_3)(K_2K_3+K_3K_1+K_1K_2)}$ (d) $\frac{E_1}{E_2} = \frac{(K_1+K_2+K_3)(K_2K_3+K_3K_1+K_1K_2)}{9K_1K_2K_3}$

15 A parallel plate capacitor is made of two square plates of side 'a' separated by a distance d (d<<a). The lower triangular portions is filled with a dielectric of dielectric constant k, as shown in the figure. Capacitance of this capacitor is **[JEE Main 2019]**

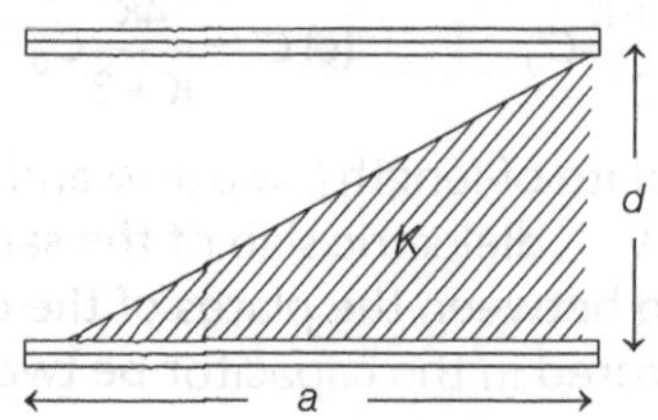

(a) $\frac{K\varepsilon_0 a^2}{d}\ln K$ (b) $\frac{K\varepsilon_0 a^2}{d(K-1)}\ln K$ (c) $\frac{K\varepsilon_0 a^2}{2d(K+1)}$ (d) $\frac{1}{2}\cdot\frac{K\varepsilon_0 a^2}{d}$

16 A parallel plate capacitor with square plates is filled with four dielectrics of dielectric constants K_1, K_2, K_3, K_4 arranged as shown in the figure. The effective dielectric constant K will be: **[JEE Main 2019]**

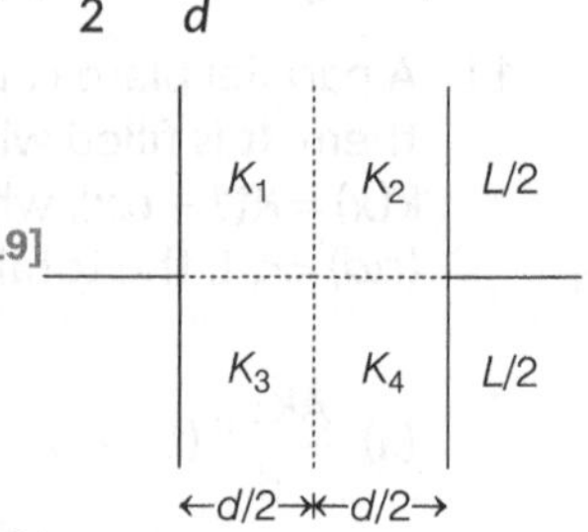

(a) $K = \frac{(K_1+K_2)(K_3+K_4)}{2(K_1+K_2+K_3+K_4)}$ (b) $K = \frac{(K_1+K_2)(K_3+K_4)}{K_1+K_2+K_3+K_4}$

(c) $K = \frac{(K_1+K_3)(K_2+K_4)}{K_1+K_2+K_3+K_4}$ (d) $K = \frac{(K_1+K_4)(K_2+K_3)}{2(K_1+K_2+K_3+K_4)}$

17 A parallel plate capacitor is of area 6 cm^2 and a separation 3 mm.
The gap is filled with three dielectric materials of equal thickness (see figure) with dielectric constants $K_1 = 10, K_2 = 12$ and $K_3 = 14$. The dielectric constant of a material which give same capacitance when fully inserted in above capacitor, would be **[JEE Main 2019]**

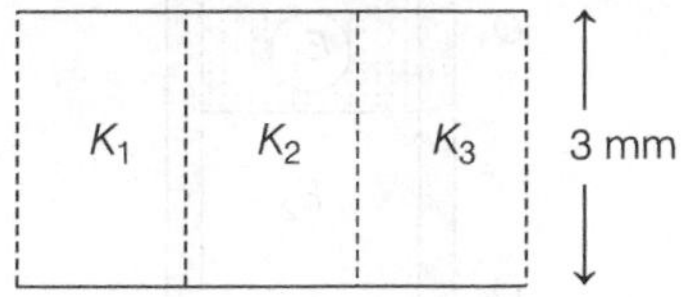

(a) 4 (b) 36 (c) 12 (d) 14

18 A parallel plate capacitor having capacitance 12 pF is charged by a battery to a potential difference of 10 V between its plates. The charging battery is now disconnected and a porcelain slab of dielectric constant 6.5 is slipped between the plates. The work done by the capacitor on the slab is **[JEE Main 2019]**

(a) 560 pJ (b) 508 pJ (c) 692 pJ (d) 600 pJ

19 A parallel plate capacitor of capacitance 90 pF is connected to a battery of emf 20 V. If a dielectric material of dielectric constant $K = \frac{5}{3}$ is inserted between the plates, the magnitude of the induced charge will be **[JEE Main 2018]**

(a) 1.2 nC (b) 0.3 nC (c) 2.4 nC (d) 0.9 nC

20 A parallel plate capacitor is made of two circular plates separated by a distance of 5 mm with a dielectric of dielectric constant 2.2 between them. When the electric field in the dielectric is 3×10^4 V/m, the charge density of the positive plate will be close to **[JEE Main 2014]**

(a) 6×10^{-7} C/m^2 (b) 3×10^{-7} C/m^2 (c) 3×10^4 C/m^2 (d) 6×10^4 C/m^2

JEE Advanced

21 Three identical capacitors C_1, C_2 and C_3 have a capacitance of 1.0 μF each and they are uncharged initially. They are connected in a circuit as shown in the figure and C_1 is then filled completely with a dielectric material of relative permittivity ε_r. The cell electromotive force (emf) $V_0 = 8$V. First the switch S_1 is closed while the switch S_2 is kept open. When the capacitor C_3 is fully charged, S_1 is opened and S_2 is closed simultaneously. When all the capacitors reach equilibrium, the charge on C_3 is found to be 5μC. The value of ε_r =... **[JEE Advanced 2018]**

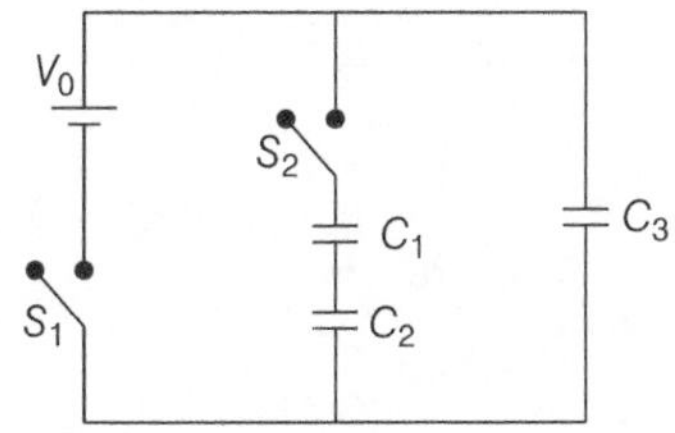

22 A parallel plate capacitor having plates of area S and plate separation d, has capacitance C_1 in air. When two dielectrics of different relative permittivities ($\varepsilon_1 = 2$ and $\varepsilon_2 = 4$) are introduced between the two plates as shown in the figure, the capacitance becomes C_2. The ratio $\frac{C_2}{C_1}$ is **[JEE Advanced 2015]**

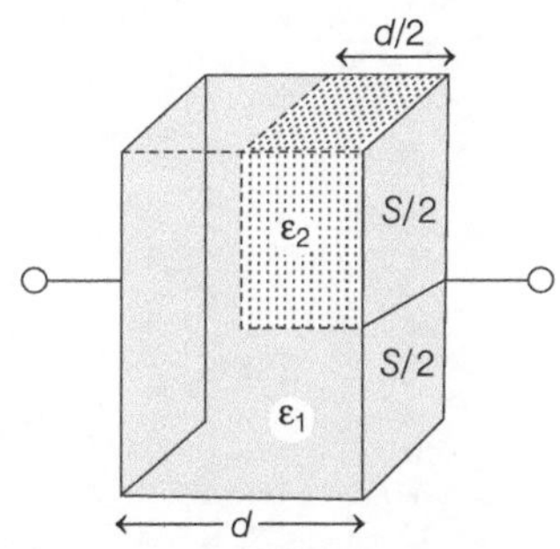

(a) $\frac{6}{5}$ (b) $\frac{5}{3}$
(c) $\frac{7}{5}$ (d) $\frac{7}{3}$

23 A parallel plate capacitor has a dielectric slab of dielectric constant K between its plates that covers 1/3 of the area of its plates, as shown in the figure. The total capacitance of the capacitor is C while that of the portion with dielectric in between is C_1. When the capacitor is

charged, the plate area covered by the dielectric gets charge Q_1 and the rest of the area gets charge Q_2. The electric field in the dielectric is E_1 and that in the other portion is E_2. Choose the correct option/options, ignoring edge effects. **[JEE Advanced 2014]**

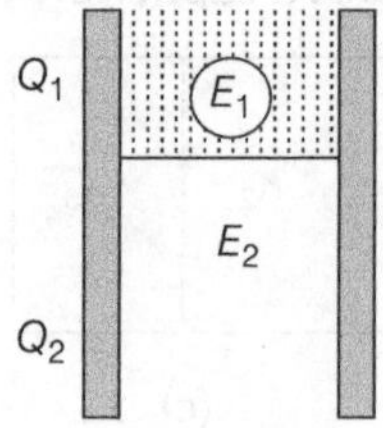

(a) $\frac{E_1}{E_2}=1$ (b) $\frac{E_1}{E_2}=\frac{1}{K}$ (c) $\frac{Q_1}{Q_2}=\frac{3}{K}$ (d) $\frac{C}{C_1}=\frac{2+K}{K}$

Answers

1. (c)	**2.** (c)	**3.** (b)	**4.** (161)	**5.** (a)	**6.** (b)
7. (864)	**8.** (3)	**9.** (c)	**10.** (b)	**11.** (c)	**12.** (b)
13. (a)	**14.** (d)	**15.** (b)	**16.** (*)	**17.** (c)	
18. (b)	**19.** (a)	**20.** (a)	**21.** (1.50)	**22.** (d)	
23. (c,d)					

For Detailed Solutions
Scan the code

CHAPTER

11

R-C CIRCUIT

LEVEL 01 \ Starter Level

1.1 R-C Circuit at t = 0 (Initial State) and t = ∞ (Final State)

1 Find out current in the circuit and charge on capacitor which is initially uncharged in the following situations.

(i) Just after the switch is closed

(ii) After a long time when switch was closed.

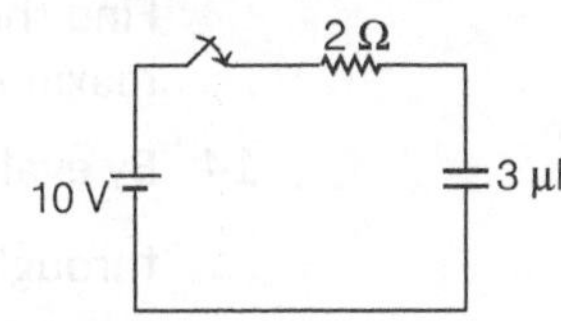

2 Find out current i_1, i_2, i_3 and charge on the capacitor in the circuit when it is initially uncharged in the following situations.

(i) Just after the switch is closed

(ii) After a long time when switch is closed.

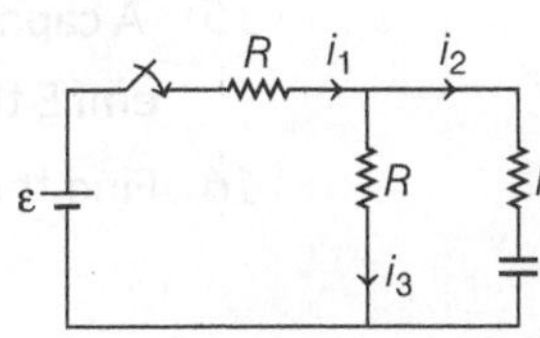

3 Find the charge on each capacitor in steady state

(i) when switch S is open

(ii) when switch S is closed.

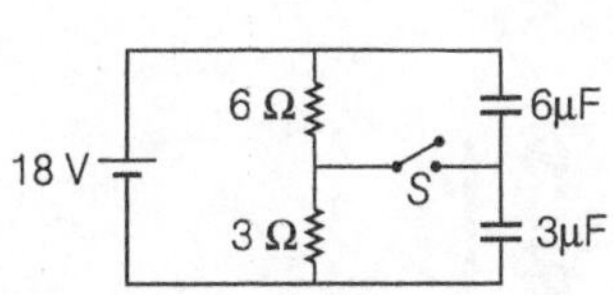

4 Find the charge on each capacitor of the given circuit in steady state.

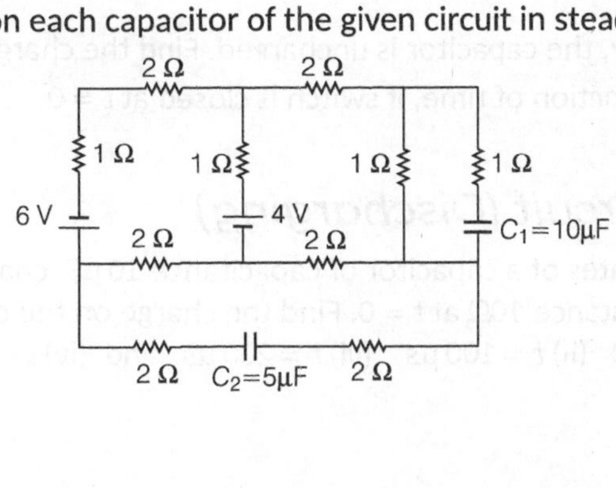

1.2 R-C Series Circuit (Charging)

5 A capacitor of capacitance 10 μF is connected to a battery of emf 2 V. It is found that it takes 50 ms for the charge on the capacitor to become 12.6 μC. Find the resistance of the circuit.

6 A 2 μF capacitor is joined to a battery of emf 50 V through a resistance of 1 kΩ. Find the charge on the capacitor 4 ms after the connections are made.

7 A capacitor is connected to a 12 V battery through a resistance of 10Ω. It is found that the potential difference across the resistor is 8 V after 1 μs. Find the capacitance of the capacitor. (Given, $\ln 3 = 1.0986$, $\ln 2 = 0.693$)

8 A parallel plate capacitor of plate area 40 cm^2 and separation between the plates 0.10 mm is connected to a battery of emf 2.0 V through a 16 Ω resistor. Find the electric field in the capacitor 10 ns after the connections are made.

9 A capacitor of capacitance 12.0μF is connected to a battery of emf 6.00 V and internal resistance 1.00 Ω through resistanceless leads. 12.0 μs after the connections are made, what will be (i) the current in the circuit, (ii) the power delivered by the battery, (iii) the power dissipated in heat and (iv) the rate at which the energy stored in the capacitor is increasing.

10 How many time constants will elapse before the charge on the capacitor in a charging *R*-*C* circuit reaches half of its steady state value?

11 How many time constants will elapse before the energy stored in the capacitor reaches half of its equilibrium value in a charging *R-C* circuit?

12 How many time constants will elapse before the power delivered by the battery drops to half of its maximum value in an *R-C* circuit?

13 A capacitor of capacitance *C* is connected to a battery of emf ε at $t = 0$ through a resistance *R*. Find the maximum rate at which energy is stored in the capacitor. When does the rate has this maximum value?

14 By evaluating $\int i^2 R dt$, show that when a capacitor is charged by connecting it to a battery through a resistor, the energy dissipated as heat equals the energy stored in the capacitor.

15 A capacitor of capacitance *C* is given a charge *q*. At $t = 0$, it is connected to an ideal battery of emf *E* through a resistance *R*. Find the charge on the capacitor at time *t*.

16 Find the charge on each of the capacitors 0.20 ms after the switch *S* is closed in figure.

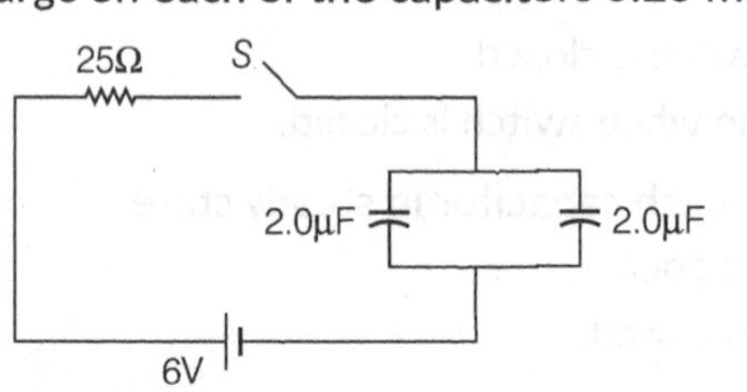

17 Initially, the capacitor is uncharged. Find the charge on capacitor as a function of time, if switch is closed at $t = 0$.

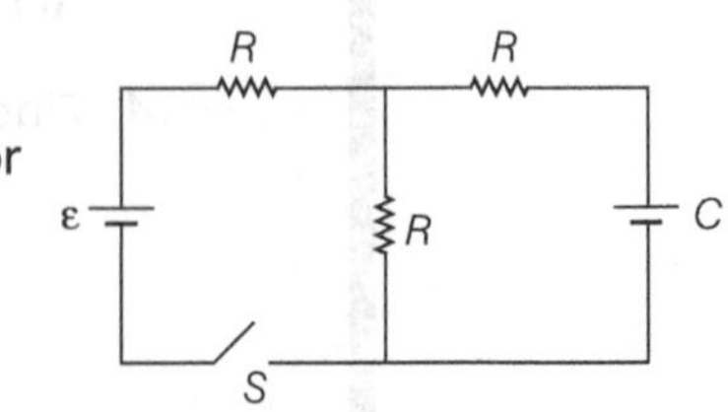

1.3 R-C Circuit (Discharging)

18 The plates of a capacitor of capacitance 10 μF, charged to 60 μC, are joined together by a wire of resistance 10Ω at $t = 0$. Find the charge on the capacitor in the circuit at (i) $t = 0$ (ii) $t = 100\ \mu s$ (iii) $t = 200\ \mu s$ and (iv) $t = 1.0$ ms.

19 The switch S shown in figure is kept closed for a long time and is then opened at $t = 0$. Find the current in the middle 10Ω resistor at $t = 1.0$ ms.

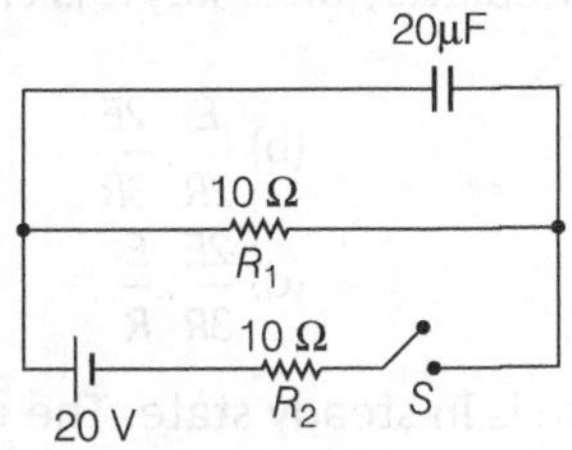

20 A capacitor of capacitance $200\mu F$ is connected across a battery of emf 10.0 V through a resistance of 40 kΩ for 16.0 s. The battery is then replaced by a thick wire. What will be the charge on the capacitor 16.0s after the battery is disconnected? (Given, $e^{-2} = 0.135$)

21 The electric field between the plates of a parallel plate capacitance $2.0\mu F$ drops to one-third of its initial value in $4.4\mu s$ when the plates are connected by a thin wire. Find the resistance of the wire. (Given, ln 3 = 1.0986)

22 A 5.0 μF capacitor having a charge of 20 μC is discharged through a wire of resistance 5.0 Ω. Find the heat dissipated in the wire between 25 μs to 50 μs after the connections are made. (Given, $e^{-2} = 0.135$)

23 A capacitor of capacitance C is given a charge Q. At $t = 0$, it is connected to an uncharged capacitor of equal capacitance through a resistance R. Find the charge on the second capacitor as a function of time.

Answers

1. (i) $q_C = 0, i = 5A$ (ii) $q_C = 30\,\mu C, i = 0$ **2.** (i) $i_1 = \frac{2\varepsilon}{3R}$ and $i_2 = i_3 = \frac{\varepsilon}{3R}, q = 0$ (ii) $i_2 = 0, i_1 = i_3 = \frac{\varepsilon}{2R}, q = \frac{C\varepsilon}{2}$

3. (i) 36μC on each capacitor (ii) $q_{6\mu F} = 72\mu C, q_{3\mu F} = 18\mu C$ **4.** $q_1 = 4\mu C$, $q_2 = 12\mu C$ **5.** 5 kΩ

6. 86.5 μC **7.** 0.25 μF **8.** 1.7×10^4 V/m **9.** (i) 2.21 A (ii) 13.2 W (iii) 4.87 W (iv) 8.37 W

10. 0.693 **11.** 1.23 **12.** 0.693 **13.** $RC\ln 2$ **14.** $\frac{1}{2}CV^2$ **15.** $CE\left(1 - e^{-\frac{t}{RC}}\right) + qe^{-\frac{t}{RC}}$

16. 10.37 μC **17.** $q = \frac{C\varepsilon}{2}\left(1 - e^{-\frac{2t}{3RC}}\right)$ **18.** (i) 60 μC, (ii) 22 μC (iii) 8.12 μC (iv) 0.003 μC

19. 6.7 mA **20.** 233.55 μC

21. 2.0 Ω **22.** 4.7 μJ **23.** $q = \frac{Q}{2}\left(1 - e^{-\frac{2t}{RC}}\right)$

For Detailed Solutions
Scan the code

LEVEL 02 \ JEE Main Level

Single Choice Correct Questions

1 The magnitude of charge in steady state on either of the plates of condenser C in the adjoining circuit is

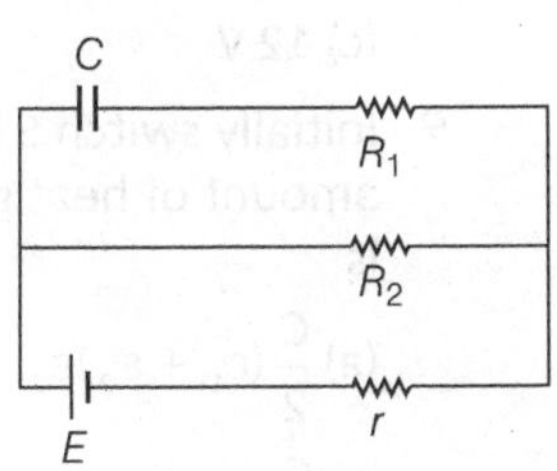

(a) CE
(b) $\frac{CER_2}{(R_1 + r)}$
(c) $\frac{CER_2}{(R_2 + r)}$
(d) $\frac{CER_1}{(R_2 + r)}$

2 A network of uncharged capacitors and resistances is as shown. Current through the battery immediately after key K is closed and after a long time interval is

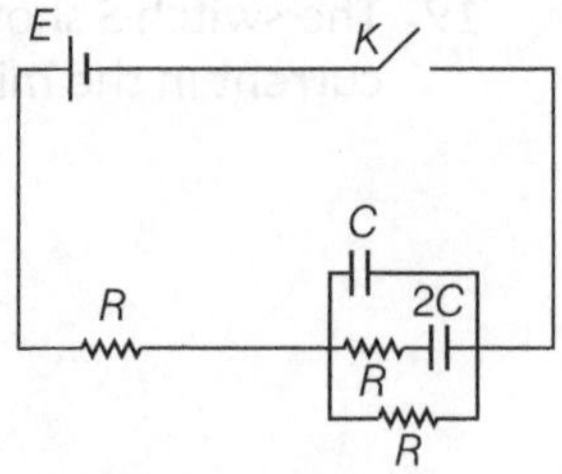

(a) $\frac{E}{R}, \frac{E}{2R}$
(b) $\frac{E}{2R}, \frac{2E}{3R}$
(c) zero, $\frac{E}{R}$
(d) $\frac{2E}{3R}, \frac{E}{R}$

3 The capacitor shown in the figure is in steady state. The energy stored in the capacitor is

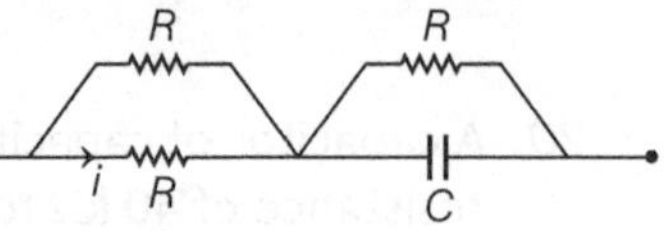

(a) Ci^2R^2
(b) $2Ci^2R^2$
(c) $4Ci^2R^2$
(d) None of these

4 In the given circuit with steady current, the potential drop across the capacitor must be

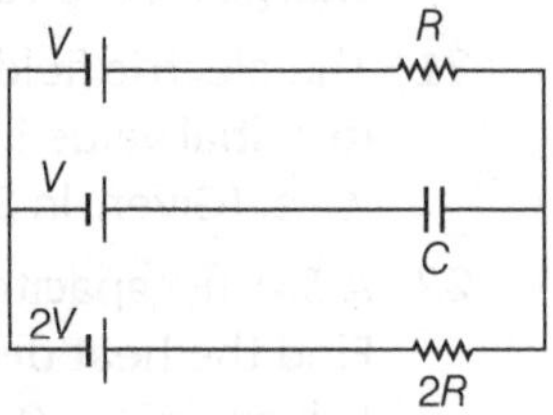

(a) V
(b) $\frac{V}{2}$
(c) $\frac{V}{3}$
(d) $\frac{2V}{3}$

5 Ratio of charges on capacitor C_1 after long time when switch S is open and closed respectively, in the given circuit, is

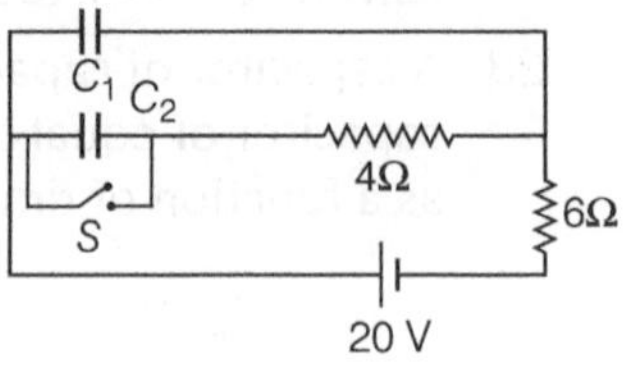

(a) 1 : 1
(b) 1 : 8
(c) 5 : 2
(d) Data insufficient

6 The charge on the capacitor in the steady state in the circuit shown is

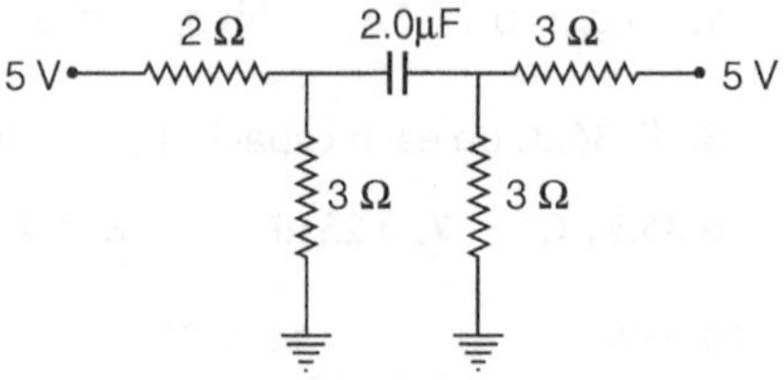

(a) 0.5 μC
(b) 1 μC
(c) 2 μC
(d) 4 μC

7 In the circuit shown, switch S_2 is closed first and is kept closed for a long time. Now, S_1 is closed. Just after that instant, the current through S_1 is

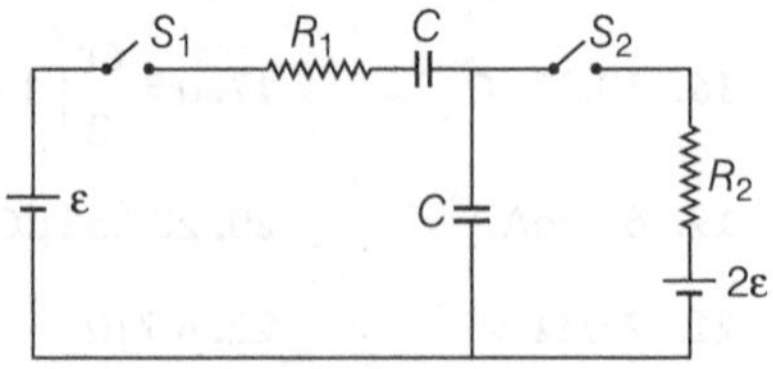

(a) $\frac{\varepsilon}{R_1}$ toward right
(b) $\frac{\varepsilon}{R_1}$ toward left
(c) $\frac{2\varepsilon}{R_1}$
(d) zero

8 In the given circuit, the potential difference across the capacitor is 12V in steady state. Each resistor have 3Ω resistance. The emf of the ideal battery is

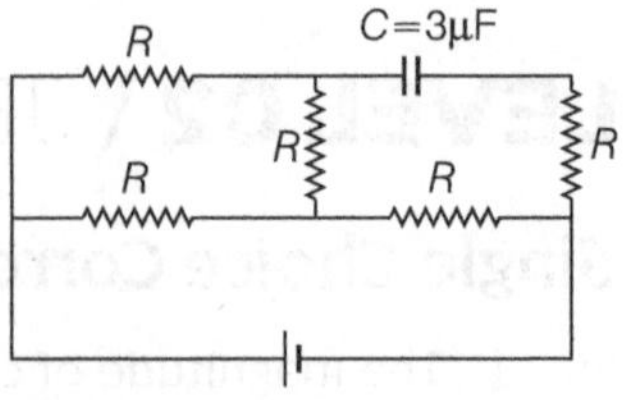

(a) 15 V
(b) 9 V
(c) 12 V
(d) 24 V

9 Initially switch S is connected to position 1 for a long time. The net amount of heat generated in the circuit after it shifted to position 2 is

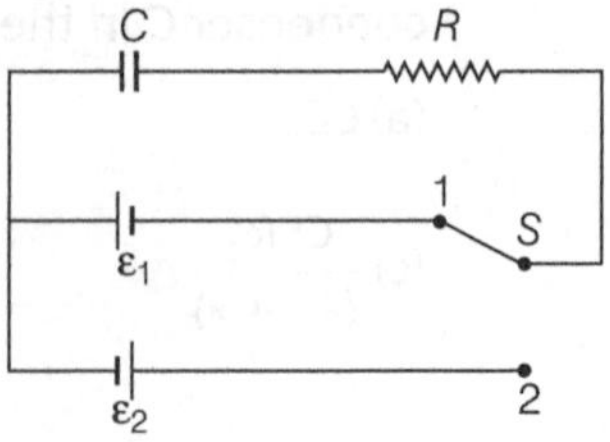

(a) $\frac{C}{2}(\varepsilon_1 + \varepsilon_2)\varepsilon_2$
(b) $C(\varepsilon_1 + \varepsilon_2)\varepsilon_2$
(c) $\frac{C}{2}(\varepsilon_1 + \varepsilon_2)^2$
(d) $C(\varepsilon_1 + \varepsilon_2)^2$

10 Heat produced in the circuit shown in the figure when switch is closed, is

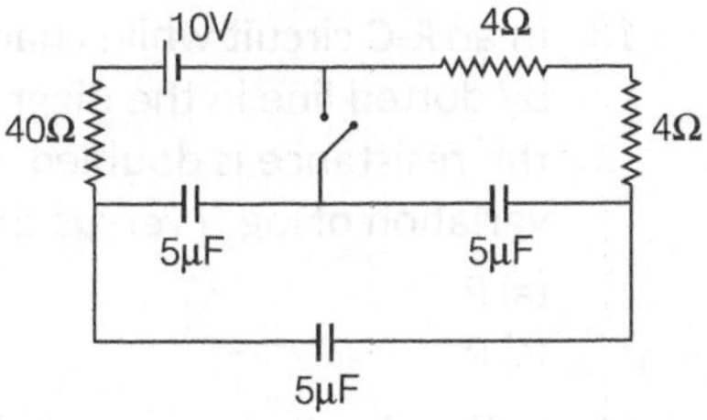

(a) 125 μJ

(b) 250 μJ

(c) 375 μJ

(d) 500 μJ

11 In the circuit, if no current flows through the galvanometer when the key K is closed, the bridge is balanced. The balancing condition for bridge is

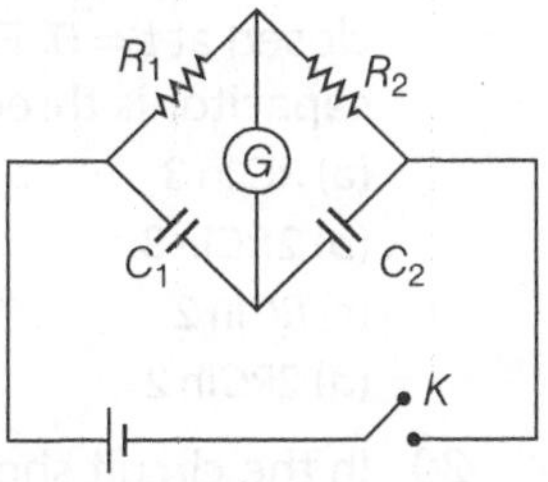

(a) $\frac{C_1}{C_2} = \frac{R_1}{R_2}$ (b) $\frac{C_1}{C_2} = \frac{R_2}{R_1}$

(c) $\frac{C_1^2}{C_2^2} = \frac{R_1^2}{R_2^2}$ (d) $\frac{C_1^2}{C_2^2} = \frac{R_2}{R_1}$

12 In given circuit, switch S is closed at $t = 0$. After what time, voltage across capacitor becomes 7 times of potential difference across R?

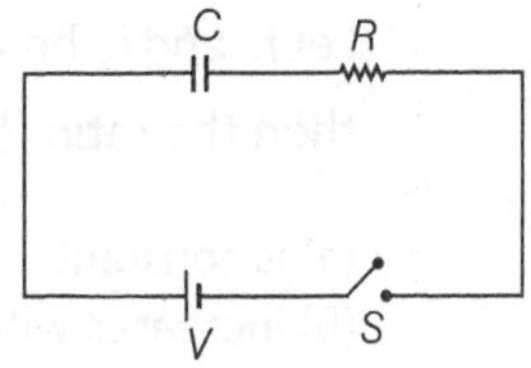

(a) $3\,CR \ln 2$

(b) $2\,CR \ln 2$

(c) $8\,CR \ln 2$

(d) $CR \ln 2$

13 A light bulb, a capacitor and a battery are connected together as shown here, with switch S initially open. When the switch S is closed, which one of the following is true?

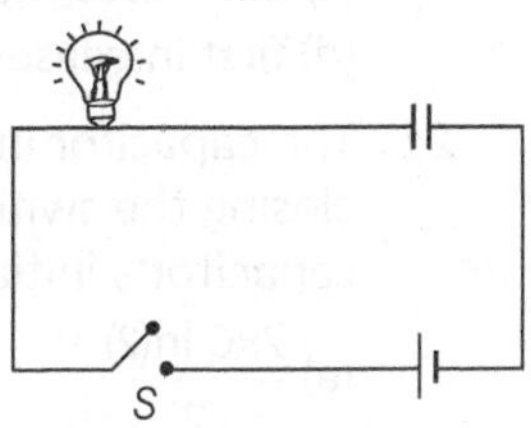

(a) The bulb will light up for some time interval after the capacitor starts charging.

(b) The bulb will light up when the capacitor is fully charged.

(c) The bulb will not light up at all.

(d) The bulb will light up and go off at regular intervals.

14 A resistor R and 2 μF capacitor in series is connected through a switch to 200 V direct supply. Across the capacitor is a neon bulb that lights up at 100 V. Calculate the value of R to make the bulb light up 6.93 s after the switch has been closed.

(a) $5 \times 10^6\ \Omega$ (b) $2 \times 10^7\ \Omega$ (c) $10 \times 10^4\ \Omega$ (d) $2 \times 10^5\ \Omega$

15 An uncharged capacitor and a resistor of big resistance are connected in series to a battery of electromotive force 4.5 V. The voltage across the capacitor is 3 V one minute after the closing the circuit. What will the voltage across the capacitor be 2 min after closing circuit?

(a) 6 V (b) 4.5 V (c) 3.5 V (d) 4 V

16 An R-C circuit with $R = 600\ \text{k}\Omega$ and $C = 10\ \mu\text{F}$, is connected to a 5 V battery until the capacitor is fully charged. Then, the battery is suddenly replaced with a new 3 V battery of opposite polarity. At what time after this replacement will the energy stored in the capacitor be zero?

(Given, $e \simeq \frac{8}{3}$)

(a) 12 s (b) 6 s (c) 3 s (d) 1.5 s

17 As situation shown in figure, the maximum value of rate of energy stored in the capacitor after the switch is closed, is

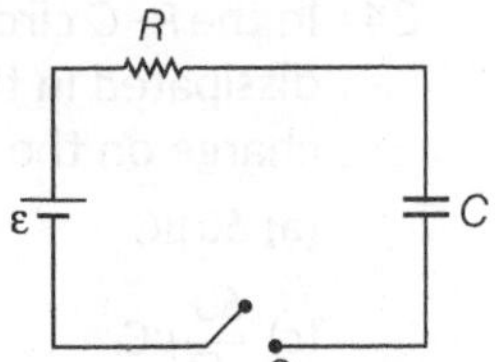

(a) $\frac{\varepsilon^2}{2R}$ (b) $\frac{\varepsilon^2}{4R}$

(c) $\frac{\varepsilon^2}{8R}$ (d) None of these

18 In an R-C circuit while charging, the graph of $\log_e i$ *versus* time is shown by dotted line in the diagram, where i is the current. When the value of the resistance is doubled, which of the solid curves best represents the variation of $\log_e i$ *versus* time?

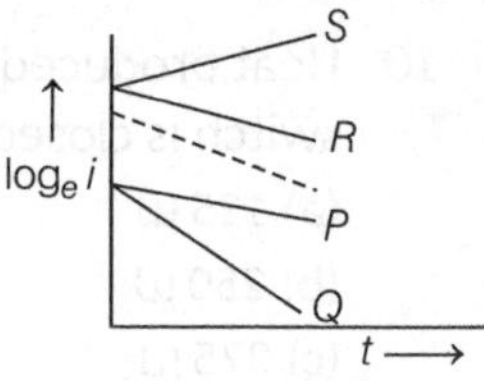

(a) P (b) Q
(c) R (d) S

19 In the circuit shown capacitor is initially uncharged and switch is closed at $t = 0$. Find the time at which the rate of energy change in capacitor is three times the power dissipated in resistor.

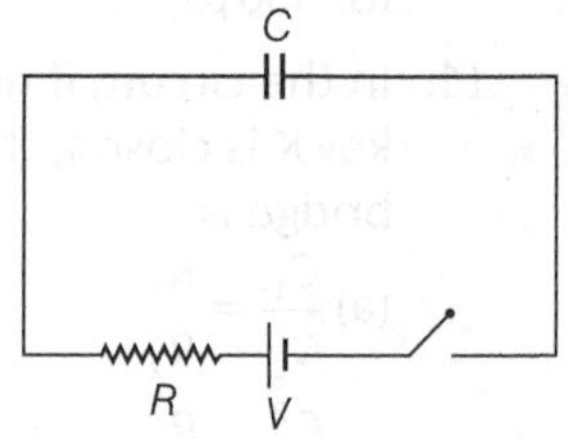

(a) $RC \ln 3$
(b) $2RC \ln 3$
(c) $RC \ln 2$
(d) $2RC \ln 2$

20 In the circuit shown in figure $C_1 = 2C_2$. Switch S is closed at time $t = 0$. Let i_1 and i_2 be the currents flowing through C_1 and C_2 at any time t, then the ratio $\frac{i_1}{i_2}$

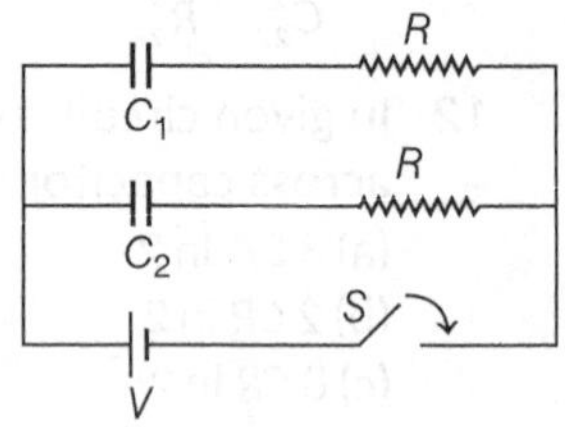

(a) is constant
(b) increases with increase in time t
(c) decreases with increase in time t
(d) first increases then decreases

21 The capacitor in the circuit shown in figure is initially charged. After closing the switch, how much time elapses until one-half of the capacitor's initial stored energy is dissipated?

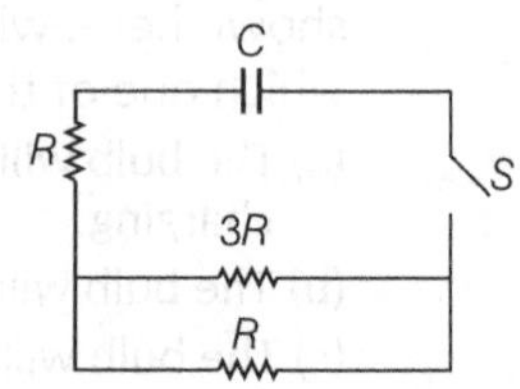

(a) $\frac{7RC \ln(2)}{2}$ (b) $\frac{7RC \ln(2)}{4}$
(c) $\frac{7RC \ln(2)}{8}$ (d) $7RC \ln(2)$

22 In a discharging R-C circuit,

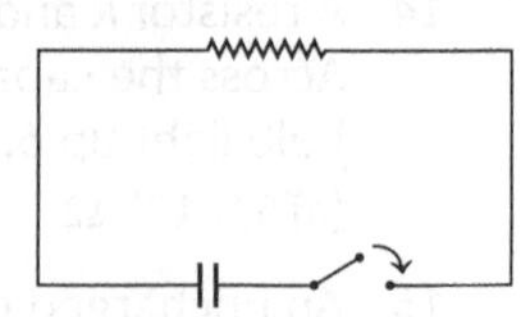

(a) ratio of charge on capacitor and current in resistance increases with time.
(b) ratio of charge on capacitor and current in resistance decreases with time.
(c) ratio of charge on capacitor and current in resistance remains constant.
(d) ratio of charge on capacitor and current in resistance first decreases, then increases with time.

23 A charged capacitor is allowed to discharge through a resistor by closing the key at the instant $t = 0$. At the instant $t = (\ln 4)\,\mu s$, the reading of the ammeter falls half the initial value. The resistance of the ammeter is equal to

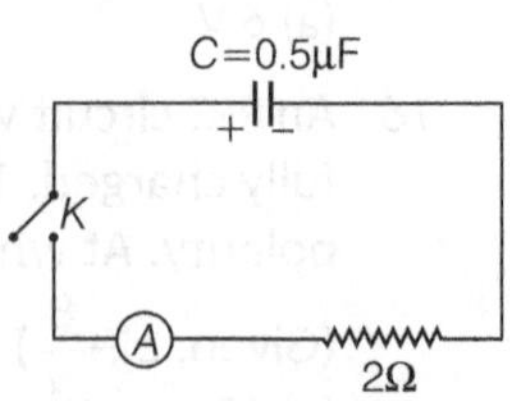

(a) 1Ω (b) 2Ω
(c) 3Ω (d) 4Ω

24 In the R-C circuit shown in the figure, the total energy of 3.6×10^{-3} J is dissipated in the 10Ω resistor when the switch S is closed. The initial charge on the capacitor is

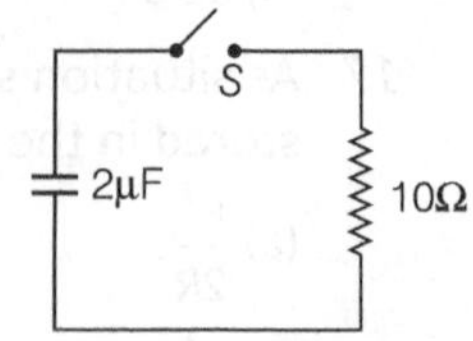

(a) 60 μC (b) $60\sqrt{2}$ μC
(c) $\frac{60}{\sqrt{2}}$ μC (d) 120 μC

25 Consider the *R*-*C* circuit shown below. What combination of values *R* and *C* will result in the slowest rate of capacitor discharge when switch *S* is closed?

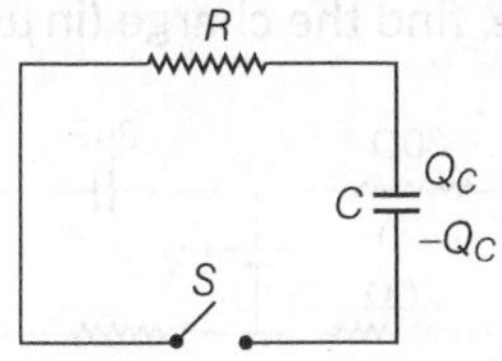

(a) 10 Ω and 50 μF (b) 100 Ω and 5 μF

(c) 10 Ω and 5 μF (d) 100 Ω and 50 μF

26 The switch *S* shown in figure is kept closed for a long time and then opened at $t = 0$, then the current in the middle 20 Ω resistor at $t = 0.25$ ms is

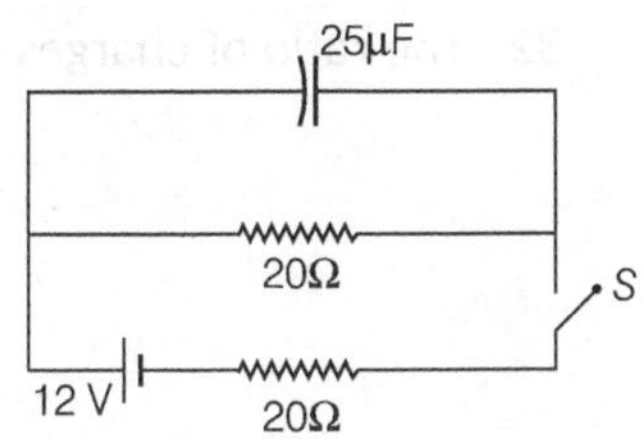

(a) 329 mA

(b) 285 mA

(c) 182 mA

(d) 28 mA

27 The figure shows an experimental plot for discharging of a capacitor in an *R*-*C* circuit. The time constant τ of this circuit lies between

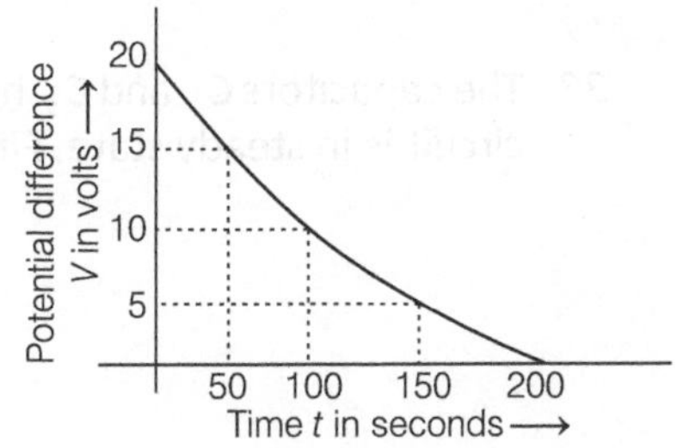

(a) 100 s and 150 s

(b) 150 s and 200 s

(c) 0 and 50 s

(d) 50 s and 100 s

28 In the transient circuit shown, the time constant of the circuit is

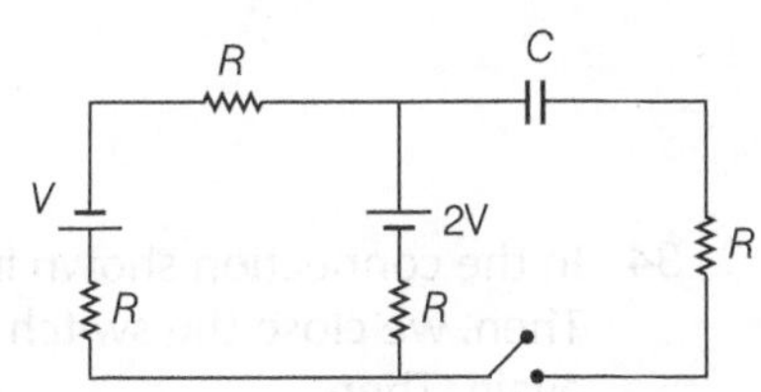

(a) $\frac{5}{3}RC$ (b) $\frac{5}{2}RC$

(c) $\frac{4}{5}RC$ (d) $\frac{7}{5}RC$

29 A capacitor $C = 100\,\mu F$ is connected to three resistors each of resistance 1 kΩ and a battery of emf 9V. The switch *S* has been closed for long time so as to charge the capacitor. When switch *S* is opened, the capacitor discharges with time constant.

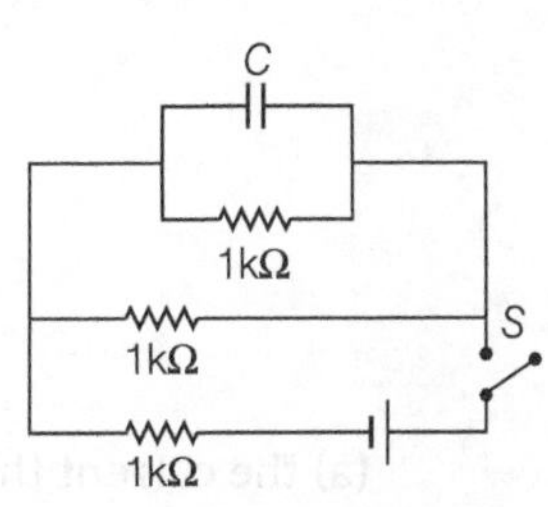

(a) 33 ms

(b) 5 ms

(c) 3.3 ms

(d) 50 ms

30 The switch *S* in the circuit diagram is closed at $t = 0$. The charge on each capacitor at any time *t* is

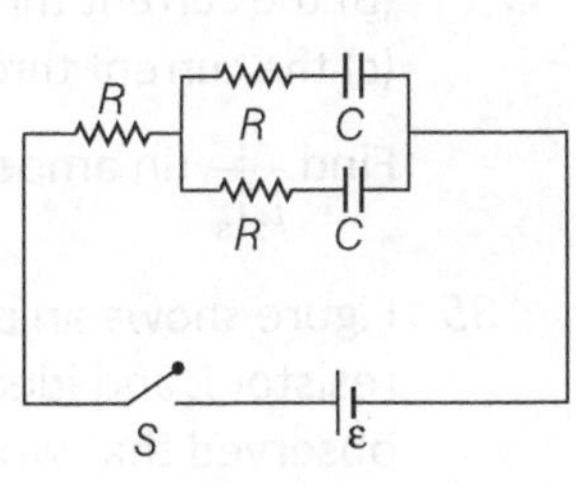

(a) $q = C\varepsilon\left(1 - e^{-\frac{2t}{3RC}}\right)$ (b) $q = C\varepsilon\left(1 - e^{-\frac{t}{2RC}}\right)$

(c) $q = C\varepsilon\left(1 - e^{-\frac{t}{3RC}}\right)$ (d) $q = C\varepsilon\left(1 - e^{-\frac{3t}{2RC}}\right)$

Numerical Value Questions

31 For the circuit shown in the figure, find the charge (in μC) on the 8 μF capacitor in the steady state.

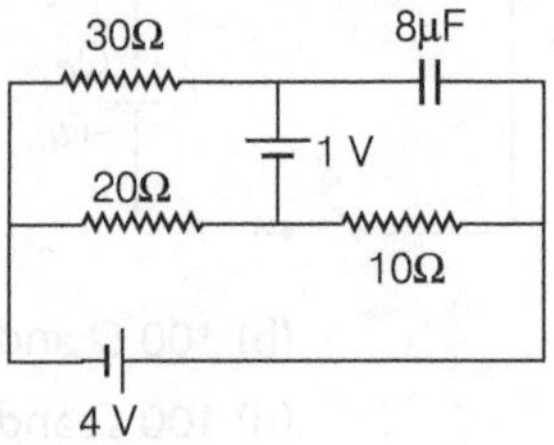

32 The ratio of charges on 5 μF and 10 μF capacitor when circuit is in steady state is $\frac{\alpha}{38}$. Find α.

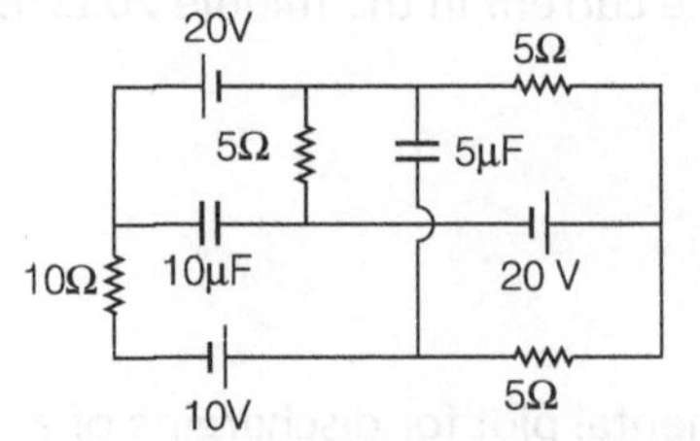

33 The capacitors C_1 and C_2 have capacitances 1 μF and 3 μF respectively in the given circuit. The circuit is in steady state. Find the total energy (in μJ) stored in the capacitors.

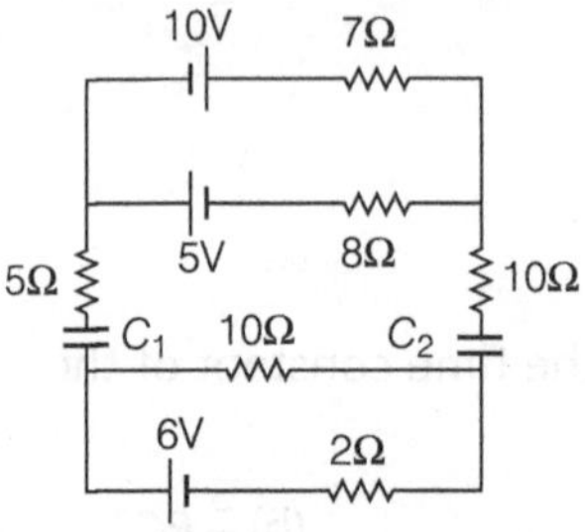

34 In the connection shown in the figure, the switch K is open and the capacitor is uncharged. Then, we close the switch and let the capacitor charge up to the maximum and open the switch again. Then,

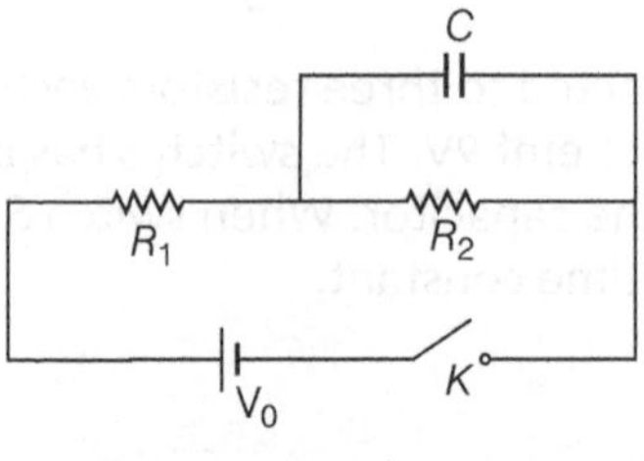

(a) the current through R_1 be i_1 immediately after closing the switch
(b) the current through R_2 be i_2 a long time after the switch was closed
(c) the current through R_2 be i_3 immediately after reopening the switch.

Find $\frac{i_1}{i_2 i_3}$ (in ampere^{-1}) (Use the following data: $V_0 = 30$ V, $R_1 = 10$ kΩ, $R_2 = 5$ kΩ)

35 Figure shows an uncharged capacitor connected in series with a resistor R and ideal battery. At $t = 0$, switch S is closed and it is observed that when charge on capacitor is 75% of maximum value, then current in the circuit is 1 A, then find the value of R (in Ω).

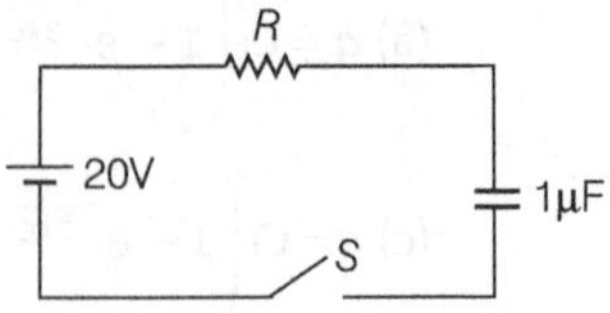

36 The switch is closed at $t = 0$. The capacitor C is uncharged but C_0 has a charge Q_0 at $t = 0$. If $R = 100\,\Omega$, $C = 2\,\mu F$, $C_0 = 1\,\mu F$, $Q_0 = 2\,\mu C$, $E = 4V$. Current in the circuit is given as $i(t) = i_0 e^{-kt}$. Find numerical value of k/I_0.

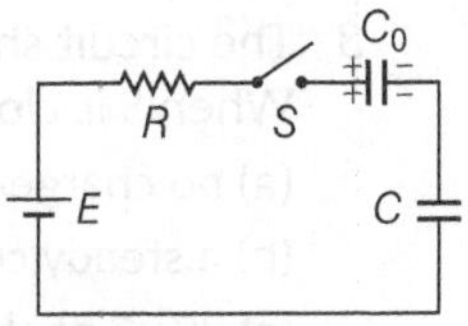

37 In the circuit shown, the switch S is closed at $t = 0$. The capacitor is initially uncharged. Then, find the current (in mA) through the resistor $R = 8\,\Omega$ at $t = 0.4$ ms. (Take, $e^{-2} = 0.135$)

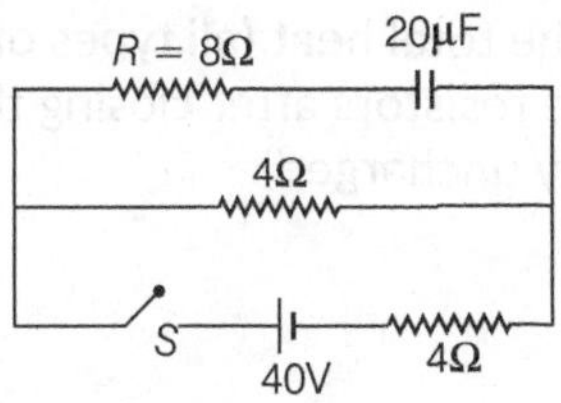

38 The figure shows two circuits with a charged capacitor that is to be discharged through a resistor as shown. The ratio of initial charge on capacitors $\frac{q_2}{q_1} = 2$. If both switches are closed at $t = 0$, the charges become equal at 10^{-4} ln 2 sec. Find the resistance R (in Ω).

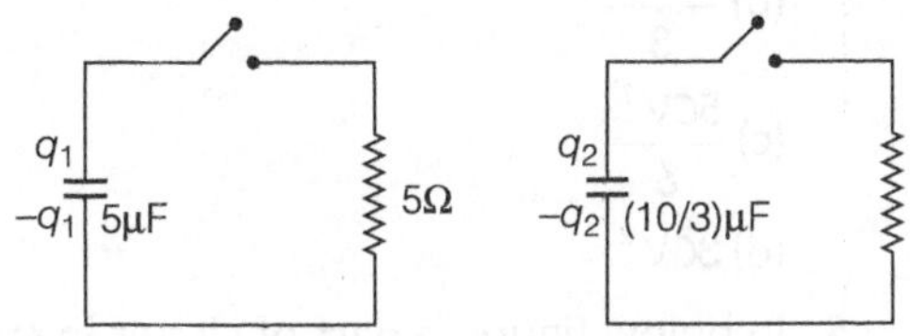

Answers

1. (c)	**2.** (a)	**3.** (b)	**4.** (c)	**5.** (c)	**6.** (b)	**7.** (b)	**8.** (a)	**9.** (c)	**10.** (a)
11. (b)	**12.** (a)	**13.** (a)	**14.** (a)	**15.** (d)	**16.** (b)	**17.** (b)	**18.** (a)	**19.** (d)	**20.** (b)
21. (c)	**22.** (c)	**23.** (b)	**24.** (d)	**25.** (d)	**26.** (c)				
27. (a)	**28.** (a)	**29.** (d)	**30.** (c)	**31.** 8	**32.** 13				
33. 24.	**34.** 750	**35.** 5	**36.** 300	**37.** 270	**38.** 6				

For Detailed Solutions
Scan the code

LEVEL 03 \ JEE Advanced Level

Single Choice Correct Questions

1 The capacitor C shown in figure is charged by closing switch S for a long time. Find the charge on it in steady state.

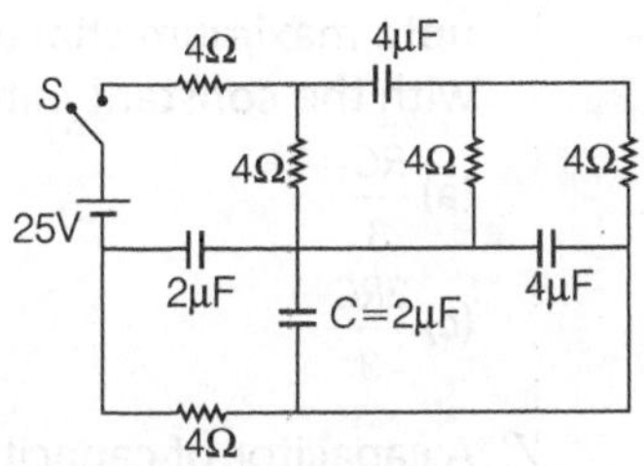

(a) $8\,\mu C$
(b) $10\,\mu C$
(c) $16\,\mu C$
(d) $20\,\mu C$

2 In the shown circuit, all three capacitors are identical and have capacitance C each. Each resistor has resistance of R. An ideal cell of emf V volts is connected as shown. Then, the magnitude of potential difference across capacitor C_3 in steady state is

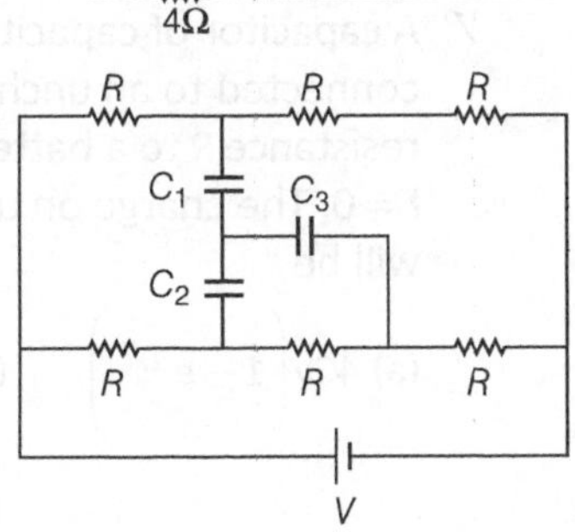

(a) $\frac{V}{3}$ (b) $\frac{V}{2}$

(c) $\frac{2V}{9}$ (d) $\frac{3V}{4}$

3 The circuit shown below shows a steady state with S open. When S is closed,

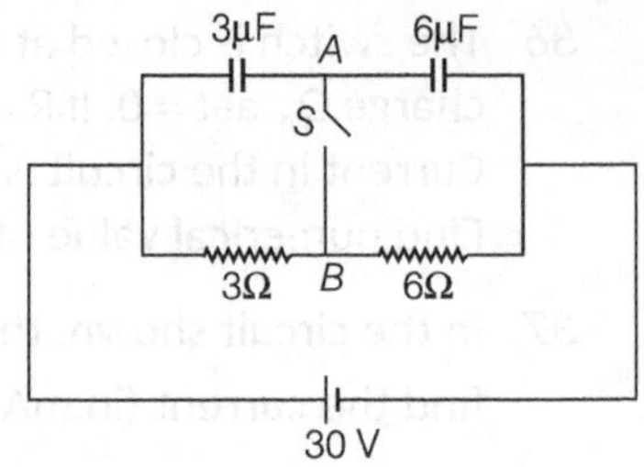

(a) no charge/current passes through S
(b) a steady current of 30 A flows through S
(c) 90 μC of charge flows from B to A
(d) 120 μC of charge flows from A to B

4 Switch S is closed at $t = 0$. Find the total heat (all types of losses included) generated in the resistors after closing the switch. (All capacitors are initially uncharged)

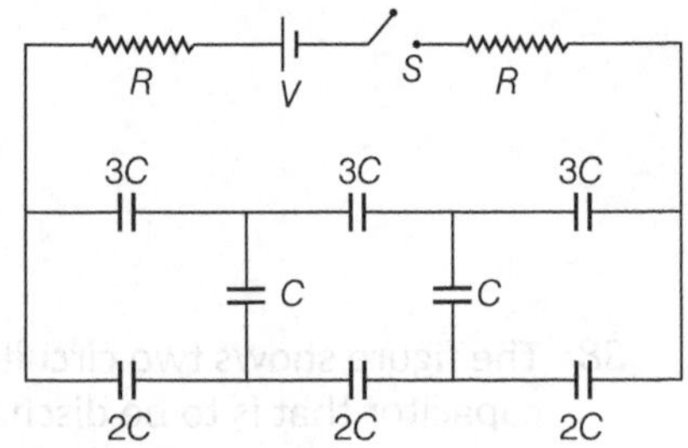

(a) $\frac{5CV^2}{3}$
(b) $\frac{6CV^2}{3}$
(c) $\frac{5CV^2}{6}$
(d) $3CV^2$

5 In below figure, a part of circuit in steady state is shown. If energy stored in the capacitor is $x \times 10^{-4}$ J, then find value of x.

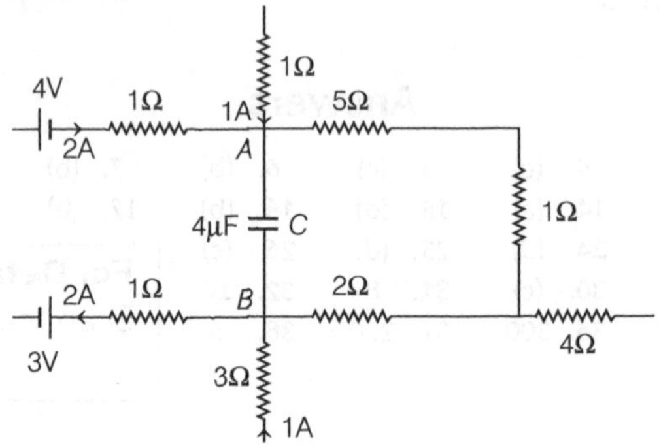

(a) 5 (b) 6
(c) 7 (d) 8

6 In the given circuit, the capacitor of capacitance C is charged by closing key K at $t = 0$. Find the time required to charge the capacitor upto maximum charge for the given circuit, if it were to be charged with the constant initial charging rate at $t = 0$ in the given circuit.

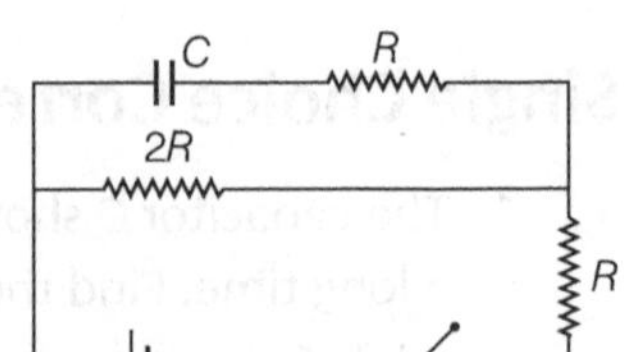

(a) $\frac{RC}{3}$ (b) $\frac{2RC}{5}$
(c) $\frac{2RC}{3}$ (d) $\frac{5RC}{3}$

7 A capacitor of capacitance $2C$ is charged to a potential $2V$ and then connected to an uncharged capacitor of capacitance C with a resistance R to a battery of emf $4V$ as shown. The key is closed at $t = 0$. The charge on uncharged capacitor as a function of time will be

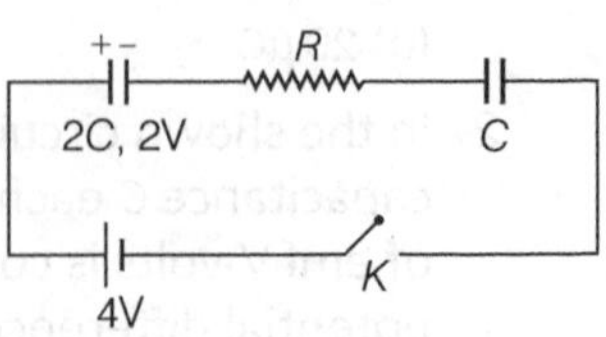

(a) $4CV\left(1 - e^{\frac{-t}{2RC}}\right)$ (b) $\frac{4CV}{3}\left(1 - e^{\frac{-t}{2RC}}\right)$ (c) $4CV\left(1 - e^{\frac{-3t}{2RC}}\right)$ (d) $\frac{4CV}{3}\left(1 - e^{\frac{-3t}{2RC}}\right)$

8 The switch in figure (a) closes when $V_C = \frac{2V}{3}$ and opens when $V_C = \frac{V}{3}$. The voltmeter reads a voltage as plotted in figure (b). What is the period T of the waveform? (Take, $R_1 = 6R$ and $R_2 = 3R$)

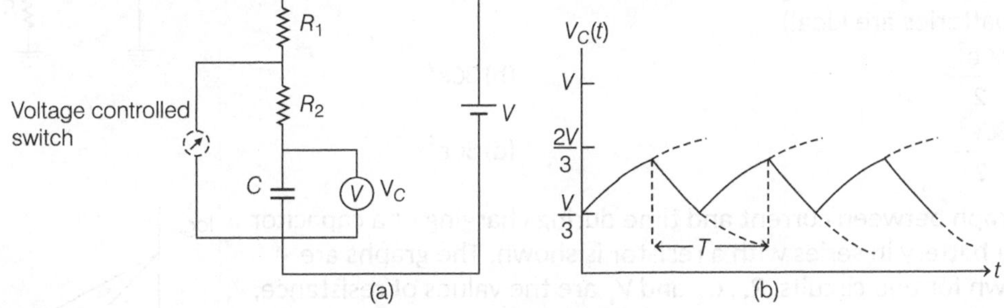

(a) $12\, RC \ln 2$
(b) $18\, RC \ln 2$
(c) $21\, RC \ln 2$
(d) None of these

9 Conducting sphere of radius a is surrounded by another spherical conducting shell of radius b. The space between them is filled with dielectric material of conductivity σ and dielectric constant K. The charges Q_1 and Q_2 are given to the inner and outer shell at time $t = 0$. Charge on outer shell at time t is

(a) $Q_1 + Q_2 \left(1 - \exp\left(-\frac{\sigma t}{\varepsilon_0 K}\right)\right)$
(b) $Q_2 - Q_1 \left(1 - \exp\left(-\frac{\sigma t}{\varepsilon_0 K}\right)\right)$
(c) $Q_1 - Q_2 \left(1 - \exp\left(-\frac{\sigma t}{\varepsilon_0 K}\right)\right)$
(d) $Q_2 + Q_1 \left(1 - \exp\left(-\frac{\sigma t}{\varepsilon_0 K}\right)\right)$

10 A metal ball of radius R is surrounded by a homogeneous medium of dielectric constant K and resistivity ρ. The ball is given a charge Q at $t = 0$, which starts leaking through the medium towards infinity. After how much time will the charge on the ball become $\frac{Q}{2}$?

(a) $[2K\rho\varepsilon_0] \ln 2$
(b) $[K\rho\varepsilon_0] \ln 2$
(c) $[K\rho\varepsilon_0 / 2] \ln 2$
(d) None of these

11 The area of the plates of the capacitor shown in figure is S. The dielectric constant and resistivity of the material of the slab inserted in the capacitor are $K = 1$ and ρ, respectively. The emf of the ideal cell is V. The current just after closing the switch is

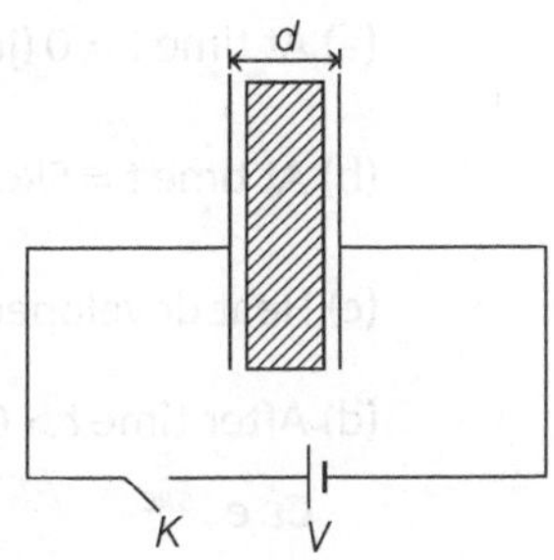

(a) $\frac{V}{\rho d} S$
(b) $\frac{2V}{\rho d} S$
(c) $\frac{V}{2\rho d} S$
(d) zero

12 In the R-C circuit shown, switch is closed at $t = 0$. Graphs showing variation of potential (V_R) across resistor and potential (V_C) across capacitor are given. Time constant of circuit is approximately equal to

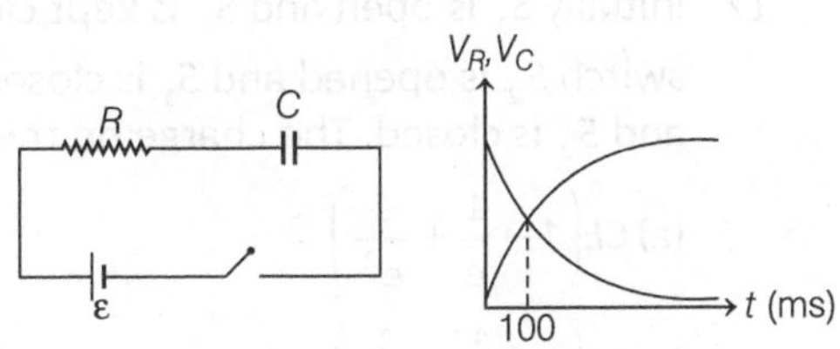

(a) 100 ms
(b) 145 ms
(c) 200 ms
(d) 300 ms

13 In the circuit arrangement shown in figure, capacitor is initially uncharged. At $t = 0$, switch is thrown to position 1. It remains closed till the current in the circuit becomes 50% of maximum current, then suddenly switch is shifted to position 2. Amount of work done by battery of emf 2ε long after switch is shifted to position 2 is (Assume all batteries are ideal)

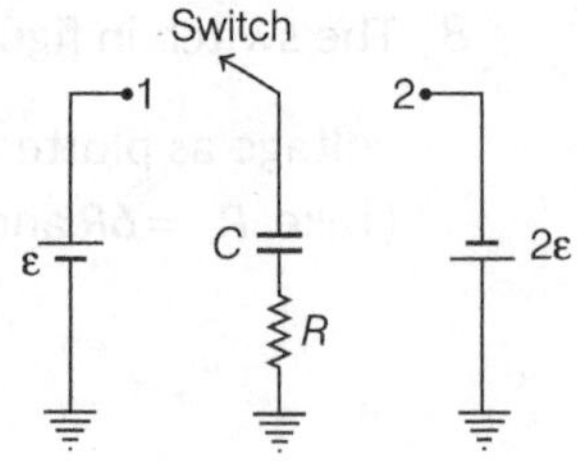

(a) $\frac{5C\varepsilon^2}{2}$ (b) $3C\varepsilon^2$

(c) $\frac{3C\varepsilon^2}{2}$ (d) $5C\varepsilon^2$

14 A graph between current and time during charging of a capacitor by a battery in series with a resistor is shown. The graphs are drawn for two circuits. R_1, C_1 and V_1 are the values of resistance, capacitance and emf of the cell in one circuit and R_2, C_2 and V_2 for another circuit, respectively. If only two parameters (out of resistance, capacitance, emf) are different in the two circuits, what is the correct option?

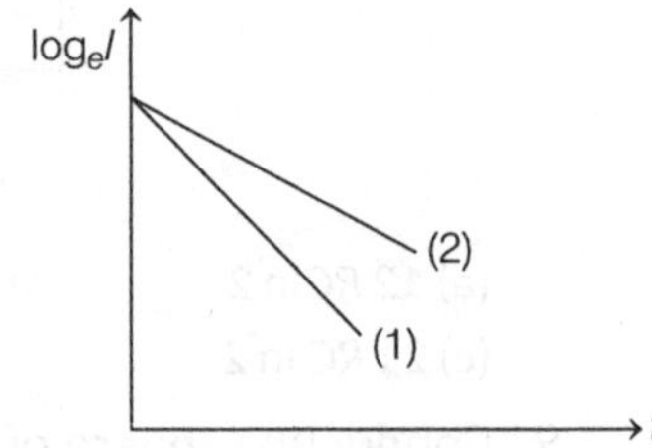

(a) $V_1 = V_2, R_1 > R_2, C_1 > C_2$ (b) $V_1 > V_2, R_1 > R_2, C_1 = C_2$

(c) $V_1 < V_2, R_1 < R_2, C_1 = C_2$ (c) $V_1 < V_2, C_1 < C_2, R_1 = R_2$

15 A charged capacitor C_1 is discharged through a resistor R. When the discharge current attains the value I_0, the key is shifted from 1 to 2. Then, the amount of heat Q liberated in the resistor starting from this moment, is

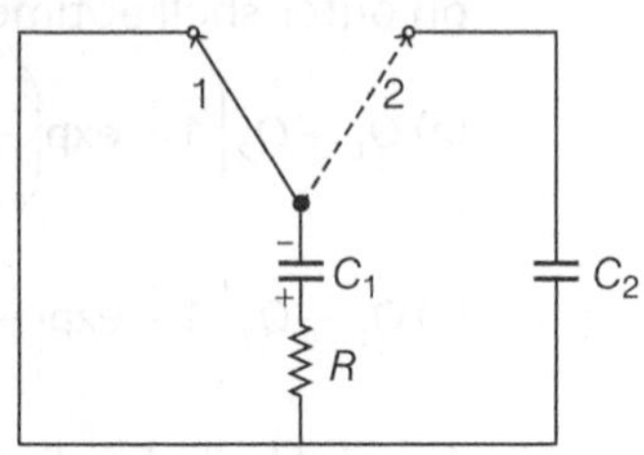

(a) $\frac{i_0^2 R^2 C_1 C_2}{2(C_1 + C_2)}$ (b) $\frac{i_0^2 R^2 C_1 C_2}{4(C_1 + C_2)}$

(c) $\frac{i_0^2 R^2}{2}\left(\frac{(C_1 + C_2)}{C_1 C_2}\right)$ (d) None of these

16 The capacitor C is initially uncharged. Switch S_1 is closed for a long time while S_2 remains open. Now, at $t = 0$, S_2 is closed while S_1 is opened. All the batteries are ideal and connecting wires are resistanceless. Find incorrect statement.

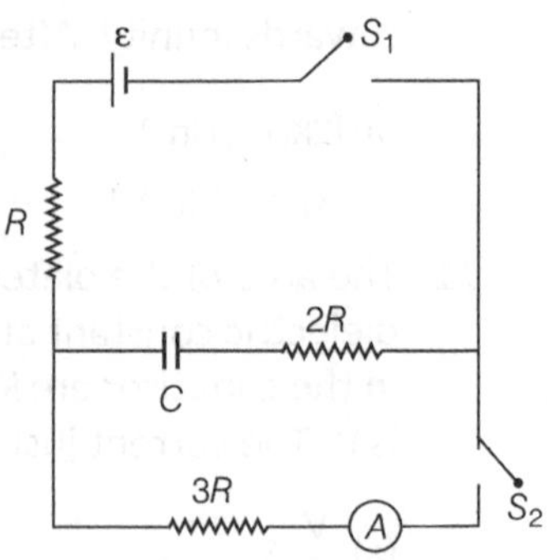

(a) At time $t = 0$ (just after S_2 is closed), reading of ammeter is $\frac{\varepsilon}{5R}$.

(b) At time $t = 5RC \ln 2$, reading of ammeter is $\frac{\varepsilon}{5eR}$.

(c) Heat developed till time $t = 5\,RC \ln 2$ in resistance $3R$ is $\frac{9}{40} C\varepsilon^2$.

(d) After time $t > 0$, charge on the capacitor follows the equation $C\varepsilon\, e^{-\frac{t}{5RC}}$

17 Initially S_1 is open and S_2 is kept closed for a long time. Now at $t = 0$, switch S_2 is opened and S_1 is closed. At $t = R_1 C$, switch S_1 is opened and S_2 is closed. The charge on the capacitor at $t = 2R_1 C + R_2 C$ is

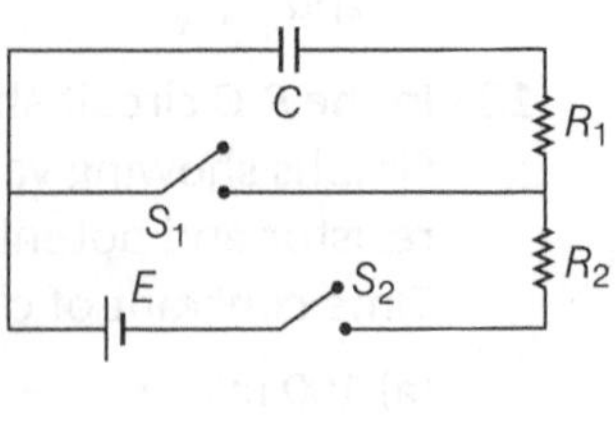

(a) $CE\left(1 - \frac{1}{e} + \frac{1}{e^2}\right)$ (b) $CE\left(\frac{1}{e} - \frac{1}{e^2}\right)$

(c) $CE\left(1 + \frac{1}{e} - \frac{1}{e^2}\right)$ (d) $CE\left(\frac{1}{e} + \frac{1}{e^2}\right)$

Multiple Choice(s) Correct Questions

18 In the circuit shown, there is steady state with the switch closed. The switch is opened at $t = 0$. Choose the correct option(s). (Given, $\varepsilon = 24$ V, $C_1 = 3$ F and $C_2 = 2$F)

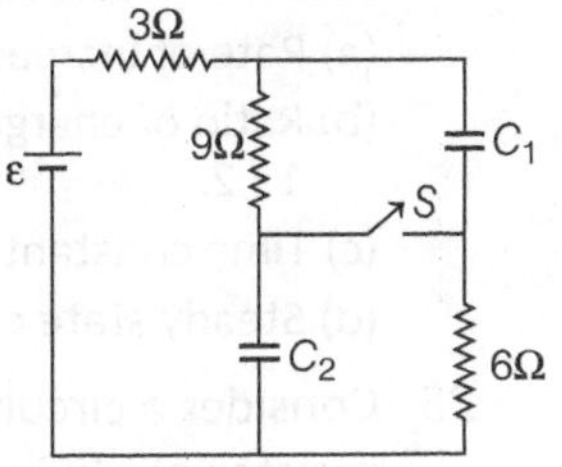

(a) The voltage across C_1 before the switch is open is 12 V.
(b) The voltage across C_2 before the switch is open is 8 V.
(c) The voltage across C_1 after a long time after the switch is open is 12 V.
(d) The voltage across C_2 after a long time after the switch is open is 24 V.

19 In the circuit shown in figure, if the galvanometer resistance is 6Ω, then in the steady state,

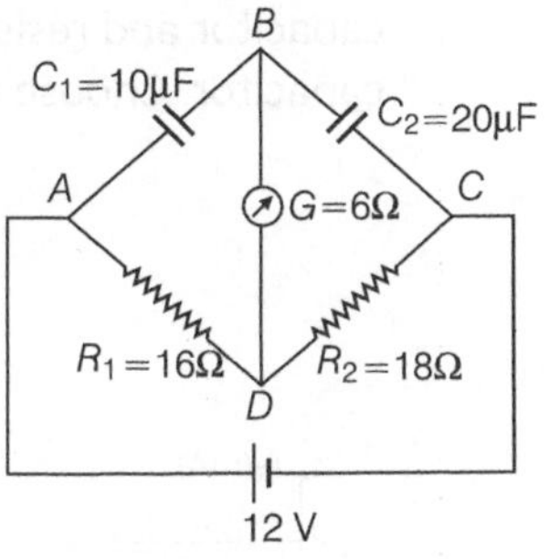

(a) no current flows through the galvanometer
(b) the current through R_2 is 4 A
(c) the charge on C_1 is 80 μC
(d) the charge on C_2 is 80 μC

20 The instantaneous charge on capacitor in two discharging *R-C* circuits is plotted with respect to time in figure. Choose the correct statement(s) (where, ε_1 and ε_2 are emfs of two DC sources in two different charging circuits and capacitors are fully charged).

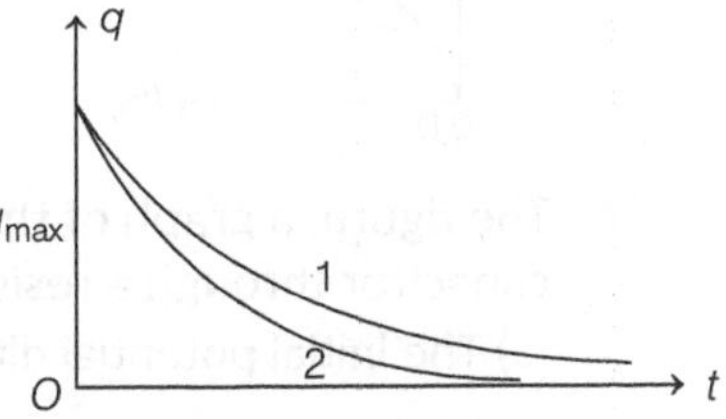

(a) $R_1C_1 > R_2C_2$
(b) $\frac{R_1}{R_2} < \frac{C_2}{C_1}$
(c) $R_1 > R_2$, if $\varepsilon_1 = \varepsilon_2$
(d) $C_2 > C_1$, if $\varepsilon_1 = \varepsilon_2$

21 Capacitors $C_1 = 1\mu$F and $C_2 = 2\mu$F are separately charged from the same battery. They are then allowed to discharge separately through equal resistors.
(a) The current in the two discharging circuits at $t = 0$ is zero
(b) The currents in the two discharging circuits at $t = 0$ are equal but not zero
(c) The currents in the two discharging circuits at $t = 0$ are unequal
(d) C_1 loses 50% of its initial charge sooner than C_2 loses 50% of its initial charge

22 A capacitor *C* charged to a value $Q = \varepsilon C$ is connected to a battery of emf ε as shown in the figure.

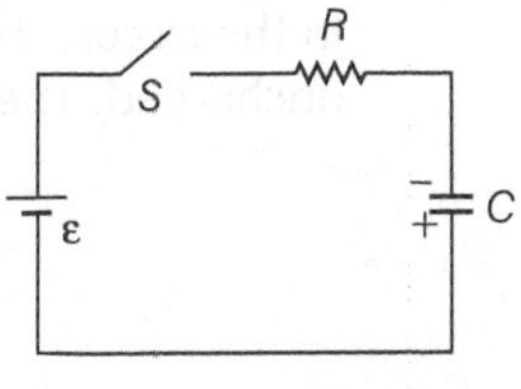

(a) Charge on the capacitor is zero at time *RC* ln 2 after the switch *S* is closed
(b) Charged on the capacitor is zero very long time after the switch *S* is closed
(c) Maximum charge on the capacitor during the process of charging is εC
(d) During the process of charging, the charging current first decreases and then increases exponentially

23 A capacitor of capacitance *C* is connected to two voltmeters *A* and *B*. *A* is an ideal voltmeter having infinite resistance, while *B* has resistance *R*. The capacitor is uncharged and then the switch *S* is closed at $t = 0$,

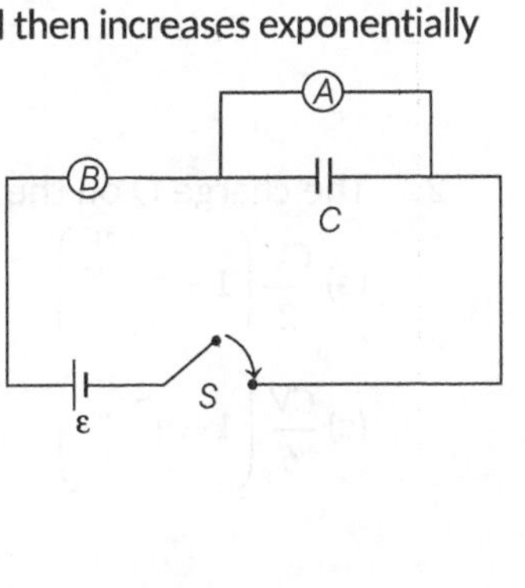

(a) Readings of *B* and *A* will be ε and zero respectively at $t = 0$
(b) During time interval $(0 \le t \le \infty)$, readings of *B* and *A* are changing
(c) Reading of *A* and *B* will be equal at $t = RC \ln 2$
(d) None of the above

24 In the circuit shown in the figure, switch S is closed at time $t = 0$. Select the correct statements.

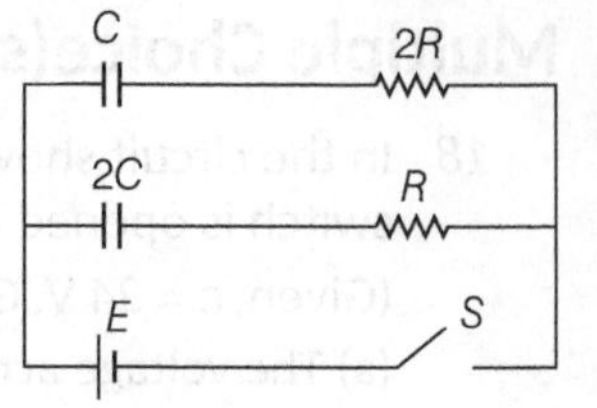

(a) Rate of increase of charge is same in both the capacitor.

(b) Ratio of charge stored in capacitors C and $2C$ at any time t would be 1 : 2.

(c) Time constant of both the capacitors are equal.

(d) Steady state charge in capacitors C and $2C$ are in the ratio 1 : 2.

25 Consider a circuit with a constant current source of $i_s = 2$ mA with a capacitor of 2 μF and resistor of 5 kΩ as shown in figure. At $t = 0$, the switch is closed. i_c and i_R denotes current in the capacitor and resistor branch respectively and V_c represents potential difference across the capacitor. Choose the correct graph(s).

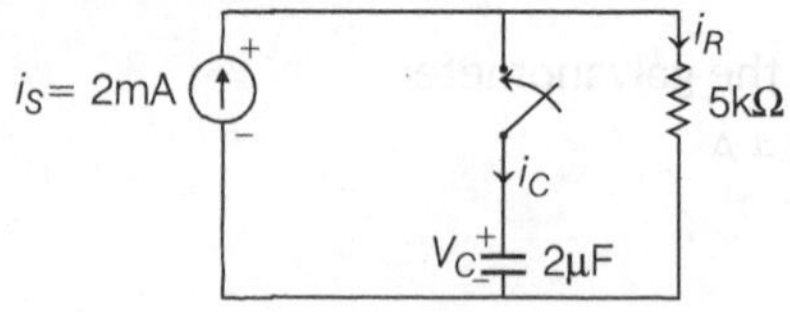

(a)
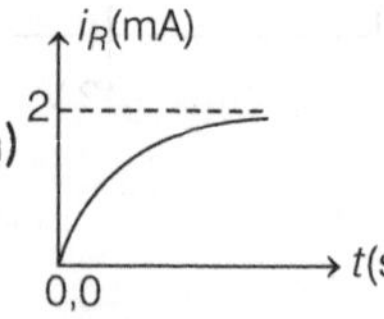

(b)
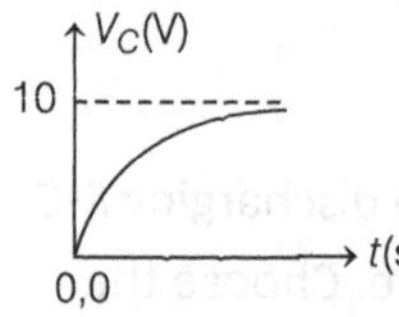

(c)
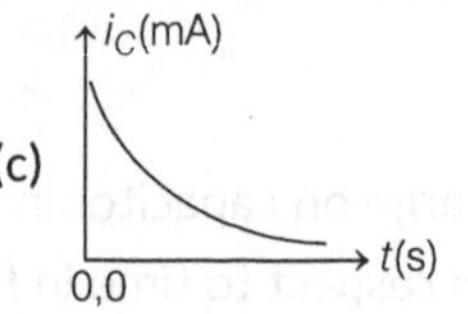

(d)
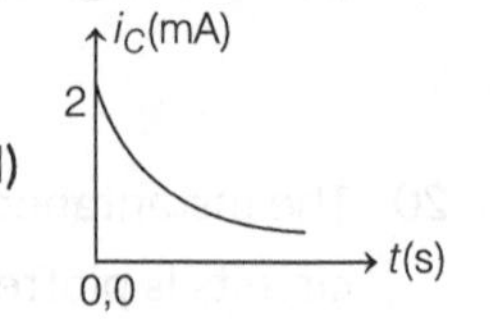

26 The figure, a graph of the current in a discharging circuit of a capacitor through a resistor of resistance 10Ω.

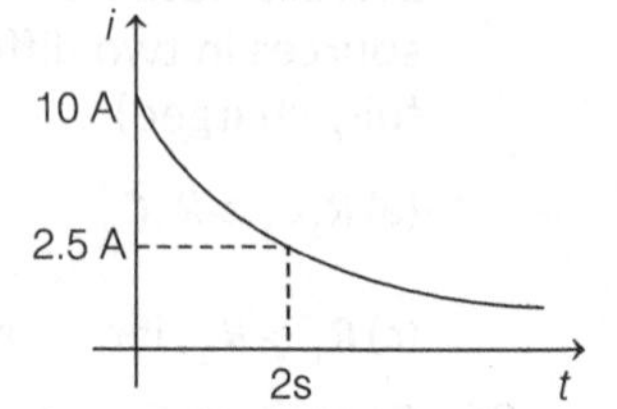

(a) The initial potential difference across the capacitor is 100 V.

(b) The capacitance of the capacitor is $\dfrac{1}{10 \ln 2}$ F.

(c) The total heat produced in the circuit will be $\dfrac{500}{\ln 2}$ J.

(d) The thermal power in the resistor will decrease with a time constant $\dfrac{1}{2 \ln 2}$ s.

Comprehension Based Questions

Comprehension-1

In the circuit shown in figure, the battery is an ideal one with emf V. The capacitor is initially uncharged. The switch S is closed at time $t = 0$.

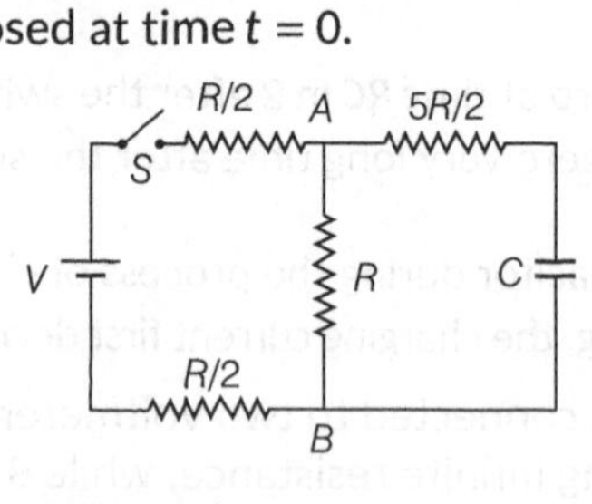

27 The charge Q on the capacitor at time t is

(a) $\dfrac{CV}{2}\left(1 - e^{-\frac{2t}{5RC}}\right)$

(b) $\dfrac{CV}{2}\left(1 - e^{-\frac{t}{3RC}}\right)$

(c) $\dfrac{CV}{6}\left(1 - e^{-\frac{2t}{5RC}}\right)$

(d) $\dfrac{CV}{6}\left(1 - e^{-\frac{t}{3RC}}\right)$

28 The current in AB at time t is

(a) $\frac{V}{2R}\left(1-e^{-\frac{2t}{5RC}}\right)$ (b) $\frac{2V}{R}\left(1-e^{-\frac{t}{3RC}}\right)$ (c) $\frac{2V}{R}\left(1-\frac{e^{-\frac{2t}{5RC}}}{6}\right)$ (d) $\frac{V}{2R}\left(1-\frac{e^{-\frac{t}{3RC}}}{6}\right)$

29 What is its limiting value i_{AB} at $t \to \infty$?

(a) $\frac{V}{2R}$ (b) $\frac{V}{6R}$ (c) $\frac{2V}{R}$ (d) $\frac{V}{3R}$

Comprehension -2

In the circuit shown involving a resistor of resistance R, capacitor of capacitance C and an ideal cell of emf ε, the capacitor is initially uncharged and the key is in position 1. At $t = 0$, the key is pushed to position 2 for $t_0 = RC$ and then key is pushed back to position 1 for $t_0 = RC$. This process is repeated again and again. Assume the time taken to push key from position 1 to 2 and *vice-versa* to be negligible.

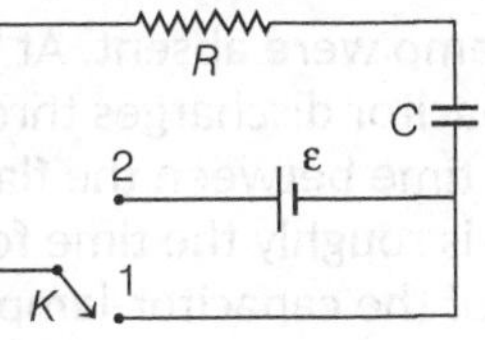

30 The charge on capacitor at $t = 2RC$ is

(a) $C\varepsilon$ (b) $C\varepsilon\left(1-\frac{1}{e}\right)$ (c) $C\varepsilon\left(\frac{1}{e}-\frac{1}{e^2}\right)$ (d) $C\varepsilon\left(1-\frac{1}{e}+\frac{1}{e^2}\right)$

31 The current through the resistance at $t = 1.5\,RC$ seconds is

(a) $\frac{\varepsilon}{e^2R}\left(1-\frac{1}{e}\right)$ (b) $\frac{\varepsilon}{eR}\left(1-\frac{1}{e}\right)$ (c) $\frac{\varepsilon}{R}\left(1-\frac{1}{e}\right)$ (d) $\frac{\varepsilon}{\sqrt{e}R}\left(1-\frac{1}{e}\right)$

32 Then, the variation of charge on capacitor with time is best represented by

(a) 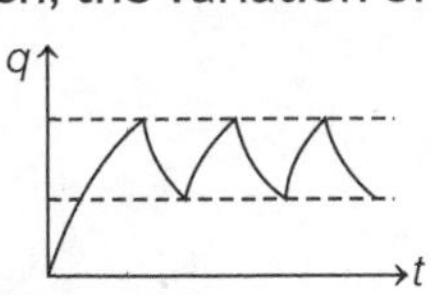

(b) 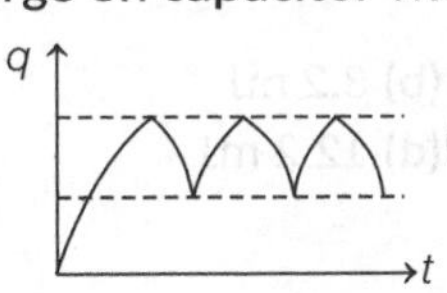

(c) 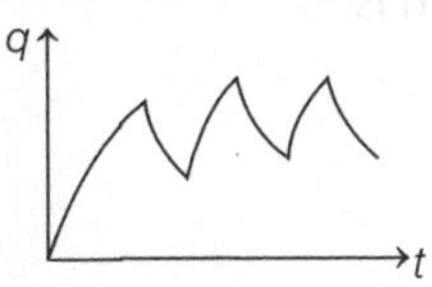

(d) 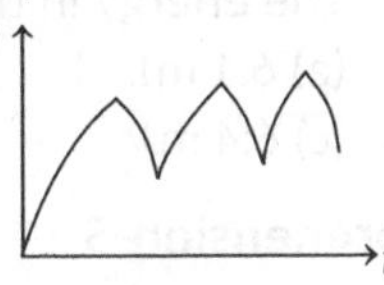

Comprehension-3

In the circuit shown in figure, switch S is closed at time $t = 0$. Thereafter the constant current source by varying its emf, maintains a constant current i out of its upper terminal.

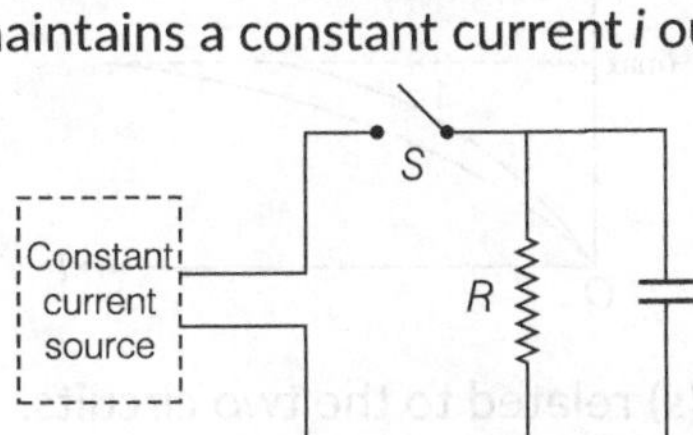

33 The expression for the charge on the capacitor as a function of time is

(a) $q = iRC\left(e^{-\frac{t}{RC}}\right)$ (b) $q = iRC\left(1-e^{-\frac{t}{RC}}\right)$ (c) $q = it\left(1-e^{-\frac{t}{RC}}\right)$ (d) $q = \frac{i}{RC}\left(1-e^{-\frac{t}{RC}}\right)$

34 Find the time at which the rate of energy storage on capacitor is equal to the rate of heat dissipation through resistor.

(a) $t = \frac{RC\ln}{2}$ (b) $t = 2RC\ln 2$ (c) $t = RC\ln 2$ (d) Not possible

Comprehension-4

A highway emergency flasher uses a 120 V battery, a 1 MΩ resistor, a 1 μF capacitor, a neon flash lamp in the circuit shown in the figure. The flash lamp has a resistance more than 10^{10} Ω when the voltage across it is less than 110 V. Above 110 V, the neon gas ionises, the lamp's resistance drops to 10 Ω, and the capacitor discharges completely. Until the capacitor voltage reaches the breakdown voltage $V_b = 110$ V, the large resistance of the flash lamp ensures that it draws a negligible current.

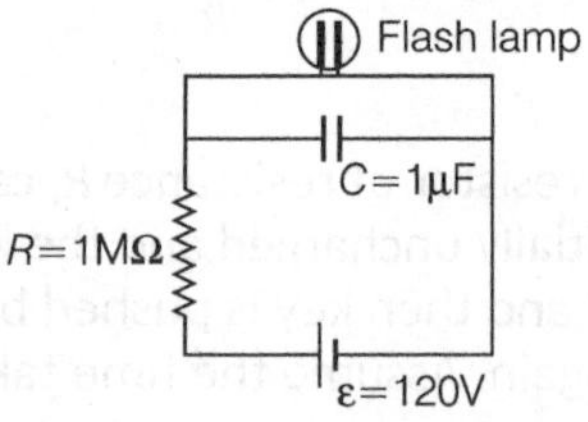

The capacitor charges as if the lamp were absent. At V_b, however the lamp resistance quickly becomes negligible, and the capacitor discharges through the lamp as if the battery and the series resistor were absent. The time between the flashes is the time for the capacitor to charge to V_b. The flash duration is roughly the time for the capacitor to discharge through the lamp, or about 3 time constant of the capacitor-lamp circuit. The flash energy is the stored energy in the capacitor at 110 V.

35 Time constant (τ_c) of the capacitor-lamp circuit is

(a) 20 μs (b) 15 μs
(c) 30 μs (d) 10 μs

36 Flash duration is

(a) 10 μs (b) 20 μs
(c) 30 μs (d) 5 μs

37 The energy in the flash is

(a) 6.1 mJ (b) 3.2 mJ
(c) 6.4 mJ (d) 12.2 mJ

Comprehension-5

The charge across the capacitor in two different *R-C* circuits 1 and 2 are plotted as shown in figure.

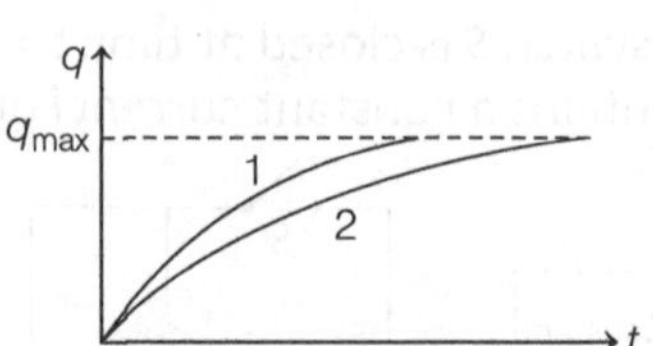

38 Choose the correct statement(s) related to the two circuits.

(a) Both the capacitors are charged to the same charge.
(b) The emf's of cells in both the circuit are equal.
(c) The emf's of the cells may be different.
(d) The emf ε_1 is more than ε_2.

39 Identify the correct statement(s) related to the R_1, R_2, C_1 and C_2 of the two *R-C* circuits.

(a) $R_1 > R_2$, if $\varepsilon_1 = \varepsilon_2$ (b) $C_1 < C_2$, if $\varepsilon_1 = \varepsilon_2$
(c) $R_1C_1 > R_2C_2$ (d) $\frac{R_1}{R_2} < \frac{C_2}{C_1}$

Comprehension-6

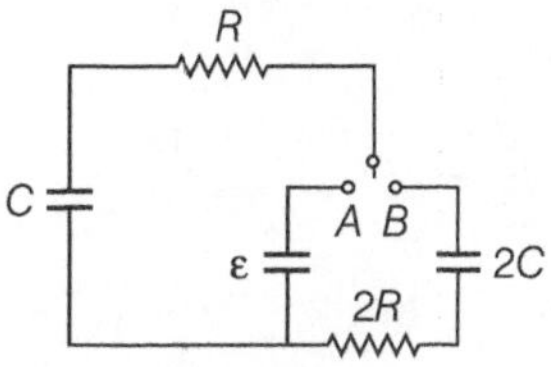

Consider the circuit in the figure. The switch is initially in position A for a sufficiently long time to establish a steady state. At $t = 0$, the switch is turned to position B. The capacitor $2C$ did not hold any charge before $t = 0$.

40 At $t \to \infty$, when a new steady state has been established, the charges held by capacitors C and $2C$ will respectively be

(a) $C\varepsilon$ and $2C\varepsilon$ (b) $\frac{C\varepsilon}{2}$ and $C\varepsilon$ (c) $\frac{C\varepsilon}{3}$ and $\frac{2C\varepsilon}{3}$ (d) $\frac{2C\varepsilon}{3}$ and $C\varepsilon$

41 The current decreases exponentially between $t = 0$ to $t = \infty$. The time constant is

(a) $9RC$ (b) $2RC$ (c) $3RC$ (d) $\frac{3RC}{2}$

42 The time, at which charge in the capacitor C is reduced to half of the value it had at $t = 0$, is

(a) $6RC \ln 2$ (b) $3RC \ln 2$ (c) $2RC \ln 2$ (d) $4RC \ln 2$

Match the Columns Questions

43 Initially capacitor was uncharged and switch was open. Switch is closed at $t = 0$. Ammeter and voltmeter are ideal. [All units are in SI]

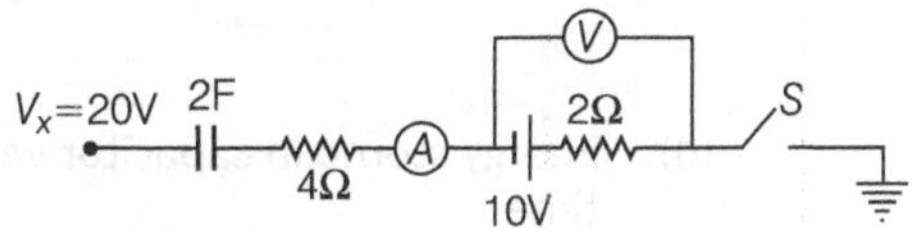

Column I		Column II	
(a)	Reading of voltmeter after long time	(p)	0
(b)	Just after closing the switch, reading of ammeter	(q)	5
(c)	Just after closing the switch, reading of voltmeter	(r)	10
(d)	Charge on capacitor after long time	(s)	20
		(t)	60

44 Consider the circuit shown in figure. There are three switches S_1, S_2, S_3. Match the following columns.

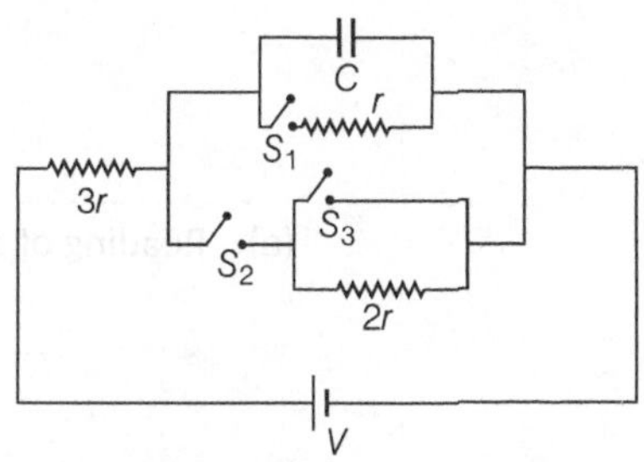

Column I		Column II	
(a)	If S_1 and S_3 are opened and S_2 is closed, then in steady state, charge on capacitor equals	(p)	$\frac{CV}{4}$
(b)	If S_2 and S_3 are opened and S_1 is closed, then in steady state, charge on capacitor equals	(q)	$\frac{2CV}{5}$
(c)	If S_1 and S_2 are closed and S_3 is opened, then in steady state, charge on capacitor equals	(r)	$\frac{CV}{3}$
(d)	If all the switches are closed, then in steady state, charge on capacitor is	(s)	$\frac{2CV}{11}$
		(t)	zero

45 In the shown circuit initially capacitor has some charge,the switch is closed at $t = 0$.

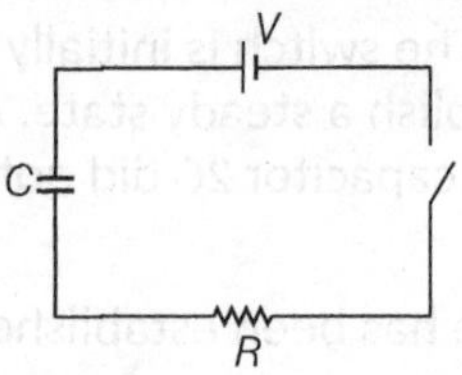

	Variables		Possible graphs
(a)	Charge on upper plate of capacitor *versus* time	(p)	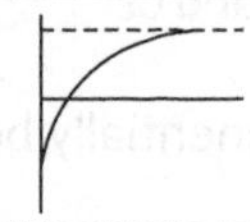
(b)	Current *versus* time	(q)	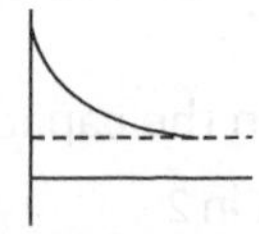
(c)	Charge *versus* current	(r)	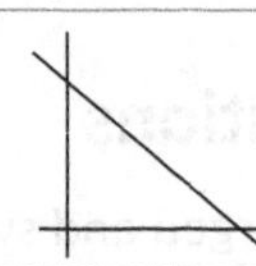
(d)	Energy stored in capacitor *versus* time	(s)	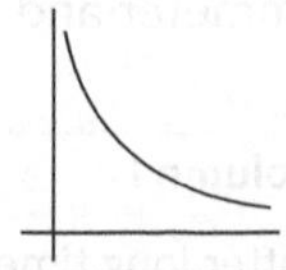

46 Column I gives some electrical circuits in steady state Column II gives some statements regarding the circuits. Voltmeter and ammeter are ideal. Match appropriately.

	Column I		Column II
(a)	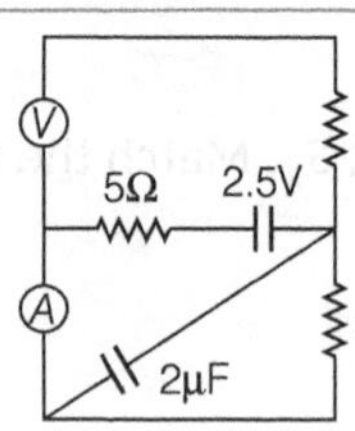	(p)	Reading of voltmeter is 2 V
(b)	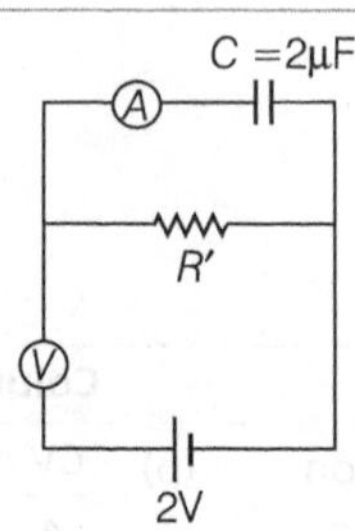	(q)	Reading of ammeter is 0.1 A
(c)	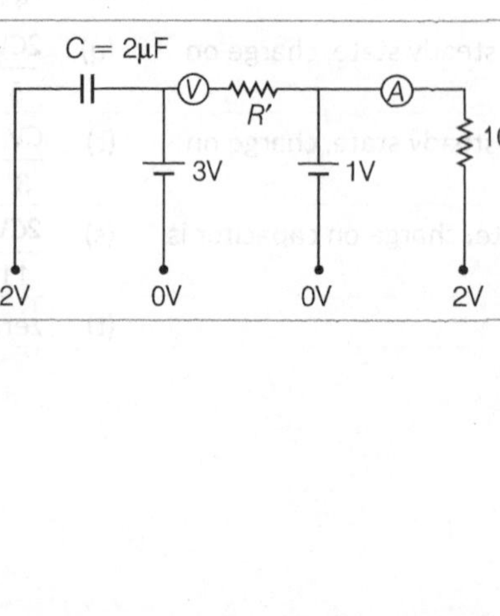	(r)	Current through R' is zero

Column I	Column II
(d) 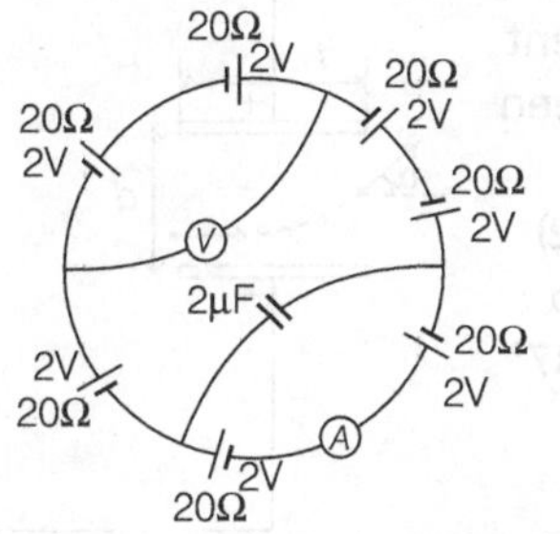	(s) Charge on capacitor is 2 μC
	(t) Current through R' is non-zero

47 In the circuit shown the figure, the capacitor is initially uncharged. At $t = 0$, the switch is closed at position (1) and remains closed for a long time. Then at $t = t'$, the switch is shifted to position (2). Match the following.

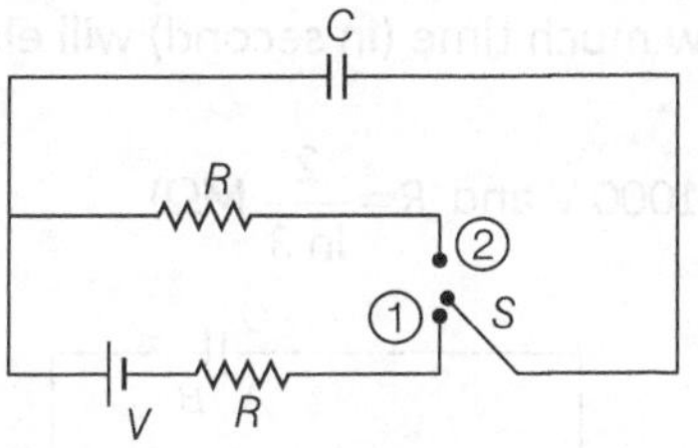

Column I		Column II	
(a)	As the capacitor charges from $t = 0$ to $t = t'$	(p)	Current in the circuit falls exponentially
(b)	As the capacitor discharges, i.e. for $t > t'$	(q)	Current in the circuit grows exponentially
(c)	Maximum current in the circuit depends on	(r)	R, C
(d)	Time to achieve 50% charging of the capacitor depends on	(s)	R
		(t)	None of these

Numerical Value Questions

48 In the circuit shown in figure, find the charge (in μC) on capacitor C_2 in steady state.

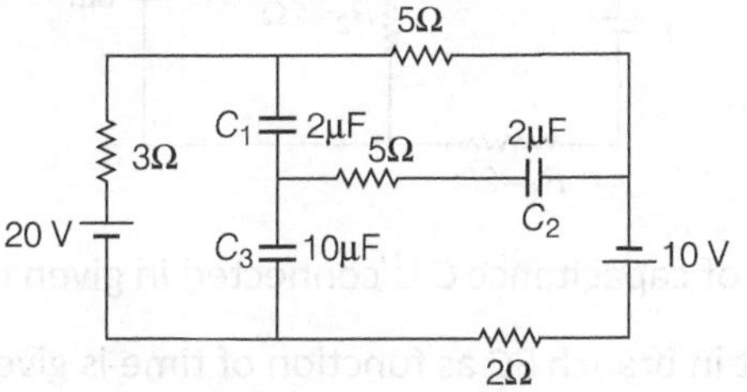

49 A system of capacitor and resistors are connected to an ideal battery of emf V as shown in figure. Initially all the capacitors are uncharged and switch is open. Now, switch S is closed at $t = 0$. Capacitance of each capacitor is C and resistance of each resistor is R and all of the symbols have their usual meanings. Current through battery just after switch is closed is $\left(\frac{nV}{3R}\right)$.

What is the value of n ?

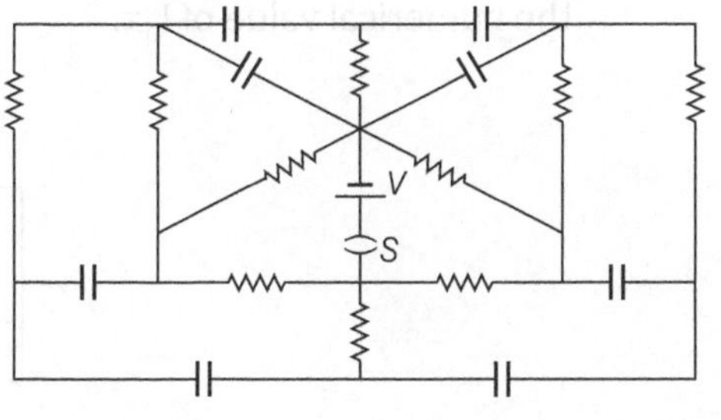

50 A parallel plate capacitor with plates of length l is included in a circuit as shown in figure. Given are the emf of the current source ε, its internal resistance r and the distance d between the plates. An electron with a velocity v_0 flies into the capacitor, parallel to the plates. Find the resistance R (in Ω) that should be connected in parallel with the capacitor, so that the electron flies out of the capacitor at an angle of 37° to the plates. Assume that circuit is in steady state and neglect gravity. (Take, l =91cm, $\varepsilon = 3$V, $r = 2\Omega$, $d = \frac{1}{3}$ mm, $v_0 = 4 \times 10^7$ m/s, $e = 1.6 \times 10^{-19}$ C, mass of electron $= 9.1 \times 10^{-31}$ kg)

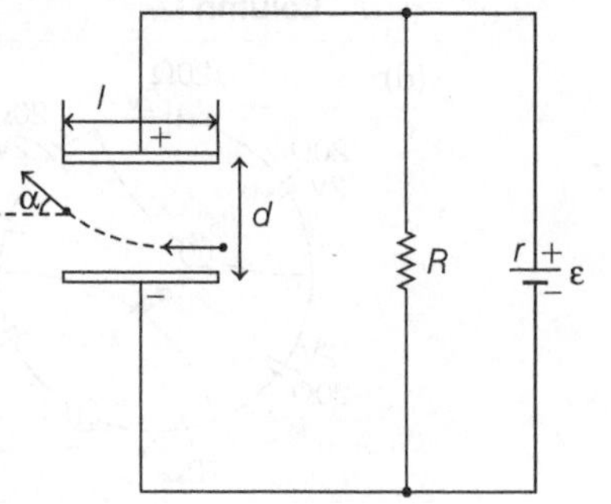

51 The figure shows a *R-C* circuit with a parallel plate capacitor. Before switching on the circuit, plate *A* of the capacitor was given a charge $-Q$. While plate *B* has no net charge. Now at $t = 0$, the circuit is switched on how much time (in second) will elapse before the net charge on plate *A* becomes zero.

(Take, $C = 1\,\mu$F, $Q = 1$ mC, $\varepsilon = 1000$ V and $R = \frac{2}{\ln 3}$ MΩ)

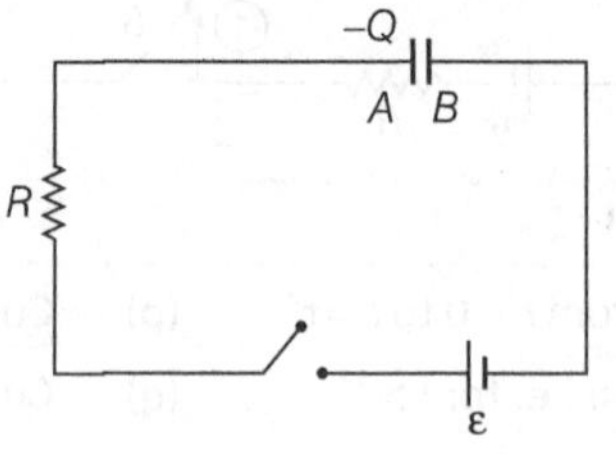

52 In the circuit shown in figure, battery is ideal. Find time (in μs) when the current in R_3 becomes $\frac{1}{e}$ time of maximum current through it.

(At time $t = 0$, the key *S* is closed, Initially capacitor is uncharged)

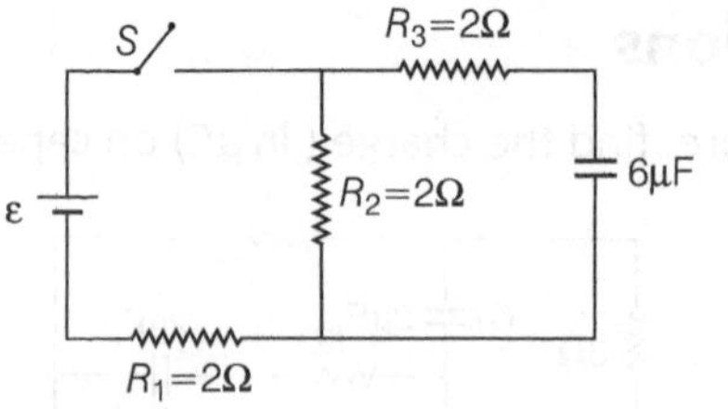

53 An uncharged capacitor of capacitance *C* is connected in given circuit diagram, switch *S* is closed at $t = 0$. If current in branch *BC* as function of time is given by $I = I_0 (Amp) e^{\frac{t}{\tau(\text{in}\,\mu s)}}$, then find the numerical value of $I_0 \tau$.

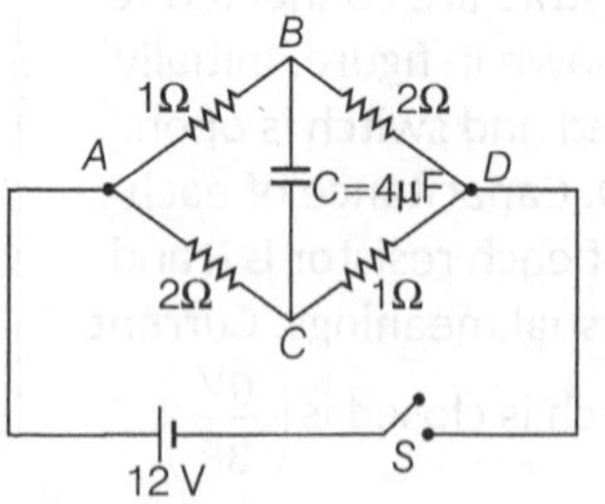

54 The horizontal plates of a capacitor are charged to a voltage of 20 V. A particle carrying a charge is in equilibrium under the influence of gravitational and electrostatic forces. Capacitor is now connected across a resistance 4 kΩ. If capacitance is 500 μF, then find the velocity (in m/s) of the particle at $t = 2$ sec upto two decimal places.
(Take, $g = 10\ \text{m/s}^2$)

55 A parallel plate capacitor has plate area A and separation d between the plates. The space between plates is filled with a dielectric of dielectric constant K and resistivity ρ. The capacitor is initially charged to Q_0 using a battery. After removing battery at $t = 0$, current density in the dielectric at any time t is $\frac{nQ_0}{\rho K \varepsilon_0 A} e^{-\left(\frac{mt}{\rho K \varepsilon_0}\right)}$. Find $n + m$.

56 Consider the shown network, the capacitor C_1 ($= 6\ \mu$F) has an initial charge $q_0 = \frac{30e}{e-1}\ \mu$C, $C_2 = 4\ \mu$F and $R = 80\Omega$. Initially C_2 is uncharged. At $t = 0$, the switch S is closed. Obtain the charge on C_2 (in μC) at $t = 192\ \mu$s.

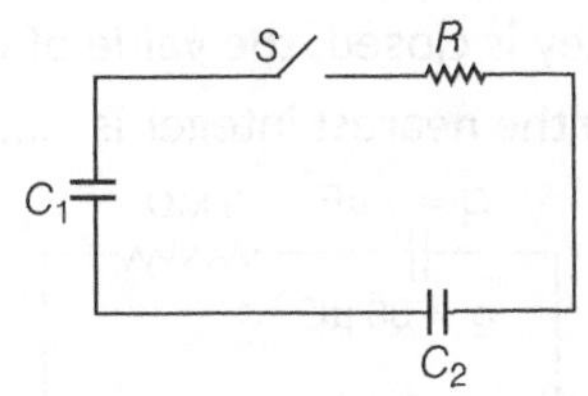

57 For the given circuit in the steady state condition, charge on the capacitor is $q_0 = 16\ \mu$C. If now the battery is removed and the nodes A and C are shorted. The time during which charge on the capacitor becomes 4 μC is X (μs) and emf of battery was Y (volt). Write value of $\frac{XY}{\ln 2}$.

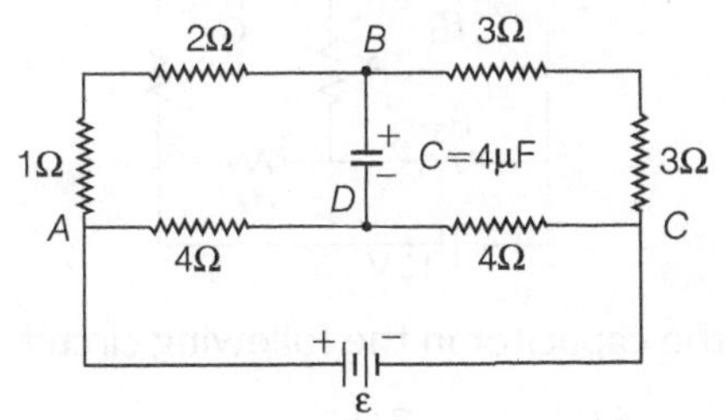

Answers

1. (d)	**2.** (c)	**3.** (c)	**4.** (c)	**5.** (d)	**6.** (d)	**7.** (d)	**8.** (a)	**9.** (d)	**10.** (b)
11. (a)	**12.** (b)	**13.** (d)	**14.** (c)	**15.** (a)	**16.** (c)	**17.** (a)	**18.** (abd)	**19.** (acd)	**20.** (ac)
21. (bd)	**22.** (ac)	**23.** (abc)	**24.** (bcd)	**25.** (abd)	**26.** (abcd)	**27.** (b)	**28.** (d)	**29.** (a)	**30.** (c)
31. (d)	**32.** (c)	**33.** (b)	**34.** (c)	**35.** (d)	**36.** (c)	**37.** (a)	**38.** (c)	**39.** (d)	**40.** (c)

41. (b) **42.** (d) **43.** a → r; b → q; c → p; d → t **44.** a → q; b → p; c → s; d → t
45. a → pq; b → s; c → r; d → q **46.** a → pqr; b → pr; c → pqrs; d → q
47. a → p; b → p; c → s; d → r
48. (10) **49.** (6) **50.** (10) **51.** (2) **52.** (18)
53. (16) **54.** (7.36) **55.** (2) **56.** (12) **57.** (768)

For Detailed Solutions
Scan the code

Previous Years' Questions

JEE Main

1 A capacitor of 50 μF is connected in a circuit as shown in figure. The charge on the upper plate of the capacitor is .... μC.

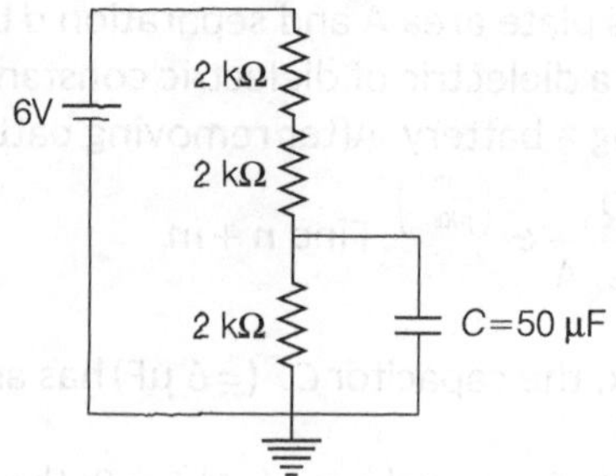

[JEE Main 2021]

2 The circuit shown in the figure consists of a charged capacitor of capacity 3 μF and a charge of 30 μC. At time $t = 0$, when the key is closed, the value of current flowing through the 5 MΩ resistor is x μA. The value of x to the nearest integer is .... [JEE Main 2021]

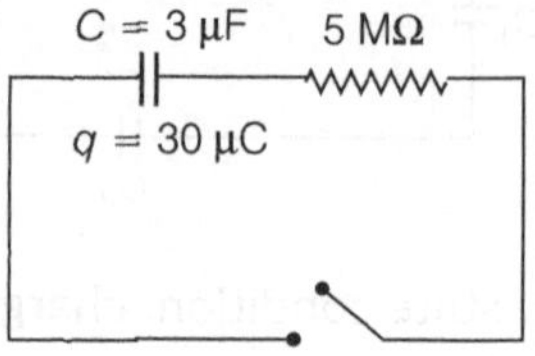

3 An ideal cell of emf 10 V is connected in circuit shown in figure. Each resistance is 2 Ω. The potential difference (in V) across the capacitor when it is fully charged is ....

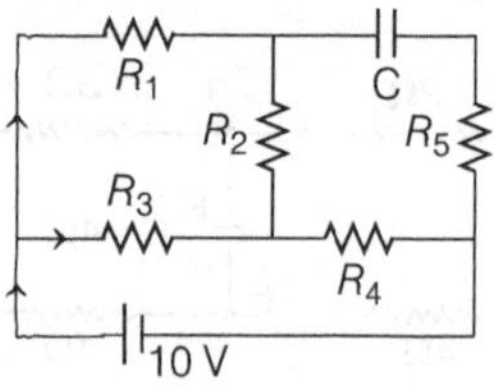

[JEE Main 2020]

4 Determine the charge on the capacitor in the following circuit [JEE Main 2019]

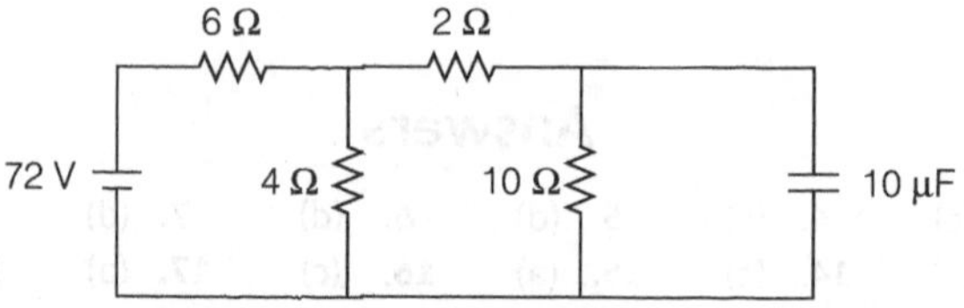

(a) 2 μC (b) 200 μC (c) 10 μC (d) 60 μC

5 In the given circuit diagram, when the current reaches steady state in the circuit, the charge on the capacitor of capacitance C will be [JEE Main 2017]

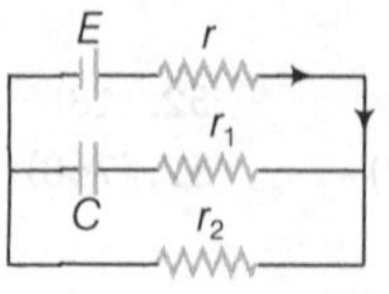

(a) $CE\dfrac{r_1}{(r_2 + r)}$ (b) $CE\dfrac{r_2}{(r + r_2)}$ (c) $CE\dfrac{r_1}{(r_1 + r)}$ (d) CE

JEE Advanced

Passage Based Questions

Passage Consider a simple R-C circuit as shown in Figure 1.

Process 1 In the circuit, the switch S is closed at $t = 0$ and the capacitor is fully charged to voltage V_0 (i.e. charging continues for time $T >> RC$). In the process, some dissipation (E_D) occurs across the resistance R. The amount of energy finally stored in the fully charged capacitor is E_C.

Process 2 In a different process the voltage is first set to $\frac{V_0}{3}$ and maintained for a charging time $T >> RC$. Then, the voltage is raised to $\frac{2V_0}{3}$ without discharging the capacitor and again maintained for a time $T >> RC$. The process is repeated one more time by raising the voltage to V_0 and the capacitor is charged to the same final voltage V_0 as in process 1.

These two processes are depicted in Figure 2. **[JEE Advanced 2017]**

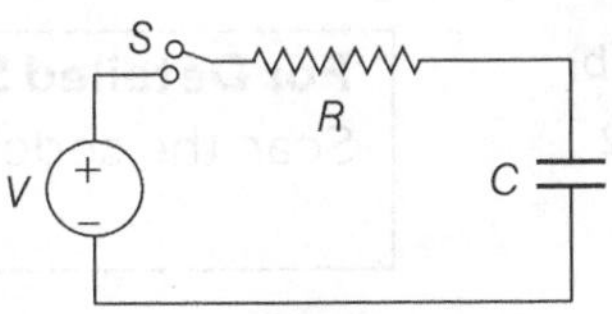

Figure 1

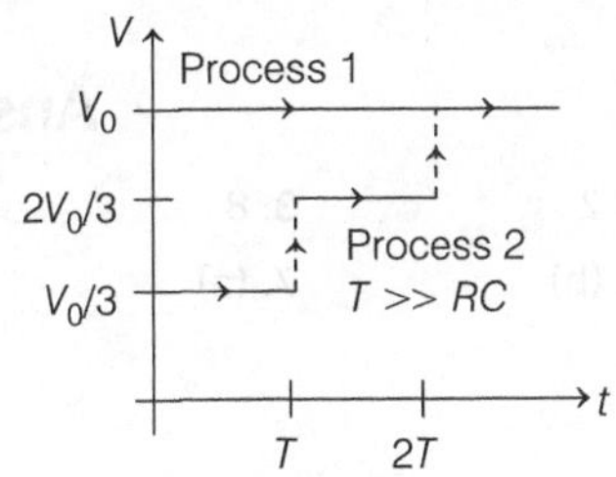

Figure 2

6 In process 1, the energy stored in the capacitor E_C and heat dissipated across resistance E_D are related by

(a) $E_C = E_D \ln 2$

(b) $E_C = E_D$

(c) $E_C = 2E_D$

(d) $E_C = \frac{1}{2}E_D$

7 In process 2, total energy dissipated across the resistance E_D is

(a) $E_D = \frac{1}{3}\left(\frac{1}{2}CV_0^2\right)$

(b) $E_D = 3\left(\frac{1}{2}CV_0^2\right)$

(c) $E_D = 3CV_0^2$

(d) $E_D = \frac{1}{2}CV_0^2$

8 At time $t = 0$, a battery of 10 V is connected across points A and B in the given circuit. If the capacitors have no charge initially, at what time (in second) does the voltage across them become 4 V? [Take : ln 5 = 1.6, ln 3 = 1.1] **[JEE Advanced 2010]**

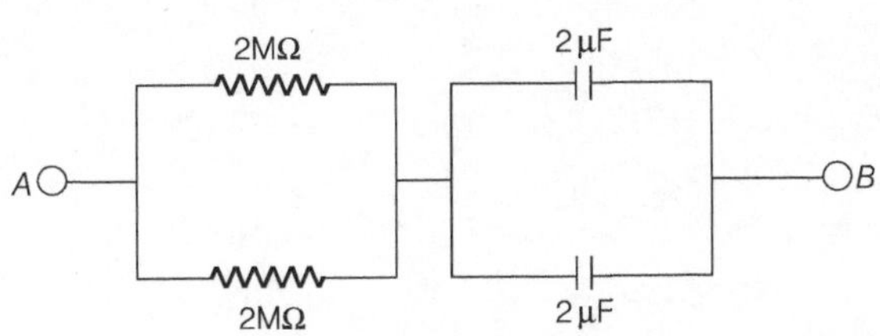

9 A parallel plate capacitor C with plates of unit area and separation d is filled with a liquid of dielectric constant $K = 2$. The level of liquid is $d/3$ initially. Suppose the liquid level decreases at a constant speed v, the time constant as a function of time t is **[JEE Advanced 2008]**

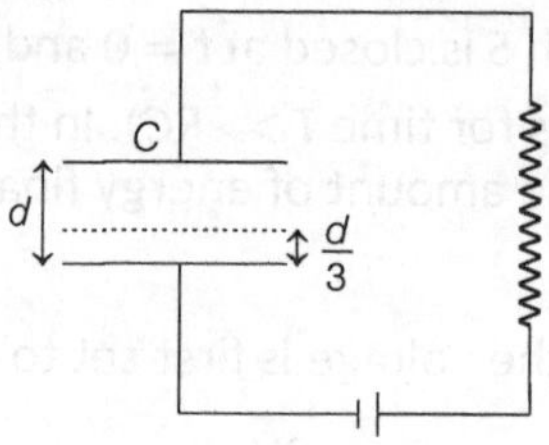

(a) $\frac{6\varepsilon_0 R}{5d + 3vt}$

(b) $\frac{(15d + 9vt)\varepsilon_0 R}{2d^2 - 3dvt - 9v^2t^2}$

(c) $\frac{6\varepsilon_0 R}{5d - 3vt}$

(d) $\frac{(15d - 9vt)\varepsilon_0 R}{2d^2 + 3dvt - 9v^2t^2}$

Answers

1. 100	**2.** 2	**3.** 8	**4.** (b)
5. (b)	**6.** (b)	**7.** (a)	**8.** 2
9. (a)			

For Detailed Solutions
Scan the code